BIOACTIVE
NATURAL PRODUCTS

Detection, Isolation, and Structural Determination

BIOACTIVE
NATURAL PRODUCTS

Detection, Isolation,
and Structural Determination

Steven M. Colegate, Ph.D.
Research Chemist
School of Veterinary Studies
Murdoch University
Perth, Western Australia

Russell J. Molyneux, Ph.D.
Research Chemist
Plant Protection Research
Western Regional Research Center
Agricultural Research Service
United States Department of Agriculture
Albany, California

CRC Press
Boca Raton   Ann Arbor   London   Tokyo

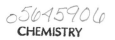

CHEMISTRY

Library of Congress Cataloging-in-Publication Data

Bioactive natural products: detection, isolation, and structural determination / [edited by] Steven M. Colegate.
    p.   cm.
Includes bibliographical references and index.
ISBN 0-8493-4372-0
1. Natural products.   2. Bioactive compounds.   I. Colegate, Steven M.
QD415.B45   1993
547.7'046—dc20   93-3582
                                                                               CIP

PREFACE

The aim of this book is to expose academics, research workers, and students to the multi-disciplinary bioactive natural products research arena. The contributors to this volume are an internationally selected group of experts with an accumulated wealth of knowledge and experience that has been tapped for this book. Thus, the book is not simply an *ad hoc* collection of treatises of barely related subjects, but a carefully selected assimilation of information describing the detection, isolation, and structural determination of bioactive natural products. The book covers all aspects of bioactive natural product research, from ethnobotanical investigations to modern, technologically assisted isolation and structural determination of active compounds.

However, the value of this book lies in the fact that it is not merely a catalog of techniques. The reader will derive the maximum benefit from the experience and expertise of the authors, since there is a strong emphasis on the rationale guiding the selection of natural resources and the isolation of bioactive compounds from these resources, and on the type and quality of structural information that can or should be obtained. The chapters describing the modern application of detection, isolation, and structural determination techniques are strongly supported by chapters detailing and reviewing research involving specific classes of bioactivity, e.g., immunomodulatory, antiviral, cytotoxic, anti-inflammatory, insect behavior modifiers, and so forth.

Steven M. Colegate, Ph.D.
Russell J. Molyneux, Ph.D.

THE EDITORS

Steven M. Colegate, Ph.D., earned a doctorate in Organic Chemistry (Synthesis, Compound Isolation, Structural Elucidation, and ESR Spectroscopy) at the University of Western Australia in 1978. The bulk of his postdoctoral work to date has been as part of a multidisciplinary team, within the School of Veterinary Studies at Murdoch University, investigating West Australian plants that are toxic to livestock.

Russell J. Molyneux, Ph.D., earned a doctorate in Organic Chemistry (Natural Product Structure Elucidation and Synthesis) at the University of Nottingham in 1963. His program as a Research Chemist at the Western Regional Research Laboratory in Albany, California has focused on the chemistry of toxic range plants, primarily in collaboration with the Poisonous Plant Research Laboratory at Logan, Utah, but also with scientists in many other parts of the world where livestock poisoning is an epidemic problem in livestock production.

THE CONTRIBUTORS

The contributors to this volume have been selected to provide a diverse range of expertise and experience in the general field of bioactive natural products. The aim has been to provide a truly multi-disciplinary approach to the problems of identifying bioactive constituents and subsequently isolating them and establishing their structures. The contributors are all experts within their particular disciplines, with well established reputations. Their contributions have been chosen to illustrate innovative approaches to the problems inherent to research on the diversity of bioactive compounds occurring in living organisms.

Yoshinori Asakawa, D.Sc.
Director
Institute of Pharmacognosy
Faculty of Pharmaceutical Sciences
Tokushima Bunri University
Tokushima, Japan

Stephen J. Bloor, Ph.D.
Scientist
Industrial Research, Ltd.
Lower Hutt, New Zealand

Lindsay T. Byrne, Ph.D.
Head
NMR Facility
Department of Chemistry
University of Western Australia
Perth, Western Australia

Shaoxing Chen, Ph.D.
Research Fellow
Department of Chemistry
National University of Singapore
Republic of Singapore

Steven M. Colegate, Ph.D.
Research Chemist
School of Veterinary Studies
Murdoch University
Perth, Western Australia

Geoffrey A. Cordell, Ph.D.
Professor
Department of Medicinal Chemistry and
    Pharmacognosy
Interim Dean
College of Pharmacy
University of Illinois at Chicago
Chicago, Illinois

Kevin D. Croft, Ph.D.
Senior Scientific Officer
Department of Medicine
University of Western Australia
Royal Perth Hospital
Perth, Western Australia

Laurent A. Décosterd, Ph.D.
Research Associate
Department of Clinical Pharmacology
University of Lausanne
Lausanne, Switzerland

Peter R. Dorling, Ph.D.
Associate Professor
School of Veterinary Studies
Murdoch University
Perth, Western Australia

William Gaffield, Ph.D.
Research Chemist
Western Regional Research Center
Agricultural Research Service
United States Department of Agriculture
Albany, California

Emilio L. Ghisalberti, Ph.D.
Associate Professor
Department of Chemistry
University of Western Australia
Nedlands, Western Australia

Achiel Haemers, Ph.D.
Professor
Department of Pharmaceutical Sciences
University of Antwerp
Antwerp, Belgium

Kurt Hostettmann, Ph.D.
Professor and Director
Institute of Pharmacognosy and
    Phytochemistry
School of Pharmacy
University of Lausanne
Lausanne, Switzerland

Clive R. R. Huxtable, Ph.D.
Associate Professor of Veterinary
    Pathology
School of Veterinary Studies
Murdoch University
Perth, Western Australia

Jinwoong Kim, Ph.D.
Associate Professor
College of Pharmacy
Seoul National University
Seoul, Korea

A. Douglas Kinghorn, Ph.D., D.Sc.
Professor
Department of Medicinal Chemistry and
    Pharmacognosy
Associate Director
Program for Collaborative Research in the
    Pharmaceutical Sciences
College of Pharmacy
University of Illinois at Chicago
Chicago, Illinois

Rudi P. Labadie, Prof. Dr.
Professor of Pharmacognosy
Department Head
Department of Pharmacognosy
Faculty of Pharmacy
Utrecht University
Utrecht, The Netherlands

Daniel Lednicer, Ph.D.
Chemist
Drug Synthesis and Chemistry Branch
Developmental Therapeutics Program
Division of Cancer Treatment
National Cancer Institute
Bethesda, Maryland

Andrew Marston, Ph.D.
Research Associate
Institute of Pharmacognosy and
    Phytochemistry
School of Pharmacy
University of Lausanne
Lausanne, Switzerland

Russell J. Molyneux, Ph.D.
Research Chemist
Plant Protection Research
Western Regional Research Center
Agricultural Research Service
United States Department of Agriculture
Albany, California

Ven L. Narayanan, Ph.D.
Chief
Drug Synthesis and Chemistry Branch
Developmental Therapeutics Program
Division of Cancer Treatment
National Cancer Institute
Bethesda, Maryland

Kip E. Panter, Ph.D.
Research Animal Scientist
Poisonous Plant Research Laboratory
Agricultural Research Service
United States Department of Agriculture
Adjunct Research Assistant Professor
Animal, Dairy, and Veterinary Science
College of Agriculture
Utah State University
Logan, Utah

John M. Pezzuto, Ph.D.
Professor and Acting Head
Department of Medicinal Chemistry and
    Pharmacognosy
College of Pharmacy
Professor and Associate Director
Specialized Cancer Center
College of Medicine
University of Illinois at Chicago
Chicago, Illinois

Lawrence J. Porter, Ph.D.
Scientist
Environmental Organic Chemistry
Institute of Environmental Health and
    Forensic Sciences, Ltd.
Lower Hutt, New Zealand

Richard J. Riopelle, M.D.
Professor
Department of Medicine (Neurology)
Queen's University
Kingston, Ontario
Canada

Teng Wah Sam, Ph.D.
Professor
School of Chemical Sciences
Universiti Sains Malaysia
Pulau Pinang, Malaysia

John K. Snyder, Ph.D.
Associate Professor
Department of Chemistry
Boston University
Boston, Massachusetts

Kenneth L. Stevens, Ph.D.
Research Leader
Department of Food Safety Research
Western Regional Research Center
Agricultural Research Service
United States Department of Agriculture
Albany, California

Shozo Takahashi, Ph.D.
Director and Professor
Faculty of Agriculture
Pesticide Research Institute
Kyoto University
Kyoto, Japan

Dirk A. R. Vanden Berghe, Ph.D.
Professor
Department of Medicine
Laboratory of Applied Medical and
    Pharmaceutical Microbiology
University of Antwerp
Antwerp, Belgium

Arnold J. Vlietinck, Ph.D.
Professor
Department of Pharmaceutical Sciences
Laboratories of Pharmacognosy and
    Phytochemistry
University of Antwerp
Antwerp, Belgium

Rosalind Y. Wong
Research Chemist
Western Regional Research Center
Agricultural Research Service
United States Department of Agriculture
Albany, California

CONTENTS

BIOACTIVE
NATURAL PRODUCTS

Detection, Isolation,
and Structural Determination

Chapter 1

INTRODUCTION AND OVERVIEW

Steven M. Colegate and Russell J. Molyneux

TABLE OF CONTENTS

0-8493-4372-0/93/$0.00+$.50
© 1993 by CRC Press, Inc.

1

I. INTRODUCTION

The investigation of bioactive natural products has, in recent years, assumed a greater sense of urgency in response to the expanding human population and its subsequent demands for food, good health, and increasing areas of land on which to live. The extinction of plant and animal species as mankind encroaches on natural habitats represents lost and irreplaceable resources, the full potential of which is unpredictable. Similarly, the loss of endemic cultures as other cultures become influential in an almost cancerous manner will result in the loss of a fount of empirical ethnobotanical knowledge that has been acquired over the course of thousands of years.

II. THE MULTIDISCIPLINARY APPROACH

Nature recognizes no artificial barriers such as those of the "academic disciplines" and thus it is no surprise to find investigators with quite different academic training studying various aspects of bioactive natural products. It is when such diversely trained investigators come together as a team or, at the very least, collaborate very closely, that the greatest benefit will arise from such studies, since the investigators approach the subject from differing perspectives that will, with a little planning, complement and stimulate each other. This multidisciplinary approach not only facilitates the solution of specific problems, such as those in plant toxicology, but may also enhance the diversity and consequent value of bioactive natural product research.

Successful multidisciplinary collaboration requires that each individual has an understanding and appreciation of the intellectual and technical contributions to the project of the other team members. The sophistication of modern scientific instrumentation is such that the human aspects of using such equipment are frequently overlooked. While almost any research problem can presently be solved by the application of the most powerful equipment on the market, it requires imagination and resourcefulness to achieve results when such techniques are unavailable or too expensive to employ. It is then essential for the collaborators to ask themselves questions concerning the most expeditious approach to be adopted; for example:

1. Can a single bioassay be used, or must more than one be employed to cover the bioactivity of interest?
2. What is the most useful technique for separation of the active compound for both structural determination and possibly large-scale biological testing? Obviously it is inefficient to develop a separation method that yields the milligram quantities necessary for structure elucidation and then have to develop an entirely different method for preparation of gram quantities for *in vivo* and *in vitro* testing.
3. What are the minimal requirements for structural determination? It is an extravagance of time and resources to employ spectroscopic techniques that provide a duplication of evidence, unless the data obtained are equivocal.
4. Can the information regarding separation and structure of the bioactive compound(s) of concern be integrated to give a useful method for detection and analysis? If so, the time required for the developement of such a method may be significantly reduced.

Such questions require that the collaborators have a basic understanding of both the potential and the limitations of the techniques and disciplines that each can bring to bear on a problem. This can only lead to good experimental design that will further the goals of the project. It is the intention of this volume to provide some measure of perception in this regard.

III. WHY ISOLATE BIOLOGICALLY ACTIVE NATURAL PRODUCTS?

The use of herbal and other naturally based medicines has a long history. However, the utilization of whole-plant or other crude preparations for therapeutic or experimental reasons can have several drawbacks, which include

1. Variation in the amount of the active constituent with geographic areas, from one season to another, with different plant parts and morphology, and with climatic and ecologic conditions.
2. Co-occurrence of undesirable compounds causing synergistic, antagonistic, or other undesirable, and possibly unpredictable, modulations of the bioactivity.
3. Losses of bioactivity due to variability in collection, storage, and preparation of the raw material.

Thus, the isolation of natural products that have biologic activity toward organisms other than the source has several advantages, including

1. Pure bioactive compounds can be administered in reproducible, accurate doses, with obvious benefits from an experimental or therapeutic point of view.
2. It can lead to the development of analytic assays for particular compounds or for classes of compounds. This is necessary, for example, in the screening of plants for potential toxicity and for quality control of food for human or animal consumption.
3. It permits the structural determination of bioactive compounds that may enable the production of synthetic material, incorporation of structural modifications, and a rationalization of mechanisms of action. This, in turn, will lead to reduced dependency on plants, for example, as sources of bioactive compounds and will enable investigations of structure/activity relationships, facilitating the development of new compounds with similar or more desirable bioactivities.

IV. DETECTION AND ISOLATION

To search for a compound that elicits a particular bioactive response, an appropriate assay is required to screen the source material and to monitor both extracts there from and subsequent purification steps. The assay could, for example, be lethality to a particular species, immunomodulatory activity, anti-inflammatory activity, or production of characteristic lesions in a species. The assay should be as simple, specific, and rapid as possible. An *in vitro* test is more desirable than a bioassay using small laboratory animals, which, in turn, is more desirable than feeding large amounts of valuable and hard-to-obtain extract to larger domestic animals. In addition, *in vivo* tests in mammals are often variable and are highly constrained by ethical considerations of animal welfare. For example, the isolation and purification of swainsonine (**1**) from *Swainsona canescens* was monitored by a simple, rapid enzyme inhibition assay developed as a consequence of a biochemical and pathologic study of the toxicosis.[1] In contrast, stypandrol (**2**), a toxin from *Stypandra imbricata* (blindgrass), was isolated using a time-consuming mouse bioassay in which the occurrence of characteristic lesions was monitored.[2]

Extraction from the plant is a trial-and-error exercise in which different solvents are tried under a variety of conditions such as time and temperature of extraction. The success or failure of the extraction process is monitored by the most appropriate assay. For example, swainsonine

(**1**) can be obtained by extraction of dried, milled *Swainsona canescens* with hot ethyl acetate for 48 h. Stypandrol (**2**), from *Stypandra imbricata*, although unstable at reflux temperatures, can be obtained by room temperature extraction of the dried or fresh plant, whereas stypandrone (**3**), a related naphthoquinone, can only be isolated from fresh plant material.[3]

Once extracted from the plant, the bioactive component then has to be separated from the coextractives. This may involve simple crystallization of the compound from the crude extract, as with the isolation of the glycoside dianellin (**4**) from *Dianella caerulea*.[4] More usually, however, it will involve further solvent partition of the coextractives and extensive chromatography, taking advantage of particular properties of the desired compound, such as acidity, polarity, and molecular size. In some cases the isolation can be assisted by prior derivatization, imparting more easily manageable properties to the desired compound. The isolation of swainsonine, as its triacetate, from *Rhizoctonia leguminicola*[5] is representative of such an approach.

Final purification, to provide compounds of suitable purity for structural analysis, may be accomplished by appropriate techniques such as recrystallization, sublimation, or distillation.

V. STRUCTURE DETERMINATION

The process of structural determination involves accumulating data from numerous sources, each of which gives some structural information, and the assimilation of these data into a chemical structure that rigorously and uniquely fits all the available structural information. There is available today a wide range of spectroscopic instrumentation, such as UV, IR, and visible absorption spectroscopies, nuclear magnetic resonance spectroscopy, and mass spectrometry, forming the backbone of modern structural analysis.

Prior to the availability of such aids to structural determination and in cases where, despite these aids, ambiguity existed, chemical modification or degradation of the unknown compound was necessary. These latter processes involve the treatment of the unknown compound with functional group-specific reagents or degradation of the compound in a predictable manner until a compound of known structure is obtained. Backtracking should then provide a structure for the unknown compound. Apart from being a time-consuming and exacting art, this method can be fraught with difficulty and ambiguity. For example, stypandrol (**2**), the toxin isolated from *Stypandra imbricata* (blindgrass), was apparently first isolated from *Hemerocallis thumbergii* and named hemerocallin, but was assigned the incorrect structure on the basis of ambiguous degradation studies.[6]

The process of spectroscopic structural determination should be closely allied to a familiarity with the scientific literature. If the compound has not already been described, it may be very similar to reported compounds that will assist in the interpretation of the data for the unknown. In this regard, an awareness of the coextractives from the plant may also be of value. For example, the facile structural elucidation of the easily obtained dianellidin (**5**) from *Stypandra imbricata* greatly expedited the structural analysis of the toxin stypandrol, which proved to be a dimer of dianellidin.

When analysis of the spectroscopic data for an unknown compound is inconclusive, and if the compound or one of its derivatives is suitably crystalline, a single crystal X-ray diffraction study should be considered. Simplistically, this involves computer-aided analysis of the diffraction pattern obtained when a single crystal is irradiated with X-rays. Correct interpretation of the data will result in a three-dimensional picture of the molecule, including the relative stereochemistry if the molecule is optically active. In some cases the absolute stereochemistry can also be determined. X-ray diffraction studies

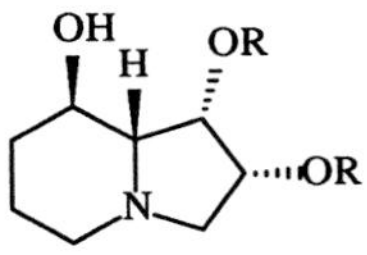

1

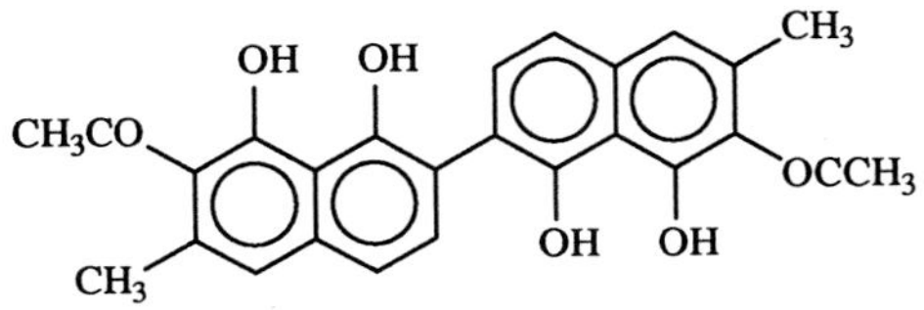

2

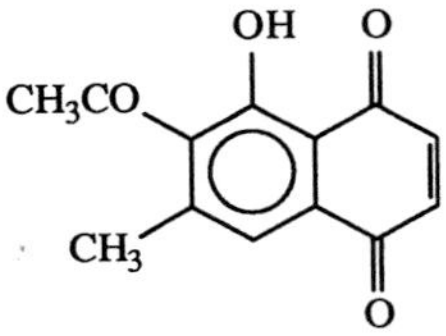

3

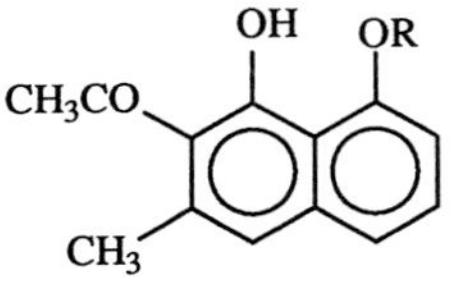

4   R = Rhamnose-Glucose

5   R = H

can give valuable information on the three-dimensional shape of the molecule, bond lengths and angles, and possible intra- and intermolecular interactions. The importance of determining the stereochemical nature of the structure is becoming more self-evident, since enantiomeric (mirror-image) structures may have entirely different biologic properties.[7] It is no longer acceptable to evaluate only the bioactivity of mixtures of such structures.

If structural analysis data are ambiguous and the compound is not amenable to an X-ray diffraction study, then chemical synthesis from precursors of known structure and stereochemistry is usually sufficient to prove or disprove a proposed structure. The chemical literature abounds with examples of how unambiguous synthesis has been vital in finally establishing a structure for an unknown compound.

VI. OVERVIEW

This book comprises chapters written by researchers from a variety of scientific disciplines, each of whom consequently approaches bioactive natural product research from a different perspective. This approach serves to highlight the multidisciplinary nature of this type of research. Despite the fact that most of the chapters in this book deal with the detection, isolation, and structural determination of bioactive compounds from terrestrial plants and fungi, the philosophy and rationale behind the research applies equally to bioactive compounds from aquatic and microbial sources. Indeed, the influence of mankind on aquatic environments parallels that on the terrestrial habitats, with consequently similar concerns for loss of species yet to be investigated or even discovered. Microbial sources of useful bioactive compounds, in conjunction with the capabilities of genetic engineering, are extremely important, especially in large-scale production of such compounds.

The chapters in the first section of this book ("Methods of Detection, Isolation, and Structural Determination") are not intended as definitive texts. Rather, they are reviews with the emphasis being on both the *philosophy and rationale* directing detection and isolation and on the *type and quality* of structural information that can be acquired from the various sources.

The second section in this book ("Specific Case Studies") comprises several chapters that deal with specific bioassays that are used to screen source material, extracts therefrom, or purified compounds, for activity. Where source material is screened, the bioassay is also used to guide the extraction of that material and the subsequent fractionation of the extracts to isolate the active constituents. Some of the chapters in this section concentrate on the bioassay details, while the others emphasize the particular isolation procedure or the structural determination rationale. Several authors in this section review the natural occurrence of compounds with specific activities and describe the rationale guiding the search for such compounds. It must be recognized that this section is not intended to provide a catalog of bioassays, but rather to illustrate the *extreme diversity* of methods and to emphasize the importance of selecting a bioassay *appropriate to the activity* sought.

A. METHODS OF DETECTION, ISOLATION, AND STRUCTURAL DETERMINATION

1. Detection and Isolation

Ghisalberti (Chapter 2) interweaves a great deal of his own extensive experience as a natural products chemist into his chapter describing the detection and isolation of bioactive natural products. This combination therefore provides a critical insight into the diversity and potential complexity of bioactive compounds and into how these compounds are pursued and discovered. Molyneux (Chapter 3) continues this theme by describing the more specific detection and isolation of highly water soluble "cryptic" alkaloids.

2. Structure Determination

Byrne (Chapter 4) describes a strategy for using nuclear magnetic resonance (NMR) spectroscopy in the structural elucidation of compounds. A brief description of NMR fundamentals and techniques is accompanied by extensive referencing to assist those readers who require more detailed information. For simplicity and so as not to detract from the NMR essence of the chapter, Byrne demonstrates this strategy, in the main, through the structural elucidation of one compound. The resulting rational and cumulative acquisition of NMR information gradually, but surely, defines the structure of the compound. This strategy, and variations of it, is observed in other chapters that concentrate on the structural-determination aspects of bioactive natural product research.

The mass spectrometry (MS) chapter by Bloor and Porter (Chapter 5) includes a brief description of MS fundamentals, but concentrates on the type and quality of MS information that can be obtained to assist in structural elucidations. The descriptions of mass analyzers, methods of ionization, methods of sample introduction, and MS/MS techniques are kept brief (but with extensive referencing), the emphasis being on providing the researcher with the alternatives available to assist in structural determination of different classes of compounds.

The determination of three-dimensional structure by single-crystal X-ray crystallography is described by Wong and Gaffield (Chapter 6). They describe the essential principles underpinning the technique, and the equipment that is available to the crystallographer. Sample requirements, the capabilities and limitations of the technique, and interpretation of experimental data are also described. This is reinforced by examples of the X-ray crystallographic structural determination of bioactive compounds from different chemical classes.

There are several methods for determination of the absolute stereochemistry of optically active compounds. One of the more recently developed, exciton chirality circular dichroism, is mentioned in several chapters in the "Specific Case Studies" section of this book. Gaffield describes the principles of this technique and the application of it in determining the absolute configuration of bioactive natural products (Chapter 7).

B. SPECIFIC CASE STUDIES

Several of the authors in this section emphasize the ethnobotanic aspects of their research programs. For example, Kinghorn and Kim (Chapter 9) describe the detection, isolation, and structural determination of potently sweet compounds from plants with recorded ethnobotanic uses. In addition, these authors highlight the importance of completely describing the absolute stereochemistry of optically active compounds, since, in many cases, only one of the enantiomers is indeed sweet. Labadie (Chapter 14) also describes an ethnobotanic approach to his work when searching for "disease-oriented" immunomodulating compounds from plants. In his chapter, Labadie briefly describes the mechanisms of the immune response and inflammation and the mode of action of immunofactors and how these relate to bioassays useful in the search for immunomodulating compounds. This is reinforced by examples of his own work in this area.

Lednicer and Narayanan (Chapter 8) outline the philosophy and the activity of the National Institutes of Health (U.S.) in the mass screening of compounds for anti-AIDS and anticancer activity. Cordell et al. (Chapter 10) expand on the bioassay, separation, and structural determination of cytotoxic natural products. This chapter describes the distribution of cytotoxic agents in plants, the rationale behind assigning a high research priority to particular plants, and the bioassays that can be used to screen for these activities. This is supported by various examples of isolation and subsequent structural elucidation, using modern NMR techniques, of cytotoxic compounds.

The chapter on antivirals by Vanden Berghe et al. (Chapter 17) includes a description of essential elements in viral replication, antiviral chemotherapy, and antiviral test methodology. The authors then review plant-derived antivirals and give examples of the detection, isolation, and structural determination of antiviral flavonoids from traditional African plants. The determination of some structure–activity relationships is illustrated by an investigation of the chemistry of 3-methoxyflavones.

Takahashi (Chapter 12) provides an intriguing insight into the behavior of insects and demonstrates how to use insect behavior observations to detect and isolate behavior modifiers.

The chapters by Asakawa (Chapter 15) and by Chen and Snyder (Chapter 16) lean more toward structural determination. Both chapters are excellent examples of rigorous NMR mapping and the determination of absolute stereochemistries of the various compounds isolated. Snyder and Chen use a molluscicidal bioassay to detect and isolate bioactive compounds from *Allium vineale*, while Asakawa subjected his compounds (from liverworts and fungi), postisolation, to various bioactivity screens.

Marston et al. (Chapter 11) illustrate the bioactivity-guided fractionation of active plant extracts, using the inhibition of replication of human colon carcinoma cells as the *in vitro* bioassay. They describe the use of modern, rapid chromatographic techniques to isolate active compounds, and emphasize the need to follow up observations of *in vitro* activity with suitable *in vivo* tests.

Sam (Chapter 18) reviews and evaluates the brine shrimp toxicity bioassay, making note of the various methods employed and providing a critical analysis of the applicability of the assay. Croft (Chapter 13) and Riopelle and Stevens (Chapter 19) give details and applications of specific, *in vitro* bioassays, i.e., anti-inflammatory and neurotoxicity bioassays, respectively. Reinforcing the value of interdisciplinary collaboration, Panter (Chapter 20) describes the use of ultrasound scanning to correlate observations on fetal activity with congenital abnormalities caused by ingestion of teratogenic plants. The extrapolation of this technique to a useful bioassay is also considered.

An interdisciplinary approach to specific problems is highlighted in the chapter by Dorling et al. (Chapter 21), which describes some investigations of plants that affect livestock. The detection, isolation, and subsequent identification of plant toxins is significantly expedited by initially developing a comprehensive description of the toxicosis from clinical and pathologic viewpoints. This knowledge then forms the basis of an assay that is used to guide the extraction and purification processes of the causative toxins. The chapter also illustrates the unpredictable "spin-off" benefits that can emerge when researchers with differing perspectives make use of the information.

REFERENCES

1. **Colegate, S. M., Dorling, P. R., and Huxtable, C. R.,** A spectroscopic investigation of swainsonine: an α-mannosidase inhibitor isolated from *Swainsona canescens, Aust. J. Chem.,* 32, 2257, 1979.
2. **Colegate, S. M., Dorling, P. R., Huxtable, C. R., Skelton, B. W., and White, A. H.,** Stypandrol, a toxic binaphthalene-tetrol isolated from *Stypandra imbricata, Aust. J. Chem.,* 38, 1233, 1985.
3. **Colegate, S. M., Dorling, P. R., and Huxtable, C. R.,** Stypandrone: a toxic naphthalene-1,4-quinone from *Stypandra imbricata* and *Dianella revoluta, Phytochemistry,* 26, 979, 1987.
4. **Batterham, T., Cooke, R. G., Duewell, H., and Sparrow, L. G.,** Colouring matters of Australian plants. VIII. Naphthalene derivatives from *Dianella* species, *Aust. J. Chem.,* 14, 637, 1961.
5. **Schneider, M. J., Ungemach, F. S., Broquist, H. P., and Harris, T. M.,** (1S,2R,8R,8aR)-1,2,8-Trihydroxyoctahydroindolizine (swainsonine), an α-mannosidase inhibitor from *Rhizoctonia leguminicola, Tetrahedron,* 39, 29, 1983.
6. **Wang, J. H., Humphreys, D. J., Stodulski, J. B. J., Middleton, D. J., Barlow, R. M., and Lee, J. B.,** Structure and distribution of a neurotoxic principle, hemerocallin, *Phytochemistry,* 28, 1825, 1989.
7. **Amato, I.,** Looking glass chemistry, *Science,* 256, 964, 1992.

Chapter 2

DETECTION AND ISOLATION OF BIOACTIVE NATURAL PRODUCTS

Emilio L. Ghisalberti

TABLE OF CONTENTS

0-8493-4372-0/93/$0.00+$.50
© 1993 by CRC Press, Inc.

I. INTRODUCTION

Natural products chemistry has always been concerned with nature and natural phenomena and, as a consequence, biologically active metabolites. Natural products research remains one of the main means of discovering bioactive compounds. Since little is known about the etiology of many human, animal, and plant diseases, it is difficult to design potentially active molecules for their treatment, and therefore leads from natural sources will continue to be sought.[1] Until recently, most natural products chemists have been more concerned with the isolation and structural elucidation of secondary metabolites than with their biologic activity. In the past it was left largely to the biologists to alert chemists to interesting interactions that might be mediated by metabolites, but there was little dialog. In 1984 Farnsworth, commenting retrospectively on the lack of success of a program aimed at discovering useful medicinal compounds from plants, stated, "a major weakness could be an almost total lack of interaction between chemists, botanists, biologists, and physicians."[2] The modern trend toward multidisciplinary research is a recognition of the necessity of collaboration if more complex problems in the life sciences are to be solved. As a consequence, the traditional separation between various disciplines is becoming less distinct. Increasingly, natural products chemists are realizing that the detection, isolation, and structural determination of a metabolite are only the first steps toward answering much broader questions.

Research into bioactive natural products has undergone a renaissance in recent years. This renewed interest parallels the development and availability of a range of new bioassays, on the one hand, and more efficient methods of isolation, separation, and purification, on the other. General screening procedures are being complemented by rational or mechanism-based screening techniques: some designed to provide an insight into old problems, others developed in response to new diseases. The impact of modern countercurrent chromatography and advances in planar chromatography and pressure liquid chromatography is beginning to be noticed. In addition, the advances in spectroscopic methods present the natural products chemist with a wide range of tools with which to tackle the problem of structure elucidation. This, in part at least, has released the chemist from the time-consuming task of chemical structure determination and has fostered a multidisciplinary approach to the development of natural products science.

Newcomers to this field need to have some working knowledge of the chemical and biologic techniques available and those being developed. The aim of this chapter is to overview the range of bioassays that can be used in the detection of bioactive compounds and to outline the various techniques for the isolation of these compounds. Given the breadth of the field, the choice of topics is, necessarily, selective. Emphasis is given to those techniques that have been well established and can conveniently be applied, and those that, although still in a state of development, appear to have potential.

II. DETECTION OF BIOLOGICALLY ACTIVE METABOLITES

The detection of biologically active metabolites is the starting point for a strategic approach in the search for potentially useful metabolites. Clues to the existence of bioactive compounds can arise from disparate sources. Folk medicine targets a selected group of plants that, if nothing else, have the weight of tradition on their side. Traditional folk remedies often have activities different from those attributed to them, but they are nevertheless useful leads. Other clues come from observations by scientists in the field who are in the best position to recognize interactions between organisms. Yet others come from screening a large number of organisms for a particular effect, from searches for a particular type of compound, from old metabolites being tested in new screens, or, not the least important, from the serendipitous discovery of new metabolites displaying pharmacologic properties. It is difficult to gauge which of these is the most efficient: the ethnobotanic route, observations, general screening, target-directed, mechanism-directed screening, or serendipity.[3] In fact, all can be considered useful approaches as long as the scientists involved keep an open mind. Thus, a search for antibiotic agents may yield a compound that has weak or no activity. In this case it is useful to venture outside the immediate area, and, given a knowledge of the structure, tests for some other activity, such as pesticidal, plant growth regulating activity, etc., may have rewarding results. Of course, the element of serendipity is always welcome, but it cannot be considered a factor in planning research.

Examples of folk medicine providing leads to bioactive natural products abound. Suffice it to point to some recent confirmations of the wealth of this source. Artemisinin (qinghaosu) (**1**) is the antimalarial sesquiterpene from a Chinese medicinal herb used in herbal remedies since ancient times.[4] Forskolin (**2**) is the antihypertensive agent from *Coleus forskohlii* Briq. (Labiatae), a plant whose use was described in ancient Hindu Ayurvedic texts.[5,6] The ginkgo tree, mentioned in Chinese medicinal books from 2800 B.C. and used in antiasthmatic and antitussive preparations, produces the ginkgolides (e.g., **3**), unusual diterpenes containing a tertiary butyl group. These compounds were isolated in 1932 and characterized in 1967, and their total synthesis was described in 1988. Their involvement in the clinical efficacy of ginkgo tree extracts was reported in 1985 when they were shown to antagonize platelet aggregation induced by platelet-activating factor (PAF).[7]

The identification of swainsonine (**4**), a potent α-mannosidase inhibitor, came from the realization that the clinical signs and pathologic effects of *Swainsona* intoxication in cattle resembled those of a hereditary condition in man and other animals known as α-mannosidosis.[8,9] This stimulated a search for other alkaloids with glycosidase-inhibiting activity and resulted in the discovery, *inter alia*, of castanospermine (**5**), a tetrahydroxyindolizidine alkaloid that inhibits replication of the human immunodeficiency virus (HIV).[10]

In the search for bioactive natural products, a common tendency is to use screening techniques that monitor biologic activity toward a problem of current interest. Following the publication of the first positive clinical data on penicillin (**6**), between 1942 and 1944, much effort was concentrated on discovering further antibiotics. Up to 1968, the same methods that would identify β-lactams were still being used, and one could have concluded that all natural β-lactams had been discovered. The introduction of new screening methods in the 1970s, which used bacterial strains supersensitive to β-lactams, tested for inhibition of β-lactamases or specifically searched for sulfur-containing metabolites, etc., resulted in the discovery of new antibiotics: the norcardicins (e.g., **7**) and carbapenems (e.g., **8**) in 1976 and the monobactams (e.g., **9**) in the 1980s.[11]

A screen designed to find metabolites from microorganisms active against parasitic infections resulted in the detection and isolation of the avermectins, e.g., avermectin$_{1a}$ (**10**). The producer organism, subsequently named *Streptomyces avermitilis* MA-4680, came from a selection of nearly 2000 cultures of soil samples collected from different environments. Screening was carried out with an *in vivo* mouse assay, using mice infected with larvae of the parasitic helminth, *Nematospiroides dubius*. The broth from different cultures was mixed with the feed given to the mice for 6 days, and the feces and intestinal contents were examined for eggs and worms. The culture broths were qualitatively assessed and graded from active to inactive, or toxic. The assay required 2 weeks,[12,13] but it is interesting to note that although such an *in vivo* test was both lengthy and expensive, it simultaneously tested for antiparasitic activity and toxicity to the host. Significantly, a deliberate choice was made in selecting microorganisms with unusual morphology and nutritional requirements.

The search for inhibitors of proteases, related to various diseases, led to the discovery of more than 100 low-molecular-weight compounds produced in microbial cultures. These compounds have both pharmacologic and immunopharmacologic activities.[14] Of particular pharmacologic value are mevinolin (**11**) and compactin (**12**), two potent inhibitors of HMG-CoA reductase, which affect cholesterol biosynthesis in humans.[15]

Cyclosporin A (**13**), a nonpolar cyclic undecapeptide, was present in an extract from the soilborne fungus, *Tolyplocladium inflatum* (formerly *Trichoderma polysporum*), which showed weak antifungal activity. The unusually low toxicity of this extract prompted its testing in a pharmacologic screening program.[16] On *in vitro* screening and in a mouse model, it was shown to inhibit antibody formation, to suppress delayed-type hypersensitivity, and to interfere with the release of inflammatory mediators. Unusually, the immunosuppression was not linked with general cytotoxic activity. Cyclosporin A was isolated from the metabolite mixture in 1973 and has since become the prototype of a new generation of immunosuppressants.[16,17]

The following illustrates that persistence can lead to unexpected results. A strain of *Penicillium cyclopium* was found to produce compactin (**12**), a hypocholestemic agent, and two known metabolites, cyclopenin (**14**) and cyclopenol (**15**).[18] Of these, cyclopenin significantly inhibited the growth of etiolated wheat seedlings. Both cylopenin and cyclopenol had been previously isolated in 1963 and tested for antibiotic activity, presumably as part of an exercise intended to discover new antibiotics for medicinal use. In any event, 20 years elapsed between the first report of the isolation of (**14**) and (**15**) and the disclosure of their relative activities in plants.

While testing cyclopenin and cyclopenol for toxicity, it was serendipitously discovered that cyclopenin produced symptoms of intoxication in animals. The picture emerging is as follows:[18] cyclopenin is active against some plants (wheat, corn, bean) and possesses tranquilizing properties similar to those of diazepam (**16**) toward vertebrates; cyclopenol is active against wheat and *Phytophthora infestans*, the pathogen responsible for late blight of potato, but does not display diazepam-like activity. This illustrates the range of actions associated with some metabolites and the selective activity of each of these two closely related compounds. Another case of serendipity is illustrated by the discovery of the now important commercial drugs vincristine (**17**) and vinblastine (**18**) in *Catharanthus roseus*. A random screening program (conducted at Eli Lilly and Company) of plants with antineoplastic activity found these anticancer agents in the 40th of 200 plants examined. Ethnomedical information attributed an anorexigenic effect to an infusion from the plant.[7,19]

Many bioactive compounds have been found in the past, but for different reasons their activity has largely not been pursued. On passing through newer screening procedures, a

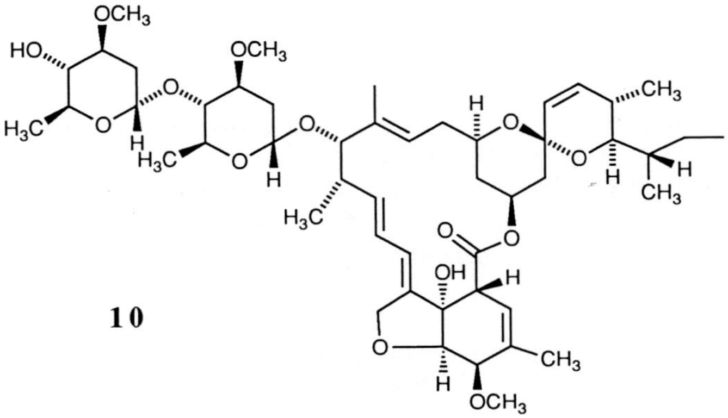

10

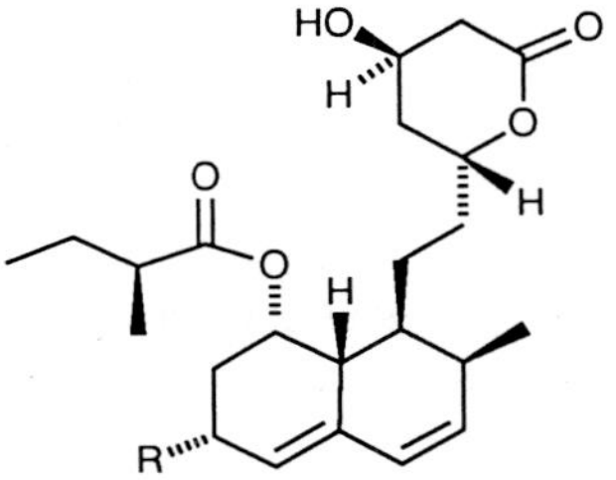

11 R = CH$_3$
12 R = H

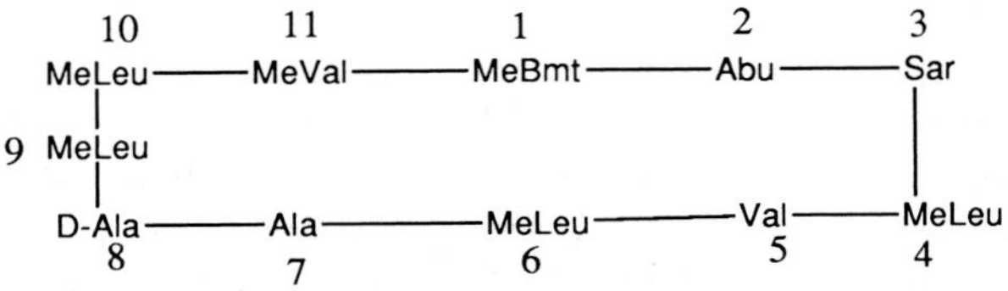

10	11	1	2	3
MeLeu	MeVal	MeBmt	Abu	Sar

9 MeLeu

D-Ala	Ala	MeLeu	Val	MeLeu
8	7	6	5	4

13

Bmt = 4R-4[(E)-2-butenyl]-
4-methyl-L-threonine

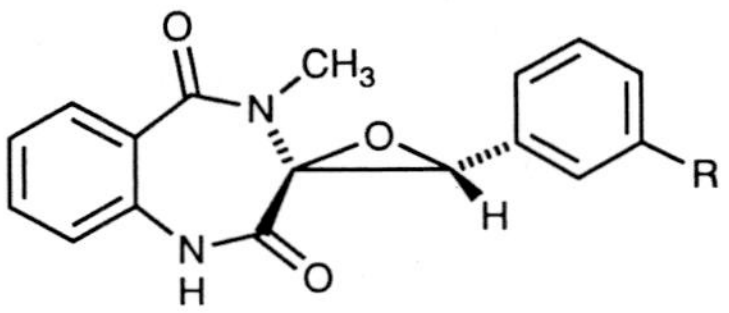

14 R = H
15 R = OH

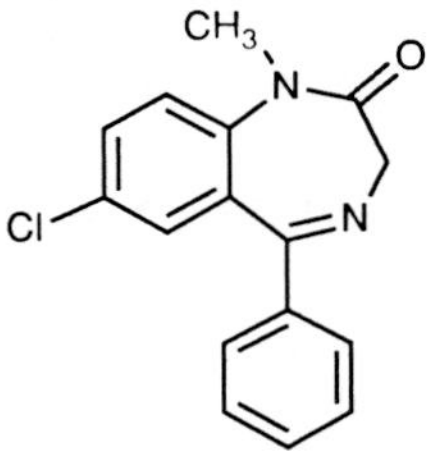

16

number of these compounds with previously unsuspected modes of action are rediscovered (e.g., the ginkgolides), and other activities become apparent.[20] For example, azidothymidine (**19**), when first synthesized and tested for anticancer activity, gave negative results. It was later found to be a reverse transcriptase inhibitor effective against the HIV-1 virus. When azidothymidine was first evaluated, the enzyme was not known.[1] In addition, the same metabolite can sometimes be detected by different screening procedures. Thus, compactin (**12**) was first discovered[21] as an antifungal compound in 1976 and found again, shortly after, in a screen for cholesterol-lowering agents.[22]

III. SCREENING FOR BIOACTIVE METABOLITES

The screening of organisms for bioactive metabolites requires a multidisciplinary approach, involving microbiologists, chemists, pharmacologists, enzymologists, biochemists, mycologists, plant pathologists, etc. This is illustrated by a recent paper summarizing a group's work on bioactive compounds from aquatic and marine sources. Sixteen authors headed the paper, reflecting contributions from chemistry, plant biology, cancer research, molecular neurobiology, and zoology departments.[23]

The primary screen, any *in vitro* or *in vivo* detecting system (bioassay), is a determining factor for a successful outcome. This might involve screens for antibiotic activity; *in vitro* inhibition tests; pharmacologic, agricultural, or veterinary screens that require diverse *in vitro* assays; and/or *in vivo* animal models. The rapid progress in the field of bioactive metabolites is due in large part to the utilization of bioassay-guided fractionation techniques. With this method most substances isolated will be those that have activity in a particular bioassay or set of bioassays, although metabolites occurring in significant quantities should not be overlooked.

Many factors can complicate matters when using bioassays or bioassay-guided fractionation. The most obvious is that the solubility of the extracts or fractions is usually limited and the selection of solvent is critical if a meaningful assay is to be carried out. Most plant extracts can be solubilized by formation of a coprecipitate with polyvinylpyrrolidone.[24] In some cases use of a powerful solvent, such as dimethylsulfoxide, is possible.[25] Other factors are not obvious and can be due to one event or a combination of events, including synergistic effects, chemical changes during extraction and manipulation of extracts, and the cancelling of activity by certain concentrations of substances.

For example, in the isolation of leurosine, a dimeric indole alkaloid related to vinblastine (**18**), the crude alkaloid fraction showed no *in vitro* activity against the P1534 leukemia system, but the pure alkaloid showed pronounced cytotoxicity in the same test.[26] On the other hand, fractionation of the extracts from *Combretum caffrum* was complicated by loss of *in vivo* activity. Modification of the extraction of a new batch of the plant eventually led to the isolation of the antineoplastic combretastatin A-1 (**20**).[27] Berberine (**21**) has a weak, but quite broad, spectrum of activity against a number of pathogenic microorganisms. The artifact (**22**), obtained on contact of ammoniacal solutions of berberine with chloroform, has greater *in vitro* antimicrobial activity.[28] When testing for bioactivity it is also useful to keep in mind the quintessence of the Paracelsian doctrine, "It's all a matter of the dose."[29]

A. PRIMARY SCREENING BIOASSAYS

There are several criteria to be met for a useful front-line screening bioassay (basic or primary screen). It must be rapid, convenient, reliable, inexpensive, sensitive, require little material, and be able to identify a broad spectrum of activities. It should also be able to be applied in-house by chemists, botanists, and others who lack the resources or expertise to carry out more elaborate bioassays.

1. Brine Shrimp Lethality Test

A general bioassay that appears capable of detecting a broad spectrum of bioactivity present in crude extracts is the brine shrimp lethality bioassay (BSLT).[30] The technique is easily mastered, costs little, and utilizes small amounts of test material. The aim of the method is to provide a front-line screen that can be backed up by more specific and more expensive bioassays once the active compounds have been isolated. It appears that BSLT is predictive of cytotoxicity and pesticidal activity. The tiny crustacean *Artemia salina* is used. The eggs are readily available and keep viable for years if refrigerated. When placed in a brine solution, the eggs hatch within 48 h and are attracted toward artificial light sources. Compounds and

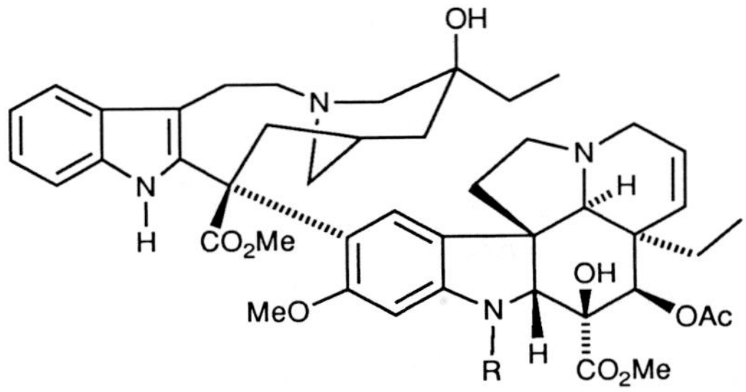

17 R = CHO
18 R = CH₃

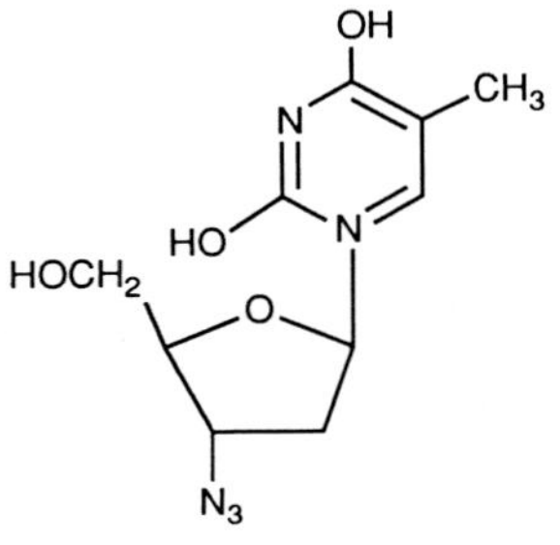

19

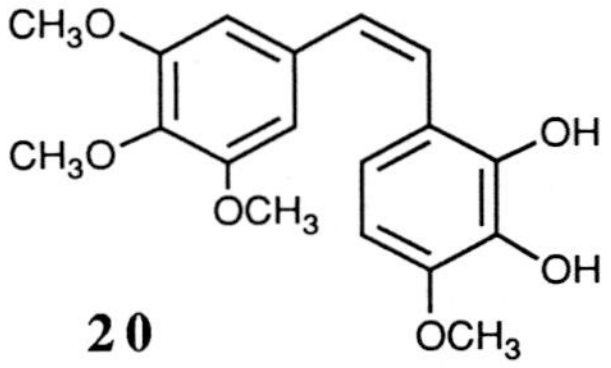

20

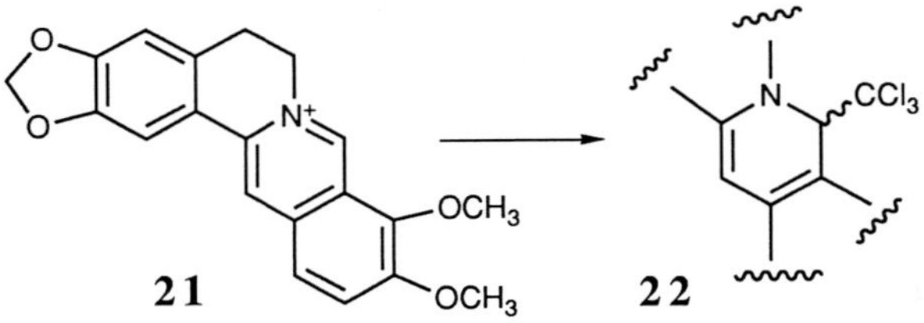

21 **22**

extracts are tested in vials containing 5 ml of brine solution and 10 brine shrimp in 3 replicates, initially at 10, 100, and 1000 ppm. Survivors are counted after 24 h with the aid of a stereoscopic microscope, and LC₅₀ values at the 95% confidence limit are calculated. Since this test was first introduced in 1982, it has been used in the isolation of *in vivo* active antitumor agents and pesticides produced by plants. Protocols for testing and for bioassay-guided fractionation have been described.[30,31] As an example, the fractionation of the bark of the paw paw plant (*Asimina triloba*) is illustrated in Figure 1.[31] A simple extraction of the bark with ethanol gave an extract with significant activity in the BSLT. A separation into lipophilic and water soluble metabolites showed the activity to be associated with the lipophilic fraction. Liquid–liquid partitioning localized the activity in the fraction of medium polarity. The active constituent (asimicin) was eventually isolated by a combination of chromatographic techniques using the bioassay to identify the most potent fractions in each

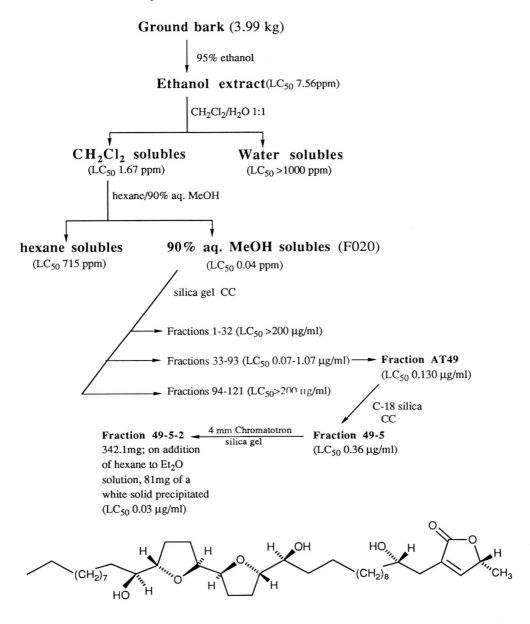

ASIMICIN
(relative configuration)

FIGURE 1. Bioassay-guided (BSLT) fractionation of extract from *Asimina triloba*.

separation. This plant and other species of the Annonaceae family have been shown to contain potently bioactive acetogenins that exhibit a broad range of activities such as cytotoxic, antitumor, antimalarial, antimicrobial, immunosuppressant, antifeedant, and pesticidal effects.[32-34]

This assay is becoming more widely used. It has been employed, *inter alia*, to evaluate plants for potential pharmacologic activity[35,36] and to characterize bioactive polyacetylenes from the sponge *Petrosia ficiformis*,[37] cytotoxic polyacetylenes from a traditional anthelmintic plant,[38] and bioactive neolignans from the leaves of *Magnolia virginiana*.[39] Minor modifications have been made to suit particular cases. The problem of solubility of extracts can be overcome by using dimethyl sulfoxide (DMSO) (1%) as a solvent or, when even this is not sufficient, the solubilizing agents Tween 80 (2% aq.) or polyvinylpyrrolidone, which is not toxic to brine shrimp at concentrations of 400 μg/ml in water.[38] The test can also be carried out in 400 μl spot plates. Sea water (50 μl) containing 10 to 40 larvae are pipetted into each spot, and 300 μl of test solution are then added, the final concentration of the sample being 8.5, 5.0, and 1.25 mg/ml. The plates are then incubated under a light source at 25 to 30°, and after 24 h the number of dead larvae are counted under a microscope. The living larvae are then killed by the addition of 50 μl of phosphate buffer (pH 1), and the total number of larvae are counted.[40]

2. Crown-Gall Tumor Bioassay

Another bench-top test that has proven useful monitors the inhibition of crown-gall tumor on potato disks.[41] This bioassay is fairly accurate in predicting 3PS (P388) *in vivo* murine antileukemic activity, giving some false positives, but few false negatives. The assay is not meant to replace the P388 assay, but it is particularly convenient for rapid screening of extracts or fractions and does not require expensive equipment or highly trained personnel. Crown-gall is a neoplastic disease induced by the Gram-negative bacterium *Agrobacterium tumefaciens*, a bacterium known to infect a number of crop plants, and is due to the transfer of a tumor-inducing plasmid from the bacterium to the plant genome. For the assay, potato tubers are surface sterilized, and a core of the tissue is extracted from each tuber. A 2-cm piece is cut from each end, and the remainder is sectioned into 0.5-cm-thick disks that are then placed onto petri dishes containing 1.5% water agar. A solution (5 μl) of the extract or compound, dissolved in a suitable solvent, is spread over a disk, and the solvent is allowed to evaporate. The disks are then inoculated with 0.1 ml of the bacterial suspension, and the plates are incubated at 27°C. The assay measures the inhibition of tumors induced by the bacterium on potato disks by various plant or other extracts. A prerequisite for this test is that the extract or substance being tested does not show antibacterial activity toward *A. tumefaciens*. This can be established by the standard agar plate-assay disk method.[42,43] The common problem of solubilizing lipophilic extracts also surfaces in this assay, although it appears that it will tolerate water/methanol mixtures up to a ratio of 1:1.[43]

3. Starfish or Sea Urchin Assay

The eggs of the starfish, *Asterina pectinifera*, have cell membranes permeable to a variety of substances. Maturation of oocytes, mediated by 1-methyladenine, occurs after 40 min, and the first polar body is released after 60 min. A second polar body is released after 100 min, while meiosis is occurring. Insemination can be achieved at any point in the maturation process, and cell division occurs after that. Exposure of the fertilized egg to different compounds will lead to different outcomes. If aphidicolin (**23**), a diterpene that selectively inhibits DNA polymerase a, is added (about 10 mg/ml), the cells die after eight or nine divisions, although this compound does not affect the maturation process. Vinblastine (**18**), an inhibitor of microtubule assembly, stops the formation of the first polar body and the first cell division. When cycloheximide (**24**), a protein synthesis inhibitor, is added, the breakdown of the germinal vesicle (meiotic division) and the first mitotic division are affected. Inhibitors of RNA synthesis show no effect on the maturation stage and allow the embryo to develop to the 64- to 128-cell stage. Cytolytic agents affect both types of divisions. This assay appears to be useful for determining which substances should be investigated for *in vivo* antineoplastic activity.[44] The fertilized sea urchin egg assay is a similar technique that can detect DNA and RNA synthesis inhibitors, microtubule assembly inhibitors, and protein synthesis inhibitors.

It is less selective than the starfish egg assay and is relatively insensitive to a series of antineoplastic agents.[45]

4. Bioassays for Antibiotic Activity

The classical agar diffusion methods have been used to isolate and identify antibiotic-producing microorganisms. It was these screening methods that helped to discover the principal antibiotics against Gram-positive bacteria and, to some extent, against Gram-negative pathogens and pathogenic fungi. The history and development of classical and modern approaches to screening for antibiotic metabolites has been discussed by several authors.[11,46,47] In screening for antibiotics the primary screen can be used not only to detect bioactivity, but also for fermentation control aimed at production of larger amounts of material.

There are a number of simple tests for antibiotic activity that can be carried out with simple equipment and a minimum of microbiologic expertise. The following example[48,49] illustrates the assay used for the detection of an antibiotic (**25**), produced by a strain of *Trichoderma koningii*, which suppressed the growth of the take-all pathogen of wheat and other cereals. A plug (7-mm diameter) taken from the growing edge of a colony of the take-all fungus (grown on one fifth potato dextrose agar, PDA) was placed at the center of a petri dish containing one fifth PDA. Samples of the ethyl acetate extract of the broth from a liquid culture of *T. koningii*, or from subsequent chromatographic fractions of the extract, were dissolved in ethanol (2 mg/ml). This solution (10 µl) was dispensed directly on top of the plug, and the solvent was allowed to evaporate in a laminar flow cabinet. The dish was incubated at 20°, and the growth of the pathogen measured (as the area of the colony) after 2, 3, 4, and 5 days, compared to a control treated only with ethanol. Inhibition of growth was obvious after 2 days: the controls showed a colony area of 110 mm^2 and the test samples an area of 32 mm^2.

In agar-dilution streak assays, up to seven different organisms can be screened simultaneously on a petri dish at a fixed concentration of extract. Weak antimicrobial agents present in low concentrations (<1%) can be detected. Representative microorganisms responsible for human infections of significance can be chosen for screening. Typical examples are *Staphylococcus aureus* (representing the Gram positives); *Escherichia coli*, *Salmonella gallinarum*, *Klebsiella pneumoniae*, and *Pseudomonas aeruginosa* (the Gram negatives); *Mycobacterium smegmatis* (the acid fasts); and *Candida albicans* (yeasts and fungi).[50] Qualitative and quantitative assays, in combination with a positive control (e.g., streptomycin sulfate), can be carried out. [51]

Bioautography is a technique that has been used to screen for antibacterial activity. The most common methods are based on the agar diffusion technique. The developed thin-layer chromatography (TLC) plate is placed in contact with an agar plate that has been inoculated with the test organism. The compound(s) diffuse(s) from the chromatographic layer to the agar plate, and after an incubation period, zones of inhibition are made visible with appropriate stains. The procedure requires the use of microbiologic equipment and suffers from the problem of differential diffusion exhibited by various classes of compounds. An improved technique, developed by Hamburger and Cordell,[52] involves application of a suspension of the microorganism in a suitable broth to the developed, dry TLC plate (silica gel). Alternatively, the plate is dipped into a staining tub containing a suspension of the test bacteria or conidia of the fungi in a nutritive broth.[53] The plate is incubated in a humid atmosphere (overnight), and zones of inhibition are detected by spraying with a reagent (the colorless p-iodonitrotetrazolium chloride) specific for dehydrogenase activity. The tetrazolium salt is converted into an intensely pink-colored formazan over a period of 4 h. The presence of antibacterial compounds is indicated by a clear spot against a colored background. The assay, using *Bacillus subtilis* and *E. coli*, proved insensitive to a number of cytotoxic compounds

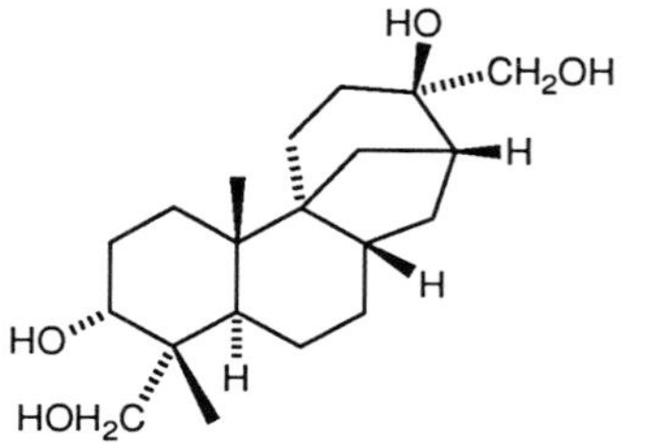

23

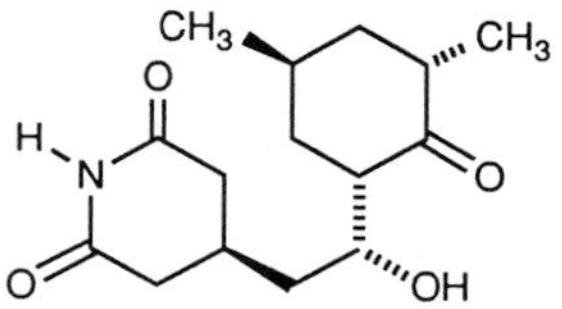

24

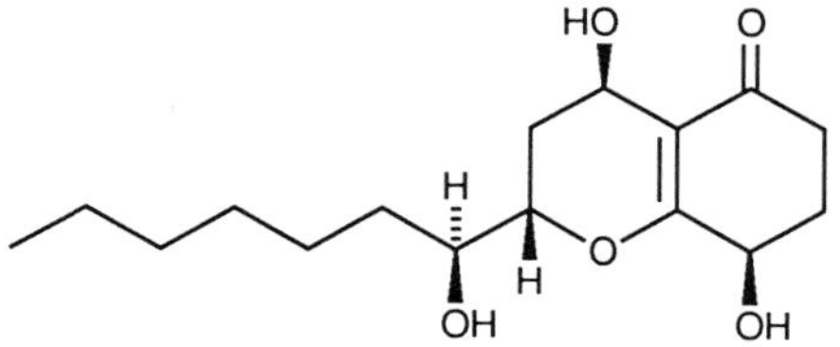

25

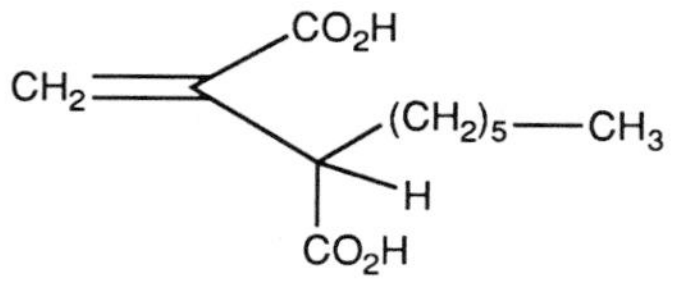

26

belonging to the camptothecin, quassinoid, and lignan series. Dispersal of bacteria can be reduced by containing the assay system in a glove bag.

While these techniques are useful as front-line screens, selective search strategies have been devised and have identified a variety of new antibiotics.[15,54-61]

5. Plant Growth Regulator Activity

Until recently, it was believed that the five major plant growth regulators, ethylene, auxins, cytokinins, gibberellins, and abscisic acid, were almost totally responsible for the differentiation and development of plants and the etiology of plant diseases. Substantial evidence now suggests that there are several natural products that can regulate the growth of plants.[62]

A simple, quick bioassay to test crude extracts or fractions utilizes lettuce seeds. A quantity of a solution of the sample (5 ml, concentration of 0.5 to 5 mg/ml) is pipetted into a petri dish. A teflon ring supporting a small screen is placed in the dish. Twenty lettuce seeds are placed on the screen and allowed to germinate under white light for 72 h. The germination rate at each concentration is recorded as a percentage of controls, and the individual root and coleoptile lengths are determined.[63-65] Interestingly, lettuce seedlings are insensitive to auxin (indoleacetic acid), but respond to gibberellin-type activity. Selective plant-growth-regulating effects of marine algal constituents on lettuce and rice seedling have been noted using a similar assay.[66]

Another assay that has been used frequently to detect plant-growth-regulating substances is the etiolated wheat coleoptile bioassay.[67,68] This assay has also detected compounds that have mycotoxic, immunosuppressant, and antifungal activity.[66]

It is important to be aware of some problems with bioassays that monitor the inhibitory effects of metabolites on seed germination. Lower phytotoxin concentrations can produce equivalent or greater inhibitory effects than can higher concentrations.[69] Also, it is well known that some compounds that are inhibitory at high concentrations can be growth promoting at lower concentrations, and vice versa. For instance, (+)-hexylitaconic acid (**26**) at 20 ppm promoted lettuce root growth by 250%, but at 100, 200, and 500 ppm, shoots and roots were inhibited.[62]

Although blue-green algae are not plants, the regulation of their growth can conveniently be considered in this section. A test for detecting substances inhibiting the growth of water-bloom-forming algae *Microcystis aeruginosa* and *Anabaena flosaquae* has been described.[70] An ethanol solution of the test sample (10 μg) is added to the culture medium of algae (5 ml). After incubation for 24 and 48 h, chlorophyll a of the algae was extracted with methanol, and its concentration was determined fluorophotometrically. This method has been used to direct the isolation of gallotannins (e.g., **27**) and flavonoids from the aquatic plant *Myriophyllum brasiliense*, which shows inhibitory activity toward the two blue-green algae.

6. Herbicidal Activity

This is an area attracting increasing interest.[71-78] There are many assays to test for activity, and they are generally target oriented. A primary herbicide screen to test for microbial production of herbicides utilizes surface-sterilized seeds of garden cress (*Lepidium sativum*), barnyard grass (*Echinochloa crusgalli*), and cucumber (*Cucumus sativum*).[73,79] As an example, seeds of garden cress and barnyard grass are placed on filter paper in petri dishes and moistened with culture broth diluted with distilled water (1:4; v/v). The dishes are incubated in darkness at 25 to 28° for 72 h. Germination and radicle growth of the seeds are then evaluated compared to controls. This assay has been used to screen 1500 soil microbial isolates, mostly actinomycetes, for herbicidal activity. Of the 4 to 12% that showed strong inhibition of seed germination or seedling growth, only 1 to 2% were strongly herbicidal in a secondary screen that involved pot trials with a greater range of test samples and observation over a 14-day period. More than 700 extracts from plant material collected for the NCI anticancer screening program have also been evaluated for plant-growth-regulating properties on tobacco and bean plants. The ethanolic extract of the stem-wood of *Camptotheca acuminata* showed selective inhibition against the tobacco plant. Separation of the active agent led to the isolation of camptothecin (**28**), an alkaloid with known antitumor activity. Apparently this compound targets the topoisomerase I enzymes, thereby strongly inhibiting the relaxation of supercoiled pBR322 DNA.[80]

Inhibition of velvetleaf germination has been used to detect compounds as models for biorational herbicides.[77,81]

Another screen, monitoring for phytotoxicity, involves placing a few microliters of the test solution over a small puncture on a detached leaf of the target plant. The leaf is then placed in a petri dish containing a paper filter saturated with water. The cover of the dish is sealed with parafilm, and the dish is incubated under controlled temperature and light conditions. Activity is normally observed by the appearance of chlorotic, necrotic, or colored spots spreading radially from the point of application, compared to the control. The effect can be seen in a period as short as 24 h.[82,83] The assay can be complicated by a number of factors. First, there is the usual problem of solubilizing an extract. Some weeds can tolerate a 1:1 mixture of methanol and water without any apparent damage to the control. Alternatively, some solubilizing agent, e.g., Tween, can be used. The more interesting problem is one of host selectivity. Some toxins affect all plant species used in the bioassay; others affect only certain species and, at reduced concentrations, even select between different cultivars. This assay was used in the isolation of maculosin (**29**), the first host-specific toxin isolated from a weed pathogen.[84,85]

Other assays involving CO_2 fixation, effects on organelles, tests on whole plants, plant parts, organelles, or tissue cultures, to monitor herbicidal activity have been described.[86]

Multiscreen procedures using simple tests are recommended in monitoring the activity of different strains. For instance, 25 strains of Streptomycetes were simultaneously tested for herbicidal, insecticidal, and acaricidal activity. Some showed activity in two or more bioassays.[87]

B. SPECIALIZED SCREENING ASSAYS

All of the assays described in the previous section have proven useful in bioassay guided (activity-directed) fractionation leading to the isolation and identification of bioactive metabolites. Several other assays are also being used, and improved and newer ones are constantly being developed; however, they are increasingly sophisticated and require the skills and expertise of biologic scientists. Some of these specialized assays are described in this section.

1. Antiviral Activity

The discovery of antiviral agents from plants and other natural sources[58] has assumed a sense of urgency. The NCI has recently implemented anti-AIDS screening and antiviral development programs in parallel with its long standing anticancer drug program.[88] Relatively few antiviral drugs are available, since they must meet a number of criteria. They must inhibit at least one of the propagation steps of the virus (inhibiting the virus completely without affecting the host cells), have a broad range of activity, and not be immunosuppressive. Most *in vitro* antiviral screening relies on plaque reduction or titer reduction assays. In the first, growing mammalian cell lines are infected with plaque-forming units of a virus (Herpes simplex type 1 [HS-1], Vesicular stomatitis [VSV], or Polio type 1), and then the cells are overlaid with a viscous base such as agar or methylcellulose. The test sample is deposited either on the layer or on paper disks, in different amounts. After 24 to 48 h the reduction of plaques formed gives a measure of the antiviral activity of the test material. Titer reduction assays are similarly performed, except that constant amounts of test substance are used with the number of viral particles being varied. In a recent modification the assay size has been reduced to fit smaller wells (5 mm in diameter), and no "thickened layer" is added. It is claimed that this leads to lower cost assays, reduction in the amount of material required, simplified processing, and elimination of interference from the thickening agent. This assay has been used to examine extracts from 61 higher plants.[89] Antiviral activity was observed in approximately 20% of the plants, which is higher than expected. It is suggested that the antiviral activity detected is due to phytoalexins produced by the plant as a protection against plant viruses. These assays simultaneously allow an estimation of cytotoxicity (loss of the cell monolayer in which the plaques are normally formed), and active extracts or compounds then become candidates for testing against tumor cell lines, e.g., P388 or L1210.

A bioautographic assay to screen for antiviral agents from insects and plants involves overlaying a TLC plate on Herpes simplex virus (HSV)-infected CV-1 (monkey kidney) cells and, after incubation, looking for areas of viral growth inhibition.[23] Antiviral testing of 40 plant species by this method suggests that terrestrial plants and marine species contain about the same proportion of antiviral compounds. Extracts from 30 species of insects were also tested, and some were found to be active. This assay also gives an indication of cytotoxicity by showing inhibition of the CV-1 cells.

Increasing attention is being focused on inhibitors of reverse transcriptase of human immunodeficiency virus type 1 (HIV-1). Recently, 156 natural products and 100 plant extracts were examined for this activity, using an assay developed for the detection of the enzyme in virions. Of the natural products, benzophenanthridine alkaloids showed potent activity. In examining plant extracts, polyphenolic substances such as tannins, themselves inhibitors of the enzyme, complicate matters. The method favored for removal of these was polyamide chromatography, carried out on 3 mg of plant extract. Solubilization of the extracts was achieved using DMSO.[25]

2. Cytotoxicity, Antitumor, and Antineoplastic Activity

Some confusion exists in the use of the terms used to describe these activities, but the National Cancer Institute (NCI) has defined these terms precisely. Cytotoxicity refers to *in vitro* toxicity to tumor cells, while antineoplastic and antitumor should refer to *in vivo* activity in experimental systems.

Most of the assays to test for these activities have been developed by the NCI. Extensive programs have been conducted by the NCI in a search for anticancer agents. Up to 1981

35,000 plant species representing 1551 genera and comprising 114,000 extracts had been screened for *in vitro* cytotoxicity and *in vivo* activity against various animal tumor systems.[7,90] Other programs have been conducted in other countries,[91,92] particularly in China and in France, but on a smaller scale. Many bioassays have been used to screen for antitumor substances in extracts and for use in fractionation. Some prove to be too sensitive (e.g., Walker carcinosarcoma 256), resulting in too many false positives, and others not sensitive enough (mouse L-1210 leukemia). More recently, the P-388 murine leukemia (for *in vitro* and *in vivo* testing) has been favored in terms of sensitivity and predictivity, while advanced *in vivo* testing included additional mouse tumors (Lewis lung carcinoma, colon 38, and CD8f1 mammary). *In vitro* prescreens for cytotoxicity, such as those utilizing 9KB human nasopharynx carcinoma or P-388 murine leukemia (9PS), have been employed to facilitate fractionation, but do not distinguish cytotoxicity from antitumor activity. Although many antitumor compounds have been isolated through these screening procedures, few of these were found to be clinically effective against slow-growing solid tumors. This is a consequence of using rapidly dividing tumors as the primary screens. On the other hand, screening with *in vivo* assays is impractical due to the time required and the high cost. The trend is to deemphasize the use of typically insensitive *in vivo* animal models and concentrate on very sensitive disease-oriented human tumor cell lines as primary screens.[58] New *in vitro* screening strategies adopted by NCI involve 80 to 100 human cell lines of major tumors. Compounds showing differential cytotoxicity for particular tumors will be further studied by *in vivo* testing with the same tumor lines (approximately 10,000 samples each year).[93,94]

The discovery of antitumor agents depends on the screening methodology used in assaying crude extracts. Also, it is important to ensure that the sample taken is representative of the crude extract. Differential solubility can be a problem, particularly when the extract is not water soluble, in which case MeOH or DMSO may need to be used. The presence of large amounts of substances that can interfere with the assay, either dominating spectrometric measurement or masking the biologic effects of smaller amounts of an active principle, must be considered and, if possible, overcome.[93] The considerable work carried out on the search for antitumor compounds from marine organisms has been reviewed.[95,96] Another recent review on bioactive compounds from marine organisms and cultivated blue-green algae places special emphasis on the diverse separation techniques applied to the isolation of novel compounds from these sources.[97]

A bioautographic assay using murine P-388 leukemia cells has been developed for the detection of antitumor compounds from natural sources. To test the sensitivity of the assay, 18 cancer chemotherapeutic drugs and natural products were assayed. All were detected by toxicity at 0.01 µg or 1.0 µg with P-388 cells. Using human HT-29 colon carcinoma cells, only 11 of the 18 were detected at 10 µg.[98]

Another strategy is to employ mechanism-based assays to detect chemopreventive agents. The search for compounds that inhibit carcinogenesis by (1) preventing the formation of carcinogens or (2) preventing binding of the carcinogen to target (blocking), or that (3) prevent development of the tumor, has led to the deployment of a number of bioassays. One of the better-characterized assays to screen for blocking agents examines the effect of compounds on the detoxifying enzyme glutathione transferase. Another is the mutation assay in *Salmonella typhimurium* (Ames test).[99] A third widely used assay monitors the metabolism of the carcinogen itself or the interaction of the carcinogen with DNA, using mammalian cell tissue culture or animals. A more recent assay is based on the metabolism of benzo[*a*]pyrene in tissue culture of hamster embryo cells to the ultimate carcinogenic diol-epoxide. Cells are plated in culture vessels and incubated for 18 to 24 h with tritiated B[*a*]P and the test substance, after which the culture medium is removed, and aliquots are extracted with a mixture of $CHCl_3$, MeOH, and water. The $CHCl_3$ layer contains starting material and the primary oxidation compound, while the more oxidized products and their conjugates are soluble in the aqueous layer. The radioactivity in each layer is measured, and the ratio is a function of the effect of the test substance on the metabolism of B[*a*]P. Further analysis by high-pressure liquid chromatography (HPLC) or ion-pair chromatography can be conducted to determine the specific metabolites produced. If the initial plant extract shows an effect on the metabolism, it is then tested for its effect on the binding of B[*a*]P to DNA, by incubating the test material with tritiated B[*a*]P and embryo cells for 24 h. The cells are harvested, the DNA isolated, and the amount of radioactivity associated with the DNA determined. An effect is considered positive if the test substance significantly inhibits binding between B[*a*]P and DNA. Application of these assays in guiding the fractionation of selected plant extracts has led to the identification of the isoflavones biochanin A and genistein and the flavones apigenin and chrysin, as chemopreventive agents. Comparative studies of these and related flavones and isoflavones have established that the presence of free hydroxyl groups at the 5 and 7 positions is required for high inhibitory activity.[93]

A rapid antimutagenic assay to screen a large number of plant extracts has been developed by Wall et al.[100] This assay evaluates the inhibition of the mutagenic activity of 2-aminoanthracene, in the presence of the Ames S-9 metabolic activation preparation,[99] by crude extracts obtained from plant materials with organic solvents. Aqueous extracts may, in some cases, contain sufficient histidine to counteract the antimutagenic effect. In these cases it may be necessary to test the extract in the absence of mutagen. This assay has been employed to evaluate more than 2000 extracts representing 39 plant families, and inhibitory activity was found in 80 samples.[100]

3. Immunostimulating Activity

This refers to compounds that stimulate the nonspecific immune system, i.e., the efficiency of granulocytes, macrophages, and complement and natural killer cells. The efficacy of compounds showing this activity fades quickly, since they do not affect immunologic memory cells. Nevertheless, they can be regarded as important adjuvants for conventional chemotherapy of infectious and tumor diseases or in autoimmune diseases when the host's immune system is suppressed. A number of *in vitro* and *in vivo* tests, such as *in vitro* phagocytosis, T-lymphocytes transformation test, leukocytes-migration test, immune-induced cytotoxicity test, and the interferon induction test, have been employed to screen plant extracts.[101,102]

4. Antimalarial Activity

Useful *in vitro* tests have become available only recently. As far back as 1947, 600 plants from 126 families were extracted, and the extracts were tested for *in vitro* activity against avian malarias. Species from over 30 genera were found to have some activity, but the extrapolation to human malarias could not be certain.[103] The continuous *in vitro* culture of the human malaria parasite *Plasmodium falciparum* represented a significant advance that led to a technique for quantitative assessment of the activity. A modification of this technique relies on measuring the ability to inhibit the incorporation of ^{3}H-hypoxanthine into plasmodia. This technique was applied to evaluate the activity of crude extracts (ethanol) of *Artemisia annua* and *A. vulgaris*. This *in vitro* test discriminated between the two extracts, with IC$_{50}$ values of 3.9 µl/ml and 250 µl/ml, respectively. *A. annua* was included in the Chinese traditional treatment of malaria, and the active principle has been shown to be artemisinin (**1**) which was absent in *A. vulgaris*. This compound has a higher chemotherapeutic index than chloroquine and is active toward strains of human malaria that are chloroquine resistant. Protocols for determination of *in vitro* anti-*Plasmodium falciparum* activity and *in vivo* anti-*Plasmodium berghei* activity in mice have been described.[103,104] It is worthwhile noticing that *in vitro* antimalarial activity does not parallel cytotoxicity against KB cells (human epidermoid carcinoma of the mouth). Thus, bruceantin (**30**) is three times as active as brusatol (**31**) in the antimalarial test, but ten times more toxic to KB cells.

27 R = GALLOYL   28

29

30 R = CH(CH$_3$)$_2$
31 R = CH$_3$

32   33

5. Amoebicidal Activity

In vitro screening methods have been developed to evaluate plant extracts and natural products with activity against *Entamoeba histolytica*, the causative agent of amoebic dysentery. Emetine (**32**) and 2,3-dehydroemetine were found to have an IC$_{50}$ value of 0.07 and 0.16 µg/ ml, respectively, whereas metronidazole, a well-tolerated drug for this infection, but known to cause tumors in experimental animals, had a value of 0.22 µg/ml. However, emetine has high cytotoxicity against guinea pig ear keratinocytes, and the cytotoxicity/amoebicidal ratio is less than 1, whereas quassin (**33**) has a more favorable ratio of 132. A number of extracts from simaroubaceous plants and their alkaloid and quassinoid components have been tested.[103,104]

6. Insecticidal Activity

In a major screening program, Alkofahi et al.[31] have submitted many species of higher plants for testing against seven indicator organisms that are agronomic pests: mosquito larvae (*Aedes aegyptii*), blowfly larvae, corn rootworm, two-spotted spider mite (*Tetranychus urticae*), southern army worm (*Spodoptera eridania*), and melon aphid (*Aphis gossypii*). As an example, an aqueous solution containing 5000 ppm of crude or fractionated extract was applied to the leaves of a squash plant, and the leaves were allowed to dry. The leaves were then removed and placed in petri dishes with army worm larvae. After 3 days the percent mortality was assessed. The tests with other organisms generally took between 1 and 3 days. The active fractions from the paw-paw fractionation scheme (F020, AT 49 and asimicin; Figure 1) were compared to the commercial pesticides pyrethrum and rotenone and proved to be of comparable activity. Kinghorn[6] has tested 20 phorbol esters, the skin irritant and cocarcinogenic diterpenes from *Euphorbia* species, for growth-inhibitory and insecticidal effects against newly hatched larvae of *Pectinophora gossypiella* Saunders (pink bollworm) and other insect pests. A close correlation was noted between the activity in the pink bollworm assay and cytotoxicity data (P-388 lymphocytic leukemia).

IV. ISOLATION AND SEPARATION

In the following sections the various techniques available for the isolation, separation, and purification of biologically active metabolites are considered. At this stage of the investigation, the chemist is now working in close collaboration with the biologic scientists. The task of the chemist is to isolate and identify the substance(s) responsible for the bioactivity observed, although often no information is available regarding the nature and class of compounds to which the metabolite belongs. There is also a question of balance involved. The identification of the biologically active metabolite is not more important than ensuring that the compound is pure and properly characterized. The natural products chemist has to ensure that adequate amounts of metabolites are isolated in as pure a form as possible, to allow proper chemical and biologic characterization. In this context, a recent example illustrates the confusion that can be introduced. Two papers, which appeared almost simultaneously, described the isolation of *ent*-kauran-3-oxo-16α,17-diol from terrestrial plants. In one, the compound is said to have mp 173 to 174°, $[\alpha]_D$ –39.2° (CHCl$_3$), and in the other, the melting point is not given and an $[\alpha]_D$ –73.1° (CHCl$_3$) is quoted. The latter sample is reported to be cytotoxic. Either the two compounds are different, in which case there has been an overreliance on interpretation of spectroscopic data, or if they are the same, the samples differ in homogeneity.

Although no one would argue that isolation and purification are essential in any work on bioactive metabolites, this is often approached in a haphazard way. The temptation to bypass careful isolation and purification processes is compounded by the fact that what might take weeks to achieve in the laboratory has to be described in one sentence in the experimental section of the paper. While structural determination has in many ways become quite routine, isolation and purification have not. This is not apparent from most papers dealing with the isolation, purification, and structural determination of natural products. Contributing to this is the attitude that purification is a trial-and-error process that, in any event, is not transposable from one problem to the next. In many papers in which long and involved separation sequences are described, the rationale for the methods chosen is often not disclosed, even *ex post facto*. In natural products chemistry, the isolation and purification of a compound is an obligatory first step that requires expertise and innovation and, as more challenging topics in this field are undertaken, can prove crucial in determining the failure or success of an investigation. It is useful to consider the following incident described by Nakanishi.[105] The structure of azadirachtin, a potent insect antifeedant, was proposed in 1975 by Nakanishi and co-workers, based largely on results obtained by applying a new ^{13}C NMR technique. The evidence was not convincing. For many years the sample of azadirachtin became a test sample for newly described NMR pulse sequences in attempts to obtain more rigorous evidence for the structure, but progress was hampered by a lack of pure material. This continued until a postdoctoral research fellow devised an isolation scheme[106] that provided 10 g of the compound and, with this, the possibility of studying its chemistry. By the time that Nakanishi's group had arrived at a revised structure (**34**) in 1986, two other groups had also arrived at the same conclusion.

The appearance of a journal, *Phytochemical Analysis*, dedicated to the publication of articles "on the application of analytical methodology in the plant sciences" will hopefully prove to be a major information source on methods of extraction, isolation, and purification of secondary metabolites.[107] There is an increasing emphasis on this topic, and various reviews and books dedicated to this field are available. Some of the earlier compilations, such as *Antibiotics: Isolation, Separation, and Purification*[108] and *Advances in Natural Products Chemistry: Extraction and Isolation of Biologically Active Compounds*[109] are still extremely useful, as is the more recent contribution from the Journal of Chromatography Library: Volume 43, *Natural Products Isolation. Separation Methods for Antimicrobials, Antivirals, and Enzyme Inhibitors.*[110] The reader is referred to these for methods of isolation for specific classes of compounds.

V. EXTRACTION

In choosing a solvent for extraction, its ability to extract components of a solute has to be considered. For instance, ionic solutes can be extracted from aqueous solutions with nonpolar solvents if neutral complexes can be generated in the aqueous phase before extraction. The more efficient the extraction step, the greater is the range of compounds present in an extract. The need to use pure solvents for extractions is obvious. Less obvious is the need to carry out the extraction under mild conditions, utilizing, whenever possible, solvents of low reactivity. The possibility of generating artifacts should never be discounted (see Section VII).

In the literature there are many basic extraction procedures that can be tried and refined if necessary, although a trial-and-error approach is often required. Solvent partition schemes have been outlined for screening plants for antitumor agents[111] or screening organisms for antitumor metabolites.[112,113] The isolation of bioactive compounds is almost "always fraught with difficulties and every step requires judgment, improvisation, and discovery."[113]

A. DRY BIOLOGIC MATERIAL

Air- or freeze-dried samples are normally extracted with a variety of solvents, and sometimes sequentially from low to high polarity, if a crude fractionation of metabolites is sought. Generally, however, a polar solvent such as ethyl acetate or methanol is used, and the separation of compounds is left until a later, chromatographic step. It is useful to consider the possibility that the choice of solvent used can determine, to some extent, whether exo- or endocellular metabolites, or a mixture of both, will be extracted. With dried material, ethyl acetate or low-polarity solvents will only rinse or leach the sample, whereas alcoholic solvents presumably rupture cell membranes and extract a greater amount of endocellular materials. For instance, washing a plant sample with ether afforded a sesquiterpene alcohol and di- and trihydroxy flavones, whereas extraction with methanol provided a different sesquiterpene alcohol and a diterpene diol, both of which, after extraction, are also soluble in ether.[114] In the extraction of dried algal samples, more lipophilic material can be extracted with dichloromethane after soaking the sample in methanol than from a dichloromethane or methanol extraction alone. A good system with high extraction potential is a 1:1 mixture of these two solvents.

Of the extraction methods used to provide fractions suitable for bioassays, the following[31] is representative. Dried, powdered material is extracted, by cold percolation or Soxhlet extractor, with 95% ethanol (Fraction 1). The ethanol extract is partitioned between chloroform and water (1:1), which normally generates three fractions: the water solubles (Fraction 2), chloroform solubles (Fraction 3), and a fraction comprising any insoluble material at the solvent interface (Fraction 4). The chloroform solubles are then partitioned between hexane and methanol, with enough water (normally 10%) added so as to generate two phases. This gives rise to two other fractions: the 90% methanol soluble (Fraction 5) and the hexane soluble (Fraction 6). If seeds or leaves coated with waxes or fatty material are used, defatting with light petroleum prior to solvent extraction is recommended. The individual fractions are then submitted for bioassay, and a qualitative and sometimes quantitative measure of their activity can be obtained. This protocol has been widely used, particularly in the detection and isolation of the biologically active acetogenins from the Annonaceae[32] and in the assaying for cytotoxicity and antitumor activity of bryophytes.[115,116]

B. FRESH MATERIAL

The water content in samples of fresh material makes the dichloromethane-methanol solvent mixture ideal for extraction purposes. Once the extract has been partitioned and the methanol removed from the aqueous fraction, the latter can be back-extracted with ethyl acetate and then with butanol, to separate the lipophilic material from the water solubles.

In the case of fermentation metabolites, it is always useful to extract the mycelial mat in this way, since often some metabolites, although they may be exuded partially into the broth, adhere to the mycelium. This is the case for the avermectins (e.g., **10**), which are quite lipophilic and are always found to adhere to the mycelial cake after broth filtration.[13] In fermentations producing the polyether antibiotics, e.g., monensin (**35**), most of the metabolites are contained in the filter cake if the fermentation broth is allowed to stand for an hour at room temperature before filtration.[117] With the Actinoplanes strain that produces the ramoplanin complex of glycolipodepsipeptide antibiotics, 85% of the activity is found in the mycelial mat. This can be extracted with water-methanol or water-acetone.[118] In any event, the mycelium may very well contain different endocellular metabolites.

For small quantities of fresh biologic material, freezing with liquid nitrogen can be a convenient step. The frozen material can be ground and extracted in the normal way. With this technique endocellular material is also extracted.

C. LIQUID CULTURE BROTH AND OTHER BIOLOGIC FLUIDS

These can be extracted sequentially with solvents of increasing polarity, from petroleum ether to butanol. The pH of the medium should be checked before extraction. In some cases extraction of the medium at varying pH ensures maximum recovery of extract and may lead to partial fractionation.

In a number of cases the organic material can be concentrated by passage of the broth through Amberlite XAD-2 or XAD-4 (polystyrene-divinylbenzene) resin columns. These are neutral resins with no ion-exchange groups, the adsorption process occurring through hydrophobic forces. To maximize adsorption, addition of 1 to 5% sodium chloride to the solution is recommended. For some compounds, e.g., the carbapenems (**8**), elution with water is sufficient; more lipophilic substances are desorbed with organic solvents (aq. acetone or methanol). Ion-pair extraction can be very useful in extracting sulfated carbapenems at neutral pH.[119] The antibiotic (**36**) and related sulfates are produced in small amounts (μg/ml) in liquid cultures of *Streptomyces olivaceus*. Direct extraction of these sulfates with organic solvents is unsatisfactory because of the acid labile carbapenem nucleus. Neutral extraction can be achieved with dichloromethane containing 10 mM of a quaternary ammonium salt such as

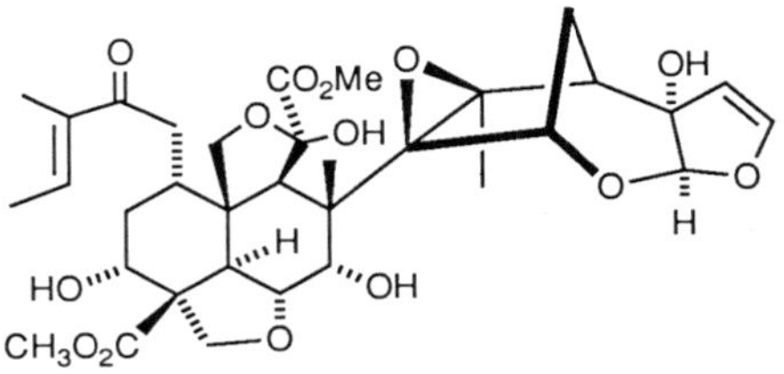

34

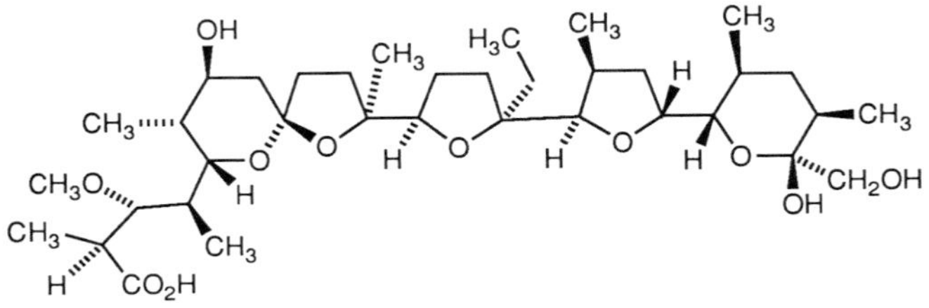

35

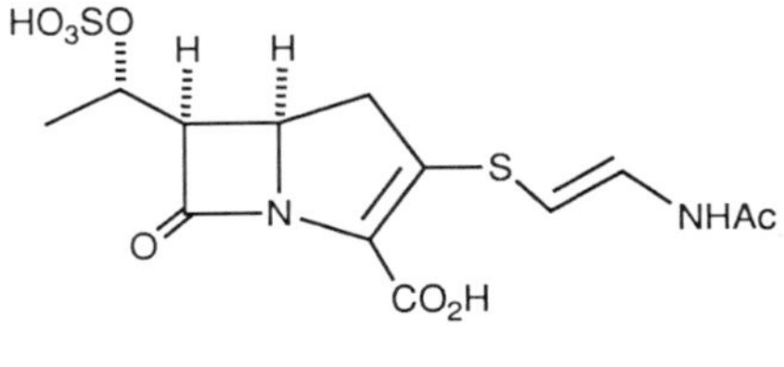

36

BDHA (benzyldimethyl-n-hexadecylammonium chloride). The organic layer is then back-extracted with 3% aqueous sodium iodide to remove the ammonium species. Another advantage of this method is that it leaves nonsulfated analogs in the aqueous solution. The zwitterionic thienamycin (**8**) can similarly be extracted; however, a more polar solvent, methylisobutyl ketone, and lower pH (about 3), necessary for ion-pair formation, are required. Contact times for the extraction and back-extraction need to be kept to a minimum to avoid decomposition of the metabolite.[119]

Centrifugal partition chromatography is a technique that potentially could be useful for the extraction of compounds from solutions (see Section IV.A.3). The solution is employed as the mobile phase, and a solvent or mixture of solvents with a high affinity for the compound of interest is chosen as the stationary phase.[120]

D. REMOVAL OF FATTY MATERIAL

Many extracts contain significant amounts of fatty material that can seriously interfere with subsequent chromatographic separation. A number of methods for partitioning such extracts are used. The extract can be dissolved in a mixture (1:1) of methanol-petroleum ether, and enough water (about 10%) is added so as to give two phases that are then separated (the origin and development of the technique has been described).[121] The aqueous methanol layer can be concentrated and the organic material recovered by back-extraction with ethyl acetate. Toluene-methanol, heptane-acetonitrile, or heptane-ethylene glycol can also be used. In the last case the ethylene glycol layer is diluted with ether, and the ethylene glycol removed by washing with water.[13]

E. SUPERCRITICAL FLUID EXTRACTION (SCF)

Although a technique still very much in a developmental phase, SCF extraction offers some advantages over conventional extraction methods, particularly when the class of compound involved is known. SCF methods are appropriate for the isolation of thermally or chemically unstable compounds. Supercritical fluids are not supersolvents, the solvent strength approaching that of liquid solvents only as their density increases.[122] Some of the advantages of SCF extraction are that it is fast, the solvent strength can be controlled by varying the pressure, many SCFs are gases at room temperature, and most are relatively inert, nontoxic, and inexpensive. Typical gases used are carbon dioxide, nitrous oxide, and ammonia, and these can be modified by the addition of small amounts of a cosolvent, e.g., methanol or water. This technique has been used in the isolation of chemotherapeutic pyrrolizidine alkaloids from plants[123] and the anti-inflammatory agents from *Calendula officinalis*.[124] The main disadvantages of the technique are that it requires pressure equipment and, at this time, instrumental costs are high. Furthermore, not enough is known about the solubility of different classes of compounds in the various solvents. Nevertheless, it is expected that this method will become more widely applicable for both analytical and preparative extractions.[125]

VI. CHROMATOGRAPHY

The separation of one or more substances from a crude extract, or fractions of an extract, can be a long and expensive process, and obtaining a pure compound often requires several separation steps involving different techniques (multistep and multidimensional[126] chromatography). This is particularly so when dealing with bioactive metabolites where the target compound(s) may be present only in trace quantities in a matrix of dozens of other constituents. There are examples in which the isolation of the bioactive metabolite is a relatively straight-forward process, especially if the fractionation is guided by a bioassay. The antibiotic agent (**25**) produced by a strain of *Trichoderma koningii* that suppressed the take-all fungus of wheat was isolated from a liquid culture in the following way.[49] Extraction of the culture broth with ethyl acetate gave a yellow oil that, from TLC, appeared to contain one major and several minor components. The bioactivity was associated with this fraction that was then subjected to flash chromatography on silicic acid. Gradient elution (light petroleum to ethyl acetate) gave six fractions, of which only the most polar exhibited bioactivity. This fraction, while giving rise to interpretable spectroscopic data, was shown by TLC to be contaminated with another compound of similar Rf. Fortunately, this impurity could be removed by chromatography on alumina (activity I) without loss of bioactivity.

Most separation procedures, however, require diverse chromatographic methods, and some of the more widely used are described in this section. In multistep chromatography the results from one chromatographic step have to be evaluated before a decision can be made on the next step. This is where expertise, intuition, trial and error, and serendipity can all play a part. The availability of high-field NMR instruments, with their greater dispersion power, can usefully be harnessed to guide the selection of separation methods. Even with mixtures of several compounds, it is often possible to infer the class to which the major components belong. An early indication of the nature of, and functionalities associated with, the compound(s) of interest can greatly facilitate the separation process.

Methods for removing large amounts of fatty material, which may hinder efficient chromatography, have been outlined in Section V.A. Filtration is often necessary to remove particulate matter or insoluble materials. This can take the form of filtration through a sintered glass funnel or a short column packed with a support of low activation, which can remove strongly polar compounds, or through millipore filters.

A. LIQUID–LIQUID CHROMATOGRAPHY

1. Countercurrent Chromatography (CCC)

Although countercurrent distribution (CCD) techniques, developed in the 1950s by Craig, provided high-resolution capabilities for separation, the method never achieved widespread use. Some reasons for this might be that the instrumentation was cumbersome and fragile, the process required a lot of solvent, and the multitube glass apparatus demanded significant time for cleaning and setting up. Nevertheless, the technique was useful in achieving separations in the actinomycin,[127] peptide antibiotic,[128] lincomycin,[129] erythromycin,[130] and streptomycin[131] classes of compounds. For instance, Gross[128] described the separation of gramicidin A, B, and C, a series of pentadecapeptides that differ at position 11, containing tryptophan, phenylalanine, and tyrosine, respectively. Erythromycin A and B, the former containing an extra tertiary hydroxyl, can also be separated by CCD.[130]

In the meantime, developments in solid-phase support chromatography, especially HPLC, attracted the natural products chemists, and for a time, CCD was overlooked.

Modern CCC owes its development to the work of Mandava and Ito[132] and its acceptance to the availability of commercial instruments (up to ten types presently).[133] The literature on this technique is voluminous, and its application to the study of bioactive natural products is increasing.[134]

2. Droplet Countercurrent Chromatography (DCCC)

Modern CCC bears some similarity to HPLC in that the solute is partitioned between different phases, but differs in the fact that the solid support is replaced by a gravitational field. CCC is a form of liquid–liquid partition chromatography involving two immiscible liquids or solutions. An apparatus keeps one phase stationary while the other liquid is pumped through it (true countercurrent techniques require both phases to be mobile). In its simplest form, DCCC, the mobile phase migrates as droplets that provide a constantly changing interface, across which the solute is continuously partitioned. Separation will depend on the differences in partition coefficients of the components of the mixture. Typically, the apparatus consists of 200 to 600 vertical columns (20- to 60-cm silanized glass tubing) of internal diameter of 2.0, 2.7, or 3.4 mm (up to 10 mm), a sample chamber, a pump, a solvent reservoir, and a detector. The choice of solvents is critical. Ternary solvent systems are used to ensure formation of suitably sized droplets of the mobile phase, and the selection of solvents can be determined using a TLC method. The major disadvantage of this technique is that it is slow. To ensure proper formation of droplets, slow flow rates are required (1 to 5 ml/h), and a chromatographic run might take several days.

The following example[135] illustrates the mildness of the technique and its efficiency in the preparative separation of crude extracts. Pyrrolizidine alkaloids tend to adsorb irreversibly on solid phases, alumina, silica, or reversed-phase supports, leading to losses of material. DCCC, therefore, was selected for the separation of this type of alkaloid from *Senecio anonymus*. For the selection of a solvent system, initial measurements of the partition coefficients of some standards, representing low-to-high-polarity alkaloids, were conducted.

These indicated that a mixture of $CHCl_3:C_6H_6:MeOH:H_2O$ (5:5:7:2) would be suitable for the moderately polar alkaloids present in the extract. The lower layer of this biphasic solvent system was selected as the stationary phase. The chromatograph used was equipped with 300 tubes of 2.7 mm inner diameter (i.d.) and 200 tubes of 3.0 mm i.d. The alkaloid fraction (3 g in 15 ml of a 1:1 mixture of the upper and lower phases) was aspirated into the instrument, followed by the mobile phase consisting of the upper aqueous layer. Flow rates up to 24 ml/h were used, and 20-ml fractions were collected and monitored by TLC. Fractions 23–26 contained **45**; fraction 30 yielded **43**; 37–43, **46**; 61–64, **39**; 68–69, **41**; 78–84, **42**. After fraction 86 the stationary phase was pumped out, giving fractions 114–115 (containing **37**) and 120 (**38**). Fractions between those cited contained mixtures of compounds. Of the ten alkaloids, eight were obtained in a pure form, and only the two minor alkaloids (**40**, **44**), exocyclic double bond isomers of **39** and **43**, were not fully separated. The more polar alkaloids (**43**, **45**, **46**) emerged first, and the least polar (**37** and **38**) were retained in the stationary phase. Although **37** and **38** formed a significant mixed fraction, 116–119 (121.5 mg), small amounts of **37** (20.5 mg) and **38** (8 mg) could be obtained in a pure form. This is remarkable considering that they differ only in double-bond geometry.

3. Other Methods

Rotation locular countercurrent chromatography (RLCC) is similar in principle to DCCC, but does not depend on formation of droplets, and in this sense, it is the most obvious extension of classical countercurrent distribution. In this case the columns (16, each separated into 37 cells by teflon disks, each with a hole in the center) are mounted cylindrically about a rotational axis (inclined 20 to 30°) and are connected in series with 1-mm teflon tubing.[136-140]

A newer development of DCCC is incorporated into the centrifugal partition chromatograph, CPC. In this instrument, columns have been replaced by channels, 400 per cartridge, drilled into a series of 12 polytetrafluoroethylene blocks arranged in a circle around the rotor of a centrifuge. The column is linked to the pump and detector through two rotary seals. The system can operate either in the ascending or descending mode. All the systems mentioned above operate under a constant gravitational field: earth's or that of a centrifugal field. Other systems that employ variable gravitational fields have been developed, and two are commercially available. These are the coil planet centrifuge countercurrent (CPC) and the high-speed countercurrent chromatographs (HSCCC).

In these cases (hydrodynamic equilibrium systems; HDES), the "column" is subjected to a gravitational field (centrifugal) augmented and varied by additional rotation (800 rpm) about its own axis, which need not be parallel. The mobile phase is pumped through the stationary phase, and once the mobile phase emerges from the coil terminus, the solute can be introduced at the head of the coil. The centrifugal force accelerates the mobile phase through the stationary phase.

Yet another modification, utilizing true countercurrent techniques, allows for introduction of the sample in the middle portion of the column. The extreme polar and nonpolar components are quickly eluted from opposite ends, followed by compounds with decreasing polarity in one phase and increasing polarity in the other phase. This seems particularly adaptable for the separation of components from crude extracts.[141]

4. General Comments on CCC

In all these cases monitoring the eluting fractions is critical. In bioassay-guided fractionation, the collected fractions should be submitted to the assay. Continuous monitoring systems, such as UV, are most convenient. TLC analysis of each fraction remains the surest method of monitoring separation, but the order of elution will not necessarily be reflected by Rf values, since in TLC, additional adsorption processes are involved.

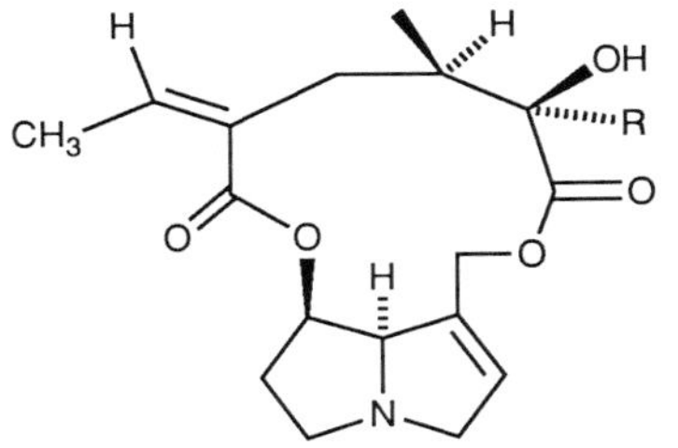

37 R = CH₃
39 R = CH₂OH

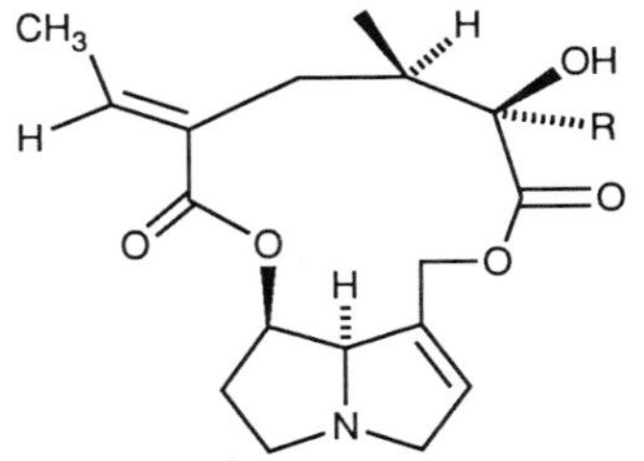

38 R = CH₃
40 R = CH₂OH

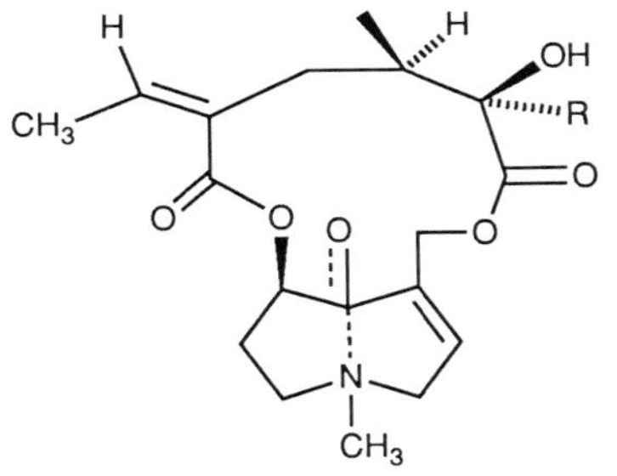

41 R = CH₃
43 R = CH₂OH

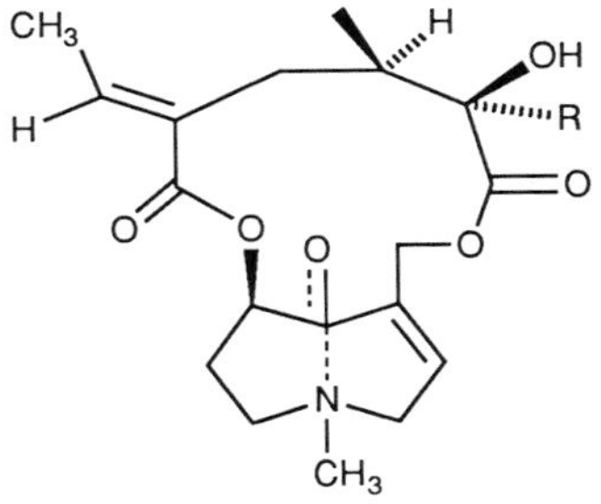

42 R = CH₃
44 R = CH₂OH

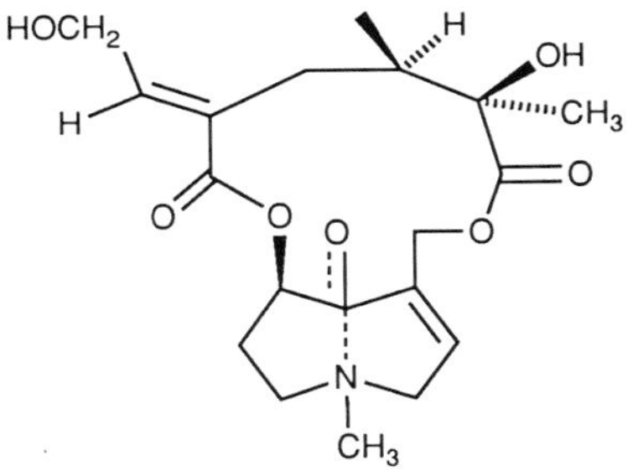

45

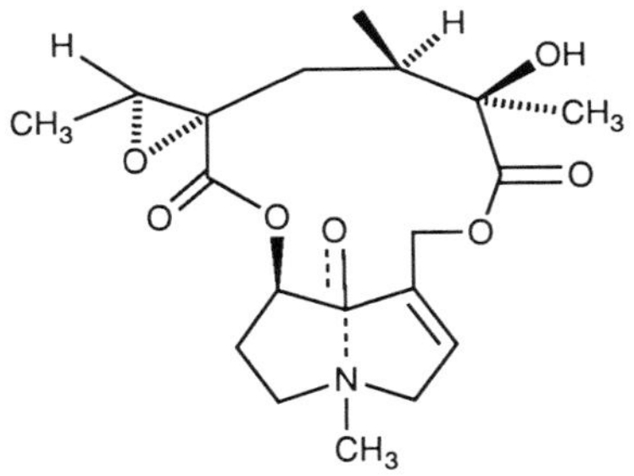

46

More detailed descriptions of the instrumentation, technique, and factors influencing resolution, together with illustrative examples, have been considered in recent reviews that emphasize application and remove the aura of mystery from the theory and nomenclature.[132-140] A chapter entitled "Getting Started in Countercurrent Chromatography" in Conway's book[134] is particularly important for the beginner. This book has the added feature of classifying the references by subject. Guides to solvent systems useful for the separation or purification of different classes of compounds have been presented by Conway,[134] Hostettmann et al.,[136,137] and McAlpine and Hochlowski.[139]

For bioactive products that can be monitored by a convenient bioassay, the selection of a suitable solvent system can be facilitated by the following method. The sample, crude extract or fraction, is shaken with the two-phase solvent system first chosen for consideration. A starting point might be the common chloroform-methanol-water combination or the less polar, hexane-ethylacetate-methanol-water mixture. The upper and lower phases are then tested for the required bioactivity, and the solvent system that gives a differential distribution of activity between the two phases should be tried initially.[139] Alternatively, and of wider applicability, preliminary studies of mobility on silica gel TLC have been used.

Modern CCC presents a number of advantages compared to adsorption and exclusion chromatography and HPLC. The first is the range of solvents that can be employed: from nonaqueous to buffered acidic and alkaline solvents. The composition of the two immiscible phases can be fine tuned to achieve the desired resolution. Compounds of unknown structure that may otherwise interact chemically with a solid phase, leading to losses by decomposition or irreversible absorption, can be handled in the mild environment of CCC. Crude extracts can easily be fractionated in a single step, preliminary to further separation. An additional advantage is the ability to adopt either a normal or reverse-phase elution mode with the same two-phase solvent system. CCC can also be adapted to a continuous extraction or enrichment mode.[120] CCC has the disadvantage of requiring an initial large capital cost, although it has low operating costs. New developments must address the question of efficiency, so that separations can be achieved in times comparable to those involved in HPLC, with the resolution increased beyond the 1000-plate scale.

B. PLANAR CHROMATOGRAPHY

Until recently, preparative thin-layer (PTLC) and paper chromatography were the only forms of this type of technique. Modern developments have led to the introduction of other planar chromatographic methods: overpressure-layer chromatography (OPLC) and centrifugal thin-layer chromatography (CTLC), methods that can provide high resolution, reproducibility, and high throughput. However, many of these new methods are still largely in a state of development, and the instrumentation is both complex and expensive. There is also a trend toward mechanization, with automatic band applicators and computerized scanning densitometers becoming available.[142]

1. Preparative Thin-Layer Chromatography (PTLC)

Traditional PTLC remains the most basic and most economical of separation techniques, since in its simplest form, it only requires a supply of precoated or home-made plates, a micropipette, a developing chamber, a sprayer, and an oven. The usefulness of PTLC as an analytic method ensures its survival and presence in every chemical laboratory. Although silica gel is the most used adsorbent material, the availability of fine-particle silica, bonded-phase and cellulose layers, as well as plates with bonded chiral phases will extend the applicability of this technique.

In the preparative mode it can be adapted for the separation of milligram to gram quantities. Although a number of TLC plates with different adsorbents are commercially available, the advantage of home-made plates is that any thickness (up to 5 mm) and many compositions (addition of silver nitrate or buffers) can be accommodated. PTLC also has a number of disadvantages. The amount of sample per plate is low (maximum of 100 mg on a 20 × 20-cm plate), and loading the sample can be haphazard when performed manually. Furthermore, recovery of the purified compound from the plate can be mechanically difficult and is potentially a hazard, as containment of fine adsorbent particles requires special precautions.

2. Centrifugal Thin-Layer Chromatography (CTLC)

A technique that is becoming more popular, albeit slowly because of the cost involved, is centrifugal thin-layer chromatography (CTLC).[138] In this method the circular preparative plate is rotated at 800 rpm while the sample is introduced in the center of the plate, followed by eluent (3 to 6 ml/s), under a nitrogen atmosphere. In the more popular version, the plate is rotated inclined to the horizontal. The concentric bands of compounds migrate to the periphery of the circular plate and are collected. The separation of UV-active substances can be monitored with the aid of a UV lamp, but in any event, the fractions collected are analyzed by analytic TLC to decide which can be pooled. This method is relatively simple, fast, does not require scraping of bands, the coated plate can be washed and regenerated, step gradient elution is possible, and the contact time between compound and stationary phase is reduced. In terms of loading capacity, a circular plate can tolerate the same amount of material as three ordinary TLC plates. The drawbacks are that the maximum thickness of the layer is 4 mm, and there is a restricted choice of stationary phases. CTLC has been useful for the separation of a wide range of substrates, from polyacetylenes to saponins and nucleotides.[138] A knowledge of good laboratory practices, an understanding of the principles involved, and a "feel" for the substances to be separated, which comes with experience, goes a long way to making PTLC a powerful separation and purification method.

3. Overpressure-Layer Chromatography (OPLC)

OPLC can be used for analytical or preparative separations. In OPLC the vapor phase of PTLC is replaced by an elastic membrane under external pressure. The mobile phase is driven by a pump through the sorbent layer, and thus the planar plate becomes a planar column. The compounds are eluted from the plate (20 or 40 cm), and fractions are individually collected. Sample sizes may vary from 50 to 500 mg. OPLC can handle increased solute loading, has higher efficiencies than PTLC, reduces separation times, and provides resolutions equivalent to those obtained with HPLC.[143,144]

The developments in this field can be followed by referring to the fundamental reviews section of *Analytical Chemistry*, a biannual review (last issue 1990[145]), to updates in the same journal,[142] to various journals (*Journal of Chromatographic Science, Chromatographia, Journal of Liquid Chromatography*, and a recent journal devoted to the topic, *Journal of Planar Chromatography — Modern TLC*).

C. COLUMN CHROMATOGRAPHY (CC)
1. Gel Filtration

Size exclusion chromatography, or gel filtration, is an important technique in the area of biopolymers. An extension of the molecular sieves exclusion principle, it also obeys the basic principles of chromatography. The support, or gel, is a neutral, porous material that allows molecules of different sizes to penetrate into the gel to different extents. This is a reversible process, so that small molecules passing into the interior of the gel can be eluted out. For any given gel, there is an exclusion limit, a molecular weight above which no penetration into the gel will occur. In general, molecules will be eluted in the reverse order of their molecular weight, although for smaller molecules, other factors, such as polarity, will play a role.[146-148]

Sephadex LH-20, obtained by alkylation of most of the hydroxyl groups of Sephadex G-25 (exclusion limit about 5000), seems to have become the most popular of the hydrophilic gels in the isolation of natural products. It is particularly useful for the removal of high-molecular-weight and polymeric material from a sample. Since these often cause problems in later chromatographic steps, gel filtration is often a prerequisite step to DCCC and pressure LC. Although primarily a fractionation technique, filtration through Sephadex LH-20 can lead to separation of compounds.

A remarkable example is illustrated by the method used for the isolation of bryostatin 1, an antineoplastic compound from the bryozoan *Bugula neritina*.[149] The isolation scheme is shown in Figure 2. The CH_2Cl_2-soluble portion of an extract was subjected to solvent partitioning, and the portion soluble in CCl_4 was collected. The extract (214 g) was divided among five large columns of Sephadex LH-20, prepared in methylene chloride-methanol. The separation was bioassay guided, and this yielded approximately 123 g of material showing activity. The separation process was repeated using a less-polar solvent combination to furnish fractions from which bryostatin 1 could be crystallized. Bryostatin 2 (deacetyl-bryostatin 1) could also be isolated in this separation.

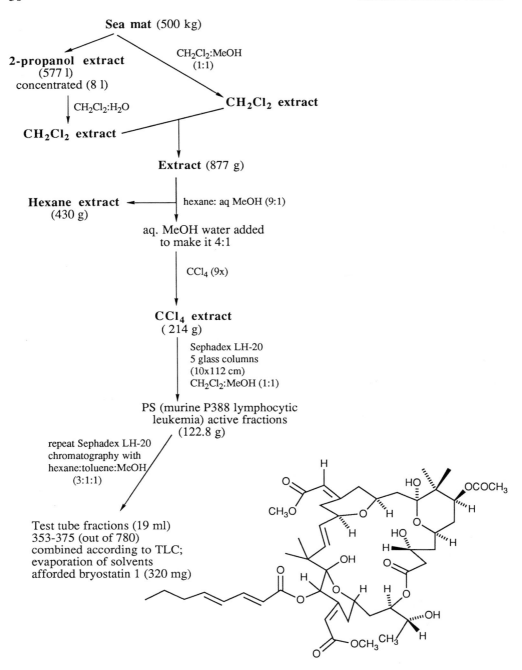

FIGURE 2. Isolation of Bryostatin 1 from *Bugula neritina.*

Partition chromatography on Sephadex LH-20 has also been used for the separation of various avermectins[13] and is the most common method for the preparative isolation of condensed tannins.[150,151] It is worthwhile noting that the separation of tannins is induced by the different adsorptivity of tannins on the gel. Because of this, higher-molecular-weight tannins are not easily recovered from the column with organic solvents. In these cases the use of a more stable vinyl polymer gel, e.g., Diaion HP-20, is indicated.[151]

2. Preparative Column Chromatography

The literature on this topic is voluminous, and reference is made only to two recent books,[148,152] a journal,[153] and reviews[154,155] that this author has found particularly helpful. The conventional gravity-driven, open-column chromatography method is still widely used in both rapid filtration and true separation modes. Sample to support ratios ranging from 1:10 to 1:300 can be used, depending on the difficulty of separation of the components of the mixture. However, it is a slow method, with the consequence that material is lost by irreversible adsorption, it has low reproducibility (the ease with which columns can be packed often leads to carelessness), and it requires large amounts of solvents that in step-gradient solvent elution are not easily recoverable. Two more recent techniques, flash chromatography and vacuum liquid chromatography, are now being widely adopted in natural products laboratories and, in terms of convenience, are superior to their forerunner, dry-column chromatography.

a. Flash Chromatography (FC)

This a very convenient and simple technique[156] that finds application in both synthetic and natural products chemistry. Briefly, the column is preferably dry-filled with adsorbent, the sample is introduced, and the solvent is forced through the column under pressure from compressed air or nitrogen (about 1 bar above atmospheric pressure). Silica gel (25 to 200 μm) is most frequently used, the sample loading depending on the ΔRf of the compounds to be separated. With ΔRf of about 0.1, the suggested column size is 50 mm (diameter) for a sample loading of 1 g.

Bonded phases, including chiral stationary phases, can also be used. Reverse-phase FC has been shown to be particularly useful in the partitioning of polar bioactive metabolites.[157] Briefly, the method involves coating the crude extract onto a reverse-phase support. This is then loaded as an aqueous slurry or a powder onto a column packed with the same support (100 g for 20 g of extract), and normal elution, step gradient from water, methanol to dichloromethane, can be carried out. Ten to twelve fractions are collected that can be tested for bioactivity and can be processed by semipreparative reverse-phase LC (RPLC). The solid support can be recycled many times.

For samples on the milligram scale, a convenient variant of this technique utilizes cartridges that function as short columns. Developed originally for "off-line" clean up of samples prior to HPLC, these cartridges are available with both normal and reversed-phase packing and are made to fit on the end of a syringe. A solution of the sample is deposited on the column normally, and elution with different solvents is achieved by pressure applied with a syringe containing the solvent. This method is useful not only for sample preparation,[158] but also for separation[159] and purification of compounds.

b. Vacuum Liquid Chromatography (VLC)

This technique rivals the method described above in its simplicity. The technique is essentially PTLC, the flow of the solvent being maintained by vacuum. The column is prepared in a sintered glass funnel, using TLC grade packing (aluminium oxide, silica gel, or reversed-phase supports). Uniform packing is achieved by initially tapping the funnel and then by application of a vacuum from below the funnel. The sample is applied uniformly at the top of the support. Step-gradient elution is used, and the column can be allowed to run dry after collection of each fraction, approximating multiple-development PTLC. All the normal stationary-phase adsorbents can be used, and the technique is applicable to large-scale separations. Sample sizes from a few milligrams to 50 g can be accommodated by choosing an appropriately sized funnel. The advantages over PTLC (reduced cost, time saving, resolution, etc.) have been listed,[160] and detailed descriptions of the simple apparatus have been given.[160-162]

D. PREPARATIVE PRESSURE LIQUID CHROMATOGRAPHY

This term covers those techniques of column chromatography in which pressure is applied by a pump operating above 2-bar pressure. Preparative, in this context, refers to amounts ranging from micrograms to kilograms. The division between low- (up to 5 bar), medium- (5–20 bar), and high-pressure (>20 bar) liquid chromatography is not simply arbitrary, but reflects the use of different columns with different size packing material and size of sample that can be fractionated.

1. Low-Pressure LC

This can be conducted with homemade glass or stainless steel columns. Ready-filled glass columns (240 × 10 mm to 440 × 37 mm) packed with silica or RP-8 support (40 to 60 μm) are commercially available. The system requires a pump capable of reaching 6-bar pressures, and an injection system. Smaller columns are suitable for sample loads up to 200 mg, whereas the larger size will tolerate up to 3 g. The selection of solvent can be extrapolated from TLC, and normally, a single solvent combination (isocratic) is used. Some representative examples have been discussed.[138] This technique is most useful for processing crude extracts into discreet smaller fractions, and while it may not necessarily provide pure compounds, the individual fractions can be submitted to chromatography with higher resolution potential.

2. Medium-Pressure LC

This mode[163] uses larger columns and higher pressures delivered by a reciprocating pump. It is a useful substitute for open-column or flash chromatography in terms of sample load, with the advantage of higher resolution and shorter separation times. Compared to these other two techniques, packings with smaller particle size (25 to 40 μm) are used. A flow rate of 100 ml/min is usual, and high loading capacity (1:25) can be achieved. The power of the technique can be illustrated with reference to the separation of the four major secoiridoid glycosides from *Gentiana lactea*. Crude extract (1.5 g) was applied to a reversed-phase column (46 cm × 26 mm i.d.; LiChroprep RP-8, 15 to 25 μm), under a maximum pressure of 36 bars, equipped with a UV detector (254 nm), and eluted with 20% and 30% aqueous methanol (flow rate 18 ml/min). The four compounds appeared between 50 and 180 min and were collected to yield **47** (8 mg), **48** (43 mg), **49** (37 mg), and **50** (14 mg), in order of elution. The resolution approached that obtainable by HPLC (see Section VI.D.3), which was used to establish optimum conditions for separation.[164]

3. High-Pressure LC

In the 20 or so years since commercial instruments became available, HPLC has had tremendous impact on separation methodology. The development in HPLC was spurred on by the discovery of DNA and the need for the separation of nanogram to microgram levels of nucleotide and nucleosides generated from hydrolysis of DNA and RNA. Commercial instruments were first referred to as "nucleic acid analyzers" or "amino acid analyzers". Developed initially for the biochemical market, HPLC has been adopted by the natural products chemist as both a preparative and an analytic technique. It has spawned the medium- and low-pressure LC methods. The development of microparticulate, chemically bonded supports added a new dimension. Reversed-phase supports were used to achieve separations not easily obtained by ion-exchange or normal adsorption or partition chromatography. HPLC has expanded to include ion-exchange, size-exclusion, affinity, immunoaffinity, ion, and chiral chromatography.[165-167] Besides the normally used silica, supports include other oxides, carbon, polymeric resins, hydroxyapatite beads, agarose, and chiral phases. Packings of various particle and pore sizes are available, and columns (stainless steel, glass-lined stainless steel, plastic cartridges) come in various sizes. Detection is still not as reliable and sensitive as one might wish. Photodiode-array (uninterrupted acquisition of UV-visible spectra) is an extension and improvement on the UV-visible detector system. The refractive index (RI) detection method is not very sensitive and is subject to variations when using gradient elution.

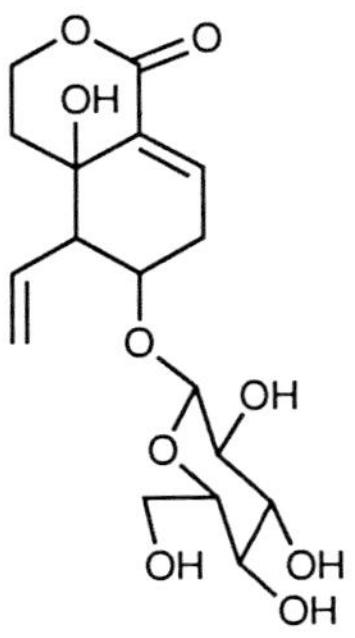

47

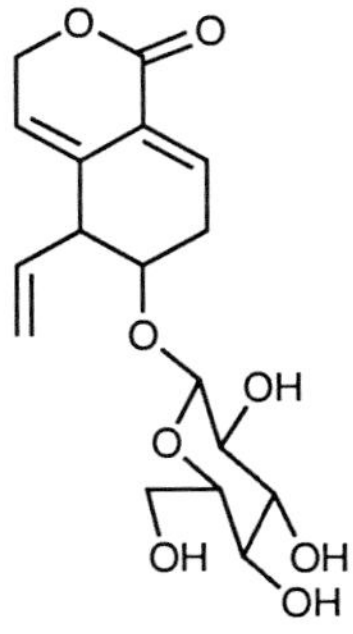

48

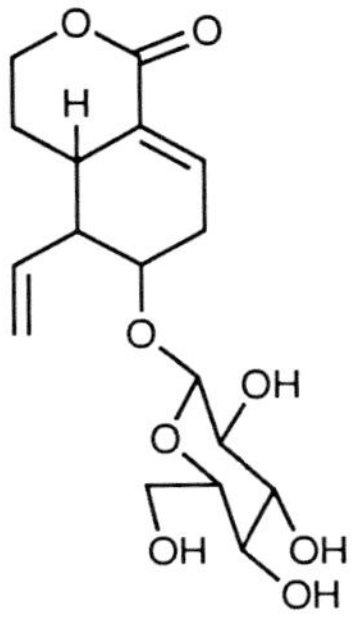

49

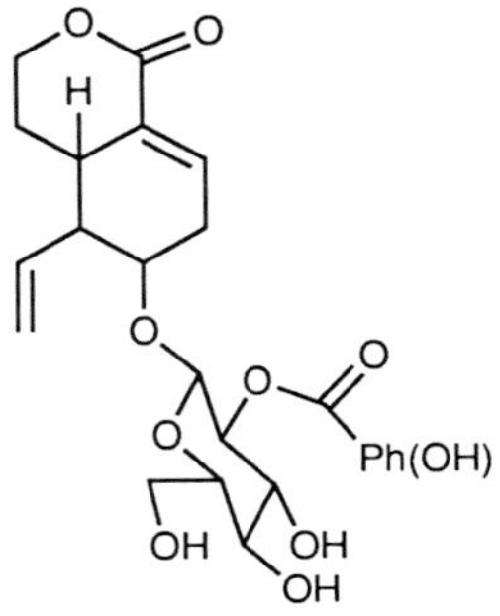

50

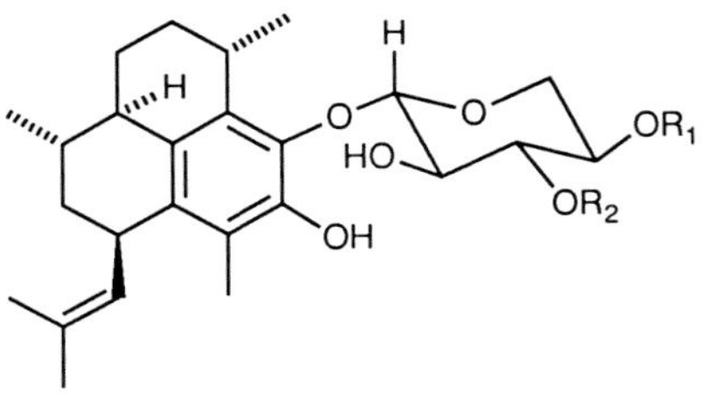

	R₁	R₂
51	Ac	H
52	H	Ac

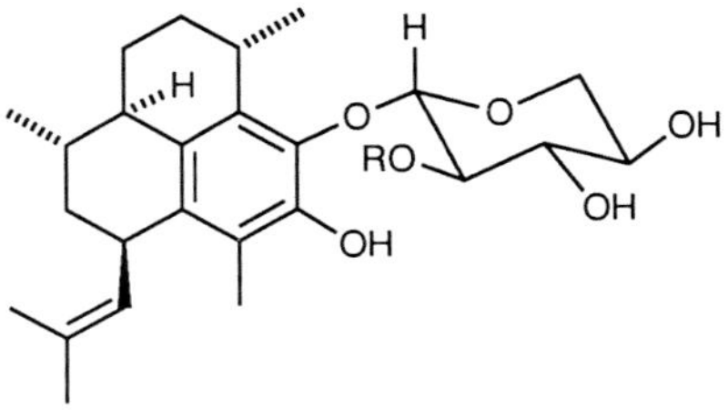

53 R = Ac
54 R = H

The advantages of HPLC are ruggedness, versatility, and separating power, particularly for hydrophilic, thermolabile compounds. The major disadvantage is the capital and maintenance costs. The excellent contribution from the *Journal of Chromatography Library,* Volume 43, contains several chapters illustrating the application of HPLC techniques.[110]

In the field of bioactive metabolites, HPLC almost always represents the final separation and purity determination step. A simple example is illustrated by the isolation and purification of four pseudopterosins: the antiinflammatory and analgesic diterpene glycosides from the

	R_1	R_2	R_3
55	CN	H	H
56	H	CN	H
57	CN	H	CH_3
58	H	CN	CH_3

sea whip *Pseudopterogorgia elisabethae*.[168] The ethyl acetate and chloroform extracts of homogenized frozen animals were combined and reextracted with chloroform. Initial separation of the extract by vacuum liquid chromatography (40 g on 300 g silica) yielded three fractions (each 0.5 l) containing the diterpenes. The first, obtained by elution with ethyl acetate-dichloromethane (1:9 to 3:7) was a mixture of **51** and **52**, the less polar monoacetate analogs. These were separated by HPLC (μ-Porasil; isooctane-ethyl acetate, 1:1). The other fractions contained **53** and **54** and were similarly separated and purified with a more polar solvent mixture (15:85). Interestingly, the monoacetate (**53**) is more polar than **51** and **52**, presumably because one possible hydrogen bonding arrangement between the glycosidic oxygen and the C-2 oxygen in the D-xylose unit is not possible in **53**.

Another example involves the isolation of the calyculins, potent antitumor metabolites from the sponge *Discodermia calyx*.[169] The sponge extract showed strong activity in the starfish egg assay[44,170] (see Section III.A.3), and so the fractionation was guided by use of this assay. The frozen sponge (1 kg) was homogenized and extracted with ethanol (3 × 5 l). The extract was partitioned between dichloromethane and water, the activity being associated with the organic fraction. This fraction (2.2 g) was subjected to low-pressure LC on silica gel, using dichloromethane-methanol as the eluent. The active fractions thus obtained were purified by reversed-phase HPLC (octadecylsilane), eluting with methanol-water, 8:2. Four active substances were obtained (about 20 to 150 mg) and designated as calyculin A–D. The major compound, calyculin A, had potent antitumor activity,[170] although it was also highly toxic to mice. The separation of these four compounds illustrates the resolution achievable by HPLC. Calyculin A (**55**) and B (**56**) differ only in the geometry of the terminal double bond, as do C (**57**) and D (**58**). Each pair is distinguished from the other by the presence or absence of a methyl group at C-32. These differences in the lipophilic portions of the molecules are sufficient to affect the degree of partitioning of each molecule into the lipophilic stationary phase.

The separation of the three phytotoxins produced by *Helminthosporium sacchari* also relied on RP-HPLC (Figure 3).[171,172] The toxin mixture was concentrated from the liquid broth on Ambersorb, then by CC on silica gel and DCCC. The final separation by HPLC (μ-Bondapak C-18) takes into account the fact that the compounds differ only in the lipophilic portion of

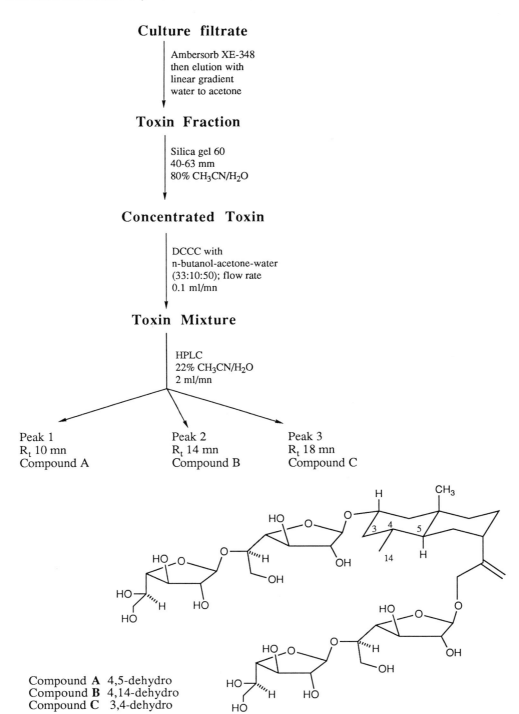

FIGURE 3. Isolation and separation of the phytotoxic metabolites from *Helminthosporium sacchari*.

the molecule. Thus, using a reversed-phase support ensures that this portion will be "recognized" by the lipophilic C-18 ends of the support. Normal-phase support would interact mainly with the hydrophilic portion of the molecule, which is identical in all three compounds.

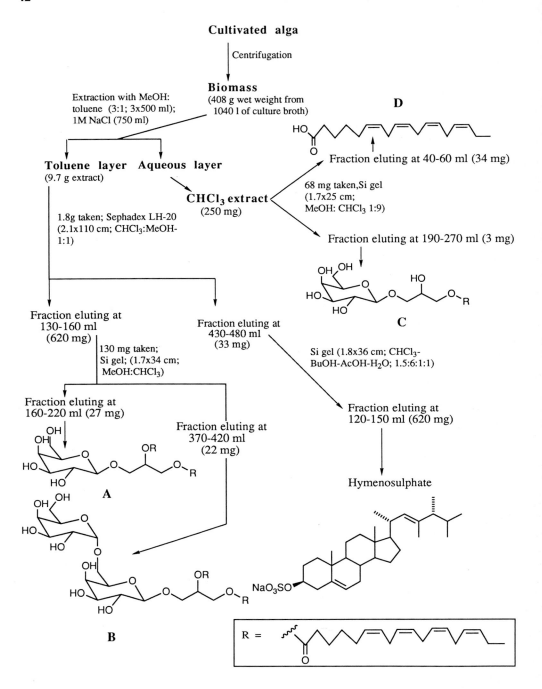

FIGURE 4. Isolation of metabolites from *Hymenomonas* sp. (Haptophyte).

4. Separation of Different Classes of Compounds

In most of the examples given so far, the focus has been on the isolation of one type of compound responsible for the bioactivity. Two cases are now considered where the fractionation of a biologically active extract led to the separation of different classes of compounds.

An extract of the microalga, *Hymenomonas* sp., showed strong Ca-releasing activity in sarcoplasmic reticulum. The isolation and separation of the active component, hymenosulfate, was carried out as shown in Figure 4.[173] There are several features of interest in this separation

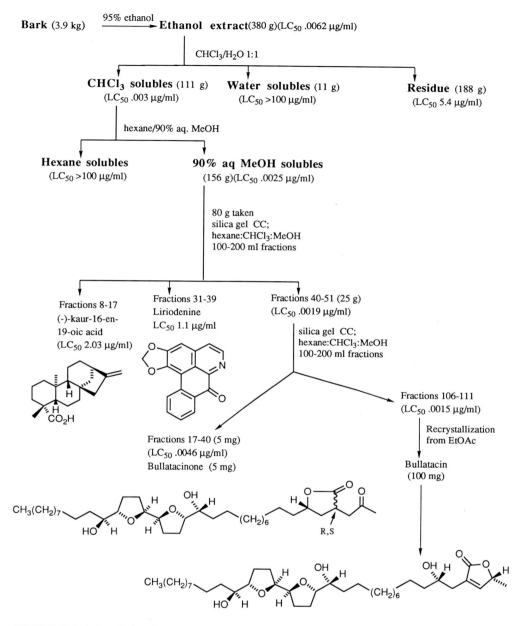

FIGURE 5. Isolation of bioactive metabolites from *Annona bullata*.

scheme. Somewhat surprisingly, the sterol sulfate is initially extracted into toluene. Given the co-occurrence of the glycolipids A, B, and C, this might be a reflection of the detergent nature of these compounds, which may enhance the solubility of the sterol sulfate in toluene. Second, the fractionation step on Sephadex LH-20 neatly separates the glycolipid components from the sterol, which is of lower molecular weight and polarity. The glycolipids A and B were separated by FC on silica gel, using a polar solvent system. Also interesting is the isolation of the fatty acid D from what was initially the aqueous methanol layer. This may be due to its detergent-like character, but since it does not appear in the toluene extract, the more likely explanation is that it is, together with C, an artifact of the extraction and arises from hydrolysis of A.

The second example concerns the separation of the bioactive metabolites from *Annona bullata*, a tree native to Cuba, extracts from which showed cytotoxic and pesticidal activities. The bioassay-guided separation scheme is illustrated in Figure 5.[174] The ethanol extract of the bark was partitioned between chloroform and water, and highest activity was found to be located in the chloroform-soluble portion. This fraction was defatted by another partitioning process, using hexane and aqueous methanol, and the bioactivity was found to be associated with the polar phase. Open CC of this fraction, monitored by TLC, separated the extract into three major fractions. The first yielded kaurenoic acid, a tetracyclic diterpene, and the second, liriodenine, a tertiary alkaloid. The less-polar nature of the diterpene can be rationalized if the hindered nature of the axial carboxylic acid is recognized. The third fraction contained a mixture of the principal active acetogenins that could be separated by further CC; the order of elution was as expected, considering the fact that bullatacinone is a dihydroxy keto lactone, whereas bullatacin is a trihydroxy lactone. All four metabolites isolated showed activity in the BSLT (see Section III.A.1), but only the acetogenins had pronounced cytotoxic effects.

Recent reviews[175,176] on modern separation methods should be consulted for extra details of techniques and solvent systems used for the separations of different classes of compounds. Information on the commercially available instrumentation is also included. For those wishing to keep up to date on the commercial suppliers and the range of accessories available, the annual "Lab Guide" of the journal *Analytical Chemistry* is a good source of information.

5. Other Chromatographic Techniques

There are many other techniques that cannot be covered due to space limitations. The most important is ion-exchange chromatography. Although superseded, to a large extent, by reversed-phase HPLC, this technique has found numerous applications in the preparative isolation of bioactive compounds representing diverse structural classes. The major references mentioned previously[108-110] contain a number of illustrative examples, often comparing ion exchange with other chromatographic techniques.

A number of chromatographic techniques appear to hold much promise, although they are still in the developmental stage. Ion-pair chromatography is a combination of ion-exchange and adsorption chromatography, employing HPLC.[140,177,178] Although its major applications have been in analytic and bioanalytic chemistry, the adaptation of this technique to the separation of charged as well as neutral molecules should prove valuable in natural products chemistry. Micellar liquid chromatography,[179,180] which relies on the partitioning of solute components into the micelles of the mobile phase, and supercritical fluid chromatography[181,182] are two emerging techniques. Enantioselective separation methods are also attracting increasing attention.[183] These methods are important in phytochemical studies where conclusive information regarding the optical purity of particular metabolites is required. Recent developments in the field of chromatography are regularly covered in the series *Advances in Chromatography*.[184]

VII. ARTIFACTS

A. ARTIFACTS FROM EXTRACTION

If the material to be extracted is not from a fresh collection, then the age and conditions of collection, drying, etc. need to be known. In the isolation of bioactive molecules, it is essential to extract under the mildest possible conditions. In certain cases this might involve carrying out the extraction under an inert atmosphere, as for the isolation of tunichrome B-1, the reducing blood pigment from the tunicate *Ascidia nigra* (Figure 6). Tunichromes are extremely sensitive to air and water, and consequently the blood from the animals was collected in test tubes under a current of dry, oxygen-free argon in the presence of tert-butylhydroxyphenyl sulfide or tert-butyl sulfide as an antioxidant.[185,186] Some algal metabo-

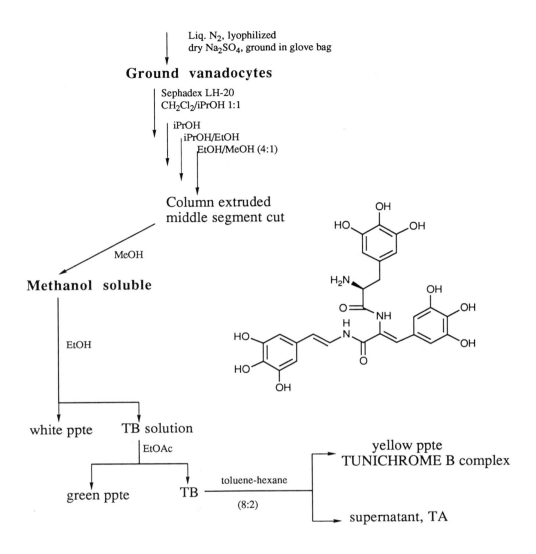

FIGURE 6. Isolation of tunichrome B (TB) complex from *Ascidia nigra* (the structure of one of the components is shown).

lites, particularly those containing bis-enol acetate (e.g., **59**) moieties and polyhalogenated metabolites (e.g., **60**) from green algae are very unstable and difficult to isolate. The widely different melting points and optical rotations quoted for samples of what are supposed to be the same compound leaves room for suspicion. Freshly collected material frozen or stored in solvent for long periods of time may decompose.[187]

Solutions of organic solvents are normally evaporated under vacuum. Aqueous solutions of an extract are better concentrated by adsorption, as mentioned in Section V.C, for large volumes or with the cartridge method[158] for small volumes. Where possible, freeze drying (lyophilization) can be a mild concentration technique, since some extracts may contain inorganic salts and acids that could catalyze reactions if the temperature is allowed to increase. As an example, weak bases such as sodium acetate and ammonia initiate epimerization at C-8' in the strained enolizable molecules of the podophyllotoxin group (**61**), forming mixtures that contain as little as 1% of the original bioactive compound.[188] It has been suggested that

59   **60**

61

62   **63**

this might occur even in saline preparations.[189] Weak acids can also induce formation of artifacts. The dehydration of **62** to **63** occurs on wetting a chloroform solution of **62** with dilute hydrochloric acid, or on contact with neutral alumina or silica gel.[190,191]

Solvents often appear as the culprits in the formation of artifacts. The isolation of the bisbenzylisoquinoline alkaloids from *Cyclea peltata* by normal methods, involving partitioning of extracts with citric acid solution, basifying with ammonia, followed by ion-exchange, alumina, and silica gel chromatography, led to the formation of artifacts.[192] These arose from involvement of ethylene glycol and dichloromethane used in extraction steps (**64** to **65** and **66**). Fractionation of berberine (**21**)-rich extracts by partitioning ammoniacal solutions with chloroform leads to the formation of the adduct **22**. On contact with acetone, **22** produces the adduct **67**.[28] The formation of acetonides on extraction of glycols with acetone, and of acetals on extraction of ketones with methanol or ethanol, can occur. The β-addition of alcoholic solvents to conjugated systems (e.g., **68** to **69**) is also possible.[193] Condensation of aryl aldehydes with acetone, the extracting solvent, has been noted.[65] The dienedioic acid, **70**, was obtained as the diethyl ester after extraction of the bark of *Phebalium nudum* with ethanol.[194] The possibility of exchanging methoxyl groups of an acetal or carboxylic ester with ethoxyl group must not be discounted.[195,196] The exchange of a carboxylic methyl ester to form the amide (**71** to **72**) on contact with aqueous ammonia has been reported.[197] Tuliposide A (**73**), isolated from *Tulipa gesneriana* L., is unstable even at –20°C and converts to the positional isomer **74**. On contact with dilute acid, **73** is hydrolyzed to give the α-methylene lactone (**75**).[28] The observation that polygodial (**76**) is generated from the diacetate (**77**) on exposure to wet ether and/or silica gel[198] suggests that polygodial may be an artifact. The presence of free polygodial in the organisms would be unlikely, given the reactivity of the dialdehyde group.

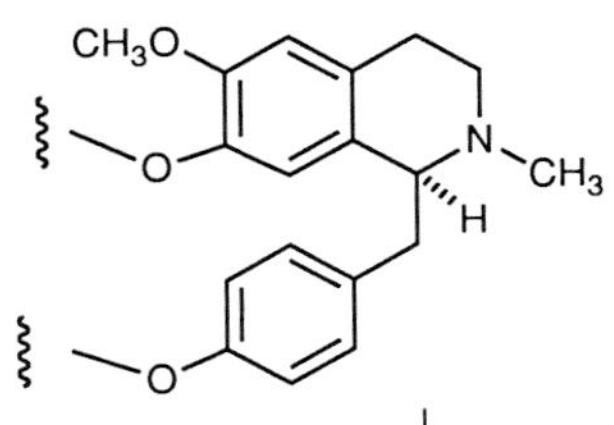

64

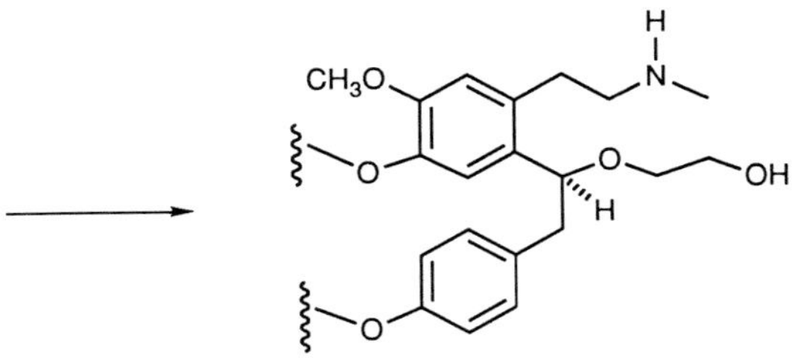

65

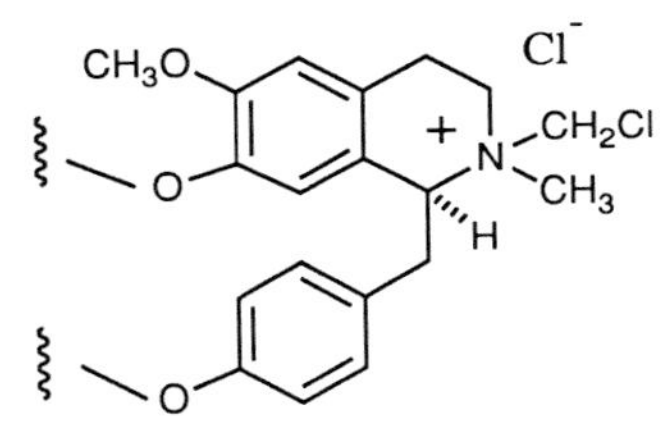

66

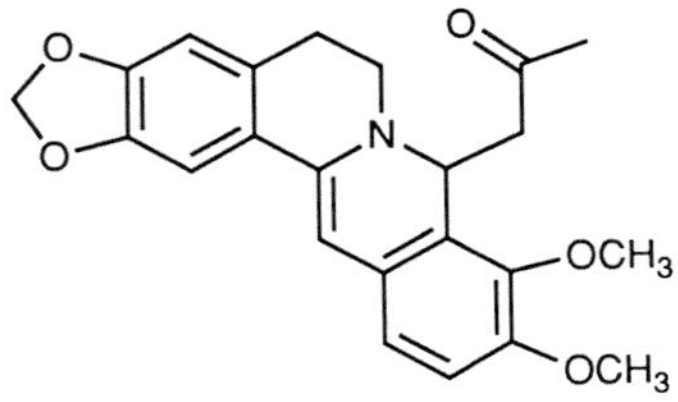

67

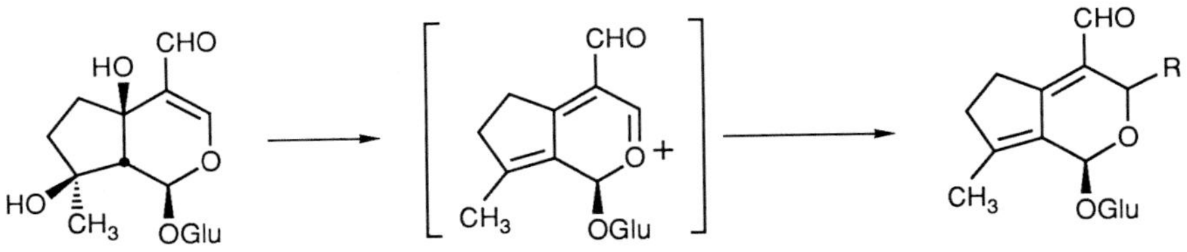

68

69   R = α-OMe
     R = β-OMe

B. ARTIFACTS FROM SEPARATIONS

Problems may also arise at the separation steps, particularly when active alumina is used.[199] A range of reactions, such as aldolization, dehydration, hydration, hydride transfer and skeletal rearrangements, can occur.[200] The occurrence of N- and O-demethylation, oxidation, rearrangements, Hoffmann β-elimination and ring expansion when isoindolobenzazepine alkaloids are chromatographed on silica gel has been noted.[201] Crude extracts can survive relatively unchanged over a period of years, but on separation, the individual components are often unstable to air, solvents, nonneutralized glass surfaces, and light. An example of the instability of a pure compound with respect to light involves the retrochalcone tepanone. Exposure of solutions of tepanone, or its methoxy derivative, to ordinary laboratory light was found to lead to an equilibrium mixture of E- and Z-isomers.[202]

Any natural product chemist who has had to work on the milligram scale knows the frustration of having to exclude or remove plasticizers. They have become the most widely

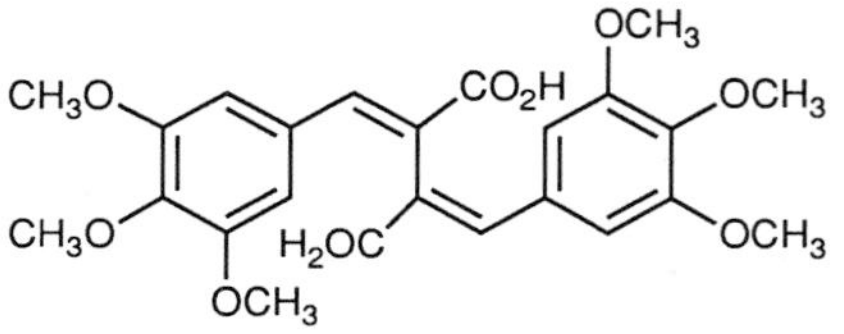

70

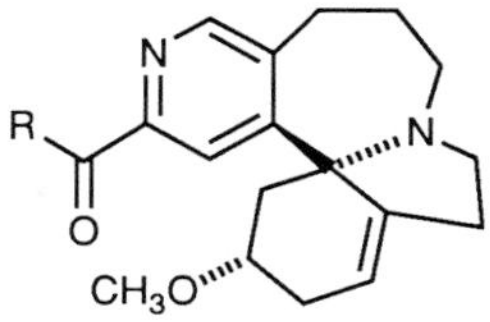

71 R = OCH₃
72 R = NH₂

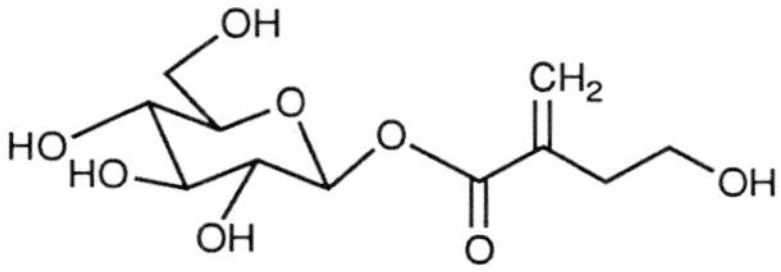

73

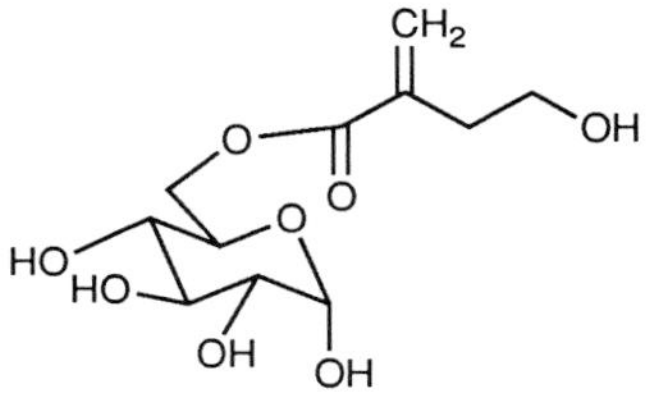

74

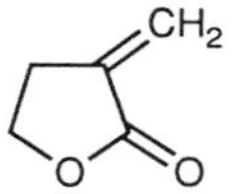

75 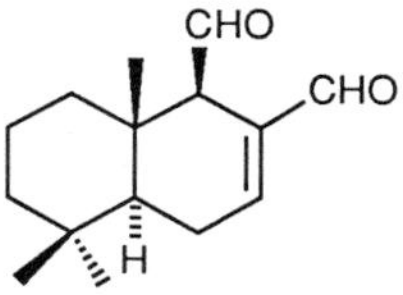   **76** 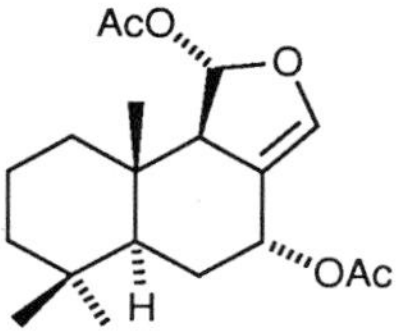   **77**

distributed isolable "unnatural products". They appear to be present in solvents and chemicals, gases (from plastic tubing), plants (stored in plastic bags), and even microorganisms (nutrient stored in plastic bags). Indeed, up to 1973 there had been 21 reports of the "natural occurrence" of phthalic acid and dialkyl phthalates,[203] and the presence of these and dimethyl terephthalates in algae has been reported.[204]

Less obvious are a number of artifacts and contaminants that plague the analytic chemist and could fool the natural products chemist. For instance, in an investigation of the metabolites that may be responsible for the plant growth promotion activity of the coprophilous fungus *Sordaria fimicola*, two major metabolites were isolated from this source.[205] The first was indole carboxaldehyde, a catabolic product of the auxin indoleacetic acid, the formation of which would explain the activity observed. The second was triacontanol, a "metabolite" that most natural products chemists would prefer to ignore. However, triacontanol is a potent plant growth stimulant, and to add a measure of uncertainty, it is a known contaminant of some filter papers. A useful compilation of such nuisance substances, together with indications of their likely origin, has been published.[206] There is much to be gained in reading these "cautionary tales".

VIII. CONCLUDING REMARKS

Even from this brief review, it is possible to gain a general idea of the activity in the areas of detection and isolation of bioactive natural products and an insight into their synergistic effects. The introduction of new, simplified bioassays can be confidently predicted. A greater choice of convenient, high-resolution separation methods will be welcomed if the costs involved do not limit accessibility. The isolation of sufficient quantities of bioactive metabolite, so that its structure can be rigorously established and its spectrum of activities adequately defined, remains a demanding and important task. A study of structure–activity relationships of promising metabolites also requires adequate supplies of material.

Research into bioactive compounds from natural sources has assumed a sense of urgency, since it has been realized that large-scale destruction of many of these reservoirs is a possibility. Much information on bioactive metabolites is being accumulated. It is inevitable that many researchers will form the impression that only a small portion of all this can be assimilated. However, this branch of natural products science is still in its infancy. As it develops, one will see a more organized and rational approach in which standardized bioassays will allow comparative studies for a broader range of bioactivities. A better understanding of the mode of action of bioactive compounds will ensure that the list of useful biopharmaceuticals will continue to grow.

REFERENCES

1. **Block, J. H.,** Progress in the design of bioactive molecules, in *Probing Bioactive Mechanisms,* Magee, P. S., Henry, D. R., and Block, J. H., Eds, ACS Symp. Ser. 413, American Chemical Society, Washington, D.C., 1989, chap. 1.
2. **Farnsworth, N. R.,** The role of medicinal plants in drug development, in *Natural Products and Drug Development,* Krogsgaard, P., Brøgger Christensen, S., and Kofod, H., Eds., Munksgaard, Copenhagen, 1984, 17.
3. **Waterman, P. G.,** Searching for bioactive compounds: various strategies, *J. Nat. Prod.,* 53, 13, 1990.
4. **Klaymann, D. L.,** Qinghaosu (artemisinin): an antimalarial drug from China, *Science,* 228, 1049, 1985.
5. **Bhat, S. V., Bajwa, B. S., Dornauer, H., de Souza, N. J., and Fehlhaber, H. W.,** Structure and stereochemistry of new labdane diterpenes from *Coleus forskohlii* Briq., *Tetrahedron Lett.,* 1669, 1977.
6. **Kinghorn, A. D.,** Biologically active compounds from plants with reputed medicinal and sweetening properties, *J. Nat. Prod.,* 50, 1009, 1987.
7. **Hamburger, M., Marston, A., and Hostettmann, K.,** Search for new drugs of plant origin, *Adv. Drug. Res.,* 20, 167, 1991.
8. **Colegate, S. M., Dorling, P. R., and Huxtable, C. R.,** A spectroscopic investigation of swainsonine: an α-mannosidase inhibitor isolated from *Swainsona canescens, Aust. J. Chem.,* 32, 2257, 1979.
9. **Colegate, S. M., Dorling, P. R., and Huxtable, C. R.,** Swainsonine: a toxic indolizidine alkaloid from the Australian *Swainsona* species, in *Toxicology of Plant and Fungal Compounds, Handbook of Natural Toxins,* Vol. 6, Keeler, R. F. and Tu, A. T., Eds., Marcel Dekker, New York, 1991, 159.
10. **Fellows, L. E. and Fleet, G. W.,** Alkaloidal glycosidase inhibitors from plants, in *Natural Products Isolation: Separation Methods for Antimicrobials, Antiviral and Enzyme Inhibitors,* Journal of Chromatography Library, Vol. 43, Wagman G. H. and Cooper R., Eds., Elsevier, Amsterdam, 1989, chap. 13.
11. **Zähner, H., Drautz, H., and Weber, W.,** Novel approaches to metabolite screening, in *Bioactive Microbial Products: Search and Discovery,* Bu'Lock, J. D., Nisbet, L. J., and Winstanley, D. J., Eds., Academic Press, London, 1982, chap. 5.
12. **Babu, J. R.,** Avermectins: biological and pesticidal activities, in *Biologically Active Natural Products. Potential Use in Agriculture,* Cutler, H. G., Ed., ACS Symp. Ser. 380, American Chemical Society, Washington, D.C., 1988, chap. 7.

13. **Miller, T. and Gullo, V. P.**, Avermectins and related compounds, in *Natural Products Isolation: Separation Methods for Antimicrobials, Antiviral and Enzyme Inhibitors,* Journal of Chromatography Library, Vol. 43, Wagman G. H. and Cooper, R., Eds., Elsevier, Amsterdam. 1989, chap. 9.

14. **Umezawa, H.**, Enzyme inhibitors produced by microorganisms, in *Natural Products Isolation: Separation Methods for Antimicrobials, Antiviral and Enzyme Inhibitors*, Journal of Chromatography Library, Vol. 43, Wagman G. H. and Cooper R., Eds., Elsevier, Amsterdam. 1989, chap. 12.

15. **Monaghan, R. L. and Tkacz, J. S.**, Bioactive microbial products: focus upon mechanism of action, *Ann. Rev. Microbiol.*, 44, 271, 1990.

16. **Borel, J.-F.**, Immunological properties of ciclosporin (Sandimmune), in *Advances in Immunopharmacology 3*, Chedid, L., Hadden, J. W., Spreafico, F., Dukor, P., and Willoughby, D., Eds., Pergamon Press, Oxford, 1986, 239.

17. **Dreyfuss, M., Harris, E., Hofman, H., Hobel, H., Pache, W., and Tscherter, H.**, Cyclosporin A and C, new metabolites from *Trichoderma polysporum* (Link ex. Pers.) Rifai, *Eur. J. Appl. Micro.*, 3, 125, 1976.

18. **Cutler, H. G., Ammermann, E., and Springer, J. P.**, Diverse but specific biological activities of four natural products from three fungi, in *Biologically Active Natural Products. Potential Use in Agriculture*, Cutler, H. G., Ed., ACS Symp. Ser. 380, American Chemical Society, Washington, D.C., 1988, chap. 6.

19. **Tyler, V. E.**, Plant drugs in the twenty-first century, *Econ. Bot.*, 40, 279, 1986.

20. **Anton, R., Beretz, A., and Haag-Berrurier, M.**, New properties for old compounds, in *Biologically Active Natural Products*, Hostettmann, K. and Lea, P. J., Eds., Clarendon Press, Oxford, 1987, 117.

21. **Brown, A. G., Smale, T. C., Hasenkamp, T. J., and Thomson, R. H.**, Crystal and molecular structure of compactin, a new antifungal metabolite from *Penicillium brevicompactum*, *J. Chem. Soc. Perkin Trans. 1*, 1165, 1976.

22. **Endo, A., Kuroda, M., and Tsujita, Y.**, ML-236A, ML-236B, and ML-236C, new inhibitors of cholesterogenesis produced by *Penicillium citrinum*, *J. Antibiot.*, 29, 1346, 1976.

23. **Rinehart, K. L., Holt, T. G., Fregeau, N. L., Keifer, P. A., Wilson, G. R., Perun, T. J., Jr., Sakai, R., Thompson, A. G., Stroh, J. G., Shield, L. S., Seigler, D. S., Li, L. H., Martin, D. G., Grimmelikhuijzen, C. J. P., and Gäde, G.**, Bioactive compounds from aquatic and terrestrial sources, *J. Nat. Prod.*, 53, 771, 1990.

24. **de Souza, N. J., Ganguli, B. N., and Reden, J.**, Strategies in the discovery of drugs from natural sources, *Ann. Rep. Med. Chem.*, 17, 301, 1982.

25. **Tan, G. T., Pezzuto, J. M., Kinghorn, A. D., and Hughes, S. H.**, Evaluation of natural products as inhibitors of human immunodeficiency virus type 1 (HIV-1) reverse transcriptase, *J. Nat. Prod.*, 54, 143, 1991.

26. **Farnsworth, N.**, Anomalous isolation of an active antitumor alkaloid from a fraction of *Catharanthus lanceus* devoid of anticancer activity, *J. Pharm. Sci.*, 61, 1840, 1972.

27. **Pettit, G. R., Singh, S. B., Niven, M. L., Hamel, E., and Schmidt, J. M.**, Isolation, structure, and synthesis of combretastatins A-1 and B-1, potent new inhibitors of microtubule assembly, derived from *Combretum caffrum*, *J. Nat. Prod.*, 50, 119, 1987.

28. **Mitscher, L. A.**, Plant-derived antibiotics, in *Antibiotics: Isolation, Separation, Purification*, Journal of Chromatography Library, Vol. 15, Weinstein, M. J. and Wagman, G. H., Eds., Elsevier, Amsterdam, 1978, 463.

29. **Albert, A.**, *Xenobiosis*, Chapman and Hall, Cambridge, 1987.

30. **Meyer, B. N., Ferrigni, N. R., Putnam, J. E., Jacobsen, L. B., Nichols, D. E., and McLaughlin, J. L.**, Brine shrimp: a convenient general bioassay for active plant constituents, *Planta Med.*, 45, 31, 1982.

31. **Alkofahi, A., Rupprecht, J. K., Anderson, J. E., McLaughlin, J. L., Mikolajczak, K. L., and Scott, B. A.**, Search for new pesticides from higher plants, in *Insecticides of Plant Origin*, Arnason, J. T., Philogène, B. J. R. and Morand, P., Eds., ACS Symp. Ser. 387, American Chemical Society, Washington, D.C.,1989, chap. 3.

32. **Rupprecht, J. K., Hui, Y. H., and McLaughlin, J. L.**, Annonaceous acetogenins: a review, *J. Nat. Prod.*, 53, 237, 1990.

33. **Rieser, M. J., Kozlowski, J. F., Wood, K. V., and McLaughlin, J. L.**, Muricatacin: a simple biologically active acetogenin derivative from the seeds of *Annona muricata* (Annonaceae), *Tetrahedron Lett.*, 32, 1137, 1991.

34. **Saad, J. M., Hui, Y. H., Rupprecht, J. K., Anderson, J. E., Kozlowski, J. F., Zhao, G.-X., Wood, K. V., and McLaughlin, J. L.**, Reticulatacin: a new bioactive acetogenin from *Annona reticulata* (Annonaceae), *Tetrahedron*, 47, 2751, 1991.

35. **Trotter, R. T., II, Logan, M. H., Rocha, J. M., and Boneta, J. L.**, Ethnography and bioassay: combined methods for a preliminary screen of home remedies for potential pharmacological activity, *J. Ethnopharmacol.*, 8, 113, 1983.

36. **McChesney, J. D. and Adams, R. P.**, Co-evaluation of plant extracts as petrochemical substitutes and for biologically active compounds, *Econ. Bot.*, 39, 74, 1985.

37. **Cimino, G., De Giulio, A., De Rosa, S., and Di Marzo, V.,** Minor bioactive polyacetylenes from *Petrosia ficiformis, J. Nat. Prod.,* 52, 345, 1990.
38. **Marles, R. J., Farnsworth, N. R., and Neill, D. A.,** Isolation of a novel cytotoxic polyacetylene from a traditional anthelmintic medicinal plant, *Minquartia guianensis, J. Nat. Prod.,* 52, 261, 1989.
39. **Nitao, J. K., Nair, M. G., Thorogood, D. L., Johnson, K. S., and Scriber, J. M.,** Bioactive neolignans from the leaves of *Magnolia virginiana, Phytochemistry,* 30, 2193, 1991.
40. **Kirivanta, J., Sivonen, K., Niemela, S. I., and Huovinen, K.,** Detection of toxins of cyanobacteria (blue-green algae) by *Artemia salina* bioassay, *Planta Med.,* December 1990.
41. **Ferrigni, N. R., Putman, J. E., Anderson, B., Jacobsen, L. B., Nichols, D. E., Moore, D. S., McLaughlin, J. L., Powell, R. G., and Smith, C. R., Jr.,** Modification and evaluation of the potato disc assay and antitumor screening of Euphorbiaceae seeds, *J. Nat. Prod.,* 45, 679, 1982.
42. **Galsky, A. G., Wilsey, J. P., and Powell, R. G.,** Crown gall tumor disc bioassay: a possible aid in the detection of compounds with antitumor activity, *Plant Physiol.,* 65, 184, 1980.
43. **Fadli, M., Aracil, J.-M., Jeanty, G., Banaigs, B., and Francisco, C.,** Novel meroterpenoids from *Cystoseira mediterranea*: use of the crown-gall bioassay as a primary screen for lipophilic antineoplastic agents, *J. Nat. Prod.,* 54, 261, 1991.
44. **Fusetani, N.,** Marine metabolites which inhibit development of echinoderm embryos, in *Biorganic Marine Chemistry,* Vol. 1, Scheuer, P. J., Ed., Springer-Verlag, Berlin, 1987, 61.
45. **Jacobs, R. S., Culver, P., Langdon, R., O'Brien, T., and White, S.,** Some pharmacological observations on marine natural products, *Tetrahedron,* 41, 981, 1985.
46. **Bêrdy, J.,** The discovery of new bioactive microbial metabolites: screening and identification, in *Bioactive Metabolites from Microorganisms, Progress in Industrial Microbiology,* Vol. 27, Bushell, M. E. and Gräfe, U., Eds., Elsevier, Amsterdam, 1989, 3.
47. **Betina, V.,** *The Chemistry and Biology of Antibiotics,* Elsevier, Bratislava, 1983.
48. **Simon, A., Dunlop, R. W., Ghisalberti, E. L., and Sivasithamparam, K.,** *Trichoderma koningii* produces a pyrone compound with antibiotic properties, *Soil Biol. Biochem.,* 20, 263, 1988.
49. **Dunlop, R. W., Simon, A., Sivasithamparam, K., and Ghisalberti, E. L.,** An antibiotic from *Trichoderma koningii* active against soilborne plant pathogens, *J. Nat. Prod.,* 52, 67, 1989.
50. **Mitscher, L. A., Drake, S., Gollapudi, S. R., and Okwute, S. K.,** A modern look at folkloric use of anti-infective agents, *J. Nat. Prod.,* 50, 1025, 1987.
51. **Jayasuriya, H., McChesney, J. D., Swanson, S. M., and Pezzuto, J. M.,** Antimicrobial and cytotoxic activity of rottlerin-type compounds from *Hypericum drummondii, J. Nat. Prod.,* 52, 325, 1989.
52. **Hamburger, M.O. and Cordell, G. A.,** A direct bioautographic tlc assay for compounds possessing antibacterial activity, *J. Nat. Prod.,* 50, 19, 1987.
53. **Baumgartner, B., Erdelmeier, C. A. J., Wright, A. D., Rali, T., and Sticher, O.,** An antimicrobial alkaloid from *Ficus septica, Phytochemistry,* 29, 3327, 1990.
54. **Fleming, I. D., Nisbet, L. J., and Brewer, S. J.,** Target directed antimicrobial screens, in *Bioactive Microbial Products: Search and Discovery,* Bu'Lock, J. D., Nisbet, L. J., and Winstanley, D. J., Eds., Academic Press, London, 1982, chap. 7.
55. **Umezawa, K., Hori, M., and Takeuchi, K.,** Microbial secondary metabolites inhibiting oncogene functions, in *Novel Microbial Products for Medicine and Agriculture, Topics in Industrial Microbiology,* Demain, A. L., Somkuti, G. A., Hunter-Cevera, J. C., and Rossmoore, H. W., Eds., Elsevier, Amsterdam, 1989, chap. 6.
56. **Hamill, R. L.,** Screens for pharmacologically active fermentation products, in *Bioactive Microbial Products: Search and Discovery,* Bu'Lock, J. D., Nisbet, L. J., and Winstanley, D. J., Eds., Academic Press, London, 1982, chap. 6.
57. **Hood, J. D.,** Inhibitors of antibiotic-inactivating enzymes, in *Bioactive Microbial Products: Search and Discovery,* Bu'Lock, J. D., Nisbet, L. J., and Winstanley, D. J., Eds., Academic Press, London, 1982, chap. 8.
58. **Jong, S.-C. and Donovick, R.,** Antitumor and antiviral substances from fungi, in *Advances in Applied Microbiology,* Vol. 34, Neidleman, S. J., Ed., Academic Press, London, 1989, 183.
59. **Nisbet, L. J. and Porter, N.,** The impact of pharmacology and molecular biology on the exploitation of microbial products, in *Microbial Products: New Approaches,* Baumberg, S., Hunter, I., and Rhodes, M., Eds., Cambridge University Press, Cambridge, 1989, 309.
60. **Gräfe, U., Dornberger, K., and Fleck, W. F.,** Approaches to new microbial metabolites with nonclassical mode of action, in *Bioactive Metabolites from Microorganisms, Progress in Industrial Microbiology,* Vol. 27, Bushell, M. E. and Gräfe, U., Eds., Elsevier, Amsterdam, 1989, 113.
61. **Demain, A. L., Somkuti, G. A., Hunter-Cevera, J. C., and Rossmoore, H. W., Eds.,** *Novel Microbial Products for Medicine and Agriculture, Topics in Industrial Microbiology,* Elsevier, Amsterdam, 1989.
62. **Cutler, H. G.,** Unusual plant-growth regulators from microorganisms, *Crit. Rev. Plant Sci.,* 6, 323, 1988.
63. **Cardellina, J. H., II, Raub, M. F., and VanWagenen, B. C.,** Plant growth regulators and insect control agents from marine organisms, in *Allelochemicals: Role in Agriculture and Forestry,* Waller, G. R., Ed., ACS Symp. Ser., 330, American Chemical Society, Washington, D.C., 1987, chap. 50.

64. **Cardellina, J. H., II,** Natural products in the search for new agrochemicals, in *Biologically Active Natural Products. Potential Use in Agriculture,* Cutler, H. G., Ed., ACS Symp. Ser., 380, American Chemical Society, Washington, D.C., 1988, chap. 20.

65. **Cardellina, J. H., II, Nigh, D., and VanWagenen, B. C.,** Plant growth regulatory indoles from the sponges *Dysidea etheria* and *Ulosa ruetzlery, J. Nat. Prod.,* 49, 1065, 1986.

66. **Kubo, I.,** Effect of a marine algal constituent on the growth of lettuce and rice seedlings, *Pure Appl. Chem.,* 61, 373, 1989.

67. **Cutler, H. G.,** *Proc. 11th Annu. Meet., Plant Growth, Regulator Society of America,* Cooke, A. R., Ed., 1984, 1.

68. **Cutler, H. G., Springer, J. P., Arrendale, R. F., Arison, B. H., Cole, P. D., and Roberts, R. G.,** Cinereain: a novel metabolite with plant growth regulating properties from *Botrytis cinerea, Agric. Biol. Chem.,* 52, 1725, 1988.

69. **Harborne, J. B.,** Recent advances in chemical ecology, *Nat. Prod. Rep.,* 6, 85, 1989.

70. **Saito, K., Matsumoto, M., Sekine, T., Murakoshi, I., Morisaki, N., and Iwasaki, S.,** Inhibitory substances from *Myriophyllum brasiliense* on growth of blue-green algae, *J. Nat. Prod.,* 52, 1221, 1989.

71. **Cutler, H. G.,** Phytotoxins of microbial origin, in *Toxicology of Plant and Fungal Compounds, Handbook of Natural Toxins,* Vol. 6, Keeler, R. F. and Tu, A. T., Eds., Marcel Dekker, New York, 1991, 411.

72. **Duke, S. O. and Lydon, J.,** Herbicides from natural compounds, *Weed Technol.,* 1, 122, 1987.

73. **Heisey, R. M., Mishra, S. K., Putnam, A. R., Miller, J. R., Whitenack, C. J., Keller, J. E., and Huang, J.,** Production of herbicidal and insecticidal metabolites from soil microorganisms, in *Biologically Active Natural Products. Potential Use in Agriculture,* Cutler, H. G., Ed., ACS Symp. Ser. 380, American Chemical Society, Washington, D.C., 1988, chap. 5.

74. **Hoagland, R. E.,** Naturally occurring carbon-phosphorous compounds as herbicides, in *Biologically Active Natural Products. Potential Use in Agriculture,* Cutler, H. G., Ed., ACS Symp. Ser. 380, American Chemical Society, Washington, D.C., 1988, chap. 13.

75. **Kenfield, D., Bunkers, G., Strobel, G., and Sugawara, F.,** Fungal phytotoxins — potential new herbicides, in *Phytotoxins and Plant Pathogenesis,* Graniti, A., Durbin, R., and Ballio, A., Eds., Springer-Verlag, Berlin, 1989, 319.

76. **Lax, A. R. and Sheperd, H. S.,** Tentoxin: a cyclic tetrapeptide having potential herbicidal use, in *Biologically Active Natural Products. Potential Use in Agriculture,* Cutler, H. G., Ed., ACS Symp. Ser. 380, American Chemical Society, Washington, D.C., 1988, chap. 2.

77. **Powell, R. G. and Spencer, G. F.,** Phytochemical inhibitors of velvetleaf (*Abutilon theophrasti*) germination as models for new biorational pesticides, in *Biologically Active Natural Products. Potential Use in Agriculture,* Cutler, H. G., Ed., ACS Symp. Ser., 380, American Chemical Society, 1988, Washington, D.C., chap. 14.

78. **Sakamura, S., Ichihara, A., and Yoshihara, T.,** Toxins of phytopathogenic microorganisms: structural diversity and physiological activity, in *Biologically Active Natural Products. Potential Use in Agriculture,* Cutler, H. G., Ed., ACS Symp. Ser. 380, American Chemical Society, Washington, D.C., 1988, chap. 4.

79. **DeFrank, J. and Putnam, A. R.,** Screening procedure to identify soil-borne actinomycetes that can produce herbicidal compounds, *Weed Sci.,* 33, 271, 1985.

80. **Buta, G. J. and Kalinski, A.,** Camptothecin and other plant growth regulators in higher plants with antitumor activity, in *Biologically Active Natural Products. Potential Use in Agriculture,* Cutler, H. G., Ed., ACS Symp. Ser. 380, American Chemical Society, Washington, D.C., 1988, chap. 19.

81. **Wolf, R. B., Spencer, G. F., and Kwolek, W. F.,** Inhibition of velvetleaf (*Abutilon theophrasti*) germination and growth by benzyl isothiocyanate, a natural toxicant, *Weed Sci.,* 32, 612, 1984.

82. **Strobel, G., Sugawara, F., and Clardy, J.,** Phytotoxins from plant pathogens of weedy plants, in *Allelochemicals: Role in Agriculture and Forestry,* Waller, G. R., Ed., ACS Symp. Ser., 330, American Chemical Society, Washington, D.C., 1987, chap. 46.

83. **Strobel, G. A., Kenfield, D., Bunkers, G., and Sugawara, F.,** Phytotoxins from fungi attacking weedy plants, in *Toxicology of Plant and Fungal Compounds, Handbook of Natural Toxins,* Vol. 6, Keeler, R. F. and Tu, A. T., Eds., Marcel Dekker, New York, 1991, 397.

84. **Stierle, A. C., Cardellina, J. H., II, and Strobel, G. A.,** Maculosin, a host-specific phytotoxin for spotted knapweed from *Alternaria alternata, Proc. Natl. Acad. Sci. U.S.A.,* 85, 8008, 1988.

85. **Stierle, A. C., Cardellina, J. H., II, and Strobel, G. A.,** Phytotoxins from *Alternaria alternata,* a pathogen of spotted knapweed, *J. Nat. Prod.,* 52, 42, 1989.

86. **Strobel, G.,** Phytotoxins, *Ann. Rev. Biochem.,* 51, 309, 1982.

87. **Blumauerová, M., Kristufek, V., Jizba, J., Sedmera, P., Beran, M., Priklilova, V., Stary, J., Kucera, M., Sajdl, P., Landa, V., Kandybin, V. N., Samoukina, G. V., Bortnik, N. I., and Barbashova, N. M.,** Research of streptomycetes producing pesticides and plant growth regulators, in *Bioactive Metabolites from Microorganisms, Progress in Industrial Microbiology,* Vol. 27, Bushell, M. E. and Gräfe, U., Eds., Elsevier, Amsterdam, 1989, 237.

88. **Boyd, M. R.,** Strategies for the identification of new agents for the treatment of AIDS: A national program to facilitate the discovery and preclinical development of new drug candidates for clinical evaluation, in *AIDS: Etiology, Diagnosis, Treatment and Prevention*, DeVita, V. T., Jr., Hellman, S., and Rosenberg, S. A., Eds., Lippincott, Philadelphia, 1988.

89. **Abou-Karam, M. and Shier, W. T.,** A simplified plaque reduction assay for antiviral agents from plants. Demonstration of frequent occurrence of antiviral activity in higher plants, *J. Nat. Prod.*, 53, 340, 1990.

90. **Suffness, M. and Douros, J.,** Current status of the NCI plant and animal program, *J. Nat. Prod.*, 45, 1, 1982.

91. **Potier, P.,** Contribution of an organic chemist to the resolution of some biological problems: consequences, *Pure Appl. Chem.*, 58, 737, 1986.

92. **Kinghorn, A. D.,** New plant-derived anticancer drugs, in *Topics in Pharmaceutical Sciences*, Breimer, D. D. and Speiser, P., Eds., Elsevier, Amsterdam, 1983, 403.

93. **Cassady, J. M., Baird, W. M., and Chang, C.-J.,** Natural products as a source of potential cancer chemotherapeutic and chemopreventive agents, *J. Nat. Prod.*, 53, 23, 1990.

94. **Suffness, M.,** New approaches to the discovery of antitumor agents, in *Biologically Active Natural Products*, Hostettmann, K. and Lea, P. J., Eds., Clarendon Press, Oxford, 1987, 85.

95. **Munro, M. H. G., Luibrand, R. T., and Blunt, J. W.,** The search for antiviral and anticancer compounds from marine organisms, in *Bioorganic Marine Chemistry*, Vol. 1, Scheuer, P. J., Ed., Springer-Verlag, Berlin, 1987, 93.

96. **Munro, M. H. G., Blunt, J. W., Barns, G., Battershill, C. N., Lake, R. J., and Perry, N. B.,** Biological activity in New Zealand marine organisms, *Pure Appl. Chem.*, 61, 529, 1989.

97. **Mynderse, J. S., Crandall, L. W., and Cardellina, J. H., II,** Bioactive compounds from marine organisms and cultivated blue-green algae, in *Natural Products Isolation: Separation Methods for Antimicrobials, Antiviral and Enzyme Inhibitors*, in Journal of Chromatography Library, Vol. 43, Wagman G. H. and Cooper, R., Eds., Elsevier, Amsterdam, 1989, chap. 10.

98. **Burres, N. S., Hunter, J. E., and Wright, A. E.,** A mammalian cell agar-diffusion assay for the detection of toxic compounds, *J. Nat. Prod.*, 52, 522, 1989.

99. **Ames, B. N., McCann, E., and Yamasaki, E.,** Methods for detecting carcinogens and mutagens with the *Salmonella*/mammalian microsome mutagenicity test, *Mut. Res.*, 31, 347, 1975.

100. **Wall, M. E., Wani, M. C., Hughes, T. J., and Taylor, H.,** Plant antimutagenic agents, I. General bioassay and isolation procedures, *J. Nat. Prod.*, 51, 866, 1988.

101. **Wagner, H.,** Search for plant derived natural products with immunostimulatory activity (recent advances), *Pure Appl. Chem.*, 62, 1217, 1990.

102. **Wagner, H.,** Immunostimulants from higher plants (recent advances), in *Biologically Active Natural Products*, Hostettmann, K. and Lea, P. J., Eds., Clarendon Press, Oxford, 1987, 127.

103. **Phillipson, J. D. and O'Neill, M. J.,** Antimalarial and amoebicidal natural products, in *Biologically Active Natural Products*, Hostettmann, K. and Lea, P. J., Eds., Clarendon Press, Oxford, 1987, p. 49.

104. **O'Neill, M. J., Bray, D. H., Boardman, P., Chan, K. L., Phillipson, J. D., Warhurst D. C., and Peters, W.,** Plants as sources of antimalarial drugs, part 4: activity of *Brucea javanica* fruits against chloroquine-resistant *Plasmodium falciparum in vitro* and against *Plasmodium berghei in vivo*, *J. Nat. Prod.*, 50, 41, 1987.

105. **Nakanishi, K.,** Natural products chemistry-past and future, in *Natural Products from Woody Plants*, Vol. 2, Rowe, J. W., Ed., Springer-Verlag, Berlin, 1989, 13.

106. **Schroeder, D. R. and Nakanishi, K.,** A simplified isolation procedure for azadirachtin, *J. Nat. Prod.*, 50, 241, 1987.

107. **Charlwood, B. V., Ed.,** Editorial, *Phytochem. Anal.*, 1, 1, 1990.

108. **Weinstein, M. J. and Wagman, G. H., Eds.,** *Antibiotics: Isolation, Separation, Purification*, in Journal of Chromatography Library, Vol. 15, Elsevier, Amsterdam, 1978.

109. **Natori, S., Ikekawa, N., and Suzuki, M.,** *Advances in Natural Products Chemistry. Extraction and Isolation of Biologically Active Compounds*, a Halstead Press Book, John Wiley & Sons, New York, 1981.

110. **Wagman G. H. and Cooper R., Eds.,** *Natural Products Isolation: Separation Methods for Antimicrobials, Antiviral and Enzyme Inhibitors*, in Journal of Chromatography Library, Vol. 43, Elsevier, Amsterdam. 1989.

111. **Mitscher, L. A., Leu, R. P., Bathala, M. S., Wu, W. N., Beal, J. L., and White, R.,** Antimicrobial agents from higher plants. I. Introduction, rationale and methodology, *Lloydia*, 35, 157, 1972.

112. **Pettit, G. R. and Cragg, G. M.,** *Biosynthetic Products for Cancer Chemotherapy*, Vol. 3, Plenum Press, New York, 1978, 47.

113. **Pettit, G. R. and Cragg, G. M.,** *Biosynthetic Products for Cancer Chemotherapy*, Vol. 2, Plenum Press, New York, 1978, 1.

114. **Ghisalberti, E. L., Jefferies, P. R., and Vu, H. T. N.,** Diterpenes from *Eremophila* species, *Phytochemistry*, 29, 316, 1990.

115. **Spjut, R. W., Cassady, J. M., McCloud, T., Suffness, M., Norris, D. H., Cragg, G. M., and Edson, C. F.,** Variation in cytotoxicity and antitumor activity among samples of the moss *Claopodium crispifolium* (Thuidiaceae), *Econ. Bot.,* 42, 62, 1988.

116. **Spjut, R. W., Suffness, M., Cragg, G. M., and Norris, D. H.,** Mosses, liverworts, and hornworts screened for antitumor agents, *Econ. Bot.,* 40, 310, 1986.

117. **Hamill, R. L. and Crandall, L. W.,** Polyether antibiotics, in *Antibiotics: Isolation, Separation, Purification,* Journal of Chromatography Library, Vol. 15, Weinstein, M. J. and Wagman, G. H., Eds., Elsevier, Amsterdam, 1978, 479.

118. **Ciabatti, R. and Cavalleri, B.,** Ramoplanin (A/16686): a new glycolipodepsipeptide antibiotic from actinoplanes, in *Bioactive Metabolites from Microorganisms, Progress in Industrial Microbiology,* Vol. 27, Bushell, M. E. and Gräfe, U., Eds., Elsevier, 1989, 205.

119. **Wilson, K. E.,** Isolation of carbapenems, in *Natural Products Isolation: Separation Methods for Antimicrobials, Antiviral and Enzyme Inhibitors,* Journal of Chromatography Library, Vol. 43, Wagman G. H. and Cooper R., Eds., Elsevier, Amsterdam, 1989, chap. 8.

120. **Nakazawa, H., Riggs, C. E., Egorin, M. J., Redwood, S. M., Bachur, N. R., Bhatnagar, R., and Ito, Y.,** Continuous extraction of urinary anthracycline antitumor antibiotics with the horizontal flow-through coil planet centrifuge, *J. Chromatogr.,* 307, 323, 1984.

121. **Pettit, G. R., Kamano, Y., Aoyagi, R., Herald, C. L., Doubek, D. L., Schmidt, J. M., and Rudloe, J. J.,** Antineoplastic agents 100. The marine bryozoan *Amathia convoluta, Tetrahedron,* 41, 985, 1985.

122. **Johnston, K. P.,** *New Directions in Supercritical Fluid Science and Technology,* ACS Symp. Ser. 406, American Chemical Society, Washington, D.C., 1989, 1

123. **Scheffer, S. T., Zalkow, L. H., and Teja, A. S.,** *Extraction and Isolation of Chemotherapeutic Pyrrolizidine Alkaloids from Plant Sources: Novel Processes Using Supercritical Fluids,* ACS Symp. Ser. 406, American Chemical Society, Washington, D.C., 1989, 416.

124. **Della Loggia, R., Becker, H., Isaac, O., and Tubaro, A.,** Topical anti-inflammatory activity of *Calendula officinalis* extracts, *Planta Med.,* December 1990.

125. **Hawthorne, S. B.,** Analytical-scale supercritical fluid extraction, *Anal. Chem.,* 62, 633A, 1990.

126. **Cortes, H. J., Ed.,** *Multidimensional Chromatography: Techniques and Applications,* Chromatography Science Series/50, Marcel Dekker, New York, 1990.

127. **Mauger, A. and Katz, E.,** Actinomycins, in *Antibiotics: Isolation, Separation, Purification,* Journal of Chromatography Library, Vol. 15, Weinstein, M. J. and Wagman, G. H., Eds., Elsevier, Amsterdam, 1978, 1.

128. **Gross, E.,** Peptide antibiotics, in *Antibiotics: Isolation, Separation, Purification,* Journal of Chromatography Library, Vol. 15, Weinstein, M. J. and Wagman, G. H., Eds., Elsevier, Amsterdam, 1978, 415.

129. **Eble, T. E.,** Lincomycin related antibiotics, in *Antibiotics: Isolation, Separation, Purification,* Journal of Chromatography Library, Vol. 15, Weinstein, M. J. and Wagman, G. H., Eds., Elsevier, Amsterdam, 1978, 231.

130. **Majer, J. P.,** Macrolide antibiotics, in *Antibiotics: Isolation, Separation, Purification,* Journal of Chromatography Library, Vol. 15, Weinstein, M. J. and Wagman, G. H., Eds., Elsevier, Amsterdam, 1978, 273.

131. **Perlman, D. and Ogawa, Y.,** Streptamine-containing antibiotics, in *Antibiotics: Isolation, Separation, Purification,* Journal of Chromatography Library, Vol. 15, Weinstein, M. J. and Wagman, G. H., Eds., Elsevier, Amsterdam, 1978, 587.

132. **Mandava, B. N. and Ito, Y., Eds.,** *Countercurrent Chromatography: Theory and Practice,* Marcel Dekker, New York, 1988.

133. **Foucault, A. P.,** Countercurrent chromatography, *Anal. Chem.,* 63, 569A, 1991.

134. **Conway, W. D.,** *Countercurrent Chromatography: Apparatus, Theory, and Practice,* VCH Publishers, New York, 1990.

135. **Zalkow, L. H., Asibal, C. F., Glinski, J. A., Bonetti, S. J., Gelbaum, L. T., VanDerveer, D., and Powis, G.,** Macrocyclic pyrrolizidine alkaloids from *Senecio anonymus*. Separation of a complex alkaloid extract using droplet countercurrent chromatography, *J. Nat. Prod.,* 51, 690, 1988.

136. **Marston, A., Slacanin, I., and Hostettmann, K.,** Centrifugal partition chromatography in the separation of natural products, *Phytochem. Anal.,* 1, 3, 1990.

137. **Hostettmann, K., Hostettmann, M., and Marston, A.,** Isolation of natural products by droplet countercurrent chromatography and related methods, *Nat. Prod. Rep.,* 1, 471, 1984.

138. **Hostettmann, K., Hostettmann, M., and Marston, A.,** *Preparative Chromatographic Techniques,* Springer-Verlag, Berlin, 1985.

139. **McAlpine, J. B. and Hochlowski, J. E.,** Countercurrent chromatography, in *Natural Products Isolation: Separation Methods for Antimicrobials, Antiviral and Enzyme Inhibitors,* Journal of Chromatography Library, Vol. 43, Wagman G. H. and Cooper R., Eds., Elsevier, Amsterdam. 1989, chap. 1.

140. **Snyder, J., Breuning, R., Derguini, F., and Nakanishi, K.,** Fractionation and proof of structure of natural products, in *Natural Products from Woody Plants,* Vol. 1, Rowe, J. W., Ed., Springer-Verlag, Berlin, 1989, chap. 2.

141. **Lee, Y. W., Fang, Q. C., Ito, Y., and Cook, C. E.,** The application of true countercurrent chromatography in the isolation of bioactive natural products, *J. Nat. Prod.,* 52, 706, 1989.

142. **Poole, C. F. and Poole, S. K.,** Modern thin-layer chromatography, *Anal. Chem.,* 61, 1257A, 1989.
143. **Erdelmeier, C. A. J., Erdelmeier, I., Kinghorn, A. D., and Farnsworth, N. R.,** Use of overpressure layer chromatography (OPLC) for the separation of natural products with antineoplastic activity, *J. Nat. Prod.,* 49, 1133, 1986.
144. **Nyiredy, Sz., Erdelmeier, C. A. J., Dallenbach-Toelke, K., Nyiredy-Mikita, K., and Sticher, O.,** Preparative on-line overpressure layer chromatography (OPLC): a new separation technique for natural products, *J. Nat. Prod.,* 49, 885, 1986.
145. **Sherma, J.,** Planar chromatography, *Anal. Chem.,* 62, 371R, 1990.
146. **Cardellina, J. H., II,** Step gradient elution in gel permeation chromatography. A new approach to natural products separation, *J. Nat. Prod.,* 46, 196, 1983.
147. **MacLeod, A. J.,** *Instrumental Methods for Food Analysis,* Elek Science, London, 1973.
148. **Poole, C. F. and Schuette, S. A.,** *Contemporary Practice of Chromatography,* Elsevier, Amsterdam, 1984.
149. **Pettit, G. R., Leet, J. E., Herald, C. L., Kamano, Y., Boettner, F. E., Baczynskyj, L., and Nieman, R. A.,** Isolation and structure of bryostatin 12 and 13, *J. Org. Chem.,* 52, 2854, 1987.
150. **Karchesy, J. J., Bae, Y., Chalker-Scott, L., Helm, R. F., and Foo, L. Y.,** Chromatography of proanthocyanidins, in *Chemistry and Significance of Condensed Tannins,* Hemingway, R. W. and Karchesy, J. J., Eds., Plenum Press, New York, 139.
151. **Okuda, T., Yoshida, T., and Hatano, T.,** New methods of analyzing tannins, *J. Nat. Prod.,* 52, 1, 1989.
152. **Verzele, M. and Dewaele, C.,** *Preparative High Performance Liquid Chromatography. A Practical Guideline,* RSL Europe, Eke, Belgium, 1986.
153. **Bidlingmeyer, B., Ed.,** *Preparative Liquid Chromatography,* Elsevier, Amsterdam, 1987.
154. **Dorsey, J. D., Foley, J. P., Cooper, W. T., Barford, R. A., and Barth, H. G.,** Liquid chromatography: theory and methodology, *Anal. Chem.,* 324R, 62, 1990.
155. **Verzele, M.,** Preparative liquid chromatography, *Anal. Chem.,* 62, 265A, 1990.
156. **Still, W. C., Kahn, M., and Mitra, A.,** Rapid chromatographic technique for preparative separations with moderate resolution, *J. Org. Chem.,* 43, 2393, 1978.
157. **Blunt, J. W., Calder, V. L., Fenwick, G. D., Lake, R. J., McCombs, J. D., Munro, M. H. G., and Perry, N. B.,** Reverse phase flash chromatography: a method for the rapid partitioning of natural product extracts, *J. Nat. Prod.,* 50, 290, 1987.
158. **Eskins, K. and Dutton, H. J.,** Sample preparation for high-performance liquid chromatography of plant pigments, *Anal. Chem.,* 51, 1885, 1979.
159. **Buchwaldt, L. and Jensen, J. S.,** HPLC purification of dextrusins produced by *Alternaria brassicae* in culture and leaves of *Brassica napus, Phytochemistry,* 30, 2311, 1991.
160. **Pelletier, S. W., Chokshi, H. P., and Desai, H. K.,** Separation of diterpenoid alkaloid mixtures using vacuum liquid chromatography, *J. Nat. Prod.,* 49, 892, 1986.
161. **Coll, J. C. and Bowden, B. F.,** The application of vacuum liquid chromatography to the separation of terpene mixtures, *J. Nat. Prod.,* 49, 934, 1986.
162. **Pieters, L. A. and Vlietinck, A. J.,** Vacuum liquid chromatography and quantitative ^{1}H nmr spectroscopy of tumor-promoting diterpene esters, *J. Nat. Prod.,* 52, 186, 1989.
163. **Leutert, T. and von Arx, E.,** Präparative Mittledruck-flüssigkeitschromatographie, *J. Chromatogr.,* 292, 333, 1984.
164. **Schaufelberger, D. and Hostettmann, K.,** Analytical and preparative reversed-phase liquid chromatography of secoiridoid glycosides, *J. Chromatogr.,* 346, 396, 1985.
165. **Synovec, R. E., Johnson, E. L., Moore, L. K., and Renn, C. N.,** Liquid chromatography: equipment and instrumentation, *Anal. Chem.,* 62, 357R, 1990.
166. **Brown, P. R.,** High performance liquid chromatography. Past developments, present status, and future trends, *Anal. Chem.,* 62, 995A, 1990.
167. **Brown, P. R. and Hartwick, R. A., Eds.,** *High Performance Liquid Chromatography,* John Wiley & Sons, New York, 1989.
168. **Look, S. A., Fenical, W., Matsumoto, G. K., and Clardy, J.,** A new class of antiinflammatory and analgesic diterpene pentosides from the marine sea whip *Pseudopterogorgia elisabethae* (Octocorallia), *J. Org. Chem.,* 51, 5140, 1986.
169. **Kato, Y., Fusetani, N., Matsunaga, S., Hashimoto, K., and Koseki, K.,** Isolation and structure elucidation of calyculins B, C, and D, novel antitumor metabolites, from the marine sponge *Discodermia calyx, J. Org. Chem.,* 53, 3930, 1988.
170. **Kato, Y., Fusetani, N., Matsunaga, S., and Hashimoto, K.,** Calyculins, potent antitumor metabolites from the marine sponge *Discodermia calyx*: biological activities, *Drugs Exp. Clin. Res.,* XIV(12), 723, 1988.
171. **Macko, V., Acklin, W., Hildenbrand, C., Weibel, F., and Arigoni, D.,** Structure of three isomeric host-specific toxins from *Helminthosporium sacchari, Experientia,* 39, 343, 1983.
172. **Macko, V., Goodfriend, K., Wachs, T., Renwick, J. A. A., Acklin, W., and Arigoni, D.,** Characterization of the host-specific toxins produced by *Helminthosporium sacchari,* the causal organism of eyespot disease of sugar cane, *Experientia,* 37, 923, 1981.

173. **Kobayashi, J., Ishibashi, M., Nakamura, H., Ohizumi, Y., and Hirata, Y.,** Hymenosulfate, a novel sterol sulfate with Ca-releasing activity from the cultured marine haptophyte *Hymenomonas* sp., *J. Chem. Soc. Perkin Trans. 1,* 101, 1989.

174. **Hui, Y.-H., Rupprecht, J. K., Liu, Y. M., Anderson, J. E., Smith, D. L., Chang, C.-J., and McLaughlin, J. L.,** Bullatacin and bullatacinone: two highly potent bioactive acetogenins from *Annona bullata, J. Nat. Prod.,* 52, 463, 1989.

175. **Marston, A. and Hostettmann, K.,** Modern separation methods, *Nat. Prod. Rep.,* 8, 391, 1991.

176. **Marston, A. and Hostettmann, K.,** New developments in the separation of natural products, in *Modern Phytochemical Methods. Recent Advances in Phytochemistry,* Vol. 25, Fischer, N. H., Isman, M. B., and Stafford, H. A., Eds., Plenum Press, New York, 1991, 1.

177. **Hearn, M. T. W.,** Ion-pair chromatography on normal- and reversed-phase systems, *Adv. Chromatogr.,* 18, 59, 1980.

178. **Hearn, M. T. W., Ed.,** *Ion-Pair Chromatography: Theory and Biological and Pharmaceutical Applications,* Marcel Dekker, New York, 1985.

179. **Yarmchuck, P., Weinberger, R., Hirsch, R. F., and Cline Love, L. J.,** Selectivity in liquid chromatography with micellar mobile phase, *Anal. Chem.,* 54, 2233, 1982.

180. **Dorsey, J. G.,** Micellar liquid chromatography, in *Advances in Chromatography,* Vol. 27, Giddings, J. C., Grushka, E., and Brown, P. R., Eds, Marcel Dekker, New York, 1987, 167.

181. **Foley, J. P. and Crow, J. A.,** Supercritical fluid chromatography for the analysis of natural products, in *Modern Phytochemical Methods. Recent Advances in Phytochemistry,* Vol. 25, Fischer, N. H., Isman, M. B., and Stafford, H. A., Eds., Plenum Press, New York, 1991, 114.

182. **White, C. M., Ed.,** *Modern Supercritical Fluid Chromatography,* Hüthig, Heidelberg, 1988.

183. **Davin, L. B., Umezawa, T., and Lewis, N. G.,** Enantioselective separations in phytochemistry, in *Modern Phytochemical Methods. Recent Advances in Phytochemistry,* Vol. 25, Fischer, N. H., Isman, M. B., and Stafford, H. A., Eds., Plenum Press, New York, 1991, 75.

184. **Giddings, J. C., Grushka, E., and Brown, P. R., Eds,** *Advances in Chromatography,* Marcel Dekker, New York, various volumes.

185. **Bruening, R. C., Oltz, E. M., Furukawa, J., Nakanishi, K., and Kustin, K.,** Isolation of tunichrome B-1, a reducing blood pigment of the sea squirt, *Ascidia nigra, J. Nat. Prod.,* 49, 193, 1986.

186. **Oltz, E. M., Bruening, R. C., Smith, M. J., Kustin, K., and Nakanishi, K.,** The tunichromes. A class of reducing blood pigments from sea squirts: isolation, structures, and vanadium chemistry, *J. Am. Chem. Soc.,* 110, 6162, 1988.

187. **Paul, V. J. and Fenical, W.,** Natural products chemistry and chemical defense in tropical marine algae of the phylum chlorophyta, in *Biorganic Marine Chemistry,* Vol. 1, Scheuer, P. J., Ed., Springer-Verlag, Berlin, 1987, 1.

188. **Hartwell, J. L. and Schrecker, A. W.,** Lignans of podophyllum, *Progr. Chem. Org. Nat. Prod.,* 15, 83, 1958.

189. **Ayres, D. C. and Loike, J. D.,** *Lignans: Chemical, Biological and Clinical Properties,* Cambridge University Press, Cambridge, 1990.

190. **Blair, G. E., Cassady, G. M., Robbers, J. E., Tyler, V. E., and Raffauf, R., F.,** Isolation of 3, 4′, 5-trimethoxy-*trans*-stilbene, otobaene, and hydroxyotobain from *Virola cuspidata, Phytochemistry,* 8, 497, 1969.

191. **Martinez, J. C., Cuca, L. E., Yoshida, M., and Gottlieb, O. R.,** Neolignans from *Virola calophylloidea, Phytochemistry,* 24, 1867, 1985.

192. **Kupchan, S. M., Liepa, A. J., Baxter, R. L., and Hintz, H. P. J.,** New alkaloids and related artifacts from *Cyclea peltata, J. Org. Chem.,* 38, 1846, 1973.

193. **Boros, C. A., Stermitz, F. R., and Harris, G. H.,** Iridoid glycosides and a pyridine monoterpene alkaloid from *Orthocarpus.* New artifactual iridoid dienals, *J. Nat. Prod.,* 53, 72, 1990.

194. **Brown, K. L., Burfitt, A. I. R., Cambie, R. C., Hall, D., and Mathai, K. P.,** The constituents of *Phebalium nudum* III. The structure of phebolin and phebalarin, *Aust. J. Chem.,* 28, 1327, 1975.

195. **Anjaneyulu, A. S. R., Rao, K. J., Rao, K. V., Row, L. R., and Subramanyan, C.,** The structures of lignans from *Gmelina arborea* Linn., *Tetrahedron,* 31, 1277, 1985.

196. **Cambie, R. C., Pang, G. T. P., Parnell, J. C., Rodrigo, R., and Weston, R. J.,** Chemistry of the Podocarpaceae LIV. Lignans from the wood of *Dacrydium intermedium, Aust. J. Chem.,* 32, 2741, 1979.

197. **Langlois, N. and Razafimbelo, J.,** Alcaloides de *Phelline* sp. aff. *Phelline lucidalucida:* origine de l'holidine et du phellinamide: configuration de la comosidine, de la lucidinine, et de leurs dérivé, *J. Nat. Prod.,* 51, 499, 1988.

198. **Cimino, G., Sodano, G., and Spinella, A.,** Occurrence of olepupuane in two mediterranean nudibranchs: a protected form of polygodial, *J. Nat. Prod.,* 51, 1010, 1988.

199. **Lederer, E. and Lederer, M.,** *Chromatography — a Review of Principles and Applications,* Elsevier, Amsterdam, 1957, 61.

200. **Posner, G. H.,** Organic reactions at alumina surfaces, *Angew. Chem. Int. Edit.,* 17, 487, 1978.
201. **Shamma, M. and Rahimizadeh, M.,** The identity of chileninone with berberrubine. The problem of true natural products vs. artifacts of isolation, *J. Nat. Prod.,* 49, 398, 1986.
202. **Colegate, S. M., Din, L. B., Ghisalberti, E., and Latiff, A.,** Tepanone: a retrochalcone isolated from *Elleipia cuneifolia, Phytochemistry,* 31, 2123, 1992.
203. **Graham, P. R.,** Phthalate ester plasticizers—why and how they are used, *Environ. Health Perspect.,* 3, 3, 1973.
204. **Güven, K. C., Reisch, J., Kizil, Z., Güvener, B., and Cevher, E.,** Dimethyl terephthalate pollution in red algae, *Phytochemistry,* 29, 3115, 1990.
205. **Ghisalberti, E. L.,** unpublished results.
206. **Middleditch, B. S.,** *Analytical Artifacts: gc, ms, hplc, tlc, and pc,* Journal of Chromatography Library, Vol. 44, Elsevier, Amsterdam, 1989.

Chapter 3

WATER SOLUBLE ALKALOIDS:
CRYPTIC BIOACTIVE NATURAL PRODUCTS

Russell J. Molyneux

TABLE OF CONTENTS

0-8493-4372-0/93/$0.00+$.50
© 1993 by CRC Press, Inc.

I. INTRODUCTION

Since the earliest days of organic chemistry, the large and diverse group of compounds known as alkaloids[1] has presented a major challenge to chemists investigating the structure and function of natural products. The enormous range of biologic activities exhibited by the alkaloids confers upon them a significant role in human affairs. They have been employed as agents of homicide and suicide and used for control of insect and mammalian pests. Their occurrence in forage and certain foodstuffs has resulted in episodic poisonings of both man and his livestock. They presently comprise a significant proportion of therapeutic remedies and, until the relatively recent development of synthetic medicinals, were a primary means for treatment of diseases. The role that they play in the production of altered states of consciousness, both in religious rituals and as drugs of abuse, is well established, and their consequent impact upon societies throughout the world is thoroughly documented.

The biologic activity of the alkaloids, in concert with their impressive structural variety, largely accounts for the enormous research effort that has been devoted to their characterization, pharmacologic evaluation, and synthesis. Alkaloids, however, present an additional significant attraction to natural products chemists. In contrast to the majority of classes of naturally occurring organic compounds, they are relatively easy to isolate in pure form. Particularly in the case of plant material, where the nonstructural organic components comprise a complex matrix dominated by chlorophyll and its degradation products, carbohydrates, and lipids, which are inherently less interesting than potentially bioactive compounds, this represents a considerable advantage. Thus, a great deal of tedious and time-consuming manipulation may be avoided, and more effort devoted to the separation and biologic testing of individual alkaloids.

While a limited number of alkaloids may occur in quite high yields in plants (or, more particularly, certain plant parts such as fruits or seeds), the great majority comprise one percent or less of the dry weight of plant material. Moreover, they rarely exist as individual compounds, but rather as mixtures of closely related structural types. Since the degree and type of bioactivity may vary significantly with quite subtle structural changes, it is essential that each alkaloid be separated, characterized, and tested. The ability to obtain a relatively pure "alkaloid complex" with minimal effort makes the task of isolating individual components far less challenging than it would be otherwise.

Modern natural products chemistry is frequently bioactivity directed. Isolation of an alkaloid fraction enables the chemist to determine whether or not the desired bioactivity is associated with the alkaloids, thus avoiding wasted efforts. If none of the activity is found in the fraction, other component classes of the total extract must be investigated. If it does occur, then it can be used to direct the separation and isolation of those alkaloids in which the highest level of activity resides.

A peculiar disadvantage of the relative ease of isolation of alkaloid fractions is that those groups that do not conform to the general isolation protocol, usually due to particular structural features, may remain undiscovered. Plants that contain such compounds may thus be reported as not elaborating alkaloids. These natural products should be regarded as "cryptic" alkaloids, since they remain concealed when standard isolation techniques are employed. The increased use of bioactivity-directed isolation methods has reduced the chance that these alkaloids will remain undiscovered and has revealed the potential existence of many additional such natural products. Nevertheless, it is the purpose of this discussion to contrast the standard isolation techniques with those that must be employed to yield the cryptic alkaloids. It is anticipated that such methods will be more routinely used in the future and, in combination with techniques yet to be developed, lead to increasing success in the search for bioactive natural products.

II. CONVENTIONAL ALKALOIDS

While alkaloids have been isolated from microorganisms, algae, insects, and marine and terrestrial animals, the great majority occur in plants. Isolation methods will therefore be discussed in relation to the latter, recognizing that the same approaches can generally be adapted to other sources. In addition to the need for isolation and purification techniques that yield primarily the alkaloid fraction, it is essential that the alkaloidal components of this fraction be recognizable as such. A number of alkaloid-specific reagents have been developed for this purpose, and these will be briefly reviewed.

A. GENERAL ISOLATION METHODS

Alkaloids frequently occur in plants in a particularly dynamic state, being transported from their site of biosynthesis and accumulating in various organs at different stages of growth.[2] They may be catabolized when no longer required for the function for which they were biosynthesized, or may accumulate in parts of the plant such as the bark or roots. It is therefore essential to examine all parts of the plant at various growth stages before reaching the conclusion that a particular species does not produce alkaloids.

The conventional approach to the extraction and isolation of alkaloids has been established for many decades and has been reviewed in detail by Cromwell[3] and Manske.[4] A general synopsis of the methodology is as follows.

Plant material is air-dried, ground to pass through a fine screen (about 1 mm), and extracted in a Soxhlet apparatus with a polar solvent such as methanol or ethanol. Pre-extraction with a nonpolar solvent, such as hexane, to remove lipid material may be advantageous, particularly when the plant material has a high fat or wax content.

The alcoholic extract is concentrated almost to dryness under vacuum, and the residue resuspended in dilute mineral acid. Hydrochloric acid is frequently used, but the natural occurrence of chlorine-containing alkaloids suggests that it is a better practice to employ dilute sulfuric acid in order to ensure that halogen-containing alkaloids are not artifacts of the isolation procedure. The inherent basicity of the majority of alkaloids ensures their ready solubility in the dilute aqueous acid, which can then be filtered to remove undissolved material and extracted with an immiscible solvent, such as ether or chloroform, to remove the great majority of neutral or acidic components.

The acid solution is then basified, usually with ammonium hydroxide, and the aqueous solution extracted with an organic solvent such as chloroform. A degree of separation of the component alkaloids, depending upon their differential basicity, may sometimes be accomplished by stepwise basification to increasingly higher pH, with organic solvent extraction at each stage. It is essential however that the ultimate pH be sufficiently high that even strongly basic alkaloids are extracted from the aqueous solution. The crude alkaloid fraction thus obtained is often sufficiently pure for crystallization of a major, or single, alkaloid component. Alternatively, a mixture of alkaloids may be separated by a considerable variety of chromatographic techniques, interference by nonalkaloidal natural products having been minimized by the above partitioning process.

It is apparent that while the majority of known alkaloids have been isolated using this general technique, certain groups are frequently overlooked because of their failure to conform to the fundamental requirements of the methodology. This deficiency is primarily due to their high water solubility and consequent partial or total insolubility in the organic solvents usually used to extract them from basified aqueous acid solution. Even when their presence is known or suspected, procedures that would ensure their detection and isolation are often not adopted. This is particularly unfortunate, since the very fact of their high solubility in aqueous solutions implies a significant bioactivity role, either within the source plant or vs. other organisms.

In fact, many alkaloids that have been adopted as drugs are not used in their free base forms, due to restricted water solubility, but are converted into derivatives, such as the hydrochloride salts, in order for them to be more readily absorbed into the tissues.

The water soluble alkaloids can be classified into three major groups, namely:

1. Quaternary alkaloid salts
2. Alkaloid N-oxides
3. Polyhydroxy alkaloids

While other examples may exist (e.g., glycoside derivatives of more conventional alkaloids), the above three classes will be discussed in some detail because they are frequently overlooked, despite their fairly common occurrence and significant biological activity.

B. DETECTION METHODS

A considerable number of general alkaloid detection reagents have been developed and their reliability attested to by decades of usage. Some of the more common are Dragendorff's reagent (potassium bismuth iodide), Mayer's reagent (potassium mercuric iodide), Sonnenschein's reagent (phosphomolybdic acid), phosphotungstic acid, ammonium reineckate, picric acid, and picrolonic acid. While not all of these are entirely specific for alkaloids, a positive test with more than one reagent provides good circumstantial evidence for the presence of an alkaloid. Details of the preparation and use of these and other less common reagents have been described,[3] and many are now commercially available.

In their earliest manifestations most of these detection methods were used as test tube reagents. While still useful as such, they are now more frequently employed as spray reagents for thin-layer chromatography. In this form their utility is more enhanced, since they give a positive reaction with individual spots and there is less chance of nonalkaloidal constituents masking the response. Furthermore, they can be used to follow the purification steps of each alkaloid in the component mixture.

While the above reagents are general for most alkaloids, specific detection methods have been devised for a great many of the particular classes of alkaloids. These are frequently structurally based and remarkably sensitive. Once an alkaloid has been established as a member of an individual group, the appropriate reagent is extremely useful for monitoring its purification or distribution within the source plant. Leading references to particular alkaloid classes can be found in the series founded by Manske and Holmes,[5] Pelletier,[6] and Saxton (The Chemical Society).[7]

III. CRYPTIC ALKALOIDS

Microscopic examination of plant tissues has shown that the alkaloids frequently occur as water soluble forms, dissolved in the cell vacuole. Staining techniques for alkaloids may fail to show their presence in fresh tissue, whereas dried tissue from the same source gives a strong positive response.[8] It has often been stated that these water-soluble constituents are salts of malic, oxalic, succinic, and other organic acids, but this often appears to be speculative, with little supporting evidence provided. The relatively large quantities of such acids present in plants makes such an assumption reasonable, while their weakly acidic nature renders the salts susceptible to decomposition on drying, grinding, and extraction of the tissues. The alkaloids are therefore generally isolated as free bases. Hydroxylic solvents such as methanol are generally sufficiently polar to extract the alkaloid salts per se if they survive the plant drying process. The subsequent acid-base partitioning process conventionally used for alkaloid purification usually generates the free-base alkaloids from these salts. Nevertheless, it is probably good practice to moisten the plant material with dilute ammonium hydroxide in order to neutralize organic acids and ensure complete extraction of the alkaloids. In spite of the simplicity of this approach, it is more often omitted than adopted.

A. QUATERNARY ALKALOIDS

Alkaloids that exist in the plant as quaternary salts present a related, but somewhat different, problem. While not extremely common, they have been found, to a greater or lesser extent, in a diversity of structural types. However, a few classes of alkaloids possess a significant subgroup of quaternary alkaloids, often co-occurring with structurally analogous tertiary bases. These include quinolines,[9] isoquinolines,[10] benzyltetrahydroisoquinolines[11] and bisbenzyltetrahydroisoquinolines,[12] aporphines,[13] benzophenanthridines,[14] indolo[2,3a] quinolizidines, which are the basis for alkaloids of the corynantheine and yohimbine types,[15] bisindoles,[16] and the curare alkaloids,[17] which may occur as mono- or diquaternary salts. It seems reasonable to suspect that the number of quaternary alkaloids identified in the above classes coincides with the intensity with which they have been investigated because of their significant pharmacological properties. An obvious corollary to this assumption is that less-well-examined classes of alkaloids may encompass many overlooked quaternary salts.

The first stage in establishing the presence of quaternary alkaloids is to remove the tertiary bases by conventional partitioning methods. The residual aqueous solution can then be screened for the presence of unextracted alkaloids, with appropriate detection reagents (e.g., Dragendorff's). If a positive response is observed, suitable methods must be adopted to isolate the quaternary salts from nonalkaloidal water-soluble constituents. Such methods include direct crystallization, precipitation as insoluble salts, extraction with polar solvents immiscible with water, and formation of pseudobases.

1. Direct Crystallization

A limited number of quaternary alkaloids are prone to crystallize directly, either from the initial extract or from the aqueous solution remaining after partitioning. This appears to be especially true for alkaloids bearing a carboxyl group, which can form intramolecular salts. Typical examples are precatorine (**1**), a constituent of the very poisonous jequirity bean (*Abrus precatorius*),[18] although it is not responsible for the toxicity, and L-stachydrine (**2**), which occurs in large amounts, together with its 3-hydroxy derivative, in the fruits of *Capparis tomentosa*.[19]

Direct crystallization may be facilitated by ion-exchange chromatographic purification of the aqueous solution. However, a number of amino acids, such as asparagine, aspartic acid, and α-aminobutyric acid, may accompany the quaternary alkaloids and contaminate them by cocrystallization.

2. Precipitation as Insoluble Salts

A variety of reagents have been employed to form insoluble salts of quaternary alkaloids in aqueous solution. Mercuric chloride and potassium mercuric iodide yield precipitates that may be decomposed by hydrogen sulfide, leaving the alkaloids in solution as hydrochlorides or hydriodides.[3] However, quaternary bases have most commonly been isolated as reineckate, picrate, or picrolonate salts. Iodides or chlorides have been directly isolated on occasion, but these are usually too soluble to precipitate from aqueous solution, although they may be rendered insoluble by addition of ethanol.

An example of isolation by precipitation is the furoquinoline alkaloid (−)-0-methylluninium (**3**), from *Lunasia quercifolia*, which has been obtained as the picrate salt.[20] The related (+)-platydesminium salt (**4**), occurring in *Skimmia japonica*, was precipitated as the reineckate salt and converted into the chloride by passage through an ion-exchange column.[21]

Analogous examples may be cited from many additional structural classes, but the occurrence and isolation of quaternary alkaloids is best illustrated by those from curare, used as

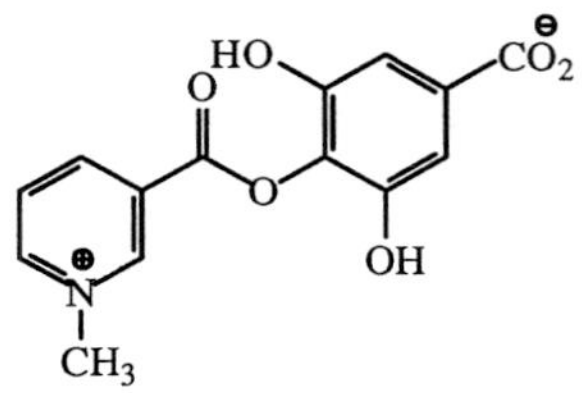

(1)

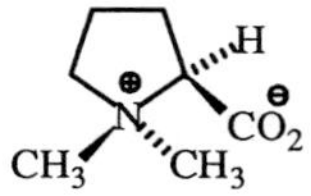

(2)

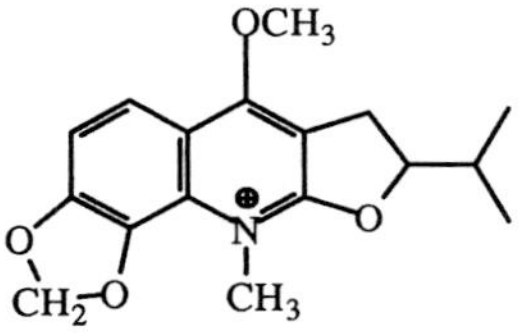

(3)

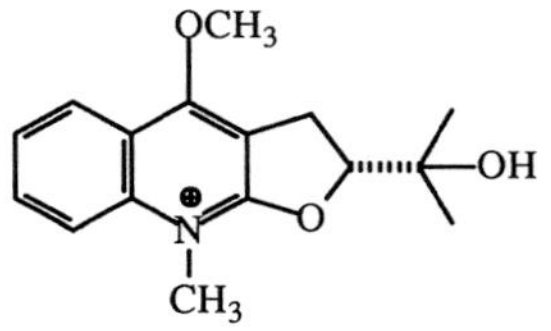

(4)

an arrow poison, in which they are particularly prevalent. Examination of calabash curare, and the bark of *Strychnos* species from which it is prepared, has shown the presence of more than 70 tertiary base and quaternary salt alkaloids. The physiologically most active constituents are the water-soluble quaternary alkaloids that remain after extraction of the tertiary bases from the basified aqueous solution with chloroform or similar solvents. After reacidification of the solution to pH 2, addition of ammonium reineckate (ammonium tetrathiocyanodiammonochromate, $NH_4[Cr(NH_3)_2(SCN)_4]$) precipitates the quaternary alkaloids as their reineckate salts, which can then be converted to their chloride or other salts for ease of handling. The latter can be separated by chromatography on cellulose or alumina to provide fractions amenable to crystallization of the pure alkaloids.

Using this technique, Battersby and co-workers[22] showed that *Strychnos toxifera* bark from British Guiana contained at least 30 quaternary alkaloids, including the double salt toxiferine-I (**5**), a potent curarizing agent.[23,24] In contrast, most of the alkaloids present in *S. toxifera* bark from Venezuela were tertiary bases.[25] This dichotomy suggests that a rather fluid relationship may exist between tertiary bases and quaternary salts in a particular plant species, possibly influenced by environmental or varietal differences. In such situations, where the isolated tertiary alkaloids have structural characteristics suggestive of the possible coexistence of quaternary bases (i.e., N-alkylation might be expected), it is a good general rule to attempt to precipitate as insoluble salts any of the latter that might be present in the residual aqueous solution.

3. Extraction with Polar Solvents

Many quaternary alkaloids, while insoluble in the relatively nonpolar solvents used to extract tertiary bases from aqueous solution, have appreciable solubility in polar, particularly hydroxylic, organic solvents. The aqueous solution may therefore be extracted with an immiscible solvent such as butanol, or the residue obtained on evaporation can be extracted with ethanol. For example, five quaternary N-methyl- and N-*iso*pentenyl dendrobinium salts [**6**; R = CH_3– or $(CH_3)_2C=CHCH_2$–] were isolated from the orchid *Dendrobium nobile* by extraction of the aqueous residue with ethanol until it showed a negative reaction to Dragendorff's reagent.[26] These were then converted to their chlorides by passage through an ion-exchange column.

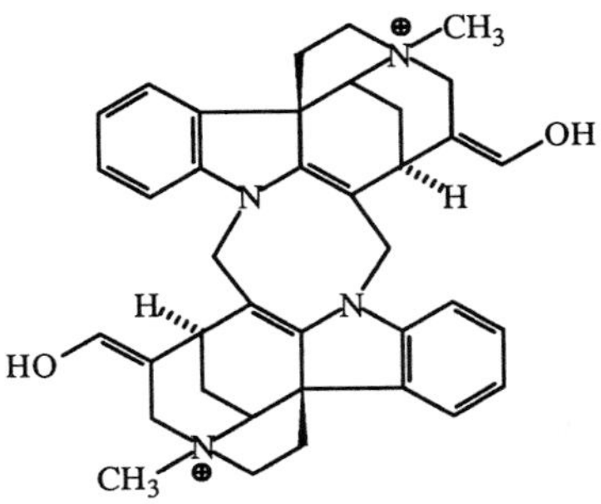

(5)

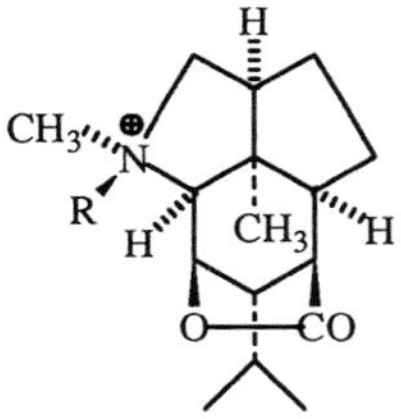

(6)

The first quaternary pyrrolizidine alkaloid to be reported was recently isolated from *Senecio integrifolius*, using a similar approach.[27] After extraction of the basified aqueous layer with methylene chloride in the conventional manner, it was exhaustively extracted with n-butanol. The latter extract yielded the N-methyldiangeloyl platynecium chloride (7). The occurrence of such a quaternary species should perhaps have been foreseen, in view of the known occurrence of N-methylated seco-pyrrolizidine alkaloids derived from otonecine, such as senkirkine (8), in *Senecio* species.[28,29] As a consequence, the genus as a whole may well warrant reinvestigation for similar quaternary alkaloids.

4. Formation of Pseudobases

Certain classes of alkaloidal quaternary bases undergo reaction with hydroxide ions to form pseudobases that are extractable into nonpolar organic solvents. These pseudobases are therefore readily formed when the acidic aqueous extract is basified during conventional alkaloid partitioning. Quaternary benzophenanthridines, such as sanguinarine (9), commonly found in the Fumariaceae and Papaveraceae,[14] have been isolated by this method, although mixtures of alkaloids must be further separated by fractional crystallization or column chromatography.

The formation of pseudobases is well illustrated by berberine, a widely distributed alkaloid with diverse biological activities.[30] Berberine was isolated from the common barberry (*Berberis vulgaris*) in 1837, and its structure elucidated in 1890.[31] The alkaloid can be obtained as a strongly basic quaternary hydroxide (10) by careful treatment of its salts with barium hydroxide. An excess of base, however, transforms the latter into the isomeric pseudobase (carbinolamine) form (11), which is a much weaker base. While the quaternary hydroxide and pseudobase forms have quite different UV absorption spectra, both give identical salts on treatment with acid. The pseudobase is in equilibrium with the amino-aldehyde (12), which is favored under strongly alkaline conditions. The potential for the amino-aldehyde to undergo condensation reactions and the possibility of dispro-

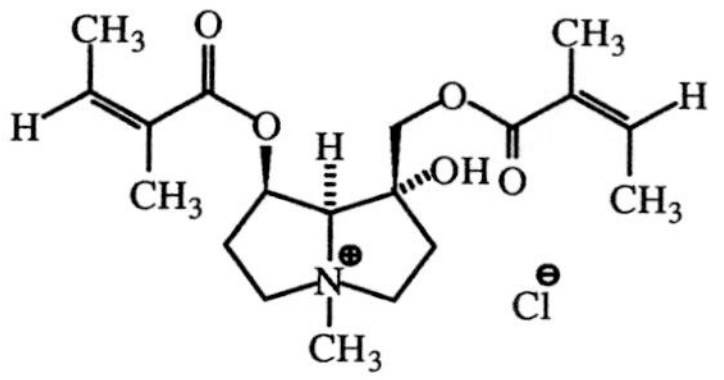

(7)

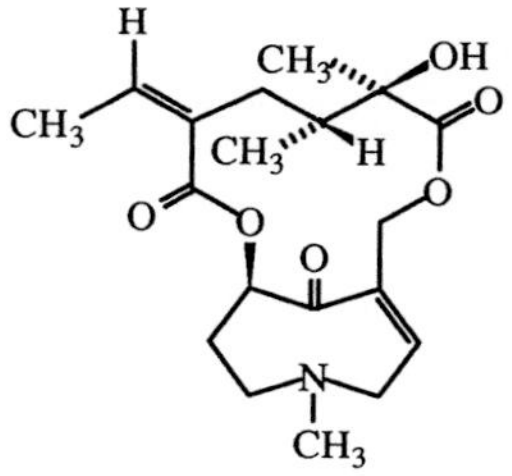

(8)

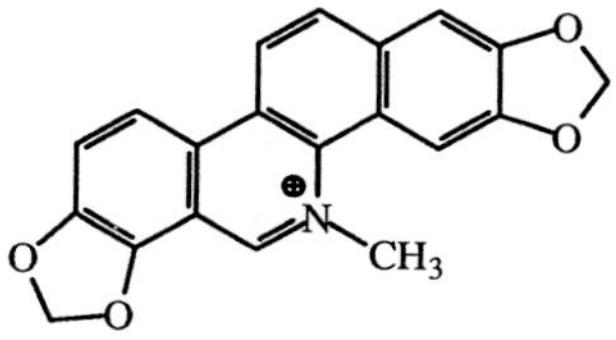

(9)

portionation of the pseudobase in alkaline solution are severe limitations on pseudobase formation as an isolation method. It is therefore frequently advantageous to convert the quaternary bases into pseudocyanides that are extractable into relatively nonpolar organic solvents.

B. ALKALOID N-OXIDES

Many alkaloid N-oxides, corresponding to a number of classes of tertiary base alkaloids, have been discovered. Their overall chemistry has been comprehensively reviewed by Phillipson and Handa.[32] Structural classes in which N-oxides are particularly well represented are the pyrrolizidine, quinolizidine, tropane, indole, and isoquinoline alkaloids. A few di-N-oxides have been isolated, corresponding to dimeric tertiary bases. As might be expected, the polar nature of the N-oxide bond confers a greatly enhanced water solubility upon these alkaloids, relative to the corresponding tertiary bases, and they are frequently not extracted into nonpolar solvents during the usual acid-base partitioning process. It is therefore probable that many alkaloid N-oxides remain undetected and that their high rate of occurrence in the above-mentioned groups is a consequence of the expectation that they should be present and are therefore deliberately sought out. In spite of such expectations, alkaloid isolation from plant genera well known to contain N-oxides is frequently performed without steps being taken to ensure the extraction of the N-oxide components. This is particularly unfortunate in the pyrrolizidine alkaloid-containing genus *Senecio*,[28,29] in which not only is the total alkaloid content often quite high, comprising several percent of the dry weight of plant material, but the greater part of this total is made up of the N-oxide rather than the tertiary base. For example, *S. riddellii* has been shown to have a peak total alkaloid content exceeding 10% of the plant dry weight, with 92% of this being riddelliine N-oxide (**13**) and 8% being riddelliine (**14**).[33]

When N-oxides have been deliberately sought, they have generally been isolated by reduction to the tertiary base and coextraction with the naturally occurring tertiary base fraction. This has led to the consequence that N-oxides are frequently regarded as artifacts or derivatives of the "parent" alkaloid, even though they have been shown to exist in the plant. This viewpoint is particularly unfortunate, since the N-oxides should be considered as

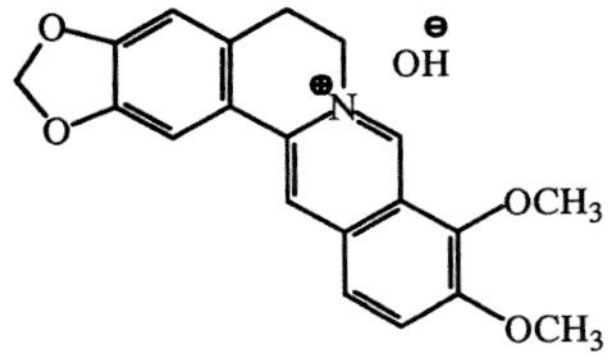

(10) 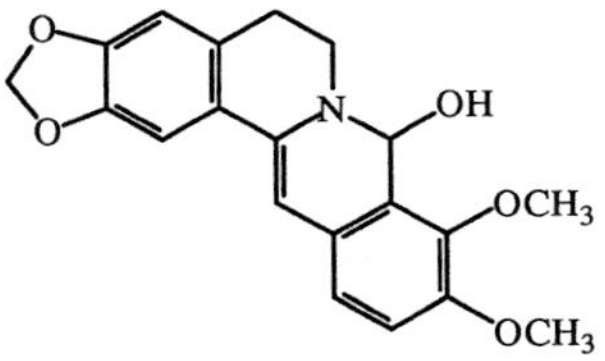 (11)

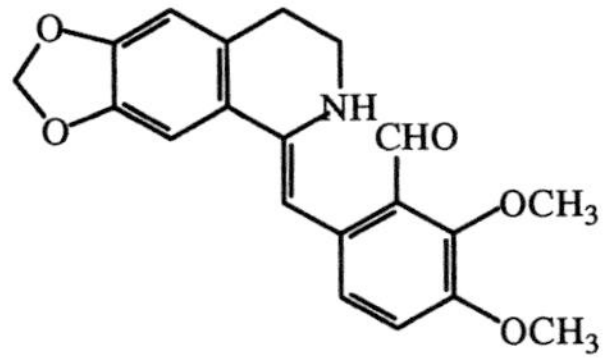

(12)

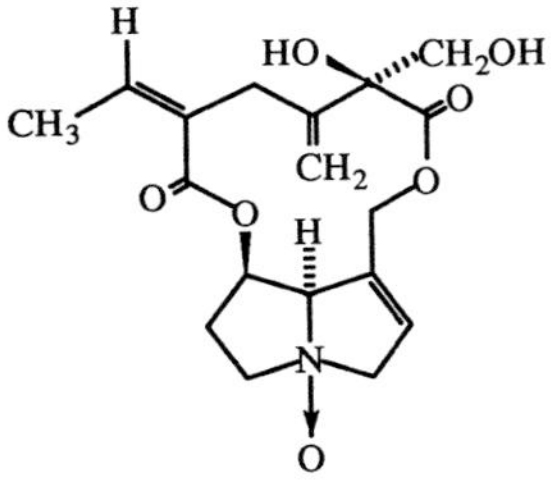

(13) 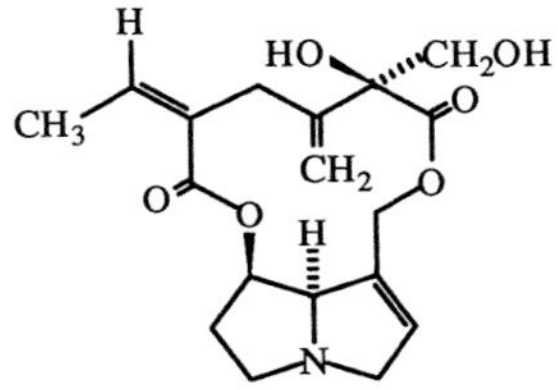 (14)

alkaloids in their own right, with quite different metabolic pathways in both plants and animals. In plants, conversion of tertiary bases to N-oxides may provide the increased solubility essential for transport of the alkaloid through the vascular system and thus account for the particularly high N-oxide content at certain growth stages.[34] As a general rule, N-oxides are less toxic than the tertiary bases in animals, and N-oxidation is a common detoxification and elimination mechanism.

Certain N-oxides are amenable to extraction from basified aqueous solution with chloroform. Usually these are alkaloids possessing a significant aliphatic moiety in their structures. Notable for their tendency to partition in this way are tropane, nicotine, and nonmacrocyclic diester pyrrolizidine alkaloids.[28,29,32] Increasing degrees of hydroxylation appear to prevent such partitioning. For example, seneciphylline N-oxide (**15**) is extractable into chloroform, whereas retrorsine N-oxide (**16**) is not.[28] However, even those N-oxides with appreciable solubility in organic solvents may not be completely extracted, and therefore it is better to isolate them via alternative methods. Moreover, N-oxides isolated by direct extraction may not be recognized as such, since the N–O bond is quite labile and the alkaloid may undergo thermal deoxygenation when analyzed by mass spectrometry, to give a spectrum virtually identical to that of the free base.

N-oxides are therefore best isolated from plant extracts by reduction to their corresponding tertiary bases that are then amenable to conventional alkaloid extraction methods. This

(15)   (16)

conversion (monitored, for example, by TLC) can be accomplished concurrently with tertiary base isolation, or the free bases can be extracted just prior to reduction of the N-oxides. The latter approach has the advantage that it provides a measurement of the free base/N-oxide ratio. Reduction can be achieved by zinc dust in acid solution, sulfurous acid, or ion-exchange chromatography on a reductive resin. The N-oxide may be regenerated by oxidation, usually with hydrogen peroxide or *m*-chloroperbenzoic acid. It must be recognized, however, that certain structural situations exist in which reoxidation may yield an N-oxide epimeric to the naturally occurring alkaloid or give a mixture of epimers. When such possibilities exist, it is essential that the synthetic N-oxide be established as identical with the natural product. This can be done by comparison of the synthetic compound with the N-oxide present in the plant extract, using thin-layer chromatography or other analytical techniques. An additional complication is that alkaloids having two potentially oxidizable N atoms may yield a di-N-oxide on reoxidation, while existing in the plant as a mono-N-oxide. Thus, five N-oxides can be synthesized from nicotine: namely nicotine-1-N-oxide (**17**), *cis-* and *trans*-nicotine-1'-N-oxides (**18**), and *cis-* and *trans*-nicotine-1,1'-di-N-oxides (**19**). However, only the 1'-N-oxide diastereomers (**18**) have been detected in *Nicotiana tabacum, N. affinis,* and *N. sylvestris.*[35] NMR shifts of the carbon atoms adjacent to the N–O functionality are particularly diagnostic of N-oxide structure in that the positive charge on the nitrogen atom causes large downfield shifts, while other atoms in the molecule are little affected.[36]

1. Zinc/Mineral Acid Reduction

The most general method employed for conversion of N-oxides to the corresponding free bases is reduction with zinc dust in mineral acid. Hydrochloric or sulfuric acids are usually used, with the former being most common. As with the conventional acid-base partitioning process, the use of hydrochloric acid always raises the possibility that chlorine-containing alkaloids may be artifacts, and sulfuric acid should, perhaps, be the reagent of choice. Treatment with barium hydroxide at a later stage removes the sulfate anions as insoluble barium sulfate.

Zinc dust reduction is a fairly mild process, complete reduction usually being achieved in 1 to 2 h at room temperature. The course of the reaction can be monitored by basification of a small portion of the solution and screening by thin-layer chromatography; the N-oxides are significantly less mobile than the free-base alkaloids.[37] Work-up of the reaction is accomplished by filtration to remove the zinc dust and basification to greater than pH 9 so that the initially precipitated zinc hydroxide redissolves. The tertiary base is then extracted with chloroform. In spite of the mildness of the process, it is essential that a retrospective view of the overall reaction be employed, since functional groups other than the N-oxide may also be reduced.

The zinc/mineral acid reduction technique has been applied to a comprehensive analysis of the variation, with growth stage, of the total alkaloid content and free base/N-oxide ratios of pyrrolizidine alkaloids in a number of *Senecio* and *Crotalaria* species.[38,39] In spite of the

(17)   (18)   (19)

potential problems mentioned above, it should be applied routinely in order to establish whether or not N-oxides are present in any given extract.

2. Sulfurous Acid Reduction

While the reduction of both alkaloidal and synthetic N-oxides with sulfur dioxide or sulfurous acid has been studied in some detail,[32] it has rarely been used as a method for the isolation of N-oxides. This is probably appropriate, since quite stable sulfamic acids may be formed that would be no more readily extractable from aqueous solution than the N-oxides themselves. A more serious defect is that N-demethylation reactions may occur, as has been demonstrated in the aporphine alkaloid series.[40]

3. Ion-Exchange Chromatography

Tertiary base and N-oxide alkaloids can be coisolated from the plant extract by adsorption onto a cation-exchange resin. While acidic and neutral components are not retained, the alkaloids are and can be eluted from the column with dilute ammonium hydroxide together with any other basic constituents. This approach has been applied to the analysis of pyrrolizidine alkaloids in tansy ragwort (*Senecio jacobaea*),[41] and to large-scale extraction of pyrrolizidine alkaloids.[42] As a means of ensuring complete isolation of both forms of the alkaloid, the method has many advantages and should be employed more frequently. The major disadvantage is that N-oxides are relatively unstable and prone to rearrangements. In addition, they may be difficult to separate from complex mixtures, thus requiring that a reduction step to the free bases be employed at some stage of the isolation.

An interesting approach to the latter problem has been the proposed use of a Serdoxit redox polymer column in combination with the ion-exchange column.[43] N-oxides are reduced to the tertiary bases by prior passage through the redox polymer, which is easily regenerated by treatment with 5% sodium dithionite solution. The greatest disadvantages of the use of such redox and ion-exchange resins are their high initial cost and the slowness of processing appreciable volumes of extract. They are probably better employed for analysis of small plant samples than for preparative purposes.

C. POLYHYDROXY ALKALOIDS

The polyhydroxy alkaloids have recently undergone a significant expansion in both their numbers and structural diversity. This increase has resulted from the discovery that many of the alkaloids are potent glycosidase inhibitors.[44] The group as a whole comprises relatively simple monocyclic pyrrolidine and piperidine alkaloids, together with bicyclic pyrrolizidine and indolizidine ring systems, bearing varying numbers of hydroxyl groups. The hydroxylation pattern confers a greater or lesser degree of water solubility on these alkaloids, which accounts for their cryptic nature. More complex polyhydroxylated ring systems also occur in nature, but the carbon skeleton and other lipophilic functionalities enable such compounds to be extracted by conventional techniques.

In addition to being water soluble, many polyhydroxy alkaloids give very weak responses to Dragendorff's and other alkaloid detection reagents. A brown color is often observed with ninhydrin,[45] quite different from the purple color generated by amino acids. Structurally

(20)   (21)   (22)

(23)   (24)

specific reagents therefore have to be devised in order to detect the alkaloids with any degree of certainty.[46] Alkaloid fractions can be analyzed for number and purity of individual alkaloids by gas chromatography of the trimethylsilyl ether derivatives.[47]

Two methods have been utilized to isolate polyhydroxy alkaloids from aqueous solution: namely direct extraction and ion-exchange chromatography. Precipitation of the alkaloids as insoluble reineckate or picrate salts has apparently not been attempted, although the technique might be expected to have distinct advantages for work up of large-scale extracts.

1. Direct Extraction

Saturated and unsaturated necine bases, such as retronecine (**20**) and platynecine (**21**), which represent the fundamental structural moieties of the pyrrolizidine alkaloid class, occur in nature primarily as mono- and diesters (e.g., riddelliine [**14**]). These esterified alkaloids are isolated in the usual manner, but the water-soluble necine bases that may be present, albeit generally in small quantities, are usually incompletely extracted by chloroform. More highly hydroxylated bases, such as rosmarinecine (**22**), are even less likely to be extracted, and the necine N-oxides are virtually insoluble in nonpolar organic solvents.

Since necine bases are the biosynthetic precursors of the more complex pyrrolizidines, it is important to determine the relative proportions of free and esterified bases in plants. This problem has been studied in depth for a number of *Heliotropium* species. Monohydroxy necines are extracted readily into chloroform, but the more highly hydroxylated bases can only be extracted from concentrated aqueous solutions with chloroform-ethanol mixtures, 3:1 or 2:1 ratios typically providing complete extraction. The method has been applied to isolation of both necine bases free in the plant and to those generated on hydrolysis of the esterified alkaloids.[48]

The degree of hydroxylation alone is not an adequate indicator of chloroform solubility. Thus, while the dihydroxyindolizidine alkaloid lentiginosine (**23**) is insoluble, its 2-epimer is sufficiently soluble for its NMR spectrum to be measured in chloroform solution.[49] Moreover, the trihydroxylated indolizidine, swainsonine (**24**), an important α-mannosidase inhibitor, was first isolated from *Swainsona canescens* by extraction into ethyl acetate, then purified by cation-exchange chromatography, extracted from active fractions with ammoniated chloroform, and crystallized from chloroform.[50] The N-oxide, with which it co-occurs in *Astragalus lentiginosus*, was found to be insoluble in this solvent and was separated from the tertiary base in this manner.[51]

Swainsonine has also been isolated, from the fungus *Rhizoctonia leguminicola*, by peracetylation of the aqueous residue, which yields derivatives having conventional solubility properties in chloroform.[52] While this technique has considerable advantages and proved

(25)   (26)

efficacious in this instance, it is burdened by the liability that the parent alkaloid must be regenerated by hydrolysis, with consequent potential for rearrangement or degradation of labile compounds.

2. Ion-Exchange Chromatography

While direct extraction has some utility for those compounds with a certain degree of solubility in organic solvents, the method of most general utility for isolation of polyhydroxy alkaloids is ion-exchange chromatography. In fact, this technique may provide a many-fold enrichment of alkaloid content even before direct extraction is attempted.

The resins most commonly used are Dowex 50 or Amberlite CG120 in their NH_4^+ forms. The crude plant extract is typically suspended in water and acidified to about pH 4 with hydrochloric acid, and insoluble material is removed by filtration. The aqueous solution is applied to the column and neutral or acidic substances eluted with water. The alkaloids and other basic substances, including a few basic amino acids, are eluted from the column with dilute ammonium hydroxide.[50] A variant of this technique is the use of Dowex 50 in its pyridinium ion form, nonbasic substances being displaced with pyridine, and the alkaloids being eluted with dilute ammonium hydroxide as before.[53] A certain degree of separation of individual alkaloids may be achieved by collection of fractions from the ion-exchange column.

Pipecolic acid-derived alkaloids are insufficiently basic for purification on the above ion-exchange resins. Preliminary purification of dihydroxypipecolic acids from *Calliandra* species has been performed on CG120 (H^+ form) resin eluted with dilute ammonium hydroxide followed by additional purification on a Dowex 50 (H^+ form) column eluted with 1-2N hydrochloric acid.[54]

Following ion-exchange chromatography, the alkaloid fraction is often sufficiently pure that major components can be obtained by crystallization, while minor constituents may be crystallizable after additional thin-layer chromatographic purification. This is well illustrated by the isolation of the potent α-glucosidase inhibitors castanospermine (25)[45] and australine (26)[55] from the seeds of *Castanospermum australe*. Other alkaloids such as 6-*epi*castanospermine[56] exist only as oils, but the hydrochloride salts are generally crystalline and suitable for X-ray structure determination.[57]

The ion-exchange purification technique suffers from the disadvantage that it is rather tedious and time consuming, especially when applied to large-scale plant extracts. While initial costs may be quite high, the resins may be reactivated and used for many separations. The disadvantages have been shown to be far outweighed by the isolation of novel biologically active compounds that have greatly enhanced the significance of polyhydroxy alkaloids in recent years.[58,59]

IV. CONCLUSIONS

Cryptic alkaloids, which are excluded from isolation by conventional techniques due to their water solubility, have too often been overlooked. This chapter has presented examples that illustrate the disadvantages of neglecting alkaloids that may have potent and useful biologic activities, merely because they do not conform to established separation and purification strategies. Techniques have been presented, based upon specific structural features, that enable such compounds to be detected and isolated. It is anticipated that the more frequent adoption of these techniques will lead to hitherto concealed alkaloids possessing novel bioactive properties.

REFERENCES

1. For a comprehensive, practical definition of the term "alkaloid" see: **Pelletier, S. W.**, The nature and definition of an alkaloid, in *Alkaloids: Chemical and Biological Perspectives*, Vol. 1, Pelletier, S. W., Ed., Wiley-Interscience, New York, 1983, 1.
2. **Waller G. R. and Nowacki, E. K.**, *Alkaloid Biology and Metabolism in Plants*, Plenum Press, New York, 1978.
3. **Cromwell, B. T.**, The alkaloids, in *Modern Methods of Plant Analysis*, Vol. 4, Paech, K. and Tracey, M. V., Eds., Springer-Verlag, Berlin, 1955, 367.
4. **Manske, R. H. F.**, Sources of alkaloids and their isolation, in *The Alkaloids. Chemistry and Physiology*, Vol. 1, Manske, R. H. F. and Holmes, H. L., Eds., Academic Press, New York, 1950, 1.
5. **Manske, R. H. F. and Holmes, H. L., Eds.**, *The Alkaloids. Chemistry and Physiology*, Vol. 1, Academic Press, New York, 1950, and subsequent volumes.
6. **Pelletier, S. W., Ed.**, *Alkaloids: Chemical and Biological Perspectives*, Vol. 1, Wiley-Interscience, New York, 1983, and subsequent volumes.
7. **Saxton, J. E., Ed.**, *Specialist Periodical Reports. The Alkaloids*, Vol 1, Chemical Society, London, 1971, and subsequent volumes.
8. **James, W. O.**, Alkaloids in the plant, in *The Alkaloids. Chemistry and Physiology*, Vol. 1, Manske, R. H. F. and Holmes, H. L., Eds., Academic Press, New York, 1950, 15.
9. **Grundon, M. F.**, Quinoline alkaloids related to anthranilic acid, in *The Alkaloids. Chemistry and Physiology*, Vol. 17, Manske, R. H. F. and Rodrigo, R. G. A., Eds., Academic Press, New York, 1979, 105.
10. **Bentley, K. W.**, *The Isoquinoline Alkaloids*, Pergamon Press, Oxford, 1965.
11. **Deulofeu, V., Comin, J., and Vernengo, M. J.**, The benzylisoquinoline alkaloids, in *The Alkaloids. Chemistry and Physiology*, Vol. 10, Manske, R. H. F., Ed., Academic Press, New York, 1968, 401.
12. **Cava, M. P., Buck, K. T., and Stuart, K. L.**, The bisbenzylisoquinoline alkaloids — occurrence, structure and pharmacology, in *The Alkaloids. Chemistry and Physiology*, Vol. 16, Manske, R. H. F., Ed., Academic Press, New York, 1977, 249.
13. **Kametani, T. and Honda, T.**, Aporphine alkaloids, in *The Alkaloids. Chemistry and Physiology*, Vol. 24, Brossi, A., Ed., Academic Press, New York, 1985, 153.
14. **Šimánik, V.**, Benzophenanthridine alkaloids, in *The Alkaloids. Chemistry and Physiology*, Vol. 26, Brossi, A., Ed., Academic Press, New York, 1985, 185.
15. **Szántay, C., Blaskó, G., Honty, K., and Dörnyei, G.**, Corynantheine, yohimbine, and related alkaloids, in *The Alkaloids. Chemistry and Physiology*, Vol. 27, Brossi, A., Ed., Academic Press, New York, 1986, 131.
16. **Cordell, G. A. and Saxton, J. E.**, Bisindole alkaloids, in *The Alkaloids. Chemistry and Physiology*, Vol. 20, Manske, R. H. F. and Rodrigo, R. G. A., Eds., Academic Press, New York, 1981, 1.
17. **Battersby, A. R. and Hodson, H. F.**, Alkaloids of calabash curare and *Strychnos* species, in *The Alkaloids. Chemistry and Physiology*, Vol. 11, Manske, R. H. F., Ed., Academic Press, New York, 1968, 189.
18. **Ghosal, S. and Dutta, S. K.**, Alkaloids of *Abrus precatorius*, *Phytochemistry*, 10, 195, 1971.
19. **Cornforth, J. W. and Henry, A. J.**, The isolation of L-stachydrine from the fruit of *Capparis tomentosa*, *J. Chem. Soc.*, 601, 1952.
20. **Hart, N. K. and Price, J. R.**, Alkaloids of the Australian Rutaceae: *Lunasia quercifolia*. III. Isolation of (–)-O-methylluninium salts, *Aust. J. Chem.*, 19, 2185, 1966.
21. **Boyd, D. R. and Grundon, M. F.**, Quinoline alkaloids. Part X. (+)-Platydesminium salt and other alkaloids from *Skimmia japonica* Thumb. The synthesis of eduline, *J. Chem. Soc. (C)*, 556, 1970.
22. **Battersby, A. R., Binks, R., Hodson, H. F., and Yeowell, D. A.**, Alkaloids of calabash-curare and *Strychnos* species. Part II. Isolation of new alkaloids, *J. Chem. Soc.*, 1848, 1960.
23. **Battersby, A. R. and Hodson, H. F.**, Alkaloids of calabash-curare and *Strychnos* species, *Q. Rev.*, 14, 77, 1960.

24. **Battersby, A. R.,** Recent researches on indole alkaloids, *Pure Appl. Chem.,* 6, 471, 1963.
25. **Asmis, H., Bächli, E., Giesbrecht, E., Kebrle, J., Schmid, H., and Karrer, P.,** Über weitere aus Calebassen isolierte quärtare Alkaloide. II. Mitteilung über Curare-Alkaloide aus Calebassen, *Helv. Chim. Acta,* 37, 1968, 1954.
26. **Hedman, K. and Leander, K.,** Studies on Orchidaceae alkaloids. XXVII. Quaternary salts of the dendrobine type from *Dendrobium nobile* Lindl., *Acta Chem. Scand.,* 26, 3177, 1972.
27. **Roeder, E. and Liu, K.,** Pyrrolizidine alkaloids from *Senecio integrifolius* var. *fauriri, Phytochemistry,* 30, 1734, 1991.
28. **Bull, L. B., Culvenor, C. C. J., and Dick, A. T.,** *The Pyrrolizidine Alkaloids. Their Chemistry, Pathogenicity and Other Biological Properties,* North-Holland, Amsterdam, 1968.
29. **Mattocks, A. R.,** *Chemistry and Toxicology of Pyrrolizidine Alkaloids,* Academic Press, London, 1986.
30. **Schiff, P. L., Jr.,** The *Thalictrum* alkaloids: chemistry and pharmacology, in *Alkaloids: Chemical and Biological Perspectives,* Vol. 5, Pelletier, S. W., Ed., Wiley-Interscience, New York, 1987, 271.
31. **Perkin, W. H., Jr.,** Berberine. Part II, *J. Chem. Soc.,* 992, 1890.
32. **Phillipson, J. D. and Handa, S. S.,** Alkaloid N-oxides. A review of recent developments, *J. Nat. Prod.,* 41, 385, 1978.
33. **Molyneux, R. J. and Johnson, A. E.,** Extraordinary levels of production of pyrrolizidine alkaloids in *Senecio riddellii, J. Nat. Prod.,* 47, 1030, 1984.
34. **Sander, H. and Hartmann, T.,** Site of synthesis, metabolism and translocation of senecionine N-oxide in cultured roots of *Senecio erucifolius, Plant Cell Tissue Organ Cult.,* 18, 19, 1989.
35. **Phillipson, J. D. and Handa, S. S.,** Nicotine N-oxides, *Phytochemistry,* 14, 2683, 1975.
36. **Molyneux, R. J., Roitman, J. N., Benson, M., and Lundin, R. E.,** ^{13}C NMR spectroscopy of pyrrolizidine alkaloids, *Phytochemistry,* 21, 439, 1982.
37. **Molyneux, R. J. and Roitman, J. N.,** Specific detection of pyrrolizidine alkaloids on thin-layer chromatograms, *J. Chromatogr.,* 195, 412, 1980.
38. **Molyneux, R. J., Johnson, A. E., Roitman, J. N., and Benson, M. E.,** Chemistry of toxic range plants. Determination of pyrrolizidine alkaloid content and composition in *Senecio* species by NMR spectroscopy, *J. Agric. Food Chem.,* 27, 494, 1979.
39. **Johnson, A. E., Molyneux, R. J., and Merrill, G. G.,** Chemistry of toxic range plants. Variation in pyrrolizidine alkaloid content of *Senecio, Amsinckia* and *Crotalaria* species, *J. Agric. Food Chem.,* 33, 50, 1985.
40. **Cava, M. P. and Srinivasan, M.,** Conversion of aporphines into N-noraporphine alkaloids, *J. Org. Chem.,* 37, 330, 1972.
41. **Ramsdell, H. S. and Buhler, D. R.,** Analytical isolation of tansy ragwort (*Senecio jacobaea*) pyrrolizidine alkaloids, in *Proc. Symp. Pyrrolizidine (Senecio) Alkaloids: Toxicity, Metabolism, and Poisonous Plant Control Measures,* Cheeke, P. R., Ed., Oregon State University, Corvallis, 1979, 19.
42. **Deagen, J. T. and Deinzer, M. L.,** Improvements in the extraction of pyrrolizidine alkaloids, *J. Nat. Prod.,* 40, 395, 1977.
43. **Huizing, H. J. and Malingré, T. M.,** Reduction of pyrrolizidine N-oxides by the use of a redox polymer, *J. Chromatogr.,* 173, 187, 1979.
44. **Elbein, A. D. and Molyneux, R. J.,** The chemistry and biochemistry of simple indolizidine and related polyhydroxy alkaloids, in *Alkaloids: Chemical and Biological Perspectives,* Vol. 5, Pelletier, S. W., Ed., Wiley-Interscience, New York, 1987, 1.
45. **Hohenschutz, L. D., Bell, E. A., Jewess, P. J., Leworthy, D. P., Pryce, R. J., Arnold, E., and Clardy, J.,** Castanospermine, a 1,6,7,8-tetrahydroxyoctahydroindolizine alkaloid, from seeds of *Castanospermum australe, Phytochemistry,* 20, 811, 1981.
46. **Molyneux, R. J., James, L. F., Panter, K. E., and Ralphs, M. H.,** Analysis and distribution of swainsonine and related polyhydroxyindolizidine alkaloids by thin layer chromatography, *Phytochem. Anal.,* 2, 125, 1991.
47. **Nash, R. J., Goldstein, W. S., Evans, S. V., and Fellows, L. E.,** Gas chromatographic method for separation of nine polyhydroxy alkaloids, *J. Chromatogr.,* 366, 431, 1986.
48. **Birecka, H., Frohlich, M. W., and Glickman, L. M.,** Free and esterified necines in *Heliotropium* species from Mexico and Texas, *Phytochemistry,* 22, 1167, 1983.
49. **Pastuszak, I., Molyneux, R. J., James, L. F., and Elbein, A. D.,** Lentiginosine, a dihydroxyindolizidine alkaloid that inhibits amyloglucosidase, *Biochemistry,* 29, 1886, 1990.
50. **Colegate, S. M., Dorling, P. R., and Huxtable, C. R.,** A spectroscopic investigation of swainsonine: An α-mannosidase inhibitor isolated from *Swainsona canescens, Aust. J. Chem.,* 32, 2257, 1979.
51. **Molyneux, R. J. and James, L. F.,** Loco intoxication. Indolizidine alkaloids of spotted locoweed (*Astragalus lentiginosus*), *Science,* 216, 109, 1982.
52. **Schneider, M. J., Ungemach, F. S., Broquist, H. P., and Harris, T. M.,** (1S, 2R, 8R, 8aR)-1,2,8-Trihydroxyoctahydroindolizine (swainsonine), an α-mannosidase inhibitor from *Rhizoctonia leguminicola, Tetrahedron,* 39, 29, 1983.

53. **Nash, R. J., Bell, E. A., and Williams, J. M.,** 2-Hydroxymethyl-3,4-dihydroxypyrrolidine in fruits of *Angylocalyx boutiqueanus, Phytochemistry,* 24, 1620, 1985.
54. **Romeo, J. T., Swain, L. E., and Bleeker, A. B.,** *cis*-4-Hydroxypipecolic acid and 2,4-*cis*-4,5-*trans*-4,5-dihydroxypipecolic acid from *Calliandra, Phytochemistry,* 22, 1615, 1983.
55. **Molyneux, R. J., Benson, M., Wong, R. Y., Tropea, J. E., and Elbein, A. D.,** Australine, a novel pyrrolizidine alkaloid glucosidase inhibitor from *Castanospermum australe, J. Nat. Prod.,* 51, 1198, 1988.
56. **Molyneux, R. J., Roitman, J. N., Dunnheim, G., Szumilo, T., and Elbein, A. D.,** 6-*Epi*castanospermine, a novel indolizidine alkaloid that inhibits α-glucosidase, *Arch. Biochem. Biophys.,* 251, 450, 1986.
57. **Nash, R. J., Fellows, L. E., Girdhar, A., Fleet, G. W. J., Peach, J. M., Watkin, D. J., and Hegarty, M. P.,** X-ray crystal structure of the hydrochloride of 6-*epi*castanospermine [(1S,6R,7R,8R,8aR)-1,6,7,8-tetrahydroxyoctahydroindolizine], *Phytochemistry,* 29, 1356, 1990.
58. **Fellows, L. E. and Fleet, G. W. J.,** Alkaloidal glycosidase inhibitors from plants, in *Natural Products Isolation,* Wagman, G. H. and Cooper, R., Eds., Elsevier, Amsterdam, 1989, 539.
59. **Molyneux, R. J.,** Polyhydroxy indolizidines and related alkaloids, in *Methods in Plant Biochemistry,* Vol. 8, Waterman, P. G., Ed., Academic Press, London, 1993, 511.

Chapter 4

NUCLEAR MAGNETIC RESONANCE SPECTROSCOPY STRATEGIES FOR STRUCTURAL DETERMINATION

Lindsay T. Byrne

TABLE OF CONTENTS

0-8493-4372-0/93/$0.00+$.50
© 1993 by CRC Press, Inc.

I. INTRODUCTION

A glance at any recent chemistry journal immediately illustrates the reliance that modern chemistry places on nuclear magnetic resonance (NMR) spectroscopy as a quick and reliable method for the elucidation of molecular structures. NMR spectroscopy has been the single most important physical method for the determination of molecular structures for more than 30 years. The power of the technique lies in that it not only defines the numbers and types of nuclei present in an organic molecule, but it also describes their individual chemical environments and, more importantly, the way they are interconnected. Driven by its potential to determine the structures of organic compounds, NMR spectroscopy has seen substantial development in the four and a half decades since the first experiments. In particular, the implementation of the pulsed Fourier transform method[1] and, subsequently, the concept of two-dimensional experiments[2] provided the seeds for vibrant growth. There are currently hundreds of multipulse experiments available to the NMR spectroscopist. However, only a small proportion of these procedures are regularly employed for the solution of molecular structures. The most useful experiments have been the subject of numerous reviews.[3-12]

Within the context of the multidisciplinary nature of bioactive natural product research, this chapter will briefly review the most utilized NMR experiments, with the aim of highlighting the types of information that each can provide. Theoretical and experimental details of each procedure will be kept to a minimum, with references to the literature for readers who require further information. The first section of this chapter briefly introduces the essential concepts. Subsequently, the most useful NMR techniques will be introduced approximately in the order that they would be applied to solve the structure of an unknown compound. Sample spectra have been chosen with the aim of clearly demonstrating each technique, without the need for detailed argument.

II. BASIC PRINCIPLES OF NMR SPECTROSCOPY

Descriptions of the fundamental aspects of NMR spectroscopy can be found, in varying degrees of detail, in most of the numerous NMR books available. However, those published before 1980 may not describe all of the concepts necessary for the understanding of modern pulsed techniques. These principles have been covered in concise,[13,14] more detailed,[15-17] and comprehensive form.[12] Two books directed at the organic chemist that describe modern spectroscopic methods are also recommended.[10,11]

TABLE 1
Properties of Some Nuclei with I = 1/2

Isotope	Natural abundance (%)	NMR frequency (MHz)		Relative sensitivity[a,b]
		at 4.70 Telsa	at 11.74 Telsa	
^1H	99.98	200.00	500.00	1.0
^{13}C	1.11	50.29	125.72	1.59×10^{-2}
^{15}N	0.37	20.26	50.66	1.04×10^{-3}
^{31}P	100.00	80.96	202.40	6.63×10^{-2}

[a] For equal number of nuclei at constant field.
[b] The expected relative strength of an NMR signal can be obtained from the product of the relative sensitivity and the natural abundance.

A. THE CONDITIONS NECESSARY FOR RESONANCE

When placed in a magnetic field, nuclei that have a nonzero spin quantum number (I) are able to absorb energy from the radio frequency range of the electromagnetic spectrum. The frequency (υ_0) at which a particular nucleus absorbs energy is dependent on the type of nucleus concerned (characterized by its gyromagnetic ratio — γ) and the strength of the magnetic field (B_0) into which it is placed. This relationship is described mathematically by the Lamor equation:

$$\upsilon_0 = \gamma B_0 / 2\pi \tag{4.1}$$

In the study of natural compounds, using NMR spectroscopy, the nuclei of prime concern are ^1H and ^{13}C, both of which have a spin quantum number of 1/2. It should be noted that both of the isotopes ^{12}C and ^{16}O have I = 0 and therefore do not give an NMR signal. The magnetic properties of some nuclei that have a spin quantum number of 1/2 are shown in Table 1.

In a magnetic field the nuclei of an isotope with I = 1/2 can occupy either of two energy states. Nuclei in the lower energy state precess about the direction of the magnetic field, while those in the higher energy state have the opposite orientation. The relative populations of the two states is described by the Boltzmann distribution, which results in only an extremely small population excess in the lower energy state. This small population excess gives rise to the net absorption of energy when the nuclei are irradiated with the correct radio frequency. The absorbed energy is subsequently lost to the surroundings (the lattice) over a period of time. This process is termed relaxation and is characterized by the spin-lattice relaxation time (T_1). The spin-lattice relaxation times of protons (^1H nuclei) are usually of the order of seconds, while those of ^{13}C nuclei may be tens of seconds. A second relaxation mechanism (spin–spin relaxation) involves the exchange of energy between nuclear spins and has an associated spin–spin relaxation time (T_2). The magnitudes of the relaxation times of the nuclei in a particular sample must be considered in the preparation of each NMR experiment.

B. THE CHEMICAL SHIFT

Soon after the first observation of nuclear magnetic resonance, it became clear that the specific environment of a nucleus slightly modifies the magnetic field it experiences.[18] The nuclei are shielded from the applied magnetic field to differing extents, depending on the electron density about each nucleus. Thus, the frequency at which a nucleus is able to absorb energy is characteristic of the environment of the particular nucleus, and it is this effect — the chemical shift — together with the phenomenon of spin–spin coupling (Section II.C.)

that makes nuclear magnetic resonance spectroscopy so useful for structural determinations. The chemical shift (δ, expressed in parts per million, ppm) is defined as in Equation 4.2.[19,20]

$$\delta(\text{ppm}) = 10^6 \left(\upsilon_{\text{sample}} - \upsilon_{\text{reference}}\right) \Big/ \upsilon_{\text{reference}} \qquad (4.2)$$

The chemical shift is thus independent of the magnetic field strength of the spectrometer used to record a spectrum. Tetramethylsilane (TMS) is used as the reference for both ^{1}H and ^{13}C spectra and is assigned the chemical shift of 0.00 ppm. In the NMR spectra of natural products, most signals for the ^{1}H nuclei occur within a range of 10 ppm, while the ^{13}C nuclei resonate over a range of 250 ppm.

C. SPIN–SPIN COUPLING

On examination of a typical ^{1}H spectrum, it is immediately obvious that most of the resonances do not appear as single lines (for example see Figure 3). Many of the signals are split into multiplets due to the effect of neighboring nuclei. This phenomenon, which is known as spin–spin coupling, occurs via the bonding electrons and thus provides information about the interconnection of coupled nuclei. In the simplest case, spin–spin coupling between two chemically different nuclei (each with spin $^1/_2$) results in the resonance of each being split into two lines: a doublet (d). The separation between the two lines is the same for each resonance and is known as the coupling constant (J), which is expressed in hertz. The magnitude of J depends on both the nature and number of bonds involved and the angular relationship of the coupled nuclei, but is independent of the magnetic field strength at which the spectrum is recorded. Each signal may show coupling to several nuclei. In documentation, J may be preceded by a superscripted number, indicating the number of bonds through which the nuclei are coupled, and followed by a subscript, indicating the nuclei involved, e.g., $^2J_{CH}$. The magnitude of three bond coupling constants are dependent on the dihedral angle between the coupled nuclei and thus provide valuable stereochemical information.[21,22] Coupling constants over four or more bonds are often too small to be observed. However, the appearance of such long-range couplings is significant, as it is usually indicative of particular stereochemical arrangements of atoms. There are many simple descriptions of the mechanism of spin–spin coupling and the use of the magnitude of the coupling constant to define the stereochemical relationship of coupled nuclei.[14,17,23] If the chemical shift separation between coupled nuclei is very small compared with their mutual coupling constant, the coupling pattern of each signal will be distorted. The separation between lines in such "second-order" signals is no longer directly related to coupling constants, and more detailed analysis is required.[15,16,23,24] Software for computer-aided analysis of such spin systems is available with most modern spectrometers. Decoupling experiments can be used to remove coupling between two nuclei by irradiating one of the spins with a second radio frequency field of suitable power while the spectrum is acquired. Experiments that facilitate the interpretation of spin–spin coupling networks are described in Section V.

D. THE NUCLEAR OVERHAUSER EFFECT

The nuclear Overhauser effect (NOE) provides a means for establishing the proximity of nuclei in space and thus supplies valuable information about molecular geometry.[25,26] The effect manifests itself as a change in the intensity of the NMR signal of a nucleus when the resonance of a second nucleus is irradiated by an additional radio frequency field. For an intensity change to be apparent, the observed nucleus must relax through the irradiated nucleus via a dipole-dipole mechanism. The magnitude of the effect falls away rapidly as the separation between the irradiated and the observed nucleus increases. The maximum possible intensity change is defined by Equation 4.3.

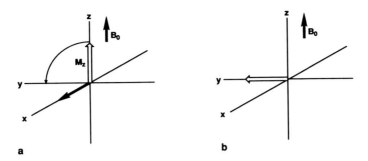

FIGURE 1. (a) The effect of a 90° pulse on the net magnetization (Mz). (b) The magnetization immediately after the pulse.

$$\text{Enhancement Factor}_{max} = \gamma_{saturated} / 2\gamma_{observed} \qquad (4.3)$$

Therefore, in homonuclear experiments, signals may be enhanced by up to 50%, while ^{13}C signals, measured with ^{1}H decoupling, may be increased by 200%.

E. PULSED FOURIER TRANSFORM NMR SPECTROSCOPY[27-31]

For nearly three decades after the first NMR experiments,[32,33] the continuous wave (CW) method was used to record spectra. Although some early experiments employed pulses,[34] the routine use of pulsed Fourier transform NMR spectroscopy was delayed by the limits of computer technology and the absence of a suitable Fourier Transform (FT) algorithm,[35] until the early 1970s. Pulsed FT NMR offered much greater sensitivity than the CW method and thereby, routine access to insensitive nuclei such as ^{13}C. In CW spectrometers each nucleus in the sample is brought into resonance by slowly sweeping the radio frequency (or magnetic field) through the chemical shift range of the isotope concerned. The sweep normally requires hundreds of seconds. Pulsed spectrometers use a short-duration radio frequency pulse to excite all of the nuclei of a particular isotope simultaneously. The resulting signal is recorded over a few seconds. The effect of the pulse is best explained by considering a group of identical nuclei in a rotating frame of reference. In this frame of reference, depicted in Figure 1, the x- and y-axes are spinning about the direction of the magnetic field (the z-axis) at the natural frequency of the nuclei. Thus, the individual nuclei appear stationary, and the excess of nuclei oriented in the direction of the field gives rise to a net magnetization in this direction (M_z). A short pulse of radio frequency applied along the x-axis (solid arrow, Figure 1a) results in the magnetization being tipped toward the y-axis (right-hand rule).[7,12] The receiver coils detect the magnetization in the x-y plane, and the resulting signal (known as a free induction decay (FID); see Figure 2a) is digitized and stored in computer memory. The pulse is described by the angle through which it tips the magnetization. Thus, after a 90° pulse, the magnetization is aligned along the y-axis, giving maximum signal (Figure 1b). A 180° pulse tips the magnetization into the –z direction. The basic pulsed experiment consists of a single pulse (usually 30 to 90°) followed by acquisition of the FID. The process is repeated with the insertion of a suitable delay before each pulse, to allow for complete relaxation, until sufficient signal has been added into computer memory. Fourier transformation of the FID (a function of time) produces the normal spectrum (see Figure 2b).

F. PULSE SEQUENCES AND TWO-DIMENSIONAL NMR EXPERIMENTS

The ability to manipulate the nuclear magnetization by combining pulses of varying lengths into sequences has led to the design of numerous experiments, each providing specific structural information. As an illustration of a pulse sequence, the COSY experiment will be described. Since this procedure is a two-dimensional (2D) experiment, the basic concepts of 2D NMR spectroscopy[36] will be introduced simultaneously.

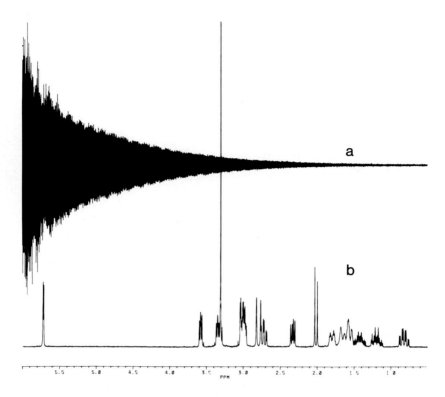

FIGURE 2. (a) A typical Free Induction Decay (FID). (b) The spectrum arising from Fourier transformation of the FID.

The COSY (COrrelation SpectroscopY) pulse sequence, in its simplest and original form,[37] is illustrated as follows :

$$\text{Relaxation Delay} - 90° - t_1 - 90° - \text{FID}\!\left(t_2\right)$$

PREPARATION : EVOLUTION : ACQUISITION

The pulse sequence begins with a delay to allow for relaxation, followed by two 90° pulses separated by a short delay (t_1). After the second 90° pulse the FID is acquired to complete the sequence. The pulse sequence of this 2D experiment is described as having three periods: the preparation period to allow the nuclear spins to reach equilibrium before each cycle, the evolution period during which the nuclear spins are allowed to evolve under the influence of the pulses and the delay, and, finally, the acquisition period. During the COSY experiment the sequence is repeated for a particular value of t_1 until sufficient signal has been acquired by the computer. The process is repeated for a number of different t_1 values (typically 256 or 512), with an FID being stored for each value. Each FID is then subjected to Fourier transformation to yield a spectrum. If these spectra are viewed, the appearance of each signal differs from spectrum to spectrum, depending on the particular value of t_1 used. Fourier transformation, with respect to t_1, gives the COSY spectrum that is a surface in 3D space. Two of the axes are frequencies (giving rise to the term "two-dimensional NMR spectroscopy"), while the third is intensity. The spectrum is best plotted as a contour map (for example see Figure 8) similar to its geographical equivalent. The COSY spectrum shows which nuclei are involved in spin–spin coupling interactions, and its interpretation is described in Section

V. The preparation, evolution, and acquisition periods are common to all 2D experiments. Some 2D experiments, e.g., the NOESY experiment (which indicates NOE interactions between nuclei), have a mixing period between the evolution and acquisition periods. The choice and use of 2D experiments for elucidating molecular structures, together with their individual attributes, have been described.[6,7,9,38]

G. THE NMR SPECTROMETER

Modern pulsed NMR spectrometers are extremely sophisticated and expensive scientific instruments (quite different from the CW spectrometers of the early 1960s). The spectrometer features a strong, stable, and homogeneous magnetic field supplied by a superconducting magnet. Modern research spectrometers have superconducting magnets with fields ranging from 4.70 Tesla to the current technologic limit of 14.09 Tesla. These magnetic field strengths correspond to ^{1}H resonance frequencies of 200 to 600 MHz (spectrometers are denoted in terms of their ^{1}H resonance frequency). The radio frequency transmitter and receiver are controlled by an advanced computer that also governs the digitization of the analog signals, the mathematical transformation of the data, and the operation of peripheral devices such as plotters, printers, and variable temperature units.

A continuing demand for spectrometers with stronger magnetic fields arises for three reasons. First, the sensitivity of the spectrometer increases significantly with the magnetic field strength ($\alpha B_0^{3/2}$). Second, at higher magnetic fields there is greater separation between signals, which results in significantly less overlap of multiplets, and therefore spectra are more easily interpreted. Finally, the increased separation of signals also results in the simplification of those "second-order" multiplets that, at lower fields, are complicated due to their close proximity. Thus, higher magnetic field strengths not only give greater access to more dilute solutions of the less sensitive nuclei, but the accompanying simplification of spectra allows the study of larger and more complex molecules.

III. STRATEGY FOR SOLVING STRUCTURES

While the exact strategy used to determine a structure is dictated by the individual demands of each sample, a typical procedure is as follows.

First, the routine ^{1}H and ^{13}C spectra are recorded and examined to identify the numbers of each type of nucleus present. This information is integrated with that obtained from other physical methods. The chemical shift of each signal describes the likely environment of the nucleus producing it. Proton coupling networks are then established, and the magnitudes of the coupling constants are used to define the angular relationship and the number of bonds between the coupled nuclei. Subsequently, the proton resonances are correlated with those of the carbon to which they are directly attached. Subunits consisting of contiguous, protonated carbon atoms terminated by heteroatoms or nonprotonated carbon atoms will now be apparent. The subunits may be linked by looking for long-range proton–proton or proton-carbon couplings between the nuclei at the ends of adjoining units. Nuclear Overhauser experiments may also indicate the proximity of nuclei on separate subunits. Usually some logical connections between the subunits can be made due to the obvious presence of well-known skeletal units (e.g., an indole residue). If sufficient sample is available, experiments using carbon-carbon coupling may be employed to assemble units connected by quaternary carbon atoms. Heavy demand on instrument time usually necessitates a careful choice of experiments in order to obtain the required information in the shortest possible time. In the following sections many of the techniques that can be used to implement this strategy are briefly described. More detailed case studies[9,38-44] and worked examples[45] as well as concise examples of the individual experiments[46] are available.

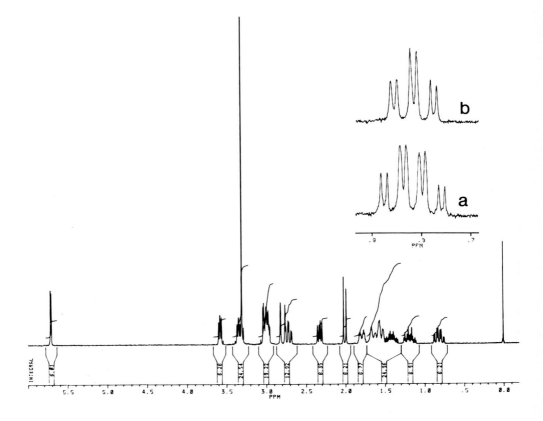

FIGURE 3. The ¹H spectrum of an alkaloid from *Margaritaria indica.* (a) Expansion of the signal at 0.82 ppm; (b) the same signal while irradiating at 3.00 ppm.

IV. OBTAINING THE BASIC INFORMATION

In order to obtain high-resolution spectra, careful preparation of the NMR sample is required. Optimum sample concentrations should be used, since halving the sample concentration increases the time required to acquire a suitable spectrum by a factor of four. The practical aspects of NMR spectroscopy have been described.[11,47]

A. THE ¹H SPECTRUM

Almost all structural studies begin with the recording of an ¹H spectrum, the ¹H nucleus being the most easily observed and, consequently, the most studied. A typical ¹H spectrum of a natural product, recorded at moderate field strength (7.05 Tesla, 300 MHz), is shown in Figure 3. The number of protons giving rise to each signal (and subsequently the number of protons in the molecule) can be determined by comparing the areas under each signal (i.e., from the integrals). The integrals in this spectrum are displayed in both digital form and the conventional form in which measurement of the vertical displacement of the integral line over a multiplet indicates its relative area. Ideally, we would like to see each multiplet completely separated from the others to allow the spectrum to be easily interpreted. In the absence of spectrometers with higher field strengths, overlap may be reduced by employing a different solvent or by using lanthanide shift reagents,[48,49] which form a complex with the molecule in question, resulting in a substantial increase in the chemical shift of nuclei close to the point of complexation.

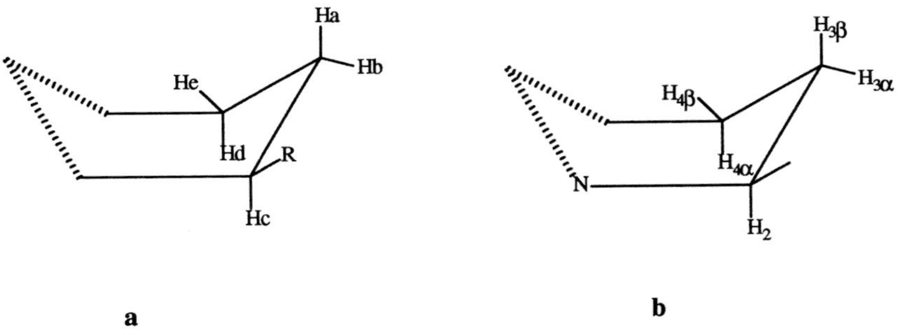

FIGURE 4. (a) The expected arrangement of protons coupled to H_a. (b) The actual arrangement in 15-α-methoxy-14,15-dihydrophyllochrysine (**1**) (i.e., $H_a = H_{3\beta}$).

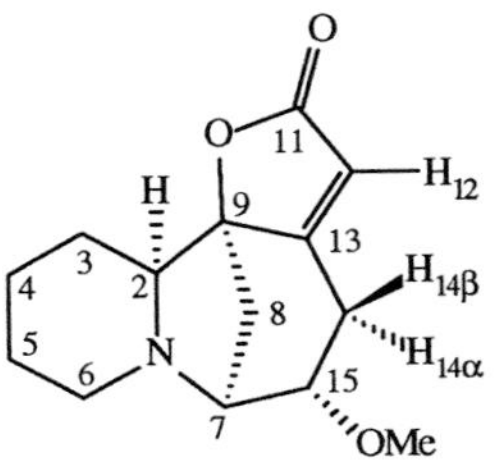

1

The chemical shift of each signal characterizes the environment of the proton from which it arises. Thus, in Figure 3, for example, the signal at 5.71 ppm is due to an olefinic proton, and the three-proton singlet at 3.30 ppm indicates the presence of a methoxyl group in the molecule.

Where the coupling pattern of a multiplet is easily interpretable, the numbers and stereochemical orientation of adjacent protons can be determined. For example, the appearance of the multiplet at 0.82 ppm (Figure 3, inset a) provides significant information about its surroundings. The coupling pattern is indicative of an axial proton on a six-membered ring (such as Ha in Figure 4a) that has two large vicinal diaxial couplings ($^3J_{HH}$ to Hc and Hd) that are of equal magnitude to its geminal coupling constant ($^2J_{HH}$ to Hb), and a smaller axial-equatorial coupling ($^3J_{HH}$ to He). The absence of a second axial-equatorial coupling indicates the likely presence of an adjacent equatorial substituent (R). Similarly, the magnitude of the sole coupling constant of the signal arising from the lone olefinic proton in the spectrum defines its immediate environment. It is typical of allylic coupling (i.e., $^4J_{HH}$ via the double bond), and therefore the carbon to which it is attached must be flanked on both sides by fully substituted carbon atoms. Furthermore, its coupling partner will be oriented at approximately 90° to the plane of the double bond, as it is in this orientation that allylic couplings are maximized.[23] The spectrum shown in Figure 3 is that of 15-α-methoxy-14,15-dihydrophyllochrysine (**1**),[50,51] and the signals just described arise from $H_{3\beta}$ (see Figure 4b) and H_{12} (which is allylically coupled to $H_{14\beta}$), respectively.

It is seldom possible to completely analyze the coupling networks from the ^1H spectrum alone. Experiments that aid the analysis of the proton spin systems are described in Section V. Resonances suspected of being due to exchangeable protons (–OH, –NH) can usually be identified by their disappearance after shaking the NMR sample with D_2O.

The free induction decay may be manipulated using one of the various window functions available on most NMR spectrometers, resulting in either increased sensitivity (useful in the case of dilute solutions) or enhanced resolution. Resolution enhancement techniques aid

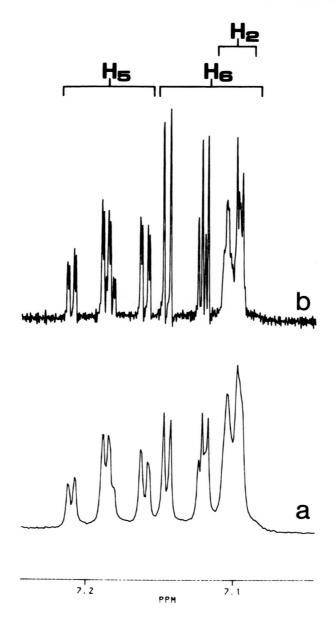

FIGURE 5. A section of the aromatic pattern of the ^{1}H spectrum of gramine (**2**): (a) normal; (b) resolution enhanced.

spectral interpretation by allowing the individual lines of a multiplet to be seen more clearly. Often small, but informative, long-range couplings may be revealed using this procedure. Figure 5 shows a part of the aromatic pattern from the spectrum of gramine (**2**). The advantages of the resolution-enhanced (b) over the normal spectrum (a) are obvious.

Comprehensive tabulated descriptions of proton chemical shifts and coupling constants are available.[23,52,53]

B. THE ^{13}C SPECTRUM

The ^{13}C spectrum[54-57] provides important structural information, since it arises from the nuclei that form the framework of the molecule, in contrast to the ^{1}H nuclei that are at the

2

periphery. It also confirms the presence of carbonyl groups and other nonprotonated carbon atoms whose presence in a molecule can only be inferred from the ^{1}H spectrum. Despite the problem of low relative sensitivity and low natural abundance, modern spectrometers can obtain a ^{13}C spectrum from a few milligrams of material in a relatively short time.

1. The ^{1}H Decoupled ^{13}C Spectrum

The normal ^{13}C spectrum is acquired with full proton decoupling either using "broad band (noise) decoupling" or the more modern "composite pulse[58,59] decoupling," which uses significantly less power, thereby minimizing heating of the sample. In the absence of coupling to ^{1}H nuclei, all of the ^{13}C signals in the spectrum appear as single lines, allowing the number of carbons in the molecule to be readily determined. By employing such decoupling methods, a great improvement in the sensitivity of the ^{13}C experiment is achieved due, in part, to the collapse of multiplets into singlets and, in part, from the nuclear Overhauser effect. The latter effect results in the protonated carbons being enhanced up to nearly three times their intensity without NOE. Nonprotonated carbons can therefore usually be identified by their lower relative intensity in the ^{1}H-decoupled ^{13}C spectrum. Figure 6a shows the composite pulse decoupled spectrum of 15-α-methoxy-14,15-dihydrophyllochrysine (**1**). All 14 carbon signals can be clearly seen. The nonproton-bearing carbons (C_9, C_{11}, and C_{13}) are readily identified, having only one half to one third the intensity of the protonated nuclei. The likely environment of each carbon can be deduced from its chemical shift. The signal at 112.99 ppm falls within the range for olefinic carbons (100 to 170 ppm), and its double-bond partner must be one of the two signals that occur at approximately 170 ppm, which is indicative of an olefinic carbon which is β to a carbonyl group. The other signal in this region must be due to the carbonyl carbon, its chemical shift being characteristic of an ester (lactone) carbonyl group. Similarly, assignments of the other resonances can be made. Knowledge of the number of ^{1}H nuclei attached to each carbon (see Section IV.B.2) is necessary for precise assignment based on chemical shift. In addition to general compilations of ^{13}C data,[54-56] specific information on natural products,[60,61] alkaloids,[62-64] steroids,[65,66] saccharides,[67-69] aromatic compounds,[70,71] and amino acids, peptides, and proteins[72-74] is available.

2. Determining the Number of Directly Attached Hydrogen Atoms

a. Fully Coupled ^{13}C Spectra[16]

Signals in a fully coupled ^{13}C spectrum are split by large $^1J_{CH}$ (typically 125 to 170 Hz, but may be as large as 250 Hz) and smaller, but significant, $^2J_{CH}$ (typically 0 to 8 Hz) and $^3J_{CH}$ (typically 3 to 13 Hz). The number of large (one bond) couplings to a particular carbon signal indicates the number of directly attached hydrogen atoms, while the smaller couplings provide information about protons on neighboring carbons.[54,56,57,70,75,76] In the coupled spectrum of **1**, as displayed in Figure 6b, the smaller two- and three-bond couplings are not resolved. The signals due to quaternary carbons (C_9, C_{11}, and C_{13}) appear as singlets, tertiary carbons (C_2, C_7, C_{12}, and C_{15}) as doublets, secondary carbons (C_3, C_4, C_5, C_6, C_8, and C_{14}) as triplets, and the primary carbon (OCH$_3$) as a quartet.

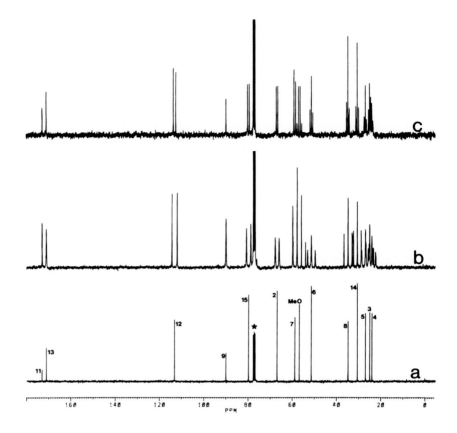

FIGURE 6. ^{13}C spectra of 15-α-methoxy-14,15-dihydrophyllochrysine (**1**): (a) composite pulse decoupled (* CDCl$_3$); (b) fully coupled; (c) single frequency off-resonance decoupled.

Even in this relatively simple spectrum, the multiplicity of some signals is not immediately obvious due to overlap with other resonances. In more complex molecules, extensive overlap of signals can severely limit the extraction of useful information from the coupled ^{13}C spectrum. For this reason, and because the time taken to acquire the data is usually prohibitive, proton-coupled ^{13}C spectra are rarely recorded.

b. Single-Frequency Off Resonance (SFOR) Decoupled Spectra

In this procedure the decoupler is set to irradiate at a single frequency (1000 to 2000 Hz) outside the ^{1}H spectral range. In the resulting spectrum (Figure 6c), one-bond couplings are reduced to 40 to 60 Hz, thus simplifying the interpretation. For example, the quartet at 56.72 ppm, which is due to the methoxyl group, is now clearly separated from the adjacent doublet arising from C$_7$. This method offers a significant sensitivity advantage over fully coupled ^{13}C spectra and was the procedure of choice prior to the introduction of spin echo (Section IV.B.2.c) and polarization transfer (Section IV.B.2.d) techniques. However, carbons attached to strongly coupled protons may be distorted in the SFOR decoupled spectrum, and overlap is still a problem in crowded regions.

c. J Modulated Spin Echo Procedures

Experiments that employ a spin echo[34] modulated by ^{1}J$_{CH}$ have been described by the above title,[10,11] but are also known by several acronyms: SEFT (Spin Echo Fourier Transform),[3] GASPE (GAted SPin Echo),[77] and APT (Attached Proton Test).[78] They are superior to the SFOR decoupling method, because signals appear as singlets, thus simplifying spectra and providing a further large gain in sensitivity. The procedure is represented schematically as follows:

$$^{13}C \qquad\qquad 90° - \tau - 180° - \tau - FID$$

$$^{1}H \text{ decoupler} \qquad OFF \qquad ON$$

The delay between pulses (τ) is set to $1/J_{CH}$, which, in the ensuing spectrum, results in signals due to CH and CH_3 carbons being of opposite phase to those of CH_2 and quaternary carbons. The phase variation arises as the spins evolve under the influence of spin–spin coupling during the period when the decoupler is off. In the J-modulated spin echo spectrum (Figure 7e) of 15-α-methoxy-14,15-dihydrophyllochrysine (**1**), the quaternary and CH_2 signals are negative, and those of CH and CH_3 carbons are positive. The same sequence, using a delay $\tau = 1/2J_{CH}$, produces a spectrum that shows only quaternary carbon resonances.

d. Methods Involving Polarization Transfer — DEPT, INEPT

Currently, the most commonly used method for determining the number of hydrogens bonded to each carbon atom is the DEPT (Distortionless Enhancement by Polarization Transfer) experiment.[79,80] This method, together with its precursor, the INEPT (Insensitive Nuclei Enhancement by Polarization Transfer) experiment,[81] involves transfer of magnetization from protons to their directly attached carbons. Thus, fully substituted carbons do not give signals in DEPT or INEPT spectra. These experiments offer a further sensitivity gain over the J-modulated spin echo procedure. The DEPT pulse sequence can be represented schematically as

$$^{1}H \qquad 90° - \tau - 180° - \tau - \Theta° - \tau - \text{decoupling}$$

$$^{13}C \qquad\qquad 90° - \tau - 180° - \tau - FID$$

The delay (τ) is set to $1/2J_{CH}$. The spectra (Figures 7b–d) of **1** were recorded with the DEPT sequence, using a final ^{1}H pulse angle (Θ) of 45°, 90°, and 135°, respectively. In the DEPT-45° spectrum (Figure 7b), all protonated carbons appear as positive singlets, while in the DEPT-90° spectrum (Figure 7c), only those resonances of carbons bearing one hydrogen can be seen. The DEPT-135° spectrum (Figure 7d) again shows all protonated carbon signals with CH_3 and CH resonances being positive, while CH_2 signals are negative. Thus, the number of hydrogens attached to each carbon in the molecule can readily be determined by comparing the ^{1}H-decoupled (Figure 7a), DEPT-90° and DEPT-135° spectra. The CH and CH_2 signals are immediately obvious from the DEPT-90° and DEPT-135° spectra, respectively. Additionally, positive signals in the DEPT-135° spectrum, which do not appear in the DEPT-90° spectrum, must arise from CH_3 groups. Resonances of quaternary carbons are those appearing in the ^{1}H-decoupled spectrum, but not in the DEPT-135° spectrum. The DEPT-45° spectrum therefore would appear to be redundant if signal multiplicity is to be determined by visual inspection alone, but must be recorded if spectral editing is required. The three DEPT spectra may be edited (added and subtracted) to produce a further three spectra, each showing only CH, CH_2, or CH_3 resonances. Modern spectrometers are capable of interpreting the DEPT spectra automatically.

The INEPT sequence produces analogous spectra to those shown for the DEPT experiment; however, a delay between pulses (rather than a pulse angle) is altered to produce the individual spectra. The sequence is more sensitive to the accuracy of setting delays than the DEPT experiment.

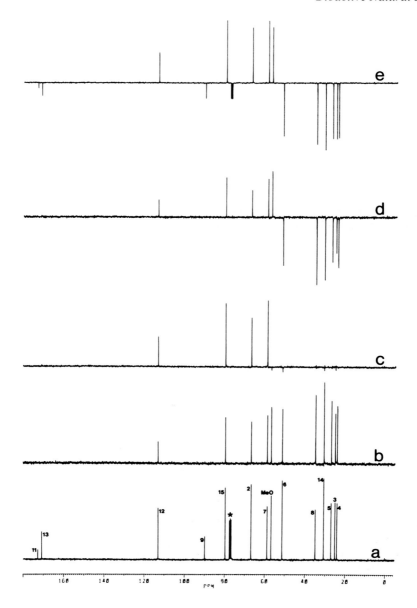

FIGURE 7. ^{13}C spectra of 15-α-methoxy-14,15-dihydrophyllochrysine (**1**): (a) composite pulse decoupled spectrum (* CDCl$_3$); (b) DEPT-45°; (c) DEPT-90°; (d) DEPT-135°; (e) J-modulated spin echo spectrum.

Variations of the above sequences, and other methods employing polarization transfer to differentiate the number of protons on each carbon atom, have been reviewed.[5]

V. ESTABLISHING THE ^{1}H COUPLING NETWORKS

A. SPIN–SPIN DECOUPLING EXPERIMENTS

Homonuclear spin–spin decoupling (double resonance) experiments are the oldest and simplest method for determining proton coupling networks. A typical experiment involves the irradiation of an individual resonance, with a second radio frequency field of suitable power, while the FID is acquired. In the resulting spectrum, the multiplets that were coupled

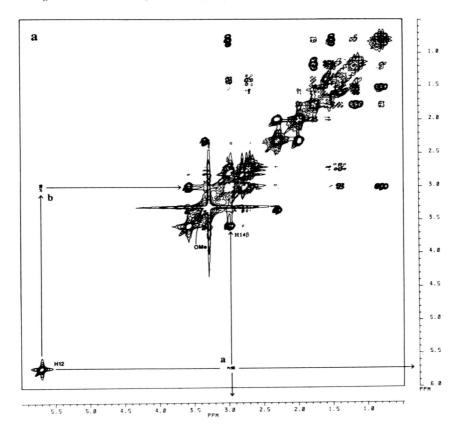

FIGURE 8a. The COSY spectrum of 15-α-methoxy-14,15-dihydrophyllochrysine (**1**).

to the irradiated nucleus are simplified, since the splitting arising from the irradiated nucleus has been removed. For example, the signal due to $H_{3\beta}$, in 15-α-methoxy-14,15-dihydrophyllochrysine (**1**) is simplified when H_2 is irradiated (Figure 3, inset b). The disappearance of a large diaxial coupling (compared with the normal spectrum; Figure 3 inset a) confirms the predicted relationship between H_2 and $H_{3\beta}$.

Modern spectrometers allow a number of decoupling experiments to be performed automatically, with each multiplet in the spectrum being irradiated in turn. Analysis of the resulting spectra allows most, if not all, of the coupling interactions in the spectrum to be defined. Problems arise when signals are very close or overlap with each other. Selective irradiations are then no longer possible, and the results in these cases may be ambiguous. The detection of changes to signals obscured in overlapping multiplets may be aided by using the spin decoupling difference technique.[82]

B. TWO-DIMENSIONAL HOMONUCLEAR CORRELATION EXPERIMENTS

The COSY experiment (Section II.F) is an extremely powerful method for tracing spin–spin coupling within an organic molecule. The normal COSY spectrum of **1** is displayed in Figure 8a. In this contour plot, signals corresponding to the one dimensional spectrum can be found along the diagonal (lower left to upper right). Cross peaks (those not on the diagonal) indicate spin–spin coupling between two nuclei, the chemical shifts of which are described by the horizontal and vertical coordinates of the cross peak. Thus, the cross peak marked (a) in Figure 8a arises from coupling between H_{12} and $H_{14\beta}$. It should be noted that the cross peaks are symmetrically disposed about the diagonal, and therefore the cross peak (b) shows the same correlation.

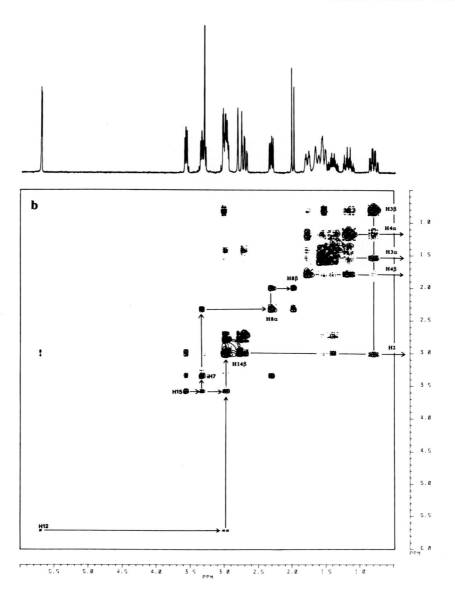

FIGURE 8b. The Double Quantum Filtered (DQF) COSY of **1**.

The analysis of a COSY spectrum begins at one or more readily identified resonances (e.g., H_{12} or $H_{3\beta}$) from which the coupling networks may be traced. The most utilized version of the experiment, the COSY with Double Quantum Filter (DQF),[83-85] is illustrated in Figure 8b (the normal 1D spectrum is shown immediately above the COSY-DQF spectrum). In this spectrum the intense resonances due to singlets (e.g., the methoxyl group) are removed, which results in significantly improved clarity. A large amount of information can be extracted from the spectrum. Starting from H_{15}, for example, cross peaks indicate coupling to H_7 and $H_{14\beta}$. In turn, H_7 is clearly coupled to $H_{8\alpha}$, which is additionally coupled to $H_{8\beta}$. Similarly, cross peaks arising from $H_{3\beta}$ allow the location of all of its coupling partners: namely $H_{4\alpha}$, $H_{3\alpha}$, $H_{4\beta}$, and H_2.

There are many other variations of the COSY experiment.[7,38] The double quantum filtered COSY, if acquired in the phase-sensitive mode,[86,87] can not only define which nuclei are coupled, but analysis of the cross peaks also shows which coupling constants arise from which nuclei (i.e., which are the active coupling constants). Insertion of a delay into the COSY sequence emphasizes cross peaks arising from small, long-range couplings.[88]

The TOCSY (TOtal Correlation SpectroscopY)[89] or HOHAHA (HOmonuclear HArtmann-HAhn) experiments[90,91] can be employed to define all of the protons within a coupling network, even though some of them may not be directly coupled. The choice of experimental mixing time in these experiments allows the selection of direct, single, double, or multiple relayed connectivities. Relay coherence transfer experiments may also be useful for identifying all of the hydrogen nuclei within a subunit.[7,11] However, these experiments are less sensitive than the TOCSY sequence.

Despite the obvious advantages of the above 2D spectra, analysis of congested regions of the spectrum may still prove to be difficult. In such cases double quantum 2D experiments[92,93] may aid interpretation, as these experiments provide an alternative presentation of the data.

C. ONE-DIMENSIONAL VERSIONS OF THE COSY EXPERIMENT

When only a small number of correlations need to be determined, a time-saving one-dimensional version of the COSY experiment[94-96] can be used. In such an experiment the first 90° pulse of the COSY sequence is usually a "soft" pulse (i.e., a low-power pulse which covers a narrow frequency range) that selectively excites a particular proton resonance. The 1D spectrum, resulting from the 1D COSY pulse sequence, shows only signals for the nuclei that are coupled to the nucleus that was selectively irradiated. Soft pulses are generated on modern spectrometers by pulse-shaping units. However, 1D COSY spectra may also be recorded using older spectrometers.[97] The use of selective pulses in NMR spectroscopy has recently been reviewed.[98]

VI. CORRELATING RESONANCES DUE TO DIRECTLY BONDED ¹H AND ¹³C NUCLEI

While experiments in Section IV were able to describe the number of hydrogens attached to each carbon atom, they were unable to show which ¹H resonances are associated with a particular carbon. Such information is a useful aid in the assignment of resonances in the individual ¹H and ¹³C spectra and, together with knowledge of the proton coupling networks, allows the complete definition of subunits of adjacent protonated carbons.

The first method used to correlate carbon resonances with those of their directly attached protons involved the acquisition of a series of ¹³C spectra. Each spectrum was acquired with decoupling at a single ¹H frequency.[99] To produce the series the frequency of the decoupler was stepped through the proton chemical shift range. By locating the carbon resonances that sharpened for a particular ¹H-decoupling frequency, one could relate the resonances due to directly bonded ¹H and ¹³C nuclei. This procedure was obviously very time consuming. Modern heteronuclear correlation techniques are considerably more efficient and may employ 2D experiments.

A. CONVENTIONAL HETERONUCLEAR CHEMICAL SHIFT CORRELATION EXPERIMENTS

1. 2D Methods

A spectrum arising from the 2D experiment[100-104] employed to correlate the chemical shifts of directly bonded ¹H and ¹³C nuclei (often referred to as H,C-COSY, or HETCOR) is illustrated in Figure 9. This spectrum, the H,C-COSY of **1** is readily analyzed, with the vertical and horizontal coordinates of each peak defining the chemical shifts of a proton and its directly attached carbon, respectively. Thus, peak (a) in Figure 9 indicates that the proton giving a resonance at 3.58 ppm (H_{15}) is directly bonded to the carbon resonating at 79.6 ppm (C_{15}).

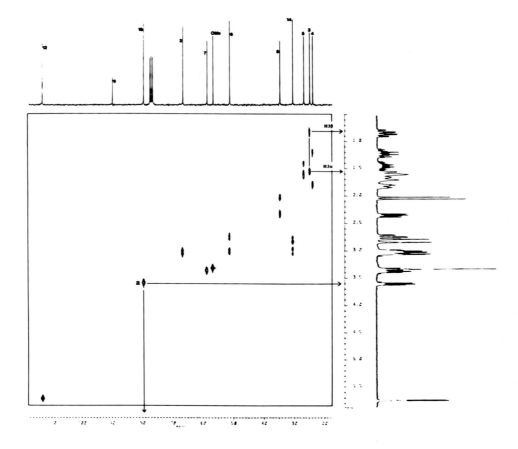

FIGURE 9. The heteronuclear chemical shift correlation spectrum (H,C-COSY) of 15-α-methoxy-14,15-dihydrophyllochrysine (**1**).

In the case of the methylene carbon atoms (namely C_3, C_4, C_5, C_6, C_8, and C_{14}), two peaks can be seen for each carbon chemical shift, thereby allowing the ready location of each of the geminal proton signals. This information can be extremely useful, as in the case of the protons on C_3. The multiplet due to $H_{3\beta}$ (0.82 ppm) has three large coupling constants of equal magnitude (see Section IV.A): two from protons on adjacent carbons, and one from its geminal partner ($H_{3\alpha}$). While decoupling experiments show the locations of the resonances of the three protons coupled to $H_{3\beta}$, they do not directly indicate which one arises from $H_{3\alpha}$ due to the equal magnitude of the coupling constants. However, the location of the $H_{3\alpha}$ signal is immediately obvious from the H,C-COSY spectrum (Figure 9). Its chemical shift is given by the second peak at the same ^{13}C chemical shift as $H_{3\beta}$. In the normal 1D spectrum the ^1H signal of $H_{3\alpha}$ is hidden within overlapping multiplets. In a similar manner, the ^1H chemical shifts of each of the other geminal pairs can also be determined.

2. 1D Methods

There are a number of one-dimensional heteronuclear shift correlation experiments[105-111] that may be useful when only a small number of correlations are of interest or if the limited amount of the sample available precludes a 2D experiment. Some of these experiments such as SHECOR[105,106] (Selective HEteronuclear CORrelation), SEPT[107] (Selective inEPT), and SDEPT[107,108] (Selective DEPT) employ selective proton pulses, while others, including SINEPT[110] and CHORTLE,[111] use nonselective proton pulses.

B. INVERSE HETERONUCLEAR CHEMICAL SHIFT CORRELATION EXPERIMENTS

Each of the heteronuclear experiments described so far has involved the excitation and/or decoupling of protons with the detection of carbon signals. In contrast, inverse spectroscopy detects the proton magnetization. Inverse (reverse) detection methods offer significantly greater sensitivity than the conventional methods. The hardware required to implement inverse techniques has only become available on the most recent generation of spectrometers. There are many techniques that have an inverse equivalent (e.g., the inverse DEPT experiment[112]).

1. 2D Methods

The 2D experiment employed for one-bond heteronuclear chemical shift correlation, using inverse detection, is known as HMQC (Heteronuclear Multiple Quantum Correlation)[113] and is eight times more sensitive than its conventional equivalent.[7] The application of this experiment, and some of its variants, in natural products chemistry has been reviewed.[9] The analysis of an HMQC spectrum is analogous to that of an H,C-COSY, except the axes of the spectrum are normally reversed. Inverse-detected 2D-heteronuclear correlation experiments have also been termed C,H-COSY experiments.[7]

2. A 1D Method

The one-dimensional analog of the HMQC sequence is the SELective INverse CORrelation (SELINCOR) experiment.[9,114] The pulse sequence uses a selective carbon pulse applied at the chemical shift of the carbon atom of interest and detects the responses of the attached protons.

VII. LONG-RANGE HETERONUCLEAR CHEMICAL SHIFT CORRELATION

Knowledge of the correlation between protons and their directly bonded carbon atoms is a prerequisite for experiments involving long-range correlations.

The experiments described in the previous sections allow the determination of all of the subunits of contiguous, protonated carbon atoms within a molecule. Thus, in the case of **1**, the fragments consisting of C_2-C_3-C_4-C_5-C_6; C_{14}-C_{15}-C_7-C_8; C_{12}, and their attached protons have now been defined. The problem of linking these units to form the correct structure now exists. Some logical connections can usually be made. For example, the chemical shifts of C_2, C_6, and C_7 indicate that they are likely to be attached to the lone nitrogen in the molecule. Evidence of long-range proton–proton coupling may show the proximity of units, as in the case of the allylic coupling between H_{12} and $H_{14\beta}$, which defines the interconnection of C_{14}, C_{13}, and C_{12}.

An extremely useful method for determining the connectivity of subunits in the molecule is by long-range heteronuclear chemical shift correlation techniques. These methods relate protons and carbons, separated by two or three bonds, via their coupling constant. Two-bond carbon-hydrogen couplings ($^2J_{CH}$; typically 0 to 8 Hz) are able to show the relationship between quaternary carbons and the protons on adjacent carbon atoms, while three-bond couplings ($^3J_{CH}$; typically 3 to 13 Hz) are able correlate resonances across a quaternary carbon or heteroatom or correlate a quaternary carbon atom with protons on the β carbon. Long-range correlations can also confirm interconnections deduced from previous experiments, within a subunit.

A. CONVENTIONAL LONG-RANGE HETERONUCLEAR CORRELATION TECHNIQUES

1. 2D Methods

Two-dimensional long-range heteronuclear chemical shift correlation methods, together with examples of their application, have been reviewed.[115] The standard H,C-COSY[103] ex-

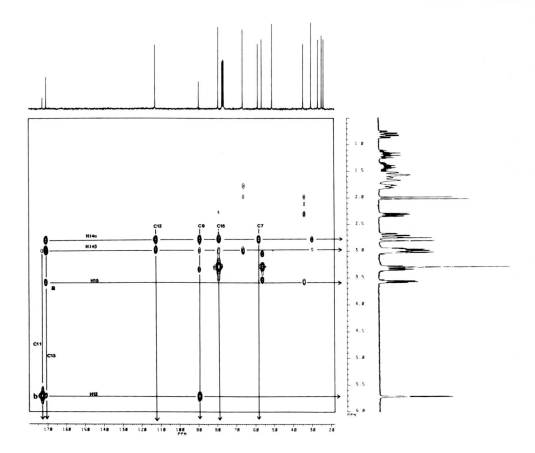

FIGURE 10. The COLOC spectrum of 15-α-methoxy-14,15-dihydrophyllochrysine (**1**) optimized for J_{CH} = 9Hz.

periment may be optimized for long-range correlations,[116,117] but relaxation during the long delays required by the small coupling constants results in a significant loss of sensitivity. Also, modulation by one-bond couplings may result in the loss of some correlation peaks. Constant evolution time experiments,[118-121] including the COLOC (COrrelation via LOng range Couplings)[118,119] and XCORFE[121] experiments, were designed to circumvent the relaxation problem, but they are still affected by one-bond modulations.

The COLOC spectrum of **1**, optimized for long-range carbon-hydrogen couplings of 9 Hz, is illustrated in Figure 10. The experiment is useful in defining the location of the nonprotonated carbon atoms (C_9, C_{11}, and C_{13}) within the molecule as well as confirming the interconnections deduced from other experiments. Peaks in the spectrum indicate long-range coupling between the nuclei at the corresponding ^{13}C (horizontal axis) and ^1H (vertical axis) chemical shifts. Hence, the peak (a) indicates a long-range interaction between H_{15} (3.58 ppm) and the nonprotonated carbon at 170.8 ppm. Some ambiguity may arise when it is unclear whether a peak arises from a two- or a three-bond coupling. In such cases logical argument based on other long-range interactions and on information from procedures described in earlier sections will normally remove the uncertainty. Thus, as most of the fragment containing H_{15} has already been defined, the peak (a) must arise via a three-bond interaction between H_{15} and the olefinic carbon attached to C_{14} (i.e., C_{13}). This carbon also shows long-range correlations with $H_{14\alpha}$, $H_{14\beta}$, and H_{12}. The extended unit C_8-C_7-C_{15}-C_{14}-C_{13} = C_{12} is now defined. In addition to the above correlation, $H_{14\alpha}$ shows interactions with C_7, C_{12}, C_{15}, and the quaternary carbon at 89.8 ppm. As H_{12} also has a long-range interaction with this quaternary carbon, then it (C_9) must be attached to C_{13}. Extension of this subunit to include C_{11} is indicated by peak (b) together with the chemical shifts of C_{12} and C_{13}, which indicate they are α and β to a carbonyl group, respectively. In some cases it may be necessary to record a second spectrum, optimized for a different long-range proton-carbon coupling constant, in order to see all of the required correlations.

The FUCOUP experiment[122] (correlation with full coupling) and the DEPT-based sequence[123-125] provide other alternatives for long-range heteronuclear correlation.

2. A 1D Method

The INAPT (Insensitive Nuclei Assigned by Polarization Transfer) experiment,[126,127] an INEPT sequence version that uses selective proton pulses, has proved useful in the structural determination of natural products[128] particularly when sample size prevents the use of 2D techniques.

B. INVERSE LONG-RANGE HETERONUCLEAR CORRELATION

1. The 2D Experiment

The significant gain in sensitivity provided by inverse detection makes the HMBC (Heteronuclear Multiple Bond Correlation) experiment[129,130] a more powerful technique for long-range heteronuclear correlation than its already important conventionally detected analogs. Analysis of the spectrum resulting from the HMBC sequence is analogous to that described for the COLOC experiment.

2. 1D Methods

1D versions of the HMBC experiment,[131-133] which employ frequency selective pulses, may be employed when connectivities to only one or two carbons need to be determined or when low solubility prevents the recording of a full HMBC spectrum.[132]

C. RELAYED COHERENCE TRANSFER EXPERIMENTS

Protons may also be correlated with neighboring carbon atoms via H-H-C relay experiments.[134-139] In these experiments magnetization is passed from one proton to a neighboring proton and subsequently to the carbon directly attached to the second proton. Such experiments, however, are less sensitive than the standard heteronuclear chemical shift correlation experiment.

VIII. ESTABLISHING MOLECULAR STRUCTURE VIA ^{13}C-^{13}C COUPLING

The 2D INADEQUATE (Incredible Natural Abundance DoublE QUAntum Transfer Experiment) sequence[140-143] offers an unambiguous method for tracing the interconnection of carbon atoms in a molecule. However, the technique suffers from extremely low sensitivity, as the experiment detects only those molecules with adjacent ^{13}C nuclei (i.e., one molecule in ten thousand). Thus, a typical experiment requires hundreds of milligrams of material. In spite of this problem, there are a number of examples of its use in the literature.[4,144]

The SELINQUATE experiment[145] is the selective 1D version of the 2D INADEQUATE procedure. Due to its low sensitivity, it is only recommended[98] when one key piece of information is required to determine a structure.

IX. ESTABLISHING THE PROXIMITY OF NUCLEI THROUGH SPACE

The nuclear Overhauser effect (NOE) allows the identification of those nuclei within a molecule that are close in space, irrespective of whether the nuclei are spin–spin coupled or not. There are two experiments commonly used to detect NOE interactions between nuclei:

the NOE difference experiment and the two-dimensional NOESY (Nuclear Overhauser Enhanced SpectroscopY) technique. The latter procedure is most useful when studying large molecules, such as proteins, while the former is able to detect small nuclear Overhauser enhancements and is usually employed in the study of small- to medium-sized molecules. The information from such experiments is valuable in defining the relative orientation of substituents and can aid the assembly of subunits into a molecular structure.

A. THE NOE DIFFERENCE EXPERIMENT[82]

The early method that was used to show NOE interactions involved measurement of the integral of a signal while another resonance was being irradiated. This integral was then compared to the value of the integral, of the same signal, obtained when the decoupler was offset from any resonance. Enhancements were noted as the percentage variation in the integration measurements (see Section II.D). The NOE difference experiment is an extension of this method.

Prior to an experiment, oxygen is removed from the sample solution,[47] thereby excluding an alternative relaxation pathway and optimizing NOE enhancements. The experiment involves irradiating a resonance for several seconds with a second radio frequency field. The irradiating field is then turned off, and an FID is acquired. The sequence is repeated for a number of scans, and the resulting FID is stored. Each resonance of interest is irradiated in turn, and a final spectrum is acquired with the decoupler offset from any resonance. In order to minimize the effects caused by spectrometer variations during the experiment, a small number of scans are used for each irradiation frequency, and the cycle of irradiations is repeated many times. Difference spectra, which are obtained by subtracting the final standard spectrum from each of the spectra in which a resonance was irradiated, show only those resonances that have NOE interactions with the irradiated signal together with a negative signal for the irradiated nucleus itself. For example, Figure 11a shows the NOE difference spectrum of **1** obtained from the irradiation of $H_{8\alpha}$. Enhancements can be seen for its geminal partner $H_{8\beta}$ (27%) and also for H_7 (6%). In addition, and more importantly, an enhancement of H_2 (8%) is observed. H_2 belongs to a separate proton coupling network, and therefore the NOE indicates the spatial proximity of $H_{8\alpha}$ and H_2 on the underside of the molecule. The NOE difference spectrum also allows the signal due to H_2 to be seen clearly, in contrast to the normal ^1H spectrum in which this resonance is partially obscured due to overlap with other resonances.

Another extremely useful interaction defined by the NOE difference experiment is that between H_{12} and $H_{14\alpha}$. Each shows an enhancement when the other is irradiated, thereby confirming their proximity. Thus, the angular relationships between H_{12} and both $H_{14\alpha}$ and $H_{14\beta}$ have been defined by NMR techniques (the latter relationship by allylic coupling; see Section IV.A).

B. THE NOESY EXPERIMENT[146-148]

The NOESY spectrum is very similar in appearance to the COSY spectrum described in Section V.B, with peaks corresponding to the one-dimensional spectrum lying along the diagonal. In the case of the NOESY spectrum, off-diagonal peaks indicate NOE interactions (exchange of magnetization) rather than spin–spin coupling interactions. A simple example of a NOESY spectrum can be seen in Figure 12. This spectrum was produced from a dilute solution of the iodo compound (**3**), a byproduct in the synthesis of a naturally occurring dibenzofuran.[149] The location of the iodine substituent, which had just been introduced, was in question. The solution to this problem was immediately obvious from the NOESY spectrum. The off-diagonal peaks corresponding to the aromatic hydrogen H_a indicate NOE interactions with both the $-CH_2O-$ protons and those of a methoxyl group. Thus, H_a is located between these two substituents. The other aromatic proton resonance (H_b) shows interactions with a methoxyl group and the aromatic methyl group, thereby defining its location and

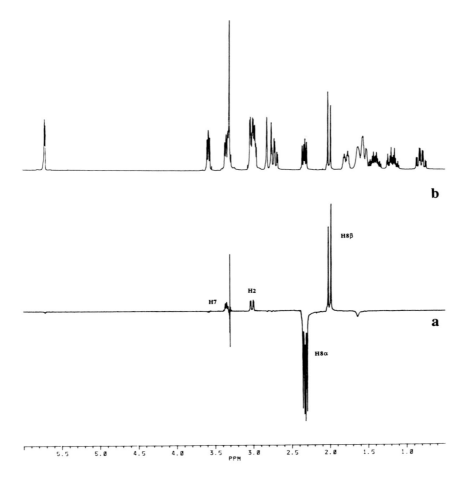

FIGURE 11. (a) The NOE difference spectrum arising from irradiation of $H_{8\alpha}$ of 15-α-methoxy-14,15-dihydrophyllochrysine (**1**). (b) The normal ^{1}H spectrum.

consequently the location of the iodine substituent. The off-diagonal peaks marked (a) in this spectrum (Figure 12) arise from chemical exchange between the hydroxyl proton and those of the water in the dilute solution.

X. CONCLUDING REMARKS

In the preceding sections only selected information arising from each spectrum has been discussed with the aim of indicating how the analysis of each spectrum is approached. Thorough analysis of each spectrum and integration of all of the available information are required in order to arrive at the correct structure. Demands on instrument time will often necessitate the selection of those experiments that are most likely to give the required information in the shortest time. The choice between a 2D technique and its 1D analog is usually based on the number of interactions to be determined, the quantity of sample, and the amount of instrument time available. While a 2D experiment is capable of showing a large number of interactions in the one spectrum, the procedure is more time consuming than its 1D alternative. As older spectrometers are replaced with the current generation of instruments, it is expected that the more sensitive inverse-detected heteronuclear correlation techniques will replace the conventionally detected procedures.

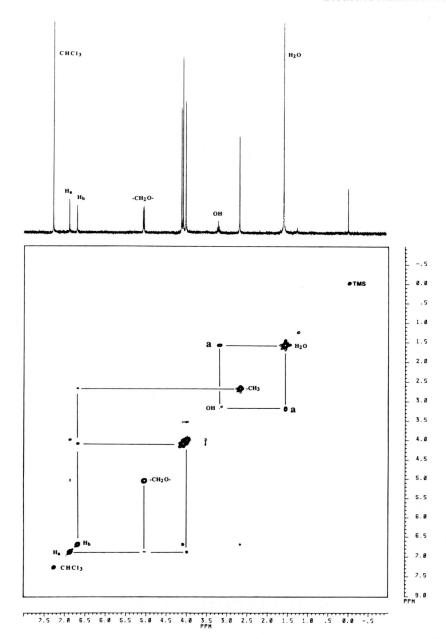

FIGURE 12. The NOESY spectrum of the iodo compound (**3**).

While x-ray crystallography has been nominated as "the ultimate arbiter of chemical structures,"[45] there are many constraints on this technique. NMR is still the first technique turned to by research workers in the quest to determine the structures of biologically active organic compounds.

REFERENCES

1. **Ernst, R. R. and Anderson, W. A.,** Application of Fourier transform spectroscopy to magnetic resonance, *Rev. Sci. Instrum.,* 37, 93, 1966.
2. **Jeener, J.,** *Ampere Summer School,* Basko Polje, Yugoslavia, 1971.
3. **Benn, R. and Gunther, H.,** Modern pulse methods in high resolution NMR spectroscopy, *Angew. Chem. Int. Ed. Engl.,* 22, 350, 1983 and references therein.
4. **Morris, G. A.,** Modern NMR techniques for structure elucidation, *Magn. Reson. Chem.,* 24, 371, 1986 and references therein.
5. **Sadler, I. H.,** The use of N.M.R. spectroscopy in the structure determination of natural products: one-dimensional methods, *Nat. Prod. Rep.,* 5, 101, 1988.
6. **Derome, A. E.,** The use of N.M.R. spectroscopy in the structure determination of natural products: two-dimensional methods, *Nat. Prod. Rep.,* 6, 111, 1989.
7. **Kessler, H., Gehrke, M., and Greisinger, C.,** Two-dimensional NMR spectroscopy: background and overview of the experiments, *Angew. Chem. Int. Ed. Engl.,* 27, 490, 1988 and references therein.
8. **Keeler, J.,** Two-dimensional nuclear magnetic resonance spectroscopy, *Chem. Soc. Rev.,* 19, 381, 1990.
9. **Martin, G. E. and Crouch, R. C.,** Inverse-detected two-dimensional methods: applications in natural products chemistry, *J. Nat. Prod.,* 54, 1, 1991.
10. **Sanders, J. K. M. and Hunter, B. K.,** *Modern NMR Spectroscopy — a Guide for Chemists,* Oxford University Press, Oxford, 1987.
11. **Derome, A. E.,** *Modern NMR Techniques for Chemical Research,* Pergamon Press, Oxford, 1987 and references therein.
12. **Ernst, R. R., Bodenhausen, G., and Wokaun, A.,** *Principles of Nuclear Magnetic Resonance in One and Two Dimensions,* Oxford University Press, Oxford, 1987.
13. **Field, L. R.,** Fundamental aspects of NMR spectroscopy, in *Analytical NMR,* Field, L. R. and Sternhell, S., Eds., John Wiley & Sons, Chichester, 1989.
14. **Williams, D. H. and Fleming, I.,** *Spectroscopic Methods in Organic Chemistry,* 3rd ed., McGraw-Hill, London, 1980, chap. 3.
15. **Becker, E. D.,** *High Resolution NMR,* 2nd ed., Academic Press, New York, 1980.
16. **Gunther, H.,** *NMR Spectroscopy,* John Wiley & Sons, New York, 1980 and references therein.
17. **Abraham, R. J. and Loftus, P.,** *Proton and Carbon-13 NMR Spectroscopy — an Integrated Approach,* Heyden, London, 1980.
18. **Andrew, E. R.,** *Nuclear Magnetic Resonance,* Cambridge University Press, London, 1955, 141 and references therein.
19. Recommendations for the presentation of NMR data for publication in chemical journals, *Pure Appl. Chem.,* 29, 627, 1972.
20. Presentation of NMR data for publication in chemical journals. B. Conventions relating to spectra from nuclei other than protons, *Pure Appl. Chem.,* 45, 217, 1976.
21. **Karplus, M.,** Contact electron-spin coupling of nuclear magnetic moments, *J. Chem. Phys.,* 30, 11, 1959.
22. **Haasnoot, C. A. G., De Leeuw, F. A. A. M., and Altona, C.,** The relationship between proton–proton NMR coupling constants and substituent electronegativities. I. An empirical generalization of the Karplus equation, *Tetrahedron,* 36, 2783, 1980.
23. **Jackman, L. M. and Sternhell, S.,** *Applications of Nuclear Magnetic Resonance Spectroscopy in Organic Chemistry,* 2nd ed., Pergamon Press, London, 1969.
24. **Roberts, J. D.,** *An Introduction to the Analysis of Spin–Spin Splitting in High Resolution Nuclear Magnetic Resonance Spectra,* Benjamin, New York, 1961.
25. **Neuhaus, D. and Williamson, M. P.,** *The Nuclear Overhauser Effect in Structural and Conformational Analysis,* VCH Publishers, New York, 1989.
26. **Noggle, J. H. and Schirmer, R. E.,** *The Nuclear Overhauser Effect,* Academic Press, New York, 1971.

27. **Macomber, R. S.,** A primer on Fourier transform NMR, *J. Chem. Educ.*, 62, 213, 1985.
28. **King, R. W. and Williams, K. R.,** The Fourier transform in chemistry. Part 1. Nuclear magnetic resonance: introduction., *J. Chem. Educ.*, 66, 213, 1989.
29. **King, R. W. and Williams, K. R.,** The Fourier transform in chemistry. Part 2. Nuclear magnetic resonance: the single pulse experiment., *J. Chem. Educ.*, 66, 243, 1989.
30. **Farrar, T. C. and Becker, E. D.,** *Pulse and Fourier Transform NMR,* Academic Press, New York, 1971.
31. **Mullen, K. and Pregosin, P. S.,** *Fourier Transform NMR Techniques: a Practical Approach,* Academic Press, London, 1976.
32. **Purcell, E. M., Torrey, H. C., and Pound, R. V.,** Resonance absorption by nuclear magnetic moments in a solid, *Phys. Rev.*, 69, 37, 1946.
33. **Bloch, F., Hansen, W. W., and Packard, M. E.,** Nuclear induction, *Phys. Rev.*, 69, 127, 1946.
34. **Hahn, E. L.,** Spin echoes, *Phys. Rev.*, 80, 580, 1950.
35. **Cooley, J. W. and Tukey, J. W.,** An algorithm for the machine calculation of complex Fourier series, *Math. Comp.*, 19, 297, 1965.
36. **Farrar, T. C.,** Two-dimensional NMR spectrometry, *Anal. Chem.*, 59, A, 1897.
37. **Aue, W. P., Bartholdi, E., and Ernst, R. R.,** Two-dimensional spectroscopy. Application to nuclear magnetic resonance, *J. Chem. Phys.*, 64, 2229, 1976.
38. **Martin, G. E. and Zektzer, A. S.,** *Two-Dimensional NMR Methods for Establishing Molecular Connectivity. A Chemist's Guide to Experimental Selection, Performance and Interpretation,* VCH Publishers, New York, 1988.
39. **van Halbeek, H.,** N.M.R. of complex carbohydrates, in *Frontiers of NMR in Molecular Biology,* Live, D., Armitige, I. M., and Patel, D., Eds., Wiley-Liss, New York, 1990, 195.
40. **Blasko, G. and Cordell, G. A.,** Proton and carbon-13 NMR assignments of biologically active natural products, in *Studies in Natural Products Chemistry, Vol. 5. Structure Elucidation (Part B),* Atta-ur-Rahman, Ed., Elsevier, Amsterdam, 1989.
41. **Gonzales-Sierra, M., Khalid, S. A., and Duddeck, H.,** Interpretation of two-dimensional NMR spectra: a pedagogical approach, *Fitoterapia,* 60, 99, 1989.
42. **Croasmun, W. R. and Carlson, R. M. K., Eds.,** *Two-Dimensional NMR Spectroscopy — Applications for Chemists and Biochemists,* VCH Publishers, New York, 1987.
43. **Kintzinger, J. P.,** Structural determination of organic compounds, in *Modern NMR Techniques and their Application in Chemistry,* Popov, A. I. and Hallenga, K., Eds., Marcel Dekker, New York, 1991.
44. **Verpoorte, R.,** Methods for structure elucidation of alkaloids, *J. Nat. Prod.,* 49, 1, 1986.
45. **Sanders, J. K. M., Constable, E. C., and Hunter, B. K.,** *Modern NMR Spectroscopy — a Workbook of Chemical Problems,* Oxford University Press, Oxford, 1989.
46. **Nakanishi, K., Ed.,** *One-dimensional and Two-dimensional NMR Spectra by Modern Pulse Techniques,* Kodansha, Tokyo, 1990.
47. **Martin, M. L, Martin, G. J., and Delpuech, J.-J.,** *Practical NMR Spectroscopy,* Heyden, London, 1980.
48. **Sievers, R. E.,** *Nuclear Magnetic Resonance Shift Reagents,* Academic Press, New York, 1973.
49. **Morrill, T. C., Ed.,** *Lanthanide Shift Reagents in Stereochemical Analysis,* VCH Publishers, Deerfield Beach, FL, 1986.
50. **Arbain, D., Byrne, L. T., Cannon, J. R., Engelhardt, L. M., and White, A. H.,** The alkaloids of *Margaritaria indica* (Euphorbiaceae). The crystal structure and absolute configuration of the hydrobromide of (+)-15α-methoxy-14,15-dihydrophyllochrysine, *Aust. J. Chem.,* 43, 439, 1990.
51. **Arbain, D., Birkbeck, A. A., Byrne, L. T., Sargent, M. V., Skelton, B. W., and White, A. H.,** The alkaloids of *Margaritaria indica.* Part 2. The structures of 4-epiphyllanthine, margaritarine and the structural revision of securinol A., *J. Chem. Soc. Perkin Trans. 1,* 1863, 1991.
52. **Bhacca, N. S.,** Nuclear magnetic resonance, in *Handbook of Spectroscopy,* Vol 2, Robinson, J. W., Ed., CRC Press, Boca Raton, FL, 1971, 329.
53. **Pretsch, E., Seibl, J., Simon, W., and Clerc, T.,** *Tables of Spectral Data for Structure Determination of Organic Compounds,* Springer-Verlag, Berlin, 1983.
54. **Stothers, J. B.,** *Carbon-13 NMR Spectroscopy,* Academic Press, New York, 1972.
55. **Levy, G. C., Lichter, R. L., and Nelson, G. L.,** *Carbon-13 Nuclear Magnetic Resonance Spectroscopy,* Wiley-Interscience, New York, 1980.
56. **Breitmaier, E. and Voelter, W.,** *¹³C NMR Spectroscopy: High Resolution Methods and Applications in Organic Chemistry and Biochemistry,* 3rd ed., VCH Publishers, New York, 1987.
57. **Wehrli, F. W. and Wirthlin, T.,** *Interpretation of Carbon-13 NMR Spectra,* Heyden, London, 1978.
58. **Shaka, A. J., Keeler, J., Frenkiel, T., and Freeman, R.,** An improved sequence for broadband decoupling: WALTZ — 16, *J. Magn. Reson.,* 52, 335, 1983.
59. **Shaka, A. J., Keeler, J., and Freeman, R.,** Evaluation of a new broadband decoupling sequence: WALTZ — 16, *J. Magn. Reson.,* 53, 313, 1983.
60. **Wehrli, F. W. and Nishida, T.,** The use of carbon-13 nuclear magnetic resonance spectroscopy in natural products chemistry, *Fortschr. Chem. Org. Naturst.,* 36, 1, 1979.

61. **Sims, J. J., Rose, A. F., and Izac, R. R.,** Application of ^{13}C nmr to marine natural products, in *Marine Natural Products — Chemical and Biological Perspectives,* Vol. 2, Scheuer, P. J., Ed., Academic Press, New York, 1978, 297.

62. **Crabb, T. A.,** Nuclear magnetic resonance of alkaloids, in *Annual Reports of NMR Spectroscopy,* Vol. 6a, Mooney, E. F., Ed., Academic Press, London, 1975, 250.

63. **Crabb, T. A.,** Nuclear magnetic resonance of alkaloids, in *Annual Reports of NMR Spectroscopy,* Vol. 8, Webb, G. A., Ed., Academic Press, London, 1978, 1.

64. **Crabb, T. A.,** Nuclear magnetic resonance of alkaloids, in *Annual Reports of NMR Spectroscopy,* Vol. 13, Webb, G. A., Ed., Academic Press, London, 1982, 1.

65. **Blunt, J. W. and Stothers, J. B.,** ^{13}C n.m.r. spectra of steroids — a survey and commentary, *Org. Magn. Reson.,* 9, 439, 1977.

66. **Smith, W. B.,** Carbon-13 NMR spectroscopy of steroids, in *Annual Reports of NMR Spectroscopy,* Vol. 8, Webb, G. A., Ed., Academic Press, London, 1978, 199.

67. **Inch, T. D.,** Nuclear magnetic resonance spectroscopy in the study of carbohydrates and related compounds, in *Annual Reports of NMR Spectroscopy,* Vol. 5A, Mooney, E. F., Ed., Academic Press, London, 1972, 305.

68. **Bock, K and Thøgerson, H.,** Nuclear magnetic resonance spectroscopy in the study of mono- and oligosaccharides, in *Annual Reports of NMR Spectroscopy,* Vol. 13, Webb, G. A., Ed., Academic Press, London, 1982, 1.

69. **Bock, K. and Pedersen, C.,** Carbon-13 nuclear magnetic resonance spectroscopy of monosaccharides, *Adv. Carbohydr. Chem. Biochem.,* 41, 27, 1983.

70. **Hansen, P. E.,** ^{13}C NMR of polycyclic aromatic compounds. A review, *Org. Magn. Reson.,* 12, 109, 1979.

71. **Memory, J. D. and Wilson, N. K.,** *NMR of Aromatic Compounds,* John Wiley & Sons, New York, 1982.

72. **Howarth, O. W. and Lilley, D. M.,** Carbon-13 NMR of peptides and proteins, *Prog. Nucl. Magn. Reson. Spectrosc.,* 12, 1, 1978.

73. **Rattle, H. W. E.,** NMR of amino acids, peptides and proteins (1977–1979), in *Annual Reports of NMR Spectroscopy,* Vol. 11A, Webb, G. A., Ed., Academic Press, London, 1981, 1.

74. **Rattle, H. W. E.,** NMR of amino acids, peptides and proteins: a brief review, 1980–1982, in *Annual Reports of NMR Spectroscopy,* Vol. 16, Webb, G. A., Ed., Academic Press, London, 1985, 1.

75. **Hansen, P. E.,** Carbon-hydrogen spin–spin coupling constants, *Prog. Nucl. Magn. Reson. Spectrosc.,* 14, 175, 1981.

76. **Marshall, J. L.,** Carbon-carbon and carbon-proton NMR couplings: application to organic stereochemistry and conformational analysis, in *Methods of Stereochemical Analysis,* Marchand, A. P., Ed., Verlag Chemie International, Weinheim, 1983.

77. **Cookson, D. J. and Smith, B. E.,** Improved method for assignment of multiplicity in ^{13}C NMR spectroscopy with application to the analysis of mixtures, *Org. Magn. Reson.,* 16, 111, 1981.

78. **Patt, S. L. and Shoolery, J. N.,** Attached proton test for carbon-13 NMR, *J. Magn. Reson.,* 46, 535, 1982.

79. **Doddrell, D. M., Pegg, D. T., and Bendall, M. R.,** Distortionless enhancement of NMR signals by polarization transfer, *J. Magn. Reson.,* 48, 323, 1982

80. **Doddrell, D. M., Pegg, D. T., and Bendall, M. R.,** Proton-polarization transfer enhancement of a heteronuclear spin multiplet with preservation of phase coherency and relative component intensities, *J. Chem. Phys.,* 77, 2745, 1982.

81. **Morris, G. A. and Freeman, R.,** Enhancement of nuclear magnetic resonance signals by polarization transfer, *J. Am. Chem. Soc.,* 101, 760, 1979.

82. **Sanders, J. K. M. and Mersh, J. D.,** Nuclear magnetic double resonance; the use of difference spectroscopy, *Progr. Nucl. Magn. Reson. Spectrosc.,* 15, 353, 1982.

83. **Wokaun, A. and Ernst, R. R.,** Selective detection of multiple quantum transitions in NMR by two-dimensional spectroscopy, *Chem. Phys. Lett.,* 52, 407, 1977.

84. **Piantini, U., Sørensen, O. W., and Ernst, R. R.,** Multiple quantum filters for elucidating NMR coupling networks, *J. Am. Chem. Soc.,* 104, 6800, 1982.

85. **Shaka, A. J. and Freeman, R.,** Simplification of NMR spectra by filtration through multiple-quantum coherence, *J. Magn. Reson.,* 51, 169, 1983.

86. **Marion, D. and Wüthrich, K.,** Application of phase sensitive two-dimensional correlated spectroscopy (COSY) for measurements of ^{1}H-^{1}H spin–spin coupling constants in proteins, *Biochem. Biophys. Res. Commun.,* 113, 967, 1983.

87. **Rance, M., Sørensen, O. W., Bodenhausen, G., Wagner, G., Ernst, R. R., and Wüthrich, K.,** Improved spectral resolution in COSY ^{1}H NMR spectra of proteins via double quantum filtering, *Biochem. Biophys. Res. Commun.,* 117, 479, 1983.

88. **Bax, A. and Freeman, R.,** Investigation of complex networks of spin–spin coupling by two-dimensional NMR, *J. Magn. Reson.,* 44, 542, 1981.

89. **Braunschweiler, L. and Ernst, R. R.,** Coherence transfer by isotropic mixing: application to proton correlation spectroscopy, *J. Magn. Reson.,* 53, 521, 1983.

90. **Davis, D. G. and Bax, A.,** Assignment of complex ^{1}H NMR spectra via two-dimensional Hartmann-Hahn spectroscopy, *J. Am. Chem, Soc.,* 107, 2820, 1985.

91. **Davis, D. G. and Bax, A.,** MLEV-17-based two-dimensional homonuclear magnetization transfer spectroscopy, *J. Magn. Reson.,* 65, 355, 1985.

92. **Mareci, T. H. and Freeman, R.,** Mapping proton–proton coupling via double quantum coherence, *J. Magn. Reson.,* 51, 531, 1983.

93. **Braunschweiler, L., Bodenhausen, G., and Ernst, R. R.,** Analysis of networks of coupled spins by multiple quantum N.M.R., *Mol. Phys.,* 48, 535, 1983.

94. **Kessler, H., Oschkinat, H., Greisinger, C., and Bermel, W.,** Transformation of homonuclear two-dimensional NMR techniques into one-dimensional techniques using Gaussian pulses, *J. Magn. Reson.,* 70, 106, 1986.

95. **Bauer, C. J., Freeman, R., Frenkiel, T., Keeler, J., and Shaka, A. J.,** Gaussian pulses, *J. Magn. Reson.,* 58, 442, 1984.

96. **Millot, C., Brondeau, J., and Canet, D.,** Determination of mutual coupling in ^{1}H NMR spectra by semiselective excitation, *J. Magn. Reson.,* 58, 143, 1984.

97. **Batta, G. and Kövér, K. E.,** Easy implementation of homonuclear 1D correlation NMR techniques. Application to oligosaccharides, *Tetrahedron,* 47, 3535, 1991.

98. **Kessler, H., Mronga, S., and Gemmecker, G.,** Multi-dimensional NMR experiments using selective pulses, *Magn. Reson. Chem.,* 29, 527, 1991.

99. **Birdsall, B. and Feeney, J.,** The ^{13}C and ^{1}H nuclear magnetic resonance spectra and methods of their assignment for nucleotides related to dihydronicotinamide adenine dinucleotide phosphate (NADPH), *J. Chem. Soc. Perkin Trans. 2,* 1643, 1972.

100. **Maudsley, A. A., Muller, L., and Ernst, R. R.,** Cross-correlation of spin-decoupled NMR spectra by heteronuclear two-dimensional spectroscopy, *J. Magn. Reson.,* 28, 463, 1977.

101. **Bodenhausen, G. and Freeman, R.,** Correlation of proton and carbon-13 NMR spectra by heteronuclear two-dimensional spectroscopy, *J. Magn. Reson.,* 28, 471, 1977.

102. **Bodenhausen, G. and Freeman, R.,** Correlation of chemical shifts of protons and carbon-13, *J. Am. Chem. Soc.,* 100, 320, 1978.

103. **Morris, G. A. and Freeman, R.,** Experimental chemical shift correlation maps in nuclear magnetic resonance spectroscopy, *J. Chem. Soc. Chem. Commun.,* 684, 1978.

104. **Bax, A. and Morris, G.,** An improved method for heteronuclear chemical shift correlation by two-dimensional NMR, *J. Magn. Reson.,* 42, 501, 1981.

105. **Cookson, D. J. and Smith, B. E.,** A selective approach to ^{13}C-^{1}H chemical shift correlation, *J. Magn. Reson.,* 54, 354, 1983.

106. **Cookson, D. J. and Smith, B. E.,** Identification of methylene groups with nonequivalent protons, *J. Magn. Reson.,* 56, 510, 1984.

107. **Doddrell, D. M., Brooks, W., Field, J., and Lynden-Bell, R. M.,** Generation of heteronuclear ^{13}C-^{1}H chemical-shift correlations using soft pulses, *J. Magn. Reson.,* 59, 384, 1984.

108. **Doddrell, D. M., Brooks, W., Field, J., and Lynden-Bell, R.,** Selective DEPT pulse sequence. A rapid one-dimensional experiment for the simultaneous determination of carbon-proton chemical shift correlations and CH$_n$ multiplicities, *J. Am. Chem. Soc.,* 105, 6973, 1983.

109. **Sarkar, S. K. and Bax, A.,** A simple and sensitive one-dimensional NMR technique for correlation of proton and carbon chemical shifts, *J. Magn. Reson.,* 62, 109, 1985.

110. **Jacobsen, H. J., Bildsøe, H., Dønstrup, S., and Sørensen, O. W.,** Simple one-dimensional NMR experiment for heteronuclear chemical shift correlation, *J. Magn. Reson.,* 57, 324, 1984.

111. **Pearson, G. A.,** High-accuracy proton-carbon chemical-shift correlations from one-dimensional polarization-transfer ^{13}C NMR spectra, *J. Magn. Reson.,* 64, 487, 1984.

112. **Bendall, M. R., Pegg, D. T., Doddrell, D. M., and Field, J.,** Inverse DEPT sequence. Polarization transfer from a spin-^{1}/2 nucleus to n spin-^{1}/2 heteronuclei via correlated motion in the doubly rotating reference frame, *J. Magn. Reson.,* 51, 520, 1983.

113. **Bax, A. and Subramanian, S.,** Sensitivity-enhanced two-dimensional heteronuclear shift correlation NMR spectroscopy, *J. Magn. Reson.,* 67, 565, 1986 and references therein.

114. **Berger, S.,** Selective inverse correlation of ^{13}C and ^{1}H NMR signals, an alternative to 2D NMR, *J. Magn. Reson.,* 81, 561, 1988.

115. **Martin, G. E. and Zektzer, A. S.,** Long-range two-dimensional heteronuclear chemical shift correlation., *Magn, Reson. Chem.,* 26, 631, 1988.

116. **Reynolds, W. F., Enriquez, R. G., Esobar, L. I., and Lozoya, X.,** Total assignment of ^{1}H and ^{13}C spectra of kauradien-9(11),16-oic acid with the aid of heteronuclear correlated 2D spectra optimised for geminal and vicinal ^{13}C-^{1}H coupling constants: or what to do when "INADEQUATE" is impossible, *Can. J. Chem.,* 62, 2421, 1984.

117. **Wernly, J. and Lauterwein, J.**, Assignment of the quaternary olefinic carbon atoms of β-carotene by 2D ^{1}H,^{13}C-chemical shift correlation via long-range couplings, *J. Chem. Soc. Chem. Commun.*, 1221, 1985.

118. **Kessler, H., Greisinger, C., Zarbock, J., and Loosli, H. R.**, Assignment of carbonyl carbons and sequence analysis in peptides by heteronuclear shift correlation via small coupling constants with broadband decoupling in t_1(COLOC), *J. Magn. Reson.*, 57, 331, 1984.

119. **Kessler, H., Greisinger, C., and Lautz, J.**, Determination of connectivities via small proton-carbon couplings with a new two-dimensional NMR technique, *Angew. Chem. Int. Ed. Engl.*, 23, 444, 1984.

120. **Bauer, C., Freeman, R., and Wimperis, S.**, Long-range carbon-proton coupling constants, *J. Magn. Reson.*, 58, 526, 1984.

121. **Reynolds, W. F., Hughes, D. W., Perpick-Dumont, M., and Enriquez, R. G.**, A pulse sequence for establishing carbon-carbon connectivities via indirect ^{13}C-^{1}H polarization transfer modulated by vicinal ^{1}H-^{1}H coupling, *J. Magn. Reson.*, 63, 413, 1985.

122. **Halterman, R. L., Nguyen, N. H., and Vollhardt, K. P. C.**, Steric hindrance to benzocyclobutene openings. First synthesis of a 1,2,3-tris(trimethylsilylated) arene by cobalt-catalyzed cyclisation and application of fully coupled two-dimensional chemical shift correlation to a structural problem, *J. Am. Chem. Soc.*, 107, 1379, 1985.

123. **Levitt, M. H., Sørensen, O. W., and Ernst, R. R.**, Multiplet-separated heteronuclear two-dimensional NMR spectroscopy, *Chem. Phys. Lett.*, 94, 540, 1983.

124. **Batta, G. and Lipták, A.**, An approach to oligosaccharide sequencing: 2D NMR-DEPT experiment for detection of interglycosidic ^{13}C-^{1}H spin–spin couplings, *J. Chem. Soc. Chem. Commun.*, 368, 1985.

125. **Batta, G. and Kövér, K. E.**, Optimized detection of small spin–spin couplings in two-dimensional chemical shift correlation experiments — COSY and DEPT, *Magn. Reson. Chem.*, 25, 125, 1987.

126. **Bax, A.**, Structure determination and spectral assignment by pulsed polarization transfer via long-range ^{1}H-^{13}C couplings, *J. Magn. Reson.*, 57, 314, 1984.

127. **Bax, A., Ferretti, J. A., Nashed, N., and Jerina, J. M.**, Complete ^{1}H and ^{13}C NMR assignment of complex polycyclic aromatic hydrocarbons, *J. Org. Chem.*, 50, 3029, 1985.

128. **Cordell, G. A. and Kinghorn, A. D.**, One-dimensional proton-carbon correlations for the structure determination of natural products, *Tetrahedron*, 47, 3521, 1991.

129. **Bax, A. and Summers, M. F.**, ^{1}H and ^{13}C assignments from sensitivity-enhanced detection of heteronuclear multiple-bond connectivity by 2D multiple quantum NMR, *J. Am. Chem. Soc.*, 108, 2093, 1986.

130. **Summers, M. F., Marzilli, L. G., and Bax, A.**, Complete ^{1}H and ^{13}C assignment of coenzyme B$_{12}$ through the use of new two-dimensional NMR experiments, *J. Am. Chem. Soc.*, 108, 4285, 1986.

131. **Bermel, W., Wagner, K., and Greisinger, C.**, Proton-detected C,H correlation via long-range couplings with soft pulses; determination of coupling constants, *J. Magn. Reson.*, 83, 223, 1989.

132. **Keniry, M. A. and Poulton, G. A.**, Assignment of quaternary carbon resonances in lambertellin by soft heteronuclear multiple bond correlation, *Magn. Reson. Chem.*, 29, 46, 1991.

133. **Crouch, R. C. and Martin, G. E.**, Selective inverse multiple bond analysis. A simple 1D experiment for the measurement of long-range heteronuclear coupling constants, *J. Magn. Reson.*, 92, 189, 1991.

134. **Bolton, P. H.**, Assignments and structural information via relayed coherence transfer spectroscopy, *J. Magn. Reson.*, 48, 336, 1982.

135. **Bolton, P. H. and Bodenhausen, G.**, Relayed coherence transfer spectroscopy of heteronuclear systems: detection of remote nuclei in NMR, *Chem. Phys. Lett.*, 89, 139, 1982.

136. **Bax, A.**, Two-dimensional heteronuclear relayed coherence transfer spectroscopy, *J. Magn. Reson.*, 53, 149, 1983.

137. **Kessler, H., Bernd, M., Kogler, H., Zarbock, J., Sørensen, O. W., Bodenhausen, G., and Ernst, R. R.**, Peptide conformations. 28. Relayed heteronuclear correlation spectroscopy and conformational analysis of cyclic hexapeptides containing the active sequence of somatostatin, *J. Am. Chem. Soc.*, 105, 6944, 1983.

138. **Bolton, P. H.**, Multiple-quantum relayed transfer spectroscopy, *J. Magn. Reson.*, 54, 333, 1983.

139. **Sarkar, S. K. and Bax, A.**, Optimization of heteronuclear relayed coherence-transfer spectroscopy, *J. Magn. Reson.*, 63, 512, 1985.

140. **Bax, A., Freeman, R., Frenkiel, T., and Levitt, M. H.**, Assignment of carbon-13 NMR spectra via double quantum coherence, *J. Magn. Reson.*, 43, 478, 1981.

141. **Bax, A., Freeman, R., and Frenkiel, T. A.**, An NMR technique for tracing out the carbon skeleton of an organic molecule, *J. Am. Chem. Soc.*, 103, 2102, 1981.

142. **Freeman, R. and Frenkiel, T. A.**, Structure of a photodimer determined by natural abundance ^{13}C-^{13}C coupling, *J. Am. Chem. Soc.*, 104, 5545, 1982.

143. **Mareci, T. H. and Freeman, R.**, Echoes and antiechoes in coherence transfer NMR: determining the signs of double quantum frequencies, *J. Magn. Reson.*, 48, 158, 1982.

144. **Müller, A., Nonnenmacher, G., Kutscher, B., and Engel, J.**, Complete proton and carbon spectral assignments of acrihellin and hellebrigenin, *Magn. Reson. Chem.*, 29, 18, 1991.

145. **Berger, S.,** Selective INADEQUATE, a farewell to 2D-NMR? *Angew. Chem. Int. Ed. Engl.,* 27, 1196, 1988.
146. **Jeener, J., Meier, B. H., Bachmann, P., and Ernst, R. R.,** Investigation of exchange processes by two-dimensional NMR spectroscopy, *J. Chem. Phys.,* 71, 4546, 1979.
147. **Kumar, A., Ernst, R. R., and Wüthrich, K.,** A two-dimensional nuclear Overhauser enhanced (2D-NOE) experiment for the elucidation of complete proton–proton cross relaxation networks in biological macro-molecules, *Biochem. Biophys. Res. Commun.,* 95, 1, 1980.
148. **Wider, G., Macura, S., Kumar, A., Ernst, R. R., and Wüthrich, K.,** Homonuclear two-dimensional ^{1}H NMR of proteins. Experimental procedures, *J. Magn. Reson.,* 56, 207, 1984.
149. **Sargent, M. V.,** Naturally occurring dibenzofurans. Part 9. A convenient synthesis of phthalides: the synthesis of methyl di-O-methylporphyrilate, *J. Chem. Soc. Perkin Trans. 1,* 231, 1987.

Chapter 5

MASS SPECTROMETRY

Stephen J. Bloor and Lawrence J. Porter

TABLE OF CONTENTS

0-8493-4372-0/93/$0.00+$.50
© 1993 by CRC Press, Inc.

I. INTRODUCTION

Organic mass spectrometry involves the generation of gas-phase positive or negative ions from molecules and their subsequent separation and detection. In the context of this chapter, the information provided by this process are used in three types of experiment:

1. To provide a molecular mass and often a molecular formula for an unknown compound by detection of a molecular ion species
2. To provide structural information regarding functional groups, glycosylation, etc., on unknown compounds from the mass of fragment ions
3. To monitor or confirm the presence of compounds (or classes of compounds) of known composition and mass spectral properties. This process may be qualitative or quantitative

Experiments 1 to 3 may be performed on a host of different instrument configurations (i.e., mass spectrometers) involving various methods of sample introduction, ionization, ion separation, and detection. The different modes of ion separation and focusing are at the heart of the method and also result in the most confusing terminology for the nonspecialist (mass spectrometry nomenclature has now been standardized[1]). These will be considered in the first section, under "mass analyzers."

In practice the various mass analyzers normally must be coupled with other devices to produce a mass spectrometer. For example, a method of sample introduction is needed, with or without prior separation of molecules, and a method of producing and detecting ions from these molecules is also required. All but the last of these components i.e., detection methods, will be discussed briefly. Obviously these topics can only be considered superficially within the confines of a single chapter. For more detailed and relatively up-to-date accounts of organic mass spectrometry, the reader may refer to books by Chapman[2] and Rose and Johnstone.[3]

II. MASS ANALYZERS

Four types of mass analyzers are in common use on currently available organic mass spectrometers. Other analyzers are used, such as time-of-flight analyzers[4] and ion cyclotron resonance analyzers (= Fourier transform mass spectrometry),[5,6] but these tend to be of a more specialist nature and normally will not be encountered in even a well-equipped laboratory. The four most common analyzers are

1. *Magnetic analyzer (B):* these act on ions with a magnetic field normal to the direction of ion travel and will bring to a common focus all ions of a given momentum with the same mass-to-charge (m/z) ratio.
2. *Electrostatic analyzer (E):* these are usually used in combination with B analyzers and act on ions with an electrostatic field normal to the direction of ion travel and will bring

to a common focus all ions of a given kinetic energy. They consist of a pair of curved parallel plates carrying a potential.

3. *Quadrupole analyzer (Q):* these mass filters use a quadrupole field with a dc component and an rf component such as to allow transmission of ions having a selected *m/z* ratio. They consist of four parallel cylindrical or elliptical rods carrying a potential.

4. *Ion-trap analyzer (IT):* these mass-resonance analyzers produce a three-dimensional, rotationally symmetrical quadrupole field capable of storing ions at selected *m/z* ratio. They produce a complex dc/rf field *via* ring and end cap electrodes.

These analyzers alone, or in combination, are used to construct various mass spectrometer configurations. The more common (and cheapest) configurations are single-focusing devices, and by far the largest number of such spectrometers are single-quadrupole instruments that are marketed by virtually all mass spectrometer manufacturers, i.e., Delsi Nermag, Extrel, Finnigan-MAT, Fisons (VG) Instruments, Hewlett-Packard, Perkin-Elmer, Varian, and Vestec. Ion-trap (IT) mass spectrometers have the potential to compete with Q spectrometers. Finnigan-MAT currently market two IT spectrometers: the ITD and ITS40.

Instruments that use B and E analyzers in combination are called double-focusing mass spectrometers, and the two fields are combined so that the B and E fields are normal to each other and each is normal to the ion path (Figure 1). This arrangement results in a high degree of focusing and mass discrimination and enables measurements to be made under medium or high resolution — a condition essential for the reliable measurement of accurate masses (see later). The more common configuration is that of an EB instrument with the so-called "normal" or Nier-Johnson geometry (strictly speaking, Nier-Johnson geometry applies only to a 90° E deflection and a 60° B deflection). Examples of such instruments are the Fisons VG70 series, the Jeol 100 and 300 series, and the Kratos MS-50, MS-80, and Concept series. BE or "reversed" geometry instruments are currently represented by the Finnigan-MAT 90 and 95 series and the Fisons VG ZAB series. More recently, Fisons has introduced an EBE instrument (triple focusing): the Autospec.

Virtually all of the EB, BE, and EBE instruments are also sold with a quadrupole attached, thus becoming "hybrid" EBQ, BEQ, and EBEQ instruments. The advantage of such instruments is that genuine MS/MS experiments (see later) may be performed. However, the problem of combining the two forms of mass spectrometry is not trivial, since the ions must be decelerated, and the image reshaped to a cylindrical beam, before entering the quadrupole. The most economic form of MS/MS is provided by triple-quadrupole (QQQ) instruments that may be constructed for about half the cost of an EB instrument. These instruments are available from a number of quadrupole manufacturers, such as Extrel, Finnigan-MAT (TSQ series), Fisons (Quattro), and Perkin-Elmer (Siex API III).

The final step in sophistication is provided by tandem mass spectrometers comprising various combinations of E and B analyzers (e.g., EBBE, EBEB, etc.). These are very expensive and are normally only encountered in specialist laboratories.

A. COMPARATIVE ADVANTAGES OF THE VARIOUS MASS ANALYZERS

Given a free choice, a research laboratory would always choose a double-focusing (EB or BE), or even a hybrid, instrument. The advantages are several: the instrument may be operated at sufficient resolution to provide accurate mass data; they are always more sensitive than Q or IT instruments, especially at higher masses (>500 Da) and for selected ion recording; they always enable some form of MS/MS experiment to be performed by linked scanning; and they are always configured so that they may accept a full range of inlet and ionization methods. Their disadvantages are the expense of purchase and maintenance (both about four times that of a good Q system) and the fact that their full potential cannot be realized without a dedicated and skilled operator.

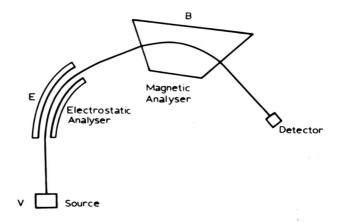

FIGURE 1. Schematic diagram of a typical double-focusing mass spectrometer. The three fields that effect the ion focusing, B, E, and V, have the following relationships in the three common forms of scanning: (1) magnetic scanning (the mode used for acquiring normal mass spectra): E and V are held constant, and B is scanned exponentially from high to low mass, resulting in a linear mass scale. (2) Selected-ion recording (the mode used for monitoring selected ions); B is held constant, and E and V are scanned together in a constant ratio and in a step-wise fashion. (3) Linked scanning (a form of tandem mass spectrometry); usually V is held constant, and B and E are scanned in a manner so that they maintain a constant mathematical relationship whose form depends on whether one wants to detect parent ions, daughter ions, etc.

The advantages of Q systems are they are relatively cheap; they are very compact (the spectrometer may be only half the size of an average GC); they are easily controlled by computer; they may be operated by less highly trained personnel; they have minimal scan rate constraints; and they perform negative-ion mass spectrometry more easily than double-focusing instruments. Their disadvantages are their low resolution that only enables accurate mass measurement under special conditions, their lack of sensitivity for higher-mass ions, and their lack of ability to perform selected ion recording experiments on low-level samples under high background conditions.

The advantages and disadvantages of IT spectrometers are similar to Q spectrometers. Additionally, they are even more compact, which makes them ideal for on-line monitoring devices in biotechnologic applications. Also, their sensitivity, which is similar to the best performing Q systems, is the same for selected-ion and full-scan spectra, which makes them ideal for applications such as drug monitoring, which require library matching. The development of IT spectrometers is still undergoing rapid changes, and it is not inconceivable that they will ultimately supplant Q spectrometers.[7] Their further great advantage is that they do not require a source; ion generation and focusing are in a common volume. This feature, however, makes them more vulnerable to sample contamination. Furthermore, they may perform MS/MS experiments within the single IT volume.[7]

B. RESOLUTION AND ACCURATE MASS MEASUREMENT

These terms are often confused by organic chemists. Commonly, a request is made for a "high-resolution" spectrum when an "accurate mass measurement" is actually required. The usual way that accurate mass measurements are performed is by bleeding in a reference compound, with accurately known mass ions, at the same time as the sample. The position of an unknown peak is then accurately determined relative to the nearest reference compound peaks of higher and lower mass. Sometimes a reference and unknown mass may have the same nominal mass, or more often, a minor peak in the reference compound spectrum will be coincident with the target peak so that the instrument must have sufficient resolution to resolve these peaks.

Resolution (R) is defined as the capacity to separate two ions of mass m and m + Δm and is usually defined in terms of "parts per million," where R = ($10^6 \times \Delta$m)/m. R may be defined in terms of one or two peaks. For double-focusing instruments, the value R will be that sufficient to separate two peaks with 10% valley definition. As this is made up of a 5% contribution from each peak if they are of equal height, then it may also be defined as the width of a single peak at 5% height. A mass spectrometer tuned to give a particular value of R is said to have a resolving power (RP) of R.

In practical terms, reference compounds are usually perfluorinated hydrocarbons that produce ions that are mass deficient (i.e., the accurate mass is less than unity). Typical examples are perfluorokerosine (PFK) and heptacosafluorotributylamine (heptacosa or PFTBA). In practice an RP of 3000 is usually sufficient for accurate mass measurements. The mass spectrometer data system will automatically generate these data from suitable combined spectra. It is considered that measured and calculated accurate masses that agree to ±5-ppm accuracy are sufficient to characterize the molecular formula of a compound.

The foregoing comments pertain to positive-ion electron impact spectra (EI⁺). The exercise is more difficult for obtaining accurate masses from positive-ion chemical ionization spectra (CI⁺), as the spectrum of PFK is very weak. This problem may be neatly overcome for ammonia CI⁺ spectra by using a low pressure of NH_3, which gives a reasonable PFK spectrum while still giving CI⁺ ions. Alternatively, modern data systems allow the sequential accumulation and addition of spectra under EI⁺ and CI⁺ conditions while using the same magnet or electrostatic analyzer scan law. However, slightly less accurate data must be tolerated under these circumstances. Accurate mass measurements with fast atom bombardment (FAB) or liquid secondary ion MS (LSIMS) (Section III.B.3) are also performed by this method where mass standard and target compound spectra are software added and processed.

The lower resolution of quadrupole mass spectrometers does not preclude obtaining accurate mass data. As stated earlier, higher resolution is needed to separate sample ions from nearby interfering ions. This problem becomes much less critical with the mass range above approximately 500 Da, due to the decreased frequency of background ions. Thus, provided a Q instrument obeys a reproducible scan law and suitable standards are available above and below the target masses, good data may often be obtained with the aid of suitable computer manipulation.[8]

The concept of resolution is also important for the monitoring and quantification of known compounds by selected-ion recording of their fragment and molecular ions. Although the inherent sensitivity of a Q spectrometer may be sufficient to detect the target ions in a gas chromatography-mass spectrometry (GC-MS) or liquid chromatography-mass spectrometry (LC-MS) run, in practice the anticipated peak in the selected-ion chromatogram may be swamped, or seriously degraded, by ionization due to the ions of material (column bleed, etc.) chromatographing at a similar retention time to the target compound. This problem may normally be completely overcome by tuning an instrument to higher resolution and utilizing a reference compound to accurately readjust the selected masses during the run so as to keep them accurately centroided. The technique also has the additional advantage of distinguishing, to a high confidence level, the target ions from interfering ions of similar mass. This approach, termed high-resolution selected-ion recording (or monitoring, HRSIR), is the cornerstone of the current protocol for monitoring chlorinated dibenzodioxins and dibenzofurans.[9] The method may lead to spectacular improvements in S/N for selected-ion chromatogram peaks compared with data obtained at low resolution or on Q instruments.

III. METHODS OF IONIZATION

A. STANDARD METHODS OF IONIZATION
1. Electron Impact Positive-Ion MS (EI⁺)

A compound is bombarded by electrons from a glowing filament at low source pressure to produce a positively charged molecular radical ion of high internal energy, which usually fragments to form positively charged fragment ions (F), neutral fragments (N), and radicals (R):

$$M + e \rightarrow M^{+\cdot} + 2e \rightarrow F^{+} + N$$

or

$$\rightarrow F^{+} + R^{\cdot}$$

It is the relative abundance of $M^{+\cdot}$, $F^{+\cdot}$, and F^{+}, together with their isotope ions, which form a characteristic EI$^+$ spectrum fragmentation pattern for a given organic molecule under defined conditions. The stability of $M^{+\cdot}$ may be so low as to be undetectable. The fragmentation pattern forms the basis for mass spectral libraries of organic compounds, which are on the disk of the data system on commercial mass spectrometers and which may be searched and matched to the spectrum of an unknown compound.

The fragment masses may have structural significance for given unknown compounds. Additionally, the mass *difference* between fragment ions may also denote the presence of commonly occurring functionalities in an organic compound. For example, a fragment ion at m/z 43 can indicate the occurrence of C_3H_7 or CH_3CO; at m/z 57, the presence of C_4H_9 or C_2H_5CO; at m/z 73, the presence of C_4H_9O, CO_2Et, or Me_3Si; and at m/z 77, the presence of C_6H_5 (phenyl). Similarly, a *loss* of m/z 60 usually denotes an aliphatic acetate (CH_3CO_2H); m/z 42, the loss of ketene; m/z 33 and 34, a thiol; m/z 28, the loss of CO; and m/z 18, the loss of H_2O. Tabulations of fragment ion and neutral loss compositions are commonly available. Once again, fragmentation ions expressed as accurate masses have more reliable diagnostic value.[10]

The interpretation of the EI$^+$ spectra of organic compounds forms an extensive literature and had great importance for the determination of the structure of unknown natural products, before the vastly improved sensitivity of NMR. These books are epitomized by the publication of Budzikiewicz et al.,[11] which gives a succinct summary of the interpretation of EI$^+$ spectra of organic compounds. More recently, two volumes have been published on biochemical applications of mass spectrometry, in which specific classes of bioactive compounds are discussed, i.e., antibiotics, hormones, vitamins, etc., together with useful chapters on the practical aspects of mass spectrometry.[12,13]

2. Positive-Ion Chemical Ionization MS (CI$^+$)

Electron impact (generally of much lower flux than for EI$^+$) is used to generate positive ions in a reagent gas at relatively high pressure (0.1 to 1 torr) inside the MS source. Ions are formed by charges exchanged between the reagent gas ions and the target molecule (M) by proton transfer, charge exchange, electrophilic addition, or anion abstraction.[14]

Reagent gases commonly used for the study of biologically active compounds are methane and ammonia. Proton addition is the dominant process for methane, i.e.,

$$M + CH_5^+ \rightarrow MH^+ + CH_4$$

whereas proton transfer or electrophilic addition may both occur, to a greater or lesser extent, for ammonia (depending on the structure of M), i.e.,

$$2M + 2NH_4^+ \rightarrow MH^+ + MNH_4^+ + NH_3$$

EI$^+$ and CI$^+$ MS form a powerful combination, as the latter will usually yield a protonated or adduct ion, i.e., MH$^+$ or MNH$_4^+$ and, hence, a molecular weight, whereas EI$^+$ will yield structural information with respect to functional groups, etc., from the mass of the fragment ions, and a characteristic "fingerprint" of the molecule for use in future reference libraries. An authoritative text on CI mass spectrometry is available.[14]

3. Electron Capture Negative-Ion MS or Negative-Ion Chemical Ionization (CI⁻)

This mode of ionization is carried out under similar conditions to CI⁺. Not only are positive ions formed, but thermal electrons are also generated that may form negative ions by electron capture: hence, the more accurate, and generally accepted, name of electron capture negative-ion MS for this method.[15] Other gases may be used, but methane has become the reagent gas of choice.[15] Although negative ions may be formed by other mechanisms, electron capture is the most important process in practical applications. The process may be pictured as

$$e^- \xrightarrow{\quad CH_4 \quad} e^- \text{ (thermal)}$$

$$e^- \text{ (thermal)} + M \rightarrow M^- \text{ (activated intermediate)}$$

$$M^- \text{ (activated intermediate)} \xrightarrow{\quad CH_4 \quad} M^-$$

Thus, methane performs two functions: (1) it provides a pool of thermal electrons, and (2) it stabilizes the resonance-activated intermediate species by collision.

In favorable cases the efficiency of electron attachment is much greater than positive-ion formation, and hence, it may provide a highly sensitive method for the detection of certain compounds. In fact, it is usually used in this mode for the low-level detection and quantification of organic molecules with good electron capture characteristics.[15] Many biologically active compounds are in this category.

4. Comparison of Ionization Methods

A good illustration of the behavior of one compound under the three ionization conditions just described is provided by the mass spectra of desacetyl doronine (Figure 2). The spectra for this compound, a pyrrolizidine alkaloid (PA) from a *Senecio* species, are from a capillary GC-MS experiment on the crude PA fraction.[16] As is often observed, the EI and CI spectra are complementary. Although the EI spectrum shows a high degree of structurally significant fragmentation, the molecular ion is of low intensity. In contrast, each of the CI spectra show an intense pseudomolecular ion. The presence of a chlorine atom is revealed by the ^{37}Cl isotope ion in the CI spectra. The marked difference in the fragmentation produced in CI⁻ with OH⁻ and NH₃ is due to the different electron affinities of the two ionizing species (OH⁻ 176.0 kJ/mol, NH_2^- 72.0 kJ/mol).

B. THE IONIZATION OF INVOLATILE OR LABILE COMPOUNDS

Many biologically active compounds are polar and, hence, involatile under normal MS conditions. Others contain very labile bonds (esters, glycosides, etc.) and, hence, do not yield molecular ion species. The following techniques are useful for providing mass spectra from such molecules.

1. Desorption Chemical Ionization (DCI) MS

This is the simplest of these techniques and involves coating a sample on an inert probe tip (usually platinum wire) and inserting the probe, which is designed to place the sample inside the source in direct contact with the CI plasma.[14] The probe is then heated very rapidly by passing a current through the wire.

2. Field Ionization (FI) and Field Desorption (FD) MS

Field desorption involves the desorption of positive ions from a surface by application of an intense electric field gradient (10^7 to 10^8 V cm⁻¹).[17] These fields are usually produced by the use of specially prepared wire emitters coated with chemically generated carbon whiskers.

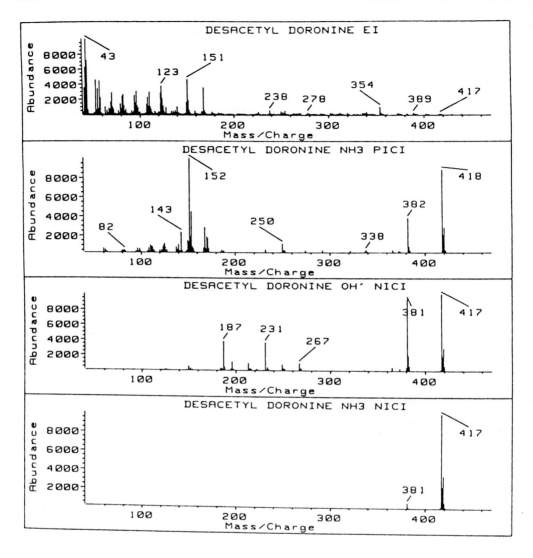

FIGURE 2. EI+, NH$_3$ CI+, OH– CI–, and NH$_3$ CI– mass spectra of desacetyl doronine. (Taken from Bicchi, C. et al., *J. Nat. Prod.*, 52, 32, 1989.)

The degree of success of the method is critically dependent on this step, and it is the lack of interlaboratory reproducibility in preparing emitters that has gained the technique a reputation for being "difficult." For an FD analysis the emitter is dipped in a solution of sample and inserted into the FD source. In the hands of experienced workers, the technique can yield excellent results.

FI differs from FD in that the sample is prevolatilized before coming into contact with the emitter surface.[17]

3. Fast-Atom Bombardment (FAB) and Liquid Secondary-Ion Mass Spectrometry (LSIMS)

These methods involve the suspension of a target compound in a liquid matrix that is then bombarded by fast argon or, preferably, xenon atoms (6 to 10 keV energy) for FAB[18] or by fast ions, generally caesium ions (8 to 20 keV or higher), for LSIMS.[19] It has been argued that FAB is a reasonable name for both methods,[20] but the fact that they do not always give equivalent results[19] means that they should probably be distinguished. There has been a more recent tendency for manufacturers to fit LSIMS, rather than FAB, capability, as it offers the advantages of avoiding the use of expensive xenon gas and because caesium ions are a more energetic species.

In either case the technique involves placing a few microliters of the sample, dispersed in a liquid matrix (usually glycerol or thioglycerol), on an inert surface (commonly stainless steel). Bombardment with fast atoms or ions displaces positive $(M + H)^+$ or negative $(M - H)^-$ ions from the surface by a sputtering mechanism; i.e., the sample is not volatilized as such. In the positive-ion mode $(M + Na)^+$ and $(M + K)^+$, ions may also be observed. It cannot be predicted *a priori* whether operation in the positive- or negative-ion mode will yield the better results, and hence, it is usual to routinely try both methods on any particular sample.

Their general applicability to very polar compounds (sugars, polysaccharides, polyphenols), labile polar compounds, and high-molecular-weight biological compounds (polypeptides, polynucleotides) has made FAB and LSIMS the methods of choice for biological organic chemistry. Figure 3 shows two FAB mass spectra of a cytotoxic saponin isolated from a *Myrsine* species.[21] The advantage of trying both positive and negative FAB are well illustrated by this example. The FAB(+) spectrum contains little structural information, whereas the FAB(−) spectrum reveals not only the pseudomolecular ion, but allows some conclusions to be made as to the nature of the sugars and the degree of branching of the oligosaccharide chain.

IV. METHODS OF SAMPLE INTRODUCTION

A. METHODS WITHOUT PRIOR SEPARATION OF COMPONENTS

The standard method of introduction of samples is as a solid or a matrix *via* a probe inserted through a vacuum lock system. For EI^+, CI^+, and CI^- mass spectrometry, samples are usually evaporated into a glass capillary, or equivalent device, that is placed on the end of a probe and brought close to the source through an inlet into the source block. The probe is usually heatable (to 750°C or more) and is also normally water cooled to ensure fast sample turnaround. The temperature gradient of the probe heating is usually under software control. FD samples are also presented on the end of a probe.

For DCI the sample is evaporated onto a small coil of wire (usually platinum), and volatilization is achieved by passing a current through the wire, whereas for FAB or LSIMS the sample matrix is smeared on the tip of a probe that is brought into the path of the atom or ion beam. The probe is not heated.

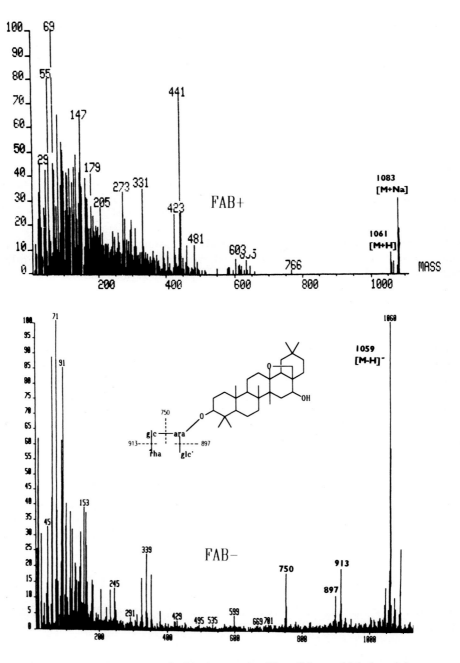

FIGURE 3. FAB(+) and FAB(−) spectra of a *Myrsine* saponin. (Bloor, S J., unpublished results).

B. METHODS WITH PRIOR SEPARATION OF COMPOUNDS
1. Gas Chromatography-Mass Spectrometry (GC-MS)

In this case organic molecules in a mixture are separated by gas chromatography, and the effluent passed into the source of a mass spectrometer or into an ion trap. In its infancy this technique had severe problems because the higher gas flows used by packed columns were not compatible with the pumping rates achieved by diffusion pumps, and so various devices, such as membrane and jet separators, were developed to deviate most of the carrier gas before the effluent entered the source. Capillary gas chromatography-mass spectrometry has, however, been called an "ideal marriage." The flow rates (0.5 to 2 ml/min) employed for 0.3 to 0.4-mm-diameter capillary columns are easily handled by turbomolecular or diffusion pumps, so that source pressures of 10^{-5} to 10^{-6} torr are readily achieved. This factor alone has ensured considerable improvements in the sensitivity of GC-MS.

Almost universally, direct coupling of the capillary column to the mass spectrometer source is used. The only useful variant is an open split interface where the end of the column is at atmospheric pressure, thereby enabling direct correlation of retention time data between GC-MS and normal GC analysis. Furthermore, it allows the solvent front to be diverted if desired.

2. Liquid Chromatography-Mass Spectrometry (LC-MS)

Methodology for LC-MS is still under rapid development and is probably just starting to emerge from an era equivalent to packed-column GC-MS. The problem is very simple: how to strip virtually all of the solvent from an LC effluent before it is presented to the high vacuum source of a mass spectrometer. Efforts to achieve such a marriage has led to several commercially available solutions each of which have their strengths and weaknesses. Games gives a good summary of the problems involved with LC-MS.[22] Currently available LC-MS interfaces are as follows.

a. Particle Beam (PB) LC-MS

The LC effluent is passed through a pneumatic nebulizer and purged with He, and the resulting aerosol is desolvated in a two-stage vacuum separator. The solute beam then passes into a normal EI/CI source where it is ionized. The advantage of particle beam LCMS is that it is capable of producing EI^+ spectra equivalent to that obtained by regular EI^+ MS and, hence, may be library searched. This makes particle beam LC-MS useful for applications such as environmental monitoring, etc. The main problem with the method is that it inexplicably fails to produce spectra for some compounds.

b. Thermospray (TS) LC-MS

Thermospray ionization is achieved by passing the effluent, containing a volatile electrolyte (ammonium acetate is normally used), at 1 to 2 ml/min through a heated stainless steel capillary that is situated diagonally opposite an auxiliary pumping line. At an appropriate temperature (250 to 300°C is typical), a supersonic jet of fine droplets is formed. As they pass through the heated source, solute ions are produced by protonation in a similar way to CI^+. The main problems of thermospray are the formation of cluster ions in the desolvation process and the fact that $(M + H)^+$ ions are virtually exclusively formed. For this reason a thermospray source is usually coupled to a triple Q MS, and fragment ions are produced by collision-activated decomposition (CAD; Section V.A) in the second Q and analyzed in the third.

c. Plasmaspray LC-MS

A variant on thermospray MS is to electrically charge the capillary tip and then induce a corona discharge inside the source, which, in turn, ionizes the solute molecules. This has the advantage that ionization does not require the presence of a buffer. The spectra produced are reported to display a significant degree of fragmentation.

d. Atmospheric Pressure Ionization (API) LC-MS

In principle, the API source is very simple. The API source region is separated from the MS analyzer by a small orifice through which the ions flow into the mass spectrometer. In practice, the LC effluent is passed through an aperture (approximately 50 μm) into one or two adjacent low-pressure regions that are normally separately pumped. The pressure drop causes supersonic, free jet expansion downstream from the aperture, with the highest intensity of gas/aerosol flow along the aperture axis. A corona discharge, or a ^{63}Ni source, ionizes the solvent, which then induces ionization of the solute by charge exchange. The solvent is pumped away in the low-pressure region. The API source produces almost solely $(M + H)^+$ ions with very little fragmentation. Mass spectra with fragmentation require collisionally activated dissociation (CAD; Section V.A) and a triple Q analyzer.

e. Electrospray Ionization (ESI) LC-MS

This is closely related to API in that the source operates at near atmospheric pressure. An electrospray is generally produced by application of a high electric field gradient to a small flow of liquid (generally 1 to 10 µl/min) from a capillary tube, and this requires stream splitting of an LC effluent. A potential difference of 3 to 6 kV is typically applied between the capillary and a counter electrode 0.3 to 2 cm away. The electric field disrupts the liquid surface and produces highly charged droplets that may be positively or negatively charged, depending on capillary polarity. Solvent evaporation is aided by pumping and using a nebulizing "curtain" of gas. Ions and ion clusters are generated that are then sampled by the mass spectrometer. The method has been applied to a wide range of solvents, and there is a small dependence of ionization efficiency on molecular mass. Ions are generally formed by proton attachment (as in TS and API).

The technique has really come into its own for the molecular mass determination of peptides and proteins.[23] These have multiple sites for proton attachment, so that the spectrum consists of a series of ions of the general form $(M_r + nH)^{n+}$, where M_r is the mass of the protein, n the number of protons, and H the proton mass. As the mass-to-charge ratio (m) of an ion peak is given by $m = (M_r + nH)/n$, then it follows that the molecular mass $[M_r = n(m - H)]$ can be found if n is known. For two consecutive peaks m_1 and m_2 (assuming that they differ by one proton attachment), $m_1 = (M_r + nH)/n$ and $m_2 = [M_r + (n + 1)H]/(n + 1)$. Then by solving these simultaneous equations, $n = (m_2 - H)/(m_2 - m_1)$, where n must be rounded to the nearest integer. If these are solved for the series of ions $m_1 \ldots m_n$ and the values of M_r averaged, then a very accurate estimate of M_r may be made. MS data systems will automatically generate these data.

In practice, the method has been very successful, as M_r correlates well with other methods of determining molecular mass. Also, the masses of the ions $m_1 \ldots m_n$ of even a large protein will be in the range of 500 to 4000 Da, so that they are amenable to measurement by good quality Q analyzers. M_r values of over 10^5 have been reported.[23]

f. Dynamic FAB/LSIMS LC-MS

This is, in principle, a very simple LC-MS method. Sample is fed continuously through a capillary that emerges at the tip of a probe that has a wick device that ensures an even and continuously renewed matrix surface being presented to the fast atom or ion beam. In many cases this offers superior results to static FAB/LSIMS experiments. Its main drawback is that the optimum flow required is about 5 µl/min, so that the effluent from conventional capillary columns must be split.

g. Direct LC-MS

The simplest method is to feed the effluent from an LC-MS directly through a capillary into the source of a mass spectrometer. Once again, flows are limited to approximately 5 µl/min (approximately equivalent to 4 ml/min helium gas flow) to accommodate the source pumping speed. The method requires splitting of column effluent and cannot be used with buffer salts, because of source contamination. It also presents difficulties in obtaining a satisfactory flow of solute molecules into the source, because of difficulties of even nebulization of the solvent stream and consistent evaporation rates (hence, the development of the other LC-MS methods).

h. Moving-Belt LC-MS

The LC effluent is deposited directly on a moving belt generally made of kaptan or polyimide. Earlier versions used an oven section to volatilize the solvent, but later, a heated nebulizer and depositor system was used that led to less thermal degradation of compounds and improved chromatographic integrity. The method is capable of producing EI and CI spectra (by volatilization of the solute off the belt into a source) or FAB spectra (by bombardment of the belt).

3. Application Examples of LC-MS

A recently described example of the application of LC-MS involves the characterization of a series of potent antitumor compounds, the ecteinascidins.[24,25] These studies involved extensive use of FAB/LCMS and FAB MS/MS. Some of the data for the most abundant of these compounds, esteinascidin 743, is reproduced in Figure 4. Figure 4a shows a reconstructed ion chromatogram [FAB(+)] for a mixture of several ecteinascidins. The column eluant [Alltech C18 microbore column (10 μm, 1 × 250 mm), MeOH-H_2O-NH_4HCO_3 (70:30:0.1), 100 μl/min] was presented to the MS via a moving-belt interface. The ion chromatogram selecting for $m/z = 744$ is shown in Figure 4b. Since the moving-belt interface provides a constantly renewed sample without the need for a matrix, the LC/FAB mass spectrum of ecteinascidin 743 (Figure 4c) is free of matrix ions. The mass spectral fragmentation scheme presented in Figure 4d is the result of a number of MS experiments, including FAB MS/MS, to clearly assign the composition of each fragment. Note that the m/z 744 ion in the FAB mass spectrum arises from dehydration. This is confirmed using HR FAB(−) MS (diethanolamine matrix), which gives an [M − H]$^-$ for ecteinascidin 743 of 760.2514.

V. MASS SPECTROMETRY/MASS SPECTROMETRY OR TANDEM MASS SPECTROMETRY

MS/MS or tandem MS are equivalent terms for a family of MS techniques (MS^n) that have found application in a variety of experiments, including fragmentation pathways, specific detection of target compounds or classes of compound, structure elucidation, and ion physics. MS/MS experiments are dependent on the observation of fragment ions that are either formed outside the source region from metastable ions, which dissociate spontaneously in flight, or are deliberately formed by collisionally activated dissociation (CAD), a process by which a gas (usually helium) is introduced in a specific region of the mass spectrometer.

Any double-focusing mass spectrometer may perform MS/MS experiments by recourse to "linked-scans" experiments in which the B and E analyzers are scanned in a fixed ratio (usually under software control). The most popular (and practical) combinations for MS/MS are, however, either triple-Q analyzers or, if the laboratory is very well off, BEQ analyzers. Ion traps are capable of MS^n experiments by using combinations of direct and rf-field pulse sequences on trapped ions in a helium reagent gas atmosphere. The viability of this approach has been demonstrated when IT and triple-Q spectrometers were directly compared for MS/MS experiments.[26] The IT offered significant advantages in terms of sensitivity over a triple Q. These aspects of IT methodology are currently under rapid development, and it is probable that IT spectrometers will play an important role in MS/MS instrumentation in the future.

A. MS/MS EXPERIMENTS

The most commonly encountered MS/MS experiments are most easily understood with reference to a triple-Q analyzer (Figure 5). In this case the type of ion to be analyzed is selected

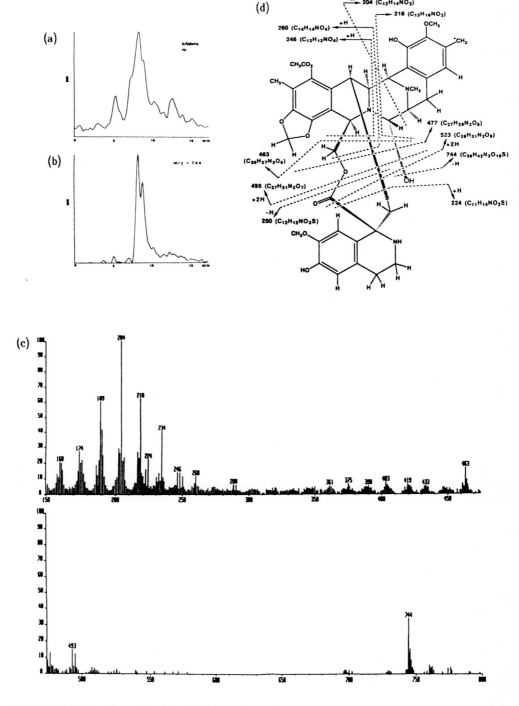

FIGURE 4. MS data for ecteinascidin 743. See text for details. (Taken from Smith, R. D. et al., *Anal. Chem.,* 62, 882, 1990; Rinehart, K. L. et al., *J. Nat. Prod.,* 53, 771, 1990.)

in Q1, the ion dissociation enhanced by collisionally activated dissociation (CAD), also called collisionally activated decomposition (also CAD) and collisionally induced dissociation (CID), in Q2, and the product ions separated in Q3.[27] The advantage of using a quadrupole, Q2, for the collision region is that it transmits ions very efficiently. The four basic experiments are

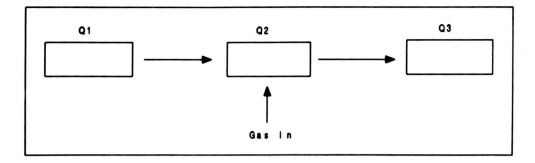

FIGURE 5. Schematic diagram of a triple-quadrupole analyzer.

1. Daughter ion spectrum: parent ions (usually a specific molecular ion) are selected in Q1 and reacted in Q2, and the fragment ions are analyzed in Q3. This is commonly used in LC-MS configurations (see previous section).
2. Parent ion spectrum: Q1 is scanned so that ions formed in the source (primary ions) enter Q2 in sequence, and Q3 is set to pass only ions of a particular m/z value (the "daughter" ion). The resulting spectrum is that of all the parent ions that give rise to the daughter ion.
3. Constant neutral loss spectrum: this experiment requires that Q1 and Q3 are scanned synchronously, but with a mass offset (difference), i.e., if $\Delta m(Q1 - Q3) = 28$, this would equal a loss of CO. The final spectrum would show only those ions with a loss of CO.
4. Multiple-reaction monitoring (MRM): this experiment is the MS/MS equivalent of selected-ion recording (SIR). The instrument is scanned to monitor only selected daughter ions formed from selected parents, hence achieving greatly increased sensitivity.

Other experiments are possible, but experiments 1 to 4 are the most commonly encountered MS/MS experiments and the most useful. Use of a BEQ analyzer will enable higher resolution of parent or daughter ions and hence greater specificity or selectivity. As stated earlier, even more complex combinations of B, E, and Q sectors are available, but these are beyond the resources of most laboratories.

B. APPLICATION EXAMPLES OF MS/MS
1. Taxol
The spectra shown in Figure 6 demonstrate the first three of the MS experiments described in the previous section. The spectra were produced during work on taxol, an anticancer agent of considerable current interest, found in the western yew tree, *Taxus brevifolia*.[28] MS/MS has proven to be a powerful tool in the search for analogs and alternative sources of bioactive compounds such as taxol. Figures 6a and 6c show the DCI mass spectra of pure taxol and the crude bark extract of another yew species, *Taxus baccata*. Under the positive ammonia DCI conditions used, taxol shows an $(M + H)^+$ ion at m/z 854 and a $(M + NH_4)^+$ ion at m/z 871. The usefulness of MS/MS techniques is shown by experiments on the clearly complex bark extract. The daughter ion spectrum of the protonated molecular ion m/z 854 for taxol itself (Figure 6b) and for the crude bark extract (Figure 6d) clearly indicate the presence of taxol in the extract. Figure 6e is an MS/MS parent ion scan of m/z 525 from the bark extract; i.e., only those parent ions which give rise to the diterpene ion at m/z 525 are observed. This experiment reveals the presence of taxol analogs with modifications in the ester side chain. Figure 6f is an MS/MS neutral loss scan of 285 from the bark extract, revealing those analogs of taxol that have the intact ester side chain. For example, the signal at m/z 836 is potentially a dehydrated analog.

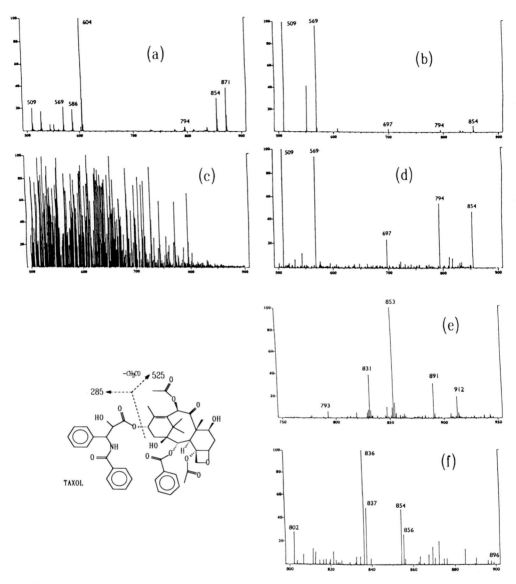

FIGURE 6. DCI MS and MS/MS spectra of taxol and crude bark extract of *Taxus baccata*, a potential alternative source of taxol. See text for details. (Reproduced from Startin, J. R., in *Applications of Mass Spectrometry in Food Science*, Gilbert, J., Ed., Elsevier, London, 1987, 289.)

2. Equisetin

A second example of the use of MS/MS is in the determination of the structure of the mycotoxin equisetin.[29] In this case the MS/MS experiments are used to increase the amount of fragmentation data available. As seen in Figure 7a, the methane CI+ spectrum of equisetin shows low fragmentation intensity. However, when the protonated molecular ion at m/z 374 is subjected to CAD (N_2) in a triple-Q mass spectrometer, the daughter ion spectrum in Figure 7c is obtained showing fragmentation to two major species at m/z 170 and 175. Further daughter ion spectra of these two fragments (Figures 7b and 7d) yield more information on the nature of these two species. The m/z 170 fragment undergoes little further fragmentation, apart from the loss of H_2O and HCHO, and exact mass measurements confirm that this

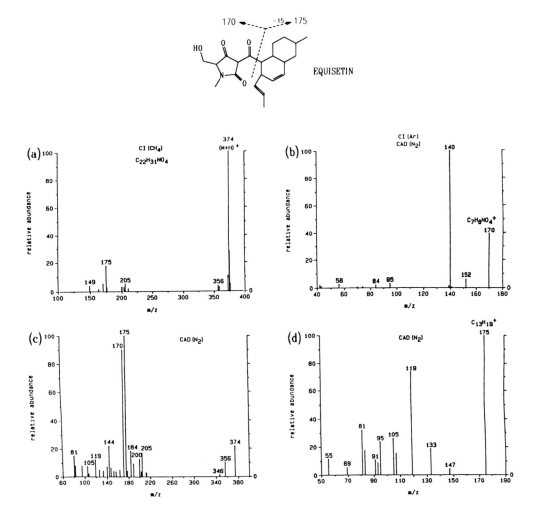

FIGURE 7. MS and MS/MS spectra of the mycotoxin equisetin. See text for details. (Taken from Wood, J. M., in *Proc. 39th ASMS Conf. on Mass Spectrometry and Allied Topics*, Nashville, TN, May 19–24, 1991.)

fragment contains all of the heteroatoms ($C_7H_4NO_4$). The m/z 175 fragment is determined to be a hydrocarbon fragment ($C_{13}H_{19}$) undergoing typical losses of CH_2 units.

VI. CONCLUSION

Mass spectrometry plays a vital role in the areas of medical, pharmaceutical, environmental, and petroleum chemistry. The cornerstone of applications in these areas is its ability to serve as a general monitoring tool. In the specific case of the analysis of target molecules, the method yields, as required, qualitative or quantitative analytic information that is unequivocal and capable of meaningful analyses of molecules at well below the picogram (i.e., femtomole) level.

Historically, mass spectrometry has played a key role in the elucidation of the structure of natural products. However, the explosive growth of NMR techniques, based on improved sensitivity, increased dispersion, and the revolution in 2D methodology, has seen a steady displacement of mass spectrometry from the area of structural elucidation of lower-molecular-weight bioactive natural products. However, mass spectrometry still remains indispensable in the area of bioactive organic molecules in two areas. First, the technique provides key information in studies of large biological molecules such as peptides, proteins, and polysaccharides, often yielding information on monomer linkages, branching, and sequencing. Second, mass spectrometry is generally required to provide a molecular mass and, ideally, a molecular formula, especially to confirm the presence or absence of heteroatoms.

A survey of several journals publishing reports of new bioactive compounds reveals the predominance of the well-established ionization methods, FAB and EI. Of 289 natural products with confirmed biological activity published in a selection of four journals over a period of 2 years, molecular ion species could best be observed for 116 compounds using EI, 27 using CI, and 99 using FAB MS. Reports of the newer MS techniques discussed in this chapter, LC-MS and MS/MS, are relatively uncommon, which, in part, can be attributed to the facts that most structure proofs rely on NMR experiments, as mentioned above, and also because of a lack of ready access to suitable equipment for LC-MS and MS/MS experiments. The work on the ecteinascidins presented earlier provides a rare example of work on bioactive compounds heavily reliant on MS.

LC-MS and MS/MS additionally provides an opportunity to study crude extracts and multicomponent fractions for the presence of specific bioactive compounds or compound types. This type of analysis is becoming increasingly important in the evaluation of alternative sources of potentially useful natural chemicals, as in the taxol example presented earlier in this chapter, and for the elimination of known classes of compounds in the search for new bioactive compounds of novel structural type.

REFERENCES

1. **Price, P.,** Standard definitions of terms relating to mass spectrometry. A report on the committee on measurements and standards of the American Society for Mass Spectrometry, *J. Am. Soc. Mass Spectrom.,* 2, 336, 1991.
2. **Chapman, J. R.,** *Practical Organic Mass Spectrometry,* John Wiley & Sons, Chichester, 1985.
3. **Rose, M. E. and Johnstone, R. A. W.,** *Mass Spectrometry for Chemists and Biochemists,* Cambridge University Press, Cambridge, 1982.
4. **Price, D. and Milnes, G. J.,** The renaissance of time-of-flight mass spectrometry, *Int. J. Mass Spectrom. Ion Process.,* 99, 1, 1990.
5. **Allison, J. and Stepnowski, R. M.,** The hows and whys of ion trapping, *Anal. Chem.,* 59, 1072A, 1987.
6. **Marshall, A. G. and Grosshans, P. B.,** Fourier transform ion cyclotron resonance mass spectrometry: the teenage years, *Anal. Chem.,* 63, 215A, 1991.
7. **Nourse, B. D. and Cooks, R. G.,** Aspects of recent developments in ion-trap mass spectrometry, *Anal. Chim. Acta,* 228, 1, 1990.
8. **Haddon, W. F., Harden, L. A., and Lieberman, A. E.,** Accurate mass measurement by quadrupole mass spectrometry; applications to LSIMS and particle-beam-EI spectra, in *Proceedings of the 39th ASMS Conference on Mass Spectrometry and Allied Topics,* Nashville, TN, May 19–24, 1991, 122.
9. **Method 1613**: tetra- through octa-chlorinated dioxins and furans by isotope dilution HRGC/HRMS. EPA Office of Water Regulations and Standards Industrial Technology Division. Office of Water Publication, revision A, 1990.
10. **Loh, S. Y. and McLafferty, F. W.,** Exact-mass probability based matching of high-resolution unknown mass spectra, *Anal. Chem.,* 63, 546, 1991.
11. **Budzikiewicz, H., Djerassi, C., and Williams, D. H.,** *Mass Spectrometry of Organic Compounds,* Holden-Day, San Francisco, 1967.
12. **Waller, G. R., Ed.,** *Biochemical Applications of Mass Spectrometry,* Wiley-Interscience, New York, 1972.
13. **Waller, G. R. and Dermer, O. C., Eds.,** *Biochemical Applications of Mass Spectrometry,* 1st Suppl. Vol., Wiley-Interscience, New York, 1980.
14. **Harrison, A. G.,** *Chemical Ionization Mass Spectrometry,* CRC Press, Boca Raton, FL, 1983.

15. **Stemmler, E. A. and Hites, R. A.,** *Electron Capture Negative Ion Mass Spectra of Environmental Contaminants and Related Compounds,* VCH Publishers, New York, 1988.

16. **Bicchi, C., Caniato, R., Tabacchi, R., and Tsoupras, G.,** Capillary gas chromatography/positive and negative ion chemical ionization mass spectrometry on pyrrolizidine alkaloids of *Senecio inaequidens* using ammonia and hydroxyl ions as the reagent species, *J. Nat. Prod.,* 52, 32, 1989.

17. **Reynolds, W. D.,** Field desorption mass spectrometry, *Anal. Chem.,* 51, 283A, 1979.

18. **Barber, M., Bordoli, R. S., Elliot, G. J., and Sedgwick, R. D.,** Fast atom bombardment mass spectrometry, *Anal. Chem.,* 59, 645A, 1982.

19. **Aberth, W., Straub, K. M., and Burlingame, A. L.,** Secondary ion mass spectrometry with cesium ion primary beam and liquid target matrix for analysis of bioorganic compounds, *Anal. Chem.,* 54, 2029, 1982.

20. **Feneslan, C. and Cotter, R. J.,** Chemical aspects of fast atom bombardment, *Chem. Rev.,* 87, 501, 1987.

21. **Bloor, S. J.,** Unpublished data, 1990.

22. **Games, D. E.,** Applications of high-performance liquid chromatography/mass spectrometry (LC/MS) in food chemistry, in *Applications of Mass Spectrometry in Food Science,* Gilbert, J., Ed., Elsevier, London, 1987, 193.

23. **Smith, R. D., Loo, J. A., Edmonds, C. G., Barinaga, C. J., and Udseth, H. R.,** New developments in biochemical mass spectrometry: electrospray ionization, *Anal. Chem.,* 62, 882, 1990.

24. **Rinehart, K. L., Holt, T. G., Fregeau, N. L., Keifer, P. A., Wilson, G. R., Perun, T. J., Jr., Sakai, R., Thompson, A. G., Stroh, J. G., Shield, L. S., Seigler, D. S., Li, L. H., Martin, D. G., Grimmelikhuijzen, C. J. P., and Gade, G.,** Bioactive compounds from aquatic and terrestrial sources, *J. Nat. Prod.,* 53, 771, 1990.

25. **Rinehart, K. L., Holt, T. G., Fregeau, N. L., Stroh, J. G., Keifer, P. G., Sun, F., Li, L. H., and Martin, D. G.,** Ecteinascidins 729, 743, 745, 759A, 759B, and 770: potent antitumour agents from the Caribbean tunicate *Ecteinascidia turbinata, J. Org. Chem.,* 55, 4512, 1990.

26. **Johnson, J. V., Yost, R. A., Kelley, P. E., and Bradford, D. C.,** Tandem-in-space and tandem-in-time mass spectrometry: triple quadrupoles and quadrupole ion traps. *Anal. Chem.,* 62, 2162, 1990.

27. **Startin, J. R.,** Applications of mass spectrometry/mass spectrometry in food analysis, in *Applications of Mass Spectrometry in Food Science,* Gilbert, J., Ed., Elsevier, London, 1987, 289.

28. **Wood, J. M., Hoke, S. H., Cooks, R. G., Chang, C.-J., and Heinstein, P. F.,** Rapid screening techniques for taxol and other taxanes using DCI-MS/MS, in *Proc. 39th ASMS Conf. on Mass Spectrometry and Allied Topics,* Nashville, TN, May 19–24, 1991, 1679.

29. **Phillips, N. J., Goodwin, J. T., Fraiman, A., Cole, R. J., and Lynn, D. G.,** Characterization of the *Fusarium* toxin equisetin: the use of phenylboronates in structure assignment, *J. Am. Chem. Soc.,* 111, 8223, 1989.

Chapter 6

DETERMINATION OF THREE-DIMENSIONAL STRUCTURE AND CONFIGURATION OF BIOACTIVE NATURAL COMPOUNDS BY X-RAY CRYSTALLOGRAPHY

Rosalind Y. Wong and William Gaffield

TABLE OF CONTENTS

0-8493-4372-0/93/$0.00+$.50

I. INTRODUCTION

The science of crystallography spans an impressive array of disciplines: chemical, biological, physical, and the material sciences. During the past several decades, X-ray crystallography has gained wide acceptance as a powerful method for the elucidation of molecular structures of bioactive components in natural products. Frequently, conventional chemical and spectroscopic techniques fail to eliminate all structures consistent with the observed data of complex naturally occurring molecules. For this reason, it is often essential to utilize the unique and unequivocal analytic method of X-ray crystallography for molecular structure determination, particularly when the amount of purified product is minute or the compound contains either new or unusual chemical linkages. The objectives of this chapter are to provide an overview of the concepts, procedures, capabilities, limitations, and practical applications of X-ray crystallography to the structural determination of bioactive natural products. The interested reader may consult additional specialized texts.[1-8]

II. X-RAY CRYSTALLOGRAPHY

X-ray crystallographic analysis was once considered a highly complicated and specialized physical technique. However, with the introduction of modern high-speed digital computers and automated crystallographic systems, it has evolved into a readily accessible analytic tool for chemical structure determination. With numerous significant advancements in instrument technology and software development, research scientists other than crystallographers may often realize the full benefit of single-crystal X-ray analysis.

The technique provides more conclusive and absolute molecular information per sample size than any other physical method. While other analytic techniques offer partial or indirect structural information, X-ray crystallography possesses the unique capability of unequivocally determining the exact three-dimensional molecular structure of crystalline substances, even in the absence of elemental analysis and spectral data. The crystal used for the experiment is usually recoverable, except for occasional minor radiation damage. In addition to providing the atomic connectivity and spatial arrangement of the atoms in the molecule, the analysis also reveals precise inter- and intramolecular bond length and bond angle information, as well as knowledge of atomic and molecular thermal vibrations and packing arrangements within the unit cell (see Section II.A). A complete and accurate structural description of the molecular geometry and stereochemistry of bioactive constituents in natural products is often vital to the rational understanding of their biological functions and activities (e.g., toxicity, carcinogenicity, plant biosynthesis, enzymatic activity, taste relationships, etc.).

A. CRYSTALLOGRAPHIC CONCEPTS AND THE DERIVATION OF CRYSTAL STRUCTURES FROM SINGLE CRYSTALS

X-rays are electromagnetic radiation of wavelength between 0.01 and 100Å that are an ideal energy source for exploring atomic and molecular dimensions. When a collimated beam of X-rays strikes a crystalline object (Figure 1), the electrons of the individual atoms act as scattering centers. Because the atoms in crystalline solids are arranged in a regular array, a characteristic diffraction pattern results that can be interpreted to provide structural information. Generally, the concept of X-ray crystallography is based on this phenomenon, which was discovered by von Laue (Nobel Prize, 1914).[9] X-ray diffraction patterns[10,11] can be observed as spots or reflections of radiation, either by photographic film or a detector in a diffractometer system (see Section II.B.2). Each reflection arises from the diffraction of X-rays by a set of planes in the crystal labeled by the ordered triplet *(hkl)* indices. The reflection angle is inversely proportional to the interplanar spacing, and it follows the Bragg equation ($n\lambda = 2d\sin\theta$, where λ is the X-ray wavelength, d is the interplanar spacing, and θ is the angle of reflection). It is

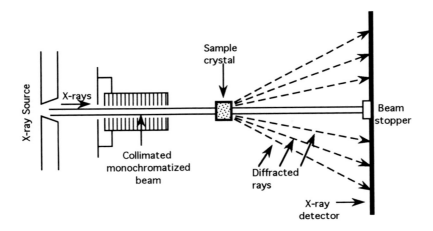

FIGURE 1. Schematic of X-ray diffraction from a single crystal.

impossible to record the entire three-dimensional *(hkl)* array on a single sheet of film in a form adequate for molecular structure analysis, because of numerous overlapping reflections. Accurate intensity data for structure determination is commonly measured by a diffractometer.

Only perfect single crystals are suitable for X-ray crystallographic studies, and it should be emphasized that not all crystalline samples are comprised of single crystals. Perfect single crystals usually possess a definite shape and distinct edges. Further, crystals of a pure chemical compound, obtained from a recrystallization, generally resemble each other in habit and color (Figure 2). By crystallographic definition, a single crystal is a body of homogenous matter composed of identical discrete units repeating indefinitely in all directions. These smallest regular repeating units are referred to as a *unit cell*. Each unit cell may contain one or more molecular clusters related to each other by specific symmetry elements. The different crystalline classes include cubic, tetragonal, rhombohedral, hexagonal, orthorhombic, monoclinic, and triclinic.[12]

A crystal structure may be compared to a wallpaper pattern, except that the former is three-dimensional and therefore more complex. A wallpaper pattern (Figure 3) has two aspects: a motif, which is the unit picture; and a symmetry element, which is the scheme of motif repetition. In a crystal structure the same two general aspects can be recognized, except that they occur in three dimensions, and apply specifically to the packing of atoms (Figure 4). The motif will be a cluster of atoms that is extended by a scheme of repetition in three dimensions, with the resultant pattern comprising the crystal structure. Therefore, structure determination generates not only the coordinates of each atom in a molecule, but also the symmetry elements that relate the molecules in the crystal to one another. The symmetry elements are translation, rotation, screw axes, reflection planes, and glide planes.[12,13] These symmetry elements, when combined with the seven crystal classes, generate 230 geometric arrangements that are referred to as *space groups*.[14]

B. X-RAY STRUCTURE DETERMINATION

X-ray structure analysis is comprised of four major procedures: (1) sample preparation; (2) intensity data collection; (3) data processing, reduction, and interpretation; and (4) structure refinement.

1. Sample Preparation

Suitable crystals of organic compounds are often grown by slow recrystallization from water, common low-boiling solvents (e.g., methanol, ethanol, hexane, benzene), or mixtures

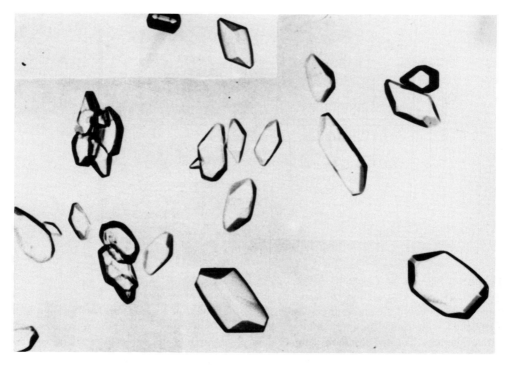

FIGURE 2. Perfect single crystals suitable for X-ray diffraction study.

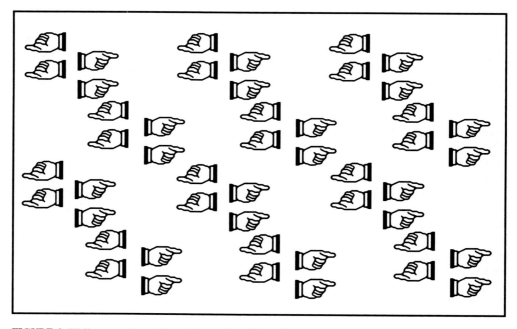

FIGURE 3. Wallpaper pattern with a scheme of motif repetition.

thereof. Solvent choice depends primarily upon the solubility of the sample compound. The solvent pair crystallization technique[15] is an alternative method for growing crystals of organic compounds that do not readily crystallize. Under a light microscope the single crystal (about 0.1 to 0.3 mm) is glued precisely to a small, thin glass fiber. Ideally, the mounting tip of

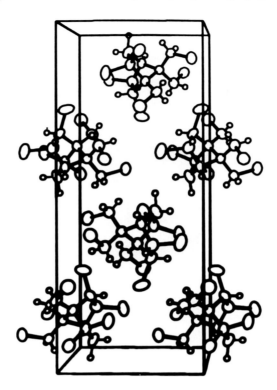

FIGURE 4. Unit cell molecular packing in a crystal structure.

the glass fiber should be smaller than one-half the dimensions of the sample crystal; otherwise, the crystal intensity data would be affected by the diffraction of the glass fiber. A crystal that is too small may not diffract strongly enough. However, one that is too large may suffer from absorption or extinction problems. A crystal of equal dimensions can be obtained by carefully cleaving a fragment of the single crystal. The mounted crystal is affixed onto a goniometer (Figure 5), a small device equipped with x- and y-translations and two perpendicular angular arcs for aligning the crystal in a desired orientation. The complete goniometer assembly is then secured onto the single-crystal X-ray diffractometer.

2. Intensity Data Collection

Accurate intensity data are measured by a computer-controlled X-ray diffractometer. A conventional automatic crystallographic system usually consists of a diffractometer module and a computer console that performs the dual functions of diffractometer control and structure calculation. Figure 6 shows a commercial diffractometer module that is equipped with a four-circle goniostat, a shielded X-ray source (combined with a beam attenuator, monochromator, and collimator), a scintillation counter detector for recording scattered radiation, and a microscope for crystal alignment. Inclusion of a low-temperature device to the diffractometer system is essential when analyzing radiation-sensitive samples. Molybdenum and copper radiation are most commonly utilized as experimental X-ray sources. Copper radiation is usually selected for organic structure analysis because of its longer wavelength that provides better resolution of the diffraction patterns generated by the analyzed crystal.

After mounting, the crystal is aligned optically in the center of the four-circle goniostat of the diffractometer. Before intensity data collection, the crystal is analyzed by the X-ray diffraction film technique. The two types of photographs most commonly employed for crystallographic studies are the *rotation photograph* (Figure 7) and the *axial photograph* (Figure 8).

FIGURE 5. Two common versions of commercial goniometers for crystal alignment (courtesy of Charles Supper Co., Natick, MA).

The size and shape of each spot of the diffraction pattern recorded on the rotation photograph reveal important clues about the quality of the single crystal. If the crystalline sample is amorphous or decomposed, a diffraction pattern will not be generated. If the sample is other than a single crystal (e.g., twinned, cracked, or with satellite subcrystals), unusual rings or spots will appear in the photographs. After the crystal quality has been confirmed, a minimum of 20 diffraction spots on the rotation photograph are selected as the starting reflection set. By using the two-dimensional film coordinate input, the crystallographic computer software automatically locates the precise orientation of these starting reflections on the diffractometer

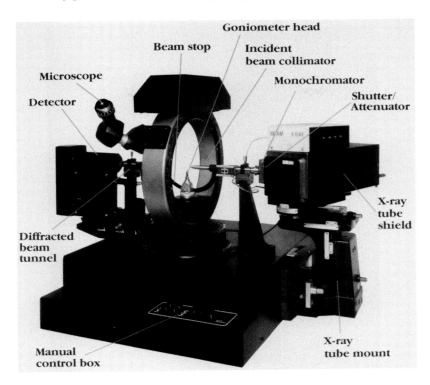

FIGURE 6. A commercial automatic X-ray single-crystal diffractometer (courtesy of Siemens Analytical X-ray Instruments, Inc., Madison, WI).

system and converts them to three-dimensional angular coordinates so that indices *(hkl)* can be assigned to each individual reflection in the starting set, from which the unit cell parameters are derived. Consequently, axial photographs of the sample are taken to confirm the crystal lattice dimensions, axial symmetry, and systematic absences in the diffraction pattern.

After careful evaluation and confirmation of preliminary structural data (unit cell class and dimensions, lattice symmetry, and space group), final data collection is begun. The intensity data are measured by moving the scintillation counter detector and crystal to a position in space where a particular *(hkl)* diffracted beam is expected to lie, and its intensity is recorded at a fixed or variable time interval. The detector and crystal are relocated to a new position, and the intensity associated with another set of *(hkl)* planes is recorded. This laborious process continues until all reflections are observed within one quadrant of the sphere of the unit cell. The minimum amount of data required to solve a typical organic structure of ten nonhydrogen atoms possessing a $10 \times 10 \times 10$ Å-unit cell size is approximately 2500 unique reflections. Thus, several continuous days of data collection are normally necessary to acquire a complete set of intensity data for structure determination. However, this period can be reduced appreciably if a fast and responsive area detector is employed rather than a conventional scintillation-counter detector. Thus, the advent of the automatic diffractometer has increased not only the accuracy of the derived-structure parameters, but has further reduced data collection time from several months to several days.

3. Data Processing, Reduction, and Interpretation

Upon completion of data collection, the diffraction data is corrected precisely for Lorentz factors, polarization effects, X-ray absorption, and decay of intensity due to crystal decomposition from radiation exposure.[16] The intensities of all unique reflections in the diffraction

FIGURE 7. X-ray diffraction pattern on a rotation photograph, displaying the crystal quality and general crystal symmetry of the crystal structure.

pattern are then analyzed by the Fourier transform method, in order to generate an electron density map of atomic peak coordinates.

Because only electrons give rise to scattered X-rays, a crystal structure examined by the X-ray diffraction method may be regarded simply as a periodic three-dimensional distribution of electron density. The resultant electron density function $[\rho_{(x,y,z)}]$ of the crystal structure is expressed mathematically by the following three-dimensional Fourier series:[17]

$$\rho_{(x,y,z)} = (1/V)\sum\sum_{+\infty}^{-\infty}\sum\left[F_{(hkl)}e^{-2\pi i(hx+ky+lz)}\right]$$

where V is the volume of the unit cell, i is $\sqrt{-1}$, and the complex number $F_{(hkl)}$ is the structure-factor amplitude. Positions of individual atoms within the unit cell are equivalent to peaks in the electron density function ρ, with their heights proportional to the atomic numbers of the relevant atoms. Consequently, the crystal structure can be solved or mapped if $F_{(hkl)}$ is known. The structure-factor amplitude is proportional to the observed diffraction intensity as follows:

$$I_{(hkl)} \approx \left|F_{(hkl)}\right|^2$$

FIGURE 8. X-ray diffraction pattern on an axial photograph, displaying the axial dimensions, symmetry, and systematic absences of the crystal structure.

The intensity (I) of a diffracted X-ray beam depends both on the scatter factor of each individual atom and its position in the unit cell, and the atomic scattering factor (F) is a measure of the efficiency of cooperation of all electrons surrounding a nucleus involved in scattering the incident radiation. Although the X-ray intensities of crystallographic reflections can be measured, the phases of the amplitudes that are necessary for the complete elucidation of the structure are not readily determined. In order to solve the phase problem (i.e., the relative time of arrival of each wave at the detector), the presence of a heavy atom (e.g., bromine or chlorine) in the molecule will greatly simplify the structure-solving process (the Patterson or *heavy-atom method*).[17] However, preparation of a heavy-atom derivative of a natural compound, without disturbing the original molecular conformation, is not always feasible. Thus, generation of a structure solution for bioactive organic molecules of considerable complexity has been an experimental challenge. The derivation of a structure solution from the diffraction data is the most crucial aspect of crystallographic structure determination, without which further structural analysis is impossible.

Hauptman and Karle (Nobel Prize, 1985) pioneered development of the *direct methods*[18] of phase determination for the elucidation of crystal structures. In order to provide a technique that was not dependent on the presence of a heavy atom in the crystal, the direct methods apply statistical routines of analysis to the measured diffraction data (lattice parameters, crystal symmetry, and unit cell contents). The direct methods generate possible structure solutions to solve the phase determination of $F_{(hkl)}$. Electron density maps for the most probable of these solutions yield recognizable molecular fragments that are refined further to reveal approximate positions for additional atoms until a preliminary structure is obtained. This approach of applying probability theory to resolve the crystallographic phase problem has been extremely effective for obtaining structure solutions, even for noncentrosymmetric (see below) organic structures containing no heavy atoms. Direct methods have tremendously simplified and revolutionized crystallographic structure determinations and have permitted the development of computer software packages (e.g., SHELX76[19]).

4. Structure Refinement

After the complete molecular structure has been determined from electron density maps, the approximate atomic coordinates are refined in order to obtain the best fit between the calculated and observed intensity data. This structure refinement process can be mathematically summarized by a conventional least-squares refinement procedure:

$$R = \left[\sum\left(\left|F_{obs}\right| - \left|F_{calc}\right|\right)\right]\Big/\sum\left|F_{obs}\right|$$

The above expression states that the discrepancy index or residual, R, is the summation of all differences between observed and calculated structure factors over the summation of all observed structure factors of all atoms in the structure. Thus, the R value is a measure of the percent discrepancy between a measurement and a calculation. This full-matrix least-squares refinement yields extremely accurate values for atomic positions, from which bond lengths and bond angles may be calculated. X-ray intensity data often are corrected further for the effects of absorption, anomalous dispersion, and secondary extinction in the refinement of crystal structures to reduce the discrepancy between the observed and calculated structure factors. All of the additional refinements result in weighted R values (R_w).[20]

In addition to the atomic positional parameters refinement, thermal vibrational parameters are calculated because atoms are not stationary in the crystal structure, but vibrate continuously about their nominal position coordinates. Although a single isotropic temperature factor per atom approximates the thermal vibration by a sphere, a more realistic set of six anisotropic temperature factors allows for motion in the form of a thermal ellipsoid. The shape of the ellipsoid represents the three-dimensional thermal vibrations of the atom. In the initial cycles of structure calculation, atomic parameters of nonhydrogen atoms are refined, first, isotropically and, subsequently, anisotropically. Often the thermal vibrational parameters of hydrogen atoms are refined isotropically. Consequently, their atomic positions are included with invariant idealized values in the structure-factor calculation, because X-ray diffraction is incapable of defining highly accurate hydrogen positions, owing to their relatively low scattering ability. Thus, if hydrogen atoms are included in a molecular display from an X-ray analysis, they are represented usually as tiny circles.

A typical X-ray crystallographic analysis generates a voluminous amount of molecular structure information. Only a brief summary of the pertinent X-ray crystal data is permitted in most publications. Supplementary structure data are deposited at the Cambridge Crystallographic Data Base,[21] Protein Data Bank,[22] or Inorganic Crystal Structure Data Base[23] and are retrievable upon request.

Attainment of a successful X-ray structure determination relies considerably on the quality of the selected crystal, the precision of the experimental procedure, and the accuracy of the structure calculation. The time span required to perform a crystallographic investigation depends upon the size of the molecule, the complexity of space group symmetry, the unit cell size, and the stability of the crystal. Approximately one week is required to accomplish the structure determination of a 30-atom molecule possessing a reasonable unit cell size and a general space group symmetry. Although this may seem undesirably laborious in comparison to other chemical or spectral methods, X-ray diffraction is the only analytical method that furnishes such extensive fundamental structural information.

Natural products whose structures were originally assigned by the application of nuclear magnetic resonance (NMR) or mass spectral techniques and that were confirmed or amended later by X-ray crystallography include the robustadials, azadirachtin, and the secoiridoid xylomollin[24] (n.b. footnote 2 therein). Conversely, although X-ray crystallographic determinations are usually held as the definitive authority for structural assignments, misinterpretation of X-ray data of the C-nucleoside pyrrolosine illustrates the pitfall of relying heavily on X-ray techniques, without confirmation from other synthetic and spectral techniques.[25]

5. Stereoscopic Viewing of Molecular Structures

X-ray crystallographic analysis is the only structure elucidation technique that yields a detailed and comprehensive pictogram as the final result. When the structure investigation is completed, interactive computer programs allow the image of the entire molecular structure to be displayed on a graphic terminal or plotter. In order to understand further the neighboring environment of the molecule, models of the structure reveal the molecular packing arrangement in the crystal and additional fundamental information, e.g., intermolecular hydrogen bonding, van der Waal or electrostatic molecular interactions, and solvent participation.

A recurring problem for the structural chemist involves the presentation of the spatial arrangement of atoms in a model that conveys three-dimensional information. Generally, stereoscopic view pairs are employed as a substitute for a solid model. *ORTEPII*,[26] the computer graphic program published by Johnson, provides scaled, two-dimensional coordinates for each member of the stereoscopic pair generated from the three-dimensional coordinates of the atoms. A coded output yields a drawing of two stereoviews, complete with tapered bonds and perspective probability ellipsoids of all atoms. Stereoscopic drawings now appear routinely in journals and textbooks, and viewing chemical structures as three-dimensional images has become a key component in reading the scientific literature.

The use of binocular vision to recover three-dimensional structure arises from the slight horizontal disparities observed by each eye upon viewing a given object or structure. Everyone that has stared either at a repeating wallpaper pattern or a periodic pattern of tiles has probably experienced the phenomenon of false depth perception due to eye mismatching of repeating elements on the pattern. Wheatstone showed that this disparity was sufficient to produce stereopsis, i.e., solid vision.[27] The two slightly different images are perceived in the brain and then synthesized into a perception of the three-dimensional image that is psychologically projected into space where it occupies the identical position as the original solid object.

Inexpensive viewers are available that permit the viewing of stereoscopic diagrams (Taylor Merchant Corp., 212 W. 35th St., New York, NY 10001). However, naked-eye stereopsis (or free viewing) can be attained with a variety of simple techniques.[28-32] An exercise that is an excellent predictor of the ability to attain naked-eye stereopsis is the *floating-fingertip* illusion where the hands are positioned to observe the merging images of the fingertips.[28]

Two of the techniques commonly employed to achieve free viewing (parallel or uncrossed viewing, i.e., left eye-left view, right eye-right view) are the *infinite stare* and the *nose to the grindstone* methods. The former method involves transferring a preestablished far convergence to a nearby stereo pair and maintaining a focus on the fused image without changing the eyes' convergence.[28,29] Thus, the goal in successful parallel free viewing is to relax the eyelines to infinity while maintaining close focus. With eyes converged at infinity, three, not two, fusion views should be visible, with the central view being that of the fused stereo image (see Figure 9). In the latter method the eyes are forced initially to unfocus, by their close proximity to the appropriate stereoview (left eye-left view, right

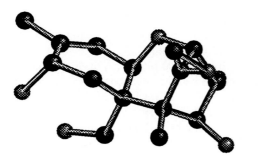 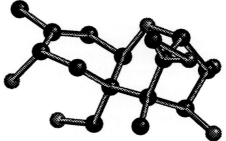

FIGURE 9. Stereoscopic views of tricothecene.

eye-right view). Upon removal of the stereoviews from close proximity to the viewer's eyes, the fused image should come into focus when the stereoviews are eventually withdrawn to reading distance.[28]

Although literature stereoviews are published usually for parallel or uncrossed viewing so that they may be viewed with standard stereoviewers, occasionally views will be reversed so that the left and right views are for the right and left eyes, respectively. An excellent technique for eye-crossed viewing (naked-eye stereopsis with inversion) employs focusing on a pencil tip that is placed between the eyes and the diagram to be viewed.[30] Upon maintaining fusion in front of the stereoview, a pseudoimage of the fused stereoview should appear that can be maintained upon removal of the pencil tip. Eye-crossed viewing is an excellent method for free viewing of poster-sized drawings, although a reflecting stereoscope is commercially available for parallel viewing of larger diagrams (VCH Publishers, 303 N.W. 12th St., Deerfield Beach, FL 33442).

III. CRYSTALLOGRAPHIC ANALYSIS
OF NATURAL PRODUCTS

Early X-ray crystallographic determinations generally verified structures of known substances (e.g., the planar, cyclic structure of benzene[33]). Later, X-ray diffraction was employed to establish molecular structures of biologically and chemically significant molecules, particularly those that had proved elusive by chemical methods. In classic studies performed in 1932, Bernal established the correct steroid skeleton structure.[34] During the 1940s and 1950s, Hodgkin (Nobel Prize, 1964) determined the structures of calciferol,[35] penicillin,[36] and vitamin B_{12}.[37] Several examples of natural product structure determination by X-ray crystallography at the Western Regional Research Center are described in the following subsections.

A. ALKALOIDS

Several rare, isomeric, dimeric quinolinone alkaloids (**1**) that differed only in their stereochemistry at C-6a and C-16a were isolated from the Brazilian tree *Euxylophora paraensis* (Rutaceae).[38,39] Determination of their structures by X-ray crystallography revealed previously unsuspected structural features that implied a new biosynthetic pathway to these compounds.

Australine (**2**), a novel pyrrolizidine alkaloid, is a toxic seed constituent of *Castanospermum australe* that has been identified as a potent inhibitor of α-glucosidase.[40] Whereas nuclear magnetic resonance spectroscopy and mass spectrometry were able to divulge only the relative stereochemical structure of australine, X-ray crystallography established its absolute configuration.

Isomeric dimeric quinolinone alkoids from Brazilian tree
Euxylophora paraensis

H 6aα,	H16aα	(dimer A)
H 6aα,	H16aβ	(dimer C)
H 6aβ,	H16aβ	(dimer E)
H 6aβ,	H16aα	(dimer F)

Paraensidimerin (1)

B. STEROIDS

The leaves of *Petunia* (Solanaceae) are toxic toward many insects, and this activity has been shown to result from the presence of a wide variety of ergostanol-type steroids. X-ray crystallography contributed to the structure determination of several toxic petuniolides (e.g., **3**)[41] and petuniasterones (e.g., **4**).[42,43]

X-ray crystallographic evidence on the solid-state conformation of 10α-cucurbitadienol (**5**) indicates that this sterol-like molecule is oriented in the crystal as a bent form rather than a flat conformation, as clearly shown by its perspective *ORTEPII* diagram, and thus may function in cellular biochemistry as the bent conformer.[44]

C. TERPENOIDS

The molecular structures and absolute configurations of several sesquiterpene lactones isolated from Russian knapweed (*Centaurea repens*), which has been implicated as the causative agent for "chewing disease" in horses, have been established by X-ray crystallography. Of these cytotoxic sesquiterpene lactones, repin (**6**) exhibits a toxicity appreciably greater than related compounds from the same plant.[45]

During an investigation of insect inhibitory substances derived from certain *Petunia* species, a related plant, *Calibrachoa parviflora*, was examined, which previously had been classified within the genus *Petunia*. A number of kauranes and pimaranes were isolated from *C. parviflora* and the structures of several new diterpenes (e.g., **7**) were established by X-ray crystallography.[46] Because none of the steroidal substances of the *Petunia* type were present, the reassignment of *Calibrachoa* into a separate genus was phytochemically supported.

D. FLAVANOID GLYCOSIDES

Determination of the molecular structure and absolute configuration of the sweetener neohesperidin dihydrochalcone (**8**) permitted comparison between the X-ray-deduced structure of the sweetener and hypothetical *active* conformations.[47] The crystallographic data lent little support to a proposal that hesperetin dihydrochalcone and other sweeteners have bent conformations responsible for conferring sweetness.

Australine - a toxic constituent from the seeds of
Castanospermum australe

HO⟍ ⟍ H   OH
 7 7a   1
6   2 ⟍ ''''''OH
 5 ─N─ 3
          8 CH₂OH

(2)

IV. DETERMINATION OF MOLECULAR CONFIGURATION

Chiral compounds are noncentrosymmetric structures; i.e., they are not superimposable, but are the inverted image of one another. It is of great importance to delineate the structural difference between two enantiomers, i.e., whether the molecule is right-handed or left-handed, because in numerous instances enantiomeric natural products exhibit significant differences in their pharmacodynamic, pharmacokinetic, and metabolic properties.[48] As a primary physical technique, X-ray crystallographic analysis is capable of providing directly the absolute configuration of a chiral molecule, without relying on comparison to other structures, if an atom is present that scatters X-rays anomalously. A link between molecular chirality (i.e., *handedness* from the Greek word for hand, χειρ)[49] and optical rotation may be attained by measuring the rotation of a solution prepared by dissolution of a crystal that has been used for X-ray measurements.[50] Determination of the absolute configuration of noncentrosymmetric crystal structures is usually attempted after the least-squares refinement of the structure has reached a satisfactory level of convergence. The X-ray approaches most frequently applied for determination of absolute configuration are those of Bijvoet pairs[51] and the R-index comparison method.[52]

The pioneering method of Bijvoet[51] utilizes the atomic anomalous dispersion of radiation by the structure and carefully compares the intensities of the Friedel pairs of a selected group of reflections in one direction *(h,k,l)* with those in the opposite direction *(−h,−k,−l)*. Some reflections are more sensitive to anomalous scattering effects than others, and their intensities may differ from those of their Friedel mates by several percent. In a noncentrosymmetric structure, each unique reflection has its Friedel mate.

(3)

Only 10 to 20 of the Friedel pair reflections most sensitive to anomalous scattering need to be compared in the complete intensity data set. However, a clear-cut majority of these carefully selected pairs should yield the same prediction for a successful assignment of absolute configuration.

Alternatively, according to the Hamilton statistical probability strategy,[52] the absolute configuration can be assigned by separately refining both of the enantiomeric structures, with the entire set of intensity data, to convergence and comparing the resulting R-indices of the observed and calculated Bijvoet differences for the two enantiomers. The atomic positional parameters for the opposite enantiomer may be generated by inverting the positional coordinates of the molecule already refined. The structure possessing the lower R-index is judged to be the correct enantiomer, although the differences in R-indices may be very small (±0.1%), and the significance of their accuracy must be carefully evaluated by statistical methods. Because both of these methods depend solely on the anomalous dispersion of radiation by the structure, they are most effective for establishing the absolute configuration of structures that contain intense scattering entities (e.g., bromine), but not for structures that are comprised predominantly of only weak anomalous scatterers. However, elements with weak dispersive ability may serve also as anomalous scatterers if the X-ray intensity data set is of high quality. Crystallographers have successfully established the absolute configuration of structures that contain no atoms heavier than oxygen.[53] An interesting caveat concerning the application of

Petuniasterones - toxic constituents from leaves of *Petunia*

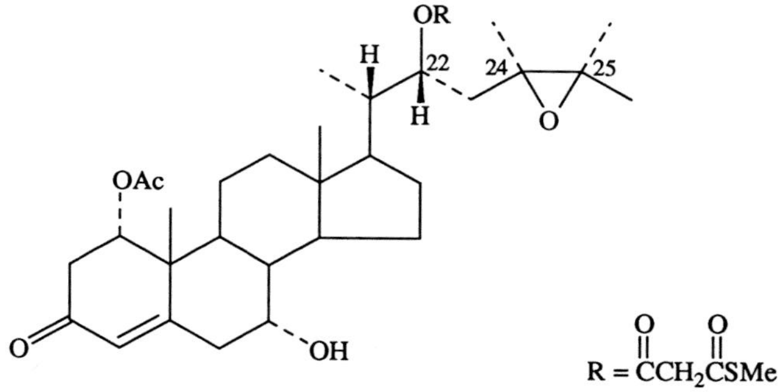

R = CCH$_2$CSMe

(4)

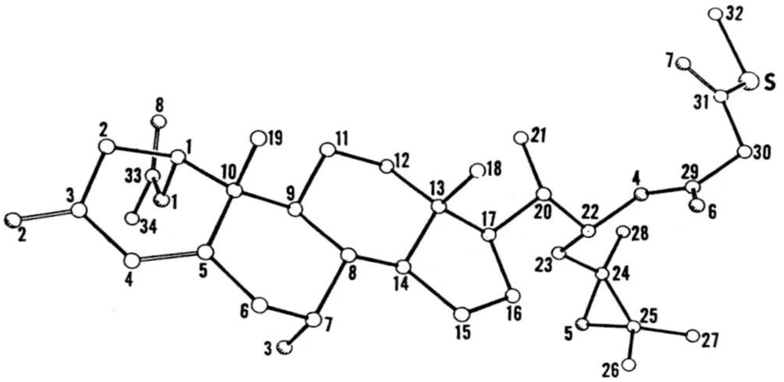

X-ray measurements to determination of absolute configuration concerns gephyrotoxin, where a configurational misassignment resulted not from misinterpretation of X-ray data, but instead from a statistical anomaly[54] (n.b. footnote 9 therein).

While chirally pure molecules can crystallize in forms that may have any phase value from 0 to 360°, racemic mixtures crystallize in space groups where phase angles are restricted to one of two values. Thus, the phase problem in the centrosymmetric space groups of racemic molecules is much simpler to solve than that for noncentrosymmetric space groups of chiral molecules. Recently, the synthesis of enantiomeric forms of natural products that contain 200 to 500 atoms (e.g., rubredoxin) has been performed, partly in order to grow centrosymmetric crystals of the racemic natural product after admixture of the synthetic with the natural enantiomer.[55] The elimination of uncertainty in phase determinations of the racemic molecular crystal should permit faster and more accurate molecular structure determination.

The assignment of *R* (rectus) or *S* (sinister) designations to chiral centers is a procedure separate from that of determining the absolute configuration and was developed in order to specify, clearly and unambiguously, the configuration of chiral atoms.[56-59] The Cahn-Ingold-Prelog sequence rule[56] involves the following steps: identification of atomic priority about a stereocenter; orientation of the molecule in space so that the stereocenter may be viewed toward the substituent of lowest priority; and, finally, allocation of the appropriate clockwise or counterclockwise path of decreasing atomic priority [e.g., the absolute configuration of australine (**2**) is (1*R*,2*R*,3*R*,7*S*,7a*R*)]. Configurational information of chiral atoms is inherent in the atomic coordinates of a chiral structure and may be derived by either of two techniques. Various computer graphic algorithms assign configurations to chiral centers, upon structural

10α -Cucurbitadienol from pumpkin seeds

$$R = \overset{\overset{\textstyle O}{\textstyle \|}}{C}\text{-}CH_3$$
$$R = H$$

(5)

input. Alternatively, the correct configurational designation for each chiral atom may be determined visually from a molecular model, by application of the sequence rule.[56]

X-ray crystallography continues to contribute greatly to our understanding of molecular structure and function. The elucidation of the three-dimensional structures of myoglobin by Kendrew (Nobel Prize, 1962),[60] hemoglobin by Perutz (Nobel Prize, 1962),[61] and DNA by Watson and Crick (Nobel Prize, 1962)[62] are landmarks in molecular biology. The evolution of supercomputers, computational techniques, and instrumental developments has led to an enormous increase in experimental flexibility and the successful solution of complex structures, especially in protein crystallography. Furthermore, the recent development of powerful three-dimensional interactive graphic computing workstations and expert system software for applications in the area of computer-aided molecular modeling and simulation techniques has added still another dimension (e.g., visualization of ligand-receptor interactions) to the already productive field of crystallographic research.

Repin - a cytotoxic sesquiterpene lactone from Russian Knapweed

Centaurea repens

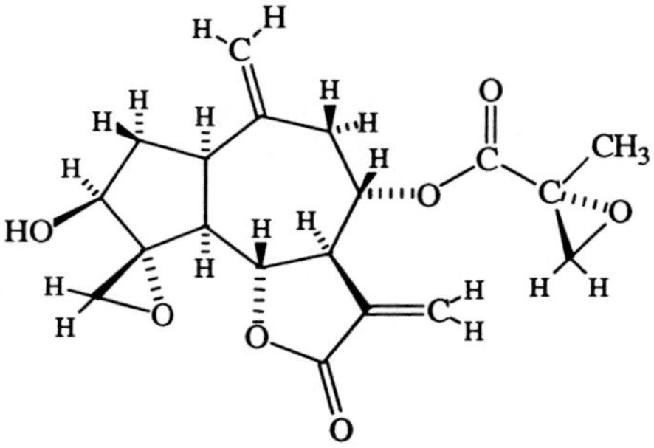

(6)

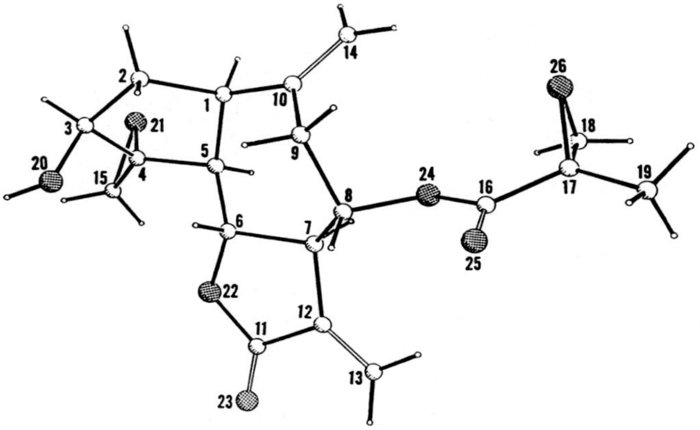

Diterpene from *Calibrachoa parviflora*

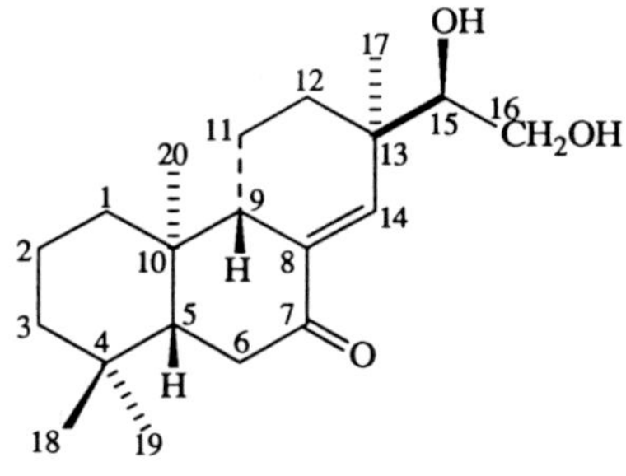

(7)

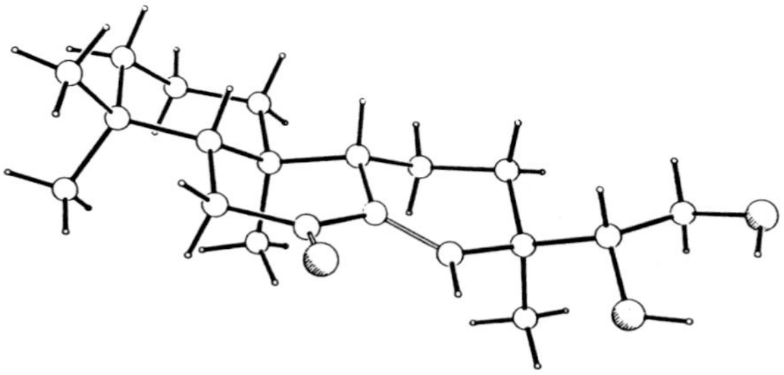

Neohesperidin dihydrochalcone - sweetener from citrus

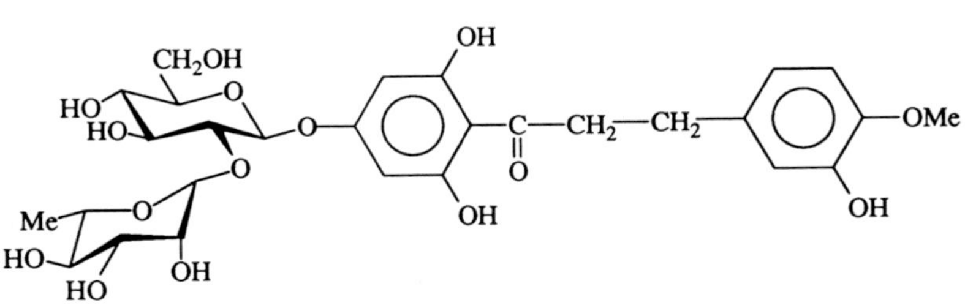

(8)

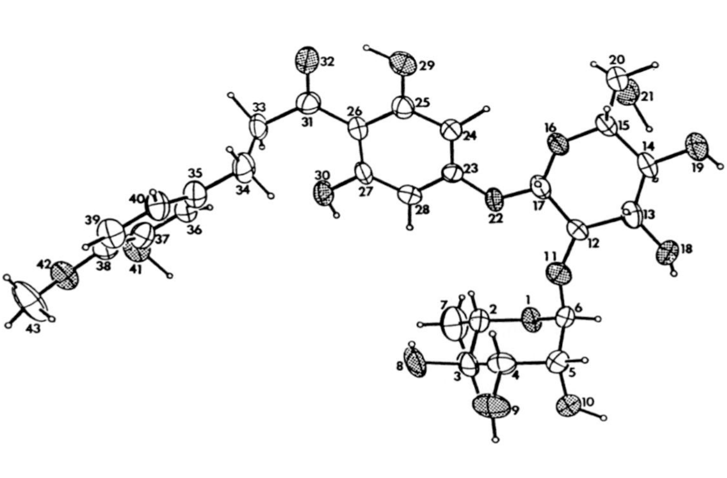

REFERENCES

1. **Bragg, W. H. and Bragg, W. L.,** *X-rays and Crystal Structure,* Bell, London, 1924.
2. **Bragg, W. H.,** *Introduction to Crystal Analysis,* Bell, London, 1928.
3. **Buerger, M. J.,** *X-ray Crystallography,* Wiley & Sons, New York, 1942; Krieger, Melbourne, FL, reprinted 1980.
4. **Buerger, M. J.,** *Crystal Structure Analysis,* Wiley & Sons, New York, 1960; Krieger, Melbourne, FL, reprinted 1980.
5. **Dunitz, J. D.,** *X-ray Analysis and The Structure of Organic Molecules,* Cornell University., Ithaca, NY, 1979.
6. **Finer-Moore, J., Arnold E., and Clardy, J.,** Some uses of X-ray diffraction in alkaloid chemistry, in *Alkaloids: Chemical and Biological Perspectives,* Vol. 2, Pelletier, S. W., Ed., John Wiley & Sons, New York, 1984, 1.
7. **Glusker, J. P. and Trueblood, K. N.,** *Crystal Structure Analysis. A Primer,* 2nd ed., Oxford University, Oxford, 1985.
8. **Stout, G. H. and Jensen, L. H.,** *X-ray Structure Determination. A Practical Guide,* Macmillan, New York, 1968.
9. **von Laue, M.,** Röntgenstrahlinterferenzen, *Physik. Z.,* 14, 1075, 1913.
10. **Bragg, W. H.,** The reflection of X-rays by crystals, *Nature,* 91, 477, 1913.
11. **Bragg, W. L.,** The diffraction of short electromagnetic waves by a crystal, *Proc. Cam. Phil. Soc.,* 17, 43, 1913.
12. **Bernal, I., Hamilton, W. C., and Ricci, J. S.,** *Symmetry: A Stereoscopic Guide for Chemists,* W. H. Freeman, New York, 1972.
13. **Wyckoff, R. W. G.,** *Structure of Crystals,* Krieger, Melbourne, FL, 1935, reprinted 1963.
14. **Hahn, T., Ed.,** Space-Group Symmetry, in *International Tables for Crystallography,* Vol. A, Reidel, Boston, 1983.
15. **Brown, J. N. and Trefonas, L. M.,** A solvent pair recrystallization apparatus for growing single crystals, *Org. Prep. Proced.,* 2, 317, 1970.
16. **Buerger, M. J. and Klein, G. E.,** Correction of diffraction amplitudes for Lorentz and polarization factors, *J. Appl. Phys.,* 17, 285, 1946.
17. **Patterson, A. L.,** A direct method for the determination of the components of interatomic distances in crystals, *Z. Kristallogr.,* 90, 517, 1935.
18. **Karle, J. and Hauptman, H.,** The phases and magnitudes of the structure factors, *Acta Cryst.,* 3, 181, 1950.
19. **Sheldrick, G. M.,** *SHELX76. A Program for Crystal Structure Determination,* Cambridge University, England, 1976.
20. **Jones, P. G.,** Crystal structure determination: a critical view, *Chem. Soc. Rev.,* 13, 157, 1984.
21. Cambridge Crystallographic Data Base, National Technical Information Service (NTIS), U.S. Dept. of Commerce, Springfield, VA.
22. Protein Data Bank, Chemistry Dept., Brookhaven National Laboratory, Upton, NY 11973.
23. Inorganic Crystal Structure Data Base, contact Paul Dellevigne, 7 Woodland Avenue, Larchmont, NY 10538.
24. **Jirousek, M. R., Mazza, S. M., and Salomon, R. G.,** Robustadials. 4. Molecular mechanics and nuclear magnetic resonance studies of conformational and configurational equilibria: 3,4-dihydrospiro[2*H*-1-benzopyran-2,2′-bicyclo[2.2.1]-heptanes], *J. Org. Chem.,* 53, 3688, 1988.
25. **Otter, B. A., Patil, S. A., Klein, R. S., and Ealick, S. E.,** A corrected structure for pyrrolosine, *J. Am. Chem. Soc.,* 114, 668, 1992.
26. **Johnson, C. K.,** *ORTEPII,* Report ORNL-5138, Oak Ridge National Laboratory, Oak Ridge, TN, 1976.
27. **Wheatstone, C.,** On some remarkable, and hitherto unobserved, phenomena of binocular vision, *Phil. Trans. R. Soc. London,* 128, 371, 1838.
28. **McKeon, T. A. and Gaffield, W.,** Viewing stereopictures in three dimensions with naked eyes, *Trends Biochem. Sci.,* 15, 412, 1990.
29. **Daniels, C. F.,** Freevision, *J. Lab. Clin. Med.,* 118, 511, 1991.
30. **Speakman, J. C.,** Visualising molecules without really trying your eyes, *New Scientist,* 78, 827, 1978.
31. **Johnstone, A. H., Letton, K. M., and Speakman, J. C.,** Stereopsis in chemistry, *Educ. Chem.,* 17, 172, 1980.
32. **Speakman, J. C.,** The key to unlock NES(s), *Chem. Br.,* 14, 107, 1978.
33. **Lonsdale, K.,** The structure of the benzene ring, *Nature,* 122, 810, 1928.
34. **Bernal, J. D.,** Crystal structures of vitamin D and related compounds, *Nature,* 129, 277, 1932.
35. **Crowfoot, D. and Dunitz, J. D.,** Structure of calciferol, *Nature,* 162, 608, 1948.
36. **Crowfoot, D., Bunn, C. W., Rogers-Low, B. W., and Turner-Jones, A.,** X-ray crystallographic investigation of the structure of penicillin, in *Chemistry of Penicillin,* Clarke, H. T., Johnson, J. R., and Robinson, R., Eds., Princeton University Press, Princeton, NJ, 1949, 310.

37. **Crowfoot, D., Pickworth, J., Robertson, J. H., Trueblood, K. N., Prosen, R. J., and White, J. G.,** Structure of vitamin B$_{12}$, *Nature,* 176, 325, 1955.

38. **Jurd, L., Wong, R. Y., and Benson, M.,** The structures of paraensidimerin A and C, two bisquinoline alkaloids from *Euxylophora paraensis, Aust. J. Chem.,* 35, 2505, 1982.

39. **Jurd, L., Benson, M., and Wong, R. Y.,** New quinolinone and bis-quinolinone alkaloids from *Euxylophora paraensis, Aust. J. Chem.,* 36, 759, 1983.

40. **Molyneux, R. J., Benson, M., Wong, R. Y., Tropea, J. E., and Elbein, A. D.,** Australine, a novel pyrrolizidine alkaloid glucosidase inhibitor from *Castanospermum australe, J. Nat. Prod.,* 51, 1198, 1988.

41. **Elliger, C. A., Wong, R. Y., Waiss, A. C., Jr., and Benson, M.,** Petuniolides. Unusual ergostanoid lactones from *Petunia* species that inhibit insect development, *J. Chem. Soc. Perkin Trans. 1,* 525, 1990.

42. **Elliger, C. A., Benson, M. E., Haddon, W. F., Lundin, R. E., Waiss, A. C., Jr., and Wong, R. Y.,** Petuniasterones, novel ergostane-type steroids of *Petunia hybridia* Vilm. (Solanaceae) having insect-inhibitory activity. X-ray molecular structure of the 22,24,25-[(methoxycarbonyl)orthoacetate] of 7α,22,24,25-tetrahydroxyergosta-1,4-dien-3-one and of 1α-acetoxy-24,25-epoxy-7α-hydroxy-22-(methylthiocarbonyl)-acetoxyergost-4-en-3-one, *J. Chem. Soc. Perkin Trans. 1,* 711, 1988.

43. **Elliger, C. A., Benson, M., Haddon, W. F., Lundin, R. E., Waiss, A. C., Jr., and Wong, R. Y.,** Petuniasterones. Part 2. Novel ergostane-type steroids from *Petunia hybrida* Vilm. (Solanaceae), *J. Chem. Soc. Perkin Trans. 1,* 143, 1989.

44. **Nes, W. D., Wong, R. Y., Benson, M., and Akihisa, T.,** Conformational analysis of 10α-cucurbitadienol, *J. Chem. Soc. Chem. Commun.,* 1272, 1991.

45. **Stevens, K. L., Riopelle, R. J., and Wong, R. Y.,** Repin, a sesquiterpene lactone from *Acroptilon repens* possessing exceptional biological activity, *J. Nat. Prod.,* 53, 218, 1990.

46. **Elliger, C. A., Wong, R. Y., Benson, M., Gaffield, W., and Waiss, A. C., Jr.,** Diterpenes of *Calibrachoa parviflora, J. Nat. Prod.,* 55, 1477, 1992.

47. **Wong, R. Y. and Horowitz, R. M.,** The X-ray crystal and molecular structure of neohesperidin dihydrochalcone sweetener, *J. Chem. Soc. Perkin Trans. 1,* 843, 1986.

48. **Gaffield, W.,** Chirality as manifested in the biological activity of natural products, in *Studies in Natural Products Chemistry,* Vol. 7, Rahman, A., Ed., Elsevier, Amsterdam, 1990, 3.

49. **Petsko, G. A.,** On the other hand..., *Science,* 256, 1403, 1992.

50. **Hulshof, L. A., Wynberg, H., van Dijk, B., and de Boer, J. L.,** Reassignment of the chirality to a series of 2,6-disubstituted spiro[3.3]heptanes by X-ray methods and implications thereof on empirical rules and theoretical models, *J. Am. Chem. Soc.,* 98, 2733, 1976.

51. **Bijvoet, J. M., Peerdeman, A. F., and van Bommel, A. J.,** Determination of the absolute configuration of optically active compounds by means of X-rays, *Nature,* 168, 271, 1951.

52. **Hamilton W. C.,** Significance tests on the crystallographic R factor, *Acta Cryst.,* 18, 502, 1965.

53. **Hope, H. and de la Camp, U.,** Determination of absolute configurations from anomalous scattering by oxygen, *Nature,* 221, 54, 1969.

54. **Fujimoto, R. and Kishi, Y.,** On the absolute configuration of gephyrotoxin, *Tetrahedron Lett.,* 22, 4197, 1981.

55. **Zawadzke, L. E. and Berg, J. M.,** A racemic protein, *J. Am. Chem. Soc.,* 114, 4002, 1992.

56. **Cahn, R. S.,** An introduction to the sequence rule, a system for the specification of absolute configuration, *J. Chem. Ed.,* 41, 116, 1964; 41, 508, 1964.

57. **Eliel, E. L.,** Recent advances in stereochemical nomenclature, *J. Chem. Ed.,* 48, 163, 1971.

58. **Eliel, E. L.,** The *R/S* system: a new method for assignment and some recent modifications, *J. Chem. Ed.,* 62, 223, 1985.

59. **Silverstein, R. M.,** Chirality in insect communication, *J. Chem. Ecol.,* 14, 1981, 1988.

60. **Kendrew, J. C.,** The three-dimensional structure of a protein molecule, *Sci. Am.,* 205(6), 96, 1961.

61. **Perutz, M. F.,** The hemoglobin molecule, *Sci. Am.,* 211(5), 64, 1964.

62. **Watson, J. D. and Crick, F. H. C.,** Molecular structure of nucleic acids. A structure for deoxyribose nucleic acid, *Nature,* 171, 737, 1953.

Chapter 7

DETERMINATION OF THE ABSOLUTE CONFIGURATION OF BIOACTIVE NATURAL COMPOUNDS, UTILIZING EXCITON CHIRALITY CIRCULAR DICHROISM

William Gaffield

TABLE OF CONTENTS

I. INTRODUCTION

Organic molecules that lack reflection symmetry (i.e., a rotation-reflection axis) are nonsuperimposable upon their mirror image and are known as enantiomers. The property of nonsuperimposability of an object on its mirror image is referred to as chirality (from the Greek word meaning "handedness"). Enantiomers, which differ only in their absolute configurations (i.e., the left- and right-handedness of their orientations), possess identical physical and chemical properties except in two important respects: they rotate the plane of polarized light in opposite directions, and they react at different rates with other chiral compounds (e.g., biologic macromolecules and DNA).

The considerable influence exerted by the shape of a molecule on its physiological properties has been recognized for many years. Although chirality is not a requirement for biologic activity, a particular biologic property is often associated with a given absolute configuration in chiral molecules. One of the earliest observations of biologic stereoselectivity was recorded over a century ago by Piutti, who isolated the two enantiomers of asparagine and observed that the dextrorotatory isomer (**1**) tasted sweet, whereas the levorotatory isomer was bland.[1]

Frequently, during the ensuing decades, the difference in both pharmacologic activity and toxicity of enantiomers and diastereomers has been stressed, e.g., the stronger potency of (–)-hyoscyamine (**2**) on terminations of the peripheral neurons in cat's eye pupils, when compared to its racemate (atropine).[2] Ample documentation has now been provided for numerous substances, demonstrating that both the pharmacokinetic and pharmacodynamic profile of enantiomers may differ due to stereospecific interaction with chiral centers on receptors. Whereas chirality is unimportant in biologic processes depending upon passive transport across membranes (e.g., absorption, distribution, and secretion), the fates and reaction rates of enantiomers can differ in systems relying upon highly chiral receptors, channels, or enzymes (e.g., active transport, protein binding, and metabolism). Thus, the impetus for determining the absolute configuration of biologically active pharmaceuticals and pesticides has resulted from the growing awareness that a biologically active racemate should be viewed not as one, but as two, different compounds.[3-6]

For over a century, from the seminal experiment of Pasteur in 1848 on the manual separation of the two enantiomorphs of sodium ammonium tartrate tetrahydrate, until 1951 when Bijvoet directly assigned the absolute configuration of crystals by anomalous X-ray scattering, only relative stereochemistry was considered. In addition to the Bijvoet method of X-ray crystallography, a nonempirical method that enables the determination of the absolute configuration of organic compounds by circular dichroism (CD) is the exciton chirality method. The basic theories underlying the exciton chirality phenomenon were provided in the 1930s by Kuhn's coupled oscillator theory[7] and Kirkwood's group polarizability theory.[8] The application of the coupled oscillator method to various natural products, notably by Nakanishi, Harada, and colleagues,[9] permits the determination of the absolute configuration of organic compounds, without reference to authentic examples.

II. EXCITON CHIRALITY METHOD

Initially, this method correlated the chirality of 1,2-glycols [e.g., rishitin dibenzoate (**3**)] with signs of the intense π-π^* Cotton effects of dibenzoates (the dibenzoate chirality rule). Extension of the technique to benzoates of phenolic hydroxy groups [e.g., derivatives of the antitumor antibiotic illudin S (**4**)], to triol benzoates of nonadjacent hydroxy groups [e.g., ajugasterone C (**5**)], and to chiral interactions between chromophores other than benzoates [e.g., (+)-glaudine (**6**) and other Rhoeadine alkaloids] led to the evolution of the broadly applicable exciton method.[10]

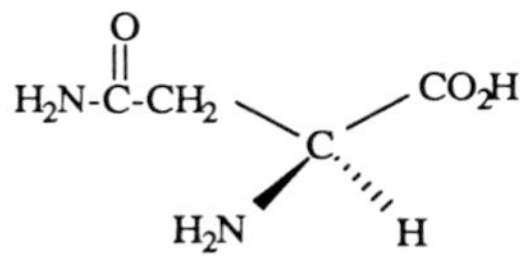

(1)

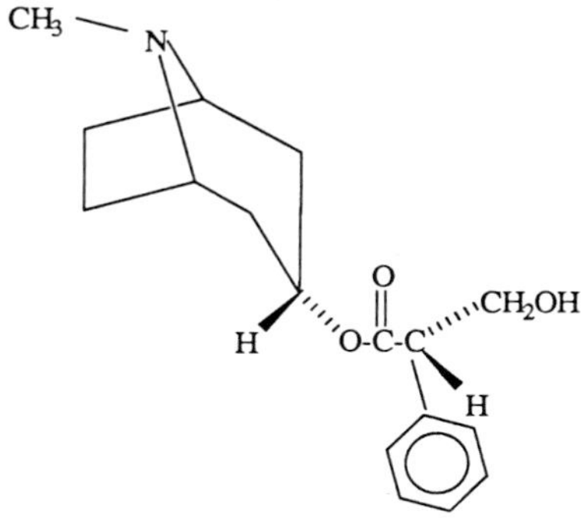

(2)

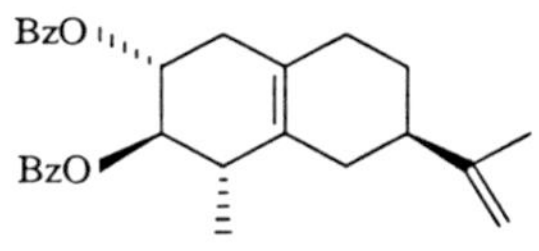

(3)

Para-substituted benzoate chromophores that are symmetric about the long axis of the chromophore and proximate to a chiral center give rise to a CD Cotton effect (CE) in the strong longitudinal L_a or intramolecular charge-transfer band. Thus, the *p*-methoxycinnamate chromophore, having a UV maximum at 311 nm, produces a single CE at 309 nm of moderate intensity. However, when two of these chromophores are chirally disposed in spatial proximity, either intra- or intermolecularly, a through-space interaction occurs that leads to a split CD with extrema of opposite signs at 287 and 322 nm (Figure 1). For a substituted dibenzoate, oppositely signed CEs at longer and shorter wavelengths (first and second CE) reflect the chirality or screwness between the two L_a electric transition moments. Despite the free rotation about the C–O bond connecting the benzoate groups to their carbon atoms, the electric transition moment (L_a) of each is approximately parallel to its C–O bond, due to the preferred *s-trans* conformation of the ester bond (Figure 2). Therefore, the chirality between the two transition moments directly reflects the chirality between the two C–O bonds. When the interaction of two (or more) chromophores is chiral, i.e., the two interacting electric transition moments are not in the same plane or do not run parallel, nondegenerate excited electronic levels result that are split by an energy gap (the Davydov splitting). In electronic spectroscopy these absorptions have only one sign (positive), which leads to a summation curve twice as intense as the integrated intensity of a single chromophore. However, in circular dichroic spectroscopy the interaction leads to slightly red- and blue-shifted CD curves of opposite signs, separated by the Davydov splitting, that define the molecular chirality. The chirality

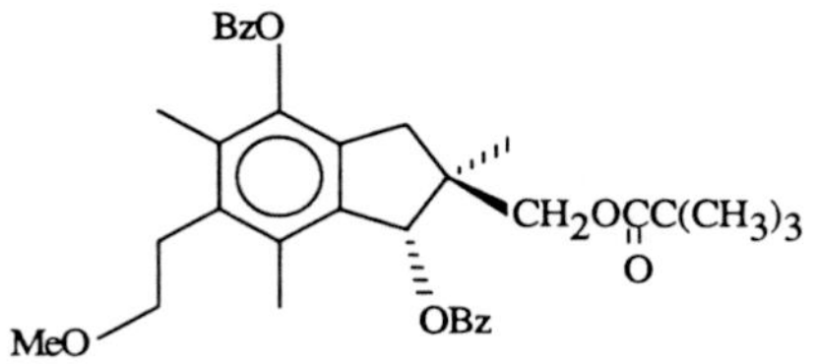

(4)

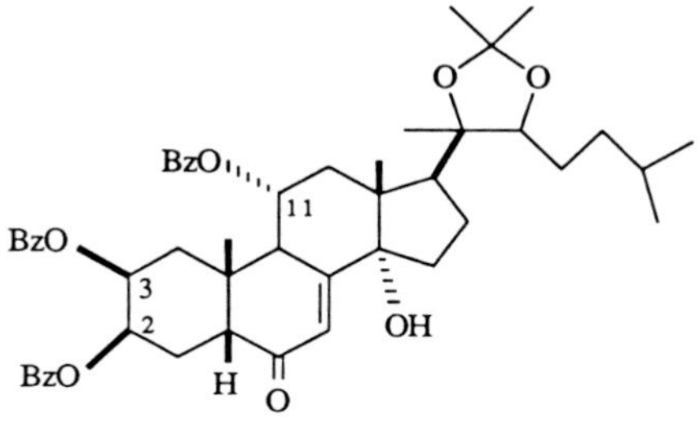

(5)

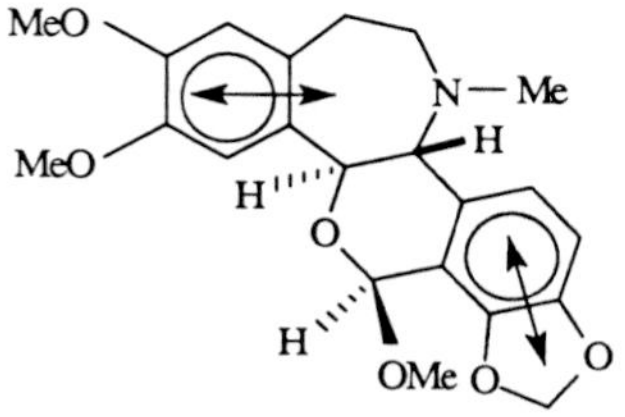

(6)

is defined as negative if the Newman projection of the two electric transition moments constitutes a left-handed screw when viewed along the line connecting the two carbon atoms, and vice versa (Figure 3). The coupled CD curves can be readily calculated from the interchromophoric distance, the oscillator strength of interacting chromophores, and the angle between transition moments. There is excellent agreement between observed and calculated exciton-split CD spectra, with the difference in intensity of the extrema of the longer-wavelength CE minus that of the shorter-wavelength CE (i.e., the CD amplitude) defined as the A value.[9]

A. SPECIAL FEATURES

Special features of the exciton chirality method include the following: (1) any chromophore can be used provided the absorption (ε) is strong and the direction of the electric transition moment (μ) is known; (2) a linear relation exists between the UV ε and CD amplitude (A), i.e., the stronger the UV ε value, the larger the A value; and (3) the extent of the coupling or amplitude of the bisignate CD curves is inversely proportional to the square of the interchromophoric distance R. Coupling of two p-bromobenzoates is observed even when the chromophores are 13 Å apart with a dihedral angle of 60°; in a vicinal dibenzoate, the A values are maximal when the dihedral angle is about 70°, but is zero at 0° or 180° (i.e., when there is no chirality). Split CD coupling is still retained when the separation in absorption maxima of the two interacting chromophores is 100 nm and the

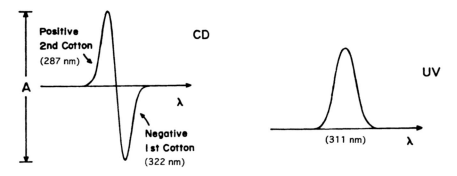

FIGURE 1. Depiction of a split CD curve (left), that arises from coupled *p*-methoxycinnamate L_a transitions (CD signs arbitrarily assigned), and the single-signed electronic spectrum (right) of summed *p*-methoxycinnamate L_a transitions.

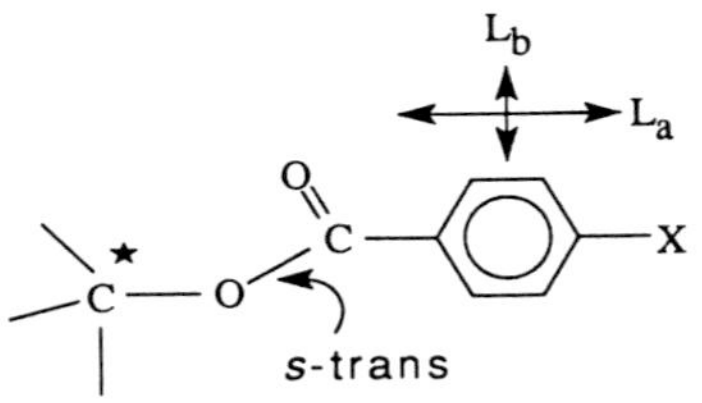

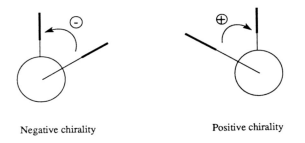

X	λ_{max} (L_a)
H	230nm
Br	244nm
N(Me)$_2$	309nm
CH$_3$O	257nm

FIGURE 2. Transitions of *p*-substituted benzoyl chromophores.

Negative chirality   Positive chirality

FIGURE 3. Definition of exciton chirality in a binary system.

total A value in multichromophoric systems is equal to the sum of interacting units (additivity relation).[11]

B. ADDITIVITY OF CD AMPLITUDES

The amplitude of a split CD curve of tri- and tetrabenzoates of monosaccharides was approximated by the sum of the three ($A_{2,3}$ + $A_{2,4}$ + $A_{3,4}$) (e.g., L-arabinose) (Figure 4) and six ($A_{2,3}$ + $A_{2,4}$ + $A_{2,6}$ + $A_{3,4}$ + $A_{3,6}$ + $A_{4,6}$) (e.g., D-glucose)-component dibenzoate units, respectively. Thus, the set of dibenzoate A values serves as constants for estimating the A value of any pyranose polybenzoate. The additivity applies not only for the single-wavelength A value, but for the entire CD curve when two chromophores are present (e.g., glycosides substituted with both *p*-bromobenzoyl and *p*-methoxycinnamoyl residues).[11] Although the strongest and most significant interaction between two benzoates results from L_a band coupling, the much weaker coupling involving higher-wavelength transversal L_b bands is relevant in defining the shape of an entire CD curve, particularly when two different chromophores are employed.[11] The additivity relation has been validated for benzoate and enone groups

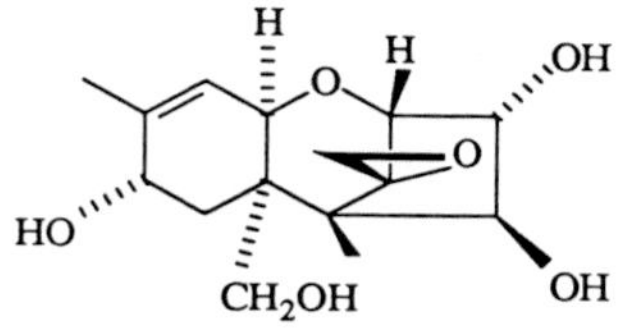

(7)

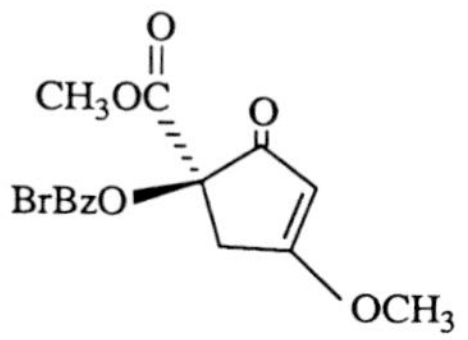

(8)

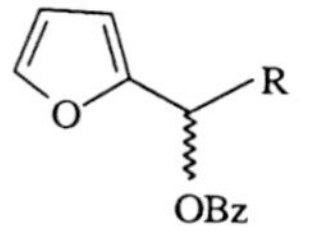

(9)

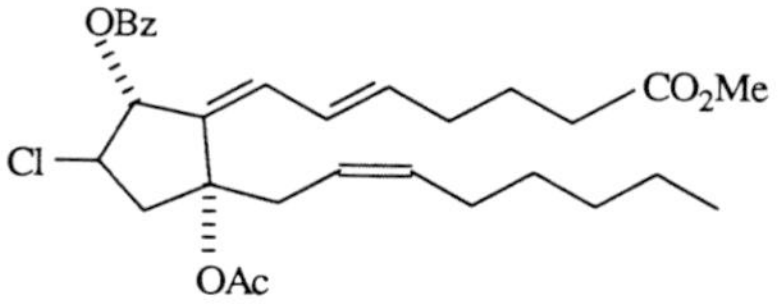

(10)

present in polyhydroxylated steroidal enones, i.e., ecdysteroids and other terpenoids. The additivity relation also holds for congested cage-structure molecules exemplified by the tricothecenes (**7**). Recent applications of the exciton chirality method to diverse coupled chromophores include (+)-kjellmanianone[12] (**8**) (enone, benzoate), 2-furylcarbinol benzoates[13] (**9**) (furan, benzoate), reduced chlorovulones[14] (**10**) (diene, benzoate), and α,β-unsaturated esters of reduced O-acetylisophotosantonin[15] (**11**) (α,β-unsaturated ester, alkene).

C. CONFIGURATION OF NATURAL PRODUCTS

Most applications of the exciton chirality method have involved conversion of the relevant hydroxy or amino group into *p*-bromobenzoates (λ_{max} 244 nm), owing to their ease of preparation, or into *p*-dimethylaminobenzoates (λ_{max} 311 nm), because of their strong absorption at longer wavelength. Although the dimethylaminobenzoate chromophore suffices for many natural products, it cannot be used when the substrate possesses UV absorption near 310 nm with a transition moment of unknown direction, because of the complicating substrate-dimethylaminobenzoyl interaction.[11]

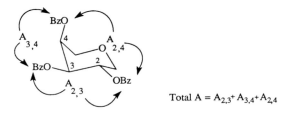

Total A = $A_{2,3} + A_{3,4} + A_{2,4}$

FIGURE 4. Additivity in amplitudes (A) of pyranose tribenzoates.

1. Mitomycin C

The antitumor agent mitomycin C (**12**) represents a biologically interesting molecule possessing strong UV absorption above 300 nm (strong bands at 245 and 309 nm and weak bands at 350 and 530 nm). To determine the absolute stereochemistry by the exciton chirality method, a new chromophore was developed, i.e., *p*-dimethylamino-cinnamate (dma-Ocin), with an intense absorption band at 361 nm that is in a transparent region of the UV spectrum of the parent compound (**12**). Either reductive or acidic activation of mitomycin C (**12**) facilitates opening of the aziridine ring with configurational retention at C-2, but with the possibility of attachment of a nucleophile at C-1 with either retention or inversion. Conversion of the authentic *trans* 1α-OH,2β-NH$_2$ mitosene into its bis *p*-dimethylaminocinnamate produced a derivative (**13**) whose positively split CD at about 360 nm (A = +46) reflected the *S*-chirality[16] of both C-1 and C-2 of mitomycin C, in accord with its absolute configuration that had been revised by X-ray analysis.[17]

2. Diterpenoid Dictyotriols

Despite numerous successful applications of the exciton chirality method as a powerful tool for the determination of the absolute configuration of both synthetic and natural compounds, extreme caution must be exercised in evaluating nonrigid structures (e.g., hydroazulenoid diterpenes) that are susceptible to either conformational distortion or unsuspected rearrangements. For example, revision of the absolute configurations of several diterpenoid dictyotriols has been shown to have resulted from the presence of an unexpected dihedral angle in dictyol B and the occurrence of an allylic rearrangement in dictyol during its conversion to a benzoate derivative.

The monobenzoates (**14**) of both 9-*epi*-dictyol B and dictyol B exhibit negative exciton chirality (allylic, benzoate coupling). Due to steric interaction between the exocyclic double bond and the allylic substituent at C-9, the dihedral angle O$_9$-C$_9$-C$_{10}$-C$_{18}$ adopts a different spatial disposition having a negative sign, as shown by molecular mechanics calculations. Thus, the most stable conformation found by molecular mechanics for the 9β-oriented dictyol B concurs with a left-handed screwness of the electric transition dipole moments of the two chromophores, in contrast to expectations derived from inspection of Dreiding models. Furthermore, mono-*p*-bromobenzoylation of dictyotadiol (**15**) appears to have resulted in an allylic rearrangement of the benzoate group from C-4 to C-2, resulting in a compound (**16**) different from that anticipated. However, application of the allylic benzoate method to the rearranged mono-*p*-bromobenzoate (**16**) leads to the correct absolute configuration.[18]

Additional caveats are provided by unexpected sign reversals of exciton couplets in both biscyanine dyes and bis-dipyrrinone systems. Sign reversal of the dyes is due not to cationic charge repulsion, but results from the adoption of a unique conformation due to steric reasons.[19] For the dipyrrinone chromophores, exciton chirality is inverted by a change from nonpolar (CHCl$_3$) to polar (DMSO) solvent.[20]

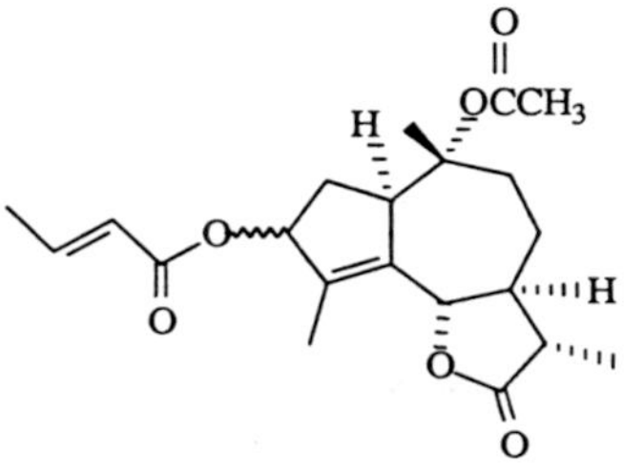

(11)

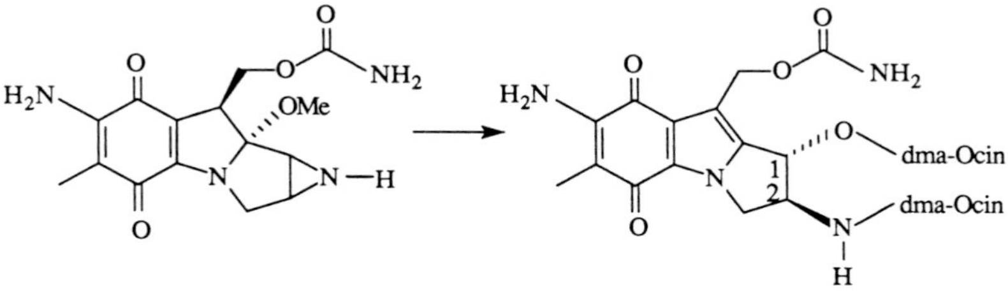

(12)   (13)

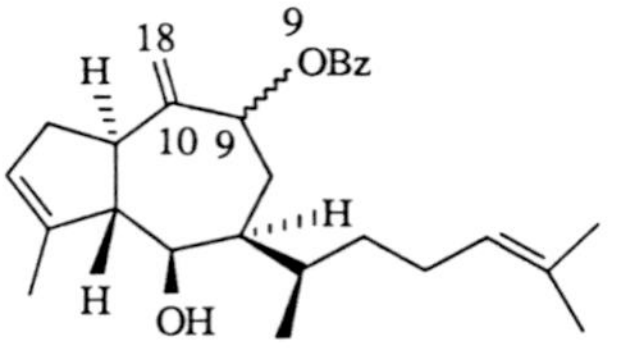

(14)

3. Acyclic Polyols and Aminopolyols

Although the exciton chirality method for CD spectroscopic determination of stereochemistry has been primarily applied to molecules with rigid ring systems, applications to simple acyclic systems have also proved successful. Polyols occur extensively in nature and are also degradatively obtained from a wide variety of natural products by treatment with either periodate or ozone [e.g., prostanoids such as punaglandin (**17**)]. Thus, the relative and absolute configurations of acyclic 1,2-, 1,2,3-, 1,2,3,4-, and 1,2,3,4,5-polyols can be established by the bichromophoric exciton chirality method after selective derivatization of primary hydroxy groups with 9-anthroyl chloride, and secondary hydroxy groups with *p*-methoxycinnamoyl chloride.[21] The use of both the degenerate and nondegenerate exciton coupling interactions of two different types of chromophores has provided a complete set of reference curves of polyols where the additive effects of all pairwise degenerate and nondegenerate interchromophoric interactions produce "fingerprint" CD curves. The dependence of exciton coupling upon interchromophoric distance has also demonstrated the presence of previously unexpected conformations that are probably stabilized by π-π aromatic ring stacking interactions (e.g., a 1,3-parallel interaction between oxygen atoms).[22] The bichromophoric CD method has been extended to configurational determination of amino polyols such as bacteriohopanoid amino tetrol (**18**).[23] As with polyols, the terminal amino group is anthroylated, and the remaining

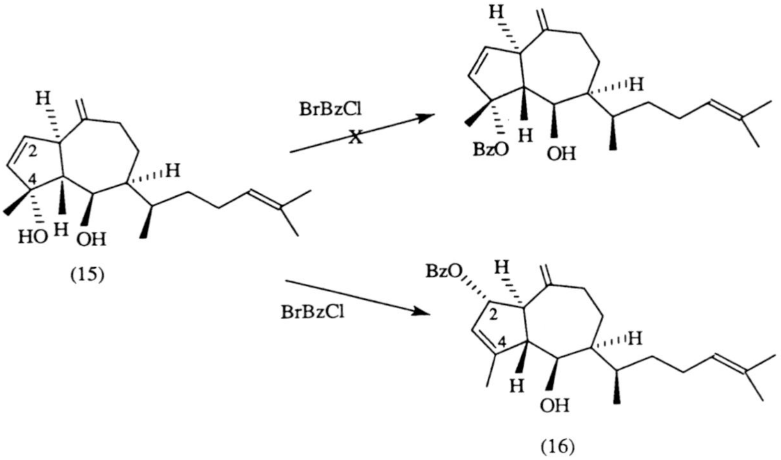

(15)

(16)

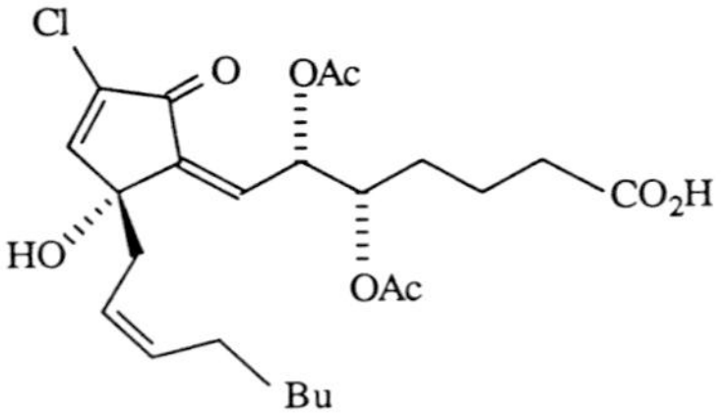

(17)

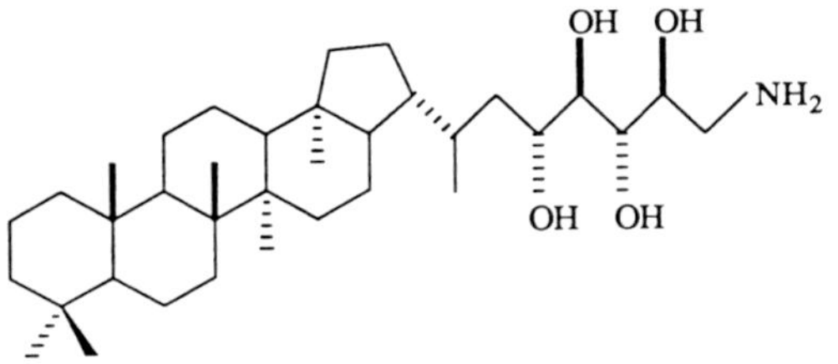

(18)

hydroxyls are *p*-methoxycinnamoylated followed by comparison of the 220- to 340-nm CD with characteristic reference spectra.[24]

D. DETERMINATION OF OLIGOSACCHARIDE LINKAGES

Not only is the CD exciton chirality method a powerful chiroptical technique for the determination of absolute configuration at stereocenters in organic molecules, but it also offers a novel strategy to elucidate oligosaccharide structure.[25] Sugars are first derivatized with exciton chromophores that differentiate free hydroxyl groups (e.g., *p*-bromobenzoate groups) from those that are involved in glycosidic linkages. The latter hydroxyl groups, liberated upon cleavage of the *p*-bromobenzoylated glycoside, are then derivatized by another chromophoric group (e.g., *p*-methoxycinnamate). The resulting sugar derivatives are separated by high-pressure liquid chromatography (HPLC) and analyzed by CD, and their spectra compared to published reference spectra in order to determine unambiguously the identity of the sugars, their linkage patterns, and absolute configurations. A new exciton chromophore (*p*-phenylbenzyl) has been developed that withstands glycoside hydrolytic conditions yet is amenable to oxidation to the more chiroptically sensitive *p*-phenylbenzoate. As with the

(19)	(20)
Benzoates	p-Phenylbenzyl ethers

substituted benzoates (**19**), the *p*-phenylbenzyl ethers (**20**) favor an *s-trans* conformation parallel to the C–O bond, and their CD spectra can be similarly interpreted in exciton chirality studies.[25]

III. EPILOG

Because CD is a special type of absorption spectroscopy, sample and solubility requirements are the same for the two techniques. Thus, similar amounts of material (several milligrams or less) are required for CD measurements as for a UV absorption spectrum. Due to the interrelationship of CD and UV, both measurements should be performed on each sample for which chiroptical data is desired. In fact, when UV extinction coefficients are known, the UV spectrum may provide an experimental determination of sample concentration. CD is a nondestructive technique, and thus, samples may be recovered after measurement. Discussions of CD measurement, instrumentation, and other aspects of chiroptical techniques may be obtained from reviews.[26,27] An experimental description of materials and methods useful in obtaining exciton-coupled CD curves is offered in Reference 28. Interested readers can consult a review on CD applications in natural product chemistry.[29]

Circular dichrographs are commercially available from the Japan Spectroscopic Company (JASCO), Tokyo, Japan; Jobin-Yvon Instruments, Longjumeau, France; and Aviv Associates, Lakewood, NJ. In addition to producing their own instruments, Aviv rebuilds and modifies Cary spectropolarimeter/dichrometers, which were the forerunner of the modern generation of spectropolarimeters.

REFERENCES

1. **Piutti, A.**, Sur une nouvelle espece d'asparagine, *Compt. Rend. Acad. Sci.,* 103, 134, 1886.
2. **Cushny, A. R.**, Atropine and the hyoscyamines; a study of the action of optical isomers, *J. Physiol.,* 30, 176, 1904.
3. **Ariens, E. J., Soudijn, W., and Timmermans, P. B. M. W. M., Eds.,** *Stereochemistry and Biological Activity of Drugs,* Blackwell Scientific, Oxford, 1983.
4. **Witiak, D. T. and Inbasekaran, M. N.,** Pharmaceuticals, optically active, in *Kirk-Othmer Encyclopedia of Chemical Technology,* Vol. 17, 3rd ed., Wiley-Interscience, New York, 1982, 311.
5. **Gaffield, W.,** Chirality as manifested in the biological activity of natural products, in *Studies in Natural Products Chemistry,* Vol. 7, Rahman, A., Ed., Elsevier, Amsterdam, 1990, 3.
6. **Wilson, K. and Walker, J.,** Chirality and its importance in drug development, *Biochem. Soc. Trans.,* 19, 443, 1991.
7. **Kuhn, W.,** The physical significance of optical rotatory power, *Trans. Faraday Soc.,* 26, 293, 1930.
8. **Kirkwood, J. G.,** On the theory of optical rotatory power, *J. Chem. Phys.,* 5, 479, 1937.
9. **Harada, N. and Nakanishi, K.,** *Circular Dichroic Spectroscopy, Exciton Coupling in Organic Chemistry,* University Science Books, Mill Valley, CA, 1983.

10. **Harada, N. and Nakanishi, K.,** The exciton chirality method and its application to configurational and conformational studies of natural products, *Acc. Chem. Res.,* 5, 257, 1972.

11. **Nakanishi, K., Kuroyanagi, M., Nambu, H., Oltz, E. M., Takeda, R., Verdine, G. L., and Zask, A.,** Recent applications of circular dichroism to structural problems, especially oligosaccharide structures, *Pure Appl. Chem.,* 56, 1031, 1984.

12. **Chen, B.-C., Weismiller, M. C., Davis, F. A., Boschelli, D., Empfield, J. R., and Smith, A. B., III,** Enantioselective synthesis of (+)-kjellmanianone, *Tetrahedron,* 47, 173, 1991.

13. **Gawronski, J., Radocki, D., Jurczak, J., Pakulski, Z., Raczko, J., Ramza, J., and Zamojski, A.,** Exciton chirality method for establishing absolute configuration of 2-furylcarbinols, *J. Org. Chem.,* 55, 1118, 1990.

14. **Nagaoka, H., Iguchi, K., Miyakoshi, T., Yamada, N., and Yamada, Y.,** Determination of absolute configuration of chlorovulones by CD measurement and by enantioselective synthesis of (–)-chlorovulone II, *Tetrahedron Lett.,* 27, 223, 1986.

15. **Lauridsen, A., Cornett, C., and Christensen, S. B.,** Exciton coupling in circular dichroic spectroscopy as a tool for establishing the absolute configuration of α,β-unsaturated esters of allylic alcohols, *Acta Chem. Scand.,* 45, 56, 1991.

16. **Verdine, G. L. and Nakanishi, K.,** *p*-Dimethylaminocinnamate, a new red-shifted chromophore, for use in the exciton chirality method. Its application to mitomycin C, *J. Chem. Soc. Chem. Commun.,* 1093, 1985.

17. **Shirahata, K. and Hirayama, N.,** Revised absolute configuration of mitomycin C. X-ray analysis of 1-N-(*p*-bromobenzoyl) mitomycin C, *J. Am. Chem. Soc.,* 105, 7199, 1983.

18. **Arroyo, P., Norte, M., Vasquez, J. T., and Nakanishi, K.,** Absolute configuration of hydroazulenoid diterpenes based on circular dichroism, *J. Org. Chem.,* 56, 2671, 1991.

19. **Gargiulo, D., Derguini, F., Berova, N., Nakanishi, K., and Harada, N.,** Unique ultraviolet-visible and circular dichroism behavior due to exciton coupling in a biscyanine dye, *J. Am. Chem. Soc.,* 113, 7046, 1991.

20. **Byun, Y.-S. and Lightner, D. A.,** Exciton coupling from dipyrrinone chromophores, *J. Org. Chem.,* 56, 6027, 1991.

21. **Wiesler, W. T. and Nakanishi, K.,** Relative and absolute configurational assignments of acyclic polyols by circular dichroism. I. Rationale for a simple procedure based on the exciton chirality method, *J. Am. Chem. Soc.,* 111, 9205, 1989.

22. **Wiesler, W. T. and Nakanishi, K.,** Relative and absolute configurational assignments of acyclic polyols by circular dichroism. II. Determination of nondegenerate exciton coupling interactions by assignment of prochiral aryloxymethylene protons for ^{1}H NMR conformational analysis, *J. Am. Chem. Soc.,* 112, 5574, 1990.

23. **Zhou, P., Berova, N., Nakanishi, K., Knani, M., and Rohmer, M.,** Microscale CD method for determining absolute configurations of acyclic amino tetrols and amino pentols. Structures of aminobacteriohopanepolyols from the methylotrophic bacterium *Methylococcus luteus, J. Am. Chem. Soc.,* 113, 4040, 1991.

24. **Zhou, P., Berova, N., Nakanishi, K., and Rohmer, M.,** Assignment of absolute stereochemistry of aminopolyols by the bichromophoric exciton chirality method, *J. Chem. Soc. Chem. Commun.,* 256, 1991.

25. **Chang, M., Meyers, H. V., Nakanishi, K., Ojika, M., Park, J. H., Park, M. H., Takeda, R., Vasquez, J. T., and Wiesler, W. T.,** Microscale structure determination of oligosaccharides by the exciton chirality method, *Pure Appl. Chem.,* 61, 1193, 1989.

26. **Woody, R. W.,** Circular dichroism of peptides, in *The Peptides,* Vol. 7, Udenfriend, S., Meienhofer, J., and Hruby, V. J., Eds., Academic Press, New York, 1985, 15.

27. **Johnson, W. C., Jr.,** Optical activity and structure of biological molecules, in *Food Analysis: Principles and Techniques,* Vol. 2, Gruenwedel, D. W. and Whitaker, J. R., Eds., Marcel Dekker, New York, 1984, 245.

28. **Wiesler, W. T., Vasquez, J. T., and Nakanishi, K.,** Pairwise additivity in exciton-coupled CD curves of multichromophoric systems, *J. Am. Chem. Soc.,* 109, 5586, 1987.

29. **Cambie, R. C., Ho, P. C., Netzke, K., Schoenfelder, W., Snatzke, F., Snatzke, G., and Schulte, J.,** Newer applications of circular dichroism in natural products chemistry, in *Natural Products Chemistry III,* Rahman, A. and Le Quesne, P. W., Eds., Springer-Verlag, Berlin, 1988, 67.

Chapter 8

ACQUISITION AND SCREENING OF NATURAL PRODUCTS AS POTENTIAL ANTICANCER AND AIDS ANTIVIRAL AGENTS

Daniel Lednicer and Ven L. Narayanan

TABLE OF CONTENTS

I. INTRODUCTION

The fruitfulness of the various approaches for identifying new chemical structures that will serve as the basis for therapeutic agents has formed the topic for a lively debate almost since the inception of medicinal chemistry. Proponents of random screening occupy one end of the spectrum, while those who would design potential new therapeutic agents on the basis of a combination of molecular biology and computer modeling occupy the other extreme. Proponents of random screening adduce considerable support from the past successes of this approach to drug discovery. It is of particular relevance to the present discussion that natural products, perhaps because of their unparalleled structural diversity, have provided a disproportionate share of new drug leads for antitumor agents. The natural products vinblastine, doxorubicin, and daunomycin, for example, are among the more important drugs in the field of cancer chemotherapy, and taxol, a compound isolated from the bark of the Western yew, has shown particular early promise in treatment of ovarian and breast cancer.

The relative paucity of basic knowledge then available led the National Cancer Institute (NCI) to place heavy reliance on random screening as the primary tool for drug discovery, starting in the early 1950s. The assays used in the screen underwent considerable evolution over the years, culminating in the high-throughput mouse leukemia P388 screen.[1] Test candidates for the screen were acquired by active solicitation of compounds from both industrial and academic laboratories. Natural products provided an important source for structurally diverse test candidates. They were, consequently, the focus of a special acquisition and screening program.[2]

A retrospective analysis of those compounds discovered by the screen that were taken into clinical trials revealed an apparent bias for agents active in leukemias, coupled with a dearth of compounds that would show activity in human solid tumors. Solid tumors, it should be noted, represent the majority of human neoplasias. After considerable effort, a high-capacity *in vitro* screen, based on tissue culture methodology utilizing cell lines derived from human solid tumors, was developed and brought on-line in late 1989. Natural products form an important sector of test candidates, for the same reason they were considered important for the earlier P388 screen, i.e., structural diversity.

In 1987 the NCI was given the mission to undertake the preclinical drug discovery phase of the development of AIDS antiviral agents, largely on the basis of its extensive experience in conducting large throughput screens and in subsequently developing active compounds. The initial effort culminated in the development of an *in vitro* screen directed at finding potential anti-HIV compounds. Natural products again form an important category of test candidates, because of the same structural diversity that makes them attractive for the antitumor screen.

II. HUMAN SOLID-TUMOR-LINE-BASED ANTICANCER SCREEN

The overall goals of the NCI drug discovery and development efforts consist of finding new and more effective drugs for the treatment of human cancers. Human neoplasms are thus considered the proper entity on which to test the effect of potential chemotherapeutic agents. Existing *in vivo* test systems, for example, xenografts on immune deficient mice, are far too slow, complex, and expensive to be used as a mass screen. Using cells derived from human cancers in an *in vitro* setting, on the other hand, is quite compatible with the desired goal. This approach has the added virtue of offering *intrinsic* activity as the end point. The effect on activity of factors such as metabolism, distribution, and elimination can be minimized. It may be noted parenthetically that the screen will of necessity turn up agents whose activity might not translate to the *in vivo* setting. In any event, the assay as it now operates involves the assay of each compound against a battery of close to 60 cell lines obtained from eight organ systems.[3,4]

The current version of the assay starts by seeding cell-line aliquots from stock solutions into individual wells of microtiter plates. The plates are then incubated for 24 h to allow the cells to stabilize. Test agents are then added to the cells at concentrations that represent 5-log dilutions, starting with a maximum concentration of one tenth millimolar. The cells are then incubated in the presence of the drug for a further 48 h. At the end of this time, the adherent cells are fixed to the plate by means of trichloroacetic acid, and after a number of washes, the cell layer is treated with the protein stain Sulforhodamine B. The optical density, which is proportional to the protein mass, is then read by automated spectrophotometric plate readers at a wavelength of 515 nm. Readings are transferred to a microcomputer, and final reports are generated using especially developed software.

Thus, for a typical compound there are no fewer than approximately 300 data points (60 cell lines × 5 concentrations) available. The data package sent to each supplier is intended to display that data in several formats that give a profile for each test agent. Figure 1 displays an excerpt from the typical raw data page sent to the suppliers of the compounds. The column headed "Panel/cell line" lists the individual human cancer cell lines used in the particular experiment, grouped by disease category (e.g., leukemia). The column headed "Time zero" presents optical densities of wells of the cell line in question, which have been put through the fixation/staining process immediately prior to treatment with drug. The "Control" column shows the same parameter for *untreated control* wells after the standard 48-h incubation period. The next five columns are the raw data, showing actual observed optical densities for each well, recorded by cell line, for each individual concentration. The increase in optical density from 0.412 to 1.047 for the cell line on the first line of the printout (CCRF-CEM) shows roughly a doubling over the period of the experiment; there is little drug effect up to a concentration of 10^{-6} molar (O.D. 0.934), after which the cell number declines precipitously. The same data, as calculated by the computer program, is shown in the next five columns under "Percent growth." Note that 100% growth indicates growth equal to untreated control. A value of 0% means that the cell mass at 48 h is the same as it was at the start of the experiment, while negative values indicate a decrease in cell mass and thus, presumably, cell kill. The last three columns are discussed below.

Figure 2 displays the same data plotted in the form of familiar dose response curves, with the plots for common disease types gathered on a single plot. It should be noted again that 100% growth indicates cell mass equal to untreated controls, and 0% indicates no increase in cell mass over time zero. Cell growth of −50 and −100% indicates the presence of only 50% of the stainable mass of that at time zero and complete cell kill, respectively. Each individual dose response curve is derived from the comparison of treated and untreated responses for those specific cell lines. The experiments are thus internally controlled so as to normalize for differences in growth rates of individual cell lines. The final panel combines all the dose responses on a single graph that permits ready visualization of sensitive cell lines. Additional points of interest in Figure 2 are three specific intercept lines: the point where a dose–response curve crosses the (1) 50% growth-inhibition line, denoted as GI50 concentration; (2) 100% inhibition, or 0% growth defined as TGI concentration; and (3) −50% growth, or 50% cell kill, defined as the LC50 concentration. The same values calculated more accurately by the computer program are shown in the last three columns in Figure 1.

The reduction of dose response data to the corresponding mean graph format, as shown in Figure 3, allows application of the COMPARE pattern recognition program. This program, in essence, compares the set of cell sensitivities to an unknown compound at all three levels (GI50, TGI, LC50) to those of standard agents or, alternatively, to those of all compounds tested to date (approximately 20,000).[5] As an example, to draw the mean graph at the GI50 level (Figure 3), the ordinary arithmetic average of the logs of the GI50s for individual cell lines are first calculated. That value (MG-MID) is then used as the vertical zero datum line. The individual log GI50 values are then plotted as proportional bars to the left (greater than) and to the right (lower than) of the datum. Groups of sensitive cell lines will be shown as strong bars to the

Panel/Cell Line	Time Zero	Ctrl	Mean Optical Densities Log10 Concentration					Percent Growth					GI50	TGI	LC50
			-8.0	-7.0	-6.0	-5.0	-4.0	-8.0	-7.0	-6.0	-5.0	-4.0			
Leukemia															
CCRF-CEM	0.412	1.047	1.056	1.011	0.934	0.370	0.306	101	94	82	-10	-26	2.23E-06	7.74E-06	>1.00E-04
HL-60(TB)	0.321	0.893	0.851	0.761	0.772	0.333	0.338	93	77	79	2	3	2.38E-06	>1.00E-04	>1.00E-04
K-562	0.335	1.403	1.366	1.331	1.238	0.576	0.326	97	93	85	23	-3	3.61E-06	7.82E-05	>1.00E-04
MOLT-4	0.507	1.316	1.318	1.229	1.151	0.400	0.338	100	89	80	-21	-33	1.96E-06	6.16E-06	>1.00E-04
RPMI-8226	0.693	1.413	1.396	1.433	1.311	0.978	0.463	98	103	86	40	-33	5.98E-06	3.50E-05	>1.00E-04
SR	0.503	1.367	1.288	1.293	1.271	0.938	0.319	91	91	89	50	-36	1.01E-05	3.80E-05	>1.00E-04

FIGURE 1. Example of *in vitro* testing results report.

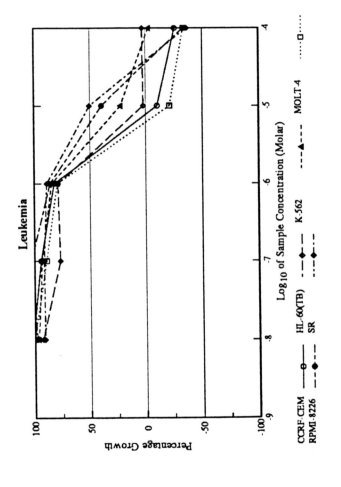

FIGURE 2. Example of dose–response curves report derived from data shown in Figure 1.

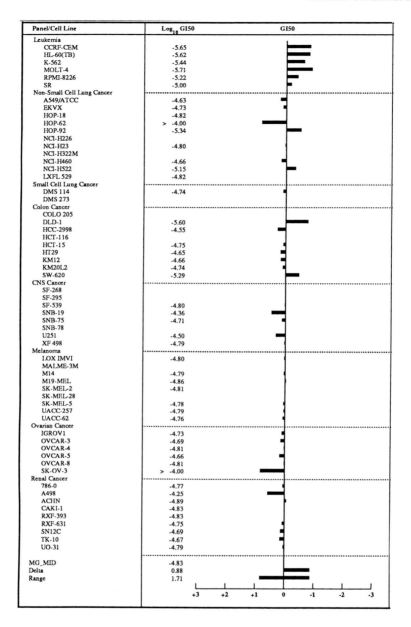

FIGURE 3. Example of mean graph format for data shown in Figure 1.

right of the datum. A slight trend in that direction is seen in leukemia lines in the case example shown in Figure 3. The same procedure is then repeated for TGI and for LC50 (not shown).

The great majority of biological assays, whether these are conducted *in vitro* or *in vivo*, allow the judgment to be made as to whether a compound is active or not. The nature of the data produced by the present assay, as well as the complexity of the goal of the program, makes that assessment difficult. A compound can be judged to be inactive if it has no effect whatsoever on cell growth, as demonstrated by a set of level-dose responses at the 100% growth level. All others are active within the narrowest meaning; whether they have promise to meet our goal constitutes a more complex judgment.

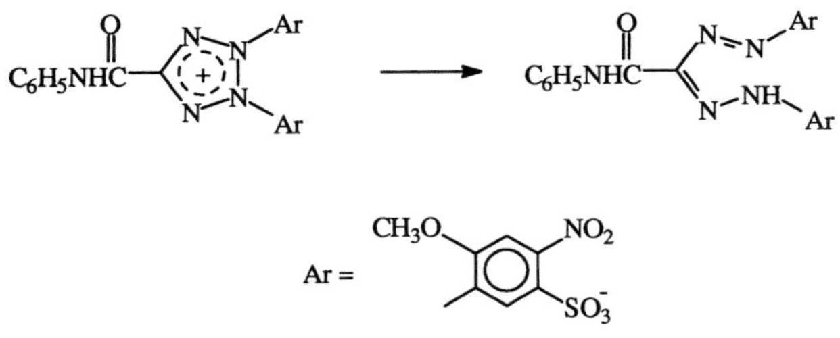

SCHEME 1. Reduction of tetrazolium group to the orange formazan entity.

III. NCI AIDS ANTIVIRAL SCREEN

That an AIDS antiviral *in vitro* screen would be required was clearly indicated for a number of reasons. First was the fact that no animal model existed for HIV infection when the program was initiated, a situation that still exists with the possible exception of the huSCID mouse (a severely immunocompromised mouse implanted with parts of the human immune system). An *in vitro* screen would have the additional advantages of high throughput capacity and modest sample size requirements. An assay was selected that would involve infecting intact cells, since this would offer less mechanistic bias than biochemical tests based on inhibition of viral enzymes such as reverse transcriptase or HIV protease. An added advantage of the whole-cell assay is the fact that such a test depends also on the cell wall penetration of potential active compounds (cell wall viral receptor blockers, of course, excepted).

The first published protocol for the assay used for the screen[6] starts with the seeding of human CEM T4-lymphocytes from stock into the wells of 96-well microtiter plates. Solutions of test compounds are then added in duplicate to two sets of wells at half-log dilutions over a 4-decade order of magnitude concentration range. HIV is added to the cells, and the plates are then incubated at 37 °C for 6 days.[7]

At the end of the incubation period, cell survival is determined by means of a vital stain. The indicator used is a soluble version (XTT; 2,3-bis[1-methoxy-4-nitro-5-sulfonylphenyl]-5-[(phenylamino)carbonyl]-2H-tetrazolium hydroxide) of the stain MTT.[8] The reduction of the tetrazolium rings, in these colorless reagents, to orange formazans (Scheme 1) is proportional to the number of viable cells, thereby permitting quantitative determination of the number of surviving cells. Thus, the assay is completed by adding a solution of XTT, containing methylphenazonium methylsulfate as the coupler, at the end of the incubation period. Dose response data are then produced automatically from the optical densities obtained on the plates in an automatic plate reader using especially developed software. The drug zidovudine (AZT) is used as a positive control for quality control purposes and as the standard for comparisons.

Typical data from the screen are displayed in Figures 4 and 5. Examining first an inactive compound (Figure 4), the dotted upper line represents the viability of uninfected CEM T-lymphocytes to increasing concentrations of the test substance: cell survival declines in a dose-related fashion as the concentration of compound increases beyond 10^{-5} molar. Cells infected with HIV show only very low survival at any level of drug, as shown by the solid lower line. Figure 5 shows typical results for an active compound. The response for uninfected cells is similar to that above, with cytotoxicity manifested quite abruptly at 100 μg/ml. The infected cells, on the other hand, show increasing levels of survival in the presence of test agent. The interval between 10 and 100 μg/ml follows the form of the well known sigmoid dose–response curve. At the higher concentration the cytotoxicity of the test agent is manifested and limits the therapeutic interval. The therapeutic index of 2-decade orders of magnitude (ratio of the

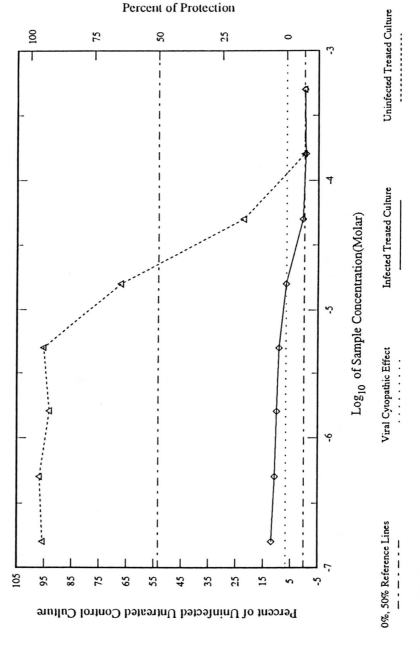

FIGURE 4. *In vitro* anti-HIV drug screening result for an inactive compound.

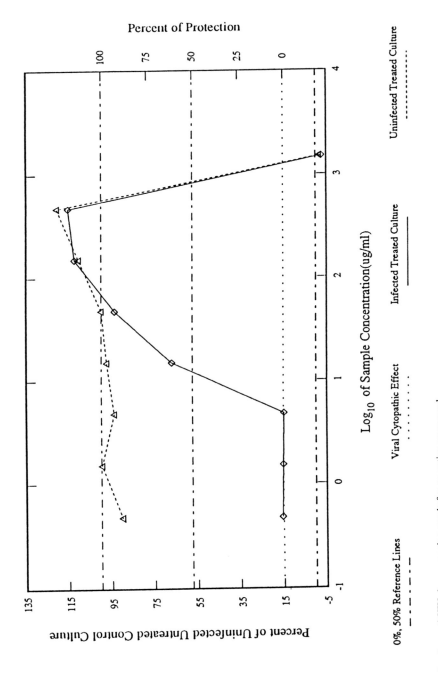

FIGURE 5. *In vitro* anti-HIV drug screening result for an active compound.

ED_{50} to the IC_{50}) would, in the usual course of events, qualify such an agent for further investigation. This particular compound was not pursued further, since it consists of a sulfated polysaccharide of indeterminate molecular weight.

IV. ACQUISITION OF COMPOUNDS FOR SCREENING

Since its inception, the NCI screening program has been viewed as a National resource that would provide important data to the chemical community. The original name for the program, Cancer Chemotherapy National Service Center (CCNSC), lives on in the prefix NSC, which is applied to all compounds that are submitted to NCI for testing in the screens. The program relies heavily on voluntary donations of compounds from academia, industry, and research institutes as a source of compounds for the screens. Donations may be made on an open or a confidential basis. In the case of the latter, not only the structure of the compounds, but also the identity of the compounds, are kept confidential within the program. NCI does not, as a consequence, reveal the names of the numerous industrial laboratories that contribute compounds to the program. It should be added that donors retain all rights to patents that may come from the screening results on their compounds, and further, the package of screening data sent out on each compound can be used to support potential patent applications. The institute has a standard "screening agreement" available that spells out the rights in more detail, and routinely provides executed copies to donors who desire a more formal agreement. Suppliers can specify whether they wish to have their compounds tested against cancer, AIDS, or both.

Recasting the screens in the *in vitro* format has resulted in a significant decrease in the minimum sample size required for initial screening. We currently ask for 25- to 50-mg samples of compound for each of the antitumor and anti-HIV screens. Larger-than-minimum amounts are appreciated if conveniently available, since this will avoid delay in follow-up tests in the event a compound is active. Compounds supplied in larger-than-minimum amounts will also have a good likelihood of being selected for testing in other mechanism-based *in vitro* biochemical screens that the NCI may develop (e.g., topoisomerase inhibition).

Natural products are a class of compounds that are of particular interest to the program because they have proven to be a rich source of biologically active compounds. They include, for example, important chemotherapeutic agents in their own right, such as doxorubicin, daunorubicin, vincristine, and bleomycin. Others, such as etoposide and teniposide, are semisynthetic analogs based on natural products. The targets for the two screens now being operated by the program represent new departures, and therefore little information about structure–activity relationships is currently available. As a result it is considered important to screen compounds that represent a wide diversity of chemical structures. The rich diversity of structural types provided by natural products thus adds to their attractiveness as potential test candidates. Special emphasis is thus placed on acquiring natural products for both the antitumor and anti-HIV screens. The program at this time accepts only isolated, pure natural products. Extracts and fractions from the extramural community are ruled out as test candidates, since those could quickly exhaust the capacity of the screens.

Modern instrumental and analytical methods have led to a marked decrease in the sample size of natural products required for characterization and structural determination. This has also led to the isolation of many compounds that form very minor constituents. The program has adjusted the sample size required for the screens to reflect this fact. The 10 mg we request for each assay will permit initial testing at the level of about 100 µg/ml. If larger amounts are available, testing can begin at correspondingly higher levels, while lesser amounts will allow a only single test without the possibility of confirmatory assays. For the above reasons, as in the case of synthetics, larger samples are appreciated if the compound is readily available.

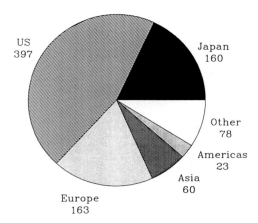

FIGURE 6. Geographic distribution of sample donors to the program.

TABLE 1
Representative Natural Products Tested in the NCI *in vitro* Screens

Structural class	Compound #
Isoquinolines	131
Steroids	63
Indolizidines	62
Pyrrolizidines	55
Benzopyranones	48
Anthracyclines	35
Indoles	32
Harmanes	17
Coumarins	16

The NCI database lists just over 10,000 compounds classed as natural products, 2500 of which have been acquired after March 1985 (NSC > 600,000). The origin of those compounds, as shown in Figure 6, reflects the international scope of the program. Just under half (397, 45%) of the 881 suppliers of compounds have addresses in the U.S. The next largest group of suppliers (163, 19%) are from Europe, while an almost identical number (160, 18%) have addresses in Japan. A surprisingly small number have addresses in countries richly endowed in the tropical rain forests often identified as sources of novel natural products. Only 60 donors have addresses in Asia, other than Japan, while only 23 have addresses in the Americas outside the U.S.

V. EVALUATION OF COMPOUNDS

As of this writing, 2536 compounds classified as natural products have been tested in the new screens, 1800 have been tested in the new antitumor screen, while 1702 have been screened against HIV. The file of compounds, stored in substructure searchable form on the NCI Drug Information System (DIS) database, is checked on a periodic basis to insure that no single group of compounds is overrepresented. Table 1 presents the results from a typical search involving some of the more common natural products classes.

This list was generated by searching for specific structural types. The presence of some large class that was not specifically sought cannot be ruled out.

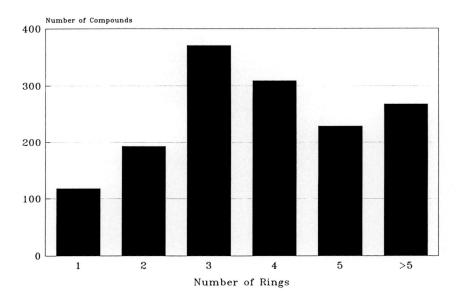

FIGURE 7. Evaluation of natural products tested in the anti-HIV program, based on number of rings.

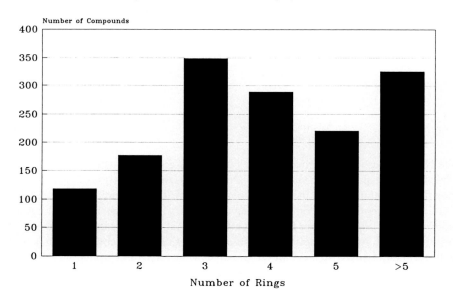

FIGURE 8. Evaluation of natural products tested in the antitumor program, based on number of rings.

A much more general search involves classifying the compounds tested for each activity by the number of rings present. Results from those searches are shown in Figures 7 and 8. It is of interest that the number of compounds with specified numbers of rings show very similar patterns in both the antitumor and anti-HIV sets; however, the number of compounds with more than five rings is significantly higher in the antitumor set.

The anti-HIV activity of a test agent is a measure of the degree of protective effect exerted by a test agent against the AIDS virus cytopathic effect on CEM T-lymphocytes. A compound is deemed *active* if it affords 100% protection at two or more concentrations ($^1/_2$ $\log_{10}$). A compound is deemed to be *moderately active* if the protective effect exceeds 50%, but never quite reaches 100%, or if it reaches 100% at only a single concentration. Table 2 summarizes

TABLE 2
Natural Products with Confirmed Anti-HIV Activity

Polyethers and derivatives	9
Sulfated polysaccharides	4
Tannins	4
Protamine derivatives	2
Lignans	2
Naphthoanthraquinone	1
Antibiotic O-sulfate	1
Phorbol ester	1
Plant protein derivative	1
Castanospermine	1
Cyclic peptide	1
Triterpene glycoside	1
Unknown structures	2

TABLE 3
Proportion of Natural Products Active in the NCI Screens

Assay	Number tested	Active[a]	Moderate[b]	Referred to BEC
Anti-HIV	2439	30	57	
%		1.2%	2.3%	
Antitumor	1867			169
%				9.1%

[a] 100% protection at 2 or more concentrations.
[b] >50% at 2 or more concentrations or <100% protection at 1 concentration.

the compounds that have been declared confirmed actives in the anti-HIV screen to date. The first entry should, however, be treated with some caution, as it includes the active natural products and four closely related derivatives prepared in an attempt to increase the therapeutic index.

Evaluation of data from the antitumor screen is somewhat more complex due to the multidimensionality of the data. Compounds that have no effect on the growth of cells, in any panel at any concentration, are easily classified as having no further interest. All others are judged by how well they match the program goals. Some of these criteria, for interest, (not necessarily in order of priority) are

- Disease-specific agents: compounds that show selectivity for cell lines from a single organ system at one of the three response levels
- Very potent, nonspecific compounds: generally cytotoxic compounds with mean LC_{50} values at or below 10^{-6} molar
- Compounds that show a pattern of responses comparable to a desired class (i.e., tubulin inhibitors) using the COMPARE program
- Compounds that show strong selectivity for cell lines from different organ systems (this may be a reflection of some new mechanism)

Compounds whose effect on cell growth in the screen meets one of those criteria are referred to the Biological Evaluation Committee for their consideration for further development. This very roughly corresponds to declaring a compound "active". The very large number of natural products found to be of interest (i.e., 99) precludes the depiction of those data in a table equivalent to Table 2.

The proportion of pure natural products found active by each of the *in vitro* screens is summarized in Table 3. The rate of compounds of interest from the antitumor screen would seem to be much higher (9.1%), at first glance, than that from the anti-HIV screen (3.5%). Those numbers should, however, be treated with caution, since test candidates for the cancer screen included a large number of compounds selected from the repository, on the basis of activity in the earlier *in vivo* screens, whereas the input for the anti-HIV screen was closer to random selection.

VI. CONCLUDING REMARKS

In summary, the drug discovery process works. The acquisition of pure natural products and the subsequent screening of these in the new NCI *in vitro* antitumor and AIDS antiviral screens is generating a variety of potential new agents worthy of further pursuit as potential therapeutic agents.

REFERENCES

1. **Driscoll, J.,** The preclinical new drug research program at the National Cancer Institute, *Cancer Treat. Rep.,* 68, 69, 1984.
2. **Suffness, M.,** Development of antitumor natural substances at the National Cancer Institute, in *Antitumor Natural Products,* Gann Monogr. Cancer Res. No. 36, Takeuchi, T., Nitta, K., and Tanaka, N., Eds., Japan Scientific Societies Press, Tokyo, 1989.
3. **Boyd, M. R.,** Status of the NCI preclinical antitumor drug discovery screen, *Prin. Pract. Onc. Updates,* 3, 1, 1989 and references therein.
4. **Monks, A., Scudiero, D., Skehan, P., Shoemaker, R., Paull, V., Vistica, D., Hose, C., and Langley, F.,** Feasibility of a high flux anticancer drug screen using a diverse panel of cultured human tumor cell lines, *J. Natl. Cancer Inst.,* 83, 757, 1991.
5. **Paull, K. D., Shoemaker, R. H., Hodes, L., Monks, A., Scudiero, D. A., Rubenstein, L., Alley, M. C., Plowman, J., and Boyd, M.R.,** Display and analysis of patterns of differential activity of drugs against human tumor cell lines: development of mean graph and COMPARE algorithm, *J. Natl. Cancer Inst.,* 81, 1088, 1989.
6. **Weislow, O. S., Kiser, R., Fine, D. L., Bader, J., Shoemaker, R. H., and Boyd, M. R.,** New soluble-formazan assay for HIV cytopathic effects: application to high-flux screening of synthetic and natural products for AIDS-antiviral activity, *J. Natl. Cancer Inst.,* 81, 577, 1989.
7. **Gulakowski, R. J., McMahon, J. B., Staley, P. G., Moran R. A., and Boyd, M. R.,** A semiautomated multiparameter approach for anti-HIV drug screening, *J. Virolog. Meth.,* 33, 87, 1991.
8. **Paull, K. D., Shoemaker, R. H., Boyd, M. R., Parsons, J. L., Risbood, P. A., Barbera, W. A., Sharma, M. N., Baker, D. C., Hand, E., Scudiero, D. A., Monks, A., Alley, M. C., and Grote, M.,** The synthesis of XTT — a new tetrazolium reagent that is bioreducible to a soluble formazan, *J. Heterocyclic Chem.,* 25, 911, 1988.

Chapter 9

POTENTLY SWEET COMPOUNDS FROM PLANTS: TECHNIQUES OF ISOLATION AND IDENTIFICATION

A. Douglas Kinghorn and Jinwoong Kim

TABLE OF CONTENTS

I. INTRODUCTION

About 50 highly sweet, naturally occurring compounds have now been discovered, and these are based on about 20 different carbon skeletons. Representatives of each of these classes of organic compounds are at least 50 to 100 times sweeter than sucrose and are often referred to as "intense" sweeteners. Not included in this category are sweet monosaccharides, disaccharides, and polyols, which are classified as "bulk" sweeteners. The potently sweet natural products that are presently known are all constituents of green plants. None have been reported so far from any species of microorganisms, insects, marine plants, or animals.[1]

This chapter has been organized so as to initially provide a brief overview of the structural types and importance of naturally occurring, highly sweet substances and then to indicate their taxonomic distribution within the plant kingdom. Next will follow a section on sweet-compound phytochemical purification procedures. Compared with investigations on natural products having other types of biological activity, there are several special considerations that must be taken into account in isolation studies directed toward the search for new natural sweetening agents. First, rather stringent efforts must be taken to identify candidate sweet plants for study, since the number of species found to biosynthesize potently sweet compounds is still extremely small. Second, it is necessary to rapidly detect high concentrations of monosaccharides, disaccharides, polyols, and phenylpropanoids that, in the vast majority of cases, tend to be responsible for the sweet taste of plants collected for laboratory examination. Third, because of the lack of suitable *in vitro* and *in vivo* experimental models capable of reliably detecting sweetness as perceived by humans, it has been necessary in the past to use human volunteer subjects to monitor chromatographic fractions for the presence or absence of sweetness. Accordingly, one should perform preliminary safety studies before isolation experiments can begin. Methodology employed in our laboratory to evaluate the sweetness intensity of pure natural products will be mentioned. However, recent progress in the use of gerbil electrophysiological and behavioral experiments to potentially substitute for the use of human subjects will be reviewed. In the final section of the chapter, the structure elucidation of novel sweeteners representing three groups of terpenoids and one type of steroidal saponin will be described.

Why search for novel, potently sweet substances from plants at all? Noncaloric and noncariogenic sucrose substitutes are in great demand in industrialized countries, for use in dietetic and diabetic foods and beverages. In most countries of North America and Western Europe, up to four principal synthetic intense sweeteners are available: namely, saccharin, aspartame, acesulfame K, and cyclamate.[2] However, the sucrose substitute profile in Japan is quite different, and natural sweeteners are very important. In this case extracts of *Stevia rebaudiana* (Bertoni) Bertoni, containing the diterpene glycosides stevioside and rebaudioside A, and extracts of *Glycyrrhiza glabra* L., the major constituent being the triterpene glycoside glycyrrhizin, are the two principal "high-intensity" sweeteners currently used in Japan.[3] In addition, three other plant constituents are commercially available as lesser-used sweeteners in Japan: the protein thaumatin, from *Thaumatococcus daniellii* (Bennett) Benth.; the dihydroisocoumarin phyllodulcin, from *Hydrangea macrophylla* Seringe var. *thunbergii* (Siebold) Makino; and a mixture of mogrosides, the sweet triterpene glycoside principles of *Siraitia grosvenorii* (Swingle) C. Jeffrey [formerly *Momordica grosvenorii* Swingle; *Thladiantha grosvenorii* (Swingle) C. Jeffrey].[3] Products made from *S. rebaudiana* have been approved recently in Brazil for the sweetening of medicines, dietetic foods and beverages, oral hygiene products, and soft drinks.[4,5] Thaumatin is an approved sweetener in Australia, Switzerland, and the U.K., in addition to Japan, and is currently being reviewed for this purpose in other countries.[6] However, it is now apparent that thaumatin will have a wider future application as a flavor enhancer and palatability improver than as a sweetener.[7]

Neohesperidin dihydrochalcone, a semisynthetic derivative of the *Citrus* flavanone, neohesperidin, is currently allowed for limited use as a sweetener in Belgium.[8]

Therefore, it seems likely that there will be a continued need for the discovery of novel high-potency sweeteners. The desired criteria for an ideal sucrose substitute are daunting and include the need for adequate sweetness intensity, solubility in water, chemical stability, and evidence of safety.[1] However, natural products, even with less than optimum taste properties, could find use in specialty markets, so that the consumer is not overexposed to any one sweetener.[9] In fact, a fairly high proportion of the known highly sweet compounds of plant origin already have commercial use, as indicated in the above paragraph. Furthermore, these compounds can also serve as lead compounds for the synthetic production and computer design of sweeteners with improved sensory or safety characteristics. Finally, potently sweet natural products may be utilized in physiologic experiments as standard sweeteners, in order to better understand the sweetness perception phenomenon.

II. STRUCTURAL TYPES OF POTENTLY SWEET COMPOUNDS AND THEIR DISTRIBUTION IN THE PLANT KINGDOM

A number of reviews describing highly sweet, naturally occurring compounds have appeared in the literature.[1,10,11] However, there have been several additional reports of potently sweet natural products published subsequent to these reviews, thereby rendering them somewhat outdated. In Table 1, examples of compounds in the major structural groups of plant-derived sweeteners are listed. The table includes several sweet compounds that are semisynthetic derivatives of natural products, such as the monoterpenoid α-*syn*-oxime derivative, perillartine, and the dihydrochalcones of naringin and neohesperidin. In general, only the prototype sweet compound in each structural class is tabulated. Examples of exceptions to this are the listing of the two sweet *ent*-kaurene glycoside constituents of *S. rebaudiana*, i.e., stevioside and rebaudioside A, since both are important sweeteners in Japan. Analysis of Table 1 shows that the major structural classes of natural products so far demonstrated as being highly sweet are terpenoids, steroidal saponins, flavonoids, and proteins. It is interesting to note, however, that the aromatic aldehyde, cinnamaldehyde, the dibenz[*b,d*]oxocin derivative hematoxylin, and two proanthocyanidins from *Cinnamomum sieboldii* Meisner are also sweet. Both the structural classes of sweet compounds from plants and the examples of such substances within these categories are continuing to increase. Aspects of sweetness–structure–activity relationships of natural sweeteners have been reviewed previously.[1]

While plants in restricted taxonomic groups often tend to biosynthesize similar classes of chemical components, there is little evidence thus far that the sweet-tasting constituents of a given plant will occur in other species of the same genus. For example, when over 100 leaf herbarium specimens of the genus *Stevia* were examined, organoleptically and phytochemically, for the occurrence of sweet-tasting *ent*-kaurene glycosides such as stevioside, such compounds were detected as expected in *S. rebaudiana*, but were found in only one additional species, i.e., *S. phebophylla* A. Gray.[43,44] A similar type of phytochemical screening study found that rubusoside, another sweet *ent*-kaurene glycoside, occurred in only one of about 40 species in the genus *Rubus*.[45] In addition, there appears to be minimal overlap in the occurrence of structurally similar sweet-tasting plant constituents at the plant family level, although sweet oleanane- and cucurbitane-type triterpene glycosides have been observed in several genera of the families Fabaceae and Cucurbitaceae, respectively.[1,10,11]

Previously, we had examined the position, in Dahlgren's classification of angiosperm (flowering plant) superorders, of the families where plants accumulating highly sweet compounds occur. No pattern was found that would enable one to predict the occurrence of sweet compounds on a taxonomic basis.[1,46] It may be pointed out that highly sweet compounds are

TABLE 1
Examples of Plant-Derived Highly Sweet Compounds

Compound class	Name	Species of origin	Ref.
Terpenes			
Monoterpene	Perillartine[a]	*Perilla frutescens* Britton	12
Sesquiterpene	Hernandulcin	*Lippia dulcis* Trev.	13
Diterpene	4b,10a-Dimethyl-1,2,3,4,5,10-hexahydrofluorene-4a,6a-dicarboxylic acid[a]		14
Diterpene Glycosides			
ent-Kaurene-types	Stevioside	*Stevia rebaudiana* (Bertoni) Bertoni	15,16
	Rebaudioside A	*S. rebaudiana*	17
	Rubusoside	*Rubus suavissimus* S. Lee	18
	Sauvioside A	*R. suavissimus*	19
Labdane-types	Baiyunoside	*Phlomis betonicoides* Diels	20
	Gaudichaudioside A	*Baccharis gaudichaudiana* DC	21
Triterpene Glycosides			
Cucurbitane-types	Mogroside V	*Siraitia grosvenorii* (Swingle) C. Jeffrey	22
	Carnosifloside V	*Hemsleya carnosiflora* C. Y. Wu et Z. L. Chen	23
Cycloartane-type	Abrusoside A	*Abrus precatorius* L.	24
Oleanane-types	Glycyrrhizin	*Glycyrrhiza glabra* L.	25
	Apioglycyrrhizin	*G. inflata* Batal	26
	Periandrin I	*Periandra dulcis* Mart	27
Steroidal Saponins	Osladin	*Polypodium vulgare* L.	28
	Polypodoside A	*P. glycyrrhiza* D.C. Eaton	29
Dihydroisocoumarin	Phyllodulcin	*Hydrangea macrophylla* Seringe var. *thunbergii* (Siebold) Makino	30
Flavonoids			
Dihydrochalcones	Glycyphyllin	*Smilax glycyphylla* Sm.	31
	Naringin dihydrochalcone[a]	*Citrus paradisi* Macfad.	32
	Neohesperidin dihydrochalcone[a]	*C. auranticum* L.	32
Dihydroflavonols	Dihydroquercetin 3-*O*-acetate 4'-(methyl ether)[a]	*Tessaria dodoneifolia* (Hook. & Arn.) Cabrera	33
	Neoastilbin	*Engelhardtia chrysolepis* Hance	34
Proteins	Thaumatins I and II	*Thaumatococcus danielli* (Bennett) Benth.	35
	Monellin	*Dioscoreophyllum cumminsii* (Stapf) Diels.	36

TABLE 1 (continued)

Compound class	Name	Species of origin	Ref.
Proteins	Mabinlin	*Capparis masaikai* Levl.	37
	Pentadin	*Pentadiplandra brazzeana* Baillon	38
	Curculin	*Curculigo latifolia* Dryand.	39
Miscellaneous			
Aromatic aldehyde	Cinnamaldehyde	*Cinnamomum osmophloeum* Kanehira	40
Dibenz[*b,d*]oxocin	Hematoxylin	*Haematoxylon campechianum* L.	41
Proanthocyanidins	Unnamed[b]	*Cinnamomum sieboldii* Meisner	42

[a] Semisynthetic derivative of a natural product.
[b] Two flavan-3-ol trimers.

already known to occur in taxonomic groups outside of the flowering plants, as in the case of the Pteridophyta (ferns), which is exemplified by the isolation of the sweet steroidal saponins, osladin, and polypodoside A.[28,29] Although no highly sweet compounds have yet been characterized from a lower plant, it appears that they occur in the moss *Bryum* (*Rhodobryum*) *giganteum*, since this species is reported to possess a saccharin-like taste.[47]

III. ISOLATION AND ASSAY PROCEDURES

A. SELECTION OF CANDIDATE SWEET PLANTS
1. Field Investigations

The most direct and probably most successful way of obtaining leads to candidate sweet-tasting plants is to make inquiries in the field or in marketplaces where medicinal plants are sold. If a plant specimen does not taste as obviously and intensely sweet as sugar cane (*Saccharum officinarum* L.), then the fact that it has a sweet taste will probably not become widely known among indigenous peoples. Plants with an extremely sweet taste that have found their way into the literature for this reason include *Stevia rebaudiana*, *Thaumatococcus daniellii*, and *Siraitia grosvenorii*.[10] In contrast, plants that possess a slight or moderate sweetness will probably not merit general attention, but may be noted, for example, by ethnobotanists, and then be appropriately documented. Alternatively, the slightly sweet taste of a particular species may be known to local populations, such as the vendors of medicinal plants, and then used to provide a colloquial name for the plant, even if the recommended use of the plant does not refer to its sweetness. To exemplify the latter point, two excellent leads were obtained from inquiries at a market for medicinal plants in Asuncion, Paraguay. First, *Tessaria dodoneifolia* (Hook. & Arn.) Cabrera was offered for sale as an emmenagogue under the Guarani name of "kaá hê-é" (sweet herb). Subsequent laboratory investigation of the plant afforded a dihydroflavonol sweetener, namely (+)-dihydroquercetin-3-acetate, which was used as a lead compound to synthesize racemic dihydroquercetin-3-acetate-4'-methylether, a compound rated as possessing 400 times the sweetness intensity of a 3% w/v sucrose solution.[33] Second, *Baccharis gaudichaudiana* DC., obtained at the same location, where it was being sold as an antidiabetic remedy under the name "chilca melosa," was found to contain a prototype labdane diterpene arabinoside sweetener, gaudichaudioside A, a compound estimated as some 55 times sweeter than 2% w/v sucrose.[21] The structure elucidation of gaudichaudioside A will be covered later in this chapter.

2. Literature Sources

Literature information on sweet-tasting plants tends to be widely dispersed, but useful material may be found in old herbals or in contemporary botanic articles. We have utilized both of these approaches to select sweet-tasting plants for laboratory investigation, as demonstrated by the following examples. First, a former graduate student, Dr. Cesar M. Compadre, searched the Mexican ethnobotanical literature in order to uncover records of plants with a sweet taste. He examined a monograph entitled *Natural History of New Spain*, written between 1570 and 1576 by the Spanish physician Francisco Hernández, in which a sweet plant was described that was known to the Aztecs by the Nahuatl name "Tzonpelic xihuitl." As a result of the description provided by Hernández and after appropriate analysis of other literature, this plant was identified as *Lippia dulcis* Trev. Subsequently, hernandulcin, a novel bisabolane sesquiterpene sweet constituent, was isolated and characterized.[13,48,49] Second, in a botanical report of a field trip to West Africa to collect sweet plants, Inglett and May pointed out that the leaves of *Abrus precatorius* L. possess an equivalent sweetness potency to sucrose.[50] When our group followed up on this lead, which was backed up by anecdotal information supplied by a botanist colleague, Dr. Julia F. Morton of the University of Miami, four novel cycloartane-type triterpene glycosides, i.e., abrusosides A–D, were isolated and characterized.[24,51] Further information about the structure elucidation of both hernandulcin and abrusosides A–D will be provided later in this chapter.

Index Kewensis is a repository of all of the Latin binomials of seed plants that have been published to date. Following the suggestion of Dr. D. Doel Soejarto, a taxonomist, this series of volumes was scanned in order to pick out specific epithets which might be indicative of sweetness. Examples of such epithets are *saccharum* (Greek, *saccharon*, sugar), *saccharifera* (sugar-bearing), *dulcis* or *dulcificum* (Latin, sweet), and *glycyrrhiza* (Greek, *glykos, glyckeros*, sweet). Among the known sweet-tasting plants listed in this manner in *Index Kewensis* are *Acer saccharum* Marsh. (sugar maple); the previously mentioned *Glycyrrhiza glabra* and *Lippia dulcis*; *Periandra dulcis* Mart. (the source of the sweet compounds, periandrins I–IV[1,10,27]); and *Richardella dulcifica* (Schumach. & Thonning) Baehni [formerly *Synsepalum dulcificum* (Schumach. & Thonning) DC] (the source of the sweetness inducer, miraculin).[1,10] By no means have all of the known plants with sweet-tasting constituents been given a specific epithet that refers to their sweet taste, but a list of about 150 potentially sweet plants was drawn up as candidates for future investigations in the field and in the laboratory, by perusal of *Index Kewensis*.[52]

B. PHYTOCHEMICAL CONSIDERATIONS

Methanol-water (4:1) is a good general solvent for plant secondary metabolites and is employed for the initial extraction of plants collected for our program on sweet natural products. The dried, 80% methanol-soluble extract of a plant is taken up in methanol-water (1:1) and then partitioned successively into petroleum ether, ethyl acetate, and butanol, in order to afford extracts of various polarities.[53] The selective partition of the sweetness, from a plant part, into one of these solvents provides a useful preliminary indication of the chemical class of the sweetener(s). For example, the petroleum ether extract might well contain sweet terpenoids or phenylpropanoids; the ethyl acetate extract, sweet flavonoids; the butanol extract, sweet di- or triterpenoid glycosides; and the aqueous methanol extract, more polar sweet glycosides. On the rare occasions when the sweetness of a plant part is due to the presence of sweet proteins, extraction would be conducted using water as the initial solvent.[38]

In an attempt to work as efficiently as possible, it is imperative to rapidly ascertain whether potently sweet constituent(s) are indeed present in a plant part under consideration. We have found that it is relatively rare for a sweet-tasting plant to contain highly sweet constituents. Rather, one of two other classes of sweet compounds, such as sugars and polyols, and sweet phenylpropanoids normally account for the perceived sweetness of an acquisition. Sugars and polyols will partition preferentially into methanol-water (1:1), or occasionally butanol, and may be concentrated by passage through a charcoal column and rapidly identified using gas chromatography/mass spectrometry (GCMS).[53] We have compared the saccharide and polyol concentration levels of several sweet-tasting plants whose sweetness is due entirely to sugars and polyols, to the constituents of the fruits of *Siraitia grosvenorii*, which contain both significant levels of sugars and polyols and over 1% w/w of the intensely sweet triterpene glycoside, mogroside V. As a result, it has been concluded that *unless the saccharide and/ or polyol content of a plant part is well over 5% w/w, then it is unlikely to exhibit an obvious sweet taste unless a highly sweet compound is also present.*[53] In addition, we have also found that sweet-tasting phenylpropanoids may occur in such high concentrations as to confer a potent sweet taste to a plant part. For example, it was demonstrated that *trans*-cinnamaldehyde was responsible for the sweet taste of *Cinnamomum osmophloeum* Kanehira leaves, in which it occurred in a yield of over 1% w/w.[40]

Another phenylpropanoid, *trans*-anethole, was determined as the constituent responsible for the sweetness of seven plants in which it proved to be the major volatile-oil constituent.[54] Since our realization that common phenylpropanoids can impart a sweet taste to a plant part, we now routinely analyze the petroleum ether extract of collected plant samples, by GCMS, for the presence and quantity of these compounds. In this manner the expense of recollecting large amounts of plant material that turn out to be sweet due to high concentrations of known phenylpropanoids can be avoided.[54]

C. ASSAY AND PRELIMINARY SAFETY STUDIES
1. Involvement of Human Subjects in Fractionation Experiments

A necessary step in the isolation of highly sweet compounds from plants is the periodic assessment of the sweetness of extracts and chromatographic fractions by volunteer human subjects.[55] There are several methods that may be used to assess the sweetness of pure, nontoxic, and nonmutagenic plant isolates, with some of these involving quite large numbers of human subjects.[13,56-58] However, in the last few years in our laboratory, we have tended to perform sensory evaluations on sweet plant constituents using a taste panel consisting of only three experienced staff personnel.[21,29,33,51] Approximate values of the sweetness intensity, relative to sucrose, of a given compound are typically determined by diluting an aqueous solution of the test compound until it exhibits an equivalent sweetness intensity to that of a 2% w/w sucrose solution. The sweetness intensity is then determined by a simple ratio of the two solution concentrations. Only a limited notion of the hedonic properties (that represent the pleasantness of taste) of a sweetener may be obtained when evaluated in this manner. However, this type of approach is sparing on the often-limited supply of a sweet compound on hand and is a useful guide to help assess the potential for commercialization of a novel sweet natural product.

Usually, there is no anecdotal or literature evidence that would indicate whether a sweet-tasting plant obtained in the field will also biosynthesize toxic substances. However, the rhizomes of *Hemsleya panicis-scandens* C.Y. Wu and Z.L. Chen (which are known to produce several cytotoxic cucurbitacins in addition to the sweet cucurbitane glycoside, scandenoside R6,[59]) and the aerial parts of *Lippia dulcis*, collected in Mexico, have been found to biosynthesize the toxic monoterpene, camphor, along with the sweet sesquiterpene, hernandulcin.[48] Therefore, to protect against the possibility of toxins cooccurring with sweet plant constituents, we have made it a practice in our laboratory to subject one or more of the initial solvent extracts of each plant acquisition to preliminary safety testing, prior to determining the presence or absence of sweetness by tasting. These safety tests comprise both an acute toxicity evaluation using mice and the use of a bacterial forward mutation assay.

In the acute toxicity experiments, single doses of plant extracts are suspended in 1% sodium carboxymethylcellulose and administered to groups of ten male Swiss-Webster mice by oral intubation up to a maximum dose of 2g/kg, depending upon the amount of material available for testing. The incidence of mortality is observed for up to 14 days after administration, and test animal body weights are recorded at intervals. Body weight variations between treated and control groups are then assessed by one-way analysis of variance.[60]

The mutation assays are also carried out on plant extracts, according to standard protocols, using *Salmonella typhimurium* strain TM677 carrying the "R-factor" plasmid pKM101, both in the presence and absence of metabolic activating systems obtained from a 9000 × g supernatant fraction derived from the livers of Aroclor 1254-pretreated rats.[61]

Attempts are made to assess the safety of the preliminary 80% methanolic extract of each plant and all subsequent extracts partitioned into solvents of different polarities. Also, purified sweet compounds that are to be evaluated by a human taste panel must be shown to be innocuous using these safety procedures. Infrequently, it has been our experience that extracts of certain candidate sweet plants are mutagenic, and in such cases these leads have not been pursued further.

2. Potential for the Use of Gerbils to Evaluate Plant Extracts and Fractions for Sweetness

The practice of submitting plant extracts to both acute toxicity tests in mice and bacterial mutagenicity evaluation, prior to being tested for the presence or absence of a sweet taste by human subjects, is both time consuming and expensive. Accordingly, we have begun to investigate an alternative approach to plant extract sweetness assessment, using electrophysiological and conditioned taste-aversion procedures on the Mongolian gerbil.[62] The electrophysiological experiment, involving stimulation of the gerbil's intact *chorda tympani* nerve, has been utilized previously for the evaluation of sweet monosaccharides, disaccharides, polyols, and many naturally occurring and synthetic intensely sweet substances.[63] In the behavioral, conditioned aversion test, the degree of similarity of taste to sucrose is determined by the amounts of experimental fluids consumed by gerbils trained to avoid sweet, salty, bitter, and sour taste qualities.[64] Data obtained on preselected extracts of varying polarities of three well-documented sweet plants, i.e., *Stevia rebaudiana*, *Siraitia grosvenorii*, and *Abrus precatorius*, showed good correlations between the results of the gerbil experiments and the presence or absence of highly sweet substances.[62]

IV. STRUCTURE ELUCIDATION PROCEDURES

A. GENERAL CONSIDERATIONS

Several of the purified sweet-tasting constituents of plants investigated in this laboratory have been found to be compounds of novel structure. In general, since we normally place heavy reliance on a combination of spectroscopic methods, particularly contemporary one- and two-dimensional NMR techniques, in our structure determinations, it has not been necessary to develop special approaches for the structure elucidation of these substances. However, in some cases we have utilized X-ray crystallography through a collaborative arrangement or have synthesized the molecule under consideration. To illustrate the variety of structural problems we have had to tackle, potently sweet substances representing the sesquiterpene (hernandulcin, Section IV.B.1), diterpene glycoside (gaudichaudioside A, Section IV.B.2), triterpene glycoside (abrusosides A–D, Section IV.B.3), and steroidal glycoside (polypodoside A, Section IV.B.4) compound classes will be discussed.

B. SPECIFIC EXAMPLES
1. Hernandulcin

As mentioned earlier, the sweet sesquiterpenoid hernandulcin (Figure 1, **1**) was isolated from the plant *Lippia dulcis*. This novel compound, which was named in honor of Francisco

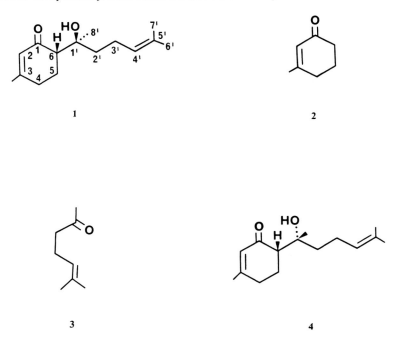

FIGURE 1. Structures of hernandulcin (**1**), 3-methyl-2-cyclohexen-1-one (**2**), 6-methyl-5-hepten-2-one (**3**), and *epi*hernandulcin (**4**).

Hernández, was found to be a minor constituent of the volatile oil of the aerial parts of the Mexican plant. The compound was preferentially soluble in petroleum ether and was purified by silica gel column chromatography and subsequent preparative thin-layer chromatography, using mixtures of hexane and acetone as the eluent and developing solutions, respectively.

The structure of hernandulcin was established on the basis of a combination of its spectral parameters, its thermal degradation profile, and its synthesis by a directed aldol condensation. The high-resolution mass spectrum of hernandulcin ($C_{15}H_{24}O_2$) exhibited fragmentation pathways characteristic of a bisabolane sesquiterpenoid. In particular, the retro Diels-Alder rupture of the cyclohexenone ring produced the fragment peak at m/z 82, and the retroaldolization of the C(6)-C(1′) bond produced the mass spectral base peak that was observed at m/z 110 (Figure 2). The IR spectrum showed evidence of the presence of a hydrogen-bonded hydroxyl group (υ_{max} 3365cm^{-1}) and an α,β-unsaturated carbonyl group (υ_{max} 1644cm^{-1}). In the ^{1}H-NMR spectrum of Compound **1**, four methyl groups were apparent, with three of these being vinylic. Also observed were two olefinic groups (δ 5.88 and 5.19 ppm), with the former signal being assigned to the α position of a conjugated carbonyl group that, along with the methyl group resonating at δ 1.97 ppm, was further assigned as constituting part of a $CH_3C=CHC=O$ unit. The second olefinic proton signal appeared as a triplet and, in association with the vinyl methyl groups that appeared at δ 1.68 and 1.63 ppm, was postulated as being part of a $(CH_3)_2C=CHCH_2$ grouping. Unambiguous assignment of the ^{1}H and ^{13}C-NMR spectra of hernandulcin was achieved by analysis of a combination of the ^{1}H-^{1}H homonuclear shift correlation (COSY) and the ^{1}H-^{13}C heteronuclear shift correlation (HETCOR) NMR spectra. Perusal of all of the spectroscopic data obtained for hernandulcin allowed the structure of this compound to be tentatively identified as 6-(1,5-dimethyl-1-hydroxy-hex-4-enyl)-3-methylcyclohex-2-enone.[13,49,65]

Confirmation of the structure of hernandulcin was obtained, in part, by the study of the thermal dissociation of the naturally occurring (+)-form of this sweetener. Distillation at 140°C under a vacuum resulted in the production of the commercially available ketones, 3-

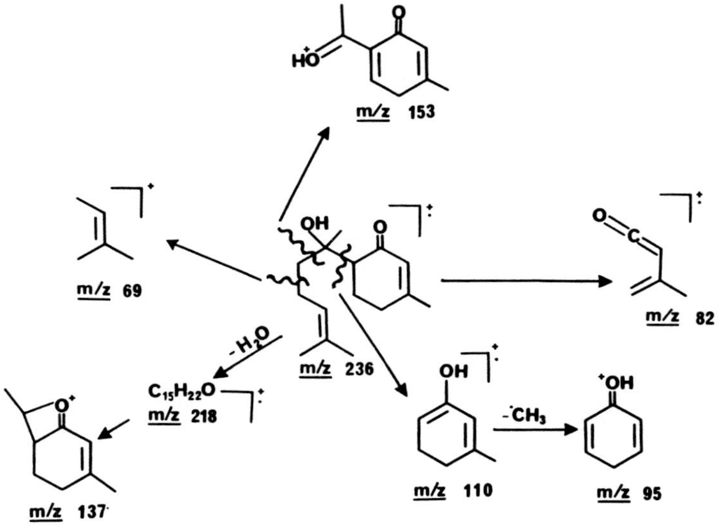

FIGURE 2. Mass spectral fragmentation pattern of hernandulcin.

methyl-2-cyclohexen-1-one (Figure 1, **2**) and 6-methyl-5-hepten-2-one (Figure 1, **3**). Furthermore, racemic hernandulcin was synthesized in about 50% yield from these same two ketones by directed-aldol condensation. The reaction occurred in a predictable stereoselective manner, with larger amounts of **1** being produced than an epimeric compound, the nonsweet *epi*hernandulcin (Figure 1, **4**). The highly stereoselective outcome of this reaction suggested that the synthetic (±)-hernandulcin produced was a mixture of the *R,R* and *S,S* enantiomers and that the (±)-*epi*hernandulcin consisted of the corresponding *R,S* and *S,R* forms.[13,49,65] The absolute stereochemistry of naturally occurring (+)-hernandulcin (Figure 1, Structure **1**) was established as 6*S*,1′*S* by Mori and Kato, who synthesized this sweet substance from (*R*)-(+)-limonene.[66]

Synthetic (±)-hernandulcin was not acutely toxic to mice and was not active as a bacterial mutagen, at the doses tested. In a preliminary sensory test using a human taste panel, natural (+)-hernandulcin was perceived to be about 1000 times sweeter than sucrose, on a molar basis. However, it was also found to exhibit undesirable hedonic effects that reduce the prospects of this compound being developed commercially.[13] A similar order of sweetness potency, relative to sucrose, was determined for (+)-hernandulcin by Mori and Kato, who also showed that the 6*S*,1′*S* diastereomer is the only one of the four isomers of this compound to exhibit a sweet taste.[66]

2. Gaudichaudioside A

Baccharis gaudichaudiana was recognized as being sweet as a result of field observations in Paraguay. Following the usual safety precautions, the sweet taste of the aerial parts of *B. gaudichaudiana* was found to concentrate preferentially into 1-butanol. Gaudichaudioside A (Figure 3, **5**) was purified by gravity and flash column chromatography over silica gel.

The molecular formula of gaudichaudioside A was established as $C_{25}H_{40}O_8$ by lithium ion-catalyzed high-resolution fast-atom bombardment (FAB) mass spectrometry. It was apparent from the IR spectrum that a conjugated aldehyde functional group (υ_{max} 1664 cm^{-1}) was present in the molecule. From the ^{13}C-NMR spectrum, three methyl, nine methylene, six methine, two quaternary, four double-bond, and one carbonyl carbon signals were evident. The chemical shifts inferred the presence of two primary alcohol groups (δ 66.9 and 59.0 ppm), one unsaturated aldehyde (δ 194.9 ppm), three methyl groups (δ 31.5, 21.5, and 16.3 ppm), and two double bonds (δ 170.1 and 131.3, and 138.7 and 125.2 ppm). The signal

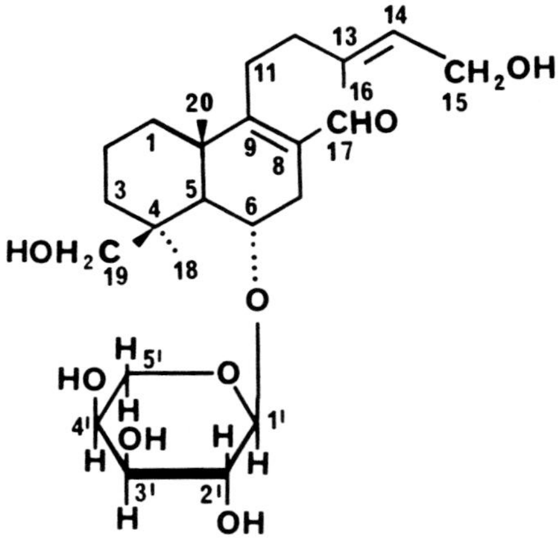

FIGURE 3. Structure of gaudichaudioside A (**5**).

resonating at δ 125.2 ppm was accorded to a protonated double-bond carbon (C-14) adjacent to a primary hydroxy group (C-15). The presence of an arabinosyl moiety was also implied from the ¹³C-NMR spectrum. Indeed, hydrolysis of **5** with 0.1 N HCl yielded L-arabinose, which was identified by comparison of the optical rotation and chromatography with an authentic sample. The coupling constant of the anomeric proton, resonating as a doublet (J = 6 Hz) at δ 4.35 ppm, indicated an α-L-arabinosyl configuration. The selective INEPT (Insensitive Nuclei Enhancement by Polarization Transfer) NMR technique[67,68] was used to decide the position of attachment of the saccharide unit to the aglycone of **5**. With this one-dimensional, *J*-modulated method, a strategically located proton is irradiated with a soft pulse, and the delay time is adjusted so as to result in the selective enhancement of a carbon atom three bonds away from the irradiated proton. Thus, irradiation of the anomeric proton signal in gaudichaudioside A enhanced C-6 of the aglycone at δ 80.1 ppm (Figure 4b). Along with other carbon atoms, C-6 was also enhanced following irradiation of H-5 (δ 1.59 ppm) and H-7α (δ 2.78 ppm) (Figure 4c and d, respectively). Other three-bond correlations were established by irradiation of H₃-16 and H₃-20 (Figure 4e and f). Attempts to characterize the aglycone, following acid or enzymatic hydrolysis of gaudichaudioside A, were unsuccessful due to its lability.[21,69]

Comparison of the spectral data of gaudichaudioside A with other diterpenoids suggested that it possessed a labdane carbon skeleton. Unambiguous ¹H- and ¹³C-NMR assignments were made for gaudichaudioside A, using a combination of ¹H-¹H COSY, ¹H-¹³C HETCOR (Figure 5), and COLOC (COrrelation spectroscopy via LOng-range Coupling) NMR measurements. The latter experiment[70] was used to support the quaternary-carbon assignments in the aglycone portion of the sweet glycoside. The relative stereochemistry at each of the chiral centers at C-4, C-5, and C-10 was determined using a 2-D nuclear Overhauser enhancement (NOESY) NMR experiment in which nOe interactions were observed between H₃-20 and H-6, thereby demonstrating the β-stereochemistry of H-6. This indicated that the L-arabinosyl unit of gaudichaudioside A was equatorial, which was supported by the large coupling constant (10 Hz) observed between H-5α and H-6β. Another nOe cross-peak was observed between H-5α and H₃-18, thereby confirming that the latter proton was in the α-equatorial position. Finally, the configuration of the C-14,15 double bond was determined as *E* from an nOe cross-peak between the terminal allylic protons and H₃-16. Therefore, the structure of gaudichaudioside A (Figure 3, **5**) was established as 15,19-dihydroxylabda-8(9)-13(14)*E*-dien-17-al-6α-*O*-α-L-arabinopyranoside.

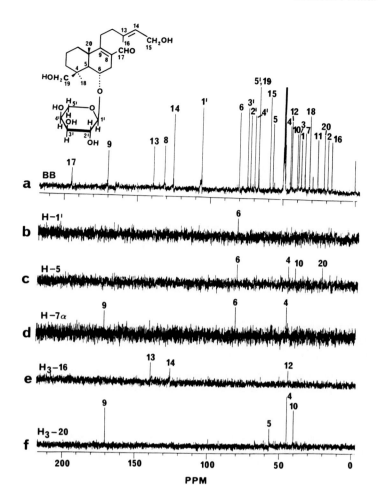

FIGURE 4. Selective INEPT NMR experiments on gaudichaudioside A.

Gaudichaudioside A was not acutely toxic to mice when administered by oral intubation at a dose of 1 g/kg body weight and was not mutagenic toward *Salmonella typhimurium* strain TM677 both in the presence and absence of a metabolic activating system. This compound was fairly soluble in water and was rated by a small taste panel as exhibiting only moderate sweetness potency, with about 55 times the intensity of a 2% w/v sucrose solution. In solution, gaudichaudioside A was found to possess a pleasant taste with a low concomitant perception of bitterness. Minor structural modification of this compound, as encountered in five labdane diterpene arabinoside analogs also isolated from the aerial parts of *B. gaudichaudiana*, led to substances found to be sweet-bitter, entirely bitter, or tasteless. However, gaudichaudioside A was the only representative of these six structurally related compounds to exhibit a pronounced sweet taste.[21,69]

3. Abrusosides A–D

The seeds of *Abrus precatorius* are well known to be toxic due to the presence of the ribosome-inactivating proteinaceous toxin, abrin.[71] In contrast, the roots and leaves of this plant do not seem to be poisonous and, in fact, have a long history of human internal consumption, particularly to substitute for licorice roots (*Glycyrrhiza glabra* and related species).[50] According to even recent literature, the sweetness of *A. precatorius* leaves is due

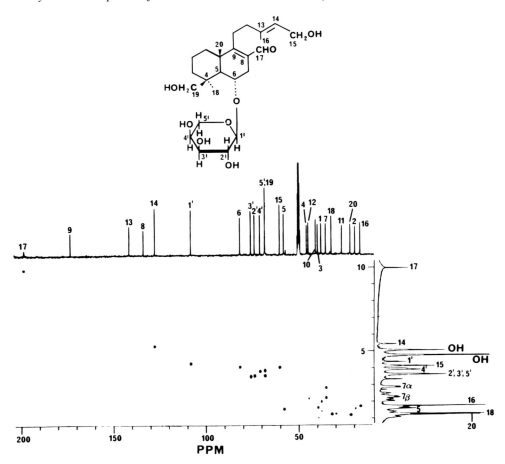

FIGURE 5. ^{1}H-^{13}C HETCOR NMR spectrum of gaudichaudioside A.

to the presence of high levels of the oleanane-type triterpene glycoside, glycyrrhizin.[72] However, when investigated in our laboratory, no glycyrrhizin was found in extracts of *A. precatorius* leaves collected in Florida. Instead, four novel, sweet, cycloartane-type triterpene glycosides, abrusosides A–D (Figure 6, **6–9**), were obtained after solvent partitioning and extensive chromatographic fractionation over silica gel. At the beginning of their purification, all four of these compounds were found to partition selectively into 1-butanol. Abrusoside A was the least polar, and abrusoside D was the most polar, of the four compounds.

Abrusoside A exhibited a molecular formula of $C_{36}H_{54}O_{10}$ by high-resolution FAB mass spectrometry and, when hydrolyzed with 1 N HCl for 4 h at 100°C, afforded D-glucose and a triterpenoid aglycone, based on a new carbon skeleton, which we called abrusogenin (Figure 6, **10**). Initial IR spectral observations showed hydroxy (υ_{max} 3430 cm^{-1}) and carbonyl group (υ_{max} 1707 cm^{-1}) absorbances in the aglycone. In the ^{1}H-NMR spectrum of abrusogenin, the presence of cyclopropyl methylene protons (δ 0.39 and 0.61 ppm) and two protons attached to carbons bearing oxygen (δ 4.09 and 4.50 ppm) was noted. Analysis of the ^{13}C-NMR spectrum of abrusogenin suggested an α,β-unsaturated δ-lactone ring, with pertinent resonances at δ 166.56, 139.65, 128.12, and 80.21 ppm. The presence of secondary alcohol and carboxylic acid functional groups in the molecule of abrusogenin was established by acetylation and methylation, respectively, using standard conditions in each case. The connectivity of the cyclopropyl ring to C-9 and C-10 was inferred by comparison of the ^{13}C-NMR and mass spectral data of abrusogenin with those of model 9,19-cyclopropyl derivatives.

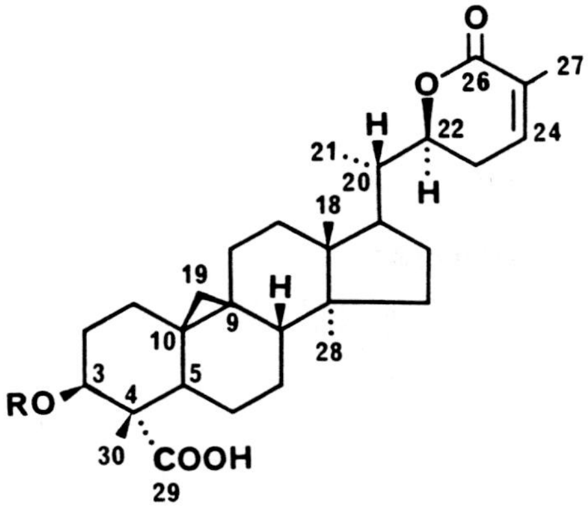

	R
6	β-D-glc
7	β-D-glcA-6-CH$_3$^{2}-β-D-glc
8	β-D-glc^2-β-glc
9	β-D-glcA-β-D-glc
10	H

Glc = glucopyranosyl; glcA = glucuronopyranosyl

FIGURE 6. Structures of abrusosides A–D (**6–9**, respectively), and abrusogenin (**10**).

Abrusogenin (Figure 6, **10**) was finally characterized as (20S,22S)-3β,22-dihydroxy-9,19-cyclolanost-24-en-26,29-dioic acid δ lactone after due consideration of its ^{1}H-^{1}H-COSY, ^{1}H-^{13}C-HETCOR, and selective INEPT NMR spectral data. The structure and stereochemistry of this aglycone were confirmed by an X-ray crystallographic analysis of abrusogenin methyl ester.[24,51,73]

The other sweet glycosides obtained from *A. precatorius* leaves, abrusosides B–D, also afforded abrusogenin on acid hydrolysis, and their molecular formulae were individually established by high-resolution FAB mass spectrometry. The configurations of the sugar moieties of these sweet glycosides were established in each case from the ^{1}H-NMR coupling constants of the anomeric proton(s) and also with reference to published ^{13}C-NMR data for model sugars. The position of the sugar attachment and the linkages of the various saccharide moieties for all four sweet compounds from this plant source were confirmed by application of the selective INEPT NMR technique on the intact heterosides. Thus, in the case of abrusoside B, which produced D-glucose and D-glucuronic acid methyl ester on hydrolysis, irradiation of H-1″ (the anomeric proton of the terminal glucose residue) (δ 5.37 ppm) led to the enhancement of C-2′ (δ 84.05 ppm) of the glucuronic acid methyl ester unit (Figure 7b). Also, similar irradiations of the 3α-H (δ 4.87 ppm) and 30-CH$_3$ (δ 1.71 ppm) protons led to, in turn, selective enhancements of C-1′ (δ 102.47 ppm) and C-3 (δ 82.59 ppm) (Figure 7c, d). In this manner the structure of abrusoside B (Figure 6, **7**) was determined as abrusogenin-3-O-β-D-glucopyranosyl-(1→2)-β-glucuronopyranoside-6′-methyl ester. The structures of abrusosides A, C, and D (Figure 6, **6, 8,** and **9**, respectively) were completed in an analogous manner.[51,73]

When separately evaluated, abrusosides A–D were found to be nonmutagenic and not acutely toxic to mice, in preliminary tests. The water soluble ammonium salts of these

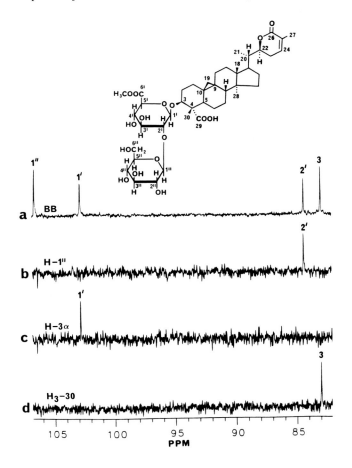

FIGURE 7. Selective INEPT NMR experiments on abrusoside B.

compounds exhibited sweetness potencies of between 30 and 100 times that of sucrose. Abrusoside B (Figure 6, 7), the sweetest of these substances, was rated by a small human taste panel at about 100 times sweeter than a 2% w/v aqueous sucrose solution. Therefore, the abrusoside sweeteners, which are based on a cycloartane skeleton, are of approximately equal sweetness potency to that of the widely used natural product sweetener glycyrrhizin, an oleanane-type triterpenoid. However, the abrusosides offer a potential advantage over the latter substance, in being easier to produce by cultivation, since they occur in the leaves of the plant rather than the roots. Additionally, and in contrast to glycyrrhizin, abrusosides A–D do not possess an unhindered α,β-unsaturated carbonyl group in their aglycone. It is the 11-oxo-12,13-dehydro- group of glycyrrhizin (Figure 8, **11**) that is responsible for its undesirable adrenocorticomimetic effects that are manifested by symptoms such as oedema and hypertension.[74] Therefore, it is hoped that these new and novel abrusoside sweeteners may not have the same type of toxic properties as glycyrrhizin and might therefore be able to be substituted for glycyrrhizin in the sweetening of foods, beverages, and medicines, especially in Japan.

4. Polypodoside A

The rhizomes of the North American species *Polypodium glycyrrhiza* D.C. Eaton exhibit a bittersweet taste and were first investigated for the chemical nature of their sweet constituent(s) some 60 years ago.[75] Our initial interest in this plant was as a potential source of osladin

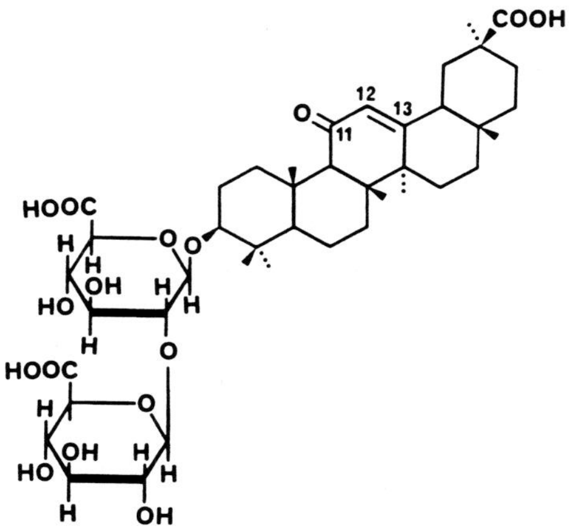

11

FIGURE 8. Structure of glycyrrhizin (**11**).

(Figure 9, **12**), a highly sweet steroidal saponin isolated in 1971, by Czechoslovakian workers, from *Polypodium vulgare* L.[28] Fractionation work on *P. glycyrrhiza* did not yield osladin. However, the closely related compound, polypodoside A (Figure 9, **13**) was isolated as a novel sweet-tasting principle.[29,76]

Polypodoside A was partitioned into 1-butanol and purified by successive chromatography over silica gel and Sephadex LH-20 columns, followed by droplet countercurrent chromatography. The solvent system used for this last step was a mixture of chloroform-methanol-isopropanol-water, and the upper layer was employed as the mobile phase in ascending-mode separation.

The elemental formula of polypodoside A was determined as $C_{45}H_{72}O_{17}$, using high-resolution FAB mass spectrometry. Hydrolytic experiments with hesperidinase enzyme produced D-glucose and L-rhamnose. In the low-resolution FAB mass spectrum, a fragment peak occurring at m/z 739 [M + H − 146]$^+$ was consistent with the loss of a terminal rhamnose unit from the intact glycoside. Another diagnostic fragment peak was observed at m/z 309, which corresponded to a dihexosyl group consisting of one unit of glucose and one of rhamnose.[29,76]

Also produced upon the hydrolysis of polypodoside A was the known aglycone, polypodogenin (Figure 9, **14**), a compound representing the $\Delta^{7,8}$ derivative of the aglycone of osladin. Since no ^{13}C-NMR assignments had been made previously for polypodogenin, this exercise was performed with the assistance of a ^1H-^{13}C HETCOR experiment. Characteristic paramagnetic glycosylation shifts for C-3 (δ 3.84 ppm) and C-26 (δ 4.45 ppm) were evident in the ^1H-NMR spectrum of polypodoside A relative to that of polypodogenin. The nature of the sugar linkages at C-3 and C-26 in polypodoside A were determined, as usual, using the selective INEPT NMR procedure. Hence, irradiation of the anomeric proton of the glucose unit (H-1′, δ 5.02 ppm) led to the selective enhancement of the C-3 signal of the aglycone moiety (Figure 10b). The anomeric resonances appearing at δ 6.26 ppm and δ 102.17 ppm in the ^1H- and ^{13}C-NMR spectra of polypodoside A were assigned to C-1″ of the rhamnopyranosyl unit attached to C-2′ of the glucose unit. The configuration of the C-3-attached disaccharide sugar (β-neohesperidose) was determined as α-L-rhamnopyranosyl-(1→2)-β-D-glucopyranose, by comparison with published ^{13}C-NMR data. When the proton signal resonating at δ 5.61 (H-1‴) was irradiated, the resonance corresponding to C-26 (δ 107.32 ppm) was enhanced (Figure 10c), thereby demonstrating the linkage of a rhamnosyl

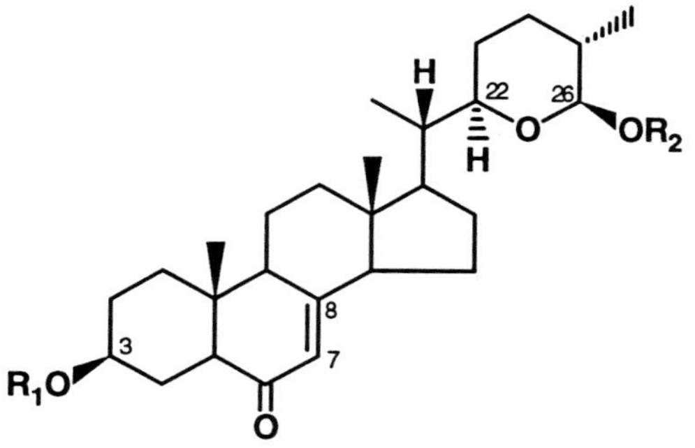

	R_1	R_2	Other
12	β-D-glc^2-α-L-rha	rha	7,8-dihydro
13	β-D-glc^2-α-L-rha	α-L-rha	--
14	H	H	H

Glc = glucopyranosyl; rha = rhamnopyranosyl

FIGURE 9. Structures of osladin (12), polypodoside A (13), and polypodogenin (14).

unit to C-26 in the molecule of polypodoside A. Comparison of the ^{13}C-NMR parameters of this C-26-attached rhamnosyl unit with those of model sugars suggested an α configuration of this sugar moiety. The stereochemistry at three chiral centers has recently been reversed in the aglycone of osladin, as a result of the application of X-ray crystallography.[77] The close comparison of the ^{13}C-NMR chemical shifts of the pyran ring of polypodoside A with those published for osladin[77] now suggests that the stereochemistry in the aglycone of polypodoside A at C-22, C-25, and C-26 is *R, S,* and *R,* respectively. In this manner the structure of polypodoside A (Figure 9, **13**) is thus 26-*O*-α-L-rhamnopyranosyl-(22*R*, 25*S*, 26*R*)-7-en-22,26-epoxy-6-oxo-5α-cholestan-3β,26-diol-3-*O*-α-L-rhamnopyranosyl-(1→2)-β-D-glucopyranoside.

Polypodoside A proved to be both nonmutagenic and not acutely toxic to mice, at the doses tested, and was rated as possessing 600 times the sweetness intensity of a 6% w/v aqueous sucrose solution. However, the compound exhibited a licorice-like off-taste and a lingering aftertaste. Polypodoside A does not seem to have good prospects for commercialization, since, in addition to its relatively unpleasant hedonic characteristics, it is rather insoluble in water and contains an α,β-unsaturated lactone functionality in its aglycone, which could conceivably produce undesirable biological effects.[29,76]

ACKNOWLEDGMENTS

The authors wish to acknowledge support of the work described herein by contract DE-02425 and grants DE-07560 and DE-08937 from the National Institute of Dental Research, Bethesda, MD, as well as from a contract awarded by General Foods Corporation, White Plains, NY. We also wish to thank several past postdoctorals and graduate students (Drs. Y. H. Choi, C. M. Compadre, F. Fullas, R. A. Hussain, Y.-M Lin, and N. P. D. Nanayakkara) for their excellent research efforts. We are grateful to Professors D. D. Soejarto and J. M. Pezzuto, University of Illinois at Chicago, and to Professor W. Jakinovich, Jr., Lehman College, City University of New York, for their ongoing collaboration on natural sweeteners. We thank the Research Resources Center, University of Illinois at Chicago, for expert assistance and

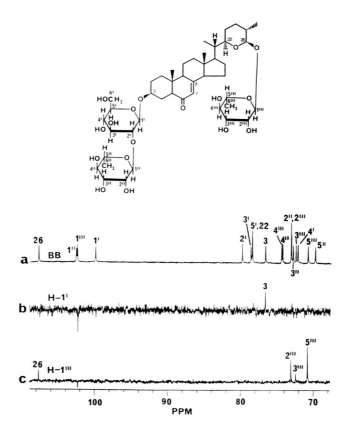

FIGURE 10. Selective INEPT NMR experiments on polypodoside A.

for the provision of NMR facilities. Dr. George Doss (formerly Asraf N. Abdel-Sayed) is acknowledged for implementation of the selective INEPT NMR pulse sequence with some of our campus NMR instrumentation.

REFERENCES

1. **Kinghorn, A. D. and Soejarto, D. D.,** Intensely sweet compounds of natural origin, *Med. Res. Rev.,* 9, 91, 1989.
2. **Grenby, T. H.,** Prospects for sugar substitutes, *Chem. Br.,* 27, 342, 1991.
3. **Anon.,** Market trends of sweeteners, *Food Chem. News (Tokyo),* June issue, 18, 1987.
4. **Anon.,** National Secretariat for Sanitary Control, Brazilian Ministry of Health, Resolution No. 14, *Diário Oficial,* January 29, 1988.
5. **Anon.,** Office of the Minister, Brazilian Ministry of Agriculture, Resolution No. 68, *Diário Oficial,* April 6, 1988.
6. **Higginbotham, J. D.,** Talin protein (thaumatin), in *Alternative Sweeteners,* O'Brien Nabors, L. and Gelardi, R. C., Eds., Marcel Dekker, New York, 1986, 103.
7. **Kinghorn, A. D. and Compadre, C. M.,** Less common high-potency sweeteners, in *Alternative Sweeteners,* 2nd ed., (revised and expanded), O'Brien Nabors, L. and Gelardi, R. C., Eds., Marcel Dekker, New York, 1991, 197.
8. **Horowitz, R. M. and Gentili, B.,** Dihydrochalcone sweeteners from *Citrus* flavanones, in *Alternative Sweeteners,* 2nd ed., (revised and expanded), O'Brien Nabors, L. and Gelardi, R. C., Eds., Marcel Dekker, New York, 1991, 97.

9. **Crosby, G. A. and Wingard, R. E., Jr.,** A survey of less common sweeteners, in *Developments in Sweeteners — 1,* Hough, C. A. M., Parker, K. J., and Vlitos, A. J., Eds., Applied Science Publishers, London, 1979, 135.

10. **Kinghorn, A. D. and Soejarto, D. D.,** Sweetening agents of plant origin, *CRC Crit. Rev. Plant Sci.,* 4, 79, 1986.

11. **Tanaka, O.,** Natural sweet principles — recent advances, *Kagaku to Kogyo (Osaka),* 61, 404, 1987.

12. **Furakawa, S. and Tomizawa, Z.,** Essential oil of *Perilla nankinensis* Decne., *J. Chem. Ind. Tokyo,* 23, 342, 1920; *Chem. Abstr.,* 14, 2839, 1920.

13. **Compadre, C. M., Pezzuto, J. M., Kinghorn, A. D., and Kamath, S. K.,** Hernandulcin: an intensely sweet compound discovered by review of ancient literature, *Science,* 227, 417, 1985.

14. **Tahara, A., Nakata, T., and Ohsuka, Y.,** New type of compound with strong sweetness, *Nature,* 233, 619, 1971.

15. **Bertoni, M. S.,** Le Kaá hâ-é. Sa nature et ses propriétés, *An. Cie. Paraguay,* 5 (Ser. 1), 1, 1905.

16. **Mosettig, E., Beglinger, U., Dolder, F., Lichti, H., Quitt, P., and Waters, J. E.,** The absolute configuration of steviol and isosteviol, *J. Am. Chem. Soc.,* 85, 2305, 1963.

17. **Kohda, H., Kasai, R., Yamasaki, K., Murakami, K., and Tanaka, O.,** New sweet diterpene glycosides from *Stevia rebaudiana, Phytochemistry,* 16, 981, 1976.

18. **Tanaka, T., Kohda, H., Tanaka, O., Chen, F.-H., Chou, W.-H., and Leu, J.-L.,** Rubusoside (β-D-glucosyl ester of 13-*O*-β-D-glucosyl-steviol), a sweet principle from *Rubus chingii* Hu (Rosaceae), *Agric. Biol. Chem.,* 45, 2165, 1981.

19. **Hirono, S., Chou, W.-H., Kasai, R., Tanaka, O., and Tada, D.,** Sweet and bitter diterpene glycosides from the leaves of *Rubus suavissimus, Chem. Pharm. Bull.,* 38, 1743, 1990.

20. **Tanaka, T., Tanaka, O., Lin, Z.-W., Zhou, J., and Ageta, H.,** Sweet and bitter glycosides of the Chinese drug, Bai-Yun-Shen (roots of *Salvia digitaloides*), *Chem. Pharm. Bull.,* 31, 780, 1983.

21. **Fullas, F., Hussain, R. A., Bordas, E., Pezzuto, J. M., Soejarto, D. D., and Kinghorn, A. D.,** Gaudichaudiosides A-E, five novel diterpene glycoside constituents from the sweet-tasting plant, *Baccharis gaudichaudiana, Tetrahedron,* 47, 8515, 1991.

22. **Takemoto, T., Arihara, S., Nakajima, T., and Okuhira, M.,** Studies on the constituents of Fructus Momordicae. III. Structures of mogrosides, *Yakugaku Zasshi,* 103, 1167, 1983.

23. **Kasai, R., Matsumoto, K., Nie, R.-L., Morita, T., Awazu, A., Zhou, J., and Tanaka, O.,** Sweet and bitter cucurbitane glycosides from *Hemsleya carnosiflora, Phytochemistry,* 26, 1371, 1987.

24. **Choi, Y.-H., Kinghorn, A. D., Shi, X., Zhang, H., and Teo, B. K.,** Abrusoside A: a new type of highly sweet triterpene glycoside, *J. Chem. Soc. Chem. Commun.,* 887, 1988.

25. **Tschirch, A. and Cederberg, H.,** Über das Glycyrrhizin, *Arch. Pharm.,* 245, 97, 1907.

26. **Kitagawa, I., Skagami, M., Hashiushi, F., Zhou, J. L., Yoshikawa, M., and Jiali, M.,** Apioglycyrrhizin and araboglycyrrhizin, two new sweet oleanene-type triterpene oligosaccharides from the root of *Glycyrrhiza inflata, Chem. Pharm. Bull.,* 37, 551, 1989.

27. **Hashimoto, Y., Ishizone, H., Suganuma, M., Ogura, M., Nakatsu, K., and Yoshioka, H.,** Periandrin I, a sweet triterpene-glycoside from *Periandra dulcis, Phytochemistry,* 22, 259, 1983.

28. **Jizba, J., Dolejs, S., Herout, V., and Sorm, F.,** The structure of osladin. The sweet principle of the rhizomes of *Polypodium vulgare* L., *Tetrahedron Lett.,* 1329, 1981.

29. **Kim, J., Pezzuto, J. M., Soejarto, D. D., Lang, F., and Kinghorn, A. D.,** Polypodoside A, an intensely sweet constituent of the rhizomes of *Polypodium glycyrrhiza, J. Nat. Prod.,* 51, 1166, 1988.

30. **Asahina, Y. and Ueno, E.,** Phyllodulcin, a chemical constituent of Amacha (*Hydrangea thunbergii* Sieb.), *J. Pharm. Soc. Jpn.,* 408, 1916; *Chem. Abstr.,* 10, 1524, 1916.

31. **Rennie, E. H.,** Glycyphyllin, the sweet principle of *Smilax glycyphylla, J. Chem. Soc.,* 49, 857, 1986.

32. **Horowitz, R. M. and Gentili, B.,** Dihydrochalcone glycosides and their use as sweetening agents, U.S. Patent 3,087,821, 1963.

33. **Nanayakkara, N. P. D., Hussain, R. A., Pezzuto, J. M., Soejarto, D. D., and Kinghorn, A. D.,** An intensely sweet dihydroflavonol derivative based on a natural product lead compound, *J. Med. Chem.,* 31, 1250, 1988.

34. **Kasai, R., Hirono, S., Chou, W.-H., Tanaka, O., and Chen, F.-H.,** Sweet dihydroflavonol rhamnoside from the leaves of *Engelhardtia chrysolepis,* a Chinese folk medicine, Huang-qi, *Chem. Pharm. Bull.,* 36, 4167, 1988.

35. **Van der Wel, H. and Loeve, K.,** Isolation and characterization of thaumatin I and II, the sweet-tasting proteins from *Thaumatococcus daniellii* Benth., *Eur. J. Biochem.,* 31, 221, 1972.

36. **Morris, J. A., Martenson, R., Debler, G., and Cagan, R. H.,** Characterization of monellin, a protein that tastes sweet, *J. Biol. Chem.,* 248, 534, 1973.

37. **Zhong, H.,** The sweet proteins from the seeds of *Capparis masaikai* Levl., in *Proc. München-Shanghai Symp. Peptide and Protein Chemistry,* June 9–10, Schloss Ringberg am Tegernsee, Germany, 1986, 109.

38. **Van der Wel, H., Larson, G., Hladik, H., Hladik, C. M., Hellekant, G., and Glaser, D.,** Isolation and characterization of pentadin, the sweet principle of *Pentadiplandra brazzeana* Baillon, *Chem. Senses,* 14, 75, 1989.

39. **Yamashita, H., Theeraslip, S., Aiuchi, T., Nakaya, K., Nakamura, Y., and Kurihara, Y.,** Purification and complete amino acid sequence of a new type of sweet protein with taste-modifying activity, curculin, *J. Biol. Chem.,* 265, 15770, 1990.

40. **Hussain, R. A., Kim, J., Hu, T.-W., Pezzuto, J. M., Soejarto, D. D., and Kinghorn, A. D.,** Isolation of a highly sweet constituent from *Cinnamomum osmophloeum* leaves, *Planta Med.,* 52, 403, 1986.

41. **Masada, H., Ohtani, K., Mizutani, K., Ogawa, S., Kasai, K., and Tanaka, O.,** Chemical study of *Haematoxylon campechianum*: a sweet principle and new dibenz[b,d]oxocin derivatives, *Chem. Pharm. Bull.,* 39, 1382, 1991.

42. **Morimoto, S., Nonaka, G.-I., and Nishioka, I.,** Tannins and related compounds. XXXV. Proanthocyanidins with a doubly linked unit from the root bark of *Cinnamomum sieboldii* Meisner, *Chem. Pharm. Bull.,* 33, 4338, 1985.

43. **Soejarto, D. D., Kinghorn, A. D., and Farnsworth, N. R.,** Potential sweetening agents of plant origin. III. Organoleptic evaluation of *Stevia* leaf herbarium samples for sweetness, *J. Nat. Prod.,* 45, 590, 1982.

44. **Kinghorn, A. D., Soejarto, D. D., Nanayakkara, N. P. D., Compadre, C. M., Makapugay, H. C., Hovanec-Brown, J. M., Medon, P. J., and Kamath, S. M.,** A phytochemical screening procedure for sweet *ent*-kaurene glycosides in the genus *Stevia, J. Nat. Prod.,* 47, 439, 1984.

45. **Seto, T., Tanaka, T., Tanaka, O., and Naruhashi, N.,** β-Glucosyl esters of 19-hydroxyursolic acid derivatives in leaves of *Rubus* species, *Phytochemistry,* 23, 2829, 1984.

46. **Kinghorn, A. D. and Soejarto, D. D.,** New highly sweet compounds from natural sources, in *Sweeteners: Discovery, Molecular Design, and Chemoreception,* Walters, D. E., Orthoefer, F. T., and DuBois, G. E., Eds., ACS Symp. Ser. 450, American Chemical Society, Washington, D.C., 1991, 14.

47. **Mizutani, M.,** On the taste of some mosses, *Miscellan. Bryol. Lichenol.,* 2, 100, 1961.

48. **Compadre, C. M., Robbins, E. F., and Kinghorn, A. D.,** The intensely sweet plant *Lippia dulcis* Trev.: historical uses, field inquiries, and constituents, *J. Ethnopharmacol.,* 15, 89, 1986.

49. **Compadre, C. M., Hussain, R. A., Lopez de Compadre, R. L., Pezzuto, J. M., and Kinghorn, A. D.,** The intensely sweet sesquiterpene, hernandulcin: isolation, synthesis, characterization, and preliminary safety evaluation, *J. Agric. Food Chem.,* 35, 273, 1987.

50. **Inglett, G. E. and May, J. F.,** Tropical plants with unusual taste properties, *Econ. Bot.,* 22, 326, 1968.

51. **Choi, Y.-H., Hussain, R. A., Pezzuto, J. M., Kinghorn, A. D., and Morton, J. M.,** Abrusosides A–D, four novel sweet-tasting triterpene glycoside constituents from the leaves of *Abrus precatorius, J. Nat. Prod.,* 52, 1118, 1989.

52. **Hussain, R. A., Kinghorn, A. D., and Soejarto, D. D.,** Sweetening agents of plant origin: literature search for candidate sweet plants, *Econ. Bot.,* 42, 267, 1988.

53. **Hussain, R. A., Lin, Y.-M., Poveda, L. J., Bordas, E., Chung, B. S., Pezzuto, J. M., Soejarto, D. D., and Kinghorn, A. D.,** Plant derived sweetening agents: saccharide and polyol constituents of some sweet-tasting plants, *J. Ethnopharmacol.,* 28, 103, 1990.

54. **Hussain, R. A., Poveda, L. J., Pezzuto, J. M., Soejarto, D. D., and Kinghorn, A. D.,** Sweetening agents of plant origin: Phenylpropanoid constituents of seven sweet-tasting plants, *Econ. Bot.,* 44, 174, 1990.

55. **Lee, C.-H.,** Intense sweetener from Lo Han Kuo (*Momordica grosvenorii*), *Experientia,* 31, 533, 1975.

56. **Spencer, H. W.,** Taste panels and the measurement of sweetness, in *Sweetness and Sweeteners,* Birch, G. G., Green, L. F., and Coulson, C. B., Eds., Applied Science, London, 1971, 112.

57. **Bartoshuk, L. M.,** The psychophysics of taste, *Am. J. Clin. Nutr.,* 31, 1068 , 1978.

58. **Schiffman, S. S.,** Comparison of taste properties of aspartame with other sweeteners, in *Aspartame: Physiology and Biochemistry,* Stegink, L. D. and Filer, L. J., Jr., Eds., Marcel Dekker, New York, 1984, 207.

59. **Kasai, R., Matsumoro, K., Nie, R.-L., Zhou, J., and Tanaka, O.,** Glycosides from Chinese medicinal plant, *Hemsleya panicis-scandens,* and structure-taste relationship of cucurbitane glycosides, *Chem. Pharm. Bull.,* 36, 234, 1988.

60. **Medon, P. J., Pezzuto, J. M., Hovanec-Brown, J. M., Nanayakkara, N. P. D., Soejarto, D. D., Kamath, S. K., and Kinghorn, A. D.,** Safety assessment of some *Stevia rebaudiana* sweet principles, *Fed. Proc.,* 41, 1568, 1982.

61. **Pezzuto, J. M., Compadre, C. M., Swanson, S. M., Nanayakkara, N. P. D., and Kinghorn, A. D.,** Metabolically activated steviol, the aglycone of stevioside, is mutagenic, *Proc. Natl. Acad. Sci. U.S.A.,* 82, 2478, 1985.

62. **Jakinovich, W., Jr., Moon, C., Choi, Y.-C., and Kinghorn, A. D.,** Evaluation of plant extracts for sweetness using the Mongolian gerbil, *J. Nat. Prod.,* 53, 190, 1990.

63. **Jakinovich, W., Jr.,** Stimulation of the gerbil's gustatory receptors by artificial sweeteners, *Brain Res.,* 210, 69, 1981.

64. **Jakinovich, W., Jr.,** Taste aversion to sugars by the gerbil, *Physiol. Behav.,* 28, 1065, 1982.

65. **Compadre, C. M.,** Studies on the Sweet Principle of *Lippia dulcis* and on Steviol, the Aglycone of Stevioside, Ph.D. dissertation, University of Illinois at Chicago, 1985.

66. **Mori, K. and Kato, M.,** Synthesis of (6*S*,1'*S*)-hernandulcin, a sweetener and its stereoisomers, *Tetrahedron,* 42, 5895, 1986.
67. **Bax, A.,** Structure determination and spectral assignment by pulsed polarization transfer via long-range ^{1}H-^{13}C couplings, *J. Magn. Reson.,* 57, 314, 1984.
68. **Cordell, G. A. and Kinghorn, A. D.,** One-dimensional proton-carbon correlations for the structure determination of natural products, *Tetrahedron,* 47, 3521, 1991.
69. **Fullas, F.,** Structure Elucidation of Bioactive Constituents of *Baccharis gaudichaudiana*, Ph.D. dissertation, University of Illinois at Chicago, 1992.
70. **Kessler, H., Bermel, G., and Griesinger, C.,** Recognition of proton spin systems of cyclosporin A via heteronuclear proton-carbon long-range couplings, *J. Am. Chem. Soc.,* 107, 1083, 1985.
71. **Olsnes, S., Heiberg, R., and Pihl, A.,** Inactivation of eucaryotic ribosomes by the toxic proteins abrin and ricin, *Mol. Biol. Rep.,* 1, 15, 1973.
72. **Akinloye, B. A. and Adalumo, L. A.,** *Abrus precatorius* leaves - a source of glycyrrhizin, *Nigerian J. Pharm.,* 12, 405, 1981.
73. **Choi, Y.-H.,** Biologically Active Constituents of *Agrostistachys hookeri* and *Abrus precatorius*, Ph.D. dissertation, University of Illinois at Chicago, 1988.
74. **Hikino, H.,** Recent research on oriental medicinal plants, in *Economic and Medicinal Plant Research*, Vol. 1, Wagner, H., Hikino, H., and Farnsworth, N. R., Eds., Academic Press, London, 1985, 53.
75. **Fischer, L. and Goodrich, F. J.,** Licorice fern and wild licorice as substitutes for licorice, *J. Am. Pharm. Assoc.,* 22, 1225, 1933.
76. **Kim, J.,** Structure Elucidation of Sweet and Bitter Principles of *Polypodium glycyrrhiza*, Ph.D. dissertation, University of Illinois at Chicago, 1988.
77. **Yamada, H., Nishizawa, M., and Katayama, C.,** Osladin, a sweet principle of *Polypodium vulgare*. Structure revision, *Tetrahedron Lett.,* 33, 4009, 1992.

Chapter 10

SEPARATION, STRUCTURE ELUCIDATION, AND BIOASSAY OF CYTOTOXIC NATURAL PRODUCTS

Geoffrey A. Cordell, A. Douglas Kinghorn, and John M. Pezzuto

TABLE OF CONTENTS

0-8493-4372-0/93/$0.00+$.50
© 1993 by CRC Press, Inc.

I. INTRODUCTION

Since the earliest recorded uses of plants as medicinal agents, mankind has been investigating terrestrial plants for their potential to treat the many forms of cancer that are so debilitating and lethal. Thus, the written records of the Egyptian, Arabic, Greek, Roman, and many other civilizations, as well as the Medieval herbals and more recent Pharmacopoeias, are replete with references to the use of plants for the treatment of cancer. Hartwell, in a series of articles organized by plant family, has reviewed these early records in considerable detail.[1]

Current estimates indicate that in the U.S., one person in three will contract cancer, and one person in five will die of cancer. Since 1954 the U.S. Federal Government, originally through the Cancer Chemotherapy National Service Center and subsequently through the National Cancer Institute (NCI), has been deeply involved in the discovery, biological evaluation, and clinical testing of anticancer agents from both synthetic and natural sources.[2] In addition, a number of the major pharmaceutical companies have embarked on drug development efforts aimed at the marketing of new agents for the treatment of cancer. Although these efforts have led to major improvements in cancer therapy, cancer deaths have remained very high, as levels of incidence have continued to increase dramatically.

For many years the National Cancer Institute conducted an extramural program aimed at discovering *in vivo* active agents. In this program, which is described in the next section, some 35,000 plants were tested for their cytotoxic activity, typically in the human carcinoma of the nasopharynx (KB) and mouse lymphocytic leukemia (P388) test systems, and most of the plant extracts were also evaluated for their activity in the *in vivo* P388 test system. Although a number of key compounds came out of this program, none of these have yet impacted the market place. Thus, there remain only three plant-derived natural products that are currently available as prescription entities, i.e., the bisindole alkaloids derived from *Catharanthus*, vinblastine, vincristine, and a semisynthetic derivative of vinblastine, vindesine. However, several compounds derived from plants are currently classified as investigational new drugs at the NCI. These include homoharringtonine, ipomeanol, pancratistatin, phyllanthoside, and taxol.[3] In addition, a number of semisynthetic derivatives of natural products are under clinical evaluation, including taxotere, 10-hydroxycamptothecin, SKF 104864, and ellipticinium.

In the years after the NCI initially discontinued their extramural drug discovery program, substantial progress was made in understanding the biology of cancer and its proliferation. These efforts revealed new opportunities for the development of bioassays that could be used for the screening of natural product extracts and synthetic compounds. Recognizing these developments, the NCI has recently introduced two drug discovery programs: an intramural program being conducted at the Frederick Cancer Research Center, and a National Cooperative Natural Product Drug Discovery Grant (NCNPDDG) program. It is, as yet, too early to expect results from either of these programs. Consequently, in this chapter we will focus on describing aspects of the distribution of the known cytotoxic agents, and some of the recent biological developments and strategies that might be employed for the discovery of agents in the future, and finally, we review some aspects of the work that has been conducted in our laboratories on the isolation of anticancer agents from plants.

II. DISTRIBUTION OF CYTOTOXIC AGENTS IN PLANTS

For more than 15 years the occurrence and chemical classes of anticancer agents from plants have been subjected to regular review, as exemplified by contributions by Kupchan,[4] Wall and Wani,[5] Cordell,[6] Cassady and Douros,[7] Suffness and Douros,[8] Sneden,[9] Suffness and Cordell,[10] Cragg and Suffness,[11] Blaskó and Cordell,[12] and Potier.[13] Thus, it is apparent that plant constituents able to kill cancer cells, and hence described as being "cytotoxic", exhibit a very large range of structural types. For example, among alkaloids, many cytotoxic representatives have been reported, such as compounds of the aconitine, acronycine, Amaryllidaceae, benzo[*c*]phenanthridine, bisindole, camptothecine, *Cephalotaxus*, colchicine, ellipticine, emetine, maytansinoid, phenanthroindolizidine, phenanthroquinolizidine, pyrrolizidine, and *Taxus* types. Other classes of plant secondary metabolites from which cytotoxic representatives have been described include acetogenins, cardenolides, coumarins, coumarinolignans, diterpenes, flavonoids, iridoids, lignans, monoterpenes, quinones, sesquiterpenes and sesquiterpene lactones, steroids, and triterpenoids (inclusive of the cucurbitacins and simaroubolides). Consequently, hundreds, if not thousands, of plant constituents have been found to be cytotoxic against one or more tumor cell types in culture.[4-13] However, it has been cautioned that not all of these classes of cytotoxic compounds are likely to possess useful antitumor activity when tested in the various *in vivo* models available, which is particularly true for substances in the cardenolide, cucurbitacin, and sesquiterpene lactone classes.[8] Even fewer compounds hold the potential of being used clinically as cancer chemotherapeutic agents.

The largest anticancer drug discovery program from plants, to date, is undoubtedly that of the NCI, Bethesda, MD. In the first phase of the NCI program, between 1955 and 1982, approximately 114,000 plant extracts from about 35,000 species were collected randomly and screened primarily against the L1210 and P388 mouse leukemia tumor systems, with nearly 5000 or 4.3% of these extracts being determined as significantly active.[8,14] Barclay and Perdue have analyzed in some detail the systematic distribution of higher plants containing antineoplastic constituents, and these include the families Cephalotaxaceae, Podocarpaceae, and Taxaceae in the gymnosperms, and families in Cronquist's orders Magnoliales (e.g., the Annonaceae), Ranunculales (e.g., the Menispermaceae), Myrtales (e.g., the Thymelaeaceae), Celastrales (e.g., the Celestraceae), Euphorbiales (e.g., the Euphorbiaceae), Sapindales (e.g., the Rutaceae and the Simaroubaceae), Gentianales (e.g., the Apocynaceae), and Liliales (e.g., the Liliaceae) among the angiosperms.[15] However, many of these taxa contain predictable chemical classes of active compounds, and by targeting them specifically, truly novel cytotoxic and antineoplastic agents may not be found. In the new plant acquisition program of the NCI, begun in 1986, some 4500 endemic plants are being collected each year from tropical rain forest regions in Africa, Southeast Asia, and Central and South America. It is noteworthy that these collections are designed to represent a broad taxonomic diversity, with ethnopharmacological information being utilized for each plant collected, if it can be reliably obtained.[14]

III. BIOLOGIC ACTIVITY-DRIVEN PROCUREMENT OF PLANT PRINCIPLES

A. SELECTION OF CANDIDATE PLANTS

Estimates indicate that there are approximately 250,000 terrestrial species of higher plants. If one imagines that a plant could yield an average of three of four chemically different plant parts (roots, leaves, bark, twig, flowers, fruits, seeds, latex, etc.), a comprehensive screening program would require the testing of a million or more samples per assay. Economically, such a plan is impractical and wasteful of precious resources when only a single biological area is being pursued. A more pragmatic approach is therefore needed that balances cost, rationale, capability, and desired biologic response. In the selection of plants for the discovery of new active agents, irrespective of the biologic area, there are five recognized strategies: the random, the taxonomic, the chemotaxonomic, the information-managed, and the ethnomedical. In the *random approach*, all available species are collected, irrespective of prior knowledge and experience, as discussed above. In the *taxonomic approach*, plants of predetermined taxa, considered to be of high interest, are sought from diverse locations. For example, given the interest in taxol derived from *Taxus* species, it would be reasonable to propose an evaluation of related *Taxus* species for higher levels of taxol and for analogs that might be available in higher yield and easily converted into taxol.[16] In the *chemotaxonomic approach*, a particular compound class, e.g., Amaryllidaceae alkaloids, may be considered as being of biologic interest, and plants thought likely to contain related compounds are collected. In the *information-managed approach*, which is almost the obverse of the chemotaxonomic approach, those plants of proven biologic activity that are unlikely to contain known active agents are collected in the anticipation of discovering unique (for patent purposes) chemical entities. In the *ethnomedical approach*, the extant oral or written information on the medicinal use of a plant is given credence, and based on the critical and weighted evaluation of this information, the plant is collected.

For each of these collection strategies, the most rational procedure is to test the material in a range of bioassays, as described below. Active leads are then arranged in order of priority, and those regarded as most promising are subjected to bioassay-directed fractionation procedures for the isolation of the active principle(s). There are both advantages and disadvantages to each of these collection strategies, depending on the particular circumstances. A sixth "approach", a nonsystematic one, is *serendipity*, where collection of the plant based on an ethnomedical use or on a particular bioactivity affords, through a combination of serendipity and acute scientific observation, the discovery of another bioactivity of commercial significance. The bisindole alkaloids of *Catharanthus* are perhaps the leading example in the whole of natural product drug discovery, where the ethnomedical use as an antidiabetic agent led to the discovery of its antileukemic properties through opportunistic observation. For the purposes of new drug discovery from plants, we believe that a combination of the ethnomedical and the information-management approaches will be found to be the most productive.

The ethnomedical approach requires both the collection and prioritization of existing ethnomedical, chemical, biologic, and clinical data so that plant collection can initially address those plants that are viewed as being most likely to yield interesting new compounds for potential development. While this may seem an unwieldy task, we can now approach this analysis through the use of the database NAPRALERT at the University of Illinois. For example, as a part of our NCNPDDG program, we have recently looked at the prioritization of plants for collection and bioassay for cancer screening. The NAPRALERT database could construct three separate files: a file of plants that have been reported ethnomedically to possess anticancer agents, a file of plant extracts that have been shown experimentally to display either *in vivo* or *in vitro* anticancer activity, and a file of compounds that possess anticancer activity. From the latter file a data set of plants that contain those compounds can be generated. Comparison of the three plant files for synonymy yields a list of plants for which there is ethnomedical or biologic activity reported, but for which there are no known active compounds. In this way an initial list of potential plants can be scored for ethnomedical, *in vitro*, *in vivo*, and clinical data to identify those plants to be collected in the initial phases of a program.[17]

Collection of species, particularly if they have been identified as being of potential interest, is a relatively unappreciated aspect of natural product drug development. The key is to have a taxonomic expert who has access to a group of experienced local collectors who are capable not only of collection, but also, with their local knowledge, of providing authoritative identification and information on the use or ecologic implications for the indigenous flora. Accurate taxonomic identification of herbarium specimens is an essential aspect of the program in the event that additional sample material is required for large-scale isolation of the active principles for more extensive testing.

B. MECHANISM-BASED BIOASSAYS

Once plants are identified, collected, and processed, it is necessary to demonstrate the ability of an extract to mediate a biological response that is consistent with the therapeutic effect of interest. For the discovery of certain categories of drugs (e.g., antimicrobial agents), clear differences between host cells and target organisms have been defined at a molecular level. Therefore, bioassays that give a strong indication of selectivity can readily be performed. Unfortunately, most biochemical characteristics of cancer cells cannot be distinguished from those of normal cells, and cancer chemotherapeutic agents that function by means of modulating discrete biochemical events that are associated with both normal and cancer cells invariably demonstrate toxicity. However, chemotherapeutic agents currently in use do mediate significant clinical responses, even though therapeutic indices may be low. The actual mechanisms leading to therapeutic efficacy are undoubtedly complex and include factors such as host metabolism, drug disposition, relative cell susceptibility, relative ability of cells or organ systems to recover from injury, etc. As a result of the complexity of this situation, there are currently no *in vitro* bioassays that can be used with confidence for the detection of a novel substance that would be predicted from the outset to serve as a clinically useful antitumor agent.

Nonetheless, mechanism-based *in vitro* assays can be designed by analogy with the types of molecular responses mediated by known (clinically effective) antitumor agents. For example, monitoring effects similar to those known to be mediated by *Vinca* alkaloids (tubulin depolymerization), taxol (tubulin stabilization), camptothecin (topoisomerase I inhibition), 2-methyl-9-hydroxyellipticinium (topoisomerase II inhibition), and bleomycin (DNA cleavage) are all reasonable avenues toward novel drug discovery. Agents that are found on the basis of these mechanism-based *in vitro* assay systems may be considered as having the potential of demonstrating biological activity that is sometimes necessary, but possibly (or probably) not sufficient to foster a clinical antitumor response. To establish the true efficacy of promising anticancer agents, subsequent evaluation in more advanced testing systems is required, followed by (pre)clinical trials. Minimally, however, such agents would be useful as molecular probes for the study of biologic processes through which modulation may lead to cancer control. Thus, utilization of these mechanism-based *in vitro* bioassay systems should aid in the procurement of unique natural products.

C. ASSAYS FOR CYTOTOXICITY

In contrast to well-defined *in vitro* test procedures, assays dealing with cell cultures have one major advantage: all potential mechanisms concerning cellular proliferation are simultaneously monitored. Obviously, it is recognized that cytotoxicity is neither necessary nor sufficient for antitumor activity. Therefore, it should be of no surprise that isolates derived from cytotoxicity-based assay procedures are often toxic toward mammalian species and do not demonstrate therapeutic efficacy. However, cytotoxicity is an activity that is consistent with antitumor activity, and virtually every known naturally occurring antitumor agent demonstrates a positive response in a cell culture system. Thus, it is likely that a previously uncharacterized plant extract that contains a novel antitumor agent would also demonstrate a positive response.

In the past, many research programs dealing with the isolation and identification of potential antitumor compounds from plants have relied on cytotoxicity for bioassay-directed fractionation. Cultured P388 or KB cells have been routinely employed. Briefly, the procedure involves treating the cells with various concentrations of the test substance and assessing cell growth after 48 or 72 h of incubation with P388 and KB cells, respectively. The results are expressed as ED_{50} values (concentration required to inhibit cell growth by 50%). As established by the NCI, ED_{50} values of ≤ 20 µg/ml for extracts and ≤ 4 µg/ml for pure compounds are considered active.[18] Although these test procedures have proven to be effective for the isolation of cytotoxic compounds that may be of novel structure, it is often the case that compounds identified as active with KB cells are not active with *in vivo* tumor models. In many instances the isolates are simply toxic.

The NCI has recently initiated a program wherein the cytotoxic potential of compounds or extracts are evaluated with a battery of human cancer cell types in culture. The objective is the discovery of agents that demonstrate selective cytotoxic activity with a cell line derived from a single type of primary human tumor.[19,20] The use of a battery of human cell lines derived from a variety of human tumor types offers great theoretical and practical advantage. The cells can be carried as solid tumors in athymic mice. Thus, more advanced testing of active principles is facilitated. Also, it is speculated that certain test materials will demonstrate cell-type specificity that will lead to isolation and identification of selectively cytotoxic agents. Based on structure and activity, such a selectively cytotoxic agent could certainly be a candidate for more advanced testing.

It is conceivable that an agent capable of demonstrating such selectivity (or a derivative thereof) could be useful in the treatment of a specific type of cancer. Irrespective of the clinical outcome, however, the concept of selective cytotoxicity implicitly suggests the presence of a cell-specific receptor that differentiates one tumor type from another. Such a discovery would be of monumental importance in terms of developing tumor-specific therapeutic strategies, and therefore a cytotoxic agent specific for one cell type would greatly aid in identifying the appropriate subcellular target. Thus, agents that demonstrate the desired selectivity will serve as important tools for probing tumor-specific receptor sites, which is an extremely rational approach toward defining tumor-specific targets for chemotherapeutic agents.

Based on these considerations, there is substantial merit associated with an approach that relies on selective cytotoxicity. Paradoxically, however, all of the well-known natural product antitumor agents that are currently utilized in clinical settings do not demonstrate selective cytotoxicity. Rather, it is typical to observe a general cytotoxic response with absolutely no cell-type specificity. Since it is not practical to assign a high priority to a plant based solely on nonspecific cytotoxic activity, it is necessary to devise a broader-based bioassay-directed drug discovery program.

At the present time we are engaged in an NCNPDDG project involving the partnership of Research Triangle Institute and Glaxo Group Research. At the University of Illinois at Chicago, cytotoxicity assays are conducted that are similar to those conducted at the NCI, although on a greatly reduced scale.[21] Additional assays are performed to evaluate antimitotic potential with cultured ASK cells and to assess the ability to reverse the multidrug-resistant phenotype with KB-VI cells. At the other research facilities involved in the NCNPDDG, extracts derived from the same plants are evaluated in mechanism-based assays. The accumulated data are then evaluated in order to decide which materials should be subjected to fractionation procedures. In an ideal case a plant extract would demonstrate activity in a mechanism-based assay and a limited number of cell lines. This type of activity profile leads to a high priority for fractionation. Alternatively, either type of activity (only mechanism-based or limited cytotoxicity) yields a high-priority rating.

General cytotoxicity alerts us to the fact that active principles are present in the plant extract, which may or may not be of interest. In general, such an extract would be placed on hold due to lack of correlative activity, but these may be reevaluated as additional knowledge is accumulated and mechanism-based assays are modified. Thus, cytotoxicity-based assays are useful for grouping plant materials into two broad categories, i.e., active or inactive. At some point in time, additional information may be accumulated to reactivate such leads. This is considered a more judicious procedure than ranking potentially useful materials as "false negatives" due to the unavoidable situation of utilizing a highly discriminatory screen.

D. BIOACTIVITY-DIRECTED ISOLATION

Following the determination that an extract is indeed active, and a strategic decision has been made to pursue the fractionation of the extract for the active principle(s), attention must be focused on the most effective methods for the continued tracking of the biologic activity so that bioactivity-directed fractionation can be used to monitor the isolation process. The monitoring of the activity is usually accomplished with the same biologic system used for the determination of activity in the original extract. However, there are exceptions, one of which occurs when multiple tests are run on the original extract and the material shows activity in more than one system. In such situations, a single assay, typically the one in which the extract shows the highest activity, is used for the fractionation. The isolates are then tested in the broad spectrum of tests that were used for the original screening of the extract. If all of the activity is not accounted for in this procedure, additional fractionation work is necessary. It is now well established that a given plant may contain two or even three classes of compounds that demonstrate cytotoxic activity. Attention should also be paid to the presence of tannins and other polyphenolic materials when enzyme-based bioassays are being used, because false-positive results are often observed. Consequently, it is usually necessary to remove these classes of compounds, for example, with the use of polyamide columns, prior to bioactivity-directed fractionation.

There are numerous techniques now available for the separation of complex mixtures of natural products, and the choice of technique is frequently guided by the anticipated nature of the active principle, based on chemotaxonomy, and the prior experience of the investigator in working with certain classes of compounds. At each fractionation step, though, both the parent fraction and the fractions derived therefrom should be subjected to bioassay. Only the active fractions are then studied in the next phase of the separation process.

IV. ISOLATION AND STRUCTURE ELUCIDATION PROCEDURES

A. FLAVONOIDS FROM *MUNTINGIA CALABURA*

As part of a collaborative research project with colleagues at Mahidol University, Bangkok, Thailand, a variety of tropical rain forest plants were screened for their cytotoxic activity against a panel of human and murine cancer cells available at the Bioassay Research Facility, College of Pharmacy, University of Illinois at Chicago. Many of the plant extracts showed activities of interest, and so these were collected in sufficient quantity to commence phytochemical isolation work. A number of publications have resulted thus far from this research collaboration.[22-26]

Muntingia calabura L. (Elaeocarpaceae) represented one of the initial plants from Thailand to exhibit preliminary cytotoxic activity. This is a tree species, indigenous to tropical America, that has been introduced into southeast Asia. The roots of *M. calabura* have had medicinal uses as an emmenagogue and as a remedy for persons with liver disease, and the flowers have been used to treat headaches and colds. However, the various parts of this plant do not seem to have been utilized previously for the treatment of cancer.[23] Prior to the start of our work, only a sparse amount of phytochemical information had been published on this species. Recollected samples of *M. calabura* obtained in Thailand and in the Philippines, however, were found to exhibit different phytochemical profiles in our laboratory. From the leaves and stems of this species obtained in Thailand, a number of known flavonoids were isolated, with the most potent cytotoxic agent being 2′,4′-dihydroxychalcone, which exhibited activities at 1.6 and 1.7 µg/ml against KB cells and a human breast cancer cell line, respectively.[27] In contrast, in an investigation on the roots of *M. calabura* collected in the Philippines, twelve novel 7,8-di-*O*-substituted flavonoids were isolated, comprised of seven flavans (Figure 1, **1–7**), three flavones (Figure 1, **8,10,12**), and two diflavans (Figure 1, **9,11**). This type of oxygenated substitution pattern does not appear to have been encountered previously among the flavonoids.

The murine lymphocytic leukemia (P388) cell line was used for activity-guided fractionation of extracts from the roots of *M. calabura* collected in the Philippines. Thus, the roots

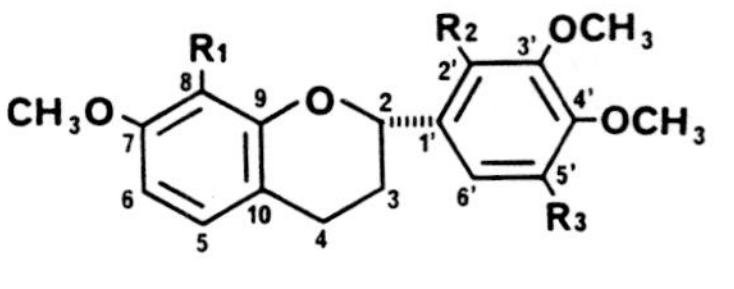

	R_1	R_2	R_3
1	H	H	OH
2	OCH_3	H	OCH_3
3	OCH_3	OH	OCH_3
4	OCH_3	H	OH
5	OH	H	OCH_3
6	OH	OH	OCH_3
7	OH	H	OH

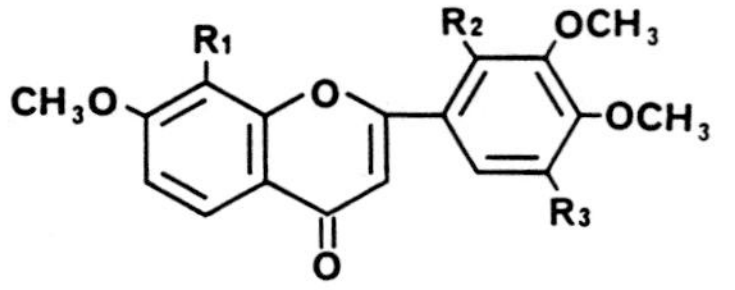

	R_1	R_2	R_3
8	OCH_3	H	OCH_3
10	OCH_3	H	OH
12	OH	H	OH

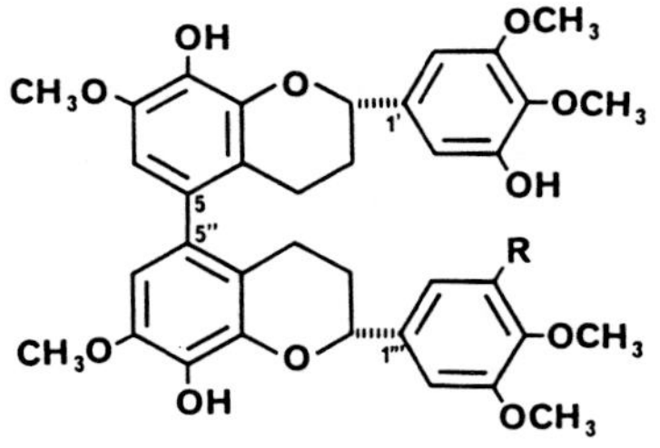

9 R = OCH_3
11 R = OH

FIGURE 1. Flavonoid derivatives **1–12** of *Muntingia calabura.*

were exhaustively extracted into methanol, and following evaporation of the methanol, the residue was partitioned between diethyl ether and water. Cytotoxic activity was found only in the organic solvent residue. The diethyl ether fraction was purified by repeated chromatography over silica gel, using chloroform-methanol and toluene-acetone mixtures as eluting solvents. As a result of this procedure, the flavans were eluted with less-polar solvents than were the biflavans and flavones. In the following paragraphs the steps involved with the structure elucidation of one example of each of these three flavonoid classes will be briefly described.

Compound **6**, 2(*S*)-8,2′-dihydroxy-7,3′,4′,5′-tetramethoxyflavan, was found to exhibit a molecular formula of $C_{19}H_{22}O_7$, from a consideration of its high-resolution mass spectral data. The presence of an unconjugated benzenoid chromophore was inferred from the observed UV and IR spectral maxima at 226 and 286 nm, and at 1500 cm^{-1}, respectively. Of the 19 carbon signals that appeared in the ^{13}C-NMR spectrum, three resonances occurring at δ 24.2 (t, C-4), 28.3 (t, C-3), and 73.0 (d, C-2) constituted the three-carbon unit of a flavan, and four methoxyl groups were also apparent (δ 56.1, 56.3, 60.6, and 61.0). In the ^{1}H-NMR spectrum of **6**, typical *ortho* coupling was apparent in ring A, with signals at δ 6.48 (d, J = 8.5 Hz, H-6) and 6.57 (d, J = 8.7 Hz, H-5) being observed. Ring B was totally substituted, with the exception of C-6′, and the proton attached to this carbon appeared in the ^{1}H-NMR spectrum as a singlet at δ 6.75. The spectrum of **6** exhibited two hydroxyl protons at δ 5.70 (s) and δ 6.00 (s). In a selective INEPT (Insensitive Nuclei Enhancement by Polarization Transfer) experiment, an NMR technique that indicates vicinal ^{1}H-^{13}C coupling,[28] the positions of these hydroxyl protons were confirmed at C-8 and C-2′, respectively (Figure 2). Irradiation ($^3J_{CH}$ = 5 Hz) of the hydroxy proton at δ 5.70 led to enhancement of carbons at δ 133.8 (C-8), 142.6 (C-9), and 145.3 (C-7) (Figure 2B). Analogous irradiation at δ 6.00

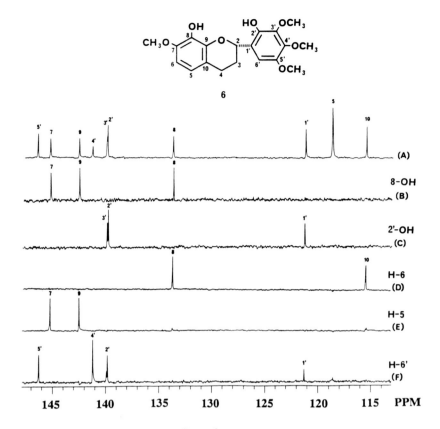

FIGURE 2. Selective INEPT experiments on flavan **6**.

selectively enhanced carbons resonating at δ 121.3 (C-1'), 139.9 (C-2'), and 140.0 (C-3') (Figure 2C). Additional selective INEPT irradiations of the unsubstituted ring A and ring B protons H-6, H-5, and H-6' (Figure 2 D–F) served to confirm the substitution pattern of the molecule. The absolute configuration of compound **6** was determined as 2S by the observation of a negative circular dichroism (CD) absorption maximum at 285 nm, which is consistent with model flavans.[23]

The flavone, **8** (Figure 1; 7,8,3',4',5'-pentamethoxyflavone), $C_{20}H_{20}O_7$, possessed an aromatic ring conjugated to a carbonyl group (UV, 241, 316 nm; IR 1640 [C = O], 1600, 1500 cm⁻¹). In the ¹H-NMR spectrum there was evidence of five methoxy proton groups (δ 3.94 – 4.05), two singlet aromatic proton systems (H-3, δ 6.72; H-2', H-6', δ 7.20), and one *ortho*-coupled aromatic proton system (H-6, δ 7.06 d, J = 8.8 Hz; H-5, δ 7.94 d, J = 8.8 Hz). The five methoxyl groups were assigned to 7,8-dimethoxy and 3',4',5'-trimethoxy ring-A and ring-B units, respectively, by analogy to literature data on known compounds containing these moieties, and by direct comparison with the identically substituted flavan, **2** (Figure 1) found to co-occur in *M. calabura* roots in this investigation.[23]

The biflavan, **11** (Figure 1), exhibited UV and IR spectral data similar to those of the seven flavans (**1–7**) isolated from *Muntingia calabura*, and the molecular ion was observed at *m/z* 662.2356 ($C_{36}H_{38}O_{12}$). The significant fragment ion at *m/z* 167, along with ¹H- and ¹³C-NMR interpretations, suggested that it was a biflavan composed of two units of compound **7** (Figure 1). As a result of a ¹H-¹³C heteronuclear shift correlation (HETCOR) NMR experiment, compound **11** was found to contain a biphenyl unit linked at C-5 and C-5″. Stereoisomerism in this molecule was therefore apparent, which may be represented as *M* and *P* in terms of helicity, according to a previous convention. The absolute configuration of **11** was determined

TABLE 1
Cytotoxicity Data of *Muntingia calabura* Flavonoids[a]

Compound	BC1	HT	Lu1	Me12	Col2	KB	KB-V	P-388
				Cell lines[b]				
1	>20	>20	>20	14.6	>20	9.4	13.3	5.9
2	>20	>20	>20	8.9	15.8	13.3	2.1	5.4
3	10.9	3.3	13.5	9.7	12.0	3.4	6.2	4.9
4	>20	>20	>20	9.0	>20	15.5	12.3	2.0
5	>20	>20	>20	9.2	>20	11.8	3.9	3.0
6	>20	>20	>20	14.5	>20	10.2	>20	2.3
7	>20	>20	>20	10.6	>20	13.8	11.1	4.1
8	—	>20	—	—	—	>20	>20	—
9	12.0	5.5	12.4	10.2	6.2	2.2	8.3	3.7
10	>20	>20	>20	>20	15.2	>20	>20	11.9
11	16.0	5.0	15.6	8.7	9.0	5.2	12.6	4.8
12	>20	>20	>20	>20	5.9	>20	>20	16.7

[a] Results are expressed as ED_{50} values (μg/ml).
[b] BC1: human breast cancer; HT (HT-1080): human fibrosarcoma; Lu1: human lung cancer; Me12: human melanoma; Co12: human colon cancer; KB: human nasopharyngeal carcinoma; KB-V: vincristine-resistant KB; P-388: murine lymphocytic leukemia.

as 2*S* and 2″*S* by the observation of a strong negative Cotton effect at 288 nm in its CD spectrum. In this manner, compound **11** was determined as a mixture of the two diastereomers [i.e., (*M*),(2*S*),(2″*S*)- and (*P*),(2*S*),(2″*S*)-] of 8,8″,5′,5‴-tetrahydroxy-7,7″,3′,3‴,4′,4‴-hexamethoxy-5,5″-biflavan.[23]

The biological test data obtained for compounds **1–12** are shown in Table 1, and it may be seen that all substances other than compound **8** demonstrated cytotoxicity against cultured P388 cells. Furthermore, the flavans **1–7** were, in general, more potently cytotoxic than the flavones, **10** and **12**, toward this murine lymphocytic leukemia cell line. There was some evidence of selective cytotoxicity by the flavans, **1–7**, toward the melanoma (Mel2) and KB cell lines. The biflavans, **9** and **11**, showed a general cytotoxicity against most of the cell lines available to the panel (Table 1). While it has been put forth that flavonoids, as a compound class, are of limited interest as potential antineoplastic agents,[8] interest in these groups of substances has been increased by the demonstrated antitumor activity of the synthetic derivative, flavone acetic acid.[29] It has been found that flavones, isoflavones, flavanones, and flavonols, which are highly methoxylated or oxygenated, tend to be more highly cytotoxic than their less-oxygenated analogs.[23,30]

B. CONSTITUENTS OF *STIZOPHYLLUM RIPARIUM*

Discussion of the cytotoxic principles of *Stizophyllum riparium* (H.B.K.) Sandw. (Bignoniaceae) is included in this chapter for two reasons. First, two separate structural classes of cytotoxic agents were isolated from this plant lead: namely, two triterpene esters based on the ursane skeleton (Figure 3, **13**, **14**), and a pregnane derivative that was accorded the trivial name stizophyllin (Figure 4, **16**).[31] Second, this latter compound exhibited somewhat unusual structural features, along with high cytotoxic potency, and so we have been able to examine its interaction with P388 cells in some depth.[32]

The plant material was initially sent to us under a contract award with the NCI. It was collected in Peru and identified by staff members of the U.S. Department of Agriculture. The genus *Stizophyllum* consists of only three species and is indigenous to tropical America, ranging from Mexico to Brazil. Prior to the commencement of our study, neither phy-

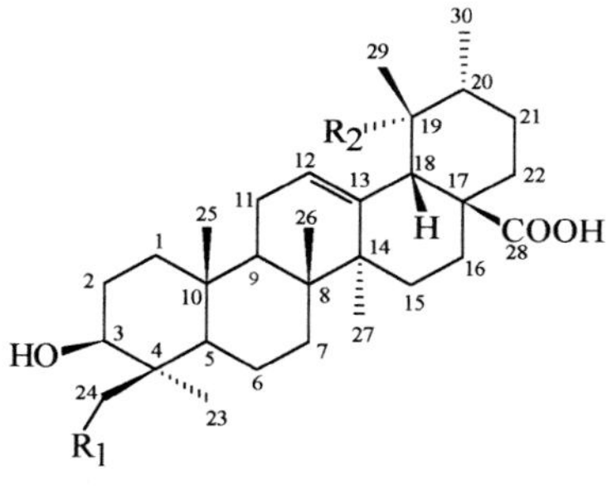

13 R_1 = 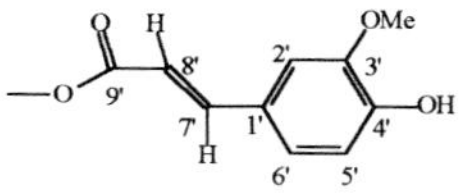 R_2 = H

14 R_1 = 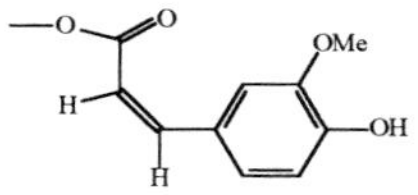 R_2 = H

15 R_1 = 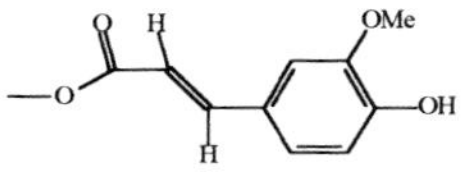 R_2 = OH

FIGURE 3. Ursane derivatives **13–15** of *Stizophyllum riparium.*

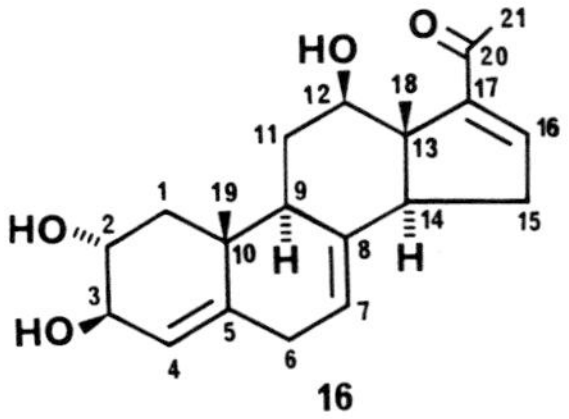

16

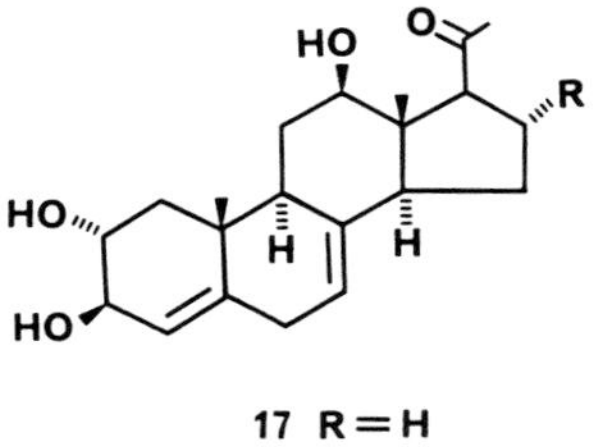

17 R=H
18 R=OCH_3

FIGURE 4. Pregnane derivatives **16–18** of *Stizophyllum riparium.*

tochemical nor biologic work had been performed on *S. riparium*.[33] The cytotoxic activity was associated with a chloroform soluble extract of the entire plant and was monitored throughout the phytochemical investigation, using the P388 cell line. Silica gel column chromatography of the chloroform soluble extract, eluted with chloroform-methanol mixtures, led to the isolation of six new compounds: namely, three ursane-type triterpenes (Figure 3, **13–15**) and three pregnane derivatives (Figure 4, **16–18**). Both the less-polar triterpenes and the more-polar pregnanes were finally purified using columns of Florisil and Sephadex L-20 and by preparative thin-layer chromatography.[31] Compound **13**, the most abundant of the three triterpenes obtained in this study, was found, by high-resolution mass spectrometry, to exhibit a molecular formula of $C_{40}H_{56}O_7$. Characteristic IR absorptions for the presence of a hydroxyl group, a carboxylic acid unit, an α,β-unsaturated ester, and an aromatic moiety were observed. The presence of significant fragment peaks at m/z 248 and 207 in the electron impact mass spectrometry (EIMS) indicated a *retro*-Diels-Alder cleavage of ring C of a triterpene unit. The skeleton of the triterpenoid part of the molecule was established by NMR data comparison with published values for Δ^{12}-ursene derivatives. In addition, the presence of a doublet at δ 2.18 ($J = 11$ Hz) for the 18β-allylic proton indicated that, apart from the presence of a C-28 carboxyl group, rings D and E were otherwise unsubstituted. It was also evident from the ^{1}H- and ^{13}C-NMR spectra of **13** that the other substituents were a secondary hydroxy group and a *trans*-ferulic acid (3-methoxy-4-hydroxycinnamic acid) unit esterified at a hydroxymethyl functionality. By comparison with model compounds, the doublet of doublets centered at δ 3.32 in the ^{1}H-NMR spectrum of compound **13** was assigned to a 3α-affixed proton, so that compound **13** was postulated as being a 3β,24-disubstituted Δ^{12}-ursene derivative. On hydrolysis with 1% KOH in methanol and subsequent methylation and acetylation, compound **13** afforded the known compound, methyl-3β,24-diacetoxyurs-12-en-28-oate. Also obtained after this hydrolysis and methylation was *trans*-ferulic acid methyl ester, and so the structural assignment for compound **13** was 3β-hydroxy-24-*trans*-feruloyloxyurs-12-en-28-oic acid.[31]

Compound **14** was an isomer of compound **13** and was obtained in much lower yield in our investigation. Analysis of its ^{1}H-NMR spectrum showed evidence for the presence of *cis*-conjugated olefinic protons (δ 6.80 and 5.78, $J = 13$ Hz, H-7$'$ and H-8$'$, respectively). Hence, this substance was structurally determined as 3β-hydroxy-24-*cis*-ferulyloxyurs-12-en-28-oic acid.[31]

The ursane-type triterpenes, **13** and **14**, were found to exhibit ED_{50} values of 1.2 and 1.0 μg/ml, respectively, when evaluated against the P388 murine lymphocytic cell line. It is of some interest to note that a further novel analog of these substances obtained from *S. riparium*, 3β,19α-dihydroxy-24-*trans*-ferulyloxyurs-12-en-28-oic acid (**15**), was deemed inactive when tested against P388 cells (activity ≤ 4 μg/ml). Compounds **13–15** were not evaluated against a panel of human cancer cell lines, due to the unavailability of the latter at the time this investigation was performed.[31]

The *S. riparium* constituents, compounds **16–18**, are the first pregnane derivatives to have been isolated from a plant in the family Bignoniaceae. The cytotoxic compound **16**, to which we have accorded the trivial name stizophyllin, exhibited a molecular formula of $C_{21}H_{28}O_4$ and possessed one or more hydroxyl groups (3390 cm^{-1}) and an α,β-unsaturated carbonyl (1643 cm^{-1}). The basic unit of stizophyllin was inferred as being 12β-hydroxypregna-16-en-20-one, by observation of ^{1}H-NMR signals at δ 0.77 (CH$_3$-18), 1.06 (CH$_3$-19), 2.37 (CH$_3$-21), 3.79 (Hα-12), and 7.00 (16-H), and by comparison with analogous data for known pregnane derivatives. From ^{13}C-NMR observations, it was apparent that three double bonds were present in the molecule of compound **16**, but there were only two additional olefinic protons observed in its ^{1}H-NMR spectrum, at δ 5.24 (H-4) and 5.35 (H-7). From the lack of a UV maximum above 241 nm, the absence of a conjugated diene system in **16** was apparent. Cross peaks observed in the ^{1}H-^{1}H homonuclear shift correlation (COSY) NMR spectrum between H-3α and H-4, and H-6β and H-7, suggested that the two extra double

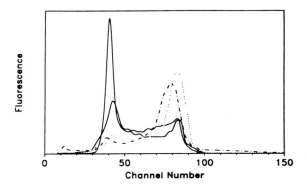

FIGURE 5. DNA fluorescence distribution of propidium iodide-stained P388 cells grown continuously in the presence of stizophyllin (4 µg/ml). Cell samples were taken from the culture at 4 (—, highest G_1/G_0), 8 (—, reduced G_1/G_0), 24 (-.-), and 48 h (···). (For additional details, see Duh, C.-Y., Kinghorn, A. D., and Pezzuto, J. M., *Chem. Biol. Interact.*, 80, 43, 1991.)

bonds of compound **16** were between carbons 4 and 5 and carbons 7 and 8.[31]

Several lines of evidence, including considerations of the coupling constants between H-1α, H-1β, and H-2β of **16**, and also between H-2β and H-3α, permitted the assignment of the stereochemistry of the two hydroxy groups in stizophyllin as 2α-equatorial and 3β-equatorial, respectively, In addition, on acetylation of the hydroxy group at C-2, selective deshielding of the 1α-proton was observed (δ 1.48 vs. 1.32 in **16**). The stereochemistry between the ring junctions was confirmed by a two-dimensional NOE experiment in which correlations were noted between H-9 and H-12α, H-9 and H-14, and between H-11β and both CH$_3$-18 and CH$_3$-19. Stizophyllin (Figure 4, **16**) was therefore assigned the structure 2α,3β,12β-trihydroxypregna-4,7,16-trien-20-one. The compound exhibited significant cytotoxicity against murine lymphocytic leukemia cells (ED$_{50}$, 0.07 µg/ml).[31] In contrast, two structurally related *S. riparium* constituents that did not bear an α,β-unsaturated carbonyl moiety, compounds **17** and **18** (Figure 4), were found to be devoid of cytotoxic activity for P388 cells.[31-33]

Stizophyllin (**16**) formed adducts with nucleophilic substances such as L-cysteine and β-mercaptoethanol. Therefore, the α,β-unsaturated carbonyl group in ring D was considered the site of the stizophyllin molecule responsible for its cytotoxic activity. The compound was found to interact with DNA, but it was suggested that this is not a covalent interaction due to the lack of mutagenicity of compound **16**. It was also found that DNA synthesis continued with cultured P388 cells in the presence of stizophyllin until a block in the cell cycle occurred in the G_2 + M phase (Figure 5), and the block appeared to be specific for the G2 phase. Therefore, it may be speculated that stizophyllin (**16**) interferes with the function of some factor akin to a sulfhydryl-containing protein that prevents P388 cells from progressing through mitosis.[32]

C. ISOLATES FROM THAI MEDICINAL PLANTS

As mentioned earlier in this chapter, over the past 10 years we have been significantly involved with research groups at Chulalongkorn University and Mahidol University, in Bangkok, Thailand. This work has resulted in a number of publications[34-52] that reflect on both the diversity of the flora and our interest in the structure elucidation of new biologically active compounds. From this program, a number of novel, interesting compounds have arisen, and while some aspects of this collaboration have been described previously,[53] it is appropriate here to summarize some of the recent developments as they relate to the discovery of plant anticancer agents and their structure elucidation.

The first compound from Thailand that we examined from a biological perspective was microminutin (**19**), the structure of which, determined through selective irradiation NMR experiments, represented the first member of a new class of substituted coumarins. This

19

20

21

compound showed weak cytotoxic activity in the P388 test system.[34] Several other compounds from Thai plants were also found to possess cytotoxic activity, including eupatorenone,[41] nimbolide and 28-deoxynimbolide,[43] 22-hydroxytingenone,[44] the iridoids of *Plumeria rubra*,[46] rotenone,[54] 6-deoxyclitoriacetal,[47] and the bisbenzylisoquinoline alkaloids of *Cyclea barbata*[50] and *Stephania erecta*.[21]

From *Wrightia tomentosa* we isolated a new type of isoflavone derivative, wrightiadione (**21**), the structure of which is deceptively simple.[48] In Thailand the bark of the plant is used as an antipyretic, and we found weak activity against HIV-1 reverse transcriptase. The IR spectrum of wrightiadione indicated the presence of two distinct carbonyl resonances (1726 and 1686 cm^{-1}), and the high resolution mass spectrum indicated a formula $C_{16}H_8O_3$. While the proton NMR spectrum showed that all of the protons were located in two *ortho*-substituted aromatic spin systems, the carbon-13 NMR data indicated only 15, rather than 16, carbon atoms. Numerous attempts were made to deduce the chemical shift of the second carbonyl carbon. An optimum pulse angle (Ernst angle) and a number of different postacquisition delays (5, 10, 20, and 30 s) to avoid signal saturation were used, but all efforts to locate this resonance failed. The UV spectrum and carbon-13 spectrum combined to eliminate a number of possible structures, and selective INEPT experiments[28,55,56] were used to assign the carbon resonances unambiguously (Figure 6). The attribution of the two aromatic rings to the sets of aromatic carbons was achieved through irradiation of the aromatic protons. For example, irradiation of H-5 at δ 7.89, using a pulse delay corresponding to $J = 8$ Hz, resulted in the enhancement of C-4 (182.53 ppm), C-7 (138.25 ppm), and C-9 (146.27 ppm), and the C-4 resonance was further enhanced when the J value was reduced to 5 Hz. Similar irradiation of H-6 (7.41 ppm) significantly enhanced C-8 (117.92 ppm) and C-10 (121.86 ppm). However, the absence of the second carbonyl signal eliminated the possibility of attributing the structure using the selective INEPT technique, even though all of the protons could be identified. This was a rather unusual situation and left us with two structures, **20** and **21**, which we could not distinguish. We consequently collaborated with our colleagues in Munich for the determination of the X-ray structure of wrightiadione, which was demonstrated to be **21**. There are no other known compounds in this class. The isolate displayed an ED$_{50}$ of 1.1 μg/ml in the P388 test system.[48]

For sheer complexity and beauty of the spectra, we have seen few examples which compare with compounds in the gambogic acid series.[49] The dried latex of *Garcinia hanburyi* is known in Thailand as "Rongthong" (gold resin) and is one of a number of exudates from *Garcinia* sp. (Guttiferae) known as gamboge. It is used internally as a drastic purgative, as well as an emetic and as a vermifuge to treat tapeworm. The extract showed moderate activity in the KB test system and the three isolates, gambogic acid (**22**), isogambogic acid (**23**), and

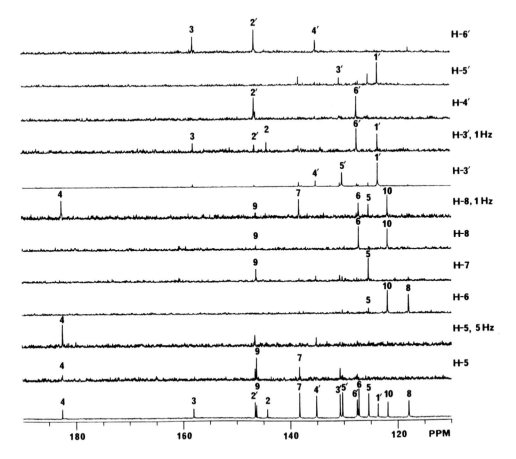

FIGURE 6. Selective INEPT experiments on wrightiadione (**21**).

isomorellinol (**24**), displayed ED_{50} values of 0.7, 0.9, and 2.3 μg/ml, and similar values in the drug-resistant KB-VI cell line.

The structure of gambogic acid, which has 38 carbons and 44 protons, was previously inferred from the X-ray structure of morellin, but no definitive evidence existed, since the proton NMR data were incomplete due to overlap at low field strength, and no carbon-13 data were available. However, the spectrum shows some rather extensive doubling of resonances due to the presence of two conformers in solution, and this situation severely challenged the interpretative power of the combination of the selective INEPT, heteronuclear multiple quantum correlation (HMQC) and heteronuclear multiple bond correlation (HMBC) spectral techniques.

The stereochemistry of the carboxyl group in the side chain was deduced through a ROESY experiment where a correlation contour was observed between H-27 and the H_3-29, confirming the *trans*-configuration of the double bond. These data served to confirm the structure of gambogic acid (**22**) and laid the groundwork for the structure determinations of the new isolates isogambogic acid (**23**) and isomorellinol (**24**).[49] A portion of the ROESY experiment for isomorellinol (**24**) is shown in Figure 7.

The vinblastine-resistant cell line KB-VI has revealed that other classes of natural product may indeed be of interest as anticancer agents, given that they show activity against this cell line, but limited or no activity against the normal KB cell line. So far we have discovered two such classes of natural products: the bisamides of *Aglaia*[52] and some lignans from *Phyllanthus amarus*.[51]

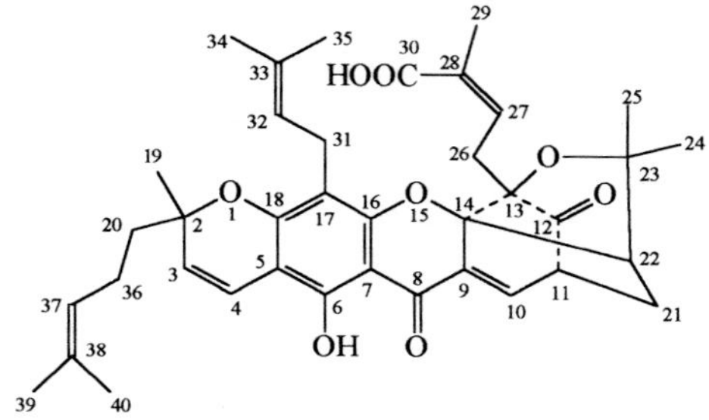

22

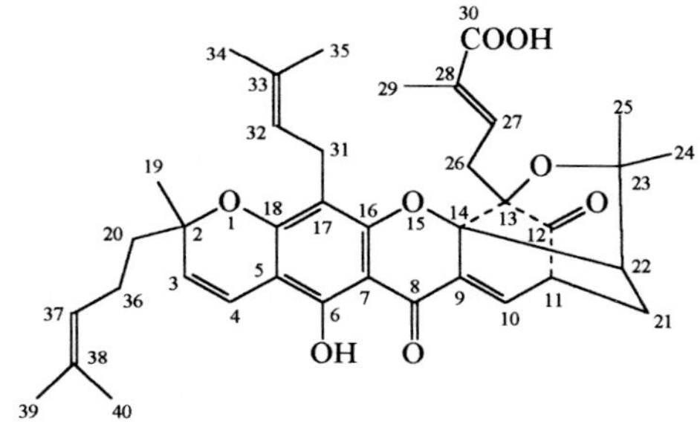

23

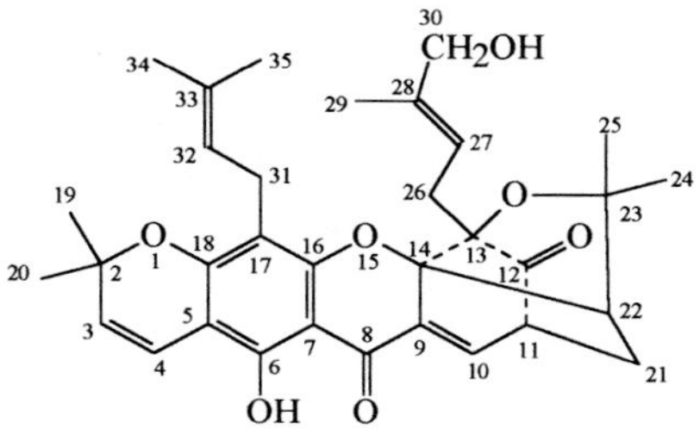

24

The genus *Aglaia* has previously afforded odorinol as an antileukemic compound,[57] and we therefore investigated the Thai species *A. pyramidata*, from which we isolated a new bisamide of spermidine, i.e., pyramidatine (**25**).[52] The structure of this compound was determined through extensive NMR studies, including homonuclear COSY, nuclear Overhauser enhancement (NOESY), attached proton test (APT), HETCOR, and selective INEPT techniques. The selective INEPT spectra (Figure 8) permitted the assignment of all of the quaternary carbons, as might be expected, and also the amide protons, when, on irradiation of the 1'-NH or the 6'-NH, the respective amide carbonyl carbon was enhanced. Surprisingly, it was also discovered that the cinnamoyl carbon resonances required revision, based on both the HETCOR and selective INEPT data. Together with pyramidatine (**25**), three other bisamides of *Aglaia* were evaluated for their cytotoxic activity against a battery of human cancer cell lines. Pyramidatine was inactive in all cell lines, but piriferine showed weak activity, and odorine (**26**) and *epi*-odorine (**27**) displayed appreciable activity against the vinblastine-resistant KB cell line.[52]

There has been substantial interest in recent years in the plant *Phyllanthus amarus* (Euphorbiaceae), which has a substantial reputation for its antihepatotoxic potential.[58] In

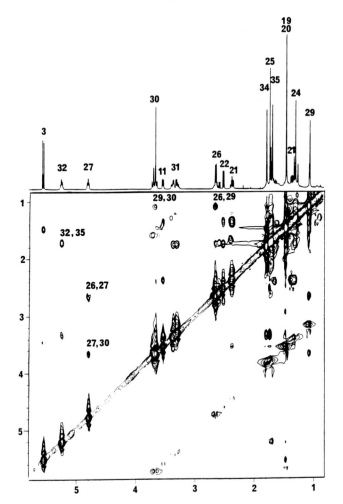

FIGURE 7. ROESY spectrum of isomorellinol (**24**).

collaboration with a research group at the Chulabhorn Research Institute in Thailand, we reinvestigated two of the principal constituents of *P. amarus*: phyllanthin (**28**) and hypophyllanthin (**29**).[51] Although the structures of these compounds appear to be straight-forward, in fact, the structure of hypophyllanthin has been revised several times, and definitive spectral data were essentially absent. Using a combination of homonuclear COSY, APT, HETCOR, and correlation spectroscopy via long-range coupling (COLOC) experiments, all of the protons and carbons of phyllanthin were assigned, and when the selective INEPT technique was applied, all of the carbons of hypophyllanthin (**29**), and consequently its structure, could be unambiguously determined. The cytotoxic potential of these compounds was then evaluated using cultured P388 cells and a battery of human tumor cell lines. Typically, these compounds did not display a cytotoxic response, including the KB-3 cell line. However, phyllanthin (**28**) demonstrated an ED_{50} value of 9.0 µg/ml with the KB-VI cells in the absence of vinblastine (VLB), and this value decreased to 2.1 µg/ml in the presence of VLB. By contrast, hypophyllanthin (**29**) did not mediate a cytotoxic response in the absence of VLB, but with VLB an ED_{50} value of 3.8 µg/ml was obtained.[51]

These effects were further studied by examining the potential of **28** and **29** to inhibit the binding of radiolabeled vinblastine with membrane vesicles derived from KB-VI cells. Hypophyllanthin (**29**) was not active, but phyllanthin (**28**) showed a dose-dependent inhibition

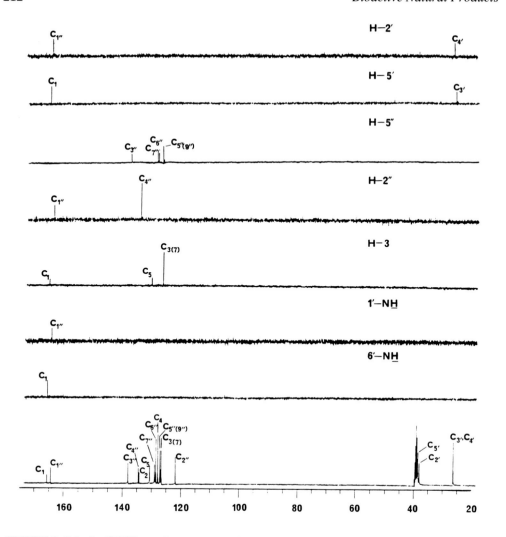

FIGURE 8. Selective INEPT experiments on pyramidatine (**25**).

of vinblastine interaction, with an IC_{50} value of 28 μg/ml, suggesting that it has an affinity for P-glycoprotein.[51]

D. BUDMUNCHIAMINES FROM *ALBIZIA AMARA*

Many, although certainly not all, clinically used anticancer agents interact either covalently or noncovalently with DNA.[7] Such interactions may or may not be quantitatively reflected in a cytotoxic response, particularly when evaluating the plant extracts for their potential to yield novel cytotoxic agents. With this in mind, we became interested in the notion that DNA could be used as an affinity probe both for the detection of extracts that contained compounds that interacted with DNA, and for their bioactivity-guided fractionation.

For our initial work we chose to use a crude preparation of DNA from calf thymus on reversed-phase C-18 high pressure liquid chromatography (HPLC). Under normal eluting conditions the DNA elutes very close to the solvent front, but in the presence of compounds that complex in some manner with DNA, the intensity of the DNA peak diminishes in direct relationship to the concentration of the ligand.[59] The potential of the technique was initially studied with a variety of compounds known to bind to DNA and with anticancer compounds

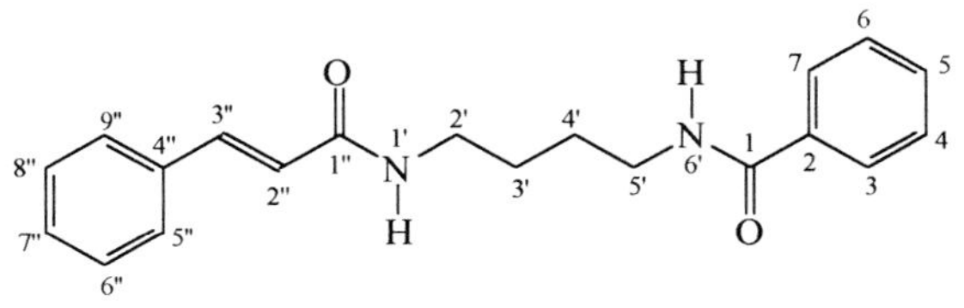

25

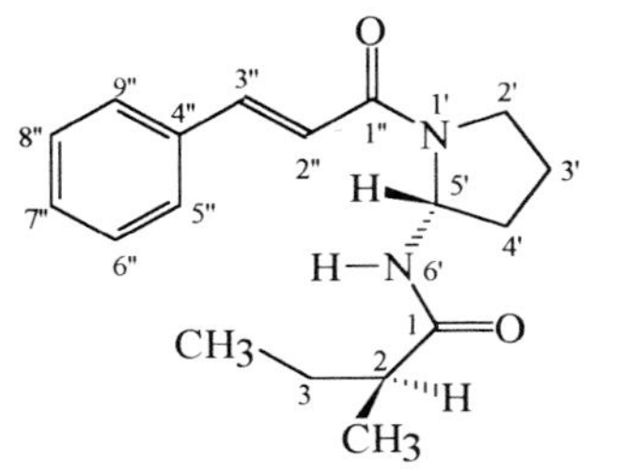

26

27 5'- epimer

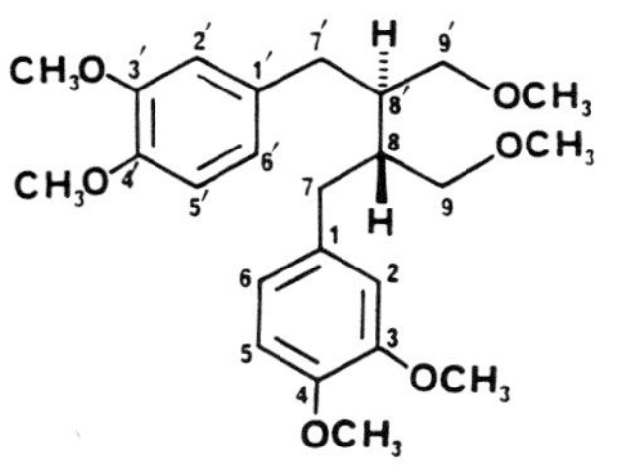

28

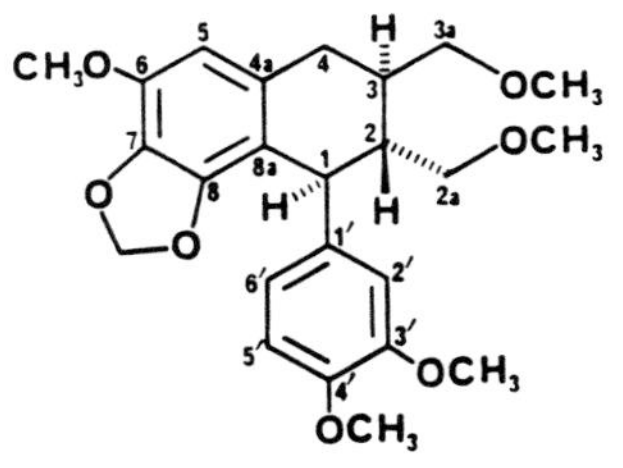

29

known not to interact with DNA. Thus, potent DNA-interacting compounds such as 9-aminoacridine, daunomycin, doxorubicin, ellipticine, and ethidium bromide gave a 70 to 100% DNA peak reduction, whereas colchicine and podophyllotoxin, which interact with tubulin rather than DNA, had little or no effect on DNA peak size.

The next step was to see if such binding agents could be detected when present in small quantities (5 to 10 mg) of crude plant extracts. When this proved successful, attention turned to the use of the bioassay for the screening of plants. We chose a group of 17 plants from our collection and assayed them for their cytotoxic activity and for their ability to affect DNA peak size. The most active of these plants was a sample of the seeds of *Albizia amara*.[60]

It was also reasoned that if, after the nonbinding plant constituents have eluted from the column, the DNA-binding materials could be selectively eluted, the technique would effectively be bioautography, for it would immediately indicate the constituents in the original extract responsible for the binding effect. In the case of *A. amara*, when the DNA-binding fraction was examined by tlc, four alkaloidal fractions were detected and their chromato-

30	Budmunchiamine D	n=6,	R_1=Me, R_2=O, R_3=H_2
31	Budmunchiamine E	n=6,	R_1=Me, R_2=H_2, R_3=O
32	Budmunchiamine F	n=4,	R_1=H, R_2=R_3=H_2
33	Budmunchiamine G	n=6,	R_1=H, R_2=R_3=H_2
34	Budmunchiamine H	n=6,	R_1=H, R_2=O, R_3=H_2
35	Budmunchiamine I	n=6,	R_1=H, R_2=H_2, R_3=O
36	Budmunchiamine A	n=4,	R_1=Me, R_2=R_3=H_2
37	Budmunchiamine B	n=2,	R_1=Me, R_2=R_3=H_2
38	Budmunchiamine C	n=6,	R_1=Me, R_2=R_3=H_2

graphic properties were used to direct, without bioassay, the isolation of these constituents on a larger scale. The savings of both bioassay capability and time should be apparent.

From these four alkaloidal fractions, the budmunchiamines A–I (30–38) were characterized in the groups A–C, D and E, F and G, and H and I, although none of them has yet been obtained in pure form. The original studies on the most abundant mixture[60] were a collaborative effort between our group in Chicago, together with groups in Munich and Budapest. The name we chose for these isolates (Bud-Mun-Chi-amine) reflects the collaborative spirit engendered between the research groups in these great cities. Details of the structural analysis of these isolates has been published elsewhere,[60,61] but some salient features will be discussed in this section.

The major constituent displayed a molecular formula of $C_{27}H_{56}N_4O$ and an IR absorption for an amide moiety. The APT and DEPT spectra displayed 24 distinct resonances, with one resonance (29.6 ppm) comprising four carbons. The structural inferences derived from this information suggested the presence of a terminal C-methyl, three N-methyl, 14 C-methylene, five N-methylene, two N-CHCH$_2$, an amide methylene, one N-methine, and an amide carbonyl carbon. In the absence of olefinic carbons, the second degree of unsaturation should therefore be a ring.

A combination of homonuclear and heteronuclear COSY spectra, together with selective INEPT experiments, indicated that the isolate was a macrocycle containing 13 carbon atoms, one amide nitrogen, three additional nitrogen atoms at positions 5, 9, and 14, and a side chain at position C-4. For example, selective INEPT irradiation of H-4 (2.83 ppm) enhanced C-2 (172.5 ppm), the N5-methyl (35.3 ppm), and C-2' (27.3 ppm), indicating that the aliphatic side chain should be located at position C-4, and the amide group at position C-3. Location

TABLE 2
Cytotoxic Activity of the MeOH Extract and Alkaloid Isolates I–IV
Obtained from the Seeds of *Albizia amara*

			Alkaloid isolate mixtures[b]		
Cell line[a]	MeOH extract[b]	I	II	III	IV
P-388	1.02	0.09	0.68	0.10	0.91
KB	2.01	0.27	1.36	0.29	1.90
Co12	5.24	0.56	2.30	1.60	1.70
BC1	10.10	0.77	4.72	1.42	4.65
Lu1	12.40	0.74	7.70	1.40	11.40
Me12	6.40	0.58	2.85	0.75	2.80
HT	2.02	0.17	1.40	0.29	1.31
KB-V1	ND[c]	0.56	ND	ND	ND

[a] For interpretation of cell line codes see Table 1.
[b] Results are expressed as ED_{50} values (μg/ml).
[c] ND = not determined.

of one of the three-carbon chains was made through the irradiation of H-6 (2.62 ppm), which enhanced C-4 (61.4 ppm) and C-8 (55.0 ppm), and a second three-carbon chain was disclosed through the irradiation of H-17. HRMS data indicated the presence of two other components, as members of a homologous series, where the differences were in the side chains.[60] These alkaloids are members of the pethecelobine group.[62-64]

The structures of the three other isolates were determined in similar fashion.[61] Isolate II showed a ketone moiety (1713 cm^{-1} and 211.31/211.42 ppm) and all the other features of isolate I, suggesting that isolate II had the same macrocyclic ring as I. As well as the two ketone carbonyl resonances, two terminal methyl resonances were noted (0.86 and 0.87 ppm, 13.76 and 13.85 ppm), which, according to the homonuclear and heteronuclear COSY spectra, were coupled to methylene resonances (1.30/22.34 ppm and 1.50/17.27 ppm). In the former case these methylene protons were coupled to the protons at 2.27 ppm attached to a carbon at 44.65 ppm, whereas in the latter case the coupled protons were at 1.35 ppm and the attached carbon at 25.96 ppm. The correlations indicated that in one isomer the carbonyl was located at C-9 in the side chain, whereas in the other isomer it was located at C-8.

Isolates III and IV were established to be mixtures, corresponding to isolates I and II, but containing one less N-methyl group. In each instance the location of the resultant new amino proton was traced to N-5 due to the absence of the resonance at 35 ppm, together with the shifts of the C-4 and C-6 resonances by about 5 ppm upfield, as well as the downfield shifts of C-3 and C-1' by about 2 ppm.

Each of the isolates I–IV was evaluated against a battery of human cancer cell lines and the P388 lymphocytic leukemia cell line. Isolate I was also tested for activity against the drug-resistant cell line KB-VI. The results are shown in Table 2. A general nonspecific response was observed, with the overall activity of the isolates I and III being very similar to, but greater than, the isolates II and IV, which possessed a carbonyl group in the side chain. It is noteworthy that the drug-resistant cell line was also susceptible to isolate I. In agreement with these relative cytotoxicity values, isolates I and II, but not isolates II and IV, displayed bactericidal responses when tested against the *Salmonella typhimurium* strain TM 677. As anticipated in view of the demonstrated absence of a covalent interaction with DNA, there were no mutagenic effects observed.

A number of other biological effects were also studied.[65] Thus, isolate I was found to inhibit the catalytic activity of DNA polymerase and RNA polymerase. For the reaction with DNA polymerase, the K_i value was 20 μM and the inhibition was competitive with respect to DNA.

In addition, isolate I was found to be a very strong, nonspecific inhibitor of PAF-induced platelet aggregation and a dose-dependent inhibitor of cyclooxygenase activity and of human lymphocyte transformation.

V. SUMMARY

This chapter has attempted to review some of the contributions that our group has made to the discovery of plant-derived cytotoxic agents. It has shown how past discoveries have demonstrated that almost every class of natural product has a member displaying some degree of cytotoxic response and how enhanced biological information, as well as the ability to construct rapid effective assays, may yield new types of active compounds. We have also shown how the selection of plants for biologic evaluation must be given careful consideration, how new cytotoxic agents from established and new classes of natural product can be isolated and characterized using the very latest in spectroscopic techniques. This chapter also highlights how new strategies, such as a DNA-binding assay, once established, can afford important new series of biologically active natural products.

This work will continue, for cancer remains a potent, extremely debilitating threat. In the future we will need to become increasingly aware of the biology of cancer and integrate these discoveries into strategies and bioassays that can be successfully used for the discovery of agents that might provide novel therapeutic modalities.

ACKNOWLEDGMENTS

This work was made possible by grants (CA 20164 and CA 33047) from the Division of Cancer Treatment, NCI, Bethesda, MD. One of us (JMP) was the recipient of a Research Career Development Award (1984–1989) from the NCI and a Research Fellowship from the Alexander von Humboldt Foundation (1990–1991). We thank the Research Resources Center at the University of Illinois at Chicago for the provision of spectroscopic facilities and Dr. George Doss for the initial establishment of the selective INEPT NMR technique at UIC. Work on the budmunchiamines was also supported by a research project between the Deutsche Forschungsgemeinschaft and the Hungarian Academy of Sciences. Finally, we thank the many collaborators whose names appear on the papers from the University of Illinois at Chicago that are cited in this chapter, and especially Dr. Norman R. Farnsworth, whose vision and leadership originally brought us together.

UPDATE

Subsequent to the preparation of this chapter, Taxol was approved by the U.S. Food and Drug Administration for use in the treatment of ovarian cancer.

REFERENCES

1. **Hartwell, J. L.,** Plants used against cancer, *Lloydia,* 30, 379, 1967; 31, 71, 1968; 32, 79, 153, 247, 1969; 33, 98, 288, 1970; 34, 103, 204, 310, 386, 1971.
2. **Suffness, M.,** Development of antitumor natural products at the National Cancer Institute, *Gann,* 36, 21, 1989.
3. NCI investigational drugs — chemical information, NIH Publication No. 92-2654, *National Institutes of Health, U.S. Department of Health and Human Services,* 1992.
4. **Kupchan, S. M.,** Novel plant-derived tumor inhibitors and their mechanism of action, *Cancer Treat. Rep.,* 60, 1115, 1976.

5. **Wall, M. E. and Wani, M. C.**, Antineoplastic agents from plants, *Ann. Rep. Pharmacol. Toxicol.*, 17, 117, 1977.
6. **Cordell, G. A.**, Anticancer agents from plants, in *Progress in Phytochemistry*, Vol. 5, Reinhold, L., Harborne, J. B., and Swain, T., Eds., Pergamon Press, Oxford, U.K., 1978, 273.
7. **Cassady, J. M. and Douros, J. D., Eds.**, *Anticancer Agents Based on Natural Product Models*, Academic Press, New York, 1980.
8. **Suffness, M. and Douros, J.**, Current status of the NCI plant and animal product program, *J. Nat. Prod.*, 45, 1, 1982.
9. **Sneden, A. T.**, Novel antitumor agents from plants, in *Anticancer and Interferon Agents: Synthesis and Properties*, Ottendrite, R. M. and Butler, G. B., Eds., Marcel Dekker, New York, 1984, 79.
10. **Suffness, M. and Cordell, G. A.**, Antitumor alkaloids, in *The Alkaloids.* Vol. XXV, Brossi, A., Ed., Academic Press, New York, 1985, 1.
11. **Cragg, G. and Suffness, M.**, Metabolism of plant-derived anticancer agents, *Pharm. Ther.*, 37, 425, 1988.
12. **Blaskó, G. and Cordell, G. A.**, Recent developments in the chemistry of plant-derived anticancer agents, in *Economic and Medicinal Plant Research*, Vol. 2, Wagner, H., Hikino, H., and Farnsworth, N. R., Eds., Academic Press, London, 1988, 119.
13. **Potier, P.**, Search and discovery of new antitumor agents (Rhône-Poulenc lecture), *Chem. Soc. Rev.*, 21, 113, 1992.
14. **Cragg, G. M., Boyd, M. R., Cardellina, J. H., II, Grever, M. R., Schepertz, S. A., Snader, K. M., and Suffness, M.**, The role of plants in the National Cancer Institute drug discovery and development program, in *Human Medicinal Agents from Plants*, Kinghorn, A. D. and Balandrin, M. F., Eds., ACS Symp. Ser. 534, American Chemical Society, Washington, D.C., in press.
15. **Barclay, A. S. and Perdue, R. E., Jr.**, Distribution of anticancer activity in higher plants, *Cancer Chemother. Rep.*, 60, 1081, 1976.
16. **Kingston, D. G. I., Samaranayake, G., and Ivey, C. A.**, The chemistry of taxol, a clinically useful anticancer agent, *J. Nat. Prod.*, 53, 1, 1990.
17. **Cordell, G. A., Beecher, C. W. W., and Pezzuto, J. M.**, Can ethnopharmacology contribute to the development of new anticancer drugs? *J. Ethnopharmacol.*, 32, 117, 1991.
18. **Geran, R. I., Greenberg, N. H., McDonald, M. M., Schumacher, A. M., and Abbott, B. J.**, Protocols for screening chemical agents and natural products against animal tumors and other biological systems (3rd ed.), *Cancer Chemother. Rep.*, 3, 1, 1972.
19. **Alley, M. C., Scudiero, D. A., Monks, A., Hursey, M. L., Czerwinski, M. J., Fine, D. L., Abbott, B. J., Mayo, J. G., Shoemaker, R. H., and Boyd, M. R.**, Feasibility of drug screening with panels of human tumor cell lines using a microculture tetrazolium assay, *Cancer Res.*, 48, 589, 1988.
20. **Scudiero, D. A., Shoemaker, R. H., Paull, K. D., Monks, A., Tierney, S., Nofziger, T. H., Curren, M. J., Seniff, D., and Boyd, M. R.**, Evaluation of a soluble tetrazolium/formazan assay for cell growth and drug sensitivity in culture using human and other tumor cell lines, *Cancer Res.*, 48, 4827, 1988.
21. **Likhitwitayawuid, K., Angerhofer, C. K., Ruangrungsi, N., Cordell, G. A., and Pezzuto, J. M.**, Cytotoxic and antimalarial bisbenzylisoquinoline alkaloids from *Stephania erecta, J. Nat. Prod.*, 56, 30, 1993.
22. **Kaneda, N., Kinghorn, A. D., Farnsworth, N. R., Tuchinda, T., Udchacon, J., Santisuk, T., and Reutrakul, V.**, Two diarylheptanoids and a lignan from *Casuarina junghuhniana, Phytochemistry*, 29, 3366, 1990.
23. **Kaneda, N., Pezzuto, J. M., Soejarto, D. D., Kinghorn, A. D., Farnsworth, N. R., Santisuk, T., Tuchinda, P., Udchachon, J., and Reutrakul, V.**, Plant anticancer agents. XLVIII. New cytotoxic flavonoids from *Muntingia calabura* roots, *J. Nat. Prod.*, 54, 196, 1991.
24. **Tuchinda, P., Udchachon, J., Reutrakul, V., Santisuk, T., Taylor, W. C., Farnsworth, N. R., Pezzuto, J. M., and Kinghorn, A. D.**, Bioactive butenolides from *Melodorum fruticosum, Phytochemistry*, 30, 2685, 1991.
25. **Kaneda, N., Pezzuto, J. M., Kinghorn, A. D., Farnsworth, N. R., Santisuk, T., Tuchinda, P., Udchachon, J., and Reutrakul, V.**, Plant anticancer agents. L. Cytotoxic triterpenes from *Sandoricum koetjape* stems, *J. Nat. Prod.*, 55, 654, 1992.
26. **Kaneda, N., Chai, H., Pezzuto, J. M., Kinghorn, A. D., Farnsworth, N. R., Tuchinda, P., Udchachon, J., Santisuk, T., and Reutrakul, V.**, Plant anticancer agents. XLIX. Cytotoxic activity of cardenolides from *Beaumontia brevituba, Planta Med.*, 58, 429, 1992.
27. **Nshimo, C. M.**, Phytochemical and Biological Studies on *Muntingia calabura* L., Ph.D. dissertation, University of Illinois at Chicago, 1991, 125.
28. **Bax, A.**, Structure determination and spectral assignment by pulsed polarisation transfer via long-range ^{1}H-^{13}C couplings, *J. Magn. Reson.*, 57, 314, 1984.
29. **Finlay, G. J., Smith, G. P., Fray, L. M., and Baguley, B. C.**, Effect of flavone acetic acid on Lewis lung carcinoma: evidence for an indirect effect, *J. Natl. Cancer Inst.*, 80, 241, 1988.

30. **Edwards, J. M., Raffauf, R. E., and Le Quesne, P. W.,** Antineoplastic activity and cytotoxicity of flavones, isoflavones, and flavanones, *J. Nat. Prod.,* 42, 85, 1979.

31. **Duh, C.-Y., Pezzuto, J. M., Kinghorn, A. D., Leung, S. L., and Farnsworth, N. R.,** Plant anticancer agents. XLIV. Cytotoxic constituents from *Stizophyllum riparium, J. Nat. Prod.,* 50, 63, 1987.

32. **Duh, C.-Y., Kinghorn, A. D., and Pezzuto, J. M.,** Cell-cycle specific cytotoxicity mediated by stizophyllin (2α,3β,12β-trihydroxypregna-4,7,16-trien-7-one), a novel electrophilic pregnane isolated from *Stizophyllum riparium, Chem. Biol. Interact.,* 80, 43, 1991.

33. **Duh, C.-Y.,** Phytochemical and Mechanistic Studies on the Constituents of *Wikstroemia elliptica* and *Stizophyllum riparium,* Ph.D dissertation, University of Illinois at Chicago, 1986, 216.

34. **Bhacca, N. S., Lankin, D. C., Tantivatana, P., Ruangrungsi, N., Valisiriraj, V., Borris, R. P., and Cordell, G. A.,** Microminutin, a novel cytotoxic coumarin from *Micromelum minutum, J. Org. Chem.,* 48, 268, 1983.

35. **Mukhopadhyay, S., Cordell, G. A., Ruangrungsi, N., Rodkird, S., Tantivatana, P., and Hylands, P. J.,** Traditional medicinal plants of Thailand IV. 3-(2′,3′-Diacetoxy-2′-methyl-butyryl)-cuauhtemone from *Pluchea indica* Less. (Compositae), *J. Nat. Prod.,* 46, 671, 1983.

36. **Ruangrungsi, N., Tippayuthpijarn, P., Tantivatana, P., Borris, R. P., and Cordell, G. A.,** Traditional medicinal plants of Thailand. VI. Isolation of cryptomeridiol from *Blumea balsamifera* (Compositae), *J. Sci. Soc. Thailand,* 11, 47, 1985.

37. **Hamburger, M. O., Cordell, G. A., Tantivatana, P., and Ruangrungsi, N.,** Traditional medicinal plants of Thailand. VIII. Isoflavonoids of *Dalbergia candenatensis, J. Nat. Prod.,* 50, 696, 1987.

38. **Meksuriyen, D., Cordell, G. A., Ruangrungsi, N., and Tantivatana, P.,** Traditional medicinal plants of Thailand. IX. 11-Hydroxy-10-methoxydracaenone and 7,11-dihydroxy-10-methoxydracaenone from *Dracaena loureiri* (Agavaceae), *J. Nat. Prod.,* 50, 1118, 1987.

39. **Meksuriyen, D. and Cordell, G. A.,** Traditional medicinal plants of Thailand. XIII. Flavonoid derivatives of *Dracaena loureiri* (Agavaceae), *J. Sci. Soc. Thailand,* 14, 3, 1988.

40. **Hamburger, M. O., Cordell, G. A., Likhitwitayawuid, K., and Ruangrungsi, N.,** 10-Demethoxykopsidasinine from *Kopsia jasminiflora, Phytochemistry,* 27, 2719, 1988.

41. **Ananvoranich, S., Likhitwitayawuid, K., Ruangrungsi, N., Blaskó, G., and Cordell, G. A.,** Determination of a new sesquiterpene skeleton through selective INEPT spectroscopy, *J. Org. Chem.,* 54, 2253, 1989.

42. **Bunyapraphatsara, N., Blaskó, G., and Cordell, G. A.,** Hortensin, an unusual flavone from *Millingtonia hortensis, Phytochemistry,* 28, 1555, 1989.

43. **Kigodi, P. G. K., Blaskó, G., Thebtaranonth, Y., and Cordell, G. A.,** Spectroscopic and biological investigation of nimbolide and 28-deoxynimbolide from *Azadirachta indica, J. Nat. Prod.,* 52, 1246, 1989.

44. **Bavovada, R., Blaskó, G., Shieh, H.-L., Pezzuto, J. M., and Cordell, G. A.,** Spectral assignment and cytotoxicity of 22-hydroxytingenone from *Glyptopetalum sclerocarpum, Planta Med.,* 56, 380, 1990.

45. **Topcu, G., Che, C.-T., Cordell, G. A., and Ruangrungsi, N.,** Iridolactones from *Alyxia reinwardti, Phytochemistry,* 29, 3197, 1990.

46. **Hamburger, M. O., Cordell, G. A., and Ruangrungsi, N.,** Traditional medicinal plants of Thailand. XVI. Biologically active constituents from *Plumeria rubra, J. Ethnopharmacol.,* 33, 289, 1991.

47. **Lin, L.-J., Ruangrungsi, N., Cordell, G. A., Shieh, H.-L., You, M., and Pezzuto, J. M.,** 6-Deoxyclitoriacetal from *Clitoria macrophylla, Phytochemistry,* 31, 4329, 1992.

48. **Lin, L.-J., Topcu, G., Lotter, H., Ruangrungsi, N., Wagner, H., and Cordell, G. A.,** Wrightiadione from *Wrightia tomentosa, Phytochemistry,* 31, 4333, 1992.

49. **Lin, L.-J., Lin, L.-Z., Pezzuto, J. M., Cordell, G. A., and Ruangrungsi, N.,** Isogambogic acid and isomorellinol from *Garcinia hanburyi, Magn. Reson. Chem.,* 31, 340, 1993.

50. **Lin, L.-Z., Xue, L., Shieh, H.-L., Angerhofer, C. K., Ruangrungsi, N., Pezzuto, J. M., Johnson, M. E., and Cordell, G. A.,** Cytotoxic and antimalarial bisbenzylisoquinoline alkaloids from *Cyclea barbata, J. Nat. Prod.,* in press.

51. **Likhitwitayawuid, K., Somanbandhu, A., Nitayangkura, S., Mahidol, C., Ruchirawat, S., Cordell, G. A., Shieh, H.-L., Chai, H., and Pezzuto, J. M.,** ^{1}H- and ^{13}C-NMR assignments of phyllanthin and hypophyllanthin: lignans that enhance cytotoxic responses with cultured multidrug-resistant cells, *J. Nat. Prod.,* 56, 233, 1993.

52. **Saifah, E., Likhitwitayawuid, K., Puripattanavong, J., Cordell, G. A., Chai, H., and Pezzuto, J. M.,** Bisamides from *Aglaia* species: structure analysis and potential to reverse resistance with cultured cells, *J. Nat. Prod.,* 56, 473, 1993.

53. **Cordell, G. A., Hamburger, M. O., Blaskó, G., Meksuriyen, D., Tantivatana, P., Ruangrungsi, N., and Bunyapraphatsara, N.,** Studies of Thai medicinal plants using one- and two-dimensional NMR techniques, in *Proc. Princess Congress I,* Vol. II, Bangkok, 1989, 507.

54. **Blaskó, G., Shieh, H.-L., Pezzuto, J. M., and Cordell, G. A.,** Carbon-13 spectral assignment and cytotoxicity study of rotenone, *J. Nat. Prod.,* 52, 1363, 1989.

55. **Cordell, G. A.,** Selective INEPT spectroscopy — a powerful tool for the spectral assignment and structure elucidation of natural products, *Phytochem. Anal.,* 2, 59, 1991.
56. **Cordell, G. A. and Kinghorn, A. D.,** One-dimensional proton-carbon correlations for the structure determination of natural products, *Tetrahedron,* 47, 3521, 1991.
57. **Hayashi, N., Lee, K. H., Hall, I. H., McPhail, A. T., and Huang, H.,** Structure and stereochemistry of (–)-odorinol, an antileukemic diamide from *Aglaia odorata, Phytochemistry,* 21, 2371, 1982.
58. **Thyagarajan, S. P., Subramanian, S., Thirunalasundari, T., Venkateswaran, P. S., and Blumberg, B. S.,** Effect of *Phyllanthus amarus* on chronic carriers of hepatitis B virus, *Lancet,* II 8614, 764, 1988.
59. **Pezzuto, J. M., Che, C.-T., McPherson, D. D., Zhu, Z.-P., Topcu, G., Erdelmeier, C. J., and Cordell, G. A.,** DNA as an affinity probe useful in the detection and isolation of biologically active natural products, *J. Nat. Prod.,* 54, 1522, 1991.
60. **Pezzuto, J. M., Mar, W., Lin, L.-Z., Cordell, G. A., Neszmelyi, A., and Wagner, H.,** DNA-based isolation and structure elucidation of the budmunchiamines, novel macrocyclic alkaloids from *Albizia amara, Heterocycles,* 32, 1961, 1991.
61. **Pezzuto, J. M., Lin, L.-Z., Mar, W., Cordell, G. A., Neszmelyi, A., and Wagner, H.,** Budmunchiamines D-I from *Albizia amara, Phytochemistry,* 31, 1795, 1992.
62. **Weisner, K., MacDonald, D. M., Valenta, Z., and Armstrong, R.,** Pithecelobine, the alkaloid of *Pithecelobium saman* Benth. I, *Can. J. Chem.,* 30, 761, 1992.
63. **Weisner, K., MacDonald, D. M., Bankiewicz, C., and Orr, D. E.,** Structure of pithecelobine. II, *Can. J. Chem.,* 46, 1881, 1968.
64. **Weisner, K., Valenta, Z., Orr, D. E., Leide, V., and Kohan, G.,** Structure of pithecelobine. III. The synthesis of the 1,5- and 1,3-deoxypithecelobines, *Can. J. Chem.,* 46, 3617, 1968.
65. **Mar, W., Tan, T.-T., Cordell, G. A., Pezzuto, J. M., Jurcic, K., Offermann, J., Redl, K., Steinke, B., and Wagner, H.,** Biological activity of novel macrocyclic alkaloids (budmunchiamines) from *Albizia amara* detected on the basis of interaction with DNA, *J. Nat. Prod.,* 54, 1531, 1991.

Chapter 11

CYTOINHIBITORY COMPOUNDS FROM HIGHER PLANTS

Andrew Marston, Laurent A. Décosterd, and Kurt Hostettmann

TABLE OF CONTENTS

I. INTRODUCTION

Cytotoxicity, or toxicity to cells in culture, can be subdivided into cytostatic activity (i.e., stopping cell growth) or cytocidal activity (i.e., killing cells).[1] Cytoinhibition is the inhibition of cell growth.

Plants have been shown to provide a useful source of natural products that are effective in the treatment of human neoplastic diseases. The search for anticancer agents from plants dates back to 1947, when Hartwell demonstrated that podophyllotoxin from *Podophyllum peltatum* (Berberidaceae) inhibited the growth of experimental tumor cells in mice.[2] The subsequent discovery of the antileukemic properties of vinblastine and vincristine, bis-indole alkaloids from *Catharanthus roseus* (Apocynaceae),[3] gave the impulse for wide-ranging investigations of plant extracts and plant-derived compounds for possible anticancer activity.

Among the species that have provided clinically useful drugs, in addition to the two mentioned above, are *Taxus brevifolia* (Taxaceae), the source of the diterpene taxol, *Ochrosia elliptica* (Apocynaceae), one of the sources of the pyridocarbazole alkaloid ellipticine, and *Camptotheca acuminata* (Nyssaceae), which contains the pyrrolo[3,4-*b*]-quinoline alkaloid camptothecin.[4] A large number of other active natural products have been detected using short-term *in vitro* growth-inhibition assays with cultured cells. These include Walker carcinosarcoma 256, mouse L-1210 leukemia, Ehrlich murine ascites tumor, sarcoma 180, and mouse P-388 leukemia cell lines.[5,6]

II. BIOASSAYS

Test systems currently used are classified into cytotoxicity-based bioassays and mechanism-based bioassays.[7] While traditional cytotoxicity-based assays are indicative for activity in leukemia, lymphoma, or a few rare tumors, their efficacy in finding products with activity in the predominantly occurring, slow growing solid tumors of humans (e.g., lung, colon, breast, skin, kidney) is strictly limited. Very recently, more representative *in vitro* protocols using cell lines derived from a wide selection of human tumor types (especially those of epithelial origin) have been introduced with the aim of finding substances that are selectively cytotoxic. This is the approach currently being implemented by the National Cancer Institute (NCI) in the U.S., for example.[1] When a plant extract shows reproducible activity against a carcinoma cell line, it is fractionated to isolate the active principles and to establish their molecular structures. The antitumor activities of the isolated products are subsequently evaluated with *in vivo* experimental models, using suitable xenografts of tumor cell lines in athymic mice.[8]

The work reported in this chapter is based on a rapid test for the screening of cytotoxic activities of plants and plant-derived natural products, using a human tumor cell line (Co-115) derived originally from an adenocarcinoma of the ascending colon.[9] In our opinion this bioassay is more representative of the vast majority of human neoplasms, which are of epithelial origin, than the various assays that involve leukemia cell lines. It should also be noted that cancers of the colon and rectum (11 to 14% of all known cancer cases) are responsible for the death of more than 50,000 people per year in the U.S. alone, and that 100,000 new cases are diagnosed each year.

The tumor cells are incubated in flat-bottomed microtiter wells with various concentrations of an ethanolic solution of a plant extract, a purified fraction, or a pure substance (Figure 1).[10] Solutions of the sample (constituting 1% of the culture medium) are tested in parallel with control experiments involving ethanol alone. After 5 days the number of living cells is measured by the method of Landegren,[11] in which the medium is treated with the chromogenic substrate, *p*-nitrophenyl-*N*-acetyl-β-ᴅ-glucosaminide (NAG). The substrate is cleaved

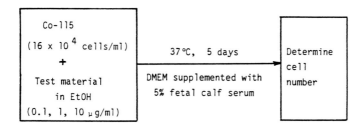

FIGURE 1. Incubation of Co-115 human colon carcinoma cells with plant extracts or pure compounds.

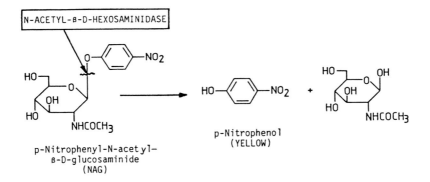

FIGURE 2. Reaction of hexosaminidase with the chromogenic substrate NAG.

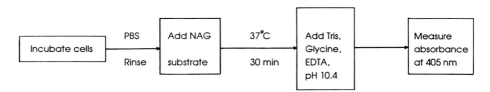

FIGURE 3. The hexosaminidase test for the determination of cell numbers.

by N-acetyl-β-D-hexosaminidase, a lysosomal enzyme ubiquitous to the cells (Figure 2). The p-nitrophenol liberated from the substrate is determined spectrophotometrically at 405 nm (Figure 3). Since there is a logarithmic relationship between the absorbance and the number of cells, the ED_{50} (concentration of extract or substance that inhibits the replication of cells by 50%) can be calculated from a semilogarithmic plot of cell number against concentration of substance tested. Determination of cell numbers by spectrophotometry means that this *in vitro* test is far more rapid than a direct counting of the cells.

When this test was applied to more than 250 extracts from 75 different plant species found mainly in the tropics, a total of 29 species with ED_{50} values lower than 10 µg/ml were discovered.[12] The most active plants were *Diospyros abyssinica, D. zombensis* (Ebenaceae), *Stillingia lineata* (Euphorbiaceae), *Hypericum annulatum, H. revolutum, H. calycinum, Mammea africana, Psorospermum febrifugum* (Guttiferae), *Erythrina berteroana, Indigofera swaziensis, Neorautanenia pseudopachyrrhiza, Swartzia madagascariensis* (Leguminosae), *Sesamum angolense* (Pedaliaceae), *Crossopteryx febrifuga* (Rubiaceae), and *Clerodendrum uncinatum* (Verbenaceae). Petroleum ether and dichloromethane extracts produced the highest activities. The systematic fractionation of these extracts has enabled the isolation of compounds that exhibit an appreciable inhibition of the growth of Co-115 cells.[13]

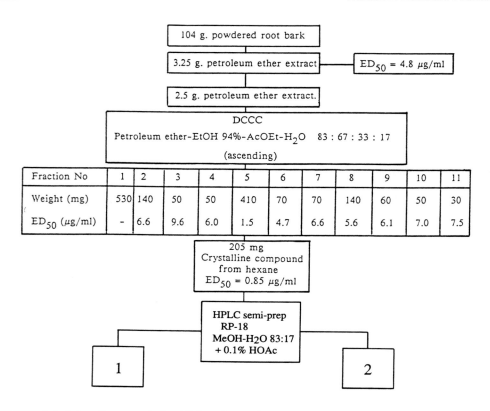

FIGURE 4. Isolation of compounds **1** and **2** from the petroleum ether extract of *Hypericum revolutum* root bark.

III. ISOLATION AND STRUCTURE DETERMINATION

Consideration of the list of plant extracts found active in the Co-115 bioassay showed that several species of the Guttiferae family gave positive results. For this reason, emphasis for the isolation of cytoinhibitory substances was placed on members of this family.

A. *HYPERICUM REVOLUTUM*

When tested in the Co-115 cell line, a petroleum ether extract of the root bark of *Hypericum revolutum*, collected on the Zomba Plateau in Malawi (central Africa), significantly inhibited cell growth (ED_{50} 4.8 µg/ml). Extraction and isolation of the active compounds were guided by the Co-115 bioassay. Since preliminary chromatography on solid (silica gel) supports led to a good deal of irreversible adsorption or decomposition of the active constituents, a portion of the extract (2.5 g) was subjected to droplet countercurrent chromatography (DCCC).[14] Using petroleum ether-94% ethanol-ethyl acetate-water (83:67:33:17) as the solvent system (the organic, upper phase constituted the mobile phase), 11 fractions were obtained (Figure 4). The most active fraction (ED_{50} 1.5 µg/ml) yielded a crystalline product (ED_{50} 0.85 µg/ml) from n-hexane. Subsequent X-ray diffraction analysis (Figure 5) established **1** for the crystals. However, the D/CI mass spectrum (NH_3, positive-ion mode) of the crystalline material revealed peaks at m/z 483 and m/z 500 in addition to the molecular ion peaks [M + NH_4]+ (m/z 486) and [M + H]+ (m/z 469) expected of **1**. This suggested the presence of a higher homolog with a molecular mass of 482. Analysis by high-pressure liquid chromatography (HPLC)-UV confirmed the presence of a minor product with the same UV spectrum as **1** (Figure 6a). Separation of the two compounds was possible by semipreparative HPLC using the same solvent (Figure 6b). Thus, from 120 mg of the original crystals, 110 mg of **1** (hyperevolutin A) and 7 mg of the homolog **2** (hyperevolutin B) were obtained.[15]

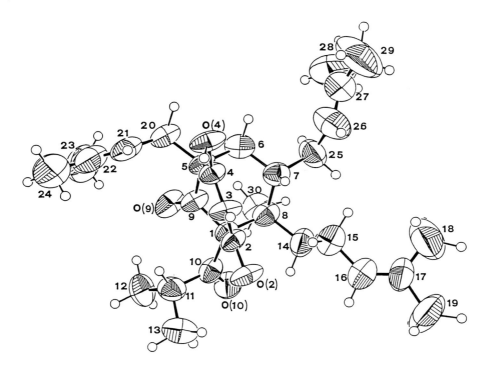

FIGURE 5. Spatial geometry and numerotation of atoms of hyperevolutin A (**1**).

The structure of **1** is similar to that of hyperforin (**3**), an antibiotic isolated from *H. perforatum*,[16] and is composed of a bicyclic skeleton. Four keto groups are arranged around the two rings, and one of these is in the enolic form. The presence of an enolic proton explains the erratic behavior on silica gel supports and the fact that irreversible adsorption occurs on the sorbent. The six-membered saturated ring has a chair conformation, while the other ring has an "envelope" conformation (Figure 7).

The structure of **2** was deduced by comparison of its ^{1}H and ^{13}C-NMR spectra with those of **1**. In the ^{13}C-NMR of **1**, the signals of the isobutyryl group appear at 41.9 ppm (CH) and at 21.5 and 20.7 ppm (2 × CH$_3$). On the other hand, hyperevolutin B (**2**) has signals

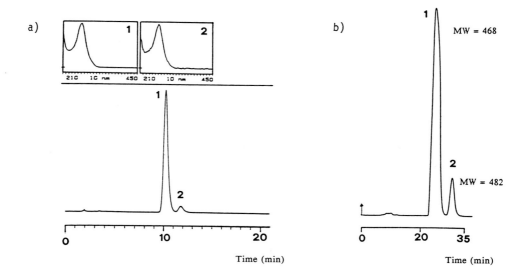

FIGURE 6. HPLC of the crystalline material obtained after DCCC. (a) Analytical HPLC. Column: μBondapak C-18 (300 × 3.9 mm); solvent: MeOH-H$_2$O 87:13 + 0.1% HOAc; flow rate: 1.5 ml/min; detection: LC-UV Hewlett-Packard HP1040, 275 nm. (b) Semipreparative HPLC. Column: μBondapak C-18 (300 × 7.8 mm); solvent: MeOH-H$_2$O 87:13 + 0.1% HOAc; flow rate: 1.5 ml/min; detection: 275 nm; sample: 120 mg in 2.4 ml MeOH (150 μl aliquots injected).

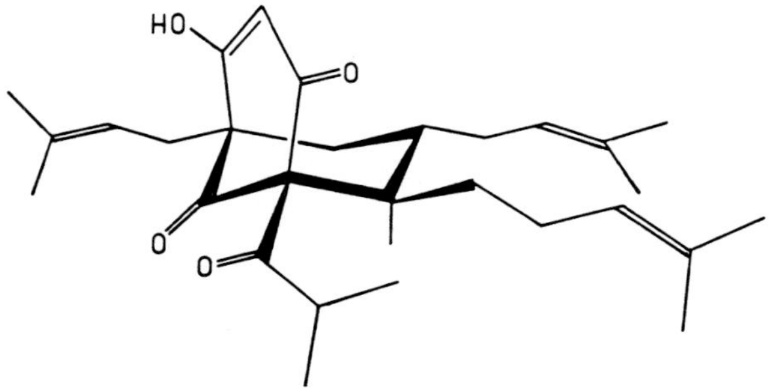

FIGURE 7. Stereochemical structure of hyperevolutin A (**1**).

characteristic of a 2-methylbutyryl group, i.e., a CH signal at 48.7 ppm (C-11), a CH$_2$ signal at 27.6 ppm (C-13), and two methyl groups at 16.6 and 11.5 ppm (C-12 and C-31, respectively).[17]

B. *HYPERICUM CALYCINUM*

This attractive plant is widespread in Europe and is used as an ornamental in borders and banks. A petroleum ether extract of the aerial parts (leaves, twigs, flowers, and buds) was cytoinhibitory in the Co-115 carcinoma cell line, with an ED$_{50}$ of 4.4 μg/ml.

1. Isolation of Active Components

In a similar fashion to *H. revolutum*, the petroleum ether extract was subjected to a preliminary fractionation, employing DCCC for quantitative recovery of injected material. The petroleum ether-94% ethanol-ethyl acetate-water (83:67:33:17 in the ascending mode) solvent system gave five fractions (3, 5, 6, 7, and 9, Figure 8) possessing higher activity than the original extract.

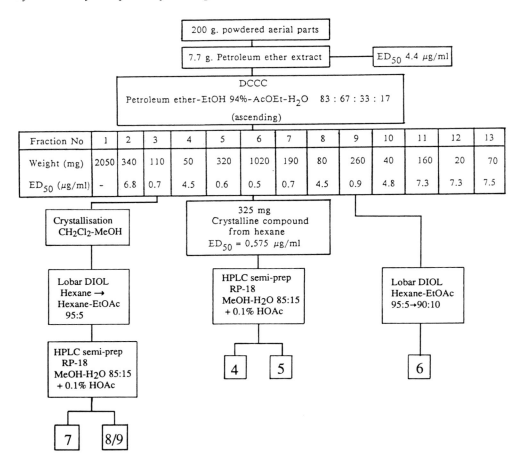

FIGURE 8. Isolation of compounds **4–9** from the petroleum ether extract of *Hypericum calycinum* aerial parts.

The major constituent of fraction 6 crystallized from n-hexane and had an ED_{50} of 0.58 µg/ml. However, the D/CI mass spectrum (NH$_3$, positive-ion mode) gave major peaks at *m/z* 445 and 431, suggesting, as observed for *H. revolutum*, the presence of homologs. HPLC analysis on octadecylsilica with methanol-water 85:15 gave a poorly resolved chromatogram (Figure 9a), but on addition of acetic acid to the solvent, well-defined peaks were produced, clearly indicating that the crystalline material was composed of two homologs (Figure 9b), each with the same UV spectrum. The two products were separated by means of semi-preparative HPLC on a C-18 column, using the same solvent system, affording 7 mg of **4** and 40 mg of **5** from 60 mg of crystalline material (Figure 10).[18]

Fraction 9 (ED_{50} 0.9 µg/ml), from DCCC of the extract, was purified by low-pressure liquid chromatography (LPLC) on a Lobar® Diol column (hexane-ethyl acetate 95:5 → 90:10), yielding 60 mg of a pale yellow product (**6**).[18]

The bioactive fraction 3, from DCCC, contained a compound that crystallized from CH$_2$Cl$_2$-MeOH. LPLC of the mother liquors on a Diol support (hexane 100% → hexane-ethyl acetate 95:5) allowed the isolation of a further product (hypercalin C). This was crystallized from acetonitrile, but, once again, was found to be a mixture of isomers (Figure 11a). Semipreparative HPLC on octadecylsilica enabled separation of **7** (MW 484) from a mixture of **8** and **9** (MW 498) (Figure 11b).[18]

Droplet countercurrent chromatography provided an excellent means for the first fractionation step of the petroleum ether extract of *H. calycinum*. Figure 12 shows how this technique

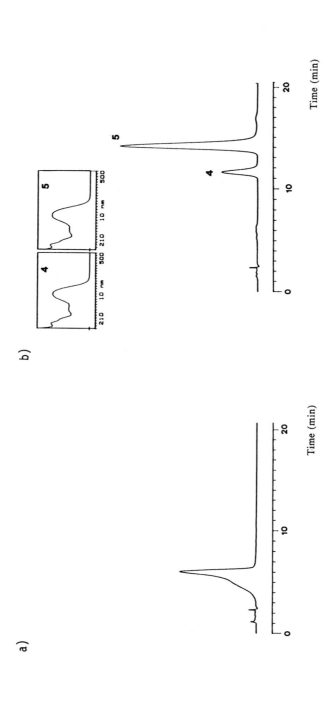

FIGURE 9. Analytical HPLC of the crystalline mixture **4/5**. Column: LiChrosorb RP-18 (250 × 4.6 mm); flow rate: 1.5 ml/min; detection: 280 nm. (a) Solvent: MeOH-H₂O 85:15. (b) Solvent: MeOH-H₂O 85:15 + 0.1% HOAc added to MeOH.

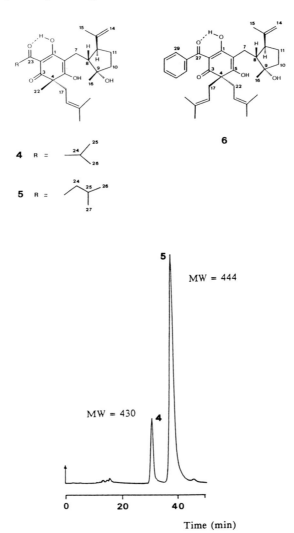

FIGURE 10. Semipreparative HPLC of the crystalline mixture **4/5**. Column: LiChrosorb RP-18 (250 × 16 mm); solvent: MeOH-H$_2$O 85:15 + 0.1% HOAc; flow rate: 10 ml/min; detection: 275 nm; sample: 60 mg.

permitted the separation of closely related compounds that differ only slightly in their R$_f$ values on thin-layer chromatography. Compound **10** was shown to be a phloroglucinol derivative that had fungicidal activity, but was not cytoinhibitory. Although DCCC has normally been used for polar compounds,[19] this example is a good demonstration of the application of the technique to lipophilic plant constituents. The tube number of the usual DCCC instrument was doubled to 580, thus allowing the separation of up to 3.5 g of extract in one run. If another 290 tubes were added, it was just possible to separate a 6 g sample, but back pressure became a limiting factor at this scale.

One disadvantage of DCCC is the length of time required to complete a separation — over 45 h in this case. For this reason, initial fractionation of the *H. calycinum* extract was attempted by centrifugal partition chromatography (CPC), a recently introduced liquid–liquid chromatography technique that permits faster run times.[20] Aliquots of the same extract were injected into two different machines, each using the same solvent system (Figure 13). The first instrument (Figure 13a) consists of 12 separation cartridges located on an enclosed rotor and has a capacity

a)   b)

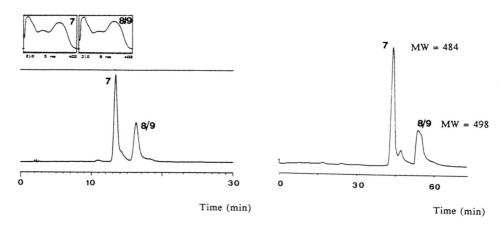

FIGURE 11. HPLC of the crystalline mixture 7/8/9. (a) Analytical HPLC. Column: LiChrosorb RP-18 (250 × 4.6 mm); solvent: MeOH-H$_2$O 85:15 + 0.1% HOAc; flow rate: 1.5 ml/min; detection: 280 nm. (b) Semipreparative HPLC. Column: LiChrosorb RP-18 (250 × 16 mm); solvent: MeOH-H$_2$O 85:15 + 0.1% HOAc; flow rate: 10 ml/min; detection: 275 nm; sample: 60 mg.

of 250 ml.[21] Although the run was complete within about 6 h, the sample load (100 mg) represents the maximum that could be injected without too much overlapping of peaks. The second instrument, the multilayer coil separator-extractor (P.C. Inc.) has at the heart of the instrument a long Teflon tube coiled around a drum (capacity 360 ml), which describes a planetary motion about a central axis.[22] Separation of the petroleum ether extract was performed in 2 h (Figure 13b), and there exists the possibility of increasing the load, especially if a two-pump modification of the instrument is employed.[23] This variant allows the quantity of stationary phase to be modified according to the loading and separation time required.

2. Structure Determination of Active Compounds

The active compounds (**4–9**) isolated from *H. calycinum* were shown to be related 2-acyl-1-hydroxy-cyclohexadien-3-ones. A characteristic of this class of molecule is the tautomerism of the cyclohexadienone ring (Figure 14).

a. Chinesin II (4)

Compound 4, from DCCC fraction 6, was an unstable colorless oil. Comparison of MS, ^{1}H-, and ^{13}C-NMR data with those of chinesin II from *H. chinense*[24] showed the compounds to be identical.

b. Hypercalin A (5)

The mass spectrum of **5** (hypercalin A) indicated a molecular weight of 444, i.e., 14 units higher than **4**, the reason being that instead of an *iso*-butyryl group, **5** possesses an *iso*-valeryl

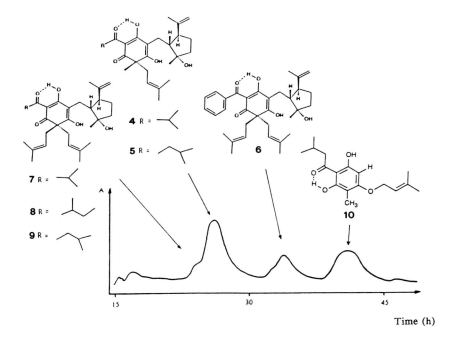

FIGURE 12. DCCC separation of a petroleum ether extract of *Hypericum calycinum* aerial parts. Instrument: Büchi DCC-Chromatograph, 580 tubes (2.7 mm I.D.); solvent system: petroleum ether-94% EtOH-EtOAc-H$_2$O 83:67:33:17 (mobile phase = upper phase); flow rate: 50 ml/h; detection: 280 nm; sample: 1.8 g extract dissolved in a mixture of the two phases.

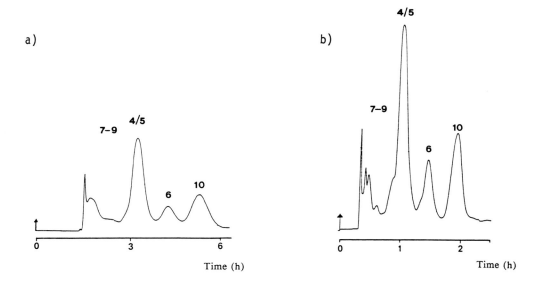

FIGURE 13. Application of centrifugal partition chromatography to the separation of a *Hypericum calycinum* petroleum ether extract. Solvent system: petroleum ether-94% EtOH-EtOAc-H$_2$O 83:67:33:17 (mobile phase = upper phase). (a) Sanki cartridge instrument (12 cartridges, 700 rpm); flow rate: 1 ml/min; detection: 275 nm; sample: 100 mg. (b) Ito multilayer coil separator-extractor (750 rpm); flow rate: 3 ml/min; detection: 275 nm; sample: 100 mg.

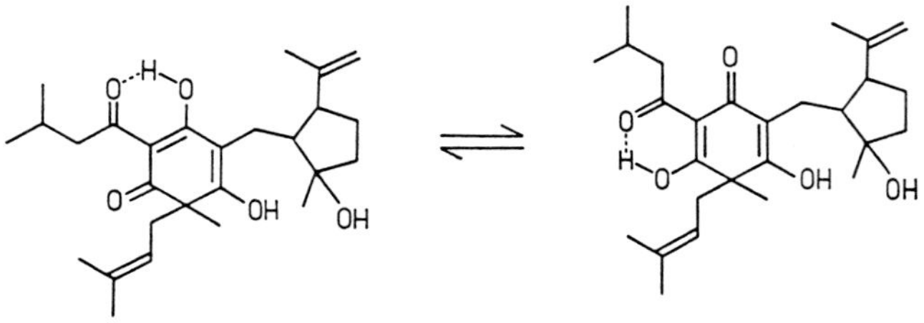

5

FIGURE 14. Tautomeric equilibrium of compound **5**.

(3-methylbutyryl) function. In the ^{1}H-NMR spectrum, the signals for protons H_2C-24 appeared as two doublet of doublets (J = 7, 14 Hz) centered at 3.04 and 2.79 ppm, while the H_3C-26 and H_3C-27 were at 0.98 and 0.96 ppm (each a doublet, J = 6.5 Hz). The ^{13}C-NMR spectrum confirmed the presence of an *iso*-valeryl group with a CH_2 at 48.2 ppm, a CH at 25.7 ppm, and two CH_3 entities at 22.8 and 22.6 ppm.

c. Hypercalin B (6)

The IR, ^{1}H, and ^{13}C-NMR spectra of **6** (hypercalin B) bear a certain resemblance to those of **4** and **5**. The enolic proton signal at 18.2 ppm, for example, is a feature of all three compounds. However, the signals due to the *iso*-valeryl chain in **5** are missing, while the presence of a phenyl group is indicated by multiplets at 7.42 and 7.33 ppm. The H_3C-22 singlet of hypercalin A (**5**) is replaced by a second prenyl chain [4.83 ppm (t), 2.56 and 2.51 ppm (2 × dd, J = 8, 14 Hz)] situated at C-4. In order to confirm the structure of **6**, recourse to different bidimensional NMR techniques was necessary. These included DQ-COSY (^{1}H-^{1}H correlation), HETCOR (^{1}H-^{13}C correlation), and LR-HETCOR (long-range ^{1}H-^{13}C correlation). As a first step it was necessary to assign the proton signals, and for this purpose, HETCOR spectra were recorded. A subsequent DQ-COSY spectrum established proton-proton connectivities. Thus, coupling between the geminal protons H_aC-7 (2.12 ppm, dd, J = 12, 15 Hz) and H_bC-7 (2.65 ppm, dd, J = 3, 15 Hz) was observed. These protons were each coupled to HC-8 (1.78 ppm, m), which, in turn, was coupled to HC-12 (2.37 ppm, ddd, J = 7, 11, 11 Hz). Coupling was observed between HC-12 and the multiplets at 1.44 (H_aC-11) and 1.74 ppm (H_bC-11). It was also possible to observe couplings between H_aC-10 and H_bC-10 (1.64 and 1.84 ppm) and between H_bC-11, H_aC-10, and H_bC-10. Long-range couplings were present in the prenyl groups as well as between H_2C-14 (4.83 and 4.86 ppm) and H_3C-15 (1.75 ppm, s).

An LR-HETCOR experiment was necessary to confirm the connectivities in the cyclohexadiene ring. This spectrum, recorded at 400 MHz, revealed a long-range correlation for C-1 and C-5 (189.0 and 176.2 ppm) with H_aC-7 and H_bC-7 (2.12 and 2.65 ppm). Coupling of C-5 and C-3 to the CH_2 moieties of the prenyl groups at C-4 was also observed.

d. Hypercalin C (7–9)

The crystalline product, hypercalin C, consisted of a mixture of three closely related substances (**7–9**). However, X-ray crystallography allowed confirmation of the structure **7** proposed after NMR experiments, since it was the major component of the mixture. As in the case of hypercalin B (**6**), compound **7** also possesses two prenyl groups, and there is a close structural resemblance between the two molecules. The only difference is the replacement of the phenyl group in **6** by the *iso*-propyl group in **7**. The six-membered ring exists in a half-chair conformation, while the five-membered ring has an "envelope" conformation, with C-12 somewhat above the plane of the other four atoms. The bond lengths in the hexacycle indicate extensive delocalization around C-2. The observation that there is an

TABLE 1
Cytoinhibition by *Psorospermum febrifugum* Root Bark Extracts in the Co-115 Human Colon Carcinoma Cell Line

Extract[a]	ED$_{50}$ (µg/ml)
Petroleum ether	4.8×10^{-1}
Chloroform	3.5×10^{-1}
Methanol	> 10

[a] Extraction with solvents of increasing polarity.

intramolecular bonding interaction between the proton on the C-5 hydroxyl group and the oxygen of the C-9 hydroxyl group suggests that this part of the molecule is fairly rigid and explains why the coupling constants of the two protons on C-7 can be so readily measured.

The two isomers **8** and **9** were not separated by semipreparative HPLC. One (**8**) has a 2-methylbutyryl group and the other (**9**) a 3-methylbutyryl group. For this reason, the ^{1}H-NMR spectrum of the mixture of **8** and **9** showed a sextet at 3.93 ppm (for the C-28 proton in **8**) and two doublet of doublets at 3.00 and 2.85 ppm (for the two C-28 protons of **9**). The difference in the acyl chains was also observed in the ^{13}C-NMR spectrum, i.e., a CH (42.0 ppm), a CH$_2$ (26.4 ppm), and two methyl groups (16.6 and 11.9 ppm) for the 2-methylbutyryl group of **8** and resonances at 48.2 (CH$_2$), 25.9 (CH), and 22.8 and 22.7 ppm (two methyl groups) for the 3-methylbutyryl group of **9**.

C. *PSOROSPERMUM FEBRIFUGUM*

Psorospermum febrifugum (Guttiferae) is a shrub with a wide distribution over southern and central Africa. It finds a variety of uses in African traditional medicine for the treatment of fever, leprosy, wounds, and skin diseases.[25] Like many members of the Guttiferae, a yellow resin can be obtained from the bark and root bark. The cytoinhibitory properties of the root bark extracts are shown in Table 1. The petroleum ether and chloroform extracts were active, but the methanol extract had no marked effect on the growth of the cells.

A previous phytochemical investigation of the plant had led to the isolation of an antileukemic xanthone from the roots, but this was obtained from the more-polar extracts.[26] Since, in the present work, the lipophilic petroleum ether extract exhibited cytoinhibitory activity, it was subjected to the bioactivity-guided fractionation procedure shown in Figure 15 for the isolation of the antiproliferative compounds.[10]

1. Isolation and Structural Determination of Lipophilic Extract Components

Flash chromatography on silica gel resulted in rapid elution of the anthraquinone 3-geranyloxyemodin (**11**), which was inactive in the Co-115 bioassay. This was followed by 3-geranyloxy-6-methyl-1,8-dihydroxyanthrone (**13**), which crystallized from fraction II and was discovered to be an artifact of the separation procedure.[10] A tetrahydroanthracene, acetylvismione D (**14**), crystallized from fraction III. This had a UV spectrum characteristic of vismiones[27] and absorptions at 1645 cm^{-1} and 1740 cm^{-1} in the IR spectrum, which indicated a chelated carbonyl group and an ester function, respectively. In the ^{1}H-NMR spectrum, the presence of two aromatic hydroxyl groups was inferred from the signals at 15.92 and 9.73 ppm. The protons at C-10, C-4, and C-2 of the vismione nucleus resonated at 6.84, 6.56, and 6.51 ppm, respectively. When compared with the equivalent spectrum of vismione D (**15**), the H$_3$C-11 signal had moved downfield to 1.70 ppm, while the two singlets for H$_2$C-5 and H$_2$C-7 had become a pair of doublets at 3.72 and 3.06 ppm and a pair of doublets at 3.20 and 2.84 ppm. Position C-3 was substituted by a geranyl moiety, as in vismione D (**15**).[27] The ^{13}C-NMR spectrum of **14** was virtually identical to that of vismione D in the regions of the aromatic ring and geranyloxy side-

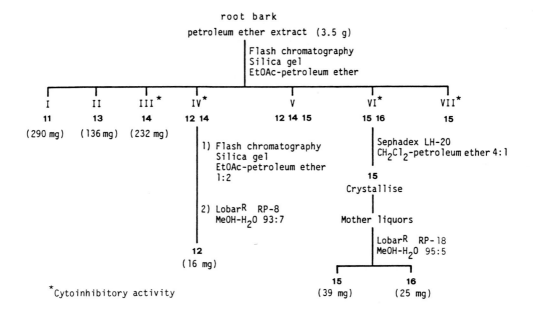

FIGURE 15. Fractionation of a petroleum ether extract of *Psorospermum febrifugum* root bark.

chain signals. The only notable differences in the carbon shifts were observed for the nonaromatic ring, the C-5 signal appearing at 39.8 ppm, C-6 at 80.7 ppm, and C-7 at 49.2 ppm. Two additional signals were also present in the spectrum: the carbonyl carbon of the acetyl group (170.4 ppm) and the corresponding methyl group (22.0 ppm).

Fraction IV also contained some **14** and further chromatography was required to isolate a second anthraquinone (**12**). This involved a second flash chromatography step and final purification by Lobar® low-pressure liquid chromatography on 40–63 μm RP-8 bonded phase. The new natural product (**12**) had a very similar UV spectrum to the emodin derivative (**11**), but the mass spectrum showed a molecular ion at *m/z* 424, representing a molecular weight 18 units higher. Instead of the broad singlet at 2.11 ppm for H_2C-16 and H_2C-17, and the multiplet for HC-18 at 5.13, found in the ¹H-NMR spectrum of **11**, compound **12** produced a triplet at 2.10 ppm and a multiplet centered at 1.52 ppm. This indicated a saturated C18 –C19 linkage, confirmed by the fact that free rotation about this bond gave rise to a singlet at 1.21 ppm for H_3C-20 and H_3C-21. Consequently, the structure of **12** is the same as **11**, except for the presence of an additional hydroxyl group at the C-19 position.[10]

A small quantity of vismione D (**15**) was isolated from fraction VI by centrifugal thin-layer chromatography[19] with ethyl acetate-petroleum ether as the eluent. However, since there was considerable adsorption to the silica-gel-coated plate, chromatography was repeated on Sephadex LH-20 in order to obtain a better yield of **15**. LPLC of the mother liquors on an octadecylsilica stationary phase, with methanol-water 95:5 as the solvent, allowed the isolation of an additional tetrahydroanthracene (**16**). The position of the geranyl side chain in this molecule was determined by ¹H-NMR spectroscopy, using NOE experiments.

The cytoinhibitory chloroform extract of the plant was shown to contain two major components. These were separated by flash chromatography on silica gel, followed by LPLC on a Lobar RP-18 column and were found to be vismione D (**15**) and vismione F (**17**) (Figure 16).[28]

2. Improved Isolation for Greater Yields

In order to obtain larger quantities of potential antiproliferative vismiones for an in-depth investigation of their *in vitro* and *in vivo* activities, a different separation strategy was adopted.

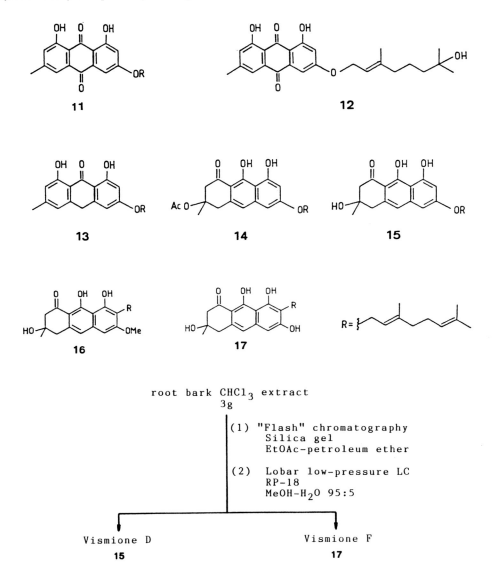

FIGURE 16. Fractionation of a chloroform extract of *Psorospermum febrifugum* root bark.

The main problem with an approach using liquid–solid chromatography was the irreversible adsorption and decomposition of the bioactive natural products on silica-gel-based supports. As a consequence, it was decided to turn to the liquid–liquid partition methods already described in the previous sections. These centrifugal techniques are rapid and permit the preparative separation of samples in a matter of hours. Using a cartridge system (Sanki), anthraquinone (**11**) and the two vismiones, **14** and **15**, were obtained in pure form directly from a 100-mg charge of the petroleum ether extract. A nonaqueous solvent system compatible with the liposolubility of these anthranoids (i.e., n-hexane-acetonitrile-methanol 40:25:10, mobile phase = upper phase) was employed.

A third vismione (**16**) eluted after **14** and **15**, but was contaminated with an impurity (Figure 17a).[20] Increasing the number of cartridges to 12 allowed a better resolution of the peaks, but lengthened the separation time to nearly 8 h (Figure 17b). However, this situation permitted an increased sample load of 500 mg of extract (Figure 17c). Similar separations

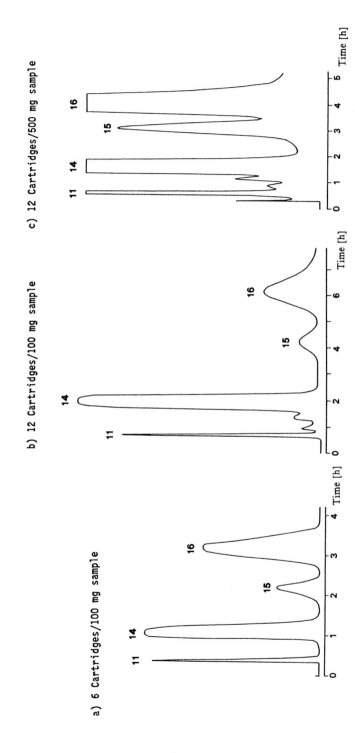

FIGURE 17. Fractionation of a *Psorospermum febrifugum* root bark petroleum ether extract by a CPC cartridge instrument. Solvent system: *n*-hexane–acetonitrile–methanol 40:25:15 (mobile phase = upper phase); detection: 254 nm. (a) Rotational speed: 1500 rpm; flow rate: 5.5 ml/min. (b) Rotational speed: 800 rpm; flow rate: 5.5 ml/min. (c) Rotational speed: 800 rpm; flow rate: 6.5 ml/min.

TABLE 2
Cytoinhibitory Activity of *Hypericum* Constituents in the Co-115 Cell Line

Substance	ED_{50} (µg/ml)
Crystals **1/2**	7.0×10^{-1}
Crystals **4/5**	3.7×10^{-1}
Compound **6**	3.4×10^{-1}
Crystals **7/8/9**	3.0×10^{-1}
5-Fluorouracil	2.3×10^{-1}
Actinomycin D	3.2×10^{-4}
Vinblastine	3.6×10^{-3}
Etoposide (VP 16-213)	2.2
Bleomycin sulfate	2.2

TABLE 3
Cytoinhibition of *Hypericum* Constituents Toward an L1210 Murine Leukemia Cell Line

Substance	L1210 Growth inhibition (IC_{50})
Crystals **1/2**	20–25 µM
Crystals **4/5**	4–6 µM
Compound **6**	10–12 µM
Navelbine	16 nM

were also possible on an Ito multilayer coil separator-extractor (P.C. Inc.) equipped with a 360 ml, 2.6 mm I.D. coil.[23]

D. BIOACTIVITY OF COMPONENTS FROM *H. REVOLUTUM*, *H. CALYCINUM*, AND *P. FEBRIFUGUM*

Compound **6**, the mixtures **1/2**, **4/5**, and the crystalline product **7/8/9** from *H. calycinum* exhibited a marked *in vitro* inhibition of the growth of Co-115 human colon carcinoma cells (Table 2). While these compounds were weaker than actinomycin D and vinblastine (from the Madagascar plant *Catharanthus roseus*, Apocynaceae), they were more effective than etoposide or bleomycin. Furthermore, hyperevolutins A (**1**) and B (**2**) and the hypercalin derivatives gave similar results in the Co-115 test to 5-fluorouracil, a synthetic molecule that is successfully used for the treatment of certain colorectal tumors.

An important step during the evaluation of potential antitumor compounds is the investigation of antiproliferative activity against other cell lines. In this way an eventual selective toxicity toward a particular cell line may be revealed. Consequently, the five cytoinhibitory compounds from *H. revolutum* and *H. calycinum* were tested in the L1210 murine leukemia cell line, and IC_{50} values were calculated (Table 3). The commercially available anticancer agent Navelbine® was used as a reference compound. The table shows that all five compounds tested exhibited reasonable growth inhibition. Further assays with human cell lines should allow conclusions to be drawn about their selectivities. Testing the active compounds *in vivo* (in athymic mice, for example) will give important information about their activity, stability, bioavailability, and toxicity.

Inhibition of Co-115 colon carcinoma cell growth by the different anthranoid pigments from *P. febrifugum* is shown in Table 4. The anthraquinones **11** and **12** were inactive, while vismiones, **14–17**, possessed a similar degree of cytoinhibition. The anthrone, **13**, had borderline activity. Cytoinhibition by vismione D (**15**) is very similar to that exhibited by the

TABLE 4
Cytoinhibitory Activity of *P. febrifugum* Root Bark Anthranoid Pigments
in the Co-115 Cell Line

Compound	ED_{50} (µg/ml)
11	> 10
12	> 10
13	4.3
14 (Acetylvismione D)	3.8×10^{-1}
15 (Vismione D)	1.5×10^{-1}
16	5.0×10^{-1}
17 (Vismione F)	5.0×10^{-1}

cancer chemotherapeutic agent 5-fluorouracil, but is less than that of the antileukemic alkaloid vinblastine from *Catharanthus roseus* (Apocynaceae). Although psorospermin and its xanthone congeners are active in the KB and P-388 bioassays, they are only found in very small quantities in the plant (less than 0.004%).[29] On the other hand, vismione D (**15**), acetylvismione D (**14**) and vismione (**16**) are the major components of the petroleum ether extract of *P. febrifugum* root bark (about 5% by weight of dry material) and are the most active of the anthranoid pigments. Preliminary experiments with vismione D (**15**) in athymic mice bearing human colon carcinoma subcutaneous xenografts have given very encouraging *in vivo* results, although the solubility of this lipophilic compound presents certain difficulties.[30]

IV. CONCLUSIONS

The plant kingdom is a huge source of very diverse pharmacologically active compounds. The potential for the discovery of anticancer drugs is, of course, immense, and the possibilities should be exploited to the full. The key to the successful separation of these substances is a strategy involving systematic bioassay-guided fractionation of active plant extracts, using as a key manipulation a suitable predictive and simple biological test. The aim of this chapter has been to illustrate this approach by means of a readily available human colon carcinoma cell line. Cytoinhibition in this cell line has been employed as a first discriminatory step in the discovery of potential antineoplastic agents. When an activity is established, recourse to a variety of modern, rapid chromatographic techniques is necessary for the efficient isolation of the active principles. Once these have been obtained, further biological testing can only be undertaken when complete structure elucidation of the compounds has been performed. The procedure may be lengthy, but when one considers the benefits gained from the antileukemic activity of the *Catharanthus* alkaloids, for example, the importance of a better understanding of plant constituents is obvious.

ACKNOWLEDGMENTS

Financial support has been provided by the Swiss National Science Foundation and the Swiss Cancer League. Bioassays were carried out at the Swiss Institute for Experimental Cancer Research (laboratory of Dr. B. Sordat) by J.-C. Chapuis, to whom we express our gratitude.

REFERENCES

1. **Suffness, M.,** New approaches to the discovery of antitumour agents, in *Proc. Phytochemical Society of Europe,* Vol. 27, *Biologically Active Natural Products,* Hostettmann, K. and Lea, P. J., Eds., Clarendon Press, Oxford, 1987, 85.
2. **Kelly, M. G. and Hartwell, J. L.,** The biological effects and chemical composition of podophyllin. I. The biological effects of podophyllin, *J. Natl. Cancer Inst.,* 14, 967, 1954.
3. **Noble, R. L., Beer, C. T., and Cutts, J. H.,** Role of chance observations in chemotherapy: *Vinca rosea, Ann. N.Y. Acad. Sci.,* 76, 893, 1958.
4. **Hamburger, M., Marston, A., and Hostettmann, K.,** Search for new drugs of plant origin, in *Advances in Drug Research,* Vol. 20, Testa, B., Ed., Academic Press, London, 1991, 167.
5. **Geran, R. T., Greenberg, M. M., MacDonald, A. M., Schumacher, A. M., and Abbott, B. J.,** Protocols for screening chemical agents and natural products against animal tumors and other biological systems, *Cancer Chemother. Rep. (Part 3),* 3, 1, 1972.
6. **Lee, K.-H., Lin, Y.-M., Wu, T.-S., Zhang, D.-C., Yamagishi, T., Hayashi, T., Hall, I. H., Chang, J. J., Wu, R.-Y., and Yang, T.-H.,** The cytotoxic principles of *Prunella vulgaris, Psychotria serpens* and *Hyptis capitata*: ursolic acid and related derivatives, *Planta Med.,* 54, 308, 1988.
7. **Cassady, J. M., Baird, W. M., and Chang, C.-J.,** Natural products as a source of potential cancer chemotherapeutic and chemopreventive agents, *J. Nat. Prod.,* 53, 23, 1990.
8. **Suffness, M. and Pezzuto, J. M.,** Assays related to cancer drug discovery, in *Methods in Plant Biochemistry* (Dey, P. M. and Harborne, J. B., Series editors), Vol. 6, *Assays for Bioactivity,* Hostettmann, K., Ed., Academic Press, London, 1991, 71.
9. **Carrel, S., Sordat, B., and Merenda, C.,** Establishment of a cell line (Co-115) from a human colon carcinoma transplanted into nude mice, *Cancer Res.,* 36, 3978, 1976.
10. **Marston, A., Chapuis, J.-C., Sordat, B., Msonthi, J. D., and Hostettmann, K.,** Anthracenic derivatives from *Psorospermum febrifugum* and their *in vitro* cytotoxicities to a human colon carcinoma cell line, *Planta Med.,* 52, 207, 1986.
11. **Landegren, U.,** Measurement of cell numbers by means of the endogenous enzyme hexosaminidase. Applications to detection of lymphokines and cell surface antigens, *J. Immunol. Methods,* 67, 379, 1984.
12. **Chapuis, J.-C., Sordat, B., and Hostettmann, K.,** Screening for cytotoxic activity of plants used in traditional medicine, *J. Ethnopharmacol.,* 23, 272, 1988.
13. **Marston, A. and Hostettmann, K.,** Antifungal, molluscicidal and cytotoxic compounds from plants used in traditional medicine, in *Proc. Phytochemical Society of Europe,* Vol. 27, *Biologically Active Natural Products,* Hostettmann, K. and Lea, P. J., Eds., Clarendon Press, Oxford, 1987, 65.
14. **Hostettmann, K.,** Droplet counter-current chromatography and its application to the preparative scale separation of natural products, *Planta Med.,* 39, 1, 1980.
15. **Décosterd, L. A.,** Isolement et Détermination de Structure de Constituents Biologiquement Actifs de Deux Espèces de la Famille des Guttifères: *Hypericum revolutum* Vahl et *Hypericum calycinum* L., Ph.D. thesis, Lausanne University, 1990.
16. **Brondz, I., Greibrokk, T., Groth, P. A., and Aasen, A. J.,** The relative stereochemistry of hyperforin — an antibiotic from *Hypericum perforatum* L., *Tetrahedron Lett.,* 24, 1299, 1982.
17. **Décosterd, L. A., Stoeckli-Evans, H., Chapuis, J.-C., Msonthi, J. D., Sordat, B., and Hostettmann, K.,** New hyperforin derivatives from *Hypericum revolutum* Vahl with growth-inhibitory activity against a human colon carcinoma cell line, *Helv. Chim. Acta,* 72, 464, 1989.
18. **Décosterd, L. A., Stoeckli-Evans, H., Chapuis, J.-C., Sordat, B., and Hostettmann, K.,** New cell growth-inhibitory cyclohexadienone derivatives from *Hypericum calycinum* L., *Helv. Chim. Acta,* 72, 1833, 1989.
19. **Hostettmann, K., Hostettmann, M., and Marston, A.,** *Preparative Chromatography Techniques: Applications in Natural Product Isolation,* Springer-Verlag, Berlin, 1986.
20. **Marston, A., Slacanin, I., and Hostettmann, K.,** Centrifugal partition chromatography in the separation of natural products, *Phytochem. Anal.,* 1, 3, 1990.
21. **Murayama, W., Kobayashi, T., Kosuge, Y., Yano, H., Nunogaki, Y., and Nunogaki, K.,** A new centrifugal countercurrent chromatograph and its application, *J. Chromatogr.,* 239, 643, 1982.
22. **Ito, Y., Sandlin, J., and Bowers, W. G.,** High-speed preparative counter-current chromatography with a coil planet centrifuge, *J. Chromatogr.,* 244, 247, 1982.

23. **Slacanin, I., Marston, A., and Hostettmann, K.,** Modifications to a high-speed counter-current chromatograph for improved separation capability, *J. Chromatogr.,* 482, 234, 1989.

24. **Nagai, M. and Tada, M.,** Antimicrobial compounds, chinesin I and II from flowers of *Hypericum chinense* L., *Chem. Lett.,* 1337, 1987.

25. **Watt, J. M. and Breyer-Brandwijk, M. G.,** *Medicinal and Poisonous Plants of Southern and Eastern Africa,* E. and S. Livingstone, Edinburgh, 1962, 498.

26. **Kupchan, S. M., Streelman, D. R., and Sneden, A. T.,** Psorospermin, a new antileukaemic xanthone from *Psorospermum febrifugum, J. Nat. Prod.,* 43, 296, 1980.

27. **Botta, B., Delle Monache, F., Delle Monache, G., Marini Bettolo, G. B., and Oguakwa, J. U.,** 3-Geranyloxy-6-methyl-1,8-dihydroxyanthraquinone and vismiones C, D and E from *Psorospermum febrifugum, Phytochemistry,* 22, 539, 1983.

28. **Botta, B., Delle Monache, F., Delle Monache, G., Marini Bettolo, G. B., and Msonthi, J. D.,** Prenylated bianthrones and vismione F from *Psorospermum febrifugum, Phytochemistry,* 24, 827, 1985.

29. **Abou-Shoer, M., Boettner, F. E., Chang, C.-J., and Cassady, J. M.,** Antitumour and cytotoxic xanthones of *Psorospermum febrifugum, Phytochemistry,* 27, 2795, 1988.

30. **Chapuis, J.-C.,** Recherche d'Activités Cyto-inhibitrices à Partir de Plantes Médicinales: Mise au Point d'un Test de Dépistage Utilisant des Cellules Carcinomateuses Coliques Humaines, Ph.D thesis, Lausanne University, 1988.

Chapter 12

INSECT BEHAVIOR MODIFIERS

Shozo Takahashi

TABLE OF CONTENTS

0-8493-4372-0/93/$0.00+$.50
© 1993 by CRC Press, Inc.

I. INTRODUCTION

Ethological studies of insects have provided important insight into the nature of environmental cues that elicit certain behavioral responses in these animals. The "key stimulus" that results in a specific behavioral response is often remarkably simple in nature. This seems to be true not only of physical stimuli, but also of chemical signals.

A number of chemicals are involved in the conveying of information in intra- and interspecific interactions between individual organisms. Chemical communication is a major means of information transfer in the insect community and is especially highly developed in locating a mate, food, and an oviposition site. Chemicals employed in this communication are termed semiochemicals and include intraspecific (pheromone) and interspecific (allelochemical) substances.[1] Allelochemicals are further subdivided into those that benefit the stimulus emitter (allomones), the stimulus receiver (kairomones), or both (synomones). The responses induced by pheromones, allomones, or kairomones may be immediate behavioral reactions or long-lasting physiological changes.

Other chemicals that modify insect behavior are classified according to the type of behavior they induce.[2] The following six categories encompass most behavior modification active chemicals:

1. A locomotory stimulant is a chemical that causes a kinetic reaction in an insect, by increasing locomotion or speeding-up the turning rate.
2. An arrestant is a chemical that causes a kinetic reaction in an insect, by decreasing locomotion or slowing the turning rate.
3. An attractant is a chemical that causes an insect to make oriented movements toward it.
4. A repellent is a chemical that causes insects to make oriented movements away from it.
5. A feeding, mating, or ovipositional stimulant is a chemical that elicits one of these three behavioral reactions.
6. A feeding, mating, or ovipositional deterrent is a chemical that inhibits (suppresses) one of these behavioral reactions.

The outstanding advances in chemical technology during the past two decades has made possible the discovery, identification, and synthesis of many such behavior-modifying chemicals. This work was accelerated by the introduction of ethological work on the role of chemical cues, and the term "chemical ethology" has been coined for such studies.[3] Identification and studies on the role of insect behavior modifiers in some of the most common insects are summarized in this chapter.

II. STIMULATION AND SUPPRESSION OF WING-RAISING ACTIVITY IN MALE *NAUPHOETA CINEREA* (LOBSTER COCKROACH)

Nauphoeta cinerea (Olivier)(Dictyoptera: Blaberidae) probably spread originally from East Africa to Madagascar, Mauritius, Indonesia, Singapore, the Philippines, and other areas of the Far East, via shipping. It now occurs in many parts of the world, including Australia, the Ryukyu Islands, the eastern part of Africa as far as Egypt, the West Indies, Brazil, Mexico, Hawaii, and Florida. Its common name, lobster cockroach, is derived from the pattern on its pronotum. Adults of this species, measuring 25 to 29 mm in length, live for about 1 year under laboratory conditions. Mating occurs after about 6 days of adulthood, and the first ootheca appears about 1 week later.

A. SEX RECOGNITION MECHANISM OF *NAUPHOETA CINEREA*

During courtship the male raises his wings and tegmina upon recognizing a sexually mature female. In contrast, when a male meets another male, antagonistic behavior with mutual aggressive antennal fencing is often observed. Many investigations into the mating and aggressive behavior of this species have been reported.[4-8] Roth and Dateo reported that the *N. cinerea* male secretes "seducin", after raising its wings and tegmina, and eventually the female mounts to lick the secretion.[5] They considered that the males raised their wings to court females and seduce them with the secretion. Recently, Sreng showed that females emit a pheromone specifically to attract males.[9] However, we have found that males raise their wings following recognition of teneral and mature females and teneral males. Thus, they raise their wings, not to court females, but as a response to sex recognition. The same behavior in males was elicited by presenting them with antennae excised from mature females. Our research provides strong evidence that *N. cinerea* males recognize other males, females, and nymphs of the species by contact chemical perception, mostly through antennal contact to the antennae and parts of body.[8,10-12]

1. Observation of Mating Behavior

Males and females were reared separately, following imaginal ecdysis, and were maintained for 2 to 4 weeks until observation began. A male and female were then confined in a glass pot (ll cm in diameter by 7 cm high) to observe their mating behavior. Observations were made in a darkroom, equipped with a red light (10 W) and maintained at an ambient temperature of $26 \pm 1°C$, during the darkphase (2:00 to 5:00 pm) of a light-dark cycle consisting of 14 h (6:00 am to 8:00 pm) of darkness and 10 h of light.

After walking for a while in the glass pot, the male touches the female's antenna(e) and/or body with his antennae. The ensuing behavior is represented schematically in Figure 1. At successful genitalia connection, he turns from beneath her, and they take an opposed (end-to-end) position. If the female is unreceptive, the male brings down his wings and tegmina, faces her again, and reexamines her by antennal contact, often accompanied by a stridulation sound.[13]

2. Observation of Aggressive Behavior

Two mature males (2 to 4 weeks old) were placed together in a glass pot, and their behavior observed. The following description of intraspecific aggressive behavior is based on observations of ten pairs.

Soon after placing the two males in the pot, they displayed aggressive antennal fencing, often butting and biting each other on the abdomen and wings. A superior-loser relationship appears after these struggles. The loser walks away or takes a subordinate posture to the superior. Once this relationship is determined between the two, the loser no longer attacks his superior. Intraspecific fighting between males, and mate selection by females, have been

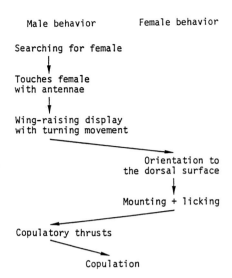

FIGURE 1. Schematic representation of the mating behavior of *N. cinerea*.

discussed by Ewing[6] and by Breed et al.[14] However, the aggressive posture observed by Ewing as part of the fighting behavior is claimed to be identical to the posture adopted when a calling male seeks to attract receptive females.[9]

3. Effect of Age on Mating and Aggressive Behavior

Each 20-day-old test male was placed in a glass pot and individually introduced to a male or female of a different age, from 3 h after imaginal ecdysis to 60 days old. The responses of the test males toward both intact *N. cinerea* and the excised antennae were observed for 2 min.

A wing-raising response was observed on more than 70% of occasions, when test males were presented with females or their excised antennae (3 h to 60 days old). When test males were presented with other males (or their excised antennae) 3 to 12 h or 1 to 2 days following imaginal ecdysis, a wing-raising response was observed on 80 to 50% of occasions, respectively. No wing-raising response was elicited in the test males when presented with males more than 3 days old, indicating a 3-day time interval for male maturation. This is in accord with a report describing the aggressive behavior of 20-day-old males toward other males older than 3 days.[7] This behavior is graphically illustrated in Figure 2.

B. EXTRACTION AND IDENTIFICATION OF WING-RAISING STIMULANTS AND SUPPRESSANTS

1. Bioassay

Since the male's wing-raising response is fully elicited upon his antennal contact with excised parts of female body, such as the antennae, a bioassay using antennae from *N. cinerea* and *Periplaneta japonica* males was devised. In order to investigate the male's response to mature males and females, antennae of both sexes (2 to 4 weeks old) were tested after excision from the body. More than 80% of the tested female antennae showed wing-raising activity, whereas the males' antennae were inactive. When the antennae were washed with ether and then brought into contact with antennae of test males, about 60% of male and female antennae initiated a wing-raising response. On the contrary, antennae excised from male *P. japonica* showed no wing-raising activity before or after the ether wash.

2. Wing-Raising Stimulants

Entire male and female cockroaches (2 to 4 weeks old) were separately immersed in hexane. The extracts were dried over anhydrous sodium sulfate, filtered, and evaporated,

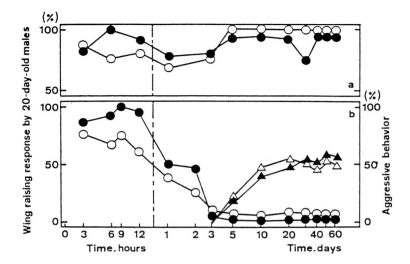

FIGURE 2. Wing-raising response by 20-day-old males; (a) to females (●) and female antennae (○) of different ages; (b) to males (●) and male antennae (○) of different ages. (b) Aggressive behavior by 20-day-old males to males (▲) and male antennae (△) of different ages.

providing crude extracts of male and female cockroaches. A solution of the crude extract (ten female or male equivalents) in 1 ml CCl_4 was used for the bioassay. Excised antennae from *N. cinerea* and *P. japonica* males were then dipped in the test solution for 1 s.

The ether-washed *N. cinerea* antennae of both sexes, after immersion in the test solution derived from the females, showed 100% wing-raising activity. Coating of the ether-washed *P. japonica* antennae with the female extract also resulted in 100% activity. The crude extract from males showed 50% wing-raising activity in this bioassay.

The crude extracts were chromatographed on silica gel. The active compounds, eliciting a 100% wing-raising response in the bioassay, were found in the hexane fraction from both sexes. GLC and gas chromatography-mass spectrometry (GC-MS) analysis of the active fractions showed that the eluate contained saturated hydrocarbons of C_{31} to C_{53} as shown in Table 1.

3. Wing-Raising Suppressants

Unmated males (a total of 7000 insects, 2 to 4 weeks old) were immersed in hexane. The hexane extract (5.31 g) was chromatographed on a silicic acid column (60 g, Polygosil 60-4063) and eluted successively with 1 l each of hexane; 5, 10, 20, and 50% ether in hexane; ether; and methanol. Of these fractions, the hexane fraction showed 100% wing-raising activity, while only the 5% ether-in-hexane fraction (0.435 g) completely suppressed the wing-raising activity of ether-washed male antennae. Further purification of the 5% ether-in-hexane fraction, on columns of silicic acid and silicic acid impregnated with 16.7% $AgNO_3$, gave a compound of mp 34 to 36°C, which was given the trivial name of nauphoetin. A solution of nauphoetin in CCl_4 (50 µg in 1 ml) suppressed the wing-raising activity of ether-washed male antennae when they were immersed in the solution for 1 s.

GC-MS analysis of nauphoetin indicated a molecular weight of 618, and the IR absorption spectrum implied an ester entity (υ_{max}, 1735 cm^{-1}). Methanolysis afforded stearyl alcohol and a compound (M^+ 380) that yielded n-pentadecanal and 8-carbomethoxyoctanal on ozonolysis. These results indicated that nauphoetin is either the (Z) or (E) isomer of octadecyl-9-tetracosenoate. Both isomers were subsequently synthesized (Figure 3) and compared to the natural compound. This established the structure of nauphoetin as octadecyl-(Z)-9-tetracosenoate (**1**). The synthetic nauphoetin (**1**) showed suppressant activity in the bioassay (Table 2).

TABLE 1
Composition of the Cuticular Hydrocarbons of *N. cinerea*

Hydrocarbon	Composition (%)
n-C_{31}	4.89
Me-C_{31}	1.10
3-Me-C_{31}	4.89
Me-C_{32}	0.34
n-C_{33}	4.54
5-, 7-, 9-, & 11-Me-C_{33}	21.32
6,10-, 8,12-, & 10,14-di-Me-C_{33}	8.58
5-, 7-, 9-, 11-, & 13-Me-C_{35}	8.58
11,15- & 13,17-di-Me-C_{35}	23.82
C_{45}	0.03
C_{47}	0.38
C_{49}	.0.74
C_{51}	2.54
C_{51}	0.37
C_{52}	3.53
C_{53}	0.40

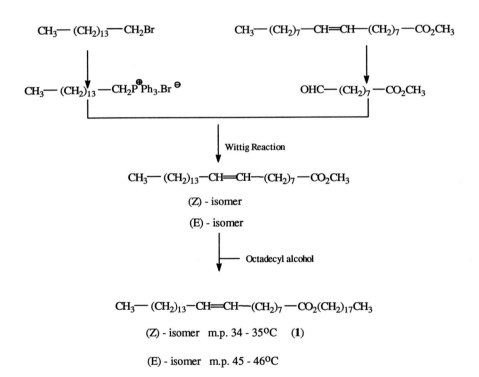

FIGURE 3. Synthesis of nauphoetin.

Mature male *N. cinerea* often raise their wings toward teneral nymphs of each instar, but not toward insects in the latter periods of nymphal stages. In fact, the last instar nymphs produce another suppressant of wing-raising stimulant, at the latter period of the stage. The suppressant has a molecular formula of $C_{51}H_{98}O$, which is different from nauphoetin.

TABLE 2
Biological Activity[a] of (Z)-Nauphoetin

Hydrocarbon (HPLC) fr. 2 +	(Z)-nauphoetin	Wing-raising (%)
10 mg/ml CCl$_4$	0	90
10 mg/ml CCl$_4$	0.0025 mg	87
10 mg/ml CCl$_4$	0.025 mg	33
10 mg/ml CCl$_4$	0.5 mg	3
10 mg/ml CCl$_4$	1.0 mg	0
10 mg/ml CCl$_4$	5.0 mg	0

[a] Substrate for the test: *P. japonica* antennae.

FIGURE 4. A general scheme of parasitization with stimuli (S_1–S_9).[15]

III. OVIPOSITION STIMULANTS IN THE COCCOID CUTICULAR WAXES OF PARASITIC WASPS

Most parasitic insects are more properly called parasitoids because offspring from eggs deposited on a host gradually eat the host as they develop. Parasitoids utilize a variety of stimuli to locate and identify their host insects, including chemical cues emanating from their host insects. They respond to host insect odor, sex pheromones, epideictic pheromones, aggregation pheromones, webbing, honeydew, body scales, and eggs. Parasitoids lay eggs after recognizing host insects by antennal contact with nonvolatile kairomones in the cuticular wax of the host insects. Various stimulants (S_1–S_9, Figure 4) present both in mobile and immobile host insects have been investigated.[15]

A. OVIPOSITION STIMULANTS IN *CEROPLASTES* SPP. TO *ANICETUS BENEFICUS*

The encyrtid parasitoid *Anicetus beneficus* Ishii and Yasumatsu was first found in Kyushu, Japan by Yasumatsu and Tachikawa in 1949.[16,17] The rate of parasitization by the parasitoid at that time was so high that the host species, the scale insect *Ceroplastes rubens* Maskell, was kept under the latent pest level in many citrus orchards in the area. Releases of *A. beneficus* were made in other areas that had a dense population of *C.rubens* and were successful in the control of the scale insect. Four *Anicetus* species have been known in Japan, and *A. beneficus* has an especially high host specificity for *C. rubens* under natural conditions.

TABLE 3
Bioassay Results of Ovipositional Behavior on a Paraffin Model[a]

Sample	No. of insects examined	No. of turnings (mean)	Total oviposition time (mean)	Drilling behavior observed (%)
Extract	12	14.4	3'23"	100
Blank	15	0	—	0

[a] M.p. of paraffin: 42 to 44°C.

However, each *Anicetus* species oviposited on *Ceroplastes* species, other than the natural host, when they were provided under laboratory conditions.[18,19]

Host recognition by the female *A. beneficus* seems to begin with perception of a chemical cue from the scale covering of the natural host, *C. rubens*. The relationship between the wax composition of the genus *Ceroplastes* and its stimulating effect on oviposition by *A. beneficus* and *A. ceroplastes* has been studied.

1. Ovipositional Behavior of *A. beneficus*

The ovipositional behavior of female *A. beneficus* toward the first instar nymph of the host scale, *C. rubens*, was observed in the laboratory.[20] The female *A. beneficus* begins by random walking, with antennal flagellation (drumming), on the host plant of the scale insect. During the walk the parasitoid perceives the scale insect via antennal contact and then examines the host.

Before ovipositor drilling the parasitoid continues to walk back and forth on the host, making several turns at the end of the scale. After this the parasitoid chooses a site for the "drilling" of the ovipositor.

2. Bioassay for Ovipositional Stimulants

To bioassay for ovipositional stimulants in the scale wax of *C. rubens*, a paraffin model was made in a hemispherical shape (3 mm in diameter and 2 mm high) and impregnated with a chloroform solution of the crude extracts of the scale wax (0.1 mg = 0.1 insect equivalent). The model was kept at room temperature overnight and then placed in a test tube (6 mm in diameter and 50 mm long) with a test female *A. beneficus*.[21] She examined the model surface with her antennae. She then walked back and forth, making several turns at the edge before drilling with her ovipositor. However, she did not deposit an egg. This ovipositional behavior was very similar to that observed toward a natural host, except for the egg deposition. Bioassay results are shown in Table 3.[22]

3. Extraction, Purification, and Identification of the Ovipositional Stimulants in the Scale Wax of the Genus *Ceroplastes*

Adult *C. rubens* (4 g) were soaked in chloroform for 15 min. The chloroform extract (3.3 g) was chromatographed on a silicic acid column (140 g) and eluted with 5 l each of hexane, benzene, ether, and methanol. Each fraction was tested for its ovipositional stimulation activity. The responses of *A. beneficus* females to the intact insect of three *Ceroplastes* species and to paraffin models loaded with the chloroform extracts are shown in Table 4. The benzene eluate (830 mg) was rechromatographed on a silicic acid column (40 g) and eluted successively with 1.2 l each of hexane and 1, 5, 10, 20, and 50% benzene in hexane. The highest activity was found in 1% benzene in hexane, and this fraction was further purified by preparative thin-layer chromatography (PTLC), using 10% ether in benzene as the developing solution. The PTLC plate was air dried and scraped off in five sections. The most active fraction was found in the section between Rf 0.4 to 0.6. The neutral compounds were extracted

TABLE 4
Ovipositional Response (%) of *A. beneficus*
to Three Species of the Genus *Ceroplastes*

Host	Intact host	Crude extract[a]	Benzene eluate[a]
C. rubens	100	100	95
C. ceriferus	80	35	75
C. japonicus	0	0	50

[a] 0.1 insect equivalent material on a paraffin model.

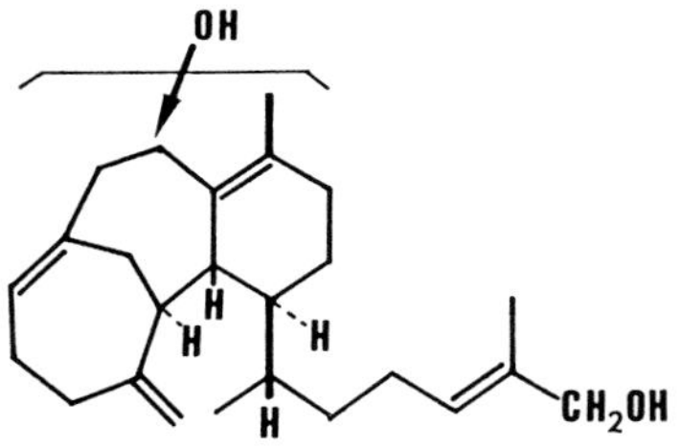

Cerorubenol-I (2)
diol-I

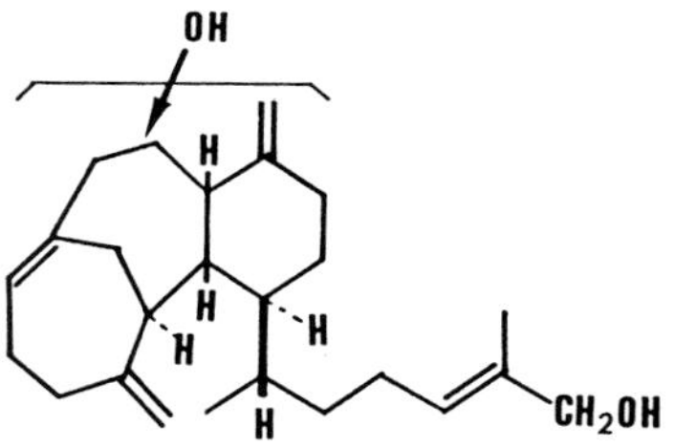

Cerorubenol-II (3)
diol-II

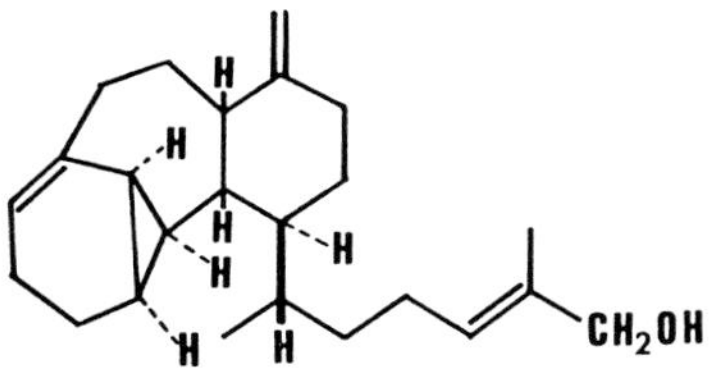

Cerorubenol-III (4)

FIGURE 5. Ovipositional stimulants in *C. rubens*.

from this section and found to be active. They were identified as the sesterterpene alcohols and diols, cerorubenol-I (2), cerorubediol-I, cerorubenol-II (3), cerorubediol-II and cerorubenol-III (4) (Figure 5). Each of the purified alcohols (2–4) and the mixture of the remaining diols exhibited an activity level of only 50%.

C. ceriferus and C. japonicus were extracted and fractionated in a similar manner. The extract from C. ceriferus showed ovipositional stimulation activity toward *Anicetus beneficus* females, but the activity was ten times less than the sesterterpenoids from C. rubens. The monocyclic sesterterpenoids ceriferol (5) and ceriferol-I (6) are the only sesterterpene alcohol components known so far in C. ceriferus.[23,24] A mixture of ceriferol and ceriferol-I elicited ovipositional behavior in about 50% of A. beneficus in a bioassay using one insect equivalent on the model (Figure 6). It is very interesting that A. beneficus responds not only to the terpenoid alcohols present in C. rubens, but also, to a certain extent, to those specific to C. ceriferus.

B. OVIPOSITION STIMULANTS IN THE COCCOID CUTICULAR WAXES OF *APHYTIS YANONENSIS*

Aphytis yanonensis De Bach and Rosen, a parasitic wasp of the arrowhead scale, *Unaspis yanonensis* (Kuwana), was introduced to Japan from China in 1980.[25] A. yanonensis is

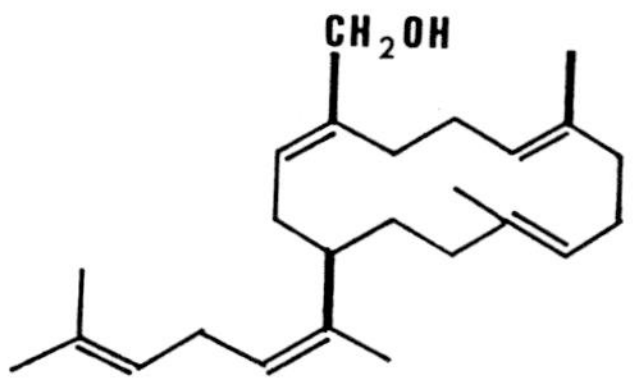

CH$_2$OH   CH$_2$OH

Ceriferol (5) 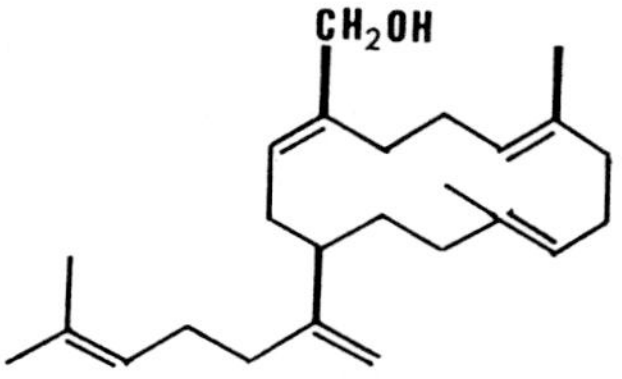   Ceriferol-I (6)

FIGURE 6. Ovipositional stimulants in *C. ceriferus.*

ectoparasitic and deposits eggs under the scales of the second instar, the immature adult and the mature female of *U. yanonensis.* Size and shape of the host scale do not appear to influence the wasp's ovipositional behavior. The average total fecundity per *A. yanonensis* female at 25°C was 17.3 progeny, but the number increased with a larger host such as *Aspidiotus nerii.*[26] Yukinari showed that *A. yanonensis* laid eggs on *Unaspis euonymi* and *Lepidosaphes cupressi* in the laboratory when it was confined with the coccoids on their host plants.[27] The localization of the ovipositional stimulant, a kairomone, in the cuticular wax of the coccoids, the possible range of the host insects for *A. yanonensis*, and the use of *Hemiberlesia lataniae* on pumpkin as a substitute host species for the rearing of *A. yanonensis* are of great interest.[28]

1. Ovipositional Behavior of *A. yanonensis*

An observation chamber was prepared by cutting a 30-mm-diameter hole in the center of a 5-mm-thick glass plate that was then covered on both sides with glass. A female *A. yanonensis*, 1 day after emergence, was introduced into the space containing a coccoid, and her behavior toward the coccoid was observed with a binocular microscope.

Typical initial behavior of the female consisted of random walking and intermittent, short flight. Antennal flagellation began when the female encountered a coccoid and moved onto it. The female then moved, with antennal flagellation, from the central part of the scale to its margin. When walking around the scale's margin, the female alternately moved backward and forward at different angles, repeating this several times and making a radial pattern. Thereafter, she began ovipositor drilling from the edge of the insect, moving backward at the same time.

Similar behavior was induced in the *A. yanonensis* female by a piece of scale cut from intact scale insects or their cuticular wax. Such behavior was not induced, however, toward scale debris after immersion in chloroform. The response of *A. yanonensis* to various coccoids is shown in Table 5. Three *Ceroplastes* species were not stimulative to *A. yanonensis*. All other coccoid cuticular wax actively stimulated the ovipositional behavior of *A. yanonensis* during the bioassay. The antennal flagellation behavior on a possible coccoid host, regardless of its shape or size, indicated that contact chemical sense is an essential cue to induce ovipositional behavior. This insect responded to *U. yanonensis* as well as *U. euonymi* and other coccoids which were subsequently shown to have a similar cuticular wax composition. Although ovipositional response to intact latania scales was less than with *U. yanonensis*, *A. yanonensis* did oviposit on this scale, and offspring emerged.

2. Extraction and Identification of Ovipositional Stimulants

Live coccoids were detached from host plants and immersed in chloroform. The chloroform solution was filtered and concentrated *in vacuo* to give a white powdery wax. The crude wax was chromatographed on silicic acid and eluted with benzene. The purified wax showed a single spot on TLC (Rf 0.5, silica gel TLC developed with benzene). Methanolysis of the wax was carried out in 5% hydrogen chloride in methanol at room temperature. The resultant fatty acid methyl esters and alcohols were analyzed by gas-liquid chromatography (Table 6).

TABLE 5
Ovipositional Behavior of *A. yanonensis* to Intact Coccoidea

Intact Coccoidea species	Ovipositional behavior observed		
	Flagellation	Drilling	Egg deposited
Diaspididae	+	+	+
Unaspis yanonensis ♀	+	+	+
Unaspis yanonensis ♂	+	+	+
Unaspis eugnymi	+	+	+
Pseudaulacaspis prunicola	+	+	?
Lepidosaphes cupressi	+	+	+
Hemiberlesia lataniae	+	+	+
Comstockaspis perniciosa	+	+	?
Pseudaocidia duplex	+	+	+
Lopholeucaspis japonica	+	+	?
Coccidae			
Ericerus pela ♂	+	+	−
Ceroplastes rubens	(+)[a]	−	−
Ceroplastes ceriferus	(+)	−	−
Ceroplastes japonica	(+)	−	−
Margaroideadae			
Icerya purchasi ♀	+	+	−

[a] A few showed brief antennal contact.

TABLE 6
Major Component of Coccoidea Cuticular Wax

Species	Wax ester (carbon number of acid–carbon number of alcohol)
Diaspidiae	
Unaspis yanonensis ♀	26–32 (5%), 28–32 (12%), 30–32 (47%), 32–32 (35%)
Unaspis yanonensis ♂	30–32 (42%), 32–32 (46%), 34–32 (12%)
Unaspis euonymi ♂♀	30–32 (25%), 32–32 (50%), 34–32 (25%)
Pseudaulacaspis prunicola ♂♀	30–32, 32–32, 34–32, 34–34 (major components); 28–28, 28–30, 28–32 (minor components)
Lepidasaphes cupressi ♂♀	30–32 (40%), 32–32 (44%), 32–34 (0.8%)
Hemiberlesia lataniae ♂♀	Mixed triglycerides (major components)
Coccidae	
Ericerus pela ♀	24–24 (3.1%), 24–26 (22.32%), 26–26 (55.2%), 28–26 (16.6%), 28–28 (3%)
Margaroidae	
Icerya purchasi ♀[a]	26–24 (13.8%), 26–26 (43.6%), 28–26 (25.2%), 28–28 (11.5%), 30–28 (6.0%)

[a] Extract from egg sac.

A mixture of wax esters isolated from *Ericerus pela*, including hexacosyl tetracosanoate (22.3%) and hexacosyl octacosanoate (16.6%), showed a high ovipositional stimulation activity. However, the major constituent, hexacosyl hexacosanoate, alone did not show the activity.

The latania scale was found to be a good substitute host for rearing *A. yanonensis* in the laboratory. Cuticular wax from *H. lataniae* (390 mg) was chromatographed on silicic acid and successively eluted with hexane, a 1:1 mixture of hexane and benzene, and 25% hexane in benzene, and then benzene alone. The fraction with high oviposition stimulation activity (207 mg, 52.5%, Rf 0.45 on silica gel TLC developed with benzene) was further separated into 16 fractions on high-pressure liquid chromatography (HPLC) (ODS; 2 × 200 mm) by eluting with 20% chloroform in acetonitrile. The major component (52.5%) of the latania

TABLE 7
Ovipositional Stimulation of Cuticular Wax to *A. yanonensis*

	Behavioral response	
Cuticular wax from	Antennal flagellation (%)	Ovipositor drilling (%)
Unaspis yanonensis	100	100
Unaspis euonymi	100	100
Pseudaulacaspis prunicola	100	100
Hemiberlesia lataniae	100	50
Lepidasaphes cupressi	50	10
Ericerus pela	100	100
Hexacosyl hexacosanoate[a]	80	0
Triolein[a]	60	30
1-Palmito-dimyristin[a]	60	50
2-Oleo-dimyristin[a]	50	40
TG mixture[b]	90	80

[a] Synthetic compound.
[b] A mixture of triolein, 1-palmito-dimyristin, 2-oleo-dimyristin, tricaprylin, tricaprin, trimyristin, and tripalmitin.

scale wax was a mixture of triglycerides with a fatty acid composition of lauric (11.83%), myritoleic (10.37%), palmitic (16.45%), oleic (9.06%), and stearic (2.10%) acids. The structures of these were confirmed by synthesis, and each synthetic triglyceride was also high in stimulation activity. It is of note that a mixture of the synthetic triglycerides was higher in activity than each of the component triglycerides individually (Table 7).

Thus, it seems apparent that *A. yanonensis* may have a strong perception of the ester group in cuticular waxes.

Two inactive compounds (48 mg, 17.5%, Rf 0.66 and Rf 0.72) identified as alpha- and delta-amyrin acetate were obtained from the cuticular waxes in the 1:1 mixture of hexane and benzene eluate of the preliminary chromatography column.

IV. EFFECT OF KAIROMONES ON THE HOST-SEARCHING BEHAVIOR OF THE WASP *APANTELES KARIYAI*

The braconid wasp *Apanteles kariyai* Watanabe is a gregarious parasitoid of the common armyworm *Pseudaletia separata*. *A. kariyai* showed a sequential oviposition behavior when observed in the laboratory. *A. kariyai* first recognizes a host habitat through antennal perception of chemical cues (arrestants) present in the feeding trace left by host larvae. The resultant, characteristic antennal contact behavior was also elicited when the female found exuvium and frass. On perceiving the arrestants, she responded in an orthokinetic manner, involving stopping, slow walking, and probing with her antennae for the host larvae. When she finds a host larva, she immediately lays her eggs.

A. BIOASSAY FOR ARRESTANTS

A bioassay using a paper disk on which was absorbed a portion of extract from the exuviae or frass showed that the arresting stimuli were present in material derived from the host larva. Female *A. kariyai*, at least 2 days old, were kept at 26°C for more than 1 h before bioassay and then confined to a petri dish (50-mm diameter). A filter paper disk (6-mm diameter) impregnated with a test solution was offered to each female. The number of arresting responses following antennal contact with the paper disk was recorded.[29]

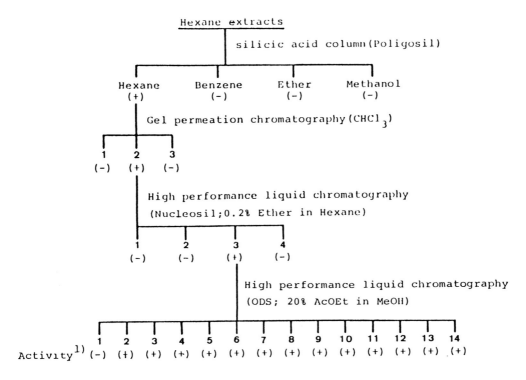

FIGURE 7. Procedure for the purification of arrestants for *A. kariyai* from hexane extracts of exuvium of *P. separata* larvae. (+) means characteristic antennal contact to a paper disk at an impregnation of 0.1 μg per disk; (−) means no response.

B. EXTRACTION AND PURIFICATION OF ARRESTANTS

Exuviae and frass were separately collected and extracted with hexane. *P. separata* larvae often bit the paper towels that covered the artificial diet in the rearing containers, leaving a buccal secretion. This secretion was extracted, with hexane, from the paper towels.

The activity-guided purification procedure of the arrestants from exuvium was carried out as shown in Figure 7.[30] The arrestants were separated into 14 fractions by HPLC on ODS (20% AcOEt in MeOH).

C. STRUCTURE DETERMINATION OF ARRESTANTS

The mass spectra of ODS-HPLC fractions 5, 8, and 11 showed similar fragmentation patterns, with molecular ions observed at m/z 478, 506, and 534, respectively. The electron impact-mass spectrum (EIMS) of fraction 8 is shown in Figure 8. The molecular formula was estimated to be $C_{35}H_{70}O$ (MW 506.5423), and the fragment ions at m/z 351, 323, 295, 281, 253, and 225 were estimated to be $C_{24}H_{47}O$, $C_{22}H_{43}O$, $C_{20}H_{39}O$, $C_{19}H_{37}O$, $C_{17}H_{33}O$, and $C_{15}H_{29}O$, respectively. Fraction 8 was considered to be a mixture of three structural isomers, each with a pair of fragment ions: m/z 225 and 351, 253 and 323, and 281 and 295. The sum of each pair is m/z 576, corresponding to $(M^+ + 70)$. A tetrahydrofuran ring (m/z 70) was assumed to be a common structural unit.

The structural elucidation of the components of fraction 8 resulted from a combination of the MS, PMR, and CMR data (Figure 9). Thus, fraction 8 consisted of a mixture of 2-undecyl-5-eicosyl-tetrahydrofuran, 2-tridecyl-5-octadecyl-tetrahydrofuran, and 2-pentadecyl-5-hexadecyl-tetrahydrofuran in a ratio of 1.2:0.8:1. Compositions of other fractions were estimated mainly by mass spectra. Fractions 5 and 11 from ODS-HPLC were found to be

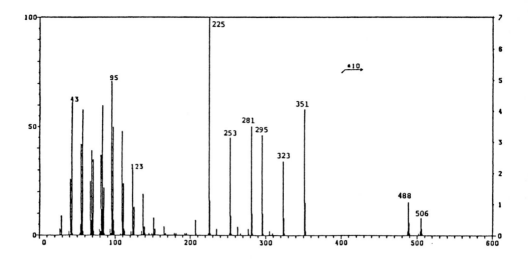

FIGURE 8. EI-mass spectrum of fraction 8 on ODS.

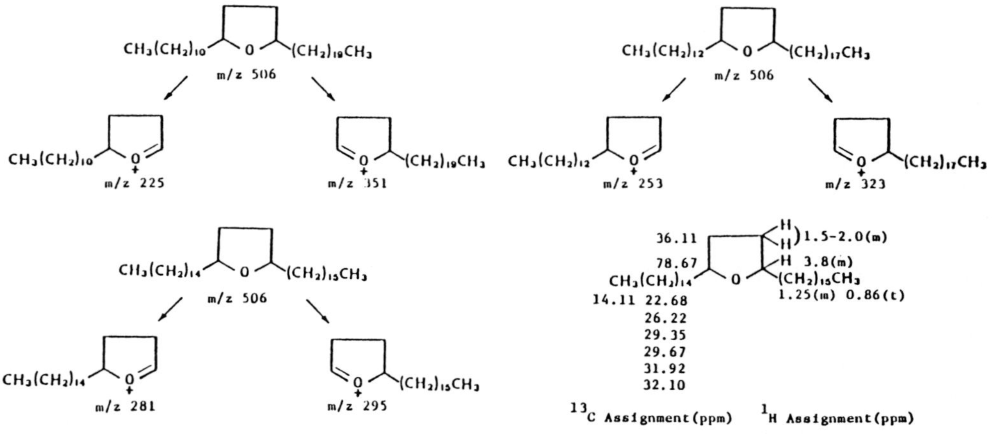

FIGURE 9. Assignment of NMR chemical shifts and mass fragments of fraction 8 on ODS.

mixtures of two and four isomers, respectively. All the compounds identified in the extract from the exuviae were also found in the buccal secretion and frass of the host larvae. It should be noted that there are four possible stereoisomers for these compounds, resulting from the two asymmetric carbons in 2,5-dialkyl-tetrahydrofurans.

D. SYNTHESIS OF ARRESTANTS AND RELATED COMPOUNDS

2,5-Dialkyl-tetrahydrofurans were synthesized either via a Wittig reaction followed by hydrogenation of the furanoid compounds or via lead tetraacetate oxidation of the appropriate alcohols, as shown in Figure 10.[31] Some MS data for the synthesized (Z)-2,5-dihexadecyl-tetrahydrofuran (**7**), (Z)-2,5-dipentadecyl-tetrahydrofuran (**8**), (Z)-2-pentadecyl-5-hexadecyl-tetrahydrofuran (**9**), (Z)-2-tetradecyl-5-heptadecyl-tetrahydrofuran (**10**), and (E)-2-tetradecyl-5-heptadecyl-tetrahydrofuran (**11**) are shown in Table 8. The retention times of **9** and **10** on ODS-HPLC were the same as that of fraction 8. Comparison of the ^{13}C-NMR chemical shifts of the methine carbons (C2 and C5) of the fractions, with those of the synthetic tetrahydro-furans, indicated that the configuration of the components in fractions 8, 9, and 11 were (E), (Z), and (E), respectively.

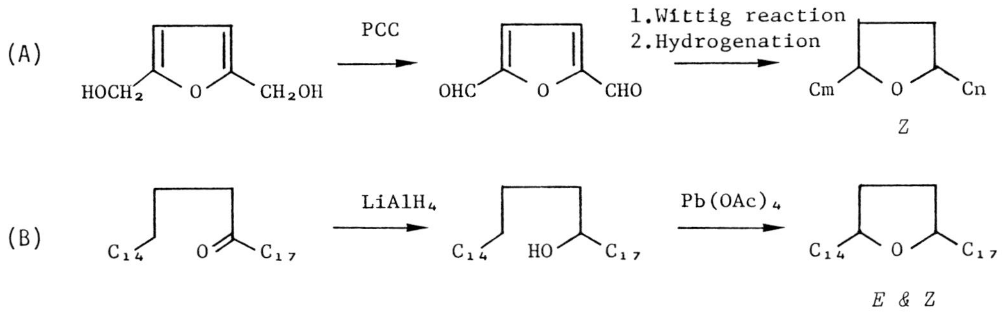

FIGURE 10. Synthetic route to 2,5-dialkyltetrahydrofurans by Wittig reaction followed (A) by hydrogenation and (B) by lead tetraacetate oxidation.

E. BIOACTIVITY OF NATURAL AND SYNTHETIC ARRESTANTS

The arrestant activity of isolated and synthetic kairomones are shown in Table 9. Fractions 8 and 9 showed higher activity than the synthetic compounds **7, 8, 9, 10**, and **11**, at a level of 10^{-4} μg per disk. Fraction 8 (E-isomer) showed a similar level of activity to fraction 9 (Z-isomer). The symmetrical, synthetic compounds **7** and **8** and the unsymmetrical synthetics **10** and **11**, which were not found among the natural arrestants, showed an activity comparable to compound **9**. Synthetic (Z)-2,5-dioctyl-tetrahydrofuran, having shorter alkyl groups than the natural arrestants, showed lower activity than fraction 9. The length of side chains at C-2 and C-5 of the tetrahydrofuran ring were considered crucial to determining the arresting activity of the compounds. The minimum length of side chains was estimated to be at least octyl or longer. No arresting activity was induced by 2-alkyltetrahydrofurans.

When different concentrations of the synthetic kairomones were arranged in patches on a filter paper, *A. kariyai* demonstrated two modes of arrestment response, which were dependent on the amount of kairomone per patch. During successive visits to the patch, walking speed increased, whereas time spent and distance walked on the patch decreased. This result indicated that an arrestant is an important factor in the host-searching behavior of *A. kariyai*. Searching behavior of parasitoids is usually stimulated by the perception of kairomones. Thus, treatment of a host habitat with arrestants would improve the use of parasitoids for pest management.

V. EFFECT OF ANTIMONES (A NEW CLASS OF ALLELOCHEMICAL) ON THE HOST-SEARCHING BEHAVIOR OF THE WASP *APANTELES KARIYAI*

A. RESPONSE OF *A. KARIYAI* TO NONHOST LARVAE

P. separata is polyphagous, and its larvae feed on leaves of many Gramineae species such as rice, wheat, millet, sugarcane, foxtail, radish, Chinese cabbage, cotton, buckwheat, etc. On the other hand, *A. kariyai* is a monophagous parasitoid of *P. separata* larvae. In the host habitat, therefore, there will probably be other nonhost noctuid larvae that feed on the same plant species. *Acantholeucania loreyi* is one of these nonhost noctuid larvae. The response of *A. kariyai* upon contact with fecal pellets from *A. loreyi* larvae, and the subsequent response upon contact with fifth-instar larvae, was observed with the insect in a petri dish.[32] Upon release, *A. kariyai* walked in the petri dish. When *A. kariyai* found the fecal pellet of *A. loreyi*, she intensively contacted it with her antennae and stayed near it, standing still or walking slowly. The wasps demonstrated this arrestment response upon contact with fecal pellets of either *A. loreyi* or *P. separata*. When a fifth-instar *A. loreyi* was then offered to the wasp, she immediately mounted it and inserted her ovipositor. However, this never resulted in any

TABLE 8
Diagnostic Ions of Natural and Synthetic Compounds

$$CH_3(CH_2)_m \quad O \quad (CH_2)_n CH_3$$

		m	*n*	M+	Diagnostic	ions
Fraction 8	ODS	10	19	506	225	351
		12	17	506	253	323
		14	15	506	281	295
Synthetic compounds						
7		15	15	520	295	
8		14	14	492	281	
9		14	15	506	281	295
10		13	16	506	267	309
11		13	16	506	267	309

TABLE 9
Response of *A. kariyai* to Arrestants and Synthetic Alkyl-Tetrahydrofurans

Compound	M+	m	n	μg per disk					
				1	10^{-1}	10^{-2}	10^{-3}	10^{-4}	10^{-5}
Isolated arrestant									
Fraction 8 on ODS (*E*–)	506	10 / 12 / 14	19 / 17 / 15	70	90	55	35	25	0 (%)[a]
Fraction 9 on ODS (*Z*–)	506	10 / 12 / 14	19 / 17 / 15	80	60	40	40	20	0
Fraction 5 on ODS (*E*–)	478	10 / 12	17 / 15	80	75	60	—	30	—
Fraction 11 on ODS (*E*–)	534	10 / 12 / 14 / 16	21 / 19 / 17 / 15	70	70	60	—	25	0
Synthetic 2,5-dialkyl-tetrahydrofuran									
Z-2,5-di-C_{16} (7)	520	15	15	65	55	50	15	0	0
Z-2,5-di-C_{15} (8)	492	14	14	75	65	60	10	0	0
Z-2-C_{15}-5-C_{16} (9)	506	14	15	70	65	40	20	5	0
Z-2-C_{14}-5-C_{17} (10)	506	13	16	80	80	45	15	0	0
E-2-C_{14}-5-C_{17} (11)	506	13	16	75	50	30	20	10	0

[a] Average value of 20 females.

TABLE 10
Chemical Structure of Major Compounds of Arrestants in Feces
of *Pseudaletia separata* and *Acantholeucania loreyi* Larvae

$$CH_3(CH_2)_m \diagup O \diagdown (CH_2)_n CH_3$$

			Arrestants isolated from feces of	
M^+	m^a	n^a	*P. separata*	*A. loreyi*
478	8	19	$-^b$	+
	10	17	+	+
	12	15	+	+
506	10	19	+	+
	12	17	+	+
	14	15	+	+
534	8	23	−	+
	10	21	+	+
	12	19	+	+
	14	17	+	+
	16	15	+	+

[a] Number of methylenes in alkyl group of 2,5-dialkyl-tetrahydrofuran.
[b] − = absence; + = presence.

offspring emergence. In contrast, when *P. separata* larvae were offered to the wasp, 70% of all ovipositor insertions resulted in offspring production.

Whitman recently proposed a term "antimone" for a new class of allelochemicals: substances produced or acquired by an organism that, when it contacts an individual of another species in the natural context, evokes in the receiver a behavioral or physiological reaction that is maladaptive to both the emitter and the receiver.[33] Thus, according to this definition, arrestants of *A. loreyi* (pseudohost) represent antimones. In contrast to kairomones present in the cuticular wax of the host insect, *P. separata*, the arrestant from *A. loreyi* evokes, in *A. kariyai*, arrestant reaction that is maladaptive to both emitter (*A. loreyi*) and receiver (*A. kariyai*).

B. CHEMICAL ANALYSIS OF ARRESTANTS ISOLATED FROM *A. LOREYI*

The same extraction and purification methods as used for the arrestants from *P. separata* larvae were applied. Feces were collected and soaked in hexane. After purification on a silicic acid column, followed by HPLC, fractions possessing arrestant activity were obtained. GC-MS analysis indicated that the arrestants from *A. loreyi* consisted of homologs with molecular weights (478, 506, 534), which were the same as those of the arrestants from *P. separata*. The chemical structures of the major homologs in the arrestant active fractions from *A. loreyi* and *P. separata* are compared in Table 10.

VI. TERMITE ANTIFEEDANTS ISOLATED FROM THE PLANTS *PHELLODENDRON AMURENSE* AND *P. CHINENSE*

The pages of the culturally important, 1200-year-old "Hyakumanto daranikyo" sutra have been well preserved at Horyuji temple in Nara. The way the paper for the sutra was made has attracted public attention because it has remained undamaged by insects and fungi for this very long period. It was recently learned that an extract from *Phellodendron amurense*

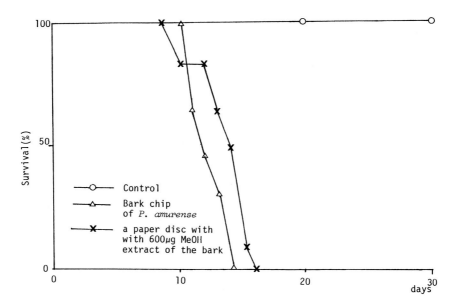

FIGURE 11. Preliminary test for antifeedant activity of *Phellodendron amurense* against *Reticulitermes speratus.*

Rupr. was added in the process of making this paper. It is believed that this addition is related to the preservation of the sutra.

P. amurense, known as "Kihada" in Japan, is an herb used medicinally as an antimalarial, febrifuge agent and also as dressing for indolent ulcers. *P. chinense* is mainly distributed in southern Asia and known as "Chinese Kihada" in Japan. The bark contains berberine, magnoflorine, phellodendrine, and palmatine. The fruits are not eaten by insects or birds. Methanol extracts of *P. amurense* bark and *P. chinense* fruits were prepared in our search for insect control agents in plants. The extracts were found to exhibit potent feeding inhibitory action against the termite *Reticulitermes speratus* (Rhinotermes).

A. BIOASSAY METHOD

Nests of *R. speratus* were collected and kept in the laboratory, and their nymphs were used in bioassay.[34] Filter papers were individually dipped for about 1 s in a solution containing Brilliant Blue FCF (San-ei Chemical Ind., Ltd.), an edible coloring material, and dried before the feeding tests. A known concentration of a methanol solution of a test sample was applied to a filter paper disk forming a test patch with a diameter of 8 mm. Untreated, dyed filter papers were fed to the termite nymphs as controls. Nymphs were released in a petri dish with a layer of moistened sea sand covering the bottom. Ten termite nymphs were used for the bioassay of each sample (except for the preliminary test in which two termites were used), and each concentration was replicated three times. The growth rate, percentage survival, and color intensity of the insect alimentary canal were compared with those of a control treatment. As shown in Figure 11, the methanol extract of *P. amurense* (600 μg per paper disk) showed strong antifeedant activity. With the methanol extract of *P. chinense* fruit (1000 μg per disk), feeding was significantly reduced (Figure 12).[35]

B. BIOACTIVITY-GUIDED FRACTIONATION OF EXTRACTS OF
P. AMURENSE BARK

The dried bark (500 g) was chipped and extracted with methanol. Evaporation of the solvent, under reduced pressure, yielded a crude extract (63.5 g) that was then sequentially extracted with water (1.5 l), chloroform (900 ml × 2), ethyl acetate (900 ml × 2), and hexane

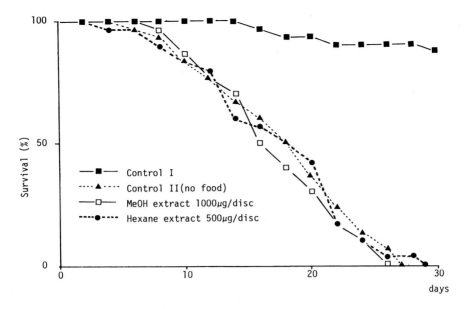

FIGURE 12. Antifeeding constituents of *P. chinense* fruits against *R. speratus.*

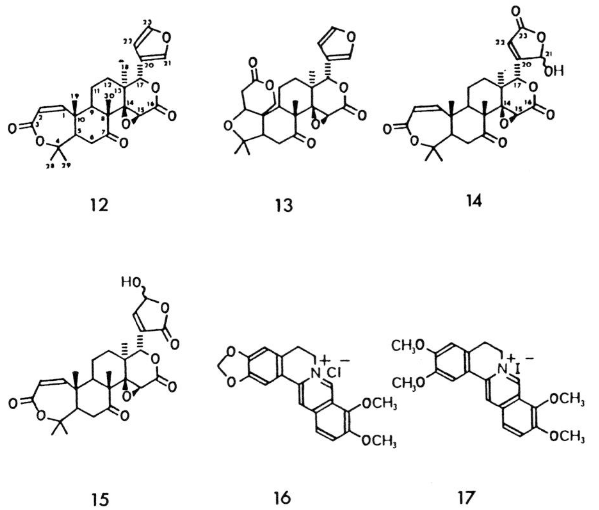

FIGURE 13. Compounds isolated from *P. amurense* extract.

(900 ml × 2). Evaporation of the solvents under reduced pressure afforded 7.9, 6.4, 0.64, and 41.8 g of extract from the hexane, chloroform, ethyl acetate, and water fractions, respectively. A bioassay showed that the activity was in the chloroform and water fractions.

A portion of the chloroform fraction (1.0 g) was first chromatographed on a silica gel column (100 g), using chloroform-methanol (20:1) as the eluent, and then on Sephadex LH-20, using methanol as the eluent. The limonoid fraction thus obtained was further purified, by using recycling HPLC on a gel filtration column (JAIGeL GS-320, methanol), to yield obacunone (**12**, 15 mg), limonin (**13**, 34 mg), kihadanin A (**14**, 3 mg), and kihadanin B (**15**, 2 mg) (Figure 13). Identification of these compounds resulted from comparison of their spectroscopic data with data reported in the literature.

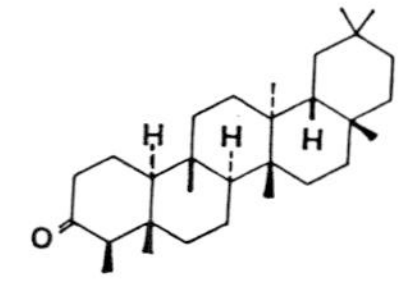

FIGURE 14. Compound isolated from *P. chinense* extract.

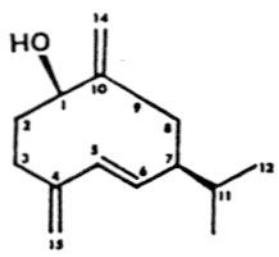

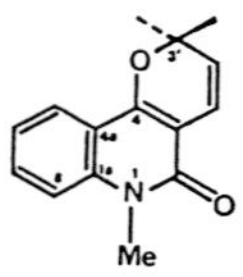

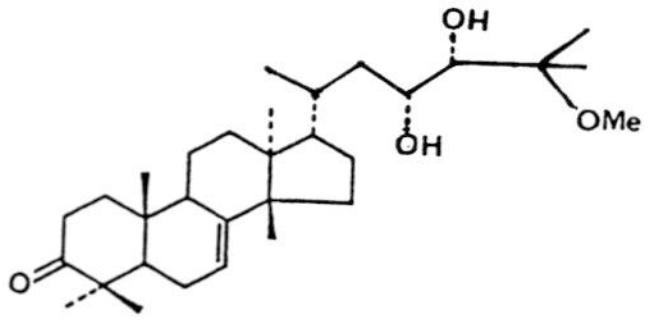

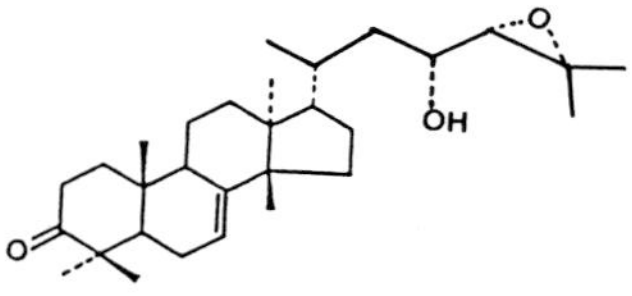

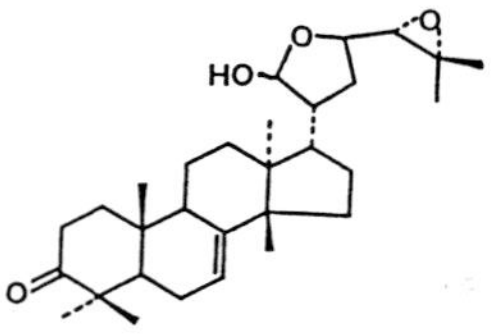

A portion of the water fraction (l0 g) was partitioned between water (300 ml) and n-butyl alcohol (250 ml × 3). When the n-butyl alcohol layer had been reduced to 60 ml by evaporation, a precipitate formed. After filtration and drying under vacuum, the residue (4.3 g) was chromatographed on a silica gel column, using a benzene-ethyl acetate-n-propyl alcohol-methanol-ethylamine solution (8:4:2:1:1) as the eluent. Fraction A (1.16 g) was dissolved in water (30 ml), and berberine chloride (**16**, 686 mg) was precipitated from the aqueous solution by adjusting the pH to 6 by addition of 1N HCl. Fraction B (1.0 g) was dissolved in water (25 ml), and palmatine iodide (**17**, 146 mg) was precipitated from the aqueous solution by adjusting the pH to 4 by addition of a 7.5% HI solution.

C. BIOACTIVITY-GUIDED FRACTIONATION OF EXTRACTS OF
P. CHINENSE FRUIT

Air-dried fruit (460 g) was steeped in methanol for 10 days. The residue obtained after evaporation of the methanol was partitioned between hexane and water. The hexane soluble part yielded a viscous yellow/brown mass (26.2 g), upon removal of the solvent *in vacuo*, which showed antifeedant activity in the bioassay. Column chromatography on silica gel, using hexane-ethyl acetate (l0:2) as the eluent, yielded friedelin (**18**), 4,10-dimethylene-7-isopropyl-5(E)-cyclodecenol (**19**), N-methylflindersine (**20**), phellochin (**21**), niloticin (**22**), and melianone (**23**) in increasing order of polarity (Figure 14).

Niloticin (**22**) had previously been isolated from *Turraea nilotica* and *P. chinense*, but its stereochemistry was not determined. The structure has now been shown by X-ray diffraction analysis to be a tirucallane-type, rather than a euphane-type, triterpenoid.[36]

D. BIOACTIVITY AND BIOGENESIS OF COMPOUNDS FROM *P. AMURENSE*
AND *P. CHINENSE*

From the bark of *P. amurense*, four limonoids and two alkaloids were isolated. They were bioassayed in a range of dosages from 100 μg to 1200 μg per disk, as shown in Table 11. Obacunone (**12**) was the most potent antifeedant, followed by berberine chloride (**16**), palmatine iodide (**17**), and kihadanin A (**14**). Kihadanin B (**15**) was inactive even at 1200 μg per disk.

Of the six compounds isolated from *P. chinenese*, only N-methylflindersine (**20**) and melianone (**23**) showed antifeedant activity. Figure 15 shows the effects of N-methylflindersine (**20**) (100 μg per disk) and melianone (**23**) (100 μg per disk) on the survival of termites. The mortality of termites treated with N-methylflindersine (l00 μg

TABLE 11
Antifeedant Activity of Limonoids and Alkaloids from *P. amurense* Against *Reticulitermes seperatus*

Compound	Dose (µg per disk)	Percent survivors (number of days)		
		10	20	30
Control	—	96.7	93.3	90.0
Obacunone (12)	100	83.3	33.3	13.0
	300	73.3	0.1	0
Kihadanin A (14)	100	96.7	83.3	80.0
	300	90.0	76.7	60.0
	600	86.7	66.7	46.7
	1200	83.3	56.7	20.0
Berberine chloride (16)	100	90.0	63.3	30.0
	300	80.0	56.7	16.7
Palmatine iodide (17)	100	93.3	70.0	43.3
	300	83.3	46.7	20.0

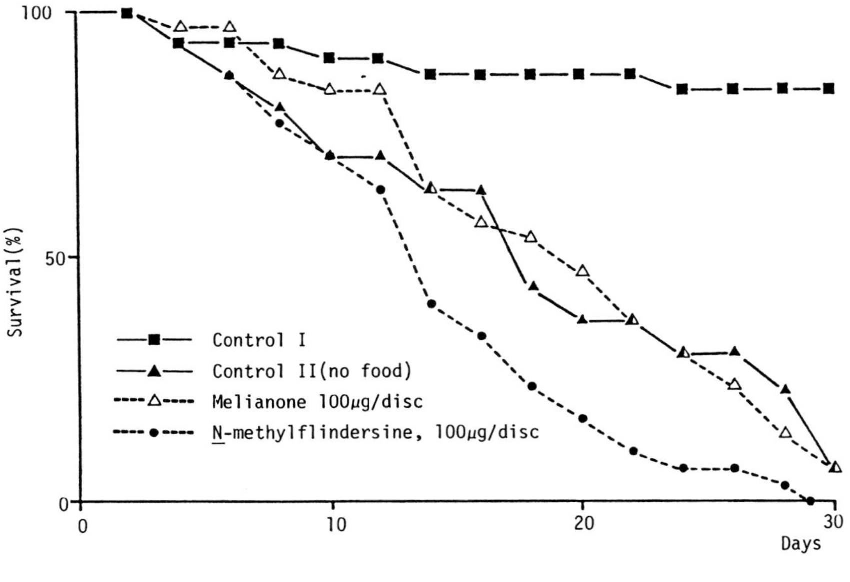

FIGURE 15. Effect of the active compounds isolated from *P. chinense* fruit on the survival rates of termite nymphs, *R. speratus*.

per disk) increased to 100% after 29 days of exposure. With melianone (**23**) (100 µg per disk) the mortality was 95% on day 30. The survival rate of termites treated with the active compounds were similar to that of control II (Figure 15). It is most probable that the two compounds had antifeeding effects.

Limonoids are known to be a group of chemically related, bitter, tetranortriterpene derivatives predominantly present in Rutaceae and Meliaceae plants. Some limonoids exhibit the antifeedant or growth disruption activity toward insects of several genera. Azadirachtin, a limonoid from *Azadirachta indica* (Meliaceae), is known to be a potent inhibitor of ecdysis of lepidopterous insects such as *Heliothis zea*, *Heliothis virescens*, *Spodoptera frugiperda*, and *Pectinophora gossypiella*.[37] Also, obacunone (**12**) and harrisonin are potent antifeedants of *Eldana saccharina* and *Maruca testulalis*. Berberine (**16**) and palmatine (**17**) are isoquinoline alkaloids associated with plants of Ranunculaceae, Berberidaceae, and Rutaceae. Berberine and palmatine are known to be principles of medicinal plants used for antibacterial and stomachic treatment.

One of the characteristic features in the structure of limonoids is the existence of a furan ring. Melianone (**23**) is an important intermediate in the furan ring formation, because of its oxidized and ring-forming carbon at C-21.[38] Niloticin (**22**), phellochin (**21**), and melianone (**23**) have different oxidation states at the side chain, but have the same stereochemical prerequisites for formation of the furanoid 21,23-oxide system, typical of limonoids. Thus, a possible biogenetic pathway to the limonoids in *P. chinense* is as follows: niloticin (**22**) $\longrightarrow$ melianone (**23**) $\longrightarrow$ limonoids.

ACKNOWLEDGMENTS

I express my gratitude to all my co-workers and colleagues who made possible the original work on insect behavior regulators. My sincere thanks to Drs. M. Fukui, J. Takabayashi, M. Kim, R. Su, M. Ishida, and T. Yamamoto, who allowed me to represent authorship of this chapter.

REFERENCES

1. **Whittaker, R. H. and Feeny, P. P.**, Allelochemicals: chemical interactions between species, *Science,* 171, 757, 1971.
2. **Dethier, V. G., Barton-Brown, L., and Smith, C. N.**, The designation of chemicals in terms of the responses they elicit from insects, *J. Econ. Entomol.,* 53, 134, 1960.
3. **Takahashi, S.**, Ethological aspects of insect pest control, in *Abstr. 22nd Int. Ethological Conf.,* Kyoto, 1991, 5.
4. **Roth, L. H. and Willis, E. R.**, The reproduction of cockroaches, *Smith. Misc. Coll.,* 122, 1, 1954.
5. **Roth, L. H. and Dateo, G. P.**, A sex pheromone produced by males of the cockroach, *Nauphoeta cinerea, J. Insect Physiol.,* 12, 255, 1966.
6. **Ewing, L. S.**, Fighting and death from stress in a cockroach, *Science,* 155, 1035, 1967.
7. **Manning, A. and Johnstone, G.**, The effect of early adult experience on the development of aggressiveness in males of the cockroach, *Nauphoeta cinerea, Rev. Comp. Animal,* 4, 12, 1970.
8. **Fukui, M. and Takahashi, S.**, Studies on the mating behavior of the cockroach, *Nauphoeta cinerea*. I. Sex discrimination by males, *Appl. Ent. Zool.,* 15, 20, 1980.
9. **Sreng, L.**, Seducin, male sex pheromone of the cockroach *Nauphoeta cinerea*: isolation, identification and bioassay, *J. Chem. Ecol.,* 16, 2899, 1990.
10. **Takahashi, S. and Fukui, M.**, Studies on the mating behavior of the cockroach, *Nauphoeta cinerea* (Olivier). II. Wing-raising stimulant in cuticular wax, *Appl. Ent. Zool.,* 15, 159, 1980.
11. **Fukui, M. and Takahashi, S.**, Studies on the mating behavior of the cockroach, *Nauphoeta cinerea* (Olivier). III. Isolation and identification of intermale recognition pheromone, *Appl. Ent. Zool.,* 18, 351, 1983.
12. **Takahashi, S. and Fukui, M.**, Studies on the mating behavior of the cockroach, *Nauphoeta cinerea* (Olivier). IV. Synthesis and biological activity of nauphoetin and related compounds. *Appl. Ent. Zool.,* 18, 357, 1983.
13. **Hartman, H. B. and Roth, L. M.**, Stridulation by a cockroach during courtship behavior, *Nature,* 213, 1243, 1967.
14. **Breed, M. D., Smith, S. K., and Gall, B. G.**, Systems of mate selection in a cockroach species with male dominance hierarchies, *Anim. Behav.,* 28, 130, 1980.
15. **Jones, R. L.**, *Semiochemicals: Their Role in Pest Control,* John Wiley & Sons, New York, 1981, Chap.12.
16. **Yasumatsu, K. and Tachikawa, T.**, Investigation on the hymenopterous parasites of *Ceroplastes rubens* Maskell in Japan, *J. Fac. Agric. Kyushu Univ.,* 9, 99, 1949.
17. **Ishii, T. and Yasumatsu, K.**, Description of a new parasitic wasp of *Ceroplastes rubens* Maskell, *Mushi Fukuoka,* 27, 69, 1954.
18. **Ohgushi, R.**, Studies on the host selection by *Anicetus beneficus,* a parasite of *Ceroplastes rubens, Mem. Coll. Sci. Univ. Kyoto,* 23(B), 55, 1956.

19. **Ohgushi, R.,** Studies on the host selection by three *Anicetus* wasps (Encyrtidae), *Mem. Coll. Sci. Univ. Kyoto,* 25(B), 31, 1958.

20. **Noda, T., Kitamura, C., Takahashi, S., Takagi, K., Kashio, T., and Tanaka, M.,** Host selection behavior of *Anicetus beneficus.* I. Ovipositional behavior for the natural host *Ceroplastes rubens, Appl. Ent. Zool.,* 17, 350, 1982.

21. **Takahashi, S. and Takabayashi, J.,** Host selection behavior of *Anicetus beneficus.* II. Bioassay of oviposition stimulants in *Ceroplastes rubens, Appl. Ent. Zool.,* 19, 117, 1984.

22. **Takabayashi, J. and Takahashi, S.,** Host selection behavior of *Anicetus beneficus.* III. Presence of oviposition of the genus *Ceroplastes, Appl. Ent. Zool.,* 20, 173, 1985.

23. **Tempesta, M. S., Iwashita, T., Miyamoto, F., Yoshihara, K., and Naya, Y.,** A new class of sesterterpenoids from the secretion of *Ceroplastes rubens, Chem. Comm.,* 1182, 1983.

24. **Pawlak, J. K., Tempesta, M. S., Iwashita, T., Nakanishi, K., and Naya, Y.,** Structures of sesterterpenoids from the scale insect *Ceroplastes ceriferus, Chem. Lett.,* 1069, 1983.

25. **Nishino, M. and Takagi, K.,** Parasite of *Unaspis yanonensis* (Kuwana) introduced from the People's Republic of China, *Plant Prot.,* 35, 253, 1981.

26. **Furuhashi, K. and Nishino, M.,** Biological control of arrowhead scale, *Unaspis yanonensis* by parasitic wasps introduced from the People's Republic of China, *Entomophaga,* 28, 277, 1983.

27. **Yukinari, M.,** Parasitization of *Aphytis yanonensis* DeBach and Rosen to some scale insects, *Proc. Assoc. Plant Prot. Shikoku,* 22, 99, 1987.

28. **Takahashi, S., Hajika, M., Takabayashi, J., and Fukui, M.,** Oviposition stimulant in the coccoid cuticular waxes of *Aphytis yanonensis* De Bach and Rosen, *J. Chem. Ecol.,* 16, 1657, 1990.

29. **Takabayashi, J., Noda, T., and Takahashi, S.,** Effect of kairomones in the host searching behavior of *Apanteles kariyai,* a parasitoid of the common armyworm, *Pseudaletia separata* Walker. I. Presence of arresting stimulants produced by the host larvae, *Appl. Ent. Zool.,* 20, 484, 1985.

30. **Takabayashi, J. and Takahashi, S.,** Effect of kairomones in the host searching behavior of *Apanteles kariyai,* a parasitoid of the common armyworm, *Pseudaletia separata* Walker. II. Isolation and identification of arrestants produced by the host larvae, *Appl. Ent. Zool.,* 21, 114, 1986.

31. **Takabayashi, J. and Takahashi, S.,** Effect of kairomones in the host searching behavior of *Apanteles kariyai,* a parasitoid of the common armyworm, *Pseudaletia separata* Walker. III. Synthesis and bioassay of arrestants and related compounds, *Appl. Ent. Zool.,* 21, 519, 1986.

32. **Takabayashi, J. and Takahashi, S.,** Effect of host-searching and postoviposition behavior of *Apanteles kariyai,* a parasitoid of *Pseudaletia separata, Entomol. Exp. Appl.,* 52, 221, 1989.

33. **Whitman, D. W.,** *Novel Aspects of Insect-Plant Interaction,* John Wiley & Sons, New York, 1988, chap.1.

34. **Kawaguchi, H., Kim, M., Ishida, M., Ahn, Y. J., Yamamoto, T., Yamaoka, R., Kozuka, M., Goto, M., and Takahashi, S.,** Several antifeedants from *Phellodendron amurense* against *Reticulitermes speratus, Agric. Biol. Chem.,* 53, 2635, 1989.

35. **Su, R., Kim, M., Yamamoto, T., and Takahashi, S.,** Antifeedants of *Phellodendron chinese* fruit against *Reticulitermes speratus, J. Pest. Sci.,* 15, 567, 1990.

36. **Su, R., Kim, M., Kawaguchi, H., Yamamoto, T., Goto, K., Taga, T., Miwa, Y., Kozuka, M., and Takahashi, S.,** Triterpenoids from the fruits of *Phellodendron chinese* Schneid.: the stereostructure of niloticin, *Chem. Pharm. Bull.,* 38, 1616, 1990.

37. **Kubo, I. and Klocke, J. A.,** Azadirachtin, insect ecdysis inhibitor, *Agric. Biol. Chem.,* 1951, 1982.

38. **Gray, A. I., Bhandari, P., and Waterman, P. G.,** New protolimonoids from the fruits of *Phellodendron chinese, Phytochemistry,* 27, 1805, 1988.

Chapter 13

ASSESSMENT OF ANTI-INFLAMMATORY ACTIVITY OF NATURAL PRODUCTS

Kevin D. Croft

TABLE OF CONTENTS

0-8493-4372-0/93/$0.00+$.50
© 1993 by CRC Press, Inc.

I. INTRODUCTION

Inflammation is a complex response to injury, involving migration of leukocytes and damage to local tissues. The vascular and cellular responses to inflammatory injury are controlled by several different classes of mediators. The inflammatory response is characterized by pain, heat, redness, swelling, cell influx, and loss of function. The various classes of chemical mediators act in concert to amplify this response. The inhibition of prostaglandin (PG) synthesis can give important anti-inflammatory effects and the nonsteroidal anti-inflammatory drugs (NSAID, e.g., aspirin, indomethacin), which inhibit the conversion of arachidonic acid (**1**) to PGs, are the most prescribed drugs for treatment of inflammatory disease.[1] The PGs not only have direct inflammatory effects, but also act synergistically with other mediators such as bradykinin. The edema induced by mediators that increase vascular permeability is enhanced by vasodilator PGs.

Leukocytes accumulate at sites of inflammation and are believed to contribute to tissue damage by releasing lysosomal enzymes and reactive oxygen species.[1] Another class of metabolites that are also important inflammatory mediators derived from leukocyte arachidonic acid are the leukotrienes (LT). LTB_4, in particular, by its potent chemotactic activity, may modulate leukocyte influx to inflammatory sites.[2] A third class of mediator that is linked biosynthetically to the prostanoids and leukotrienes is platelet-activating factor (PAF), a phospholipid that activates most inflammatory cells and has a variety of biological activities related to inflammation.[3] Interestingly, several natural products isolated from medicinal plants used in the treatment of inflammatory disease have been shown to be potent receptor antagonists for PAF, with the ability to block the inflammatory effects induced by PAF. Among these are the neolignan kadsurenone, isolated from *Piper futokadsura*,[4] and the ginkgolides, isolated from *Ginkgo biloba*.[5]

The prostanoids, leukotrienes, and PAF are not stored in tissues, but are synthesized from precursor fatty acids and phospholipids, upon cell stimulation. A large amount of arachidonic acid is stored as an ester, predominantly by reaction at the 2 position of cell membrane phospholipids. The initial step in the biosynthesis of the eicosanoids (i.e., PGs and LTs) is the enzymatic release of free arachidonic acid from phospholipid stores, predominantly by the action of phospholipase A_2. Free arachidonic acid can be metabolized by the cyclooxygenase enzyme complex, present in most animal cells, to unstable prostaglandin endoperoxides that are pivotal in the formation of several other products. The endoperoxides can be converted enzymatically to prostacyclin (PGI_2) and thromboxane (TXA_2) or to the primary PGs such as PGE_2 or $PGF_{2\alpha}$. PGI_2 and TXA_2 are unstable at physiologic pH and temperature and have short biologic half-lives and are usually measured as the more stable hydrolysis products 6-keto-$PGF_{1\alpha}$ and TXB_2, respectively. The biosynthesis of prostanoids has been reviewed in detail by Samuelsson and colleagues.[6] An alternative pathway of arachidonate metabolism is oxidation by the lipoxygenase enzyme. In leukocytes this is predominantly a 5-lipoxygenase that generates an epoxide at the 5,6-position of arachidonic acid and is intermediate to the formation of the biologically active LTB_4, LTC_4, etc.[7]

Membranes of a number of inflammatory cells contain a significant proportion of C-1 ether-linked phospholipids that also have an arachidonyl group at the C-2 position and thus may serve as a common source of the two substrates, arachidonic acid and lyso-PAF, for eicosanoid and PAF biosynthesis, respectively. The formation of PAF from the lyso precursor is via a specific acetyltransferase, to produce 1-O-alkyl-2-O-acetyl-*sn*-glycero-3-phosphorylcholine.

II. ANIMAL MODELS OF INFLAMMATION

A number of *in vivo* animal models have been used to assess anti-inflammatory activity. For example, the carrageenin-induced paw edema method[8] measures the change in volume of a rat paw 3 h after injection of 50 µl of 1% carrageenin solution into the plantar surface. Paw volume is measured using a plethysmometer, and test drugs can be given orally 30 min prior to the carrageenin challenge. Control animals should be given an oral dose of the vehicle

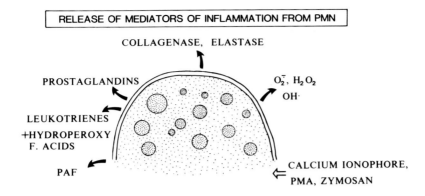

FIGURE 1. Release of mediators of inflammation from polymorphonuclear leukocytes (PMN, neutrophils).

used for the test drugs. Dose response cures can be constructed for each test compound, using percent inhibition of edema volume compared to the control.

Carrageenin-induced pleurisy can be used as an alternative model that measures exudate volume and cell infiltration.[9] In this case, 30 min after treatment of rats with doses of test compound or vehicle (for control), 150 µl of 1% carrageenin solution is injected intrapleurally. After 3 h, animals are killed and pleural exudates collected for measurement of volume and leukocyte count. This model has the advantage that exudates can also be used to measure the formation of eicosanoids, for example by radioimmunoassay (RIA).

Antipyretic tests can be carried out in rats that have hyperthermia, induced by subcutaneous injection of a 10 to 12% yeast suspension.[10] Test compounds and vehicle can be administered 10 h after induction of hyperthermia, and the rectal temperature is measured at various time points.

An interleukin-1-dependent model of inflammation mediated by neutrophils, which involves unilateral injection of 300 U of purified recombinant human interleukin-1 (IL-1) in mouse ears, has been described.[11] The inflammatory response results in seven- to tenfold increases in ear thickness within 24 h after treatment. This model appears to be insensitive to NSAIDs, suggesting that prostanoids do not play a role in mediating IL-1-induced inflammation.

Leukocyte chemiluminescence can be measured as an indication of the formation of reactive oxygen species. This technique has been used to detect compounds with possible anti-inflammatory activity.[12] Using this system, inflammatory leukocytes can be isolated and compounds tested *in vitro* for inhibition of opsonized zymosan-induced chemiluminescence. Alternatively, test compounds may be administered to the animal prior to cell isolation and stimulation. The *in vitro* technique suffers from the disadvantage that some compounds may quench the chemiluminescence and give a false positive result.

The *in vivo* inflammatory models have the advantage that they measure a biological endpoint or response and reflect the effectiveness of anti-inflammatory compounds, regardless of the possible mechanisms that may be involved. On the other hand, they have the disadvantage of variable biological response and inaccuracies in measurements such as paw volume. In addition, reasonably large amounts of pure compound may be necessary for administration to a large number of animals, in order to construct dose response curves.

III. *IN VITRO* LEUKOCYTE ASSAY

Often, natural products are isolated with only a few milligrams of pure material being available for biological testing. Since isolated rat inflammatory leukocytes or human peripheral leukocytes, in response to stimuli such as calcium ionophore, produce leukotrienes, prostanoids, and PAF, they are good model cells to study the effects of test compounds on the synthesis of the various lipid mediators (Figure 1). Small amounts of compounds may

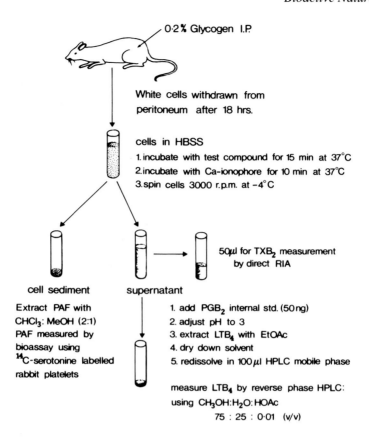

FIGURE 2. Summary of *in vitro* leukocyte assay.

be tested *in vitro*, and information on the possible mechanism of anti-inflammatory activity may be obtained. However, compounds that show potential activity *in vitro* should also be assessed using an *in vivo* inflammatory model.

The *in vitro* leukocyte assay is summarized in Figure 2.

A. PREPARATION OF LEUKOCYTES[13]

Rats are injected intraperitoneally with 5 ml oyster glycogen (0.2%) in phosphate-buffered saline, and then 18 h later the peritoneum is lavaged with heparinized (20 U/ml) Hepes-buffered Hanks solution (HBHS) containing 0.1% bovine serum albumin, pH 7.4 (25 ml). Exudate cells, which yield more than 85% neutrophils, can then be assessed by differential staining and light microscopy. Blood contamination of the exudate fluid should be avoided. Leukocytes are washed and resuspended in HBHS to a concentration of between 0.5 to 1×10^7 cells per milliliter, which can be determined using a coulter counter.

Leukocytes can also be isolated from human venous blood collected into a one-tenth volume of 3.8% trisodium citrate.[14] The citrated blood is layered over mono-poly resolving medium (Flow Laboratories) and centrifuged at $500 \times g$ for 30 min. The top layer of cells contains monocytes and platelets, while the second layer, rich in neutrophils, is removed and diluted with HBHS, centrifuged at $150 \times g$ for 10 min, and then the cell pellet treated with lysing buffer[15] at 4°C for 5 min to remove red cells. After washing twice with HBHS, cells are resuspended at a concentration of 0.5 to 1×10^7 cells per milliliter.

B. INCUBATION OF CELLS WITH TEST COMPOUNDS

Cells are incubated with test compounds in doses ranging from 1 to 200 μM for 15 min at 37°C prior to cell stimulation with the calcium ionophore A23187 (1 to 5 µg/ml) for 10 min. Control incubations are carried out with vehicle only (10 µl DMSO per 1 ml of incubation mixture). Other, more physiological cell stimulants, such as opsonized zymosan or the peptide formyl methionyl leucyl phenylalanine (FMLP), can be used in place of calcium ionophore. However, much lower levels of eicosanoids are generated, which makes HPLC analysis (of LTB_4, for example) more difficult, and RIA methods must therefore be employed.

C. ANALYSIS OF PROSTANOIDS, LEUKOTRIENES, AND PAF

The major products of 5-lipoxygenase in stimulated leukocytes are LTB_4 and 5-hydroxyeicosatetraenoic acid (5-HETE). These products can be conveniently measured by reverse-phase HPLC, using a methanol:water:acetic acid (70:30:0.01 v/v) mobile phase with UV absorbance monitored at 270 nm for LTB_4 and at 234 nm for 5-HETE. After the incubation with ionophore, the cells are centrifuged, and PGB_2 (50 ng) is added to the supernatant as an internal standard. The pH is adjusted to 3 with formic acid, and the mixture extracted with ethyl acetate. The solvent is evaporated to dryness under nitrogen, and the residue reconstituted in mobile phase prior to analysis. The leukotrienes can also be measured by RIA at very low concentrations, using commercially available kits.

The prostanoid metabolites PGE_2, TXB_2, and 6-keto $PGF_{1\alpha}$ can all be measured directly in supernatant from cell incubates by specific RIA.[16] Samples are assayed neat and after serial dilution with phosphate-buffered saline.

Since most PAF formed in stimulated leukocytes remains cell bound, measurements of this substance are made on organic extracts of the cell pellet. The pellet is extracted with 2 ml of $CHCl_3$:CH_3OH (2:1), and after removal of solvent, the lipid residue is dissolved in ethanol (50 µl) and reconstituted in Hepes-buffered Tyrodes solution, containing 0.25% bovine serum albumin, to a total volume of 1 ml. The extracted PAF is quantitated by bioassay using ^{14}C-labeled rabbit platelets,[17] with standard curves constructed using authentic PAF (Sigma Chemicals). Full details of this assay have been described.[18,19]

IV. CASE STUDIES

A. DIFFERENTIAL INHIBITION OF TXA₂ AND LTB₄ BIOSYNTHESIS BY TWO NATURALLY OCCURRING ACETYLENIC FATTY ACIDS

A number of acetylenic fatty acids are known to interfere with fatty acid metabolism and inhibit cyclo-oxygenase and lipoxygenase enzymes in a variety of tissues. The seed oil of two Australian species of plant were examined, and two acetylenic fatty acids isolated, by preparative HPLC. Crepenynic acid (**2**) was isolated from *Ixiolaena brevicomta*,[20] a plant linked with extensive sheep mortalities in eastern Australia, and ximenynic acid (**3**) was obtained from *Santalum accuminatum*,[21] a plant reputed to have medicinal properties by the Australian aborigines.[22] The effects of these products on LTB_4 and TXB_2 production in rat peritoneal leukocytes was studied and compared with the nonacetylenic fatty acids linoleic (**4**) and ricinoleic (**5**) acids.[23] In concentrations ranging from 10 to 100 μM, linoleic acid and ricinoleic acid had only minimal effects on LTB_4 and TXB_2 production in ionophore stimulated cells (Figure 3). Ximenynic acid (**3**) demonstrated dose-dependent inhibition of LTB_4, TXB_2, and 6-keto $PGF_{1\alpha}$ production, with an IC_{50} of 60 μM for LTB_4. Crepenynic acid (**2**), on the other hand, was a more effective inhibitor of the cyclo-oxygenase products having an IC_{50} for TXB_2 of less than 10 μM (Figure 4). The acetylenic bond is a necessary feature for inhibition of eicosanoid production, since other unsaturated fatty acids of a similar chain length show no, or only slight, inhibitory effects. The acetylenic bond is likely to act via an intermediate allene or vinylhydroperoxide to irreversibly bind to and deactivate the enzyme,

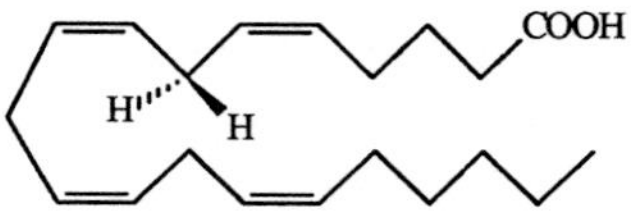

1

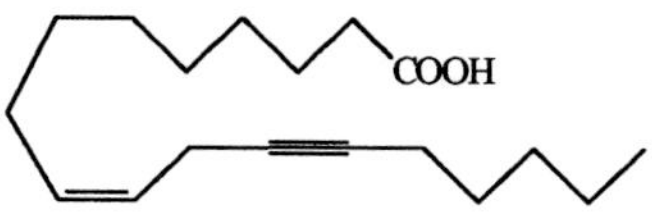

2

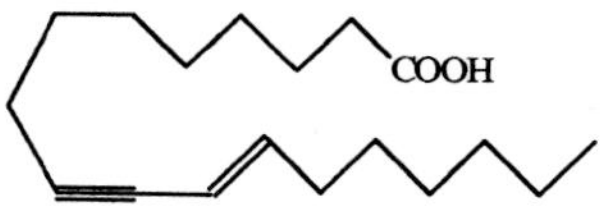

3

although in this case the two acetylenic compounds differentially inhibit the cyclo-oxygenase and lipoxygenase products of stimulated leukocytes.

B. INHIBITION OF LTB$_4$ AND PAF SYNTHESIS IN LEUKOCYTES BY A SESQUITERPENE LACTONE

In a screening of compounds from Philippine medicinal plants for anti-inflammatory activity, the sesquiterpene lactone scandenolide (**6**) was isolated from *Mikania cordata*. Scandenolide was found to inhibit completely the chemiluminescence of whole blood exposed to opsonized zymosan or phorbol myristate acetate (PMA), at a concentration of 100 μ*M* (Figure 5).[24] In concentrations ranging from 10 to 200 μ*M*, scandenolide (**6**) showed a dose-dependent inhibition of the lipoxygenase products LTB$_4$ (IC$_{50}$ 15 μ*M*) and 5-HETE (IC$_{50}$ 30 μ*M*), as shown in Figure 6. Figure 7 shows a representative HPLC trace following a control incubation and an incubation carried out in the presence of 50 μ*M* scandenolide, demonstrating the lowered levels of LTB$_4$ and 5-HETE. In contrast, there was only slight inhibition of the cyclo-oxygenase product TXB$_2$ over the entire concentration range tested. The effect of this compound on phospholipase A$_2$ activity was tested by using leukocytes prelabeled with ^{3}H-arachidonic acid. At scandenolide concentrations of 100 μ*M*, phospho-

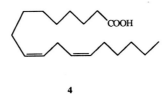

4

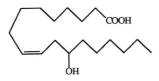

5

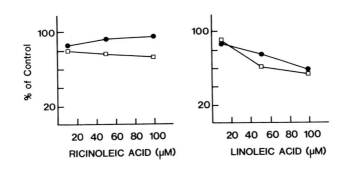

FIGURE 3. The effect of ricinoleic acid (5) and linoleic acid (4) on the formation of LTB_4 (●) and TXB_2 (□) by ionophore-stimulated rat leukocytes. Results are expressed as percent of control and are the mean of duplicate experiments.

lipase activity remained at 60% of control values, and it is therefore unlikely that effects on this enzyme are responsible for the inhibition seen. It is presumed that suppression of LTB_4 and 5-HETE is due to direct effects on the 5-lipoxygenase, since any effect acting through inhibition of phospholipase A_2 would cause a decrease in all arachidonate metabolites, including TXB_2.

Scandenolide showed strong inhibition of PAF formation in stimulated leukocytes, with an IC_{50} of less than 20 μM and nearly complete inhibition at concentrations above 100 μM (Figure 8). This is the first activity of this type attributed to a sesquiterpene lactone, and although the point at which PAF synthesis is inhibited remains to be elucidated, such activity is very interesting and may be of biologic relevance for anti-inflammatory activity.

Similar experiments were conducted with the sesquiterpene lactone coronopilin (7), which at the same concentrations showed no inhibition of either LTB_4 or PAF production.

Plants containing sesquiterpene lactones have been used as anti-inflammatory remedies,[25] and the mode of action of a number of individual compounds has been studied.[26,27] Sesquiterpenes such as helenalin (8) have been shown to uncouple oxidative phosphorylation of

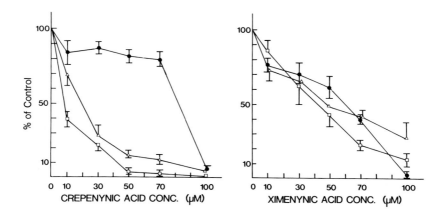

FIGURE 4. The effect of crepenynic acid (**2**) and ximenynic acid (**3**) on the formation of LTB$_4$ (●), TXB$_2$ (□), and 6-ketoPGF$_{1\alpha}$ (△) by ionophore-stimulated rat peritoneal leukocytes. Results are expressed as percent of control and represent the mean and standard error of four experiments.

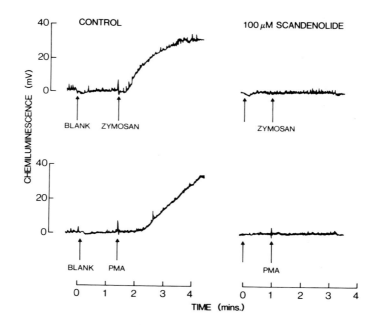

FIGURE 5. Inhibition of whole blood chemiluminescence by 100 µ*M* of scandenolide (**6**).

neutrophils and inhibit lysosomal enzyme activity and neutrophil chemotactic migration. At relatively high concentrations (in the millimolar range) cyclo-oxygenase activity has been suppressed. The α-methylene-γ-lactone moiety and the β-unsaturated cyclopentanone ring, as found in helenalin (**8**), are considered necessary for anti-inflammatory activity. Scandenolide (**6**) possesses some of these structural features, and our studies suggest it is a potent inhibitor of lipoxygenase and PAF synthesis, but not cyclo-oxygenase activity at micromolar concentrations.

The link between the biochemical activities displayed by scandenolide and anti-inflammatory activity of the plant were confirmed in two *in vivo* test systems.[28]

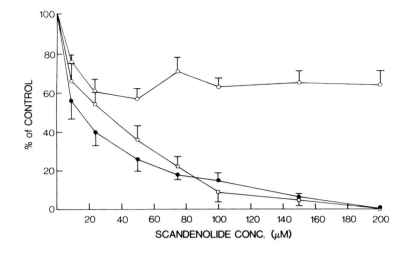

FIGURE 6. The effect of scandenolide (**6**) on the formation of LTB$_4$ (●), 5-HETE (□), and TXB$_2$ (○) by ionophore-stimulated rat peritoneal leukocytes. Results are expressed as percent of control and represent the mean and standard error of six experiments.

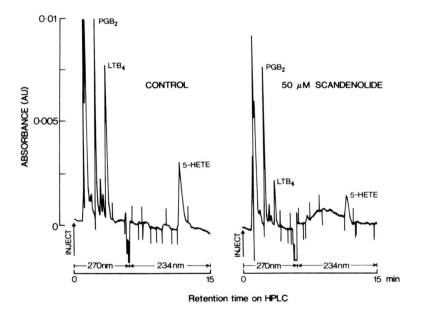

FIGURE 7. Representative reverse-phase HPLC analysis of LTB$_4$ and 5-HETE from ionophore-stimulated rat leukocytes from control and cells incubated with 50 μM scandenolide (**6**).

Purified extracts of the plant *M. cordata* significantly suppressed exudate volume and leukocyte accumulation in the carrageenin-induced pleurisy model, at an oral dose of 200 mg/kg. A dose-dependent inhibition of mouse paw edema was observed, with an ID$_{50}$ of 100 mg/kg. Thus, the biochemical activity described for one of the constituents of this plant, i.e., scandenolide, may explain, at least in part, some of its medicinal properties.

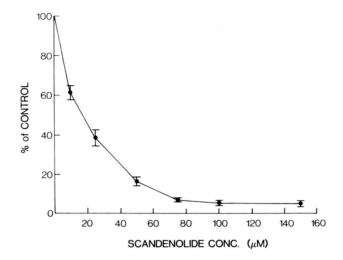

FIGURE 8. The effect of scandenolide (**6**) on the formation of platelet-activating factor (PAF) by ionophore-stimulated rat peritoneal leukocytes. Results are expressed as percent of control and represent the mean and standard error of 4 experiments.

6

7

8

C. EFFECTS OF DIETARY n-3 FATTY ACIDS ON EICOSANOID AND PAF FORMATION IN RAT INFLAMMATORY LEUKOCYTES

There has been considerable interest over the past decade in the effects of dietary fish oils, which are rich in n-3 fatty acids such as eicosapentaenoic acid (EPA), on human health and disease.[29] While most interest has centered on effects on the cardiovascular system, there is evidence that fish oil supplementation is effective in the treatment of inflammatory diseases such as active rheumatoid arthritis.[30] The alleviation of rheumatoid arthritis in human subjects has been linked with a reduction in neutrophil LTB_4 production. We have used the rat model to study the effect of dietary EPA on the formation of LTB_4, TXB_2, and PAF by ionophore-stimulated leukocytes.[19] Rats were fed for 3 weeks on a diet supplemented with either 10% by weight of "Max EPA" fish oil or a coconut oil/safflower oil control diet. At the end of this period, peritoneal leukocytes were harvested as previously described. The EPA-rich diet significantly increased the EPA content of cell phospholipids and reduced the arachidonic acid (**1**) levels by 35% compared to the control diet. LTB_4 production, as measured by reverse-phase HPLC, was decreased by 50% with the EPA-rich diet. There was concomitant formation of the biologically less-active LTB_5, derived from the EPA substrate, which was identified by HPLC and showed cross-reactivity with the LTB_4 antibody (Figure 9). The amount of LTB_4 and LTB_5 produced by stimulated leukocytes closely resembled the changes in arachidonic acid and EPA content of leukocyte phospholipids.

TXB_2 production in stimulated leukocytes from the EPA-fed animals was also significantly reduced compared with the control group. PAF production, on the other hand, was not altered by dietary treatment. Despite the lack of effect on PAF formation, the ability of an EPA-rich diet to decrease LTB_4 and TXB_2 production suggests that these diets may attenuate leukocyte activity and have useful anti-inflammatory properties.

V. CONCLUSION

The *in vitro* inflammatory leukocyte model provides a useful and convenient system for the study of the effects of potential anti-inflammatory compounds on important classes of lipid mediators. Tests can be carried out with small amounts of purified compounds (e.g., 2 to 5 mg), or alternatively, the model can be used to evaluate the effects of dietary components on leukocyte activity and eicosanoid synthesis. The use of *in vivo* inflammatory models must also be considered for the complete assessment of the anti-inflammatory potential of natural products.

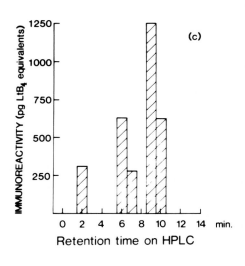

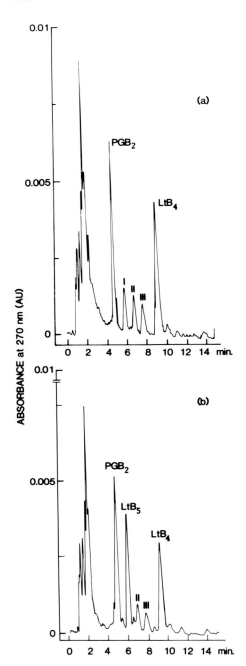

FIGURE 9. Formation of LTB$_4$ and LTB$_5$ and their corresponding stereoisomers by calcium ionophore-stimulated rat peritoneal cells (1×10^7 cells per milliliter). (a) HPLC profile of products obtained from rats fed the coconut oil containing 3% safflower oil control diet: II and III correspond to 6-*trans* isomers of LTB$_4$; I corresponds to LTB$_5$. (b) HPLC profile of products from rats fed the Max EPA oil diet. (c) Immunoreactivity of 1-ml fractions collected after HPLC separation of products obtained from rats fed Max EPA oil [i.e., HPLC profile (b)]. Immunoreactivity was assessed using LTB$_4$ antibody obtained from Wellcome Diagnostics.

REFERENCES

1. **Salmon, J. A. and Higgs, G. A.,** Prostaglandins and leukotrienes as inflammatory mediators, *Br. Med. Bull.,* 43(2), 285, 1987.
2. **Bray, M. A.,** The pharmacology and pathophysiology of leukotriene B$_4$, *Br. Med. Bull.,* 39, 249, 1983.
3. **Braquet, P., Touqui, L., Shen, T. Y., and Vargaftig, B. B.,** Perspectives in platelet activating factor research, *Pharmacol. Rev.,* 39, 97, 1987.
4. **Shen, T. Y., Hwang, S. B., Chang, M. N., Doebber, T. W., Lam, M. T., Wu, M. S., Wang, X., Han, G. Q., and Li, R. Z.,** Characterization of a platelet-activating factor receptor antagonist isolated from haifenteng (*Piper futokadsura*): specific inhibition of *in vitro* and *in vivo* platelet activating factor induced effects, *Proc. Natl. Acad. Sci. U.S.A.,* 82, 672, 1985.
5. **Braquet, P.,** Involvement of PAF in various immune disorders using BN 52021 (ginkolide B): a powerful PAF antagonist isolated from *Ginkgo biloba,* in *Advances in Prostaglandin, Thromboxane and Leukotriene Research,* Vol. 16, Raven Press, New York, 1986, 179.
6. **Samuelsson, B., Goldyne, M., Granstrom, E., Hamberg, M., Hammarstrom, S., and Malmsten, C.,** Prostaglandins and thromboxanes, *Ann. Rev. Biochem.,* 47, 997, 1978.
7. **Samuelsson, B.,** Leukotrienes: mediators of immediate hypersensitivity reactions and inflammation, *Science,* 220, 568, 1983.
8. **Winter, C. A., Risley, E. A., and Nuss, G. W.,** Carrageenin induced edema in hind paw of the rat as an assay for anti-inflammatory drugs, *Proc. Soc. Exp. Biol. Med.,* 111, 544, 1962.
9. **Ku, E. C., Raychaudhuri, A., Ghai, G., Kimble, E. F., Lee, W. H., Colombo, C., Dotson, R., Oglesby, T. D., and Wasley, J. W. F.,** Characterization of CGS 8515 as a selective 5-lipoxygenase inhibitor using *in vitro* and *in vivo* models, *Biochim. Biophys. Acta,* 959, 332, 1988.
10. **Teotino, U. M., Friz, L. P., Gandini, A., and Bella, D. D.,** Thio derivatives of 2,3-dihydro-4H-1,3-benzoxazin-4-one. Synthesis and pharmacological properties, *J. Med. Chem.,* 6, 248, 1963.
11. **Maloff, B. L., Shaw, J. E., and DiMeo, T. M.,** IL-1 dependent models of inflammation mediated by neutrophils, *J. Pharmacol. Meth.,* 22, 133, 1989.
12. **Bird, J. and Giroud, J. P.,** An appraisal of the technique of polymorphonuclear leukocyte chemiluminescence as a means to detect compounds with anti-inflammatory activity, *J. Pharmacol. Meth.,* 14, 305, 1985.
13. **Terano, T., Salmon, J. A., and Moncada, S.,** Effect of orally administered eicosapentaenoic acid on the formation of leukotriene B4 and leukotriene B5 by rat leukocytes, *Biochem. Pharmacol.,* 33, 3071, 1984.
14. **McCulloch, R. K., Croft, K. D., and Vandongen, R.,** Enhancement of platelet 12-HETE production in the presence of polymorphonuclear leukocytes during calcium ionophore stimulation, *Biochim. Biophys. Acta,* 1133, 142, 1991.
15. **Boyum, A.,** Separation of leukocytes from blood and bone marrow, *Scand. J. Clin. Invest.,* 21, 77, 1968.
16. **Mahoney, D. P., Barden, A., Beilin, L. J., and Vandongen, R.,** Radioimmunoassay of prostaglandins and thromboxane B2 in extracted and unextracted urine and serum using an iodinated ligand, *Prost. Leuk. Med.,* 12, 11, 1983.
17. **Henson, P. M.,** Activation and desensitization of platelets by platelet activating factor derived from IgE-sensitized basophils, *J. Exp. Med.,* 143, 937, 1976.
18. **Croft, K. D., Sturm, M. J., Codde, J. P., Vandongen, R., and Beilin, L. J.,** Dietary fish oils reduce plasma levels of platelet activating factor precursor (lyso-PAF) in rats, *Life Sci.,* 38, 1875, 1986.
19. **Croft, K. D., Codde, J. P., Barden, A., Vandongen, R., and Beilin L. J.,** Effect of dietary fish oils on the formation of leukotriene B4 and B5, thromboxane and platelet activating factor by rat leukocytes, *Clin. Exp. Pharm. Phys.,* 15, 517, 1988.
20. **Ford, G. L.,** Semi-preparative isolation of crepenynic acid, a potential inhibitor of essential fatty acid metabolism, *J. Chromatogr.,* 346, 431, 1985.
21. **Hatt, H. H. and Schoenfeld, R.,** Some seed fats of the Santalaceae family, *J. Sci. Food Agric.,* 7, 130, 1956.
22. **Cribb, A. B. and Cribb, J. W.,** in *Wild Medicine in Australia,* Fontana Books, Sydney, 1983, 51.
23. **Croft, K. D., Beilin, L. J., and Ford, G. L.,** Differential inhibition of thromboxane B2 and leukotriene B4 biosynthesis by two naturally occurring acetylenic fatty acids, *Biochim. Biophys. Acta,* 921, 621, 1987.
24. **Ysrael, M. C. and Croft, K. D.,** Inhibition of leukotriene and platelet activating factor synthesis in leukocytes by the sesquiterpene lactone scandenolide, *Planta Med.,* 56, 268, 1990.
25. **Hall, I. H., Lee, K. H., Starnes, C. O., Sumida, Y., Wu, R. W., Waddell, T. G., Cochran, J. W., and Gerhart, K. G.,** Anti-inflammatory activity of sesquiterpene lactones and related compounds, *J. Pharm. Sci.,* 68, 537, 1979.
26. **Hall, I. H., Lee, K. H., and Sykes, H. C.,** Anti-inflammatory agents. IV. Structure activity relationships of sesquiterpene lactone esters derived from helenalin, *Planta Med.,* 53, 153, 1987.

27. **Hall, I. H., Starnes, C. O., Lee, K. H., and Waddell, T.G.,** Mode of action of sesquiterpene lactones as anti-inflammatory agents, *J. Pharm. Sci.,* 69, 537, 1980.
28. **Ysrael, M. C. and Croft, K. D.,** An evaluation of the anti-inflammatory properties of *Mikania cordata, Acta Manilana,* 38, 75, 1990.
29. **Nelson, G. J.,** *Health Effects of Dietary Fatty Acids,* American Oil Chemists' Society, Champaign, Illinois, 1991.
30. **Kremer, J. M., Jubiz, W., Michalek, A., Rynes, R. I., Bartholomew, L. E., Bigaouette, J., Timchalk, M., Beeler, D., and Lininger, L.,** Fish oil fatty acid supplementation in active rheumatoid arthritis, *Ann. Int. Med.,* 106, 497, 1987.

Chapter 14

IMMUNOMODULATORY COMPOUNDS

Rudi P. Labadie

TABLE OF CONTENTS

0-8493-4372-0/93/$0.00+$.50
© 1993 by CRC Press, Inc.

I. INTRODUCTION

"Immunomodulating activity" is a collective term indicating biologic or pharmacologic effects on humoral or cellular factors functioning in the immune response. Each factor and each functional system involved in the immune response may be influenced in various ways. The actual effects may be specific or nonspecific. Hence, mechanistic studies have to be performed to distinguish between these categories. Some agents may combine specific and nonspecific effects. Because of regulatory interactions between humoral and cellular immunofactors in the course of functional processes of the immune response, the *in vivo* net effect of an immunomodulator determines whether a stimulatory or a suppressive action will result. Negative-feedback mechanisms appear to be frequent in the immune system. Thus "immunosuppression" may result from stimulation of inhibitory cells or humoral factors, as well as from inhibition of effector cells or activating humoral factors. On the other hand, "immunostimulation" arises from stimulation of effector cells or the production of their metabolic inducers, and possibly also from inhibition of factors that limit immunogenicity.

Since, depending on the dose range, one and the same agent is able to exert immunostimulatory as well as immunosuppressive effects, it is preferred in this chapter to deal with the selected active compounds under the general term immunomodulators. Whether an agent is termed to be immunostimulatory or immunosuppressive should also depend on the amount and the extent of evidence based on *in vitro*, *in vivo*, and clinical studies.

The relevance of immunomodulatory effects to drug therapy is basically connected with the question of whether the bioassays used are functional. For example, not all compounds that show enhancement of macrophage phagocytosis *in vitro* and an increased carbon clearance in animals are capable of a curing effect on animals with bacterial infections.[1,2] A curative effect is only produced by those immunomodulators that not only enhance macrophage phagocytosis, but also stimulate the secretion of monokines like CSF (colony-stimulating factors) from macrophages. The experimental models to be used in the search and the development of immunomodulators should advisedly be functional and disease oriented.[2] As for the latter, understanding of pathophysiological processes and the role of immunofactors involved is important in a research approach focused on new immunomodulators.

It has become clear now that raw natural products represent prominent and promising sources for molecules with interesting immunomodulating properties. This line of research in a programmed form is relatively young, but fast growing. Before the second half of the 1970s, immunomodulation was mainly of interest within the field of immunology. In the years toward the end of that decade, however, a mutual interest in immunomodulatory activities arose among an international group of pharmacologists and immunologists. The early advances in immunopharmacology were reviewed in 1982.[3] The 1970s and 1980s were also marked by the appearance of papers describing immunomodulatory properties of compounds isolated from fungi, mushrooms, and plants.

A picture of advances in this field is given in some reviews by Lindequist and Teuscher,[4] Wagner and Proksch,[5] Wagner,[6] and Labadie et al.[7] The studies reviewed in these publications were all carried out on natural products that have a reputation in ethnomedical practices. Leads derived from traditional medicine through scientific studies appear to be influential for the discovery of new immunomodulatory compounds and novel mechanisms of action. Such compounds may serve innovative drug development. However, in a broader perspective, studies on ethnomedical resources can be of great impact on drug therapy as a whole.

The scope of this chapter is to outline the assaying of immunomodulatory activity, the approaches to the selection of crude natural product sources, and the experimental strategies aimed at the detection and isolation of immunomodulatory compounds. We also discuss a selected group of immunomodulatory compounds representing diverse molecular structural classes.

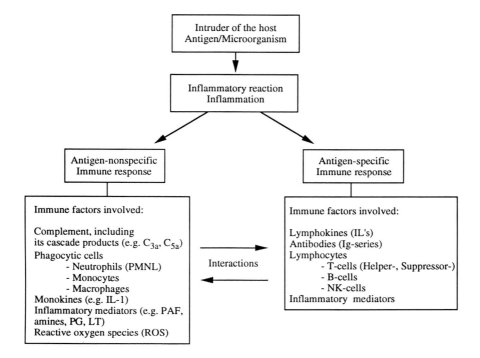

II. THE ASSAYING OF IMMUNOMODULATORY ACTIVITY

A. IMMUNE RESPONSE, INFLAMMATION, AND IMMUNOFACTORS

The basic function of the immune system is recognition and destruction of the "nonself" that has entered the host. The anatomic and physiologic components involved in the immune response can be divided into molecular factors (or humoral factors) and cellular factors. The immune response includes both antigen-nonspecific and antigen-specific processes (Scheme 1).

The immunofactors considered here also function in inflammation reactions; therefore, considerations concerning modulation of the immune response and inflammatory processes overlap. Hence, certain immunomodulatory properties coincide with anti-inflammatory effects.

Prominent immunofactors functioning in the antigen-nonspecific immune response are the complement system, including the peptides (e.g., C3a and C5a) formed during activation of the complement cascade; the phagocytic cells [i.e., polymorphonuclear neutrophilic leukocytes (PMNL), mononuclear phagocytes (or monocytes), and macrophages]; the production of reactive oxygen species (ROS) by activated PMNL, in particular, but also by monocytes and macrophages, constituting an important mechanism; among other mediators, activated monocytes and macrophages producing monokines that interact with lymphocytes; and the so-called mediator cells (i.e., mast cells, basophils, eosinophils, and platelets) participating in the immune response, in that they produce and release soluble mediators [such as vasoactive amines, heparin, chemotactic factors, platelet-activating factor (PAF), and arachidonic acid metabolites (prostaglandins and leukotrienes)], which enhance inflammatory reactions.

Prominent immunofactors of the antigen-specific immune response are T-lymphocytes (including T-helper cells, T-suppressor cells, and cytotoxic T-cells) and B-lymphocytes, and other types of lymphocytes such as natural killer cells (NK-cells); the immunoglobulins (IgM, IgG, IgA, IgD, IgE) and their subclasses produced by B-cells functioning as antibodies; and the lymphokines or interleukins (ILs) representing typical lymphocyte metabolic products that function as messengers to other lymphocytes, phagocytes, or nonimmune cells.

The overall immune responsiveness is brought about through functional interactions of the separate cellular and humoral immunofactors. Antigen-nonspecific and antigen-specific processes are closely and interdependently linked in a regulatory network of the immune response. A detailed treatise of this immunoregulatory network goes far beyond the scope of this chapter; hence, only an outline of major interactions is given here.[3,8] Functional activation of both pathways of complement leads to lysis of cells foreign to the host. In the terminal phase of the complement cascade, the molecular complex C5b67 binds to the invading cell membranes and mediates lysis together with C8 and C9. In itself, this is an antigen-nonspecific defense mechanism. During the complement cascade, however, extremely important products are formed. These peptide substances functionally link complement with mast cells, basophils, phagocytes (PMN, monocytes, etc.),[9,10] and B-cells.[11] The compounds C3a and C5a are known to be anaphylotoxins. Both these cationic peptides induce degranulation of mast cells and basophils and provoke the release of histamine, increased capillary permeability, and smooth muscle contraction. In addition C5a is a very potent chemotactic factor. Very small concentrations of C5a (5 to 10 ng/ml) are able to provoke direct migration of PMN, monocytes, and eosinophils.[9,10] The adhesiveness and aggregation of human polymorphonuclear leukocytes (PMN) have been found to increase under the effect of C5a.[9,12,13] The cellular adhesion processes, e.g., those in which leukocytes adhere to endothelial cells of blood vessels before their migration through the vessel wall, have recently been recognized to be important in inflammation.[14-16] Another complement cascade product that interacts with phagocytic cell function, in the process called immunophagocytosis, is C3b. This fairly large molecule (180 kDa) affixes to bacteria, fungi, or other cells foreign to the host and renders such particles able to be recognized and bound by phagocytic leukocytes that bear the so-called C3b- and iC3b-receptors at their surfaces.[11,16] The phagocytic leukocytes are activated through this process to perform phagocytosis. Although C3b is known to serve these phagocytic purposes primarily, C4b can also bind particles that are then recognized by PMN and monocytes. It is not known, however, to what extent this phenomenon leads to phagocytosis.[9]

The functional interaction between the complement system and antibodies is indicated by the initial activation process of the classical pathway of complement. In particular, IgG or IgM attached to an antigen (immune complexes) is needed to activate the complement cascade through the classical pathway. In addition, microorganisms opsonized with specific antibody work together with alternative-pathway C3-fragments to be recognized by phagocytic leukocytes through specific receptors (for C3b or a fragment of C3b) on their membranes. A small fragment of C3, with molecular weight of 10 kDa and designated C3e, was found to mobilize leukocytes *in vitro* and *in vivo*. This factor is also indicated as leukocytosis-promoting factor.[17,18]

It is clear that the functional interactions (including feedback mechanisms) between antigen-nonspecific humoral and cellular factors are manifold. However, interactions between antigen-nonspecific and antigen-specific immunofactors are also intensive and intimate. For instance, macrophages and lymphocytes interact through the phenomenon of antigen presentation. Macrophages, having processed the antigen that has been taken up, present it (in a processed form) to T-lymphocytes. In this process, cell-to-cell contact takes place, and the macrophage and lymphocyte share histocompatibility-linked gene products. The macrophage seems to select discrete regions of the antigen molecule for recognition by sensitized T-cells.[3,19-22] Besides this direct cell-cell contact, macrophages and lymphocytes may influence each others' functions indirectly. Such interactions are mediated through secretion of cytokines (i.e., monokines and lymphokines) that provoke specific activities. Monokines (e.g., IL-1) produced by monocytes and macrophages effectuate proliferation of lymphocytes or enhance the stimulation of other agents such as antigens or mitogens. Lymphokines produced by T-lymphocytes (e.g., IL-2) activate macrophages. Such activation may include enhanced phagocytosis, increased microbial killing, increased adherence to foreign surfaces, increased secretory activity, and attraction and retention activities at sites of inflammation. The latter activities of lymphokines are not restricted to monocytes or macrophages, since lymphokines such as IL-5 are known to have chemotactic effects on eosinophils, basophils, and neutrophils. But the leukocyte inhibition factor (LIF) impairs the migration of these cells in the same manner as MIF (macrophage/monocyte migration inhibition factor) does in the case of monocytes/macrophages.

Activated T-lymphocytes also produce IL-4, a lymphokine that stimulates B-cells to produce antibodies. In cases of an allergic or inflammatory reaction, the antibody (IgE) produced would trigger mast cells to release inflammatory mediators, including histamine, prostaglandin D_2 (PGD$_2$), and leukotriene C_4 (LTC$_4$). The lymphokines IL-4 and IL-3 also represent growth factors for mast cells. Together with basophils, mast cells play a key role in immediate hypersensitivity and other allergic reactions.

Generally, besides the cytokines, specific functions of communication within the immune response network are performed by other groups of mediators. In inflammatory conditions the following mediators play key roles: vasoactive amines (e.g., histamine), prostaglandins (PGD$_2$, PGE$_2$, PGF$_2$), thromboxane, leukotrienes (LTB$_4$, LTC$_4$, LTD$_4$, LTE$_4$), platelet-activating factor (PAF), peptides (substance P), adenosine, and bradykinin.

The picture given above of the functional interactions between cellular and humoral immunofactors is far from complete. It illustrates, however, the interdependency of regulatory and effector cells, their molecular messengers, and the mediators in the antigen-specific and antigen-nonspecific immune response.

Research focused on the development of immunomodulators is directed toward activities that can be expressed in terms of stimulation or inhibition of immunofactor function, and its interaction with functions of other immunofactors and relevant nonimmune cells. In order to explore immunomodulatory properties and potencies, the selection and development of relevant and functional bioassays is of primary importance.

B. BIOASSAYS FOR IMMUNOMODULATORY ACTIVITY

The literature concerned with assaying of immunomodulatory activity is very extensive. It comprises many methodologically diverse approaches and techniques. An important reason underlying this situation is the bewildering complexity of the immune system. Understanding of its functions and mechanisms is thus of primary importance. The reader is referred to authoritative sources in immunology for current concepts, immunologic principles that form the basis for experimental methods, and human diseases discussed from an immunological point of view.[23-25] For the purpose of assaying T- and B-cell functions *in vitro* and *in vivo*, and assays concerned with antibodies, cytokines, nonlymphoid cells, and cell sorting, a very useful compilation has been written by Coligan et al.[26] An extensive and authoritative treatise on the methods for studying mononuclear phagocytes, by Adams et al., represents a broad and practical orientation in the field.[27] As for measurements concerned with cells involved in inflammation, inflammatory mediators, chemotactic properties, and cells and enzymes involved in the production of free oxygen radicals and hydrogen peroxide, reference is made to Higgs and Williams,[28] Wilkinson,[29] Fantone and Ward,[30] Fehér et al.,[31] and Van Dyke and Castranova.[32]

In practice, immunomodulatory activity *in vitro* is recorded on different levels, ranging from effects on parts of organs (e.g., trachea rings), whole-cell preparations (e.g., polymorphonuclear leukocytes, monocytes, macrophages, lymphocytes), enzyme systems, and receptor binding. Although, depending on the type of test performed, specific effects can be measured, each category of *in vitro* testing has its limitations in predicting capacity for *in vivo* conditions. In addition, positive results of activity in animals do not guarantee a similar effect in humans. The strategy followed in screening programs is to select a battery of *in vitro* and *in vivo* assays that permit a fair chance of scoring in clinical conditions. The criteria applied in the selection of bioassays are the relevancy and the significance of immune effector functions or dysfunctions under normal, health-threatening, and pathologic conditions. Thus, by choice of relevant and functional bioassays, the search for, and development of, immunomodulators can, and should be, disease oriented.

Recently, a selection of bioassays used to screen plant extracts and plant compounds for immunostimulatory and anti-inflammatory effects was described by Wagner and Jurcic.[33] Simultaneously, Beretz and Cazenave dealt with assays to be used in screening natural products for effects on platelet functions *in vitro* and *in vivo*.[34] The same chapter also deals with platelet function assays using receptor-binding models and platelet enzymes. Using immunologic parameters (i.e., complement- and macrophage-mediated cytotoxicity for antihepatotoxic effects), Kiso and Hikino described *in vitro* assays with hepatocytes and *in vivo* hepatitis models using mice.[35] Also, in a very recent article, assays based on receptor (for IL-1 and IL-2) binding, enzyme (extracellular phospholipase) activity, cellular adhesion, and release of inflammatory mediators (e.g., leukotriene B$_4$) were described by Devlin et al.[36]

III. NATURAL PRODUCT RESOURCES AND THEIR SELECTION

The use and selection of natural products in the search for, and development of, immunomodulators can be based on two different sources of information. The first is generated from taxonomic, chemotaxonomic, phytochemical, and immunopharmacologic studies and possibly from clinical data on single plant constituents and standardized plant extracts. The second is derived from classical texts on traditional medicine, oral interviews and consultations with contemporary practitioners, and practices of traditional and folkloric medicinal systems. A rational balance between both these sources of information is of utmost significance. The majority of studies carried out on natural-product-derived immunomodulators so far represent approaches that have a poor basis in traditional medicinal knowledge or ethnomedical practices. The decision making in the selection of natural products, and the strategy of the experimental approach, may be strongly influenced by basic knowledge and evaluation of traditional ethnomedical information. Already, at such an early stage, the research program can be disease oriented.

For fruitful research on natural-product-derived immunomodulatory compounds, we feel that the empirical knowledge that forms the basis of ethnomedical practices is as important as pure botanic, biologic, phytochemical, and (immuno)pharmacologic data. To take full advantage of the potentials of natural product resources and the empirical knowledge in traditional medicine, we found that an ethnopharmacognostic inquiry, preceding the immunopharmacognostic and immunopharmacologic experimental studies, was most effective.[7,37] With such an approach, literature data and observations in ethnopharmacognostic field work are combined and evaluated. Ethnopharmacognostic field inquiries utilize questionnaires, including questions through which four areas are covered. These are ethnobotany, ethnopharmacy, ethnopharmacology, and ethnomedicine. These areas appear to be strongly integrated in traditional and folk medicine. Moreover, many aspects of data in these areas are presented and practiced in an intermingled form. This may lead to confusion and mistakes during the interpretation and evaluation of data collected. Pitfalls and errors may also arise from mistakes of a linguistic or philologic nature.

To overcome such problems in the selection of natural products, the proper experts in the fields of linguistics, philology, and anthropology should collaborate and be consulted. In the end, the systematic handling of the questionnaire designed for an ethnopharmacognostic inquiry leads to a better chance to arrive at

1. The plant species that deserve priority in the investigation
2. The type of ethnomedicinal preparation that contains and conditions, respectively, the active constituents
3. A more appropriate basis for the selection of relevant (immuno)pharmacologic tests
4. The proper pharmacotherapeutic category or subcategory, or indications of medical importance

An evaluative ethnopharmacognostic inquiry is concluded with a list of ethnomedical preparations, the medicinal plant species, and the crude drugs that serve as raw material. Such a listing represents the most relevant candidates, within the ethnic area under investigation, for assaying biologic activity. In the case of immunomodulatory activity, the selection is specifically based on data of ethnopharmacologic claims and ethnomedical uses that were evaluated to be closely or basically connected with immunologic disorders, immune response, and host-compromising conditions, or dysfunctions of immunofactors. Thus, in the field study the questions are concerned with traditional preparations and therapies in cases of the full diversity of inflammatory and autoimmune diseases (e.g., rheumatoid arthritis, systemic lupus erythematosus), infections, skin diseases, burns, and wounds. Within the context of immunomodulation, it is also relevant to focus field study questions on traditional therapies that aim at vitalization, strengthening, and stimulation of body functions and enhancement of resistance against diseases. Last but not least, one may find good natural product candidates for immunomodulators among plants that are used specifically in convalescence stages and in diets.

IV. LABORATORY EXPERIMENTATION AND STRATEGY

The general approach is to keep track of scientifically evaluated data resulting from the field inquiry. Initially, this refers to the identity of the plant species, the plant part to be used, the processing of the plant material, and the conditions and procedures applied to prepare the traditional medicine in relation to the disorder for which it is applied. Such information is taken into account to design controlled experimentation in laboratory screening programs. Initial screening for activity is thus performed with crude extracts that are closest in nature to the traditionally used medicinal products. Since water is most frequently and commonly used as the solvent to prepare traditional medicines, laboratory-prepared aqueous extracts can often be the starting point of screening for activity. It is worthwhile giving attention to the phenomenon that aqueous extracts usually contain a mixture of constituents that range from normal water soluble compounds to poorly soluble, nonsoluble, and colloidally dispersed molecular entities. Thus, although an immunomodulatory active compound is extractable with water and can be found in the crude aqueous extract, when isolated in pure form it may turn out to be insoluble or poorly soluble in water or other polar solvents.

In traditional medicinal practices, differential biologic oils and fats are used in preparation procedures. Since the major part of the water originally present is evaporated during such preparation processes, one may expect an extraction in favor of lipophilic active plant constituents. Depending on specific processing conditions, the nature of the raw plant material, and the chemical nature of the plant compounds involved, such medicinal oils may be pure solutions or dispersal systems. A rational examination of such traditional systems may provide hints for the type of immunomodulatory compounds that are involved.

Another approach to the nature of crude extracts is based on the fact that they are usually composed of constituents displaying a wide range of molecular weights. A separation into fractions on the basis of molecular weights easily leads to an insight into the molecular weight range in which immunomodulators can be found systematically.

V. DETECTION AND ISOLATION OF IMMUNOMODULATORY COMPOUNDS

The assessment of the presence of one (or more) immunomodulatory compounds in plant material, or its medicinal preparation, is not straightforward. False positive results may occur due to effects that are nonselective and nonspecific. Some plant constituents may bind Ca^{2+} or Mg^{2+} and, in effect, show a nonspecific anticomplementary activity. Other specifically active compounds, if present, may then be masked. Another example of disturbances in crude extract screening is caused by certain polyphenolic plant constituents that may deactivate proteins nonselectively. In the presence of such polyphenolic compounds, which interfere in enzyme- and receptor-based assays, specific activity of other constituents is difficult to detect. False results are also obtained when activity shown is actually based on cytotoxic effects on immune cells. The diversity of molecular structures in crude plant extracts poses a widely unpredictable situation in the initial screening phase. However, crude extracts represent the most logical starting point for initial screening studies.

Besides aqueous extracts, one may choose differently prepared crude extracts, e.g., using ethanol or methanol as solvents in the initial screening phase. Crude extracts containing mainly lipophilic constituents, which is the case when nonpolar solvents are used for extraction, generally present problems in *in vitro* test systems, because of poor solubility in hydrophilic buffers. In such cases the samples to be tested in the bioassay can be dissolved in ethanol or dimethylsulfoxide (DMSO).

Immunomodulatory activity in the screening stage of crude extracts is preferably established through a wide concentration range (e.g., 0.1, 1, 10, 100, and 1000 µg/ml). This is useful, first, to detect the concentration range where activity is expressed, and second, a wide range of concentrations may indicate differential types of effects or the presence of more than one (major) active compound. Once the presence of activity is established in a dose–response relationship, certain bioassays allow tentative conclusions as to the selectivity and/or the specificity of the effect. For example, a crude extract that exhibits a potent inhibitory effect on the classical pathway of complement activation and no activity at all on the alternative pathway of complement suggests interference with C1, C2, C3, or C4, or IgG or IgM, or the aggregated forms of these immunoglobulins. It is clear that mechanistic studies will focus selectively on each of these immunofactors to find out the specific interaction. In instances of so-called "high-capacity" screening programs[36] where receptor binding or enzyme activity assays are used, results also allow tentative conclusions concerning the specific mechanisms under study. So in the initial screening, directions are indicated for further immunomodulatory testing.

In fact, crude extract assaying is followed up by so-called "activity-guided" fractionation and isolation procedures. In principle, any fractionation or separation method can be applied provided no significant loss or decomposition of constituents occurs. Fractionation methods of choice are those based on molecular weight (or size) and on extractability in an eluotropic series of solvents. Qualitative and quantitative evaluation of immunomodulatory activity of each fraction determines the decision as to which fractions are taken further in the process of purifying and isolating the active compound(s). This is illustrated in the experimental data (Tables 1 and 2) published by Kroes.[38] Through these studies it was found that the fraction containing the compounds with molecular weight lower than 10 kDa, from an aqueous macerate of *Woodfordia fruticosa* flowers, almost quantitatively represented the inhibitory activity of luminol-enhanced chemiluminescence effected by zymosan-stimulated polymorphonuclear leukocytes (PMN). Further separation of the <10-kDa fraction was effected through successive percolations with petroleum ether (40 to 60°), diethyl ether, and ethyl acetate, respectively. The data in Table 1 show that the most active constituents are found in the diethyl ether extract ($IC_{50} = 1.9$ µg/ml), and this fraction represents 56% of the inhibitory activity of the <10-kDa fraction. Although the compounds remaining in the aqueous fraction

TABLE 1
Inhibition of Luminol-Enhanced Chemiluminescence (CL_{lz})
Generated by Zymosan-Stimulated PMNs by Fraction <10 kDa (1-Mr)
and by Subfractions Obtained by Perforation

Fraction	% of 1-Mr fraction	CL_{lz} inhibition IC_{50} in µg/ml	Conc. at IC_{50} of 1-Mr fraction in µg/ml	Contribution of fractions to activity of 1-Mr[a]
Fraction <10 kDa (1Mr)	100	4.8 ± 0.2		50
Petroleum ether (PE)	0.26	>100	0.01	
Diethyl ether (DE)	13.0	1.9 ± 0.1	0.62	28
Ethyl acetate (EA)	2.0	3.9 ± 0.2	0.10	0
Residue (AQ)	85.1	11.1 ± 0.8	4.2	20

[a] Inhibition (%) corresponding to the concentration of the fraction present at the IC_{50} of the 1-Mr.

TABLE 2
Inhibition of Luminol-Dependent Chemiluminescence (CL_{lz})
Generated by Zymosan-Stimulated Polymorphonuclear Leukocytes
by Subfractions of the Diethyl Ether (DE) Fraction

Fraction	% of DE fraction	CL_{lz} inhibition IC_{50} in µg/ml	Conc. at IC_{50} of DE fraction in µg/ml	Contribution of fractions to activity of DE[a]
Diethyl ether (DE)	100	1.9 ± 0.1		50
DEa	67.2	2.3 ± 0.2	1.28	42
DEw	2.6	2.4 ± 0.2	0.05	0
DEwm	10.1	4.2 ± 0.4	0.19	8
DEm	0.7	5.8 ± 0.4	0.01	0

[a] Inhibition (%) corresponding to the concentration of the fraction present at the IC_{50} of DE.

(AQ residue) show lower activity, the whole fraction still accounts for 40% of the total activity. Such data are important in defining the rationale underlying complex traditional phytomedicines and for guiding the isolation of the most active compound and indicating in which fraction(s) to find other active constituents. Further separation of the diethyl ether fraction, using an Amberlite XAD-2 column, was performed by successive elution with acidified water (DEa), water (DEw), methanol-water (1:1) (DEwm), and methanol (DEm). The results shown in Table 2 indicate that the acidified water elute (DEa) contains the major part of the original diethyl ether extract mass (DE) and represents 84% of its activity; 99% of the DEa fraction consisted of only one compound, and this was identified as gallic acid. This sequence of experimentally controlled extraction, fractionation, and isolation procedures, guided by a discriminatory outcome of a bioassay, led to the detection and the isolation of pure active compounds. Of course, these experimental procedures can be guided by differential bioassays. The decision on which bioassay is selected is determined by disease-oriented considerations, i.e., the type of immunomodulatory activity that seems relevant for the drug therapy in perspective and under study.

Although the disease-oriented approach and the activity-guided isolation strategy appear to be efficient and fruitful in laboratory experimentation, not all recently isolated biogenic immunomodulators were found by such a pathway. Some researchers still follow the rule, "Chemistry first, then biologic activity." However, the biogenic compounds selected to be dealt with in the next section were not discriminated on this basis.

VI. IMMUNOMODULATORY COMPOUNDS
AND THEIR STRUCTURES

As described at the beginning of this chapter, immunomodulatory activity can be expressed through many different sites and mechanisms. From a molecular point of view, the compounds that are known so far to modulate immune responses can belong to almost any structure class. In a way, this reflects the diverse and various sites, mechanisms, and levels of interference with the immune response. Natural-product-derived immunomodulators can be classified as low and high molecular weight compounds. A more specific classification reveals, however, that active compounds were found belonging to any of the following substance classes: carbohydrates, terpenes, steroids, phenolics of different biogenetic classes, coumarins, amino acids, peptides, proteins, glycoproteins, alkaloids, and other N-containing organic compounds.

The immunomodulatory compounds discussed in this section are molecular representatives of the substance classes mentioned above. Attempts have not been made to be complete in this review.

A. CARBOHYDRATES

Biogenic carbohydrates represent a very interesting group of immunomodulating agents. In purely immunologically oriented studies, several polysaccharides (e.g., zymosan, inulin), sulfated polysaccharides (e.g., dextransulfate, heparin), protein-bound lipopolysaccharides (e.g., endotoxin), protein-free lipopolysaccharides (LPS), and glycoproteins (e.g., lectins) have been of interest for many years. Research work has also been carried out on monosaccharides and sugar alcohols. L-fucose as well as L-rhamnose are capable of inhibiting lymphokine activity *in vitro*, and both these sugars were shown to be effective in suppressing *in vivo* manifestations of cellular immunity.[39] In addition, Nair and Schwartz demonstrated that L-fucose, L-rhamnose, and α-methyl-D-mannoside are able to block the SSF (soluble suppressor factor[s]) activity of lymphocytes.[40] Also Koszinowski and Kramer reported that methyl-D-mannose counteracted T-suppressor cells in their regulatory immune functions.[41] More recently, however, investigations on immunomodulatory carbohydrates have focused mainly on polysaccharides from fungi, lichens, algae, and higher plants. The reader is referred to review articles and chapters in books for general information on chemical structures and immunomodulatory activity.[4-6,42-44]

The general procedure for polysaccharide fractionation, after initial extraction with hot water or dilute alkaline solution, includes alternate steps of precipitation and solubilization using differential agents. Further purification steps may involve separation techniques like ion-exchange, gel permeation, and affinity chromatography. Molecular weights of polysaccharides are determined on columns of Sepharose CL-2B or Sepharose CL-4B, or by gel permeation using Sephacryl 16-200 or Sephadex G-150 and G-200.[43]

The structural determination of polysaccharides is usually based on data derived from conventional chemical procedures such as complete and partial acidic hydrolysis, alkaline degradation, methylation and methanolysis, and specific reduction reactions on (uronic acid-)carboxyl and hemi-acetal groups to sugar alcohols (alditols). The reaction products of these procedures are analyzed by means of gas chromatography-mass spectrometry (GC-MS), usually after acetylation. ¹H-NMR, but especially ¹³C-NMR, spectroscopic analysis also represent a powerful method for structure studies in the field of polysaccharides. Systematic performance of these procedures and methods leads to the basic and specific structure features of the sometimes most complicated polysaccharides.

Complete acidic hydrolysis of purified polysaccharide, using trifluoroacetic acid (TFA) stepwise in different concentrations, leads to a component sugars mixture that can be analyzed using a sugar analyzer or thin-layer chromatography and chemical reactions. A part of the monosaccharide mixture can be converted into the corresponding alditol acetates through reduction with 1-cyclohexyl-3-(2-morpholinoethyl)-carbodiimide, methyl-p-toluenesulfonate, and sodium borohydride. The corresponding alditol acetates are prepared by reaction with acetic anhydride in pyridine and are analyzed qualitatively and quantitatively by means of GC and GC-MS. The latter procedure is also performed using deuterated sodium borohydride ($NaBD_4$). This helps to differentiate between the mass spectra of isomeric alditols and isomeric methylated glycosides. For example, by reduction with $NaBD_4$, the mass spectra of the alditols from 2,3- and 3,4-di-O-methylpentoses, as well as those of 2,4- and 3,5-di-O-methylhexoses, can be distinguished. Another advantage in using deutero-borohydride is that for each substitution pattern of a sugar component, a unique mass spectrum is obtained. This means that in methylation analysis of polysaccharides, the mass spectrum of a partially methylated alditol acetate obtained from a methylated sugar, in combination with the sugar composition of the original polysaccharide and relative retention time (T) values of alditol acetates, will generally give evidence for the characterization of the methylated sugar. A requirement is that the mass spectrum of a methylated alditol acetate having the same substitution pattern should be known. Mass spectra and T-values of known methylated alditol acetates can be found in the literature. Basic data were already provided by Björndal et al.[45,46] Additional and more recent data on newly identified polysaccharides can be found in the literature.[47-56,70] Methylation of polysaccharides is performed in a dimethylsulfoxide solution with methylsulfinyl carbanion and methyl iodide, according to the conventional method of Hakamori (see reference 45). To determine positional links between monosaccharides in polysaccharides, methylated polysaccharides are hydrolyzed to obtain the partially methylated monosaccharides. The latter mixture is reduced with sodium borodeuteride and acetylated to the corresponding alditol acetate and analyzed by GC and GC-MS. Methylated polysaccharides can be hydrolyzed either by refluxing their methanol solution with 3% hydrochloric (methanolysis) or with dilute sulfuric acid in acetic acid.

Proton-NMR has very limited value because the molecular size and complexity of polysaccharides do not allow detailed molecular structure information. Proton spectra of polysaccharides suffer from line broadening because the molecular motion of the polymer is less than isotropic, so there is incomplete averaging of the localized magnetic fields.[58] On the other hand, ^{13}C-NMR, allowing a greater dispersion of chemical shifts, is a powerful method in structure elucidation studies of polysaccharides. Valuable information is obtained on structural sequences in simple polymers from ^{13}C-NMR spectra of polysaccharides. J_{C-H} values usually provide significant data with regard to glycosidic configuration. In addition, molecular ratios and chemical structure may be probed by increasing the spin lattice relaxation time.[43,58]

Assignments of signals in a polysaccharide can be made by comparing the chemical shifts with those of the component monosaccharides, methylglycosides, or oligosaccharides, having the appropriate configurations. A strong downfield shift of the O-glycosylated carbon in the polysaccharide is found (α effect) in comparison with the corresponding carbon in the monosaccharide. A smaller upfield shift of the adjacent carbon signal (β effect) is found in polysaccharides in comparison with corresponding monosaccharide carbon atoms. In ultrastructure studies of polysaccharides, solid-state cross-polarization magic angle spinning (CP/MAS) ^{13}C-NMR spectroscopy has been found useful. The chemical shifts of several carbons in a polysaccharide in the solid state are different from those in solution, because of conformational differences. In such studies, for example, the C-3 signal of a β-glucosyl moiety in the helical form appears at δ 89, whereas in the random coil structure, the C-3 signal resonated at a higher field, i.e., δ 86.[59] Thus, some polysaccharides of the six-branched $\beta(1 \rightarrow 3)$-glucan type were demonstrated to appear in two types of conformations: a helical and a random coil form. Depending on the solvents used to dissolve a polysaccharide, one of the two conformations is favored.[43] At this point it is noted that conformational transition of polysaccharides during extraction or purification may have consequent alterations in the biological activity.

1. Fungal

The polysaccharides, mostly glucans and some mannans, isolated from fungi have been shown to display interesting immunomodulating properties. The molecular weights of the different compounds range from 5 to 1000 kDa. A major chemical feature of these immunomodulating fungi polysaccharides is the presence of $\beta 1 \rightarrow$ 3-linkage of D-glucose units in the main chain. Side-chain D-glucose units are linked via $\beta 1 \rightarrow 6$ to the main chain. Compounds typical of this group are lentinan, isolated from *Lentinus edodes* Berk., schizophyllan from *Schizophyllum commune,* and polysaccharide H11 from mycelia of *Poria cocos.*[60-63] These neutral glucans show antitumor activity.

Lentinan now represents the classical example of a nontoxic immunomodulating fungal polysaccharide. Its antitumor activity is based on an immunostimulating effect and not on a direct effect on tumor cells. As early as 1969 it was shown by Chihara et al. that two polysaccharide fractions from *Lentinus edodes* effected complete regression or growth inhibition of tumors developed in mice following subcutaneous injection of sarcoma 180 ascites.[64] Further studies on the molecular structure of the active polysaccharides resulted in establishment of a single compound (i.e., lentinan) with a molecular weight around 1000 kDa. It was shown to be a branched molecule having a backbone of $(1 \rightarrow 3)$-β-D glucan and side chains of both β-D$(1 \rightarrow 3)$- and β-D$(1 \rightarrow 6)$-linked D-glucose residues, together with a few internal β-D$(1 \rightarrow 6)$ linkages.[60] Interestingly, a hydrolysis product of lentinan, called "small lentinan," showing an entirely identical IR-spectrum and having a molecular weight of 16,200, was demonstrated to be as active as lentinan (MW 1,000,000) itself.

The immunomodulating action of lentinan was first proved to be associated with lymphocyte-mediated immune response in mice, by Maeda and Chihara.[65] In continuation of this, Dennert and Tucker demonstrated that lentinan stimulated T-helper cell priming, which amplified antibody synthesis.[66] Okuda et al. showed that lentinan and other antitumor fungal polysaccharides inhibit complement activation through the alternative pathway.[67] *In vivo* experiments with lentinan in combination with OK432 and bacterial lipopolysaccharide were carried out by Abe et al.[68] These three agents together were able to effect almost complete regression of the solid-type tumor MH134 in mice. Clinical studies demonstrated the usefulness of lentinan in different cancer forms.[68]

Neutral polysaccharides, such as lentinan, are generally extracted with hot water. They remain in the nondialyzable fraction and are precipitated with alcohol. A fractional precipitation of polysaccharides can be performed by treating the original hot water extract with different amounts of ethanol. The water/ethanol proportions range from 1:0.5 to 1:5 volume units. Alternatively, zinc chloride solutions have been used to extract $\beta(1 \rightarrow 3)$-D glucans branched at C6.[43] A purification step for these polysaccharides, including lentinan, is performed by precipitation with cetyl trimethyl ammonium hydroxide (CTA-OH). Uronic acids containing polysaccharides are usually extracted with dilute aqueous sodium hydroxide or sodium carbonate solution. Fractional solubilization with acetic acid is used to free the polysaccharide from its salt. From such solutions, acidic polysaccharides can also be precipitated with ethanol. Copper salts or boric acid are also utilized to precipitate and purify polysaccharide fractions.[43]

2. Higher Plants

Recently, many higher plants have been subjected to research efforts focused on immunomodulatory polysaccharides. Table 3 summarizes the plant species and the nature of immunomodulatory activity effectuated by the polysaccharide fractions or identified moieties isolated from the plant materials. The reader is referred to the references provided for experimental details.

Unlike the fungal [six-branched $(1 \rightarrow 3)$ β-D glucans] immunomodulatory polysaccharides, those isolated from higher plants so far have been shown to be neutral or acidic heteroglycans.

TABLE 3
Higher Plant Species Containing Polysaccharides with Immunomodulatory Properties

Plant species	Immunomodulatory activity[a]	Refs.
Echinacea purpurea	PhE; CCE; MPhE; IL-1; LAF	5, 6, 42, 69, 70
Echinacea angustifolia	PhE; CCE	5, 6, 42
Eupatorium perfoliatum	PhE; CCE	5, 6
Acanthopanax senticosus (= *Eleutherococcus senticosus*)	PhE; CCE	5, 71
Sabal serrulata	PhE; CCE	5, 6
Calendula officinalis	PhE; CCE	5, 6
Baptisia tinctoria	PhE; CCE	5, 6
Arnica montana	PhE; CCE	5, 6
Achyrocline satureioides	PhE; CCE; MPhE; ACCA	5, 6, 69
Carthamus tinctorius	PhE; CCE	6
Chamomilla recutita	PhE; CCE	6
Althaea officinalis	PhE; CCE	5, 6, 72
Plantago major	PhE; CCE	6
Abelmoschus esculentus (= *Hibiscus esculentus*)	PhE; CCE	6
Symphtum officinale	PhE; CCE	6
Tilia officinalis	PhE; CCE	6
Tilia argentea	PhE; CCE	6
Eupatorium cannabinum	PhE; CCE	5
Angelica acutiloba	ACCA; Ifs; Mit	47, 48, 57, 73
Aloe vera	ACCA; AdA; DTHind	74, 75
Viscum album	NKct; LEC/TuCCon	76–78
Azadirachta indica	ACCA; DTHind; AdA	79–81
Picrorhiza kurroa	ACC; ACCA; AdA	82–84
Curcuma longa	AC; CCE; AlP	85, 86
Astragalus mongholicus	MPhE	87
Benincasa cerifera	sBcP; Mact; IgMe; IgGe; DTHind	88
Artemisia princeps	AC	49, 89
Panax ginseng	AC	50
Abelmoschus manihot	AC	51, 52, 72
Hydrangea paniculata	AC	53, 72
Althaea rosea	AC	54, 72
Plantago asiatica	AC	55, 56, 72
Zizyphus jujuba var. *inermis*	AC	72, 90
Cassia angustifolia	Mit; CMind; MPhE	91
Hibiscus sabdariffa	Mit; CMind; MPhE	91

[a] Abbreviations: PhE = granulocyte phagocytosis enhancement; CCE = carbon clearance enhancement; sBcP = stimulation of B-cell proliferation and differentiation; Mact = macrophage activation into antitumor macrophages; IgMe = enhancement of IgM antibody response; IgGe = enhancement of IgG antibody response; MPhE = macrophage phagocytosis enhancement; ACCA = anticomplementary activity in classical and alternative pathways; ACC = anticomplementary activity on the classical pathway selectively; AC = anticomplementary activity; Ifs = stimulation of interferon production; Mit = mitogenic activity; AdA = adjuvant activity on specific antibody production; DTHind = induction of delayed-type hypersensitivity in mice footpad; NKct = natural killer cell cytotoxicity; LEC/TucCon = lytic effector cell/K562 tumor cell conjugates formation; AlP = alkaline-phosphatase inducing activity; CMind = induction of cytotoxic macrophages.

Usually these active compounds are highly branched and represent complex stereochemical molecules. The main neutral monosaccharides found after hydrolysis are xylose, arabinose, rhamnose, galactose, and glucose. The uronic acids, which are found most frequently in the acidic polysaccharides, are galacturonic, glucuronic, and 4-O-methylglucuronic acids. Some

→4)-ß-Xyl p-(1→4)-ß-Xyl p-(1→4)-ß-Xyl p-(1→
2   n
↑
R   n=2.0-2.5

R = →3)-α-4-O-Me-GluA p-(1→

→4)-ß-Xyl p-(1→

ß-Xyl p-(1→

R = →5)-α-Ara f-(1→

α-Ara f-(1→

Xyl p = Xylopyranosyl

GluA p = Glucuronic acid pyranosyl

Ara f = Arabinofuranosyl

FIGURE 1. Molecular structure of polysaccharide PSI from *Echinacea purpurea.*

immunomodulatory plant polysaccharides carry associated peptide or protein moieties. In many studies carried out so far, it is not yet clear whether these peptide/protein moieties are covalently or noncovalently bound to the carbohydrate moiety, or to what extent they contribute to the immunomodulatory effects. In this discussion we restrict ourselves to pure polysaccharides and those polysaccharides that are associated with a protein moiety that is present in less than 10% of the substance.

a. Echinacea purpurea[5,6,42,69,70]

From the aerial parts of this plant, two immunomodulatory polysaccharides were isolated and coded PSI and PSII.

The first (PSI) was found to be a heteroxylan with a mean molecular weight of 35,000 Da. The monosaccharide components of this substance are arabinose, xylose, and 4-O-methylglucuronic acid, in a molar ratio of 1:4.9:0.9. The backbone of the structure consists of (1 → 4) β-bonded xylopyranose. Branching occurs at about each fifth xylose unit. The side-chain sugar units are attached at C-2 and C-3 of xylopyranose units of the main chain. These and other binding features of this 4-O-methylglucuronoarabinoxylan are portrayed in Figure 1. The arabinofuranosyl residues were suggested to be attached to the backbone at position C-3 as α(1 → 5)-linked side chains and as nonreducing terminal groups. The β-linked 4-O-methylglucuronic acid occupies a nonreducing terminal group at position C-2 of the main-chain xylopyranose units.

The second immunomodulatory polysaccharide (PSII) isolated from *Echinacea purpurea* was characterized as an acidic arabino-rhamnogalactan, having a molecular weight of 450,000 Da. This polysaccharide is composed of rhamnose, arabinose, galactose, and glucuronic acid, in a molar ratio of 0.8:0.6:1:0.6. The molecular structure represents a highly branched moiety. There are chains consisting of rhamnose and galactose in equimolar ratio. Rhamnose units are α(1 → 2)-linked to each other, and galactose units are linked to each other by an α(1 → 4) bond. Galactose units are α(1 → 2)-coupled to rhamnose units. Branches are attached to the rhamnose-galactose chain at the C4-O position of rhamnose. Figure 2 portrays the different linkages typical for PSII from *Echinacea purpurea.*

→2)-α-Rha p-(1→2)-α-Rha p-(1→4)-α-Gal p-(1→4)-α-Gal p-(1→

4

↑

R

R = →3)-ß-GluA p-(1→   R = →3)-ß-Glu p-(1→

→5)-α-Ara f-(1→   →4)-ß-Xyl f-(1→

→4)-α-Gal p-(1→   →2)-α-Rha p-(1→

                              ß-Gal p-(1→

FIGURE 2. Molecular structure of polysaccharide PSII from *Echinacea purpurea.*

The crude polysaccharide fraction containing PSII and PSI was obtained through a sequence of extraction and precipitation steps. The powdered plant material was first refluxed with methanol, and the extract obtained was discarded. The dried plant material was then extracted twice with 0.5 M aqueous sodium hydroxide (5 ml/g plant material), and allowed to stand overnight. After separation from the plant material, the supernatant was treated with three volumes of ethanol and was stirred for 24 h. The brownish precipitate was collected, resuspended in water, and treated with TCA (15%) while stirred at 0°. The mixture, in which a precipitate had formed, was kept for 1 h at room temperature.

To the supernatant, separated from the precipitate, three volumes of ethanol were added. After storage for 24 h and centrifugation, the precipitate collected was resuspended in 2% aqueous sodium acetate solution. Insoluble material was separated by centrifugation, and the supernatant mixed with an equal volume of ethanol. After standing for 62 h, the supernatant was decanted and the precipitate collected by centrifugation. This precipitate containing the crude polysaccharide mixture was dissolved in water and subjected to dialysis against water for 3 days. The solid, crude polysaccharide fraction was obtained after lyophilization. The yield of crude polysaccharide through this method is 8.93 g out of 1 kg powdered plant material.[70]

The isolation and purification of the immunomodulatory-active 4-O-methyl-glucurono-arabinoxylan from the crude polysaccharide was carried out by means of dialysis against 0.02 M phosphate buffer solution (pH 7.5) and subsequent fractionation by ion-exchange chromatography on a DEAE Sepharose Cl-6B column equilibrated with the same buffer system. Elution was performed starting with the buffer and continued with a linear gradient of 0.0 to 0.1 M sodium chloride solution. The fractionation was monitored by polarimetric detection. Fraction II was purified further by rechromatography on DEAE Sepharose Cl-6B and on Biogel P-60, using water as eluent. The homogeneity of the purified polysaccharide was checked by high-pressure liquid chromatography (HPLC) on μ-Bondagel E-250, E-500, and E-100 columns equilibrated with 0.2 M Pi buffer (pH 7). The same columns can be used to estimate the molecular weight of the purified polysaccharide, using a standard mixture of dextrans of known molecular weights. The yield of the purified heteroxylan is about one fifth of the crude polysaccharide fraction.

Table 4 illustrates the results that lead to structure determination of 4-O-methylglucuronoarabinoxylan when it was subjected to methylation analysis of the alditol acetates derived from the compound. The relative retention time (T)-value, i.e., relative to the retention time of 1,5-di-O-acetyl-2,3,4,6-tetra-O-methyl-glucitol on the GC-analysis column, indicates the identity of the partially methylated alditol acetate and its corresponding sugar. The substitution pattern of the methylated alditol acetate indicates the links of the component sugars

TABLE 4
Methylation Analysis of 4-O-Methylglucuronoarabinoxylan from Aerial Parts of *Echinacea purpurea*[70]

Methylated alditol acetates	Deduced linkages	Relative retention time[a] T	Molar ratio[b] A	B
2,3,5-Me$_3$-Arabinitol-Ac	Ara f-(1→	0.42	1	0.3
2,3-Me$_2$-Arabinitol-Ac	→5)-Ara f-(1→	1.07	2.3	0.8
2,3,4-Me$_3$-Xylitol-Ac	Xyl p-(1→	0.53	0.4	0.1
2,3-Me$_2$-Xylitol-Ac	→4)-Xyl p-(1→	1.19	10.0	10.0
2-Me-Xylitol-Ac	→4)-Xyl p-(1→ 3	2.14	2.3	1.2
3-Me-Xylitol-Ac	→4)-Xyl p-(1→ 2			
2,3,4,6-Me$_4$-Glucitol-Ac	Glc p-(1→	1.00	—	0.6
2,4,6-Me$_3$-Glucitol-Ac	→3)-Glu p-(1→	1.81	—	0.1

[a] T is relative to 1,5-di-O-acetyl-2,3,4,6-tetra-O-methylglucitol. A = 4-O-methylglucuronoarabinoxylan; B = the reduced arabinoxylan.

[b] Estimated by GC of alditol acetates.

involved and also whether they are present as pyranosyl or furanosyl rings. The relative abundance of one or more sugars in specific links indicates main-chain structures. Lower methylated alditol residues indicate branching locations. Molar ratios of the different alditols and their corresponding sugars indicate the type of polysaccharide under investigation. From the ratios found, it was clear that the polysaccharide PSI is an arabinoxylan. The presence of 2,3,5-tri-O-methylarabinitol indicated the existence of nonreducing terminal arabinofuranosyl side-chain units present in the ratio 1:12.3 to xylosyl residues. The presence of 2,3-di-O-methylarabinitol indicated the existence of 1,5-linked arabinofuranosyl residues in the side chain of the polysaccharide. Its molar ratios to xylosyl residues and to arabinofuranosyl end groups are 1:5.3 and 2.3:1, respectively. The presence of 2,3,4-tri-O-methylxylitol in a molar ratio of 1:30.7 to xylosyl residues of the main chain indicated the existence of xylopyranosyl residues as nonreducing terminal units. The relatively high abundance of 2,3-di-O-methylxylitol indicates the (1 → 4)-linked xylopyranosyl residues constituting the backbone of the polysaccharide. The presence of 2-O-methylxylitol and 3-O-methylxylitol indicated that the main chain is branched at positions C-3 and C-2, respectively, of the xylopyranosyl moieties. Generally, the presence of uronic acids can be demonstrated and estimated by analysis of the total hydrolyzed native polysaccharide in an automatic sugar analyzer, or chemically by means of the *m*-hydroxydiphenyl reagent. However, the GC-MS analysis of methylated alditol acetates resulted in establishment of the presence of 2,3,4,6-tetra-O-methylglucose and 2,4,6-tri-O-methylglucose indicating that glucuronic acid is present as a nonreducing terminal sugar in (1 → 2)-linkage to the xylose units of the backbone.

The assignments of chemical shifts for 4-O-methylglucuronoarabinoxylan in D$_2$O solution are given in Table 5. The chemical shifts of anomeric carbons of polysaccharides are generally found in the region between 110 and 95 ppm. In the case of 4-O-methylglucuronoarabinoxylan, the chemical shifts at 101.78, 101.42, and 101.05 have been assigned to the anomeric carbon signals of the β-(1 → 4)-linked xylopyranose residues. The signals at 109.38 and 107.62 ppm have been assigned to the arabinofuranosyl residues, whereas the signal at 97.64 ppm represents the chemical shift of the anomeric carbon of the α-linked 4-O-methylglucuronic acid residue. Its methyl group carbon atom resonates at 59.83 ppm. As can be seen from Table 5, chemical shifts of other carbons (C-1 to C-5) are found at a significantly higher field than the signals of the anomeric carbons.

TABLE 5
^{13}C-NMR Chemical Shifts of 4-O-Methylglucuronoarabinoxylan
(from *Echinacea purpurea*) Dissolved in D$_2$O^{70}

Compound linkage groups before acidic degradation	Chemical shifts (ppm)						
	C-1	C-2	C-3	C-4	C-5	C-6	-OMe
→4)-β-Xyl p-(1→	101.78	72.82	73.83	76.52	63.09		
→2,4)-β-Xyl p-(1→	101.42	76.23	73.60	76.67	63.02		
→3,4)-β-Xyl p-(1→	101.05	71.44	(79.48)[a]	(75.9)[a]	62.88		
α-Ara f-(1→	109.38	81.45	76.96	84.1	61.44		
→5)-α-Ara f-(1→	107.62	81.43	76.94	82.42		176.78	59.83
4-O-Me-β-GluA p-(1→	97.64	71.44	72.41	82.49	70.03		

[a] Assignments not definitive.

In conclusion, a point of discussion is whether this or any other pure polysaccharide is, at least qualitatively, representative of immunomodulatory properties for which it was isolated from the active crude polysaccharide fraction or the crude plant extract, and characterized. In this case it was found that the methylglucuronoarabinoxylan showed strong phagocytosis-enhancing effect *in vitro*, but unlike the crude polysaccharide fraction from which it was isolated, no enhancement of phagocytosis *in vivo*, judged from the carbon clearance in animals, could be demonstrated. In such cases the significance of action-guided isolation, fractionation, and purification becomes clear.

b. Angelica acutiloba

Several extensive studies have been carried out on the immunomodulatory properties and the molecular structure of polysaccharides isolated from the roots of *Angelica acutiloba*.[47,48,57,73] The anticomplementary arabinogalactan (coded AR-4IIa-1) was cleanly separated on the basis of activity from other less interesting fractions, as far as anticomplementary activity is concerned.[47,73] The structure of the further purified AR-arabinogalactan was determined to be composed mainly of arabinose and galactose, in a molar ratio of 1.2:1.0; a small amount of galacturonic acid (2.8%); and traces of glucose and mannose. The polysaccharide contains 5.7% protein. The backbone of AR-arabinogalactan consists of (1 → 6)-linked galactopyranosyl residues. Arabinose is mostly present as α-L-arabinofuranosyl residues in nonreducing side chains and terminals. The galactopyranosyl residues might be linked through (3 → 1) glycosyl bonds to the arabinosyl residues and side chains. The possible structure of AR-arabinogalactan is pictured in Figure 3. This AR-arabinogalactan, when subjected to the activity of exo-α-L-arabinofuranosidase, is 50% hydrolyzed. The only sugar detected was arabinose.

In the same experimental sequence of activity-guided fractionation of crude polysaccharides, a fraction coded AR-4IIc, which showed interferon inducing activity, was isolated and characterized. However, besides the carbohydrate moiety, which consisted of 31.5% uronic acid, 12.1% arabinose, and smaller percentages of galactose, glucose, and rhamnose, 30.2% of the fraction was protein.

In another series of investigations, Yamada, Kiyohara, and co-workers described a pectic arabinogalactan with anticomplementary activity.[48,57,92-94] This polysaccharide, coded AGIIb-1, has a very complex, highly branched structure. It is water soluble and contains approximately 92.7% neutral sugar, 8.7 to 12.2% uronic acid, and 2.3% protein. The component sugars, arabinose, galactose, rhamnose, galacturonic acid, and glucuronic acid, are present in the molar ratios 1.8~2.2:1.0:0.2~0.3:0.2~0.4:0.1. AGIIb-1 is mainly built up of an arabino-3,6-galactan moiety. Most of the arabinose is present as α-L-arabinofuranosyl residues in the

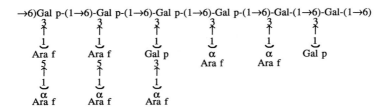

FIGURE 3. Possible structure of an anticomplementary AR-arabinogalactan from the roots of *Angelica acutiloba*.[47,73]

nonreducing terminals, while the highly polymerized and branched side chains, which are attached to positions 3 and 6, are (1 → 6)- and (1 → 3)-linked galactose. Some arabinose containing chains are also attached to (1 → 4)-linked galactose residues. (1 → 4)-linked arabinopyranosyl residues are also present at the reducing terminal. A neutral arabinogalactan that is obtained on mild acid hydrolysis contains (1 → 3)-, (1 → 4)-, and (1 → 6)-linked and 3,6-di-O-substituted galactose. The acid arabinogalactans that are found on mild acid hydrolysis, in addition to the neutral fraction, carry nonreducing terminal glucuronic acid, (1 → 4)-linked galacturonic acid, and 2,4-di-O-substituted rhamnose.

In relation to the anticomplementary activity of AGIIb-1, it is of interest that treatment with exo-α-L-arabinofuranosidase markedly increased the activity of the remaining part of the polysaccharide. Mild acid hydrolysis of AGIIb-1 yielded four polysaccharide fractions: a neutral arabinogalactan (N-I), two acidic arabinogalactans, and a neutral arabinan. The most potent anticomplementary activity was observed with the neutral arabinogalactan (N-I) and was superior to AGIIb-1. The product (AF-N-I) obtained after treatment of N-I with exo-α-L-arabinofuranosidase is even more active than N-I.

These findings indicate that anticomplementary polysaccharides, apart from specific structural features important for activity, need not be acidic (anionic) in nature. In addition, it was found that AF-N-I had a marked increase of activity through the alternative pathway of complement. This is also true for the enzyme-treated product of AGIIb-I.[57]

c. Aloe vera

A high molecular weight, nondialyzable fraction that showed anticomplementary activity was obtained from the water extract of *Aloe vera* gel.[74,75] Activity-guided purification using DEAE-Sephacel and Sephacryl S-300 yielded two anticomplementary, protein-free polysaccharides (B-I and B-II) with molecular weights of 320,000 Da and 200,000 Da, respectively. The anticomplementary activity was shown to be based on consumption of C3, the crucial component of the alternative pathway activation. In addition to anticomplementary activity, these polysaccharides were shown to inhibit the opsonization of zymosan in human pooled serum (HPS) and to display adjuvant activity in mice by an increase of specific antibody production. Although to a lesser extent than was the case with antibody enhancement, the induction of delayed-type hypersensitivity (DTH) was also significantly increased. Structural investigation of these polysaccharides demonstrated that both lack uronic acids and that both contain only neutral sugars, with mannose as the main sugar.

Thus, B-I consists of 92.1% mannose, 3.2% glucose, 3.8% galactose, and 0.9% arabinose, and B-II consists of 83.7% mannose, 8.8% galactose, 3.9% glucose, and 3.6% arabinose.

d. Picrorhiza kurroa

The immunomodulatory active, polysaccharide-containing macromolecular fraction from *Picrorhiza kurroa*[82-84] roots has shown particularly interesting properties. One of the macromolecular entities (coded CM1Sr1Sd2), which was isolated and purified on the basis of its anticomplementary activity, is discussed here. It is a very potent anticomplementary substance, with an IC_{50} of 0.2 μg/ml, acting selectively on the classical pathway activation.

TABLE 6
Quantitation of Anticomplementary Effects of Subfractions
of the Methanol-Extract from *Picrorhiza kurroa* Root Material

Extract and subfractions	Dry-weight % of the starting Me-OH-extract	Inhibitory activity on complement activation			
		Classical pathway		Alternative pathway	
		IC_{50}[a] (μg/ml)	Recovery (%)	IC_{50} (μg/ml)	Recovery (%)
Me-OH-extract	—	20 ± 2	—	2230 ± 110	—
H₂O-soluble part	84	20 ± 3	92	>1500	—
H₂O-insolube part	16	290 ± 14	1	>1500	—
Fraction CM1	1.8	0.6 ± 0.07	66	>250	—
Fraction CM2	82	125 ± 8	14	>1500	—
Fraction CM1Sr1	0.9	0.3 ± 0.04	66	>250	—
Fraction CM1Sr2	0.9	>100	—	>250	—
Fraction CM1Sr1Sd1	0.4	9 ± 1	1	>100	—
Fraction CM1Sr1Sd2	0.5	0.2 ± 0.04	55	>100	—

[a] Mean ± 1 SEM (n =3); fractions CM1 and CM2 were eluted from a Sephadex G-200 column; fractions CM1Sr1 and CM1Sr2 were eluted from a Sepharose-2B column; fractions CM1Sr1Sd1 and CM1Sr1Sd2 were eluted from a Sephadex G-200 column with SDS added to the eluent.

The substance is thermostable, i.e., its potent classical pathway inhibitory activity is retained after heating an aqueous solution at 100°C for 30 min. However, the substance can be extracted effectively with methanol or water at 4°C during 24 h of stirring. In aqueous solution the compound tends to exist in a complex state with a molecular weight more than 10^7. In the presence of a detergent, however, a basic molecular unit with a molecular weight of 50,000 Da is obtained. Activity-guided fractionation and purification is performed by gel permeation on Sephadex G-200 and Sepharose 2B columns, using 0.05 *M* sodium phosphate buffer (pH 7.0) as the eluting solvent. To isolate the basic molecular unit, 0.1% sodium dodecyl sulfate (SDS) is added to the eluent. Table 6 shows quantified data on subfractions and purified substances, illustrating recovery and enrichment of active compounds through this activity-guided fractionation procedure.

The high potency of anticomplementary activity prompted us to scrutinize the isolated polysaccharide-containing macromolecule for the presence of bacterial lipopolysaccharide (LPS). Thus, applying the LAL-gelation assay,[95] it was shown that the content of LPS was almost negligible,[84] and hence, its interference has been ruled out. Moreover, the presence of long-chain fatty acids in CM1Sr1Sd2 could not be detected.

Further chemical characterization of CM1Sr1Sd2 showed a content of about 1% of protein and 40% of carbohydrate, in which the component sugars glucose, galactose, and mannose were present in the molar percentages 72:17:11. Further research on structural details of this highly and selectively active compound is in progress.

e. *Viscum album*

Interesting advances have been made recently in the field of immunomodulatory agents from *Viscum album*[76-78] and their mechanism of action. The first studies referred to here involve active fractions in commercial preparations (Iscador-M and Iscador-Q) derived from *Viscum album* growing on apple trees and oak trees, respectively, and in freshly prepared extracts of *Viscum album* growing on oaks.[76,77] These commercial preparations include a step involving fermentation of the fresh extracts. It was found, however, that the fresh extracts, compared to Iscador-M and Iscador-Q, showed similar immunomodulatory activity, and the fermentation procedure seemingly does not affect the activity.[76] The immunomodulatory activities under consideration here are enhancement of human NK cytotoxicity and increase of the formation of NK-cell/K562 tumor cell conjugates. As far as carbohydrates are concerned, two immunomodulatory fractions were investigated. No detailed structural data are available to date, but the active fractions were characterized by using biochemical agents.

First, the nondialyzable fraction (MW > 10,000) of the *V. album* extract was found to enhance NK cytotoxicity of human peripheral blood monocytes and to increase the formation of NK-cell/K562 tumor cell conjugation. The active component in this fraction was shown to be a rhamnogalacturonan. This was based on the finding that treatment with poly-α-D-galacturonidase and α-rhamnosidase resulted in loss of both immunomodulatory effects. In addition, the activities shown by mistletoe rhamnogalacturonan were inhibited in the presence of galacturonic acid and acetylated rhamnose and also the structurally related acetylated mannose and acetylated mannonic acid γ-lactone.

On the basis of results obtained by using a specially adapted agglutination assay, Mueller and Anderer postulated an interesting mechanism of action.[77] This mechanism involves bridging of the NK cells and the tumor cells by the active rhamnogalacturonan moiety. It was shown that terminal acetylated rhamnose or acetylated mannose binds to NK cells in a dose-dependent manner, but not to K562 tumor cells. On the other hand, terminal galacturonic acid is bound to the K562 tumor cells, but not to NK cells.

Another active fraction, isolated from *V. album* extract, Helixor® M, which shows enhancement of NK cytotoxicity of human blood monocytes, is a low molecular component that was obtained as a dialyzable fraction (<1000 Da). To obtain the active component, the dialyzable fraction was lyophilized and subjected to chromatography on a Biogel P2 column with a bed volume of 180 ml. Elution was performed with water at a flow rate of 18 ml/h at room temperature. The fraction between the void volume and the salt fraction, coded HM-BP,[78] contained the immunomodulatory material and was free of cytotoxic constituents. Interestingly, this NK-cytotoxicity-enhancing *V. album* constituent was biochemically characterized as an oligosaccharide. It was only found in the commercial Helixor® M preparation from *V. album* growing on apple trees and not in Iscador® M or Iscador® Q, nor in *V. album* extracts derived from mistletoe of firs and pines. Its NK-cytotoxicity-enhancing effect also differed from that of the "rhamnogalacturonan" from the Iscador® M preparation, in the sense that the oligosaccharide showed a lymphokine-like activity. Only after preincubation, for many hours, of the human monocytes with fraction HM-BP was a strong NK cytotoxicity expressed against differential tumor cell lines. The maximum expression of NK cytotoxicity was observed after 72 h of preincubation.

The presumed oligosaccharide nature of the active component, in fraction HM-BP, was based, first, on the fact that treatment with α-glucosidase or endoglycosidase F abrogated the NK-cytotoxicity-enhancing activity completely and, second, on the low molecular weight (≤ 1000 Da). Partial inactivation of the HM-BP fraction was observed after treatment with endoglycosidase D or hemicellulase. Treatment of HM-BP with carboxypeptidase, various proteases, and phospholipases and ribonuclease had no effect on its NK-cytotoxicity-enhancing activity. This practically rules out active compounds with a peptide, phospholipid, or RNA nature.

In vivo experiments showed that subcutaneously and intraperitoneally administered HM-BP reduced tumor take incidence and development in mice, when applied before and after tumor challenge.

B. PROTEINS, PEPTIDES, AND GLYCOPROTEINS

Proteins, in crude or pure form, have been known for a long time to be agents that evoke immunological reactions. In fact, any medicine of the peptide or polypeptide class should be suspected of antigenic activity or provoking immunogenicity. Taking into consideration that many molecular factors and mediators involved in the immune response are proteins, peptides, or glycoproteins, it is understandable that natural products of such chemical nature might easily interfere or interact with functional immunofactors and immunosystems. Indeed, many different peptides can bind to major histocompatibility complex (MHC) Class I and Class II molecules both *in vitro* and *in vivo*. Through such bindings on the surface of antigen-presenting cells, peptides serve as ligands for T-cell receptors of CD8[+] and CD4[+] T-cells, respectively.[96] Modern research focuses on selectivity and specificity of natural-product-derived peptides in their interaction with differential functional processes of the immune response. For recent information on immunomodulating peptides, we refer the reader to earlier reviews.[3,4,97,98]

In this section we focus on a few advances made with molecular entities isolated from natural products.

1. *Viscum album* Lectin (MLI)

MLI is a β-galactoside-specific lectin isolated (in addition to the immunomodulatory active carbohydrates, Section VI.A.2.e) from the *Viscum album* preparation, Iscador®.[99] Iscador is a proprietary extract that has approval for clinical application in tumor therapy in Germany. An extensive background literature on *Viscum album* lectins is found within the report by Hajto et al.[99] The summarized immunomodulatory properties of MLI include significant increases *in vivo* in rabbits of (1) natural killer cytotoxicity, (2) the frequency of large granular lymphocytes (LGL), and (3) phagocytic activity of granulocytes.

The relevance of these enhancements (in rabbits) for clinical situations was studied. In such clinical experiments, breast cancer patients were treated intravenously and subcutaneously with Iscador dosages corresponding to 1.65 (±0.35) ng/kg MLI and 12 ng/kg MLI, respectively. Compared to results obtained in rabbits, a qualitatively similar profile of changes was measured in the patients with breast cancer. These were augmentation in phagocytic activity of PMN leukocytes and frequency of large granular lymphocytes (LGL) and an increase in the number of neutrophils and lymphocytes was observed.

The isolation of the β-galactoside-specific lectin MLI from commercial Iscador was performed through application of affinity chromatography, using a support with immobilized lactose. For this purpose, the lactose was coupled to divinyl sulfone-activated Sepharose 4B. Thus, the lectin (MLI) is then bound to the affinity ligand on the matrix by batchwise incubation of the mistletoe extract (Iscador) in a sample rotator overnight at 4°C. The resultant slurry is then poured into a column (4 ml resin for 3 ml extract), and the "flow through" (i.e., the nonbinding fraction) is similarly treated with D-galactosamine-Sepharose 4B. The lectins, retained by both of these columns, are eluted by the specific sugars and, additionally, by 0.2 *M* glycine (pH 2.6) in the case of the second column. To avoid lectin inactivation, the glycine eluent is immediately neutralized. All eluents are extensively dialyzed against water to remove sugar and salt.

The carbohydrate-binding subunit of the active lectin MLI has been proved to display the same immunomodulatory properties in rabbits, as described above. MLI is composed of two subunits having different functional domains. The toxic subunit has an apparent molecular weight of 29,000 and is termed chain A. The carbohydrate-binding subunit has an apparent molecular weight of 34,000 and is termed chain B. The *in situ* cleavage of the disulfide bond of MLI is achieved with β-mercaptoethanol, and the subunits can be separated by means of polyacrylamide gel electrophoresis in the presence of 0.1% sodium dodecyl sulfate.

The quantitative determination of the lectin content in mistletoe extract is based on application of specific antibodies raised against MLI in rabbits. The determination is carried out by applying an optimized ELLA technique. This is based on lectin binding to an immobilized ligand (asialofetuin) and subsequent binding of specific antibody to the bound lectin. The specific binding of rabbit antibodies was then quantitatively assessed using goat anti-rabbit peroxidase and subsequent measurement, in an enzyme-linked immunosorbent assay plate reader at 492 nm, of the generation of a colored product from the substrate phenylenediamine hydrochloride.

Two matters are of particular interest in this assay. First, the carbohydrate chains of the lectin, generally representing the main immunological epitopes, had been destroyed by preceding periodate treatment, thereby restricting antibody binding to antigenic determinants established by peptide sequences. Second, to focus evaluation of lectin on functionally active, and not just immunologically reactive, lectins in the extract, binding of the lectins to a natural glycoprotein as ligand, immobilized on microtiter plates, precedes the immunological detection.

Further molecular characterization of this lectin is needed to clarify mechanistic aspects of the immunomodulatory properties demonstrated.

2. *Jatropha multifida* Cyclic Peptides

Among the immunomodulators of different structural classes found in the latex of *Jatropha multifida*,[100-103] the cyclic peptides represent a special group. Their discovery as plant compounds with specific structural features, as well as their selective immunomodulatory activity, have opened new lines for the development of novel immunoactive drugs. These cyclic peptides, named labaditin and biobollein, selectively inhibit the classical pathway of human complement activation. In mechanistic studies an apparent inactivation of C1, but not C4 and C2, was demonstrated. Subsequently, it was found, using a solid-phase C1q-binding assay, that the peptides interfered with the binding of aggregated IgG to C1q. In the functional complement assay, it has been shown that the peptides bonded to the antibody-coated target cells and thereby inhibited classical pathway activation. These findings indicate that these peptides bind to aggregated and antigen-bound IgG and most probably block the antibody C1q-acceptor site. This binding property of the peptides is restricted to IgG and does not extend to IgM and IgA. The selectivity of the binding property appears to be even more pronounced when the subclasses of IgG are considered. Labaditin and biobollein selectively bind the most prominent, and complement activating, IgG subclass, i.e., IgG1.

To isolate labaditin and biobollein, on the basis of activity-guided fractionation from the latex of *Jatropha multifida*, the crude latex is first mixed with five times its volume of demineralized water. The resultant precipitate is separated from the supernatant, and the latter is extracted with n-hexane. The aqueous fraction is lyophilized, and the solid residue subjected to chromatographic separation on a polyamide column. By performing a gradient elution using equal volumes of water, methanol/water (1:4), methanol/water (2:3), methanol/water (3:2), and methanol successively, a series of fractions is collected. The extracts and fractions were all tested for modulatory effects on both classical and alternative pathways of human complement activation and on the production of reactive oxygen species (ROS) by zymosan-stimulated human PMN leukocytes. The methanol/water (3:2) eluent contained the cyclic peptides. These combined fractions, in a solution of 1% $NaHCO_3$, were exhaustively extracted with ethyl acetate. The solid residue obtained after evaporation of the ethyl acetate under reduced pressure was dissolved in methanol and applied to a Sephadex LH-20 column. This column was eluted with methanol, and the fractions were monitored by thin-layer chromatography (TLC) and by activity on the classical pathway of human complement activation. Thin-layer chromatographic analysis was performed on silica gel 60 F254 plates, using $CHCl_3/MeOH/H_2O$ (=13:10:2) as the developing solvent in a saturated chamber. The cyclic peptides were detected by spraying with vanillin-sulfuric acid reagent, followed by heating at 110°C for 5 min. The peptides show a purple to purple-blue color. The Rf values are 0.69 for labaditin and 0.74 for biobollein.

Structural analysis established a cyclic decapeptide structure for labaditin and a cyclic nonapeptide structure for biobollein. Using an automatic amino acid analyzer, after general hydrolysis in 6 N HCl at 110°C for 48 h, and for tryptophan determination, in 6 N HCl with

TABLE 7
Amino Acid Composition and Molecular Masses of the Cyclic Peptides Labaditin and Biobollein from *Jatropha multifida* Latex

Amino acids	HPLC measurement values[c] Results[a]		Mass Results[b]	Molecular mass by MS-FAB
In labaditin:				
Thr	1.9 (2)	-.- (2)	238	1070
Gly	2.0 (2)	2.0 (2)	150	
Ala	1.1 (1)	1.0 (1)	89	
Val	2.0 (2)	2.0 (2)	234	
Ile	1.0 (1)	1.0 (1)	131	
Trp		1.5 (2)	408	
			$1250 - (10 \times 18) \rightarrow$ 1070	
In biobollein:				
Ser	1.0 (1)	0.7 (1)	105	868
Gly	2.0 (2)	2.1 (2)	150	
Ala	2.1 (2)	2.2 (2)	178	
Ile	0.9 (1)	1.0 (1)	131	
Leu	2.0 (2)	2.1 (2)	262	
Trp		0.8 (1)	204	
			$1030 - (9 \times 18) \rightarrow$ 868	

[a] Results after hydrolysis with 6N HCl at 110°C for 48 h.
[b] Results after hydrolysis with 6N HCl + 4% thioglycolic acid at 110°C for 24 h.
[c] The numbers in parentheses are deducted assuming labaditin and biobollein are single peptides.

4% thioglycolic acid at 110°C for 24 h, the amino acids were identified, and their molar ratios determined. Mass spectrometric analysis using FAB experiments resulted in molecular weights of 1070 and 868 for the decapeptide and the nonapeptide, respectively. The results of amino acid analysis and the molecular masses found by FAB-MS, summarized in Table 7, unambiguously proved the cyclic structure of both these peptides. The molecular fragmentation in the FAB mass spectra is shown in Figure 4.

Determination of the amino acid sequence in the peptides was facilitated by the use of two-dimensional homonuclear shift correlation (COSY) and nuclear Overhauser enhancement (NOESY) ^{1}H-NMR techniques. For practical reasons, only the work done on labaditin is discussed here. First, a COSY spectrum was used for the assignments of ^{1}H-chemical shifts of individual amino acid residues. In such a spectrum, correlations can be seen between the J-coupled protons of each amino acid residue. Thus, one spin system corresponding to a valine was identified by cross-peaks between the NH and the Cα proton, between Cα and Cβ protons, between the Cβ and the Cτ and the Cτ1 protons. In this way the spin systems corresponding to one Ala, two Gly, one Ile, two Thr, two Val, and the aliphatic part of two Trp systems were obtained. In the aromatic region of ^{1}H-NMR spectrum, two aromatic spin systems corresponding to Trp could be identified. On the basis of NOEs, observed in the NOESY spectrum, between the aromatic ring resonances and the Cβ protons, the complete Trp spin system was identified. These results displayed complete accordance between the amino acid composition determined by HPLC analysis of the hydrolysis products and by the spin system analysis. Table 8 shows the chemical shifts of the individual amino acid residues of labaditin in DMSO-d6 at different temperatures. Regarding the ten amide NH chemical shifts, the five between 8.2 and 9.0 are at a normal field and are temperature dependent (>0.2 ppm difference with the 338 K spectrum). The other five shifts appear at a relatively high field and are not, or almost not, temperature dependent, probably indicating hydrogen bonds. This is especially so for 7Gly, 9Gly, and 10Thr.

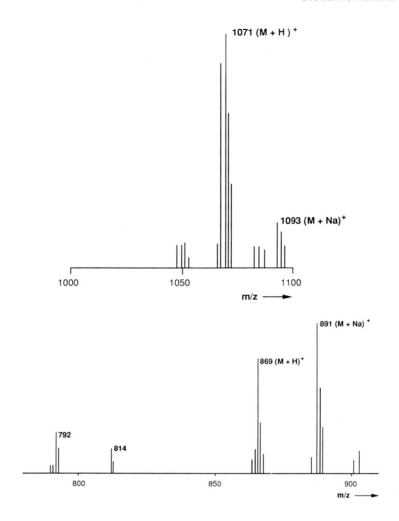

FIGURE 4. FAB-mass spectra of labaditin (top) and biobollein (bottom).

The sequence of the individual amino acids in the cyclic decapeptide was determined by means of sequence-specific NOE assignments in the NOESY spectrum,[104] in which cross-peaks can be observed between protons that are spatially close to each other (<4 Å). Thus, apart from NOEs between protons belonging to the same amino acids, NOEs between protons belonging to adjacent amino acids can also be observed. Depending on the exact structure, strong $d_{\alpha N}(i,i + 1)$ connectivities (for extended structures) or relatively strong $d_{NN}(i,i + 1)$ connectivities are almost always found between neighboring amino acids.[105] These connectivities form the basis for sequential assignment methods of protein ¹H-NMR spectra. For the amino acid sequence determination of labaditin, the sequential connectivities were used in a reversed way, in the sense that the $d_{\alpha N}$ and d_{NN} connectivities were used to establish the amino acid sequence of the cyclic decapeptide.[101]

All amide protons and most α protons showed separate chemical shifts in the NOESY spectrum. Only the α protons of the 4Ile and 8Val have the same chemical shifts in this spectrum. Assignments of the connectivities with these protons, however, could be made through the cooccurrence of NOEs with other protons of these particular amino acids. In this way connectivities were assigned between 1Ala-NH and 4Ile-αH and between 2Trp-NH and 8Val-αH. A combined COSY and NOESY spectrum (high field region only), plotted at the same scale, showing the

TABLE 8
Chemical Shifts of the Individual Amino Acid Moieties of Labaditin in DMSO-d$_6$ (Conc. 30 mg/ml) at 303 K and 338 K, and their Differences[a]

Proton	δ 303 K (ppm)	δ 338 K (high) (ppm)	$\Delta\delta$ 303 K/I (ppm)	$\Delta\delta$ 303/338 K (ppm)	Proton	δ 303 K (ppm)	δ 338 K (high) (ppm)	$\Delta\delta$ 303 K/I (ppm)	$\Delta\delta$ 303/338 K (ppm)
Ala 01					Trp 02				
Hα	3.78	3.92	−0.6	−0.1	CH2	7.15	7.14	0.0	0.0
Hβ	1.27	1.29	0.0	0.0	CH4	7.44	7.46	−0.2	0.0
NH	8.97	8.59	0.1	0.4	CH5	6.94	6.94	−0.1	0.0
					CH6	7.02	7.03	−0.1	0.0
Gly 07					CH7	7.27	7.28	−0.1	0.0
Hα1	3.29	3.44	−0.5	−0.2	CNH	10.77		0.0	
Hα2	3.85	3.90	0.1	0.0	Hα	4.54	0.00	0.0	0.0
NH	7.78	7.67	−0.4	0.1	Hβ1	2.96	3.00	0.0	0.0
					Hβ2	3.07		−0.1	
Gly 09					NH	8.78	8.47	0.7	0.3
Hα1	3.44	3.48	−0.3	0.0					
Hα2	4.44		0.7		Trp 06				
NH	7.52	7.60	−0.6	−0.1	CH2	7.02	7.03	−0.2	0.0
					CH4	7.50	7.51	−0.2	0.0
Ile 04					CH5	6.97	6.96	0.0	0.0
Hα	4.22	0.00	0.0	0.0	CH6	7.04	7.04	−0.1	0.0
Hβ	1.64	1.73	−0.3	−0.1	CH7	7.28	7.30	−0.1	0.0
Hγ1a	0.90	1.05	−0.3	−0.1	CNH	10.77		0.0	
Hγ1b	1.49	1.50	0.0	0.0	Hα	4.51		−0.1	
Hγ2	0.80		−0.1		Hβ1	2.69	2.79	−0.2	−0.1
Hδ	0.79		0.0		Hβ2	3.32	3.32	0.3	0.0
NH	8.43	8.09	0.2	0.4	NH	7.83	7.76	−0.3	0.1
					Val 05				
Thr 03					Hα	3.67	3.81	−0.6	−0.1
Hα	4.66	4.58	0.4	0.1	Hβ	1.82	1.86	−0.2	0.0
Hβ	3.83	3.94	−0.1	−0.1	Hγ1	0.32	0.51	−0.6	−0.2
Hγ	0.91	0.95	−0.2	0.0	Hγ2	0.63	0.67	−0.3	−0.1
NH	8.70	8.24	0.9	0.5	NH	8.20	7.98	0.3	0.2
					Val 08				
Thr 10					Hα	4.22		0.0	
Hα	4.71	4.64	0.6	0.1	Hβ	2.06	2.07	0.1	0.0
Hβ	4.35	0.00	0.4	0.0	Hγ1	0.82	0.83	−0.1	0.0
Hγ	1.16	1.13	0.1	0.0	Hγ2	0.83	0.83	−0.2	0.0
NH	7.30	7.34	−0.5	0.0	NH	7.68	7.63	−0.2	0.0

[a] The ^{1}H-NMR spectra were recorded on a Bruker WM-300 spectrometer. The differences recorded under $\Delta\delta$ 303 K/I refer to comparisons with chemical shifts given by Kessler et al.[104] for the same amino acid moiety in protected small linear peptides. The numbering of individual amino acids is according to their amide proton chemical shift; the amide proton chemical shift at the lowest field being number 1.

sequential d$_{\alpha N}$ connectivities of the ten amide NH protons, is presented in Figure 5. Thus, starting at the NH frequency of the unique 1Ala residue, one observes a d$_{\alpha N}$ connectivity to the unique 4Ile spin system in addition to the internal αN cross-peak of Ala. Thus, the 4Ile is likely to be adjacent in the sequence to the 1Ala residue. Now one continues on to the NH frequency of 4Ile and identifies a d$_{\alpha N}$ connectivity to 3Thr. On the 3Thr-NH frequency, we observe two connectivities to the next Gly. On the 9NH frequency, we observe a weak d$_{\alpha N}$ connectivity to 6Trp. Similarly, one finds d$_{\alpha N}$ connectivities from 6Trp-NH to 5Val-αH, from 5Val-NH to 10Thr-αH, from 10Thr-NH to 2Trp-αH, from 2Trp-NH to 8Val-αH, from 8Val-NH to two 7Gly-α protons, and from 7Gly-NH to 1Ala-αH.

The sequence consistent with these observations is given in Figure 6 (with observed d$_{NN}$ and d$_{\alpha N}$ values), and the structural formula consistent with this sequence is shown in Figure 7. The J$_{NH\alpha H}$ of 5 Val (J = 4.1Hz) is rather unusual, indicating a *cis-* rather than a *trans*-configuration between the 5Val-NH and 5Val-αH. Whereas both the α protons of 7Gly have

¹H-NMR

NOESY-COSY conectivity diagram; sequential assigments via d$_{\alpha N}$

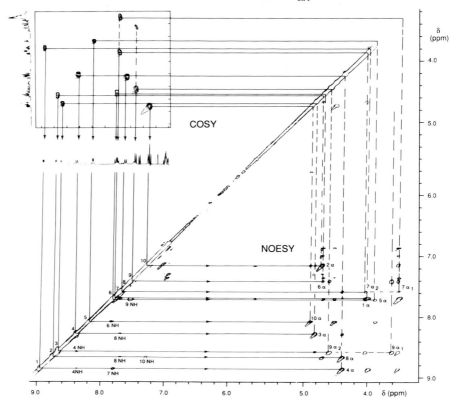

FIGURE 5. COSY-NOESY connectivity diagram. Indicated are the sequential assignments of labaditin via d$_{\alpha N}$. A NOE indicated as 4α on the line of 1 identifies the interaction of the amide proton of 1 with the α proton of 4.

almost the same $J_{NH\alpha H}$ (~6Hz), those of 9Gly are quite different (J = 2.7 and 9.1Hz), indicating a *cis-* and a *trans-*configuration, respectively, and thus a more rigid conformation. According to these observations, the molecule probably contains two turns containing 5Val and 7Gly, respectively. This is in agreement with the two observed d$_{NN}$ connectivities at these residues. The extended structure between the two turns probably forms an antiparallel β structure stabilized by hydrogen bonds. The presence of these H-bonds explains the low dependency of the chemical shift on temperature for a number of the NH protons.

A more complete description of the structure of this cyclic decapeptide is possible by extending the quantitative analysis of all NOEs from the NOESY spectrum. Also, other studies are in progress.

The structure of biobollein has been determined in a similar fashion, and the amino acid sequential structure is shown in Figure 8.

C. ALKALOIDS

Alkaloids do not, so far, play a major role within the diverse group of immunomodulating natural product molecular classes. Older data on well-known compounds like aristolochic acid, colchicine, demecolcine, vincristine, vinblastine, tylophorine, emetine, and cepharanthine can be found elsewhere.[4-6,69] A systematic screening of alkaloid-rich medicinal plants for

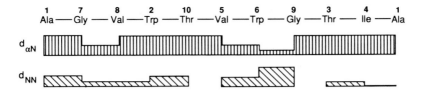

FIGURE 6. Summary of the sequential $d_{\alpha N}$ and d_{NN} connectivities in the NOESY. The height of the bars indicates the intensity of the NOE cross peaks.

FIGURE 7. Amino acid sequential structure of the cyclic decapeptide labaditin from *Jatropha multifida* latex.

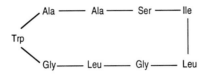

FIGURE 8. Amino acid sequential structure of the cyclic nonapeptide biobollein from *Jatropha multifida* latex.

immunomodulatory activity is still lacking in modern research. In this section we deal only with oxindole alkaloids from *Uncaria tomentosa* roots.[106]

A traditional preparation of this plant material is known to be taken orally against cancer. A patent application (WO 8201.130) submitted by Keplinger describes the efficacy of differential extracts from *U. tomentosa* roots for the treatment of several forms of malignancies. A hydrolyzed extract from *U. tomentosa* root powder, packed in capsules, is marketed (as ImmunAct) for inflammatory and immune deficiency diseases.

Some of the oxindole alkaloids from *U. tomentosa* roots have been found to enhance phagocytosis capacity of PMN leukocytes *in vitro* and to stimulate phagocytosis *in vivo*, as judged by the carbon clearance in mice. The crude ethanol and aqueous extracts of the roots display these activities. The alkaloid-enriched fraction was found to have an enhanced *in vitro* activity, but the purified alkaloid mixture failed to show *in vivo* activity. However, on addition of a 10% aqueous catechin macerate to the purified alkaloid mixture, *in vivo* activity was displayed. This synergistic phenomenon has to be considered when rationalizing biological activity of compounds from natural products and phytomedicines.

Among the oxindole alkaloids (from the *U. tomentosa* roots) tested for phagocytic activity, isopteropodine-HCl was shown to be the most potent compound. Pteropodine, isomitraphylline, and isorynchophylline displayed much weaker activity, and mitraphylline and rynchophylline did not demonstrate activity. The difference in phagocytic activity between these structurally closely related oxindole alkaloids provides interesting structure/activity relationship aspects for study.

The isolation of the crude alkaloid mixture proceeds by mixing the root powder with 10% ammonia and drying the mixture by means of infra-red light, prior to maceration in ethyl acetate. The ethyl acetate solution of the free bases is extracted with 2% sulfuric acid, which is subsequently treated with solid Na_2CO_3. The ethyl acetate extract of the resultant alkaline

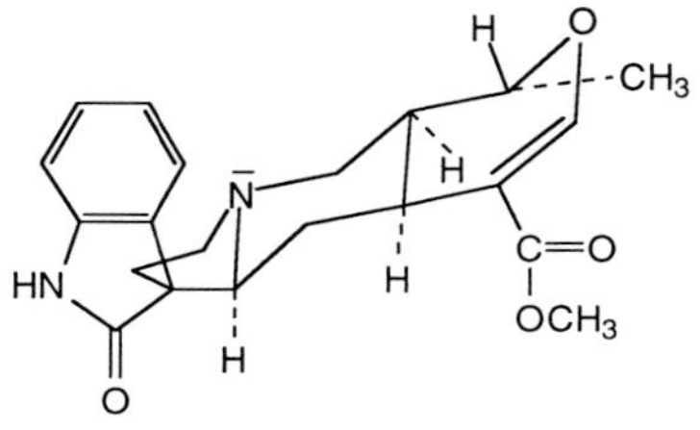

isopteropodin

FIGURE 9. Structural formula of isopteropodine.

(pH 8 to 9) solution is dried over Na_2SO_4, and the enriched alkaloid mixture is obtained as a light brown powder after distilling off the organic solvent.

The preparative isolation of the different alkaloids is performed by successive column chromatography and preparative thin-layer chromatography, using silica gel 60 (63 to 200 μm) and silica gel 60 F254 precoated TLC plates (0.5 mm), respectively. The silica gel column is eluted with a chloroform/acetone (6:3) mixture. The most potent immunomodulatory alkaloid, isopteropodin (Figure 9), is detected in the eluates by means of analytical TLC, using silica gel 60 F254, precoated TLC plates (0.25), and chloroform/acetone (5:4), chloroform/ethanol (95:5), and ethyl acetate/isopropanol/conc. ammonia (100:2:1) as solvent systems. The Rf values of isopteropodin in the three TLC solvent systems are 0.73, 0.48, and 0.83, respectively. The eluates containing isopteropodin are combined and subjected to preparative TLC. Thus, the fraction dissolved in methanol/ethyl acetate (1:1) is applied to preparative thin-layer plates, and the chromatogram developed with n-hexane/chloroform/acetone (1:9:1) (5 cm) and twice (7.4 and 10 cm) with chloroform/acetone (9:1). The band of isopteropodin is visualized under UV light (254 nm) and is then eluted with methanol/ethyl acetate (1:1) from the adsorbent.

The oxindole alkaloid mixture extracted from the crude plant material can also be analyzed by means of ion-pair HPLC. This method provides a direct possibility to compare the alkaloid mixtures accumulating in different varieties of *U. tomentosa* and related *Uncaria* species. The column material used is a μ-Bondapak C_{18} (particle diameter 10 μm). For the separation of penta- and tetracyclic oxindole alkaloids, acetone/water (25:75) plus 2.2% PIC B7 (corresponding to 5 mM heptane sulfonic acid-sodium salt and 15 mM acetic acid) is used as the eluent.

Typical pentacyclic oxindole mass fragmentation peaks are found at m/e 223, 208, 180, and 69. Further physical and analytic data of isopteropodine and related oxindole alkaloids have been reported elsewhere.[106]

D. TERPENOIDS

Earlier work on immunomodulatory terpenoids has been reviewed by Lindequist and Teuscher[4] and by Wagner and Proksch.[5] Within the subclass of sesquiterpenoids, several substances, including zexbrevin from *Zexmennia brevifolia*, helenalin from *Arnica montana*, alantolacton and 11,13-dihydroalantolacton from *Inula helenium*, diketocoriolin B from *Coriolus consor*, and ovalicin from the culture medium of *Pseudorotium ovalis*, have been demonstrated to modulate the immune response. Although no structure/activity relationships can be established yet, these compounds are characterized by a strongly oxygenated molecular structure pattern containing lactone-, ester-, ketone-, epoxy-, cyclic ether-, hydroxy-, and methoxy-groups.

Within the subclass of the diterpenoids, the phorbol derivatives from *Euphorbia* species are well known for their immunomodulatory activity. For example, phorbol-myristate-acetate

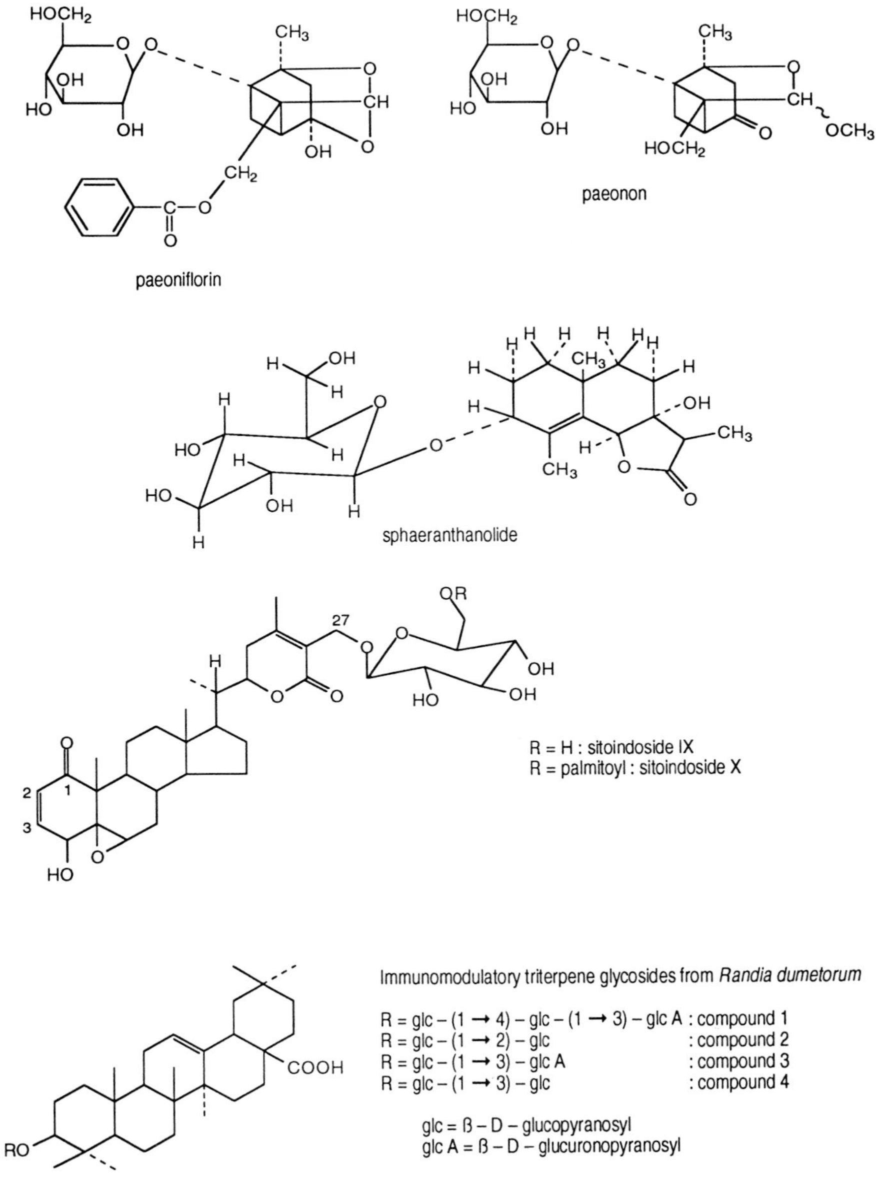

FIGURE 10. Immunomodulatory terpenoids.

(PMA) is currently used as an inducer of metabolic activation processes in immunological studies. In the 1970s, withaferin A, within the subclass of steroids, was shown to inhibit rejection of transplants in chicks and to suppress adjuvant arthritis in rats. Immunosuppressive activity is also known to be displayed by the *Solanum nigrum* steroidal alkaloid, solasodine.

Recently, other terpenoids showing immunomodulatory activity have been identified (Figure 10). Some of these compounds are discussed here. Paeoniflorin, a benzoylated monoterpene glycoside from the root bark of *Paeonia albiflora,* has been found to inhibit experimental (picryl-chloride-induced) contact hypersensitivity in mice and passive cutaneous anaphylaxis reaction, induced by BSA plus complete Freund's adjuvant, in guinea pigs. Paeonon, a debenzoylated keto derivative prepared from paeoniflorin was proved to be a more potent anti-anaphylactic agent. In addition, paeonon, in contrast to paeoniflorin, is able to inhibit paw tissue swelling in the adjuvant arthritis, induced by *Mycobacterium butyrica*, in rats.[107]

An example of a newly identified immunostimulating sesquiterpene glycoside is sphaeranthanolide, isolated from the flowers of *Sphaeranthus indicus*.[108] Using the Jerne plaque assay, the authors reported that this compound enhanced antibody production in mice challenged by sheep erythrocytes. The compound was isolated by extraction with methanol, followed by chromatographic purification on Sephadex LH-20 and silica gel columns. The isolation was monitored by TLC, the compound giving a dark blue color with orcinol sugar reagent. Final purification was performed by crystallization from methanol. The molecular structure was established on the basis of detailed data from ^{1}H-NMR, ^{13}C-NMR, two-dimensional spectroscopic analysis, applying homodecoupling experiments, COSY-45, NOESY, and NOE difference measurements.

Another interesting development in immunomodulatory terpenoids involves the results obtained with two newly isolated glycowithanolides, sitoindoside IX and sitoindoside X.[109] These C_{27}-O-glycosidic derivatives of withaferin A were found to mobilize and activate peritoneal macrophages, increase phagocytosis, and increase the activity of the lysosomal enzymes secreted by the activated macrophages. Concerning effects on the central nervous system, both these compounds produced significant antistress activity in albino mice and rats and augmented learning acquisition and memory retention in both young and old rats. With respect to the ethnomedical use of *Withania somnifera* (Ashwagandha in Sanskrit) in Ayurveda, these findings seem to be consistent with the claimed attenuation of cerebral function deficits in geriatric patients and nonspecific host defense. An additional interesting aspect of these investigations is that the immunostimulatory properties of the sitoindosides, being glycosidic forms of withaferin, contrast with the immunosuppressive properties of their aglycone.

In the subclass of the pentacyclic triterpenoids, four glycosidic compounds isolated from the fruits of *Randia dumetorum* have been found to enhance the proliferation of human lymphocytes *in vitro*.[110] They are oleanolic acid glycosides and are pictured in Figure 10.

Recent investigations have shown that when cannabidiol (CBD), one of the well-known constituents of *Cannabis sativa,* is orally administered to mice (10 mg/kg), blood LTB_4 production is inhibited. Cannabidiol is just as effective as the dual lipoxygenase/cyclo-oxygenase inhibitor BW755C and the lipoxygenase inhibitor BWA4C, used at a dose of 50 mg/kg. In the same blood samples, TXB_2 production is stimulated by cannabidiol. A23187 stimulation of LTB_4 synthesis in human PMN leukocytes *in vitro* is inhibited by cannabidiol in a dose-related manner (IC_{50} 5.4 μM). In lower-dose ranges, CBD stimulated TXB_2 production in PMN cells, but at higher doses TXB_2 synthesis was inhibited.[111]

Since cannabidiol is comprised of monoterpene and acetogenic phenolic ring-system entities, it bridges the terpenoid and phenolic types of immunomodulator.

E. PHENOLICS

Many representatives of different structural classes of phenolic compounds have been found to modulate functional factors of the immune response. Apart from specific conformational characteristics of particular phenolic compounds interfering with immunologic sites of action, phenolic compounds, in general, can interact chemically and biochemically with electron-transferring systems or (oxidizing and reducing) enzymes. In terms of immunomodulating mechanisms, phenolics may interact with systems regulating and producing oxygen-derived free radical species. On the other hand, phenolics of certain configurations and molecular size may bind biochemically functional metal ions and form complexes with enzymes. Although one has to be aware of nonspecific interactions, the search for useful immunomodulatory phenolic compounds is aimed at selectivity and specificity of action. The variability and diversity in natural-product-derived phenolic constituents is of such an extent, and hitherto so unpredictable, that extensive research efforts will be undertaken for many

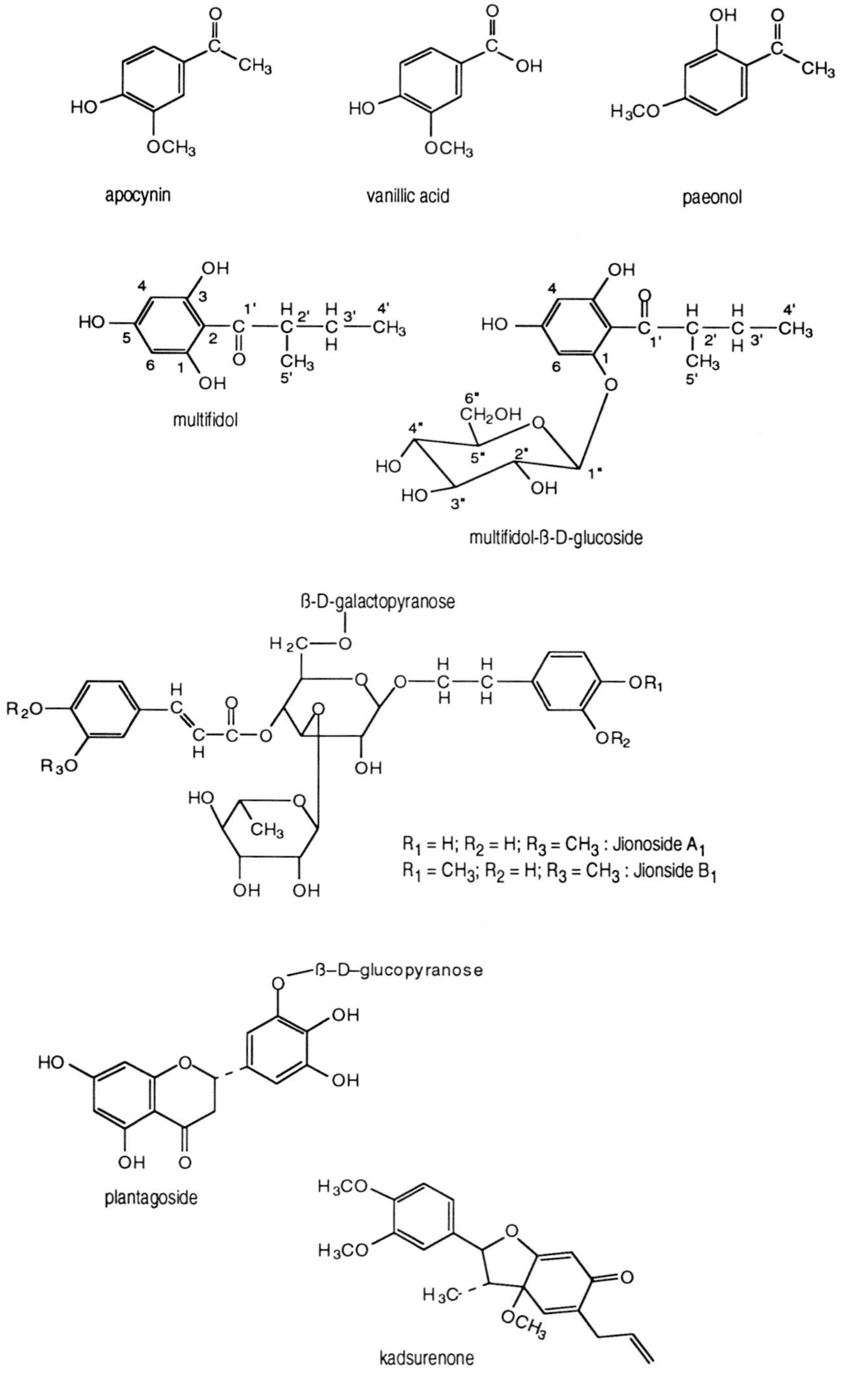

FIGURE 11. Immunomodulatory plant phenolics of different structure classes.

years. Interesting immunomodulatory compounds have been found within the different structure classes of phenolics. Earlier data have already been referred to.[4-6] A selection of recent advances achieved with phenolic immunomodulators is treated in this section, ranging from simple monocyclic phenolics to multiring phenolic molecules (Figure 11).

The simple molecule of apocynin represents basic features of a *p*-acetophenol associated with potent and very interesting immunomodulatory activity.[83,84,112,113] Although its molecular structure has been known since 1971, the immunomodulatory properties were studied only recently. *In vitro*, apocynin is a potent inhibitor of neutrophil superoxide anion (O_2^-) release. This inhibitory effect of apocynin is restricted to cells that are able to release peroxidase and ROS. Peroxidase-deficient cells are insensitive to apocynin. It was shown that myeloperoxidase (MPO) is required to activate apocynin and other *o*-methoxycatechols, such as vanillic acid and ferulic acid, to become inhibitors of the cellular respiratory burst.[84,112] *In vivo* experiments using rats in which arthritis was induced by type II collagen (CII) demonstrated that apocynin administered orally in drinking water was a potent antiarthritic compound.[113] The anti-inflammatory properties of apocynin were further explored in terms of effects on arachidonic-acid-derived mediators. It turned out that apocynin inhibits the formation of thromboxane A_2, whereas the release of prostaglandins E_2 and $F_{2\alpha}$ is stimulated by apocynin. In addition, it was shown that apocynin is a potent inhibitor of arachidonic-acid-induced aggregation of bovine platelets, possibly through inhibition of thromboxane formation.[114] These results suggested that apocynin might become a valuable therapeutic agent in the treatment of inflammatory (including arthritic) diseases.

Vanillic acid, ferulic acid, picroside II, picroside III, and picroside IV, although less potent *in vitro* than apocynin and having an *ortho*-hydroxy-methoxy-phenyl grouping in common, show similar inhibition of ROS-production by activated PMN leukocytes *in vitro*. As all of these compounds, like apocynin, are constituents of *Picrorhiza kurroa* roots, its ethnomedical use against inflammatory disorders seems consistent with the above-mentioned findings. Interestingly, an isomer of apocynin, called paeonol, isolated from *Paeonia moutan* root bark, was also shown to inhibit the formation of thromboxane B_2 and platelet aggregation.[115]

Another simple phenolic compound that inhibits luminol-dependent chemiluminescence by zymosan-stimulated PMN leukocytes is gallic acid. Subsequently, it was shown in *in vivo* experiments that gallic acid inhibits both acute (zymosan-induced), and delayed-type hyper-sensitivity-mediated, footpad swelling in mice. To effect this activity, gallic acid works on different levels. First, gallic acid is a scavenger of superoxide anions (O_2^-). Second, gallic acid interferes with the release and the activity of MPO, and third, it influences the assembly of active NADPH-oxidase. Each of these effects, working individually or all together, may contribute to the anti-inflammatory activity of gallic acid.[38]

Multifidol and multifidol glucoside, recently isolated from *Jatropha multifida* latex, were also shown to inhibit the production of ROS released from activated PMN leukocytes. These compounds also displayed an inhibitory effect on the classical pathway activation of complement *in vitro*.[7,100,116] The compounds were isolated as follows.

The latex was mixed with an equal volume of water, resulting in a precipitate that was separated by centrifugation. The supernatant was lyophilized, and the water soluble lyophilizate was chromatographed on a polyamide column (grain size 0.16 μm) that was eluted successively with equal volumes of water, methanol/water (1:4), methanol/water (2:3), methanol/water (3:2), and methanol. Multifidol and its glucoside were collected in the fractions eluted in the range of methanol/water (1:4). The two compounds were then separated effectively from each other by preparative TLC on silica gel 60, F254 (1.0 mm) plates, using chloroform/methanol/water (13:10:2) as the developing solvent system. Multifidol and its glucoside (Rf 7.1 and Rf 5.4, respectively) were detected under UV 254 nm and subsequently eluted with methanol. The structures of the (2-methylbutyryl)phloroglucinol and its $(1 \rightarrow 1'')$-β-D-glucopyranoside were determined from the ^{1}H-NMR, COSY, ^{13}C-NMR, and 2D-spectra.[116]

Comparative analysis of NMR chemical shifts strongly indicated the attachment of the glucosyl moiety at C_1-O of the phloroglucinol moiety. In the aglycone the chemical shifts of C_4 and C_6 in the ^{13}C-NMR spectrum, and those of C_4-H and C_6-H in the ^{1}H-NMR spectrum, have the same δ value, i.e., δ 94.7 and δ 5.81, respectively, due to symmetry of the aromatic moiety. In the glucoside, however, these carbons and protons were no longer identical, occurring at ^{13}C:δ 95.0 and 97.2 and at ^1H:δ 5.85 and 6.07. Further proof of the attachment of the glucosyl to C_1-O was provided by ^1H-NOESY experiments. A correlated cross-relaxation was observed between the anomeric $C_{1''}$-H and the aromatic C_6-H, but not with C_4-H. The CD spectrum of multifidol exhibited a positive Cotton effect with only one maximum at 280 nm, indicating that the substance isolated possesses the (2S)-(+)-configuration. The CD spectrum of the multifidol-glucoside displayed a positive Cotton effect at 280 nm but, additionally, a negative Cotton effect at 310 nm. This indicated that the aglycone part of the glycoside has the same absolute configuration (2S) as multifidol.[116]

Two new phenylethyl alcohol glycosides, jionoside A1 and jionoside B1, isolated from the roots of *Rehmannia glutinosa* var. *hueichingensis* were recently found to be immunosuppressive. This activity was established by recording inhibitory effects on hemolytic plaque-forming cells from mice immunized with sheep red blood cells (SRBC) injected intravenously. The isolation and structure determination of these compounds was performed by Sasaki et al.[117] The activity-guided fractionation showed that fractions other than that containing jionosides A1 and B1 displayed similar immunosuppressive activity. Briefly, the active compounds were isolated from an n-butanol-soluble fraction of the residual ethanol extract from the *Rehmannia* roots. Subfractionation of this butanol fraction was performed on a Diaion HP-20 column using water, 50% aqueous methanol, and methanol successively as eluents. The subfraction obtained by eluting with 50% aqueous methanol was further separated on Sephadex LH-20 [eluted with increasing amounts of methanol in water (0:1) → (1:0)], MCI gel CHP2OP [eluted with water/methanol (9:1) → (4:6)], and Fuji gel ODSG3 [eluted with water/methanol (5:95) → (30:70)] columns. Eight compounds, including jionosides A1 and B1, were isolated, all giving green colorations on spraying with ferrichloride reagent on TLC. Both jionosides A1 and B1 contain a trisaccharide moiety composed of rhamnose, glucose, and galactose. In the molecule of jionoside A1, the β-D-glucopyranosyl moiety is glycosidically bound at position C_1-O to a 3,4-dihydroxy-phenylethyl moiety and at position C_4-O, by ester-binding to a feroyl moiety. Jionoside B1 differs in that the β-D-glycopyranosyl moiety is glycosidically bound to a 3-hydroxy-4-methoxy-phenylethyl moiety.

Flavonoids, another group of phenolic natural products, have been shown to exhibit immunomodulatory activity.[7,81,100,118,119] In a comparative study it was found that most of the flavonoids investigated decreased luminol-dependent chemiluminescence by inhibition of ROS production by activated neutrophils. Some flavonoids inhibited MPO release, while others, in addition, also inhibited MPO activity. Experimental evidence indicated that the OH groups in the B-ring of flavonoid molecules were essential determinants for the inhibition of O_2^- release by activated neutrophils. Flavonoids methoxylated in the B-ring are only inhibitory on ROS-producing neutrophils in the presence of myeloperoxidase.[119] Highly or completely methoxylated flavonoids, e.g., tangeretin, exhibit strong anticomplementary activity.[120]

Recently, a flavanone glucoside named plantagoside (Figure 11), isolated from *Plantago asiatica* seeds, was found to suppress antibody response to sheep red blood cells, and concanavalin-A-induced lymphocyte proliferation as measured by [^{3}H]-thymidine incorporation. Plantagoside is a specific inhibitor of jack-bean α-mannosidase and mouse liver α-mannosidase activity. The isolation, structure determination, and immunosuppressive activity of plantagoside are described by Yamada et al.[121]

Oligomeric flavonoids have also been found to modulate immune reactions. For example, Kosasi et al. describe a proanthocyanidin isolated from *Jatropha multifida* latex, which inhibits complement activation through the classical pathway.[122] The alternative pathway activation of complement is relatively insensitive to this proanthocyanidin. This selectivity is due to the selective depletion of Ca^{2+}, needed for the classical pathway activation. The proanthocyanidin does not bind Mg^{2+}.

As an example of a lignan exhibiting antagonistic activity toward PAF, the compound kadsurenone is mentioned. Kadsurenone, isolated from *Piper futokadsura,* a Chinese medicinal plant, is a specific and competitive inhibitor of PAF binding to its receptor. The compound exerts its antagonistic activity both *in vitro* and *in vivo* after oral administration.[123]

VII. CONCLUSION

Natural product resources provide excellent raw material for the discovery and development of novel immunomodulatory compounds. Along with the discovery of new and active molecular entities, new mechanisms of immunomodulatory activity can be found. Consequently, innovative approaches to therapy of immunologic disorders may be developed. However, the selection of an appropriate conceptual and experimental approach in the search for active molecular entities is crucial to hit the targets.

As a program forming part of innovative drug development, the search for, and the research on, biogenic immunomodulatory compounds should be disease oriented. This will be a guideline to the selection of basic and functional bioassays. A selected pathologic disorder or medical indication should also play an essential role in the design of the ethnopharmacognostic investigations. These studies scrutinize empirical knowledge and traditional practices in order to select candidates for laboratory experimental work. It has been shown that an activity-guided isolation is an appropriate strategy to discover natural-product-derived immunomodulatory compounds.

ACKNOWLEDGMENT

I am grateful to my secretary, Mrs. Sun des Tombe, for her skilled assistance and her dedication to quality.

REFERENCES

1. **Drews, J.,** Experimental models relevant for therapy, in *Decision Making in Drug Research,* Gross, F., Ed., Raven Press, New York, 1983, 49.
2. **Mayer, P., Hamberger, H., and Drews, J.,** Differential effects of ubiquinone Q7 and ubiquinone analogs on macrophage activation and experimental infections in granulocytopenic mice, *Infection,* 8, 256, 1980.
3. **Sirois, P. and Rola-Pleszczynski, M.,** *Immunopharmacology,* Vol. 4 of Research Monographs in Immunology, Turk, J. L., Ed., Elsevier, Amsterdam, 1982.
4. **Lindequist, U. and Teuscher, E.,** Pflanzliche und mikrobielle Wirkstoffe als Immunmodulatoren, *Pharmazie,* 40, H1, 10, 1985.
5. **Wagner, H. and Proksch, A.,** Immunostimulatory drugs of fungi and higher plants, in *Economic and Medicinal Plant Research,* Vol. 1, Wagner, H., Hikino, H., and Farnsworth, N. R., Eds., Academic Press, New York, 1985, 113.
6. **Wagner, H.,** Immunostimulants from higher plants (recent advances), in *Biologically Active Natural Products,* Hostettmann, K. and Lea, P. J., Eds., Clarendon Press, Oxford, 1987, 127.
7. **Labadie, R. P., Van der Nat, J. M., Simons, J. M., Kroes, B. H., Kosasi, S., Van den Berg, A. J. J., 't Hart, L. A., Van der Sluis, W. G., Abeysekera, A., Bamunuarachchi, A., and De Silva, K. T. D.,** An ethnopharmacognostic approach to the search for immunomodulators of plant origin, *Planta Med.,* 55, 339, 1989.
8. **Rola-Pleszczynski, M. and Stankova, J.,** Cytokine gene regulation by PGE$_2$, LTB$_4$ and PAF, *Mediators Inflammation,* 1, 5, 1992.
9. **Goldstein, I. M.,** *Complement in Infectious Disease,* Current Concepts™, A Scope Publication, Upjohn Company, Kalamazoo, MI, 1980.

10. **Fernandez, H. N., Henson, P. M., Otani, A., and Hugli, T. E.,** Chemotactic response to human C3a and C5a anaphylatoxins. I. Evaluation of C3a and C5a leukotaxis *in vitro* and under simulated *in vivo* conditions, *J. Immunol.,* 120, 109, 1978.

11. **Brown, E. J.,** Complement receptors and phagocytosis, *Curr. Opin. Immunol.,* 3, 76, 1991.

12. **Craddock, P. R., Fehr, J., Brigham, K. L., Kronenberg, R. S., and Jacob, H. S.,** Complement and leukocyte-mediated pulmonary dysfunction in hemodialysis, *N. Engl. J. Med.,* 296, 769, 1977.

13. **Craddock, P. R., Hammerschmidt, D., White, J. G., Dalmasso, A. P., and Jacob, H. S.,** Complement (C5a)-induced granulocyte aggregation *in vitro.* A possible mechanism of complement-mediated leukostasis and leukopenia, *J. Clin. Invest.,* 60, 260, 1977.

14. **Summers, J. B.,** Anti-inflammatory patents highlights from the first half of 1991, *Curr. Opin. Ther. Patents,* 1(10), 1460, 1991.

15. **Ward, P. A. and Marks, R. M.,** The acute inflammatory reaction, *Curr. Opin. Immunol.,* 2, 5, 1989.

16. **Detmers, P. A. and Wright, S. D.,** Adhesion-promoting receptors on leukocytes, *Curr. Opin. Immunol.,* 1, 10, 1988.

17. **Ghebrehiwet, B. and Müller-Eberhard, H. J.,** Description of an acidic fragment (C3e) of human C3 having leukocytosis-producing activity, *J. Immunol.,* 120, 1774, 1978.

18. **Rother, K.,** Leukocyte mobilizing factor, a new biological activity derived from the third component of complement, *Eur. J. Immunol.,* 2, 550, 1972.

19. **Erb, P. and Feldman, M.,** The role of macrophages in the generation of T-helper cells. II. The genetic control of macrophage-T-cell interaction for helper cell induction with soluble antigens, *J. Exp. Med.,* 142, 460, 1975.

20. **Lipsky, P. E. and Rosenthal, A. S.,** Macrophage-lymphocyte interaction. II. Antigen-mediated physical interactions between immune guinea pig lymph node lymphocytes and syngeneic macrophages, *J. Exp. Med.,* 141, 138, 1975.

21. **Rosenthal, A. S., Blake, J. T., Kahn, C. R., Mann, D., and Galloway, J.,** Genetic control of the immune response to insulin: a clinical study of adverse immunologic reactions to insulin, in *Insulin, Chemistry, Structure and Function of Insulin and Related Hormones, Proc. 2nd. Int. Insulin Symp.,* Walter de Gruyter & Co., Berlin, 1980, 585.

22. **Rola-Pleszczynski, M.,** Anatomy of the immune response, in *Immunopharmacology,* Sirois, P. and Rola-Pleszczynski, M., Eds., Elsevier, Amsterdam, 1982, 1.

23. **Roitt, I. M.,** *Essential Immunology,* 6th ed., Blackwell Scientific, Oxford, 1988.

24. **Paul, W. E.,** *Fundamental Immunology,* 2nd ed., Raven Press, New York, 1989.

25. **Sites, D. P. and Terr, A. I.,** *Human Immunology,* Appleton & Lange, East Norwalk, CT, 1991.

26. **Coligan, J. E., Kruisbeek, A. M., Margulies, D. H., Shevach, E. M., and Strober, W.,** *Current Protocols in Immunology,* Greene Publishing Associates and Wiley-Interscience, New York, 1991.

27. **Adams, D. O., Edelson, P. J., and Koren, H. S.,** *Methods for Studying Mononuclear Phagocytes,* Academic Press, New York, 1981.

28. **Higgs, G. A. and Williams, T. J., Eds.,** *Inflammatory Mediators,* VCH Publishers, Deerfield Beach, FL, 1985.

29. **Wilkinson, P. C.,** *Chemotaxis and Inflammation,* 2nd ed., Churchill Livingstone, Edinburgh, 1982.

30. **Fantone, J. C. and Ward, P. A.,** *Oxygen-Derived Radicals and Their Metabolites: Relationship to Tissue Injury, Current Concepts^{TM},* A Scope Publication, Upjohn Company, Kalamazoo, MI, 1985.

31. **Fehér, J., Csomós, G., and Vereckei, A.,** *Free Radical Reactions in Medicine,* Springer-Verlag, Berlin, 1987.

32. **Van Dyke, K. and Castranova, V., Eds.,** *Cellular Chemiluminescence,* Vol. I–III, CRC Press, Boca Raton, FL, 1987.

33. **Wagner, H. and Jurcic, K.,** Assays for immunomodulation and effects on mediators of inflammation, in *Methods in Plant Biochemistry,* Vol. 6, *Assays for Bioactivity,* Hostettmann, K., Ed., Academic Press, London, 1991, 195.

34. **Beretz, A. and Cazenave, J.-P.,** Assays for platelet aggregation and related enzyme activities useful for the analysis of plant material, in *Methods in Plant Biochemistry,* Vol. 6, *Assays for Bioactivity,* Hostettmann, K., Ed., Academic Press, London, 1991, 235.

35. **Kiso, Y. and Hikino, H.,** Assay methods for antihepatotoxic activity, in *Methods in Plant Biochemistry,* Vol. 6, *Assays for Bioactivity,* Hostettmann, K., Ed., Academic Press, London, 1991, 219.

36. **Devlin, J. P., Carter, C., and Homon, C. A.,** High capacity screening: natural products as immune regulators, in *Economic and Medical Plant Research,* Vol. 5, *Plants and Traditional Medicine,* Wagner, H. and Farnsworth, N. R., Eds., Academic Press, London, 1991, 83.

37. **Labadie, R. P.,** Problems and possibilities in the use of traditional drugs, *J. Ethnopharmacol.,* 15, 221, 1986.

38. **Kroes, B.H.,** Nimba Arishta — Impact of the Preparation Process on Chemical Parameters and Immunomodulatory Activity, Ph.D. thesis, Utrecht University, Utrecht, 1990.

39. **Baba, T., Yoshida, T., Yoshida, T., and Cohen, S.,** Suppression of cell-mediated immune reactions by monosaccharides, *J. Immunol.,* 122, 838, 1979.

40. **Nair, M. P. N. and Schwartz, S. A.,** Suppression of human natural and antibody-dependent cytotoxicity by soluble factors from unstimulated normal lymphocytes, *J. Immunol.,* 129, 2511, 1982.

41. **Koszinowski, U. H. and Kramer, M.,** Selective inhibition of T-suppressor-cell function by a monosaccharide, *Nature,* 289, 184, 1981.

42. **Bauer, R. and Wagner, H.,** *Echinacea*-species as potential immunostimulatory drugs, in *Economic and Medicinal Plant Research,* Vol. 5., *Plants and Traditional Medicine,* Wagner, H. and Farnsworth, N. R., Eds., Academic Press, London, 1991, 253.

43. **Srivastava, R. and Kulshreshtha, D. K.,** Bioactive polysaccharides from plants, *Phytochemistry,* 28, 2877, 1989.

44. **Yamada, H.,** Natural products of commercial potential as medicines, *Curr. Opin. Biotechnol.,* 2, 203, 1991.

45. **Björndal, H., Lindberg, B., and Svensson, S.,** Gas-liquid chromatography of partially methylated alditols as their acetates, *Acta Chem. Scand.,* 21, 1801, 1967.

46. **Björndal, H., Lindberg, B., and Svensson, S.,** Mass spectrometry of partially methylated alditol acetates, *Carbohydr. Res.,* 5, 433, 1967.

47. **Kiyohara, H. and Yamada, H.,** Structure of anti-complementary arabinogalactan from the root of *Angelica acutiloba* Kitagawa, *Carbohydr. Res.,* 193, 173, 1989.

48. **Yamada, H., Komiyama, K., Kiyohara, H., Cyong, J.-Ch., Hirakawa, Y., and Otsuka, Y.,** Structural characterization and antitumor activity of a pectic polysaccharide from the roots of *Angelica acutiloba, Planta Med.,* 56, 182, 1990.

49. **Yamada, H., Otsuka, Y., and Omura, S.,** Structural characterization of anti-complementary polysaccharides from the leaves of *Artemisia princeps, Planta Med.,* 52, 311, 1986.

50. **Gao, Q.-P., Kiyohara, H., Cyong, J.-C., and Yamada, H.,** Characterization of anti-complementary acidic heteroglycans from the leaves of *Panax ginseng* C. A. Meyer, *Carbohydr. Res.,* 181, 175, 1988.

51. **Tomoda, M. and Suzuki, Y.,** Plant Mucilages. XVIII. Isolation and characterization of a mucilage, "*Abelmoschus*-mucilage M," from the roots of *Abelmoschus manihot, Chem. Pharm. Bull.,* 25, 3061, 1977.

52. **Tomoda, M., Suzuki, Y., and Satoh, N.,** Plant Mucilages. XXIII. Partial hydrolysis of *Abelmoschus*-mucilage M. and the structural features of its polysaccharide moiety, *Chem. Pharm. Bull.,* 27, 1651, 1979.

53. **Tomoda, M. and Satoh, N.,** Plant Mucilages. XVII. Partial hydrolysis and a possible structure of paniculatan, *Chem. Pharm. Bull.,* 25, 2910, 1977.

54. **Tomoda, M., Shimada, K., and Shimizu, N.,** Plant Mucilages. XXXII. A representative mucilage, "*Althaea*-mucilage R," from the roots of *Althaea rosea, Chem. Pharm. Bull.,* 31, 2677, 1983.

55. **Tomoda, M., Yokoi, M., and Ishikawa, K.,** Plant Mucilages. XXIX. Isolation and characterization of a mucous polysaccharide, "*Plantago*-mucilage A," from the seeds of *Plantago major* var. *asiatica, Chem. Pharm. Bull.,* 29, 2877, 1981.

56. **Tomoda, M., Shimizu, N., Shimada, K., Gonda, R., and Sakabe, H.,** Plant Mucilages. XXXIV. The location of O-acetyl groups and the structural features of *Plantago*-Mucilage A, the mucous polysaccharide from the seeds of *Plantago major* var. *asiatica, Chem. Pharm. Bull.,* 32, 2182, 1984.

57. **Kiyohara, H., Cyong, J.-Ch., and Yamada, H.,** Relationship between structure and activity of an anti-complementary arabinogalactan from the roots of *Angelica acutiloba* Kitagawa, *Carbohydr. Res.,* 193, 193, 1989.

58. **Selvendran, R. R. and Ryden, P.,** Isolation and analysis of plant cell walls, in *Methods in Plant Biochemistry,* Vol. 2, *Carbohydrates,* Dey, P. M., Ed., Academic Press, London, 1990, 549.

59. **Ohno, N., Adachi, Y., Suzuki, I., Oikawa, S., Sato, K., Suzuki, Y., Ohsawa, M., and Yadomae, T.,** Two different conformations of antitumor glucans obtained from *Grifola frondosa, Chem. Pharm. Bull.,* 34, 2555, 1986.

60. **Sasaki, T. and Takasuka, N.,** Further study on the structure of lentinan, an antitumor polysaccharide from *Lentinus edodes, Carbohydr. Res.,* 47, 99, 1976.

61. **Mitani, M., Ariga, T., Matsuo, T., Asano, T., and Saito, G.,** Anti-tumor effect of schizophyllan, an immunomodulator on syngeneic tumors, *Int. J. Immunopharmacol.,* 2, 174, 1980.

62. **Kozima, T., Tabata, K., Kawabata, S., and Misaki, A.,** Chemical and physical properties of schizophyllan, a therapeutically useful fungal polysaccharide, *Int. J. Immunopharmacol.,* 2, 174, 1980.

63. **Kanayama, H., Adachi, N., and Togami, M.,** A new antitumor polysaccharide from the mycelia of *Poria cocos* Wolf., *Chem. Pharm. Bull.,* 31, 1115, 1983.

64. **Chihara, G., Maeda, Y., Hamuro, J., Sasaki, T., and Fukuoka, F.,** Inhibition of mouse sarcoma 180 by polysaccharides from *Lentinus edodes* (Berk.) Sing., *Nature,* 222, 687, 1969.

65. **Maeda, Y.Y. and Chihara, G.,** Lentinan, a new immuno-accelerator of cell-mediated responses, *Nature,* 229, 634, 1971.

66. **Dennert, G. and Tucker, D.,** Antitumor polysaccharide lentinan — a T cell adjuvant, *J. Nat. Cancer Inst.,* 51, 1727, 1973.

67. **Okuda, T., Yoshioka, Y., Ikekawa, T., Chihara, G., and Nishioka, K.,** Anticomplementary activity of antitumour polysaccharides, *Nature New Biol.,* 238, 59, 1972.

68. **Abe, S., Takahashi, K., Tsubouchi, J., Yamazaki, M., and Mizuno, D.,** Combination therapy of murine tumors with lentinan, bacterial lipopolysaccharide and a *Streptococcus* preparation, OK 432, *Gann* (Japanese Journal of Cancer Research, Tokyo), 74, 273, 1983.

69. **Wagner, H.,** Search for plant derived natural products with immunostimulatory activity (recent advances), *Pure Appl. Chem.*, 62, 1217, 1990.

70. **Proksch, A. and Wagner, H.,** Structural analysis of a 4-O-methylglucuronoarabinoxylan with immunostimulating activity from *Echinacea purpurea, Phytochemistry,* 26, 1989, 1987.

71. **Fang, J., Proksch, A., and Wagner, H.,** Immunologically active polysaccharides of *Acanthopanax senticosus, Phytochemistry,* 24, 2619, 1985.

72. **Yamada, H., Nagai, T., Cyong, J.-C., and Otsuka, Y.,** Relationship between chemical structure and anti-complementary activity of plant polysaccharides, *Carbohydr. Res.,* 144, 101, 1985.

73. **Yamada, H., Kiyohara, H., Cyong, J.-Ch., Kojima, Y., Kumazawa, Y., and Otsuka, Y.,** Studies on polysaccharides from *Angelica acutiloba, Planta Med.,* 48, 163, 1984.

74. **'t Hart, L. A., Van Enckevoort, P. H., Van Dijk, H., Zaat, R., De Silva, K. T. D., and Labadie, R. P.,** Two functionally and chemically distinct immunomodulatory compounds in the gel of *Aloe vera, J. Ethnopharmacol.,* 23, 61, 1988.

75. **'t Hart, L. A., Van den Berg, A. J. J., Kuis, L., Van Dijk, H., and Labadie, R. P.,** An anti-complementary polysaccharide with immunological adjuvant activity from the leaf parenchyma gel of *Aloe vera, Planta Med.,* 55, 509, 1989.

76. **Mueller, E. A., Hamprecht, K., and Anderer, F. A.,** Biochemical characterization of a component in extracts of *Viscum album* enhancing human NK toxicity, *Immunopharmacology,* 17, 11, 1989.

77. **Mueller, E. A. and Anderer, F. A.,** Chemical specificity of effector cell/tumor cell bridging by a *Viscum album* rhamnogalacturonan enhancing cytotoxicity of human NK cells, *Immunopharmacology,* 19, 69, 1990.

78. **Klett, C. Y. and Anderer, F. A.,** Activation of natural killer cell cytotoxicity of human blood monocytes by a low molecular weight component from *Viscum album* extract, *Arzneimittelforschung,* 39(II), 1580, 1989.

79. **Van der Nat, J. M., Klerx, J. P. A. M., Van Dijk, H., De Silva, K. T. D., and Labadie, R. P.,** Immunomodulatory activity of an aqueous extract of *Azadirachta indica* stem bark, *J. Ethnopharmacol.,* 19, 125, 1987.

80. **Van der Nat, J. M., 't Hart, L. A., Van der Sluis, W. G., Van Dijk, H., Van den Berg, A. J. J., De Silva, K. T. D., and Labadie, R. P.,** Characterization of anti-complement compounds from *Azadirachta indica, J. Ethnopharmacol.,* 27, 15, 1989.

81. **Van der Nat, J. M.,** *Azadirachta Indica* Bark: An Immunopharmacognostical Study of its Traditional Use in Inflammatory Disease, Ph.D. thesis, Utrecht University, Utrecht, 1989.

82. **Simons, J. M., 't Hart, L. A., Van Dijk, H., Fischer, F. C., De Silva, K. T. D., and Labadie, R. P.,** Immunomodulatory compounds from *Picrorhiza kurroa*: isolation and characterization of two anti-complementary polymeric fractions from an aqueous root extract, *J. Ethnopharmacol.,* 26, 169, 1989.

83. **Simons, J. M., 't Hart, L. A., Labadie, R. P., Van Dijk, H., and De Silva, K. T. D.,** Modulation of human complement activation and the human neutrophil oxidative burst by different root extracts of *Picrorhiza kurroa, Phytother. Res.,* 4, 207, 1990.

84. **Simons, J. M.,** Immunomodulation by *Picrorhiza kurroa* — basis for its Ethnomedical Use? Ph.D. thesis, Utrecht University, Utrecht, 1989.

85. **Gonda, R., Tomoda, M., Shimizu, N., and Kanari, M.,** Characterization of polysaccharides having activity on the reticuloendothelial system from the rhizome of *Curcuma longa, Chem. Pharm. Bull.,* 38, 482, 1990.

86. **Gonda, R., Tomoda, M., and Takada, K.,** The core structure and immunological activities of ukonan B, the representative polysaccharide from turmeric and its degradation products, *Pharm. Pharmacol. Lett.,* 2, 50, 1992.

87. **Fang, S., Chen, Y., Xu, X., Ye, C., Zhai, S., and Shen, M.,** Studies of the active principles of *Astragalus mongholicus* Bunge. I. Isolation, characterization and biological effect of its polysaccharides, *Youji Huaxue,* 1, 26, 1982; *Chemical Abstracts,* 96, 177941r, 1982.

88. **Kumazawa, Y., Nakatsuru, Y., Yamada, A., Yadomae, T., Nishimura, C., Otsuka, Y., and Nomoto, K.,** Immunopotentiator separated from hot water extract of the seed of *Benincasa cerifera* Savi (Tongashi), *Cancer Immunol. Immunother.,* 19, 79, 1985.

89. **Yamada, H., Ohtani, K., Kiyohara, H., Cyong, J.-C., Otsuka, Y., Ueno, Y., and Omura, S.,** Purification and chemical properties of anti-complementary polysaccharide from the leaves of *Artemisia princeps, Planta Med.,* 51, 121, 1985.

90. **Tomoda, M., Takahashi, M., and Nakatsuka, S.,** Water-soluble carbohydrates of *Zizyphi* Fructus. II. Isolation of two polysaccharides and structure of an arabinan, *Chem. Pharm. Bull.,* 21, 707, 1973.

91. **Kraus, J. and Roszkopf, F.,** Relationship between immunological activity *in vitro* and antitumor effect *in vivo* of various polysaccharides, *Pharm. Pharmacol. Lett.,* 1, 11, 1991.

92. **Kiyohara, H., Yamada, H., Cyong, J.-C., and Otsuka, Y.,** Studies on polysaccharides from *Angelica acutiloba.* V. Molecular aggregation and anti-complementary activity of arabinogalactan from *Angelica acutiloba, J. Pharmacobiodyn.,* 9, 339, 1986.

93. **Kiyohara, H., Yamada, H., and Otsuka, Y.,** Unit structure of the anti-complementary arabinogalactan from *Angelica acutiloba* Kitagawa, *Carbohydr. Res., 167,* 221, 1987.

94. **Yamada, H., Kiyohara, H., Cyong, J.-C., and Otsuka, Y.,** Structural characterization of an anti-complementary arabinogalactan from the roots of *Angelica acutiloba* Kitagawa, *Carbohydr. Res., 159,* 275, 1987.

95. **Van Noordwijk, J., Hillen, F. C., and Hagelen, F.,** The relative activity of endotoxins of different origins in the LAL gelation method and in the LAL chromogenic substrate method, in *Detection of Bacterial Endotoxins with the Limulus Amebocyte Lysate Test,* Watson, S. W., Levin, J., and Novitsky, T. J., Eds., Alan R. Liss, New York, 1987, 103.

96. **Adorini, L. and Nagy, Z. A.,** Peptide competition for antigen presentation, *Immunol. Today,* 11, 21, 1990.

97. **Werner, G. H., Floc'h, F., Migliore-Samour, D., and Jollès, P.,** Immunomodulating peptides, *Experientia,* 42, 521, 1986.

98. **Floc'h, F.,** Natural or synthetic peptides as immunomodulating drugs, in *Peptides: a Target for New Drug Development,* Bloom, S. R. and Burnstock, G., Eds., IBC Technical Services, London, 1991, 109.

99. **Hajto, T., Hostanska, K., and Gabius, H.-J.,** Modulatory potency of the β-galactoside-specific lectin from mistletoe extract (Iscador) on the host defence system *in vivo* in rabbits and patients, *Cancer Res.,* 49, 4803, 1989.

100. **Kosasi, S.,** Immunomodulators from the Latex of *Jatropha multifida* L., Ph.D. thesis, Utrecht University, Utrecht, 1990.

101. **Kosasi, S., Van der Sluis, W. G., Boelens, R., 't Hart, L. A., and Labadie, R. P.,** Labaditin, a novel cyclic decapeptide from the latex of *Jatropha multifida* L. (Euphorbiaceae) — Isolation and sequence determination by means of two-dimensional NMR, *FEBS Lett.,* 256, 91, 1989.

102. **Kosasi, S., Quarles van Ufford, L., Van den Berg, A. J. J., Van Dijk, H., and Labadie, R. P.,** Selective inhibition of the classical complement pathway by two cyclic peptides from *Jatropha multifida* L. (Euphorbiaceae), *Complement and Inflammation, Laboratory and Clinical Research, Abstr. 3rd European Meet. on Complement in Human Disease,* Starsia, Z. and Nydegger, U., Eds., S. Karger, Basel, 7, 155, 1990.

103. **Labadie, R. P. and Van Dijk, H.,** International Patent application, WO 91/16345, 1991.

104. **Kessler, H., Steuer Nagel, S., Gillessen, D., and Kamyama, T.,** Complete sequence determination and localisation of one imino and three sulfide bridges of the nonadecapeptide Ro 09-0198 by homonuclear 2D-NMR spectroscopy. The DQF-Relayed-NOESY-experiment, *Helv. Chim. Acta,* 70, 726, 1987.

105. **Wüthrich, K.,** *NMR of Proteins and Nucleic Acids,* John Wiley & Sons, New York, 1986.

106. **Wagner, H., Kreutzkamp, B., and Jurcic, K.,** Die alkaloide von *Uncaria tomentosa* und ihre Phagozytose-steigernde Wirkung, *Planta Med.,* 419, 1985.

107. **Yamahara, J., Yamada, T., Kimura, H., Sawada, T., and Fujimura, H.,** Biologically active principles of crude drugs. II. Anti-allergic principles in "Shoseiryu-to", anti-inflammatory properties of paeoniflorin and its derivatives, *J. Pharmacobiodyn.,* 5, 921, 1982.

108. **Shekhani, M. S., Shah, P. M., Yasmin, A., Siddiqui, R., Perveen, S., Kahn, K. M., Kazmi, S. U., and Rahman, A.-U.,** An immunostimulant sesquiterpene glycoside from *Sphaeranthus indicus, Phytochemistry,* 29, 2573, 1990.

109. **Ghosal, S., Lal, J., Srivastava, R., Bhattacharya, S. K., Upadhyay, S. N., Jaiswal, A. K., and Chattopadhyay, U.,** Immunomodulatory and CNS effects of sitoindosides IX and X, two new glycowithanolides from *Withania somnifera, Phytother. Res.,* 3, 201, 1989.

110. **Dubois, M.-A., Benze, S., and Wagner, H.,** New biologically active triterpene-saponins from *Randia dumetorum, Planta Med.,* 56, 451, 1990.

111. **Formukong, E. A., Evans, A. T., Evans, F. J., and Garland, L. G.,** Inhibition of A23187-induced release of leukotriene B_4 in mouse whole blood *ex vivo* and human polymorphonuclear cells *in vitro* by the cannabinoid analgesic cannabidiol, *Phytother. Res.,* 5, 258, 1991.

112. **Simons, J. M., 't Hart, B. A., Ip Vai Ching, T. R. A. M., Van Dijk, H., and Labadie, R. P.,** Metabolic activation of natural phenols into selective oxidative burst agonists by activated human neutrophils, *Free Radic. Biol. Med.,* 8, 251, 1990.

113. **'t Hart, B. A., Simons, J. M., Knaan-Shanzer, S., Bakker, N. P. M., and Labadie, R. P.,** Antiarthritic activity of the newly developed neutrophil oxidative burst antagonist apocynin, *Free Radic. Biol. Med.,* 9, 127, 1990.

114. **Engels, F., Renirie, B. F., 't Hart, B. A., Labadie, R. P., and Nijkamp, F. P.,** Effects of apocynin, a drug isolated from the roots of *Picrorhiza kurroa,* on arachidonic acid metabolism, *FEBS Lett.,* 305, 254, 1992.

115. **Hirai, A., Terano, T., Hamazaki, T., Sajiki, J., Saito, H., Tahara, K., Tamura, Y., and Kumagai, A.,** Studies on the mechanism of antiaggregatory effect of Moutan Cortex, *Thromb. Res.,* 31, 29, 1983.

116. **Kosasi, S., Van der Sluis, W. G., and Labadie, R. P.,** Multifidol and multifidol glucoside from the latex of *Jatropha multifida, Phytochemistry,* 28, 2439, 1989.

117. **Sasaki, H., Nishimura, H., Morota, T., Chin, M., Mitsuhashi, H., Komatsu, Y., Maruyama, H., Guorui, T., Wei, H., and Yu-lang, X.,** Immunosuppressive principles of *Rehmannia glutinosa* var. *hueichingensis, Planta Med.,* 55, 458, 1989.

118. **Wagner, H., Kreutzkamp, B., and Jurcic, K.,** Inhaltsstoffe und Pharmakologie der *Okoubaka aubreville-*Rinde, *Planta Med.,* 404, 1985.

119. **'t Hart, B. A., Ip Vai Ching, T. R. A. M., Van Dijk, H., and Labadie, R. P.,** How flavonoids inhibit the generation of luminol-dependent chemiluminescence by activated human neutrophils, *Chem. Biol. Inter.,* 73, 323, 1990.

120. **Kroes, B. H. and Labadie, R. P.,** unpublished data.

121. **Yamada, H., Nagai, T., Takemoto, N., Endoh, H., Kiyohara, H., Kawamura, H., and Otsuka, Y.,** Plantagoside, a novel α-mannosidase inhibitor isolated from the seeds of *Plantago asiatica,* suppresses immune response, *Biochem. Biophys. Res. Commun.,* 165, 1292, 1989.

122. **Kosasi, S., 't Hart, L. A., Van Dijk, H., and Labadie, R. P.,** Inhibitory activity of *Jatropha multifida* latex on classical complement pathway activity in human serum mediated by a calcium-binding proanthocyanidin, *J. Ethnopharmacol.,* 27, 81, 1989.

123. **Shen, T. Y., Hwang, S.-B., Chang, M. N., Doebber, T. W., Lam, M.-H. T., Wu, M. S., Wang, X., Han, G. Q., and Li, R. Z.,** Characterization of a platelet-activating factor receptor antagonist isolated from haifenteng (*Piper futokadsura*): specific inhibition of *in vitro* and *in vivo* platelet-activating factor-induced effects, *Proc. Natl. Acad. Sci. U.S.A.,* 82, 672, 1985.

Chapter 15

BIOLOGICALLY ACTIVE TERPENOIDS AND AROMATIC COMPOUNDS FROM LIVERWORTS AND THE INEDIBLE MUSHROOM *CRYPTOPORUS VOLVATUS*

Yoshinori Asakawa

TABLE OF CONTENTS

0-8493-4372-0/93/$0.00+$.50

319

I. INTRODUCTION

The bryophytes are taxonomically placed between algae and pteridophytes, and there are about 24,000 species in the world. They are divided into three classes: Hepaticae (liverwort), Musci (moss), and Anthocerotae (hornwort). A number of bryophytes have been used as medicinal plants in North America, China, and Europe to treat burns, bruises, external wounds, etc. Some thalloid liverworts (for example, *Conocephalum conicum* and *Marchantia polymorpha*) contain diuretic, antimicrobial, antifungal, and anticancer active substances.[1,2]

It is known that some bryophytes produce intensely hot, bitter, and saccharine-like substances. Generally, bryophytes are not damaged by insect larvae, snails, slugs, and other small animals. Furthermore, some liverworts cause strong allergenic contact dermatitis and allelopathy. It has been established that the previously known biologic activity of the Hepaticae is associated with the oil bodies that are composed of terpenoids and lipophilic, aromatic compounds. We have been interested in the biologically active substances present in bryophytes and have studied the chemistry and pharmacology of about 750 species, as well as their application as a source of cosmetics and medicinal or agricultural drugs.

Little attention has been paid to the chemical constituents of inedible mushrooms in Japan, except for the poisonous species. Recently, we found that the fungus *Cryptoporus volvatus* (Polyporaceae) contained a large amount of drimane-type sesquiterpenoids that showed strong inhibition of superoxide anion radical release.[3-5] The separation, isolation, structure characterization, and biologic activity of several unique terpenoids and aromatic compounds from some liverworts and *Cryptoporus volvatus* have been summarized in this chapter.

II. MARCHANTINS, CYCLIC BIS(BIBENZYLS), AND THEIR RELATED COMPOUNDS FROM LIVERWORTS

Marchantia polymorpha, a thalloid liverwort belonging to the Marchantiales, is one of the common liverworts and grows on wet soil. It has been known that the extract of this species can cause allergenic contact dermatitis, shows inhibitory activity against Gram-positive bacteria, and has diuretic activity.[1,2]

A. ISOLATION

M. polymorpha, collected in Japan, was dried and ground mechanically. The powder (8 kg) was extracted with methanol for 1 month. After evaporation of the solvent, a small portion of the residue was checked on thin-layer chromatography (TLC). A large pink spot appeared after spraying the TLC plate with 30% sulfuric acid, heating at 100 to 120°C for a few minutes, and then being allowed to stand for 48 h. The remaining extract (100 g) was chromatographed on silica gel (n-hexane-EtOAc gradient), and each resultant fraction was rechromatographed on Sephadex LH-20 (CHCl$_3$-MeOH, 1:1) to give marchantins A (MA, **1**, 30 g), B (MB, **2**, 1 g), C (MC, **3**, 560 mg), D (MD, **4**, 100 mg), E (ME, **5**, 86 mg), F (MF, **6**, 5 mg), and G (MG, **7**, 5 mg). MA (**1**) and its related compounds (**2–7**, **21–23**) have also been isolated from the German, Indian, and South African *M. polymorpha*; from Japanese *M. paleacea* var. *diptera*, and *M. tosana*;[1,2] and from South American *M. plicata*.[6] The yield of MA is dependent upon the species. For example, 100 to 120 g of MA have been isolated from 2 kg of dried Japanese *M. paleacea* var. *diptera*.

B. STRUCTURE DETERMINATION
1. Marchantin A

The molecular formula of MA (**1**), C$_{28}$H$_{24}$O$_5$ ([M]$^+$ 440.1617) was established by high-resolution mass spectrometry. The UV and IR spectra indicated the presence of an aromatic ring [λ_{max} 217 nm (log ε, 4.91), 275 (4.10), 279 (4.07); υ_{max} 1620, 1605, 1585 cm^{-1}] and

1 $R^1 = R^2 = R^3 = OH, R^4 = R^5 = R^6 = H$

2 $R^1 = R^2 = R^3 = R^4 = OH, R^5 = R^6 = H$

3 $R^1 = R^3 = OH, R^2 = R^4 = R^5 = R^6 = H$

4 $R^1 = R^2 = R^3 = R^5 = OH, R^4 = R^6 = H$

5 $R^1 = R^2 = R^3 = OH, R^4 = R^6 = H, R^5 = OMe$

6 $R^1 = R^2 = R^3 = R^4 = R^5 = OH, R^6 = H$

a hydroxyl group (υ_{max} 3560 cm^{-1}). The ^{1}H NMR spectrum contained the signals of four benzylic methylenes (δ ppm 2.77, 4H, m; 3.00, 4H, m), three protons that disappeared on addition of D$_2$O, 12 protons (δ 6.41–7.15, m) on benzene rings, and one methine proton (δ 5.13, 1H, d, J = 1.8 Hz). Chemical modification of **1** (Scheme 1) afforded invaluable structural information. That MA possessed three phenolic hydroxyl groups was confirmed by methylation of **1** with methyl iodide, which gave a trimethyl ether (**8**, [M]$^+$ 482; δ 3.62, 3.84, 3.86 each s, 3H), and by acetylation of **1**, which gave a triacetate (**9**, [M]$^+$ 566; υ_{max} 1770 cm^{-1}; δ 1.63, 2.24, 2.28, each 3H, s). Treatment of **1** with methylene iodide in dimethylsulfoxide in the presence of cupric oxide gave a methylene dioxide (**10**, [M]$^+$ 452; δ 5.93, 2H, s), indicating the presence of the *ortho*-diol in the structure of MA. The remaining two oxygen atoms were ether functionalities, since neither carbonyl nor hydroxyl absorption bands were observed in the IR spectrum of **8**. On the basis of the above data, the structure of MA was suggested to be a cyclic bis(bibenzyl) in which two bibenzyls were linked by two ether oxygens. Hydrogenation of **8**, in the presence of platinum oxide, gave a hydrogenolized product (**11**), which was subsequently acetylated or methylated to afford the monoacetate (**12**) or the tetramethyl ether (**13**), respectively. Furthermore, **11** was methylated with trideuteriomethyl iodide to yield a mono-trideuteriomethyl ether (**14**). In the mass spectrum of each product (**11–14**), the base peaks were observed at *m/z* 167, 167, 181, and 184, respectively, together with an intense fragment ion at *m/z* 91. The base peak in the mass spectrum of **12** was due to a deacetyl fragment ion. The above data implied that the cleavage of one of the ether linkages occurred at C-1. Treatment of **13** with sodium in liquid ammonia furnished a monomethoxy bibenzyl (**15**, [M]$^+$ 212; δ 3.83, 3H, s) and a monohydroxy-trimethoxybibenzyl (**16**, [M]$^+$ 288; υ_{max} 3520 cm^{-1}; δ 3.76, 3.84, 3.85, each 3H, s). In the ^{1}H NMR spectrum of **15**, the lack of an A$_2$B$_2$ multiplicity pattern indicated that the methoxyl group was placed at C-13 or C-14. On the other hand, *meta*-coupled protons were observed at δ 6.20 and 6.40 in **16**, suggesting that C-1′, C-2′, and C-6′ were substituted. The proposed structures of the two bibenzyls were confirmed by synthesis, as shown in Scheme 2. Birch reduction of **8** gave two bibenzyl derivatives (**17** and **18**), the structures of which were established by their spectral data and by synthesis as shown in Scheme 3. These results showed that one of the ether oxygens was linked between C-1

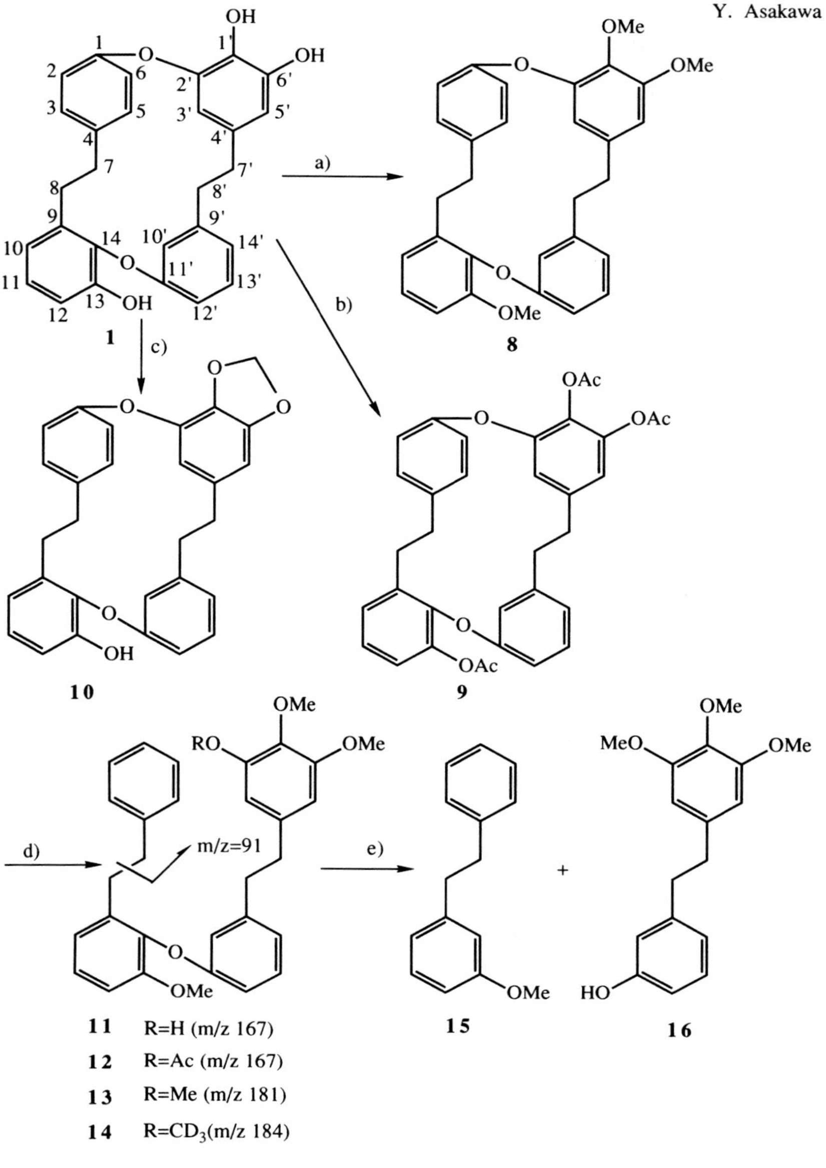

a) CH₃I/OH⁻   b) Ac₂O/Py   c) CH₂I₂/CuO/DMSO   d)H₂/PtO₂   e)liq.NH₃/Na

SCHEME 1. Chemical transformation of marchantin A (**1**).

and C-2′ and the other between C-11′ and the B-ring. That the second ether oxygen was linked between C-14 and C-11′ was established by spin decoupling of the ¹H NMR spectrum of MA (**1**) as well as by the ¹³C-¹H (Figure 1) and long range ¹³C-¹H 2D COSY NMR (Figure 2) spectra and NOE experiments (Figure 3) of **8**.[7,8]

Thus the substitution patterns of the four benzene rings of MA were established and the structure was therefore represented as **1**. Although MA and its derivatives are very viscous materials, MA-trimethyl ether (**8**) was obtained as a single crystal when the chloroform solution of **8** was passed through a column packed with a mixture of silica gel and dried magnesium sulfate (1:1). The result of an X-ray diffraction crystallographic analysis of **8** is

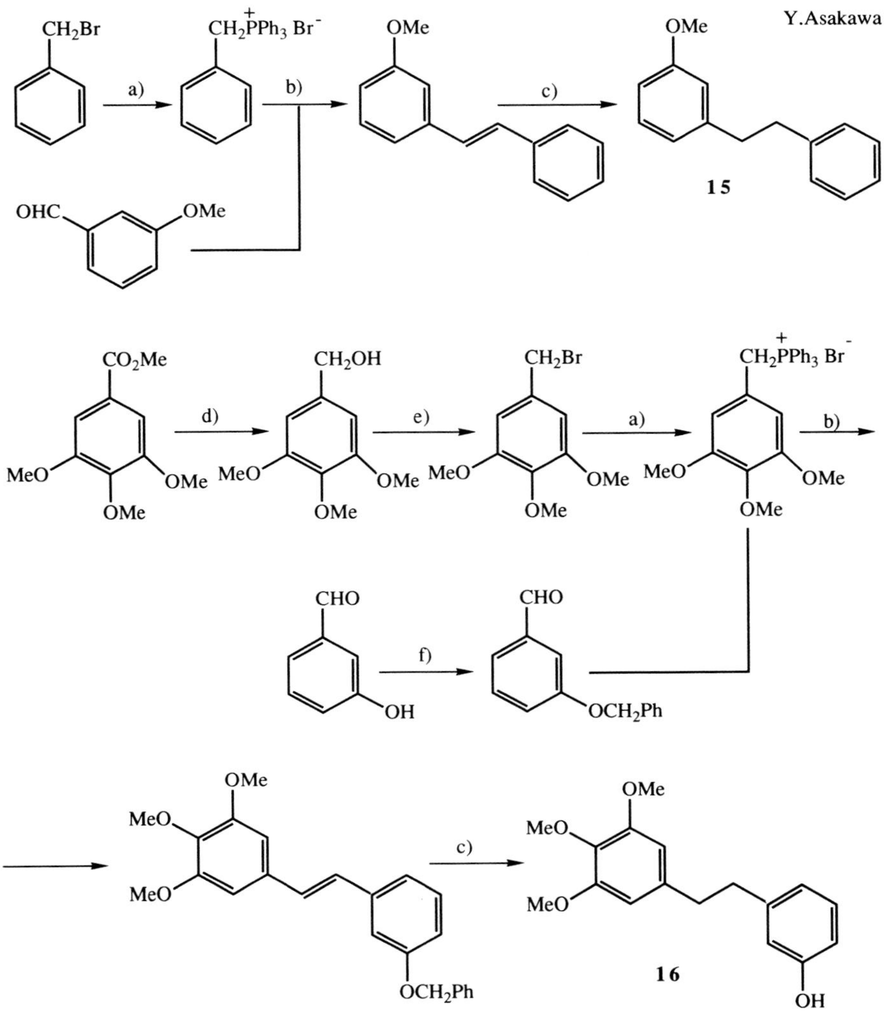

a) PPh$_3$/DMF b) MeONa/EtOH c) H$_2$/Pd-C/EtOH d) LiAlH$_4$/Et$_2$O e) HBr/C$_6$H$_6$

f) PhCH$_2$Br/Me$_2$CO/OH$^-$

SCHEME 2. Syntheses of bibenzyls **15** and **16**.

shown in Figure 4. In the ^{1}H NMR spectra of **1** and **8**, H-3′ appeared at remarkably high field (δ 5.13). This chemical shift is caused by the paramagnetic effect of the two benzene rings A and D that sandwich H-3′, as shown in Figure 4.

The total synthesis of MA (**1**) has been accomplished in 12 steps by Kodama et al.[9]

2. Other Marchantins and Related Compounds

The structures of marchantins B–G (**2–7**) were established in a manner similar to that described for marchantin A.[7] Marchantins H (**19**) and I (**20**) have been isolated from *Plagiochasma intermedium* (Marchantiales) and *Riccardia multifida* (Metzgeriales).[7]

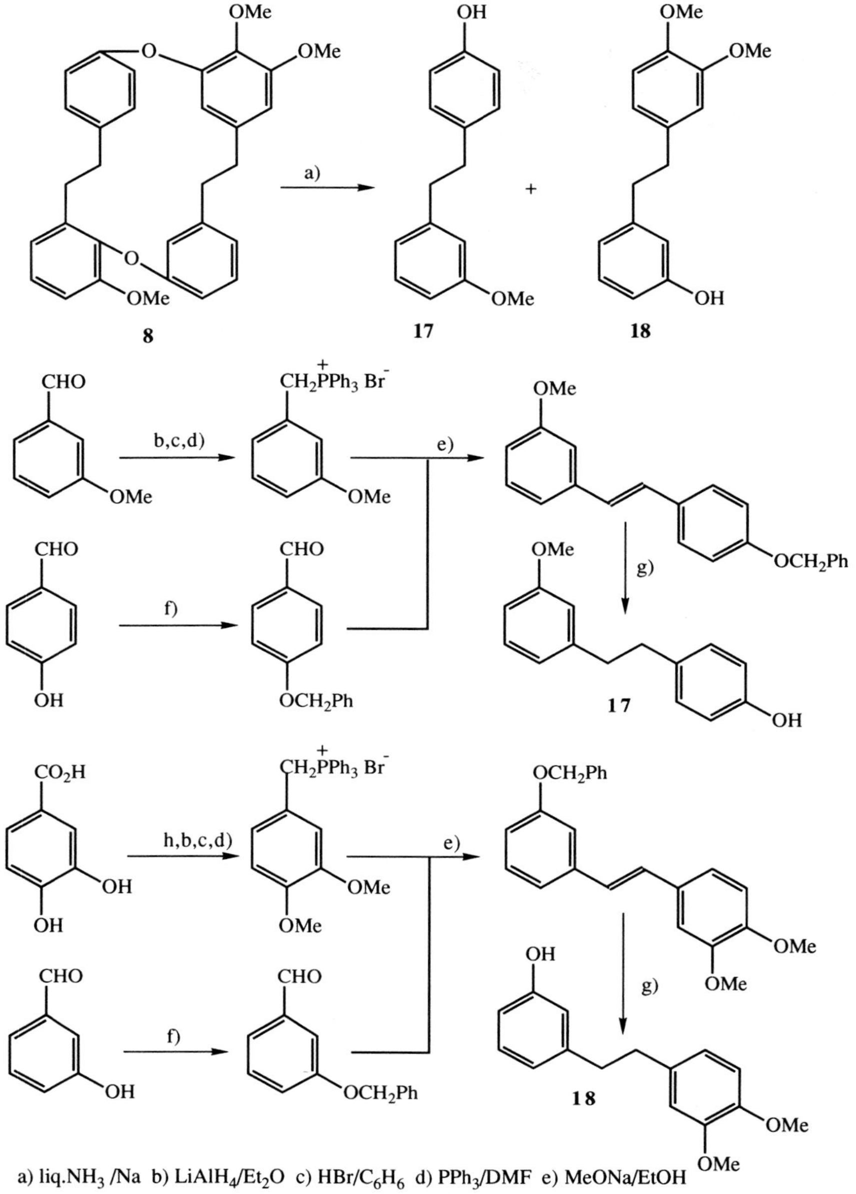

a) liq.NH$_3$ /Na b) LiAlH$_4$/Et$_2$O c) HBr/C$_6$H$_6$ d) PPh$_3$/DMF e) MeONa/EtOH
f) PhCH$_2$Br/Me$_2$CO/OH$^-$ g) H$_2$/Pd-C/EtOH h) MeI/Me$_2$CO/OH$^-$

SCHEME 3. Degradation of marchantin A trimethyl ether (**8**) and syntheses of bibenzyls **17** and **18**.

The similar macrocyclic bis(bibenzyls) (**21–23**) have been isolated from German *M. polymorpha*.[10] The riccardins (**24–29**) are closely related to the marchantins, having a biphenyl ether and biphenyl linkage rather than the bis(biphenyl ether) linkages. They have been found in *Marchantia, Reboulia* (Marchantiales), *Riccardia,* and *Monoclea* (Monocleales) species.[1,2] *R. multifida* produced not only riccardin-type cyclic bis(bibenzyls), but also a marchantin-like cyclic bis(bibenzyl) (**25**). The Indian *M. polymorpha* and *M. palmata* contained isomarchantin C (**30**) and isoriccardin C (**31**). *Radula* species (Jungermanniales) elaborate linear bis(bibenzyl) ethers (**32–34**), which

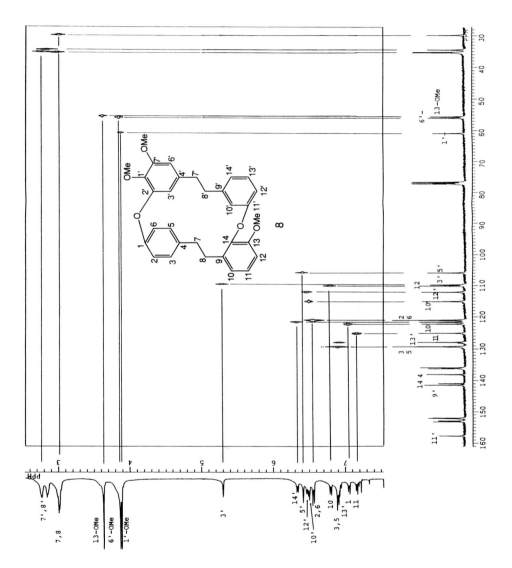

FIGURE 1. ^{13}C-^1H 2D COSY NMR spectrum of marchantin A trimethyl ether (**8**).

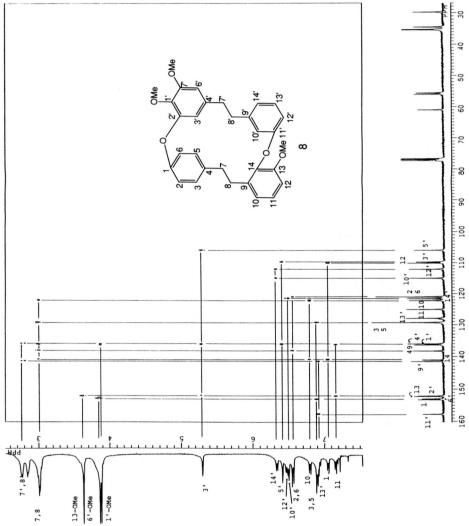

FIGURE 2. Long-range ^{13}C-^{1}H 2D COSY NMR spectrum of marchantin A trimethyl ether (**8**).

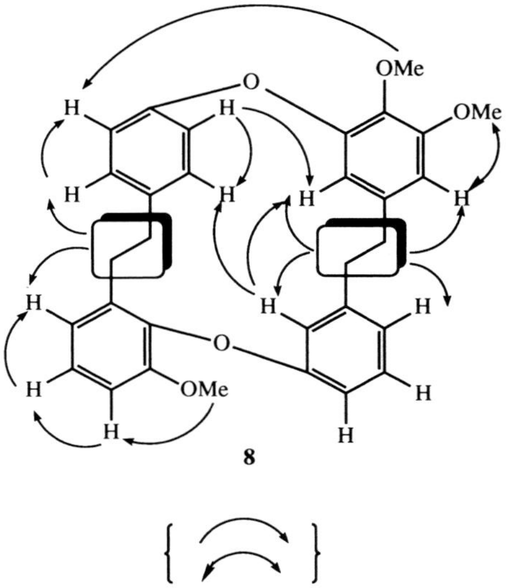

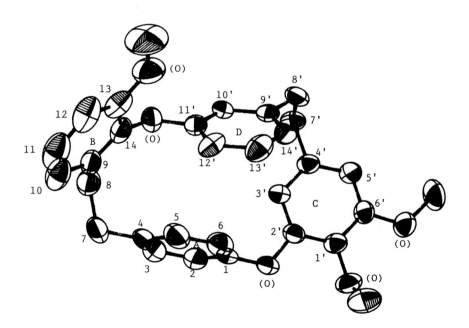

NOE

FIGURE 3. NOEs of marchantin A trimethyl ether (**8**).

FIGURE 4. Molecular structure of marchantin A trimethyl ether (**8**).

may be precursors of the macrocyclic bis(bibenzyls) found in the other liverworts. *Plagiochila sciophila* (*P. acanthophylla* subsp. *japonica*) (Jungermanniales) biosynthesizes plagiochins A–D (**35–38**), which are cyclic bis(bibenzyls) possessing a biphenyl and a biphenyl ether linkage.

Almost all *Radula* species so far examined produce not only linear bis(bibenzyls), but also prenylated bibenzyls (**39–47**).

19 $R^1 = R^3 = R^4 = OH$, $R^2 = R^5 = R^6 = H$

21 $R^1 = R^2 = R^3 = OH$, $R^4 = R^6 = H$, $R^5 = OEt$

22 $R^1 = R^2 = R^3 = R^4 = OH$, $R^5 = OMe$, $R^6 = H$

23 $R^1 = R^2 = R^3 = R^6 = OH$, $R^4 = R^5 = H$

C. BIOACTIVITY

Marchantin A (MA, **1**) shows various biological activities: cytotoxic (ED_{50} 8.39 µg/ml for KB cells); cardiotonic [increases coronary blood flow (2.5 ml/min at 0.1 mg)]; antimicrobial (Table 1); antifungal (Table 2); 5-lipoxygenase and calmodulin inhibitory activity (Table 3). MB (**2**), MC (**3**), riccardin A (**24**), riccardin B (**25**), and perrottetin E (**32**) also showed cytotoxic activity (ED_{50} 10.0, 10.0, 10.0, 10.0, and 12.5 mg/ml for KB cells). MD (**4**), ME (**5**), **24**, and **32** exhibited 5-lipoxygenase and calmodulin inhibitory activity (Table 3). The prenylated bibenzyls (**39**, **40**, **42**, **43**, **46**) also showed the 5-lipoxygenase and calmodulin inhibitory activity (Table 3).

Marchantins are structurally similar to bis(bibenzyl)isoquinoline alkaloids, such as d-tubocurarine (curare, **48**, Figure 5), which are pharmacologically important drugs that induce muscle relaxation. Surprisingly, MA (**1**) and its methyl ether (**8**) showed muscle relaxation activity.[11] The dose–response curves of **8** and d-tubocurarine (**48**) in rectus abdominis of frog (RAF) are shown in Figures 6 and 7. Nicotine in frog Ringer's solution shows the maximum contraction of RAF at a concentration of 10^{-6} M. After preincubation with **8** (at concentrations of 2×10^{-7} to 2×10^{-4} M) in Ringer's solution, nicotine (10^{-8} to 10^{-4} M) was added. At a concentration of 10^{-6} M, the contraction of RAF decreased about 30%, and each curve showed parallel shifts from left to right. This phenomenon has also been observed with d-tubocurarine (**48**). Using acetylcholine as the agonist, instead of nicotine, the same results were observed (Figure 7). MA (**1**) and its trimethyl ether (**8**) also showed muscle relaxation activity against mouse *in vivo*. It is thus noteworthy that some nonnitrogenous, cyclic bis(bibenzyls) isolated from liverworts cause a concentration-dependent decrease of contraction of RAF. The mechanism of action of MA (**1**) and its trimethyl ether (**8**) in the process of muscle relaxation is still unknown. However, it has been determined by MM2 calculations that the conformation of MA and **8** and the presence of an *ortho*-hydroxyl group in **1** and *ortho*-methoxy group in **8** are necessary for muscle relaxation activity.

III. MASTIGOPHORENES: ISOCUPARENE-TYPE SESQUITERPENE DIMERS FROM THE LIVERWORT *MASTIGOPHORA DICLADOS*

Mastigophora diclados is a rather primitive liverwort and is commonly found in tropical Asiatic areas.

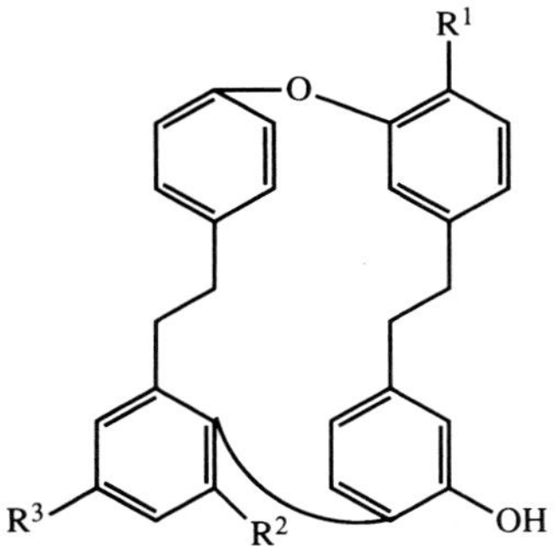

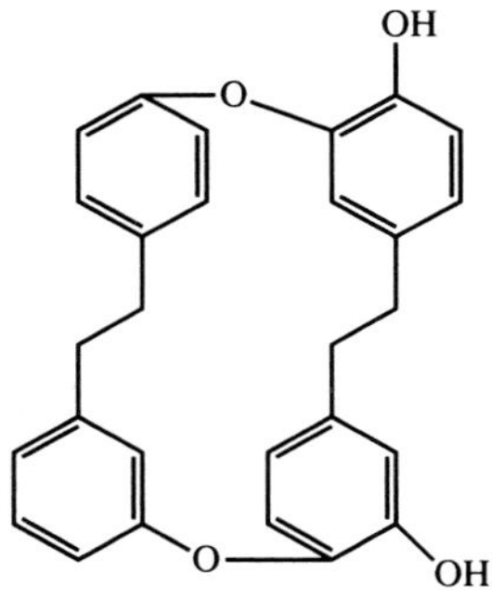

24 R^1 =OH, R^2 =H, R^3 =OMe
26 R^1 =R^3 =OH, R^2 =H
27 R^1 =OMe, R^2 =H, R^3 =OH
28 R^1 =R^2 =OH, R^3 =H
29 R^1 =OH, R^2 =OMe, R^3 =H

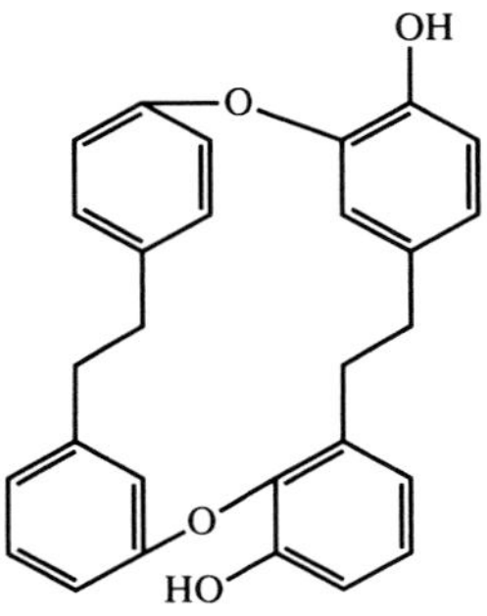

25

30

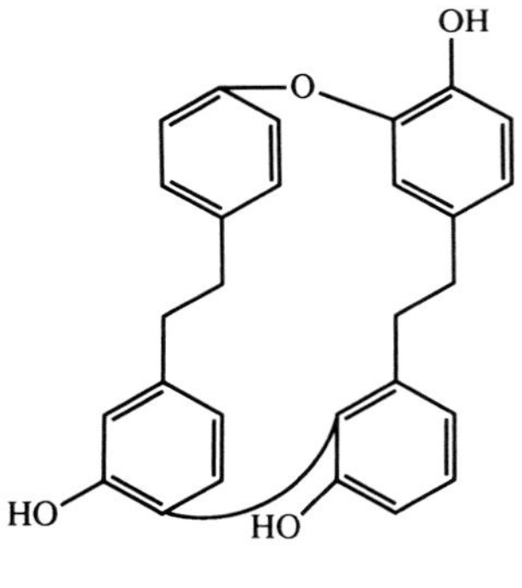

31

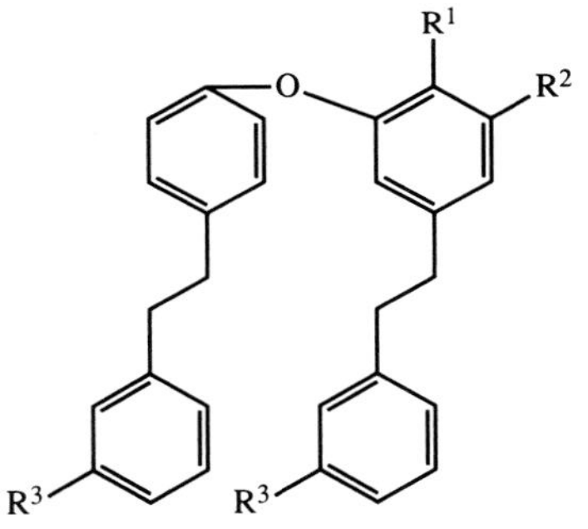

32 R^1 =R^3 =OH, R^2 =H
33 R^1=R^2 =R^3 =OH
34 R^1 =OMe, R^2 =R^3 =OH

A. ISOLATION

The dried, whole plant (220 g) was extracted with diethyl ether for 1 month. The crude extract (6.0 g) was first chromatographed on silica gel (n-hexane-EtOAc gradient), and then each fraction was further purified on Sephadex LH-20 (MeOH-CHCl$_3$, 7:3) to give mastigophorene A (**49**, 36 mg), mastigophorene B (**50**, 61 mg), (–)-herbertenediol (**53**, 300 mg), β-herbertenol (**54**, 1.5 g), 3,4-isocuparenediol (**51**, 20 mg), mastigophorene C (**55**, 30 mg), and mastigophorene D (**57**, 70 mg).

B. STRUCTURE DETERMINATION
1. Mastigophorenes A and B

High-resolution mass spectrometry showed that mastigophorenes A (**49**) and B (**50**) had the same molecular formula: C$_{30}$H$_{42}$O$_4$. IR and UV absorption spectroscopy indicated the

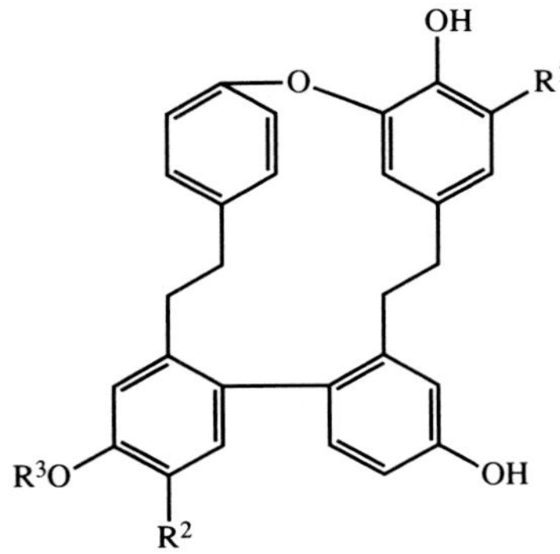

35 $R^1 = R^2 = OH, R^3 = Me$

36 $R^1 = OH, R^2 = R^3 = Me$

37 $R^1 = H, R^2 = OH, R^3 = Me$

38 $R^1 = R^2 = H, R^3 = Me$

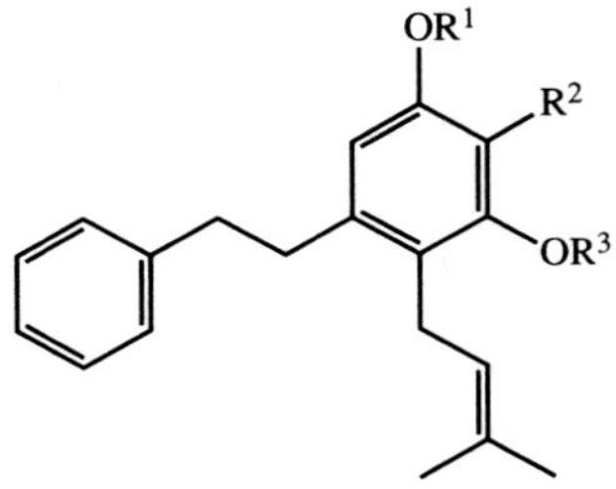

39 $R^1 = R^3 = H, R^2 = OH$

40 $R^1 = R^2 = R^3 = H$

41 $R^1 = Me, R^2 = R^3 = H$

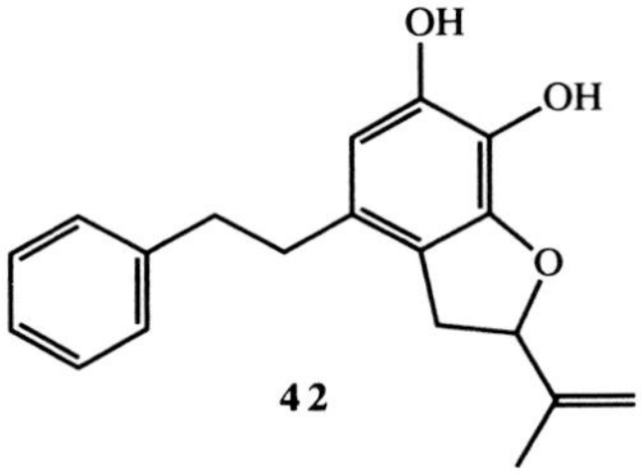

42

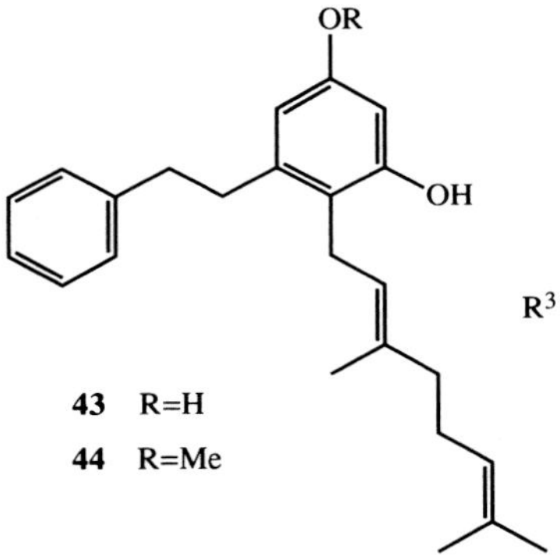

43 R=H

44 R=Me

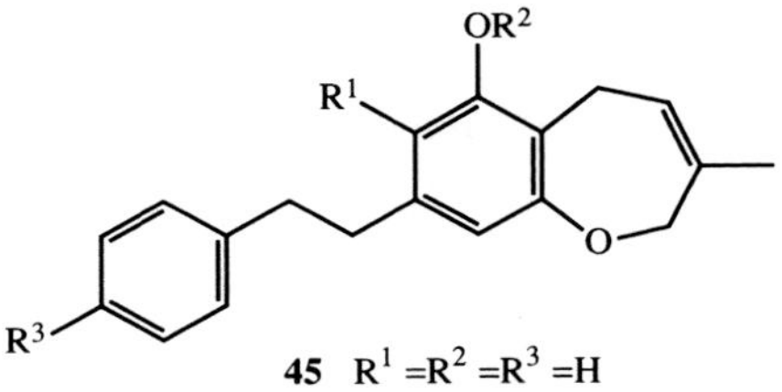

45 $R^1 = R^2 = R^3 = H$

46 $R^1 = CO_2H, R^2 = R^3 = H$

47 $R^1 = CO_2H, R^2 = H, R^3 = OH$

presence of a hydroxyl group (υ_{max} 3550 cm^{-1}) and a benzene ring [λ_{max} 213 nm (log ε 4.49), 387 (3.57)]. The ^{1}H and ^{13}C NMR spectra for mastigophorene A were almost identical to those for mastigophorene B. It was also noted that the NMR data were closely related to those of herbertenediol (**53**). The main differences were the absence of a pair of *meta*-coupled aromatic protons in **49** and **50** and replacement of the aromatic methine carbon signal at δ 113.4 in **53** with the aromatic quaternary carbon signal at δ 117.1 and 117.0 for **49** and **50**, respectively. These spectral data suggested that A and B were symmetrical dimers of herbertenediol (**53**), presumably linked through an aryl-aryl bond at the C-1 or C-3 position of **53**. The ^{13}C NMR signals for the quaternary carbons of the biphenyl bond were observed at relatively high field (δ 117 ppm), indicating that the dimerization of **53** occurs at C-3. In addition, these proposed structures for **49** and **50** were supported by an observed NOE between the sole aromatic proton [δ 6.86 (**49**) and 6.85 (**50**)] and the H-14 methyl signal, and by the long-range ^{13}C-^{1}H 2D COSY NMR experiments shown in Table 4. On the basis of the above spectral data, mastigophorenes A and B must be atropisomers at the biphenyl axis, leading to diastereoisomeric nonequivalence.

The absolute configurations at the aryl-aryl axis of **49** and **50** have been established by the circular dichroism (CD) exciton chirality rule. The CD spectrum of **49** showed the first

TABLE 1
Antimicrobial Activity of Marchantin A (1)

Microorganism	MIC (μg/ml)
Acinetobacter calcoaceticus	6.25
Alcaligenes faecalis	100
Bacillus cereus	12.5
B. megaterium	2.5
B. subtilis	2.5
Cryptococcus neoformans	12.5
Enterobacter cloacae	100
Escherichia coli	100
Proteus mirabilis	100
Pseudomonas aeruginosa	100
Salmonella typhimurium	100
Staphylococcus aureus	3.13–25

TABLE 2
Antifungal Activity of Marchantin A (1)

Fungi	MIC (μg/ml)
Alternaria kikuchiana	100
Aspergillus fumigatus	100
A. niger	25–100
Candida albicans	100
Microsporum gypseum	100
Penicillium chrysogenum	100
Piricularia oryzae	12.5
Rhizoctonia solani	50
Saccharomyces cerevisiae	100
Sporothrix schenckii	100
Trichophyton mentagrophytes	3.13
T. rubrum	100

TABLE 3
5-Lipoxygenase and Calmodulin Inhibitory Activity of Bis(Bibenzyls) and Prenylbibenzyls Isolated from Liverworts

Compound	Inhibition	
	5-Lipoxygenase 10^{-6} mol	Calmodulin ID_{50} μg/ml
Marchantin A (**1**)	94.4	1.85
Marchantin D (**4**)	40	6.0
Marchantin E (**5**)	36	7.0
Riccardin A (**24**)	4	20.0
Perrottetin A (**39**)	76	3.5
Prenylbibenzyl (**40**)	50	4.9
Perrottetin D (**42**)	40	2.0
Prenylbibenzyl (**43**)	11	4.0
Prenylbibenzyl (**45**)	×[a]	95.0
Radulanin H (**46**)	15	17.0
Prenylbibenzyl (**47**)	×	18.5

[a] × = not tested.

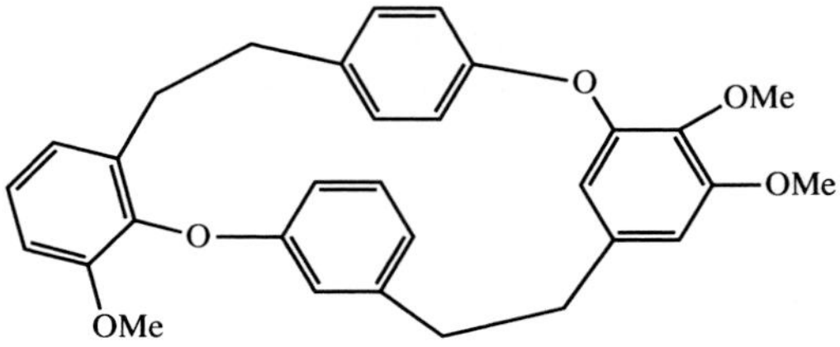

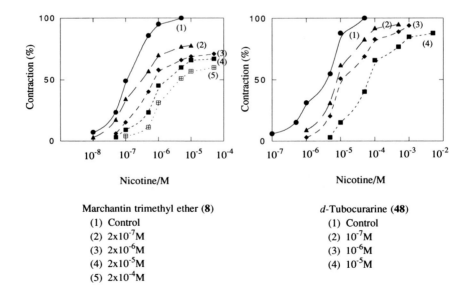

Marchantin trimethyl ether (**8**)

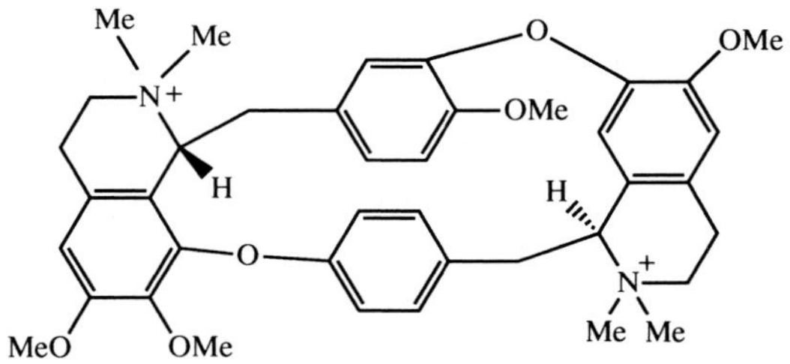

d-Tubocurarine (O,O',N-trimethyl-) (**48**)

FIGURE 5. Structural similarity between marchantin A trimethyl ether (**8**) and *d*-tubocurarine (**48**).

Marchantin trimethyl ether (**8**)
  (1) Control
  (2) 2x10⁻⁷M
  (3) 2x10⁻⁶M
  (4) 2x10⁻⁵M
  (5) 2x10⁻⁴M

d-Tubocurarine (**48**)
  (1) Control
  (2) 10⁻⁷M
  (3) 10⁻⁶M
  (4) 10⁻⁵M

FIGURE 6. Dose–response curves in rectus abdominis of frog for marchantin A trimethyl ether (**8**) and *d*-tubocurarine (**48**) in nicotine.

positive Cotton effect at 222 nm and the second negative Cotton effect at 202 nm, whereas **50** showed the first negative and second positive Cotton effects at 215 and 202 nm, respectively, indicating that **49** has an (S)-configuration and **50** has an (R)-configuration at the biaryl axis.

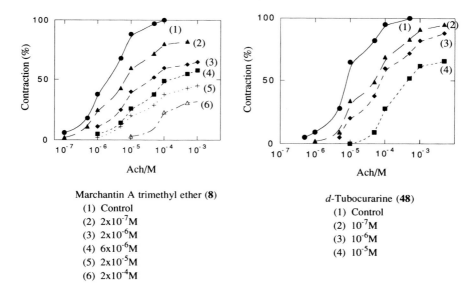

Marchantin A trimethyl ether (**8**)
(1) Control
(2) 2×10^{-7}M
(3) 2×10^{-6}M
(4) 6×10^{-6}M
(5) 2×10^{-5}M
(6) 2×10^{-4}M

d-Tubocurarine (**48**)
(1) Control
(2) 10^{-7}M
(3) 10^{-6}M
(4) 10^{-5}M

FIGURE 7. Dose–response curves in rectus abdominis of frog for marchantin A trimethyl ether (**8**) and *d*-tubocurarine (**48**) in acetyl choline.

Accordingly, the structures of mastigophorenes A (**49**) and B (**50**) were assigned as (S)-3,3′-biherbertenediol and (R)-3,3′-biherbertenediol, respectively.

2. Mastigophorenes C and D

The structures of these two additional dimers (**55** and **57**) were established by a combination of chemical modification (acetylation) and extensive NMR spectral examination.

C. BIOSYNTHESIS AND BIOACTIVITY

Mastigophorenes might be biosynthesized via phenoxy radicals produced by one electron oxidation of (–)-herbertenediol (**53**), which is a cometabolite in *M. diclados*. The essential features of this proposed biosynthetic pathway are shown in Scheme 4.

Mastigophorenes A, B, and D exhibited neurotrophic properties at 10^{-5} to 10^{-7} *M*, causing greatly accelerated neuritic sprouting and network formation in the primary neuritic cell culture derived from the fetal rat hemisphere (Table 5).[12] Mastigophorene C and the monomeric isocuparenes, **51**, **53**, and **54**, on the other hand, suppressed neuritic differentiation.

IV. CRYPTOPORIC ACIDS: DRIMANE-TYPE SESQUITERPENES AND THEIR DIMERS FROM A FUNGUS *CRYPTOPORUS VOLVATUS*

The fungus *Cryptoporus volvatus* grows on decayed pine trees, and its fruit body emits a resinous odor that attracts adult insects such as *Parabolitophagus felix* and *Ischnodactylus loripes*. In China this fruit body has been used in the ablactation (weaning) of infants, because it contains powerfully bitter substances. Previously, ergosterol,[13] and an antitumor-active protein-polysaccharide,[14] and volatile components including 1-(E,Z)-3,5- and 1-(E,E)-3,5-undecatrienes and a few mono- and sesquiterpenoids,[15] have been found in the fruit body of *C. volvatus*. We were also aware that *C. volvatus* contained a large amount of bitter sesquiterpenoids, and subsequently, we isolated and identified cryptoporic acids (CAs) A–

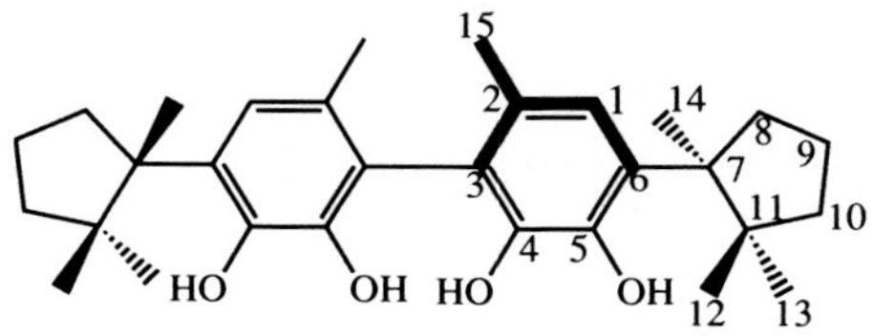

49

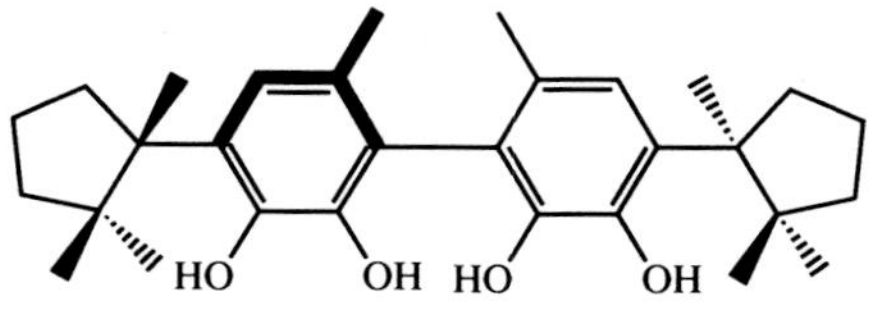

50

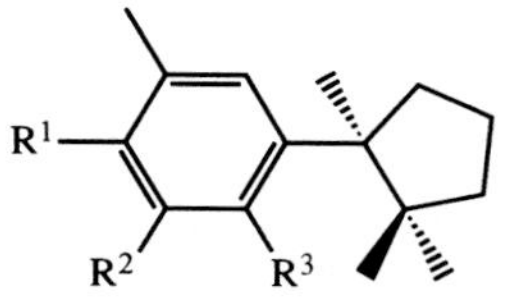

5 1	$R^1=R^2=OH, R^3=H$
5 2	$R^1=R^2=OAc, R^3=H$
5 3	$R^1=H, R^2=R^3=OH$
5 4	$R^1=OH, R^2=R^3=H$

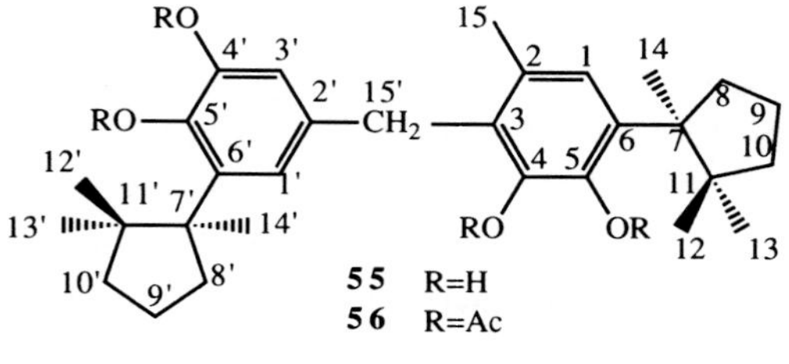

5 5 R=H
5 6 R=Ac

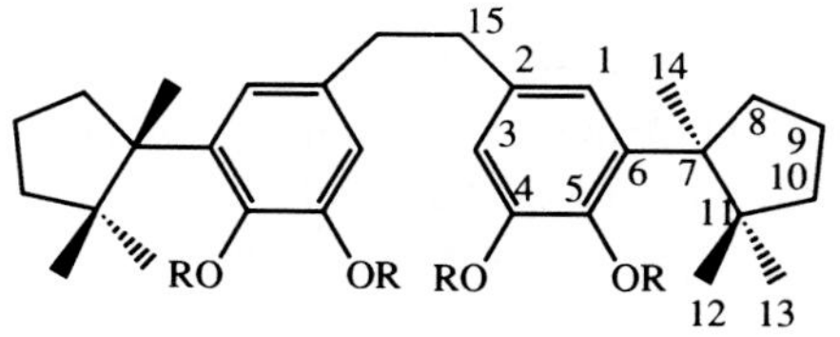

5 7 R=H
5 8 R=Ac

G (**59, 71, 77, 80, 84, 88, 90**). Recently, cryptoporic acid H (**92**) was isolated from the cultured broth of *C. volvatus*, and cryptoporic acid I (**93**), in conjunction with CA-H, has been found in *Ganoderma* neo-*japonicum* culture.[16] None of the CAs described in this chapter have been detected in the cultured broth of either fungi.[16]

A. ISOLATION

C. volvatus (1.16 kg) was macerated and extracted with ethyl acetate (3 l) overnight. After filtration and evaporation of the solvent, a viscous oil (75.95 g) was obtained. The residue was further extracted with ethyl acetate (3 l) for 4 days, which yielded a further 21.86 g of the oil. The two extracts were combined, and a part (75.60 g) of this oil was chromatographed on silica gel (CHCl₃-EtOH gradient). Each fraction was then rechromatographed on silica

TABLE 4

^{13}C-^{1}H Correlation in the Long-Range ^{13}C-^{1}H COSY for 49, 50, 55, and 58

C	Correlated H in 49 and 50	Correlated H in 55	Correlated H in 58
1	H–15	H–15	H–15, 3
1′		H–15′, 3′	
2	H–15	H–15, 15′	H–15
2′		H–15′	
3	H–1, 15	H–1, 15, 15′	H–1, 15
3′		H–1′, 15′	
4	H–1	H–15′	
5	H–1	H–1	H–1, 3
5′		H–1′, 3′	
6	H–14	H–14	H–14
6′		H–14′	
7	H–1, 12, 13, 14	H–1, 12′, 13′	H–1, 12, 13
7′		H–1′, 12′, 13′	
8	H–14	H–14	H–14
8′		H–14′	
10	H–12, 13	H–12, 13	H–12, 13
10′		H–12′, 13′	
11	H-12, 13, 14	H–12, 13, 14	H-12, 13, 14
11′		H–12′, 13′, 14′	
12	H–13	H–13	H–13
12′		H–13′	
13	H–12	H–12	H–12
13′		H–12′	
15	H–1	H–1	H–1, 3
15′		H–1′, 3′	

gel (CHCl$_3$-Me$_2$CO-AcOH, 1:1:0.l) to give ergosterol (2.351 g), cryptoporic acid A (CA-A, **59**, 3.291 g), CA-B (**71**, 1.772 g), CA-C (**77**, 5.063 g), CA-D (**80**, 5.692 g), CA-E (**84**, 5.95 g), CA-F (**88**, 875 mg), and CA-G (**90**, 2.018 g). An alternative isolation method involved partitioning the ethyl acetate extract between water and ethyl acetate. The organic layer was subsequently washed with saturated sodium bicarbonate, followed by neutralization with 1 N-hydrochloric acid, and then chromatographed on Sephadex LH-20 (CHCl$_3$-MeOH, 1:1) to furnish CA-A to CA-G. By this method, 700 g of CA-E (**84**) was rapidly obtained from 40 kg of the fresh fruit body of *C. volvatus*.

B. STRUCTURE DETERMINATION

1. Cryptoporic Acid A

The NMR and IR spectra of cryptoporic acid A (CA-A, **59**), C$_{23}$H$_{36}$O$_7$ ([M]$^+$ 424.2458), showed the presence of a carboxylic acid group (υ_{max} 3450–2500 and 1715 cm^{-1}; δ_C 170.9 s) and two carbomethoxyl groups (υ_{max} 1735, 1740 cm^{-1}; δ_C 170.9, 171.3, each s; δ_H 3.68, 3.75 each s). The ^{1}H and ^{13}C NMR spectra implied three tertiary methyls, an exocyclic methylene, a methylene, and a methine carbon, each bearing an ether oxygen, six methylenes, three methines, and two quaternary carbons. The double quantum filter (DQF) ^{1}H-^{1}H 2D COSY NMR spectrum (Table 6) indicated partial structures A and B (Figure 8). The ^{1}H and ^{13}C NMR spectral patterns of CA-A were similar, in part, to those of a drimane-type sesquiterpene primary alcohol, albicanol (**61**) isolated from the liverworts *Diplophyllum albicans*[17] and *Bazzania* species[18]. This indicated that **59** might be composed of albicanol linked through the primary alcohol to the dimethyl ester of isocitric acid (partial structure B). The positions of the ether linkage between C-11 and C-1′ and the carboxylic acid group at C-3′ of isocitric acid dimethyl ester were confirmed by the analysis of the ^{13}C-^{1}H and long-range ^{13}C-^{1}H 2D

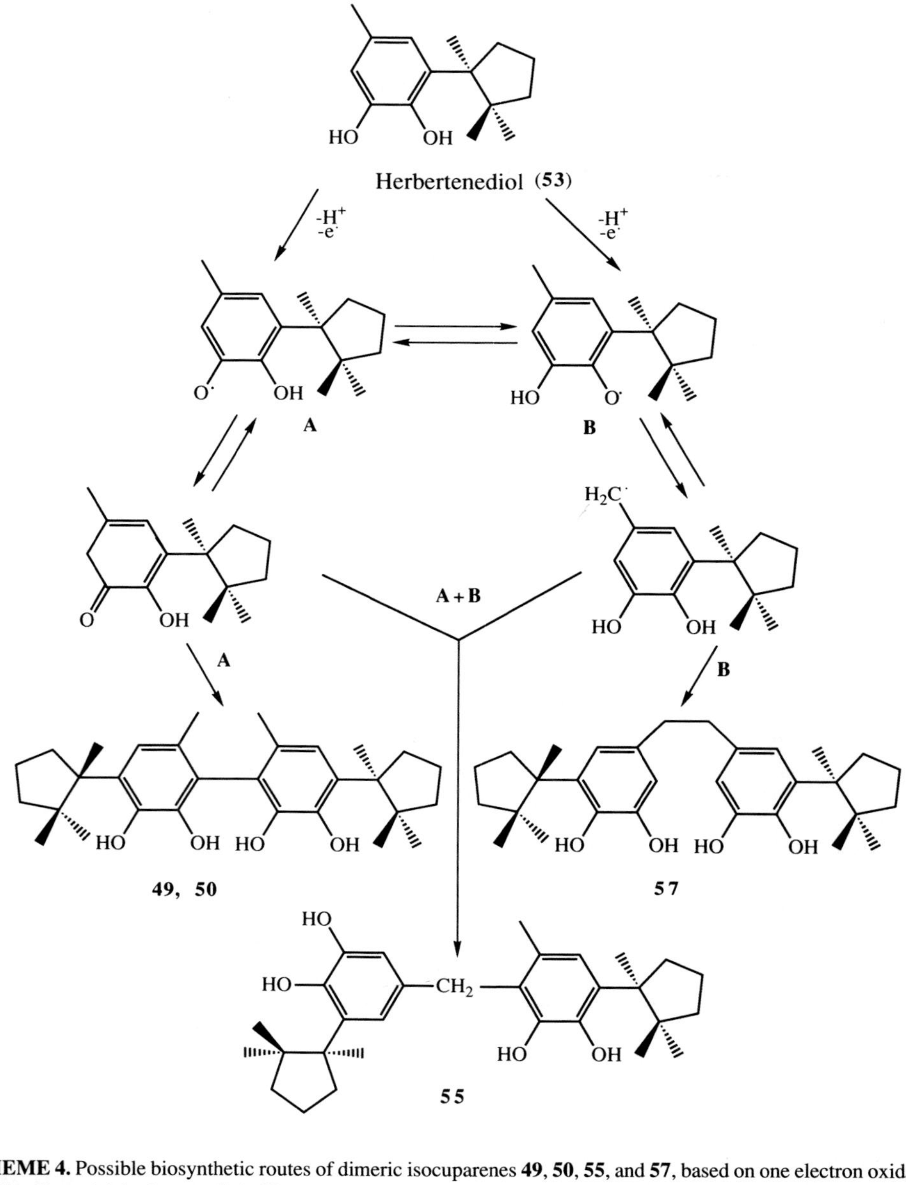

SCHEME 4. Possible biosynthetic routes of dimeric isocuparenes **49**, **50**, **55**, and **57**, based on one electron oxidative coupling from (–)-herbertenediol (**53**).

COSY NMR spectra of **59** (Tables 7 and 8). The relative stereochemistry of the sesquiterpene part of **59** was established to be the same as that of **61** by observation of NOEs between (1) H-13 and H-14, (2) H-11 and H-13, (3) H-5 and H-15, and (4) H-5 and H-9.

The absolute configuration of **59** was also determined by reference to that of **61**. Thus, the absolute stereochemistry of the ketone **62**, derived from ozonolysis of **61**, was established by observation of a negative Cotton effect at 307 nm in the CD spectrum.[17] An analogous ketone, **63**, was derived from the ozonolysis of the trimethyl ester (**60**), $C_{24}H_{38}O_7$ ([M]$^+$ 438.2631), of **59**. The CD spectrum of **63** also showed the negative Cotton effect at 274 nm, indicating that the sesquiterpene part possessed the same absolute configuration as that of **62**.

TABLE 5
Effect of Mastigophorenes (49, 50, 55, and 57) on Cell Morphology
in Primary Cell Culture of Fetal Rat Cerebral Hemisphere

Compounds 0.5% EtOH	Conc. (M)	4 day	7 day	10 day
Mastigophorene A (**49**)	10^{-5}	NS (+)[a]	NS (+)	NS (+)
	10^{-6}	NS (+)	NS (+)	NS (+)
	10^{-7}	NS (+)	NS (+)	NS (+)
Mastigophorene B (**50**)	10^{-5}	Dec[b]		
	10^{-6}	NS (+)	NS (+)	NS (±)
	10^{-7}			
Mastigophorene C (**55**)	10^{-5}	Dec		
	10^{-6}	NS (+)		
	10^{-7}	NS (+)		
Mastigophorene D (**57**)	10^{-5}	Dec		
	10^{-6}	NS (+)	NS (+)	NS (+)
	10^{-7}	NS (+)	NS (+)	NS (±)

[a] NS = neuritic sprouting.
[b] Dec = death of neuron.

TABLE 6
¹H–¹H 2D COSY NMR Spectral Data for CA–A (59)

H	Correlated H
5	H–6
6	H–5, H–7a, H–7b
7a	H–6, H–7b, H–12a, H–12b
7b	H–6, H–7a, H–12a, H–12b
9	H–11a, H–11b, H–12a, H–12b
11a	H–9, H–11b
11b	H–9, H–11a
12a	H–7a, H–7b, H–9, H–12b
12b	H–7a, H–7b, H–9, H–12a
1′	H–2′
2′	H–1′, H–3′a, H–3′b
3′a	H–2′, H–3′b
3′b	H–2′, H–3′a

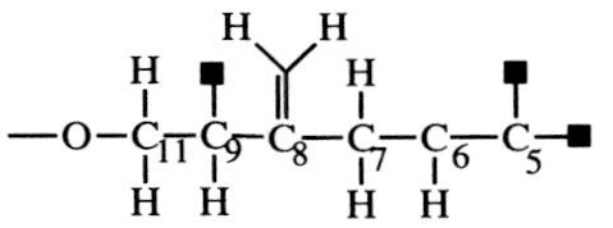

A

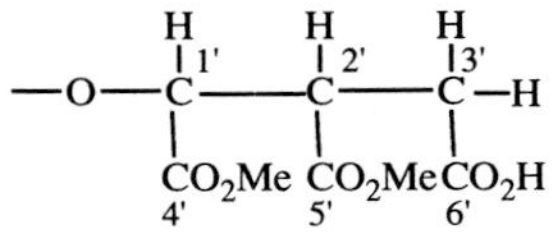

B

FIGURE 8. Partial structures (A and B) of cryptoporic acid A (**59**).

The absolute configuration of the isocitric acid moiety of CA-A (**59**) was established by chemical transformation and comparison of the optical rotations of the products with those of known compounds. Thus, Kaneko et al.[19] had previously reported that the (+)-tetraacetate (**66**) was obtained from (–)-isocitric acid (**64**) via reduction of (+)-isocitric acid lactone (**65**) by lithium aluminium hydride (LiAlH₄) and subsequent acetylation (Scheme 5). In a similar manner, (+)-alloisocitric acid lactone (**68**), derived from (+)-alloisocitric acid (**67**), was converted to the (+)-tetraacetate (**69**). To produce such a tetraacetate from **59**, CA-A trimethyl ester (**60**) was reduced by LiAlH₄ and acetylated to give a triacetate (**70**), $C_{27}H_{44}O_7$ ([M]⁺ 480.3087; δ 2.04, 2.06, 2.08, each 3H, s). Treatment of **70** with boron trifluoride etherate in Ac₂O afforded the (–)-tetraacetate (**66′**), the spectral data of which were identical to those of the (±)-tetraacetate (**66″**) prepared from the commercially available (±)-isocitric acid lactone (**65″**). Thus, the absolute configuration of the isocitric acid moiety of CA-A was established as 1′(R) and 2′(S), and therefore, the absolute stereochemistry of CA-A is represented as **59**.

2. Cryptoporic Acid B

The ¹H and ¹³C NMR spectral data of CA-B (**71**), $C_{23}H_{36}O_8$ ([M]⁺ 440.2405) resembled those of CA-A (**59**), except that one of the three tertiary methyl groups of **59** (δ$_H$ 0.87, 3H,

TABLE 7
^{1}H–^{1}H 2D COSY NMR Spectral Data for CA–A (59)

C	Correlated H
1	H–1
2	H–2
3	H–3
5	H–5
6	H–6
7	H–7a, H–7b
9	H–9
11	H–11a, H–11b
12	H–12a, H–12b
13	H–13
14	H–14
15	H–15
1′	H–1′
2′	H–2′
3′	H–3′a, H–3′b

TABLE 8
C–H Correlation in the ^{13}C-^{1}H Long-Range 2D COSY NMR Spectrum of CA-A (59)

C	Correlated H
1	H–2, H–3, H–13
2	H–1, H–3
3	H–1, H–2, H–14, H–15
4	H–5, H–6, H–14, H–15
5	H–3, H–14, H–15
6	H–5, H–7
7	H–9, H–12
8	H–6, H–7, H–9, H–11
9	H–5, H–7, H–12
10	H–1, H–2, H–5, H–9, H–13
11	H–9, H–1′
13	H–1, H–5, H–9
14	H–3, H–5, H–15
15	H–3, H–5, H–14
1′	H–11, H–2′, H–3′
2′	H–1′, H–3′
3′	H–1′
4′	H–1′, H–2′, –OMe
5′	H–1′, H–2′, H–3′, –OMe
6′	H–3′

s; δ_C 33.7, q) was replaced by a hydroxymethyl group (δ_H 3.04, 3.39 each 1H, d, J = 11.0 Hz; δ_C 71.3, t). The ^{1}H–^{1}H 2D COSY NMR spectrum of CA-B confirmed the presence of the same partial structures (A and B) as seen in CA-A (**59**). The carboxylic acid group was confirmed to be at C-3′, by analysis of the ^{13}C-^{1}H and long-range ^{13}C-^{1}H 2D COSY NMR spectra of CA-B. The presence of the hydroxyl substituent was further substantiated by chemical evidence. Thus, treatment of CA-B with CH_2N_2 gave a trimethyl ester (**72**), $C_{24}H_{38}O_8$ ([M]$^+$ 454.2551; υ_{max} 3550cm^{-1}; δ 3.68, 6H, s; 3.75, 3H, s), which was acetylated to afford a monoacetate (**73**, δ 2.08, 3H, s). Reduction of **72** with LiAlH$_4$, followed by acetylation, gave a tetraacetate (**74**), $C_{29}H_{46}O_9$ ([M]$^+$ 538.3139). The position of the hydroxyl group at C-15 was established by observation of an NOE between H-5 and H-15. This was supported

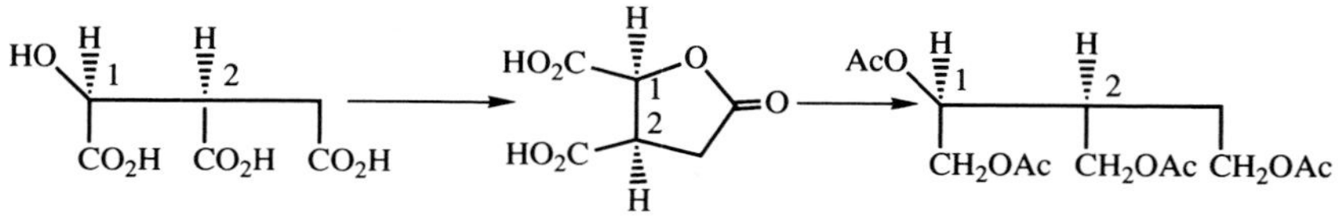

64 (1*S*,2*R*) **65** (1*S*,2*R*) **66** (1*S*,2*R*)

66' (1*R*,2*S*)

65'' (1*S*,2*R*+1*R*,2*S*) **66''** (1*S*,2*R*+1*R*,2*S*)

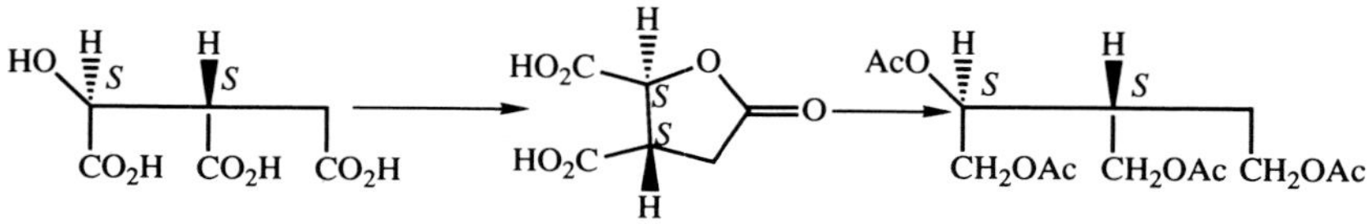

67 **68** **69**

SCHEME 5. Chemical transformation of (–)-isocitric acid (**64**) and (+)-alloisocitric acid (**67**).

by the fact that the carbon signals for C-3, C-4, and C-14, in CA-B, appeared at higher field than those of **59**, due to the expected γ shielding effect.

The absolute stereochemistry of CA-B was established in a similar manner to CA-A. Thus, ozonolysis of **73** gave a monoketone (**75**), $C_{25}H_{38}O_{10}$ ($[M]^+$ 498.2468), which demonstrated a negative Cotton effect at 276 nm in the CD spectrum, showing that the absolute configuration of the sesquiterpene part was identical to that of **59**. The absolute configuration of the isocitric acid entity was established by chemical correlation with CA-A (**59**). Thus, CA-B trimethyl ester (**72**) was treated with *p*-toluenesulfonyl chloride in pyridine to give a monotosylate (**76**, δ 2.46, 3H, s; 7.36, 7.78, each 2H, d, J = 8.3 Hz). This was then reduced by sodium iodide-zinc in hexamethylphosphoramide (HMPA) to afford a detosylated product, the spectral data and optical rotation of which were identical to those of the trimethyl ester (**60**) derived from CA-A. Therefore, the structure of CA-B was elucidated as 15-hydroxycryptoporic acid A (**71**).

3. Cryptoporic Acid C

The NMR data for cryptoporic acid C (CA-C, **77**), $C_{45}H_{68}O_{14}$ (elemental analysis), indicated five tertiary methyls, three methoxyls, two exocyclic methylenes, two methylenes bearing an ether oxygen, four ester carbonyls, and two carboxylic acid groups. The ^{1}H-^{1}H and ^{13}C-^{1}H 2D COSY NMR spectra indicated that CA-C was a conjugate of CA-A and CA-B, esterified between the hydroxyl group at C-15 of **71** and one of the carboxylic acid groups in the isocitric acid moiety of CA-A. This deduction was supported by chemical modifications. Methylation of CA-C with CH_2N_2 or MeI gave a pentamethyl ester (**78**, $[M]^+$ 860; δ 3.67, 3.68, 3.69, 3.75, and 3.76, each 3H, s). Subsequent reduction by $LiAlH_4$ and then acetylation yielded two acetates (**70** and **74**) that were identical to those prepared from CA-A (**59**) and CA-B (**71**), respectively. The position of two carboxylic acids at C-3′ and C-3‴ and of dimerization at C-5′ with the primary alcohol of CA-B were further established by the ^{13}C-^{1}H and long range ^{13}C-^{1}H 2D COSY NMR spectra of **77**. Observation of a negative Cotton effect (275 nm) in the CD spectrum of diketone **79**, prepared from CA-C by ozonolysis, and correlation with CA-A, established the absolute stereochemistry of CA-C, which is therefore represented as the dimeric drimane-type sesquiterpenoid **77**.

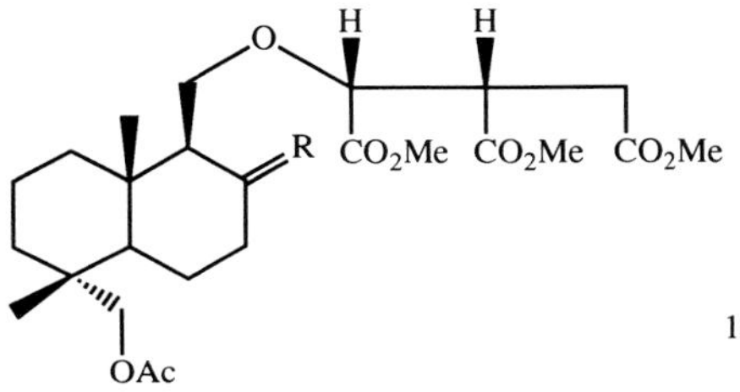

73 R =CH$_2$

75 R =O

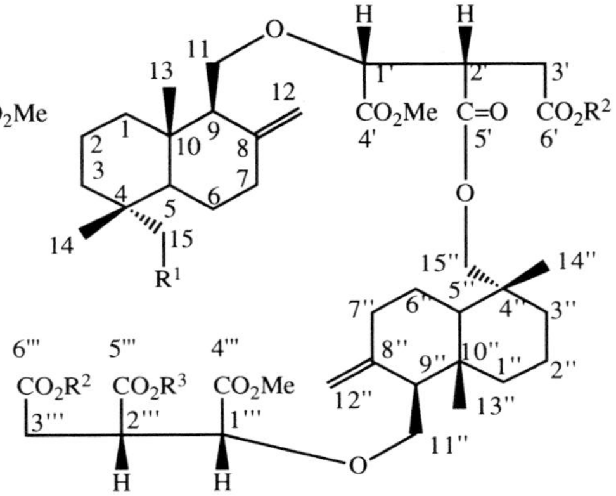

77 R^1= R^2 =H, R^3 =Me

78 R^1 =H, R^2 =R^3 =Me

84 R^1 =OH, R^2 =H, R^3 =Me

85 R^1 =OH, R^2 =R^3 =Me

86 R^1 =OAc, R^2 =R^3 =Me

88 R^1 =R^2 =R^3 =H

89 R^1 =H, R^2 =R^3 = -CH$_2$COC$_6$H$_4$Br(p)

90 R^1 =OH, R ^{2}=R^3 =H

91 R^1 =H, R^2 =R^3 = -CH$_2$COC$_6$H$_4$Br(p)

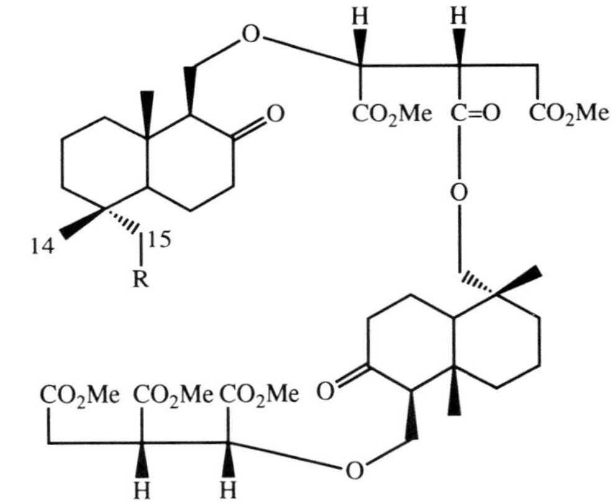

79 R=H

87 R=OAc

4. Cryptoporic Acid D

The molecular formula for CA-D (**80**) was established as C$_{44}$H$_{64}$O$_{14}$ by elemental analysis and the observation of a molecular ion peak at *m/z* 844 in the mass spectrum of its methyl ester (**81**). The ^{1}H NMR spectrum of CA-D showed the presence of two tertiary methyls, one methoxyl group, an exocyclic methylene, and a methylene bearing an ether oxygen. IR

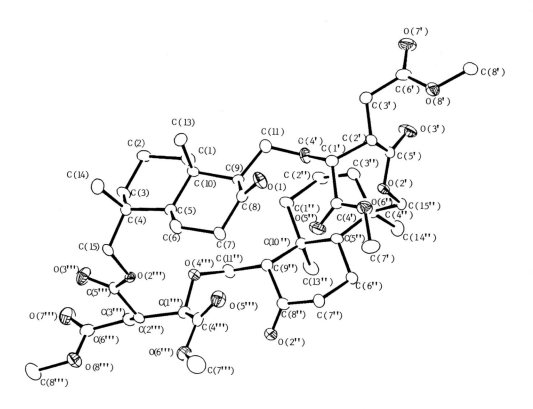

FIGURE 9. Molecular structure of diketone (**83**) prepared from cryptoporic acid D (**80**).

absorption spectroscopy indicated the presence of a carboxylic acid (v_{max} 3450–2500, 1725 cm^{-1}) and an ester group (1740 cm^{-1}). In the ^{13}C NMR spectrum, 22 carbon signals were observed, including two ester carbonyls and a carboxylic acid carbonyl. The ^{13}C and the ^{1}H-^{1}H 2D COSY NMR spectra of CA-D were quite similar to those of CA-B (**71**), suggesting that it might be a symmetrical dimer of **71**, esterified at C-5′ and C-15″, and C-5‴ and C-15. This was supported by reduction of the CA-D tetramethyl ester (**81**) with LiAlH$_4$, followed by acetylation to give a tetraacetate (**74**). The optical rotation and spectral data of **74** were identical to those of the tetraacetate prepared from CA-B (**71**). Further support was obtained by the formation of the di-*p*-bromophenacyl ester (**82**) from **80**, by reaction with *p*-bromophenacyl bromide. The location of the carboxylic acids and the ester linkages were confirmed to be C-3′ (C-3‴) and C-5′ (C-5‴) by analysis of the long-range ^{13}C-^{1}H 2D COSY NMR spectra of **80** and **82**. The suggested structure was confirmed by an X-ray crystallographic analysis (Figure 9) of the diketone (**83**), C$_{44}$H$_{64}$O$_{16}$, derived from CA-D by ozonolysis of its methyl ester (**81**).

The absolute configuration of CA-D was established to be the same as CA-A, by observation of a negative Cotton effect at 276 nm in the CD spectrum of the diketone **83** and by structural correlation to CA-A through CA-B.

5. Cryptoporic Acid E

The molecular formula, C$_{45}$H$_{68}$O$_{15}$, of cryptoporic acid E (CA-E, **84**) was established by a combination of positive ([M + Na]$^+$ 871) and negative ([M–1]$^-$ 847) ion FAB mass spectrometry and elemental analysis. The IR spectrum showed a broad band at 3400–2450 cm^{-1} and an intense band at 1715 cm^{-1}, consistent with a carboxylic acid group, and an absorption at 1740 cm^{-1} attributable to an ester carbonyl group. The ^{1}H and ^{13}C NMR spectra

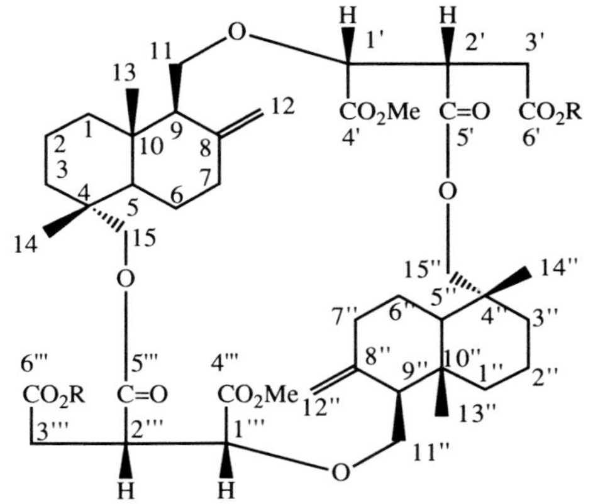

80   R=H

81   R=Me

82   R=-CH$_2$OCC$_6$H$_4$Br (p)

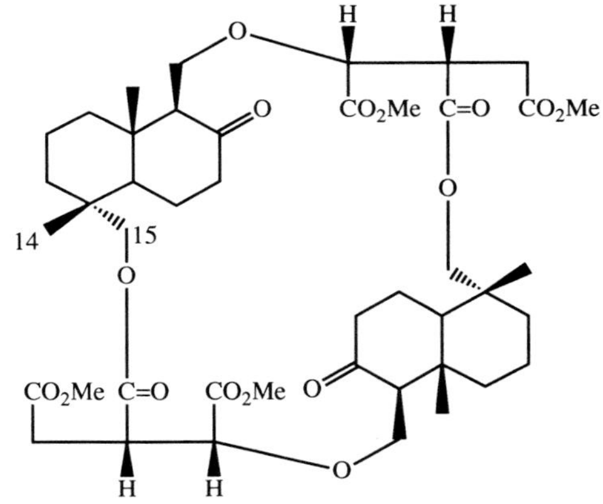

83

of CA-E resembled those of CA-C (**77**), except for the presence of a hydroxymethyl group in place of one of the four tertiary methyl groups, indicating that CA-E was a dimer of CA-B in which the carboxyl group at C-2′ was esterified with the hydroxyl group at C-15″. This was supported by the ^{1}H-^{1}H, ^{13}C-^{1}H and long-range ^{13}C-^{1}H 2D COSY NMR spectra of CA-E. Furthermore, analysis of the 2D COSY spectra established the position of two carboxylic groups at C-3′ and C-3‴ and the ester linkage between C-5′ and C-15″.

The absolute stereochemistry of CA-E was determined by the chemical correlation with CA-B. Confirmation was obtained by acetylating the pentamethyl ester (**85**) of CA-E, followed by ozonolysis to afford a diketone (**87**). The CD spectrum of **87** showed the negative Cotton effect at 275 nm, indicating that the absolute configuration of the sesquiterpene part of CA-E was identical to that of CA-A, B, C, and D. Therefore the structure of CA-E was characterized to be **84**.

6. Monodemethylcryptoporic Acids

The spectroscopic data for cryptoporic acid F (CA-F, **88**), C$_{44}$H$_{66}$O$_{14}$ (elemental analysis), and cryptoporic acid G (CA-G, **90**), C$_{44}$H$_{66}$O$_{15}$ (elemental analysis), indicated that they were probably monodemethyl analogs of CA-C and CA-E, respectively. The presence of three

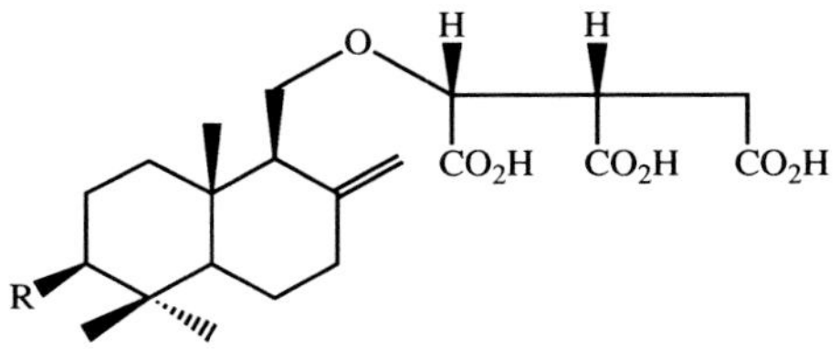

92   R=H

93   R=OH

TABLE 9
Inhibitory Effects of Cryptoporic Acids A–G (59, 71, 77, 80, 84, 88, and 90) on Superoxide Anion Radical Release from Guinea Pig Peritoneal Macrophages Induced by O_2^- Stimulant FMLP[a]

Cryptoporic acids	ID_{50} µg/ml
59	13
71	25
77	0.07
80	0.1
84	0.05
88	0.3
90	0.15

[a] FMLP = formyl methionyl leucyl phenyalanine.

carboxylic acid groups in each was confirmed by the formation of the tri-*p*-bromophenacyl esters (**89** and **91**). Methylation of CA-F and CA-G afforded pentamethyl esters identical to those formed from CA-C and CA-E. Thus, CA-F and CA-G were assigned the absolute stereochemical structures **88** and **90**, respectively.

C. BIOACTIVITY

All CAs isolated from the fruit body of *C. volvatus* were intensely bitter to the taste and completely inhibited elongation of the second coleoptile and germination of rice in husk at a concentration of 200 ppm.

Superoxide anion radical (O_2^-), which is produced by inflammation and injury due to radiation, etc., is now recognized to play an important role in hepatic, cardiac, renal, and pancreatic insufficiency and in injury to other organs. Inhibitors of superoxide anion radical release and radical scavengers are necessary to prevent human diseases caused by ischemia and inflammation. CAs strongly inhibited the release of superoxide anion radical from guinea pig peritoneal macrophages, induced by the O_2^- stimulant FMLP (formyl methionyl leucyl phenylalanine: 10^{-7} M), at concentrations from 0.05 to 25 µg/ml (IC_{50}), as shown in Table 9. CA-C (**77**) also inhibited the release of O_2^- from rabbit polymorphonuclear leukocytes, induced by O_2^- stimulant FMLP (10^{-7} M), at a concentration of 2 µg/ml (IC_{50}), as shown in Table 10.

In addition, CA-C inhibited the release of lysosomal enzymes from rat peritoneal neutrophil cells, at a concentration (IC_{50}) of 40 µg/ml (Table 11).

CA-E (**84**) inhibited the tumor-promotion activity of okadaic acid, in two stage carcinogenesis experiments on mouse skin, at a concentration of 1.2 to 5.9 M, while CA-D (**80**) slightly enhanced tumor promotion at similar concentrations (Figure 10). The activation effect of CA-D on protein kinase C and other protein kinases, at concentrations of up to 100 µM, is shown in Figure 11.[20]

TABLE 10

Inhibitory Effect of Cryptoporic Acid C (77) on Superoxide Anion Radical Release from Rabbit Polymorphonuclear Leukocytes Induced by $O_2^-\cdot$ Stimulant FMLP

Compound	O_2^- release inhibition (%)	
	6 μg/ml	88
77	3 μg/ml	70
	1.5 μg/ml	20
	IC_{50}:2 μg/ml	

TABLE 11

Inhibitory Effects of Cryptoporic Acids A–G (59, 71, 77, 80, 84, 88, and 90) on Lysosomal Enzyme Release from Rat Peritoneal Neutrophil Cells[a]

Cryptoporic acids	IC_{50} μg/ml
59	>50
71	>50
77	40
80	>50
84	>50
88	>50
90	>50

[a] Stimulant: FMLP ($10^{-6}M$) and cytochalasin B (5 μg/ml).

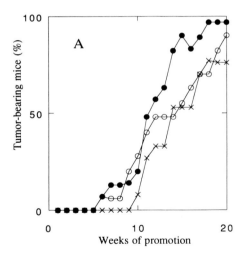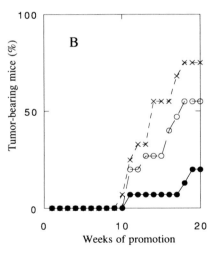

FIGURE 10. Effects of cryptoporic acids D (**80**) (A) and E (**84**) (B) on the percentage of tumor-bearing mice of the group treated with 7,12-dimethylbenz[*a*]anthracene (DMBA) plus okadaic acid. The skin of the back of female CDS-1 mice was initiated by a single application of 100 μg of DMBA. From 1 week after initiation, 1 μg of okadaic acid dissolved in 0.1 ml of Me$_2$CO was applied to the same area on the back of mice, twice a week, until week 20 of tumor promotion. Each group consisted of 15 mice. Cryptoporic acids dissolved in 0.1 ml Me$_2$CO were applied on the same area 10 min before okadaic acid. (A) Groups treated with DMBA and okadaic acid (✕), DMBA and okadaic acid plus 1 mg cryptoporic acid E per application (○), and DMBA and okadaic acid plus 5 mg cryptoporic acid D per application (●). (B) Groups treated with DMBA and okadaic acid (✕), DMBA and okadaic acid plus 1 mg cryptoporic acid D per application (○), and DMBA and okadaic acid plus 5 mg cryptoporic acid E per application (●).

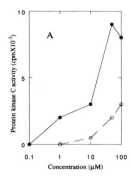
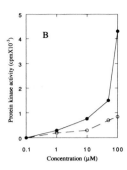

FIGURE 11. Effects of cryptoporic acids D (**80**) and E (**84**) on activation of protein kinases. Protein kinase C was partially purified from mouse brain by DEAE-cellulose column chromatography. Protein kinase C activity was determined by incorporation of ^{32}P histone III-S. The activity of the other protein kinases was measured by incorporation of ^{32}P into histone H1. (A) Activation of protein kinase C by cryptoporic acids D (●) and E (○). (B) Stimulation of ^{32}P incorporation into histone H1 by other protein kinases in the presence of cryptoporic acids D (●) and E (○).

V. CONCLUSION

Some species of the Hepaticae (liverworts) and the Polyporaceae produce a large amount of peculiar terpenoid and aromatic compounds that show a range of biological activities. The isolation and structural determination of the compounds described in this chapter represents only a very small proportion of the bioactive compounds that have been isolated from bryophytes and inedible mushrooms.

These spore-forming plants will continue to be a valuable source of bioactive compounds.

REFERENCES

1. **Asakawa, Y.,** Biologically active substances from bryophytes, in *Bryophyte Development: Physiology and Biochemistry,* Chopra, R. N. and Bhatla, S. C., Eds., CRC Press, FL, l990, 259.
2. **Asakawa, Y.,** Terpenoids and aromatic compounds with pharmacological activity from bryophytes, in *Bryophytes: Their Chemistry and Chemical Taxonomy,* Zinsmeister, H. D. and Mues, R., Eds., Oxford University Press, Oxford, l990, 369.
3. **Hashimoto, T., Tori, M., Mizuno, Y., and Asakawa, Y.,** Cryptoporic acids A and B, novel bitter drimane sesquiterpenoid ethers of isocitric acid, from the fungus *Cryptoporous volvatus, Tetrahedron Lett.,* 28, 6303, 1987.
4. **Hashimoto, T., Tori, M., Mizuno, Y., Asakawa, Y., and Fukazawa, Y.,** The superoxide release inhibitors, cryptoporic acids C, D, and E: dimeric drimane sesquiterpenoid ethers of isocitric acid from the fungus *Cryptoporus volvatus, J. Chem. Soc. Chem. Commun.,* 258, l989.
5. **Asakawa, Y., Hashimoto, T., Mizuno, Y., Tori, M., and Fukazawa, Y.,** Cryptoporic acids A–G, drimane-type sesquiterpenoid ethers of isocitric acid from the fungus *Cryptoporus volvatus, Phytochemistry,* 31, 1992 (in press).
6. **Nagashima, F., Nishioka, E., Kameo, K., Nakagawa, C., and Asakawa, Y.,** Terpenoids and aromatic compounds from selected Ecuadorian liverworts, *Phytochemistry,* 30, 215, 1991.
7. **Tori, M., Toyota, M., Harrison, L. J., Takikawa, K., and Asakawa, Y.,** Total assignment of ¹H and ¹³C NMR spectra of marchantins isolated from liverworts and its application to structure determination of two macrocyclic bis(bibenzyls) from *Plagiochasma intermedium* and *Riccardia multifida, Tetrahedron Lett.,* 26, 4735, 1985.
8. **Tori, M.,** Application of 2D NMR techniques to structure determination of natural products, in *Studies in Natural Products Chemistry,* Vol.2, Atta-ur-Rahman, Ed., Elsevier, Amsterdam, 8l, 1988.

9. **Kodama, M., Shiobara, Y., Sumitomo, H., Matsumura, K., Tsukamoto, M., and Harada, C.,** Total syntheses of marchantin A and riccardin B, cytotoxic bis(bibenzyls) from liverwort, *J. Org. Chem.,* 53, 72, 1988.

10. **Asakawa, Y., Tori, M., Masuya, T., and Frahm, J.-P.,** Ent-sesquiterpenoids and cyclic bis(bibenzyls) from the German liverwort *Marchantia polymorpha, Phytochemistry,* 29, 1577, 1990.

11. **Taira, Z., Hashimoto, T., Takei, M., Endo, K., Sakiya, Y., and Asakawa, Y.,** Structure–activity relationship of non-nitrogen macrocyclic aromatic compounds, marchantin A and its analogues from some liverworts, possessing d-tubocurarine-like activity, in Proc. 11th Symp. on Medicinal Chemistry, Tokushima, Japan, December 4–5, 1990, 124 (Japanese), 182 (English).

12. **Asou, H., Iwasaki, N., Hirano, S., and Dahl, D.,** Mitotic neuroblasts in dissociated cell cultures from embryonic rat cerebral hemispheres express neurofilament protein, *Brain Res.,* 332, 355, 1985.

13. **Lee, C.-O., Chung, J.-W., and Kim, B.-K.,** Studies on the constituents of higher fungi of Korea. XXXII, *Kor. J. Mycol.,* 9, 153, 1981.

14. **Kim, B.-K., Robbers, J. E., Chung, K.-S., Chung, H.-S., and Choi, E.-C.,** Antitumor components of *Cryptoporus volvatus, Kor. J. Mycol.,* 10, 111, 1982.

15. **Hayashi, N., Mikata, K., Yasuda, S., and Komae, H.,** Volatile component of *Cryptoporus volvatus,* in Proc. 31st Symp. on the Chemistry of Terpenes, Essential Oils and Aromatics, Kyoto, Japan, September 10–12, 1987, 59.

16. **Hirotani, M., Furuya, T., and Shiro, M.,** Cryptoporic acids H and I, drimane sesquiterpenes from *Ganoderma* neo-*japonicum* and *Cryptoporus volvatus, Phytochemistry,* 30, 1555, 1991.

17. **Ohta, Y., Andersen, N. H., and Liu, C.-B.,** Sesquiterpene constituents of two liverworts of genus *Diplophyllum.* Novel eudesmanolides and cytotoxicity studies for enantiomeric methylene lactones, *Tetrahedron,* 33, 617, 1977.

18. **Toyota, M., Asakawa, Y., and Takemoto, T.,** Sesquiterpenes from Japanese liverworts, *Phytochemistry,* 20, 2359, 1981.

19. **Kaneko, T., Katsura, H., and Wakabayashi, K.,** Correlation of the β-asymmetric carbon atoms of isocitric acids to methylsuccinic acid, *Chem. Ind.,* 1187, 1960.

20. **Matsunaga, S., Furuya-Suguri, H., Nishiwaki, S., Yoshizawa, S., Suganuma, M., Hashimoto, T., Asakawa, Y., and Fujiki, H.,** Differential effects of cryptoporic acids D and E, inhibitors of superoxide anion radical release, on tumor promotion of okadaic acid in mouse skin, *Carcinogenesis,* 12, 1129, 1991.

Chapter 16

GENERAL STRATEGY FOR THE STRUCTURE DETERMINATION OF SAPONINS: MOLLUSCICIDAL SAPONINS FROM *ALLIUM VINEALE*

Shaoxing Chen and John K. Snyder

TABLE OF CONTENTS

0-8493-4372-0/93/$0.00+$.50
© 1993 by CRC Press, Inc.

I. INTRODUCTION

The purpose of this chapter is to describe a general strategy for the isolation and structure determination of saponins, then to illustrate this strategy with specific work on the saponins of *Allium vineale* (Liliaceae).[1,2]

By the original definition, saponins are compounds that foam when shaken in water.[3,4] Plants or parts of plants that contained these compounds had often been used as soaps, from which the term "saponin" derived. Compounds that are capable of inducing this frothing behavior are, more often than not, characterized structurally by the presence of an oligosaccharide unit(s) appended to a steroid or triterpenoid aglycone. However, since the original term "saponin" only refers to a physical property in aqueous solution, strictly speaking, there were no specific structural "requirements" that a compound must meet in order to be knighted a "saponin." Nevertheless, the evolution of the word "saponin" is now accepted to be a triterpenoid or steroidal glycoside. Compounds of other structural types that, upon agitation, induce foaming in water are simply referred to by a structurally more descriptive name. The term saponin is used by different workers to be, more or less, structurally restrictive.[3,4] For example, some prefer to restrict the use primarily for spirostane steroidal glycosides.

Saponins remain of interest for various reasons, but as with all natural products, biological activity is of primary concern.[3-5] The pronounced hemolytic activity of saponins has led some workers to propose that saponins be defined on the basis of this biological property. Another biologic property commonly associated with saponins is molluscicidal (snail killing) activity.[6] Indeed, the use of the molluscicidal bioassay to guide isolations to biologically active natural products frequently leads to saponins. The potential of this activity in controlling schistosomiasis by eliminating the snail vector in the life cycle of the parasite schistosome is an oft-quoted rationale for employing this bioassay as a guide in natural products isolations.[6,7] The value of this activity to the plant, as protection against herbivorous snails, is equally obvious.[8,9] While it had been reported that 79% of all plant species examined contained saponins, for a long time their role in the life history of plants was undetermined.[10] Whatever other purposes these natural products serve, it is now fairly well recognized that saponins are important defense compounds with a variety of antimicrobial and antifungal activities.

An example of this defensive role, which also illustrates the continuously evolving natural war between prey and predator, is the group of saponins known as the avenacins, found in the roots of oats. These saponins, which have triglucosidic saccharide chains, are the source of the resistance of oats to a variant of the Take-All fungus, *Gaeumannomyces graminis* (Sacc) Arx et Olivier (var. *tritici* Walker), which attacks the stem base.[11] A second variant of the Take-All fungus (var. *avena*) tolerates the avenacins, and it has been suggested that this tolerance stems from the ability of this variant to deglucosidate the most potently antifungal avenacin, avenacin A_1, though this may not be the only mechanism employed by this fungal variant to circumvent the chemical defenses of oats.[12,13] Indeed, the importance of the nature of the saccharide unit of other saponins for antifungal activity has been noted.[14,15]

Saponins from *Allium* species have been the target of intense investigations for some time, particularly since World War II. A large number of these investigations were undertaken in an effort to find an alternative source of diosgenin, which until recently served as a critical starting material for the pharmaceutical industry, accounting for roughly two thirds of all synthetic steroids.[16] Diosgenin was reported from *A. albidum* in 1970,[17] followed by similar findings of diosgenin from most other *Allium* species examined. Not only were the important agricultural *Allium* species investigated, such as *A. cepa* (onion), *A. sativum* (garlic), and others, but the search for new diosgenin sources, as well as investigations into biologically active saponins focusing on *Allium* species, was especially intense in the Soviet Union, where a wide range of *Alliums* native to central Asia were examined.[18] In addition to known diosgenin-bearing saponins, numerous new saponins, including several with previously unreported aglycones, such as alliogenin[19,20] and gantogenin[21] from *A. giganteum*, neoalliogenin from *A. turcomanicum*,[22,23] and karatavigenin C from *A. karataviense*,[24] were discovered.

The basis for choosing *A. vineale* for investigation in the present work was the absence of noticeable predation and excellent durability of this species in the environs of southeastern Pennsylvania (in other words, it was a common weed that was difficult to control!). Reports of applications of *A. vineale* in traditional medicines are rare, but confusion over nomenclature may be part of the problem. While common field guides[25] identify *A. vineale* by the common name "Field Garlic," it has been noted that this species is often called crow garlic.[26,27] Indeed, Britton and Brown identify *A. vineale* by the common name of either crow or field garlic in their treatise.[28] Crow garlic, which is identified in Gerard's herbal as *A. sylvestre*, has found use as a garlic substitute in both flavoring and medicine.[29] *A. vineale* has probably also found use in this respect, though perhaps under the name of crow garlic. Crow garlic has also been referred to as *A. oleraceum*, as distinct from field garlic, *A. vineale*.[26]

Though the crude extract of *A. vineale* showed both molluscicidal and antifungal activities, the molluscicidal, rather than the fungicidal, bioassay was used to guide the isolation to the active compounds, which proved to be saponins, for two reasons. First, the molluscicidal test is extremely easy to perform in a chemistry research laboratory, with a minimum amount of effort.[6] Furthermore, it was felt that the antifungal test would lead to many sulfur-bearing natural products that are very common in the genus *Allium* and are typically found as fungicidal components.[30-32] Since much of the chemistry of these sulfur-bearing natural products had already been revealed in excellent detail by Block et al. of SUNY-Albany,[33-35] assays that did not lead to these compounds were thought to be the most productive. The antifungal tests were subsequently performed on the saponins after purification. As a result of this work on the *A. vineale* saponins, nine saponins (**1–9**), seven of which were previously unreported, were isolated. The main aglycone of these compounds was diosgenin, with the largest saponin having a hexasaccharide unit. As must also be pointed out, however, since a bioassay-guided isolation of **1–9** was employed, other saponins that were present in inactive fractions would not have been detected. Thus, the isolation of **1–9** by no means is intended to represent all of the saponins present in *A. vineale*.

II. GENERAL STRATEGY FOR THE ISOLATION OF SAPONINS

A. EXTRACTION

Various approaches to the purification of saponins from crude plant extracts have been presented in several reviews.[3,36] In this laboratory the general procedure followed for the initial extraction of the crude plant material depends upon the state of the plant source.

Fresh plant material is washed after collection and immediately ground with methanol for extraction on the day of collection to prevent fungal growth, and the sample is subjected to exhaustive methanol extraction (usually three extractions). Due to the water content of the plant, this amounts to an aqueous methanol extraction for the first extraction. The plant residue that remains after the methanol extraction may be subsequently extracted with water, but most saponins will be removed in the methanol extract.

B. PARTITIONING OF EXTRACTS

With the crude extract(s) in hand, an initial partitioning scheme separates the components on the basis of solubilities, with subsequent chromatography ultimately providing the purified compound. In the case of bioassay-guided isolations, each fraction obtained during the course of the isolation procedure is assayed for activity, using any one or more bioassays. The partitioning sequence used in the isolation of the *A. vineale* saponins provided a solubility-based separation of the crude methanol extract into petroleum ether, 5% aqueous methanol, *n*-butanol, and water soluble fractions, as presented in Section IV. The bulk of the saponins were found in the *n*-butanol soluble fraction, though less-polar saponins (i.e., fewer sugars in the saccharide unit) had also partitioned into the 5% aqueous methanol fraction to a small extent, and some of the more-polar saponins (more sugars) also appeared in limited amounts in the water soluble fraction.

At times, raw plant material may be obtained only in a dried state. In this case the procedure used in these laboratories is to extract the material sequentially with petroleum ether, methanol, and 50% aqueous methanol. The latter two extracts are combined and then partitioned as described above for fresh collection, with the petroleum ether fraction (which usually contains relatively little material) being combined with the original petroleum ether extract. Figure 1 illustrates the initial processing of plant material to produce the preliminary subfractions.

C. PURIFICATION OF ACTIVE COMPONENTS

Purification of the compounds of interest then proceeds through a combination of modern liquid–liquid partition chromatography[37-40] and adsorption (usually silica gel) and/or exclusion (LH-20) chromatography, depending upon the polarity of the active fraction. Two liquid–liquid partition chromatographic techniques incorporated into the isolation scheme in this work that are particularly effective in isolating polar compounds such as saponins are rotation locular countercurrent chromatography (RLCC)[41,42] and droplet countercurrent chromatography (DCCC).[43-47] In addition, Ito's planetary coil countercurrent chromatograph[38,40,48-50] has also found significant applications in the isolation of saponins and related compounds.[51,52] As described in detail elsewhere, there are several advantages of liquid–liquid partition chromatography, especially as applied to polar compounds.[53] Liquid–liquid partition chromatography, which does not utilize a solid support, avoids irreversible adsorption onto the stationary phase and stationary-phase-catalyzed rearrangements and decomposition. This can be a troublesome problem when using extensive silica gel chromatography for the separation of polar compounds. Equally important, incorporation of a liquid–liquid partitioning technique into the isolation scheme for natural products enables both differential adsorptive behavior as well as differential partitioning behavior to be exploited for purification. Compounds that do not separate well by adsorption chromatography may often be easily resolved by liquid–liquid partition chromatography.

Of the two types of liquid–liquid partition chromatography without solid support employed in this work, the RLCC has the greatest loading capacity (up to 10 g in this work), but with lower resolution (about 300 theoretical plates, under common operating procedures). The only limitation in the choice of a solvent system is that it must be biphasic. This makes the RLCC ideal for early fractionation steps and nicely complements flash chromatography in this regard. The DCCC provides much better resolution (more than 1000 theoretical plates, under common operation procedures), but with a limited loading capacity (500 mg in this work),[54] and thus is more commonly employed as a final or penultimate purification step. The choice of solvent systems for use in the DCCC, however, is limited by the ability of the biphasic system to form droplets of suitable size and with sufficient density difference to enable the mobile phase to pass through the stationary phase as droplets without displacement of the stationary phase in a reasonable amount of time. This limits the applications of the DCCC, but fortunately, the polarity of most saponins is ideal for the solvent systems in common use for DCCC separations, and there are many reports of the purification of saponins using the DCCC.[40,43,44,55,56] While the final purification of the natural products is usually accomplished with HPLC, DCCC can be better for isolating polar compounds, such as saponins, often encountered in the *n*-butanol soluble fraction. Final, reversed-phase high-performance liquid chromatography (HPLC) purification of saponins may still be necessary after using DCCC, but the separation is usually much easier.

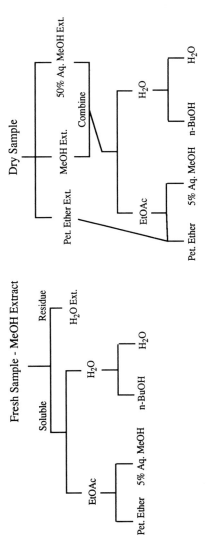

FIGURE 1. General extraction and initial partitioning sequence of crude samples for both fresh collections and dried materials.

III. GENERAL STRATEGY FOR THE STRUCTURE DETERMINATION OF SAPONINS

Since saponins may be regarded as possessing two different structural subunits (i.e., an aglycone and a saccharide), newly discovered saponins will typically vary from previously reported structures in only one of these two subunits. While new aglycones are continuously appearing in the literature, the majority of new saponins vary only in the nature of the saccharide. The structural novelty of the saccharide moiety is, in turn, usually only a matter of sugar sequence (new individual sugars are quite rare and often the cause for celebration on the part of the discoverer). Therefore, the structure assignment of saponins often breaks down into the identification of the aglycone and determination of the sugar sequence, including linkage sites between the individual monosaccharide units of the oligosaccharide.

Traditionally, the structure determination of saponins has required extensive derivatization and degradation studies.[3] Typically, a complete hydrolysis of the isolated natural compound was used to identify the aglycone as well as the individual sugars of the oligosaccharide. Sequencing of the sugars was subsequently accomplished by permethylation of the saponin, with subsequent complete hydrolysis, the nonmethylated sites on each sugar thereby locating the linkage sites. The sequence finally was deduced from the partial hydrolyses resulting in compounds with smaller oligosaccharides still attached to the aglycone. Unambiguous sequence assignment, however, required that each hydrolysis product corresponding to sequential loss of single sugar residues be successfully isolated from these partial hydrolysis experiments.

In the past decade, advances in spectroscopy, particularly nuclear magnetic resonance (NMR) and mass spectroscopy, have all but eliminated the need for such laborious efforts. A complete structure determination can be routinely done on the intact saponin, even with relatively large oligosaccharide units. Such an approach has the well-discussed advantages of reducing the amount of sample necessary for structure determination, as well as avoiding undesired chemistry, such as rearrangements and decomposition, during the degradation and derivatization procedures. The advent of multidimensional NMR has also enabled structural studies to proceed spectroscopically, even when severe overlap exists in the one dimensional spectra.[57]

The strategy for applying NMR spectroscopy to the full structure determination of saponins is virtually the same as with any newly isolated natural product (Figure 2). In order to avoid structure assignment based upon chemical shift rationales alone, which can lead to misinterpretation, it was important in this work to confirm all assignments by directly observed interactions (scalar or dipolar couplings).

Recording of the simple one-dimensional proton and carbon spectra, including the determination of the carbon multiplicities by use of the DEPT[58](Distortionless Enhancement by Polarization Transfer) or APT[59,60] (Attached Proton Test) sequences, often gives clues as to the nature of the aglycone as well as the saccharide subunits. In particular, it is often possible to identify the aglycone[61] if it is known and to count the number of sugars present from the ^{13}C-NMR spectrum. The number of sugars can usually be determined simply by counting the number of anomeric carbons present, though caution must be taken not to confuse a possible aglycone acetal carbon with a sugar anomeric carbon. Confirmation of the sugar count is obtained from the mass spectrum. The proton-coupling networks can be subsequently delineated by use of various two-dimensional techniques that may vary from instrument to instrument depending upon the software and hardware available. In the work presented in this chapter, homonuclear correlation spectroscopy (COSY)[62-64] and double-quantum-filtered/ phase-sensitive COSY (DQFCOSY, run in phase-sensitive mode)[65,66] served to reveal the geminal and vicinal proton couplings. An important advantage of the DQFCOSY technique relative to COSY (or any of the simple variations of COSY, such as COSY-45[64]) is reduction in the intensity of the diagonal peaks without sacrificing the intensity of the off-diagonal peaks

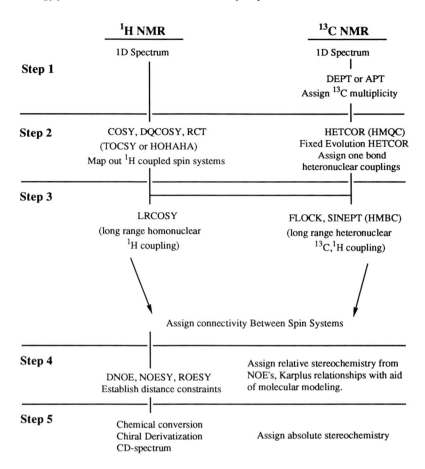

FIGURE 2. General strategy for the application of NMR in structure elucidation of new natural products. Pulse sequences listed in parentheses were not used in this work, but would be used as method of choice if available.

that arise from coherence transfer.[67,68] Though more time consuming to acquire than a normal COSY spectrum, the spectral overlap in the region of the glycosidic protons often requires analysis of off-diagonal elements that lie close to the diagonal itself. Thus, one is able to "work" closer to the diagonal with the DQFCOSY spectrum (i.e., reduction in the intensity of diagonal elements reduces their overlap with close-lying off-diagonal peaks).

As the proton spectrum becomes increasingly complex, severe signal overlap may make it impossible to trace completely each individual proton-coupled spin system when the spectral congestion becomes too extreme even among the off-diagonal elements of the two dimensional DQFCOSY spectrum. Relayed coherence transfer (RCT),[69,70] to a great extent, can overcome this problem. Depending upon availability, TOCSY,[71] also known as HOHAHA,[72] could be substituted for RCT. Introduced by Ernst in 1982,[69] the RCT sequence relays the coherence transfer generated in the basic COSY sequence, to spins sharing a coupled partner. Coherence will be relayed, and therefore an off-diagonal element will appear, between spins that are not directly coupled but have a common "coupled partner." For example, coherence will be relayed between H1 and H3 in glucose via their mutually coupled partner H2, giving rise to an off-diagonal peak linking H1 and H3, even though these two protons are not coupled.

In early examples, Schmitz et al. had demonstrated the advantage of this experiment in mapping out sugar spin systems of glycosides.[73,74] While glycosidic protons may be overlapped in the NMR spectrum, the anomeric protons are usually fairly well resolved. Therefore, the relayed coherences appear as off-diagonal peaks cleanly read from the horizontal or vertical cross sections through the anomeric resonance that is usually resolved from any other signals, with the exception of the direct coherence transfer between H1/H2.

It had been our intention to increase the number of relay steps, thereby enabling the individual sugar spin systems to be observed by virtue of the sequential appearance of H1/H2, H1/H3, H1/H4, H1/H5, and possibly even H1/H6 off-diagonal elements. This would be accomplished by an initial coherence transfer from the DQFCOSY experiment, followed by single, double, triple, and, if possible, quadruple relayed coherence transfer experiments, respectively. If successful, these off-diagonal peaks would appear in relatively clean horizontal (or vertical) cross sections at the anomeric proton frequencies. The number of relays possible would, of course, be restricted by loss of signal due to relaxation. As discussed in Section VI.B, triple relays were indeed quite feasible and were applied to saponin **9**, though the fourth relay did not succeed due to signal loss. It nevertheless must be kept in mind that during the RCT experiment, *all* coherence transfers are being relayed. Thus, while the H1/H2 coherence is being relayed to H3 by the single RCT experiment, H6/H5 coherence is also being relayed to H4. When resolved, these additional coherence relays become useful.

With the proton-coupling networks resolved, a heteronuclear correlation experiment (for example, HETCOR,[75,76] now commonly referred to as heteronuclear COSY, was used in this work, though other variations also exist[57]) then enables assignment of each proton to the carbon resonance corresponding to its directly bonded carbon. Heteronuclear multiple quantum correlation (HMQC[77]) would be substituted on instruments equipped for inverse detection.[78] One difficulty with the standard HETCOR experiment, frequently encountered in natural product structure elucidations, is the relatively poor signal intensity of methylene carbons with magnetically nonequivalent protons. At times, these correlations may even be impossible to see. To overcome this limitation, Reynolds has introduced the fixed evolution HETCOR experiment, which proved to be particularly useful in working with saponins.[79]

In the basic HETCOR pulse sequence (Figure 3A, which includes carbon decoupling during the evolution period), the evolution period is incremented from 0 to t_{max} to form the first time domain (t_1), where t_{max} depends upon the desired spectral window (SW1) in the F_1 dimension (the ^{1}H dimension) and the number of increments of the evolution period (NI), which also sets the F_1 digital resolution, i.e.,

$$t_{max} = NI/SW1 \qquad DR_1 = SW1/NI = 1/t_{max}$$

A typical setup for a HETCOR experiment in this work utilized 512 increments (NI = 512), with a spectral window of 4000 Hz in the proton dimension (SW1 = 4000 Hz), giving $t_{max} = 0.128$ s. Although this provides adequate resolution in the proton dimension ($DR_1 = 7.8$ Hz), the relative length of the evolution period can result in significant relaxation prior to signal detection for samples with relatively long correlation times, resulting in considerable loss in signal intensity for these carbons. Such was the case with the saponins in this work; e.g., as shown in later sections, the relaxation times of the anomeric carbons in saponin **6** varied from 0.159 to 0.247 s.

In addition to signal loss due to relaxation, ^{1}H/^{1}H coupling will also reduce the intensity of the ^{1}H/^{13}C cross peaks. By introduction of a BIlinear Rotation Decoupling (BIRD) pulse into the original HETCOR sequence,[80] Bax was able to decouple vicinal and other smaller couplings (Figure 3B).[81] This modified HETCOR sequence of Bax is the standard experiment available in most software supplied with commercial spectrometers and is the HETCOR sequence used in this work. Stronger geminal couplings, however, are not completely decoupled by this sequence, further reducing the intensity of methylene signals in the "normal" HETCOR experiment.

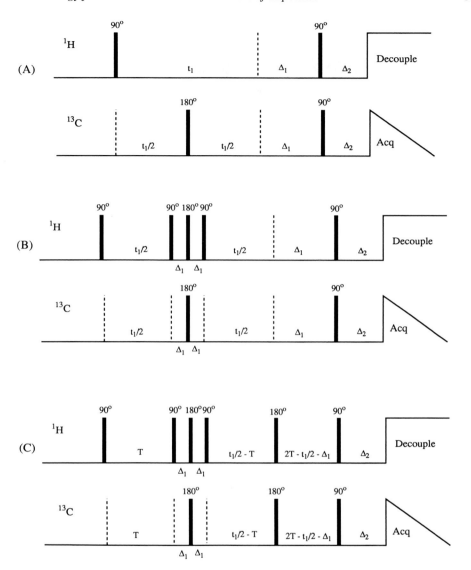

FIGURE 3. Pulse sequences for one-bond heteronuclear correlation (heteronuclear COSY). (A) Basic HETCOR pulse sequence with carbon decoupling in the evolution period. (B) Standard HETCOR pulse sequence with BIRD pulse for proton–proton decoupling. (C) Reynolds' fixed evolution HETCOR sequence. In typical operations, Δ_1 = $1/2J_{CH}$, Δ_2 = $1/4J_{CH}$, and T is a fixed interval, set to 18 ms in this work.

This loss of intensity of the methylene $^1H/^{13}C$ cross peaks due to $^1H/^1H$ coupling can be understood by considering the relative peak height of a singlet in comparison to one line of a doublet, with both singlet and doublet integrating for the same number of nuclei. Clearly the height of the singlet will be, in principle, twice that of the individual lines of the doublet, barring line-width differences. Similarly, should the $^1H/^{13}C$ cross peaks show proton multiplicity in the cross section through the carbon frequency, the height of the lines of this multiplet would be significantly less than those cross peaks that are singlets in the proton dimension. Since one typically views the HETCOR experiment in a contour plot, observation of these proton-nondecoupled cross peaks would require a threshold that goes relatively deep into the contour. In many cases, however, this loss of intensity due to proton coupling reduces the signal intensity of the $^1H/^{13}C$ cross peaks to the level of the noise, and they are not seen.

The fixed-evolution HETCOR experiment reduces signal loss due to relaxation and also enhances the sensitivity of the methylene cross peaks by fully decoupling all proton/proton coupling. Developed by modifying Kessler's correlation spectroscopy via long-range coupling (COLOC) experiment[82] for one-bond couplings, this sequence "marches" a pair of 180° proton and carbon pulses across an evolution period, fixed at a relatively small value to minimize relaxation, to create the first time domain. While the initially reported sequence[83] suffered from a lack of resolution in the proton dimension (typically on the order of 15 Hz), a subsequent modification[84] increased the resolution to that obtained in the standard HETCOR experiment (7 to 8 Hz) by allowing the pair of 180° proton and carbon pulses to traverse the evolution period as well as the delay Δ, thereby increasing the number of t_1 increments (Figure 3C). Thus with the fixed-evolution HETCOR sequence, the loss of intensity of the methylene signals due to relaxation and proton/proton coupling is minimal.

With the carbon signals identified, subsequent linking of the individual proton spin systems can then be accomplished either through long-range homonuclear correlation via scalar coupling (i.e., $^4J_{H,H}$ or longer), as detected by long-range COSY (LRCOSY),[64] or by long-range heteronuclear correlation of scalar coupling. In our work a variation of a long-range HETCOR, known as FLOCK, since the pulse sequence incorporates a number of BIRD pulses,[85] has been utilized for the two-dimensional spectroscopy, while selective INEPT (SINEPT),[86,87] a more sensitive one-dimensional sequence, has also been employed. Other sequences for detecting long-range heteronuclear couplings, such as COLOC[82] and heteronuclear multiple bond correlation (HMBC)[88] would also serve this purpose. Nuclear Overhauser enhancements (NOEs), observed either with the two-dimensional 2D-NOE sequence[89,90] or by difference NOE (DNOE), may also be needed to link spin systems.

With the spin systems linked, the remaining problem is the stereochemical assignment. Typically, relative stereochemical assignments are solved by a combination of Karplus-type relationships deduced from the vicinal coupling constants and distance constraints imposed by NOEs. Absolute stereochemical assignments can be accomplished in a variety of ways, though for saponins this is usually straightforward since the absolute stereochemistries of the sugars and aglycones are almost always known, and a direct comparison of optical rotations with literature values is usually sufficient. When new aglycones or sugars are discovered, conversion to a structure of known absolute chirality is often routine. When appropriate, other methods can be used to assign absolute stereochemistry, including circular dichroism spectroscopy[91-93] and NMR analysis of systematic shifts using Mosher's acid[94-98] or mandelic acid[99-101] derivatives (mainly for secondary alcohols). X-ray analysis is rarely applied to saponins since they are usually not obtained in crystalline form.

For saponins, where the primary concern is usually the sugar sequence, each individual sugar forms an isolated, proton-coupled spin system. Once the proton spin system for each sugar has been delineated, the carbon assignments can then be made from a HETCOR experiment, which often gives considerable insight to the identity of the sugars, including distinguishing the furanose and pyranose forms when necessary, the sites of substitution, as well as the anomeric stereochemistry, simply from the chemical shifts.[61,102-108] Confirmation of the sugar identity then becomes a matter of stereochemical assignment. For example, if a sugar is a hexopyranose and all *trans*-diaxial coupling is observed between the pyranose ring protons, then the sugar is β-glucose. 1,3-Diaxial NOEs between H1/H3, H1/H5, H2/H4, and H3/H5 would confirm such an assignment, and the $^1J_{H1/C1}$ coupling constant, routinely measured from the heteronuclear 2D-J spectrum (HET2D-J),[109,110] would confirm the anomeric stereochemistry.

With the individual sugars assigned, determining their sequence and linkage sites would be only a matter of linking the individual spin systems (in this chapter, the term "sequencing" means the order of the individual sugars in the saccharide chain, without indicating substitution patterns, which is referred to as "linkage"). While the ^{13}C chemical shifts of the sugar carbons would give a clue as to substitution sites on each individual sugar from the well-known glycosidic downfield shift, direct observation of dipolar or scalar interactions that crossed the glycosidic bonds is imperative. Since the proton spin systems would be previously known, these interresidue couplings would therefore reveal the sequence and linkage sites. Clearly, this approach to sequencing requires that all the sugar protons be unambiguously assigned. In the work presented in this chapter on the *A. vineale* saponins, interresidue NOEs were used to assign the linkage sites and confirm the sequence. Applications of long-range heteronuclear interresidue scalar interactions to assign linkage sites and sequences have also been reported.[111-113] Moreover, supporting evidence for the saccharide structure could also come from a measurement of the ^{13}C T_1 relaxation times, to distinguish the terminal and interior sugars of the saccharide chain,[114,115] as well as analysis of the fragmentation pattern in the mass spectra.[116,117]

The value of the T_1 measurements stems from the dependence of this relaxation time on the correlation time (τ_c) of the magnetic moment of the carbon nucleus, which, in turn, is dependent upon both the isotropic motion of the molecule and the local freedom of the nucleus itself.[118-120] Since the overwhelmingly dominant relaxation pathway for methylene and methine carbons is dipolar coupling with the directly bonded protons, all other factors being the same, differential internal motion within a molecule will be the prime cause of variation of T_1 relaxation times for carbons of the same multiplicity. The application of this to the determination of a sugar sequence relies upon the shorter τ_c for terminal sugars, resulting in longer T_1 relaxation times, in comparison to interior sugars of the saccharide chain that have less local freedom. For the sake of comparison between sugar residues, however, it is important that all other structural factors be as similar as possible. Thus, in the work on the *A. vineale* saponins, it was felt that the most accurate comparisons would be made between the T_1 relaxation times of the anomeric carbons, since these positions represented substituted methine sites.

For the *A. vineale* saponins, negative-ion fast-atom bombardment (FAB) MS[121] was employed to record the mass spectra. In the past decade, FAB-MS has become a prominent technique to determine the molecular weight and oligosaccharide sequence of underivatized glycosides.[122,123] In the FAB-MS spectrum, the fragmentation pattern often shows sequential cleavage of the glycosidic bonds, similar to the pattern seen in the classic, partial acidic hydrolyses. While the precise mechanism of the fragmentation is unknown, particularly with negative-ion MS,[124] studies comparing the FAB-MS fragmentations with partial acidic hydrolysis have shown that the relative yields of the pseudoglycosides in the partial hydrolysis correlate well with the relative abundance of their corresponding ions in FAB-MS.[125] In essence, a "spectroscopic partial hydrolysis" is obtained with FAB-MS, with the fragmentation occurring in the sample matrix prior to ionization. With negative-ion FAB, the molecular ion [M-H]⁻ as well as fragmentation ions retaining the aglycone [(M-H)-sugar(s)]⁻ are usually observed. When available, daughter-ion fragmentation patterns, as obtained from MS-MS, can be particularly helpful.[126,127]

IV. ISOLATION OF SAPONINS FROM *A. VINEALE*

A. vineale, collected in the Pottstown region of southeastern Pennsylvania in June 1986, was thoroughly washed and divided into three parts: flowers, stems/leaves, and bulbs. Each part was then immediately extracted with methanol, and the crude methanol extracts screened for molluscicidal activity, using the South American snail *Biomphalaria pfeifferei*, a known schistosome carrier. At 1000 ppm, only the extract of the bulbs showed activity after 24 h, so only this extract was investigated, using the molluscicidal assay to guide the isolation to the active constituents.

The crude methanol extract was divided into four fractions, by standard partitioning, to separate the components on the basis of solubility. Thus, petroleum ether, 5% aqueous

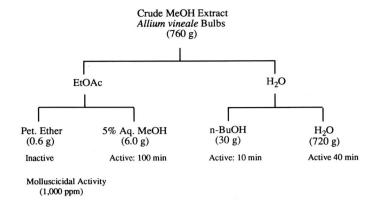

FIGURE 4. Initial partitioning of the crude methanol extract of *A. vineale* bulbs and activity in the molluskicidal bioassay (1000 ppm).

methanol, *n*-butanol, and water soluble fractions were obtained (Figure 4). With the exception of the petroleum ether fraction, the remaining three fractions all showed molluscicidal activity, though the *n*-butanol fraction was considerably more active, and our attention turned to this fraction. Given the well-known propensity for saponins to be found in the *n*-butanol fraction as well as the established record of saponins as excellent molluscicides,[6] it was thought that the active compounds sought were most probably saponins. Furthermore, if saponins were responsible for the molluscicidal activity of the 5% aqueous methanol and water soluble fractions as well, then the strong possibility existed that activity of the aqueous methanol fraction resulted from the partitioning of some of the less-polar saponins present mainly in the *n*-butanol fraction, while the activity of the water soluble fraction was a consequence of partial partitioning of the more-polar saponins of the *n*-butanol fraction. This was subsequently shown to be the case, as described later in this section.

The *n*-butanol fraction was subjected to RLCC liquid–liquid partition chromatography using a solvent system of *i*-butanol/*n*-propanol/water (5:3:10), selected as previously described,[41] in the ascending mode of operation ("organic" phase as the mobile phase). With this system, the less-polar saponins had the shorter retention times. A total of 28 g of the *n*-butanol fraction was chromatographed using this system, with seven separate runs through the instrument at 4 g per run. Four active fractions (500 ppm, molluscicidal in less than 24 h) were obtained, i.e., "Fractions A–D" in order of increasing polarity. Flash chromatography on each fraction using chloroform/methanol/water solvent systems of varying ratios produced a single active saponin fraction from each. Thus, flash chromatography of Fraction A, using $CHCl_3/CH_3OH/H_2O$ (78:20:2), produced saponin fraction "a", flash chromatography of Fractions B and C, using $CHCl_3/CH_3OH/H_2O$ (70:30:10, lower layer), produced saponin fractions "b" and "c", respectively, and flash chromatography of Fraction D, using $CHCl_3/CH_3OH/H_2O$ (65:35:10, lower layer), produced saponin fraction "d" (Figure 5). The molluscicidal activities of these saponin fractions were established at 300 ppm, while other fractions from each of these chromatographic runs were inactive in the molluscicidal test at this concentration. The saponic nature of the components of each of these fractions (a–d) was confirmed by ^{1}H-NMR, which also indicated that each fraction was still a mixture.

Final purifications of the saponin fractions "a–c" were accomplished by reversed-phase C_{18} HPLC, using aqueous methanol systems for elution (Figure 5). Three pure saponins (**1–3**) were isolated from fraction "a", two (**4** and **5**) from fraction "b", and a single saponin (**6**) was the main component of fraction "c". Purification of the component saponins from fraction "d" by HPLC was exhaustively examined, but was not possible under any conditions. Application of DCCC to this fraction, using a chloroform/methanol/water system

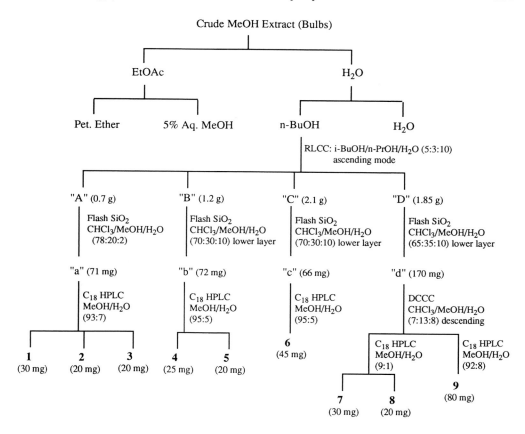

FIGURE 5. Isolation scheme leading to *A. vineale* saponins with molluskicidal activity. Scheme only shows active fractions from 28 g of the *n*-butanol soluble material.

(7:13:8, descending mode), gave two different subfractions that had identical thin-layer chromatography (TLC) behavior on both normal and reversed-phase plates. Reversed-phase C_{18} HPLC on the subfraction that eluted first from the DCCC gave two additional saponins (**7** and **8**), while the second subfraction yielded a single pure saponin (**9**). Thus, nine pure saponins were obtained from the *n*-butanol soluble fraction of the methanol extract of the bulbs of *A. vineale*.

The use of the RLCC technique early in the fractionation of the crude extract, as well as in the final purification of saponins **7–9** (DCCC) greatly facilitated the isolation procedure. In such a fashion, both the relative partitioning behavior as well as the relative adsorption behavior of these saponins were exploited for their purification. Indeed, the use of the DCCC was critical in separating saponins **7–9**.

The earlier suggestion that some of the less-polar saponins had partitioned into the 5% aqueous methanol soluble fraction, and some of the more-polar saponins into the water soluble fraction, thereby accounting for the activity of these crude fractions, was confirmed. A TLC analysis of the aqueous methanol fraction did indeed show the presence of the same saponins as found in saponin fractions "a" and "b", while the water soluble fraction contained the saponins of saponin fraction "d". Rapid flash column separation obtained saponin fractions "a" and "b" from the crude aqueous methanol soluble fraction, and saponin fraction "d" from a small portion of the crude water soluble fraction. No other fraction from these flash columns had molluscicidal activity, so the nine saponins isolated from the *n*-butanol soluble fraction appeared to be the only molluscicidal compounds present in the bulbs of *A. vineale*.

V. STRUCTURE ELUCIDATION OF SAPONINS 1–4

With the saponins in hand and the purity indicated by HPLC confirmed by NMR, the next task was to clarify the structures by identifying the aglycone and saccharide units using the general strategy outlined in Section III. The NMR spectra were run in pyridine-d_5 since no saponin-derived aromatic signals were present and thus no overlap with solvent resonances would occur in the ^{1}H and ^{13}C-NMR spectra. The saponins were also soluble in DMSO-d_6, but sample recovery from the NMR solution was easier from pyridine. In order to minimize signal overlap in the ^{1}H-NMR spectra, as well as to simplify the spectra, the hydroxyl protons were exchanged for deuterium by a minimum of three D_2O suspension/lyophilization cycles. In addition, D_2O was also added to the pyridine-d_5 NMR solution to eliminate any residual hydroxyl proton resonances. The amount of D_2O added, usually one to two drops, was adjusted to maximize signal dispersion of the remaining saccharide resonances. The addition of excessive D_2O led to precipitation of the saponins.

From the ^{1}H- and ^{13}C-NMR spectra, as well as the high-resolution mass spectroscopy (HRMS), it was apparent that saponins **1–3** each contained the same disaccharide, but with different steroidal aglycones. Several key features in the initial spectra of each saponin provided important clues to the structures. A steroidal aglycone was indicated by the presence of only four methyl group signals (three singlets and a doublet for saponins **1** and **2**, and two singlets and two doublets for saponin **3**) and 27 nonsaccharide carbons (two of which were sp^2 hybridized) in the ^{1}H- and ^{13}C-NMR spectra, respectively. A vinyl proton also appeared in each ^{1}H-NMR spectrum as a broadened doublet, typical for H-6 of phytosteroids. The subsequent HETCOR spectrum confirmed the olefinic nature of this resonance, by one-bond coupling with the sp^2 hybridized C-6. Tables 3–5 in Section IX give useful ^{1}H and ^{13}C chemical shift assignments for all saponins.

The disaccharide nature of the sugar moiety in saponins **1–3** was also apparent from simple integration of the ^{1}H resonances and a count of the carbon signals, with the two anomeric carbons occurring at δ 100 and 102 ppm. The characteristic methyl doublet, δ 1.7–1.8 ppm, along with the broadened singlets for H-1 and H-2 (δ 6.3–6.4, and 4.8 ppm, respectively) suggested that one of the two sugars was α-rhamnose. The second sugar was suggested to be β-glucose in the following manner, as exemplified for saponin **1**. A COSY spectrum quickly delineated the two separate sugar proton spin systems. From the one-dimensional ^{1}H-NMR spectrum of saponin **1** (Figure 6A), two resonances of the second sugar were well resolved, i.e., a doublet at δ 5.01 (J = 7.3 Hz), which could be assigned to the anomeric proton with *trans*-diaxial coupling with H-2, and a triplet at δ 4.13 (J = 9.4 Hz) which, from the COSY spectrum, could be assigned to H-4. Since the coupling constants for this resonance also required *trans*-diaxial coupling, a complete *trans*-diaxial relationship was required, as is only possible in β-glucose. The ^{13}C resonances (Figures 6B and C) indicated that the rhamnose was attached to the C-2 position of the glucose since the glucose C-2 resonance was shifted downfield roughly 4.5 ppm (saponin **1**: δ 79.6; saponin **2**: δ 79.7; saponin **3**: δ 79.7) from the normal position of an unsubstituted C-2 shift (the unsubstituted C-2 chemical shift of terminal glucose residues in the larger saponins was approximately δ 75 ppm). This conclusion was subsequently confirmed for each saponin (**1–3**) by a 2D-NOE spectrum that revealed an NOE between H-2 of the glucose residue and the rhamnose anomeric proton. Also observed in these spectra were NOEs between the glucose anomeric protons and a proton on the respective aglycones, subsequently assigned to the steroidal H-3 protons.

With the sugar spin systems resolved and the saccharide carbons subsequently assigned by a HETCOR spectrum, the assignment of the aglycone for saponins **1–3** was routine. The ^{13}C-NMR spectrum for each saponin had a resonance for a nonprotonated ketal carbon (saponin **1**: δ 109.3; saponin **2**: δ 121.7; saponin **3**: δ 109.6), suggesting that the aglycones of saponins **1** and **3** were spirostane steroids. With the carbon spectra in hand, comparison

Structures 1 - 9

Isonuatigenin: **I**   Nuatigenin: **N**   Diosgenin: **D**

with literature values[61] enabled the assignment of the aglycones of saponins **1–3** as isonuatigenin, nuatigenin, and diosgenin, respectively. As mentioned in Section I, this latter aglycone has been frequently found in the saponins isolated from other *Allium* species. Thus, saponins **1–3** were assigned as 3-O-[α-L-rhamnopyranosyl-(1 → 2)-O-β-D-glucopyranosyl]-isonuatigenin (**1**), 3-O-[α-L-rhamnopyranosyl-(1 → 2)-O-β-D-glucopyranosyl]-nuatigenin (**2**), and 3-O-[α-L-rhamnopyranosyl-(1 → 2)-O-β-D-glucopyranosyl]-diosgenin (**3**), in full agreement with the ¹H, ¹³C-NMR, and HRMS data. A search of the literature revealed that saponin **3** had been previously reported from *Ophiopogon planiscapus*, also of the Liliaceae family.[128] Comparison of the ¹³C chemical shifts reported in that work with those recorded for saponin **3** confirmed the identity. Saponins **1** and **2** were, to the best of our knowledge, previously unreported.

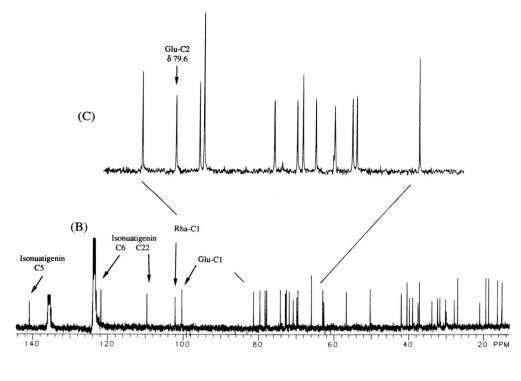

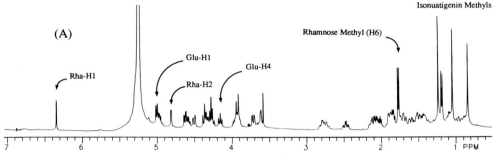

FIGURE 6. NMR spectra of saponin 1 (93.94 kG). (A) ^{1}H-NMR spectrum (400 MHz); (B) ^{13}C-NMR spectrum (100 MHz); (C) expansion of ^{13}C-NMR spectrum from δ 64–84. Significant peaks discussed in text are labeled.

With the structures of these saponins established, the crude 5% aqueous methanol soluble fraction (Figure 4) was reexamined for the presence of the parent steroidal aglycones. Using ^{1}H-NMR as a guide, flash chromatography followed by HPLC led to the isolation of (−)-(25 S)-nuatigenin and (−)-(25 S)-isonuatigenin. Nonglycosidated diosgenin was not detected. As shown by Tschesche, nuatigenin and isonuatigenin interconvert under acidic conditions.[3,129] Since the isolation of these two aglycones, as well as the isolation of saponins **1** and **2**, employed silica gel chromatography, it cannot be unambiguously stated that both are true natural products, rather than one being an artifact of the isolation procedure.

VI. STRUCTURE ELUCIDATION OF SAPONINS 4–9

Examination of the ^{1}H and ^{13}C-NMR spectra of the remaining saponins (**4–9**) indicated that they all were diosgenin saponins with increasing numbers of sugars in the saccharide

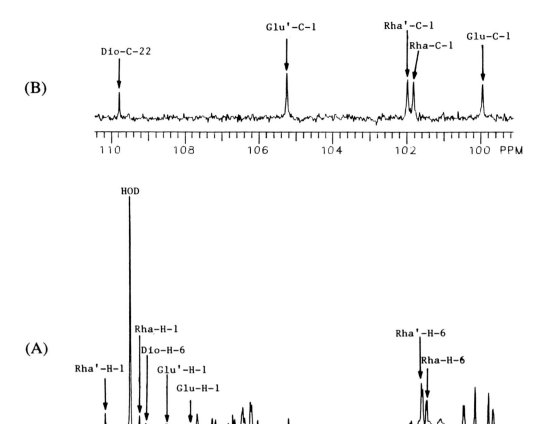

FIGURE 7. NMR spectra of saponin 6 (93.94 kG). (A) ^{1}H-NMR spectrum (400 MHz); (B) Sugar anomeric region of ^{13}C-NMR spectrum (100 MHz). Significant peaks discussed in text are labeled.

unit. Furthermore, the diosgenin triglycoside (**4**) was also identical to a saponin isolated from *O. planiscapus*. The general NMR strategy described in Section III was then applied to the structure determination of the remaining five previously unreported saponins **5–9**, as well as to saponin **4**, which verified the structure of this latter compound. This section details this approach for two of the saponins: the diosgenin tetrasaccharide (**6**) and the hexasaccaharide (**9**). A similar analysis of the other saponins (**5,7,** and **8**) led to the reported structures, and only unique features of these structure determinations will be presented (Section VI.C). As the number of sugar units in the saccharide chain increased, there was increasing overlap in all ^{1}H dimensions of the NMR spectra, which necessitated more particular attention.

A. THE STRUCTURE OF SAPONIN 6

Saponin **6** was actually the first saponin studied after saponin **1–4**, even though it had one more sugar in the saccharide unit than saponin **5**, because it was isolated in relatively large amounts (30 mg). Since the presence of the diosgenin aglycone was determined from the ^{13}C-NMR spectrum, the task at hand was to assign the oligosaccharide unit.

The presence of four sugars was initially apparent from the signals for four anomeric carbons (δ 99.9, 101.8, 102.0, 105.2 ppm, Figure 7B). Two of these sugars were thought to

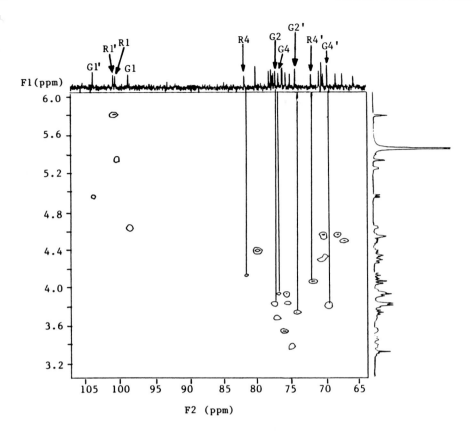

FIGURE 8. Sugar carbon region of HETCOR spectrum of saponin 6 (93.94 kG, ¹H-NMR: 400 MHz; ¹³C: 100 MHz). Anomeric carbons are labeled, though lines are not drawn to avoid congestion. Significant peaks discussed in text are labeled.

be α-rhamnose residues due to the presence of two distinct methyl doublets at δ 1.58 and 1.60 ppm (J = 6.0 Hz for each), and two broad singlets at δ 5.36 and 5.83 ppm in the ¹H-NMR spectrum (Figure 7A), typical α-rhamnose resonances. These latter two resonances were confirmed as anomeric protons by the HETCOR spectrum (Figure 8), which established the one-bond coupling with the anomeric carbon resonances at δ 101.8 (coupled with proton δ 5.36) and δ 102.0 (coupled with proton δ 5.83).

1. Sugar Identification

In order to map out the spin systems of each individual sugar, a DQFCOSY was recorded. The DQFCOSY spectrum (Figure 9) outlined the remaining resonances from the C6 methyl group through H5 and H4 to H3 of each rhamnose residue. *Trans*-diaxial couplings between H3/H4, and H4/H5 ($J_{3,4}$ = 9.0 Hz, $J_{4,5}$ = 9.5 Hz), as measured by a homonuclear 2D-J (HOMO2D-J) spectrum,[130,131] as well as the typical $J_{5,6}$ of 6.0 Hz (which were identical for both rhamnose residues) confirmed the rhamnose stereochemistry.

Due to the weak couplings between the rhamnose H1/H2 and H2/H3 vicinal proton pairs, coherence transfer within these two sets of protons was not observed in the DQFCOSY spectrum. The H1/H2 and H2/H3 correlations were observed via dipolar coupling in the 2D-NOE spectrum (Figure 10, the H2/H3 distance, an ax/eq relationship, is 2.3 Å, while that of H1/H2, an eq/eq relationship, is about 2.5 Å), which completed the assignment of each rhamnose spin system.

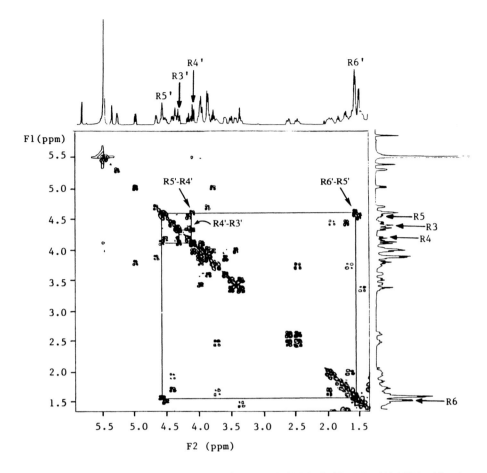

FIGURE 9. DQFCOSY spectrum (phase sensitive) of saponin **6** (93.94 kG, ¹H-NMR: 400 MHz) with coherence transfers between rhamnose Rha′ H3/H4, H4/H5, and H5/H6, delineated. Similar connectivity for rhamnose Rha is not drawn, but these protons are labeled on vertical axis.

Once the proton spin system of each rhamnose had been delineated, the HETCOR experiment then enabled assignment of the carbon signals of these sugars (Figure 8). These carbon assignments suggested that one of the rhamnose residues was a terminal sugar, while the other was interior, linked at C4 to another sugar. This was suggested by the C4 chemical shift of the latter rhamnose (δ 83.0 ppm), which was almost 10 ppm further downfield than C4 of the terminal (and therefore unsubstituted) rhamnose (δ 73.2 ppm).

The proton spin systems of the two remaining sugars were also delineated in the DQFCOSY spectrum (Figure 11), with the subsequent HETCOR spectrum then enabling carbon assignments. The all *trans*-diaxial couplings between the pyranose ring protons (^3J = 8 to 9 Hz, with the exception of $^3J_{1,2}$, which tends to be slightly less, 7 to 8 Hz for β-glucose) confirmed the β-glucose stereochemistry. These coupling constants were obtained either from the ¹H-NMR spectrum or from the HOMO2D-J spectrum.

2. Anomeric Orientation

The α-orientation of the rhamnose anomeric positions, which to the best of our knowledge is characteristic of all natural rhamnose residues, and the β-orientation of the glucose anomeric positions were confirmed by measuring the one-bond coupling constants with a heteronuclear 2D-J (HET2D-J) spectrum. For the rhamnose residues, a $^1J_{C,H}$ of 167 Hz for each was

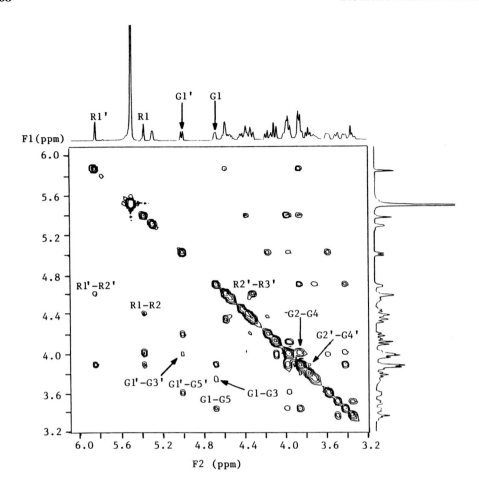

FIGURE 10. 2D-NOE spectrum of saponin **6** (93.94 kG, ^{1}H-NMR: 400 MHz), with intraresidue NOEs indicated by arrows.

recorded. The $^1J_{C,H}$ coupling constant for the anomeric center of α-rhamnosides has been established as 164 to 168 Hz, while that of β-rhamnosides is approximately 10 Hz less (152 to 158 Hz)[132] due to the anomeric effect.[102] In the same HET2D-J spectrum, the β-anomeric stereochemistry was also confirmed for the two glucose units by the $^1J_{C,H}$ values of 159 Hz for both residues. In contrast, α-glucose anomers are reported to have $^1J_{C,H}$ values about 10 Hz greater (i.e., about 170 Hz).[133]

3. Sequencing and Linkage of Sugar Units

Of the two β-glucose sugars, one was thought to be a terminal residue, while the other was thought to be a branch point residue with substitution at C2 and C4, based upon the chemical shifts observed in the ^{13}C-NMR spectrum. Thus, the C2 carbon was shifted downfield by 3 ppm, while C4 was downfield by 7 ppm (δ 78.7 and 77.9, respectively) relative to the terminal glucose residue (δ 75.7 and 70.8, respectively; see HETCOR spectrum, Figure 8).

The suggestion of a branched tetrasaccharide with a terminal glucose and a terminal rhamnose was supported by measurement of the T_1 relaxation times. These measurements were recorded in a sample maintained at 30°C (to eliminate ambient temperature fluctuations during the course of the experiment) and previously degassed by purging with argon. These relaxation times clearly indicated that one glucose (Glu′, T_1 0.234 s) and one rhamnose (Rha′,

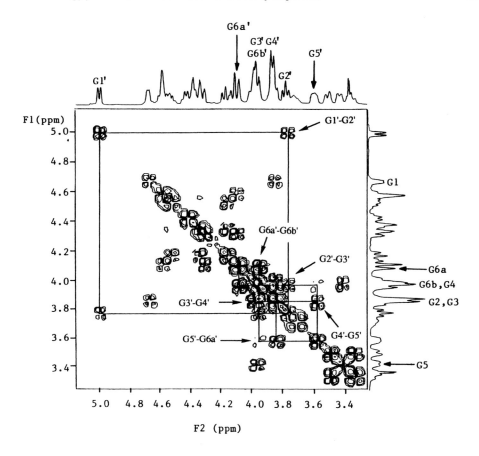

FIGURE 11. DQFCOSY spectrum (phase sensitive) of saponin **6** (93.94 kG, ¹H-NMR: 400 MHz), with coherence transfers between glucose Glu′ protons delineated. Similar connectivity for glucose Glu is not drawn, but these protons are labeled on vertical axis.

T₁ 0.247 s) had significantly longer relaxation times than those of their respective counterparts Glu (T₁ 0.159 s) and Rha (T₁ 0.193 s), suggesting that Glu′ and Rha′ were terminal sugars. It should also be noted that in this particular case, the anomeric carbon with the shortest relaxation time was that of Glu, the sugar residue that was both attached to the aglycone and that served as a branching residue in the oligosaccharide subunit. While a first-order approach would conclude that such a sugar residue should indeed have the shortest relaxation time since its mobility would be the most restricted (longest τ_c), this conclusion did not stand up upon examination of the anomeric carbon relaxation times in the larger saponins (*vide infra*, anomeric carbon relaxation times for saponins **6–9** are compiled in Table 6, Section IX), and therefore, structural interpretation of T₁ relaxation times was limited to differentiating terminal and interior sugars.

The sequence of the sugars was resolved, and the linkage sites, suggested by the ¹³C-NMR chemical shifts, confirmed by the 2D-NOE experiment (Figure 12). As described in Section III, NOEs that "crossed" the glycosidic bonds between the anomeric protons and protons at the site of substitution were crucial in establishing the sequence and linkage sites. Thus, NOEs between Glu-H1 and the diosgenin H-3, Rha′-H1 and Glu-H2, Rha-H1 and Glu-H4, as well as Glu′-H1 and Rha-H4 were observed. These NOEs required that the saccharide be described as, [β-glucopyranosyl(1 → 4)-α-rhamnopyranosyl(1 → 4)]-[α-rhamnopyranosyl(1 → 2)]-β-glucopyranoside.

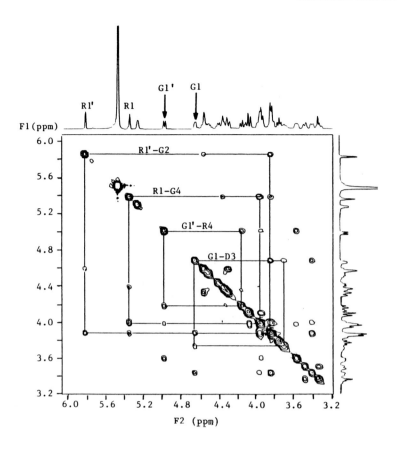

FIGURE 12. 2D-NOE spectrum of saponin **6** (93.94 kG, ¹H-NMR: 400 MHz), with interresidue NOEs indicated, establishing the linkage sites.

The sequence established by the NOEs was supported by the negative-ion FAB-MS (Figure 13). In addition to the molecular ion (m/z 1029, [M-H]⁻), the FAB-MS also showed four clear fragmentation ions for glycosidic bond cleavage. Ions at m/z 883 and 867, for loss of the terminal rhamnose ([(M-H)-146]⁻) and the terminal glucose ([(M-H)-162]⁻), respectively, were observed, as were ions corresponding to the loss of the terminal disaccharide Glu′-Rha (m/z 721) and loss of the full tetrasaccharide (m/z 431) corresponding to [diosgenin]⁻.

B. THE STRUCTURE OF SAPONIN 9

With the increased number of sugars, the ¹H-NMR spectra of saponin **9**, as well as the ¹H dimensions of all 2D-NMR spectra of saponin **9** were very congested. Application of DQFCOSY was no longer sufficient to resolve the individual proton spin systems of each sugar, as the off-diagonal resonances were themselves severely overlapped. The relayed coherence transfer (RCT) experiment was utilized to overcome this problem of signal overlap. This is illustrated initially with the glucose residues of saponin **9**.

1. Identification of Sugar Units

From the DQFCOSY spectrum (Figure 14), the glucose H2 positions were easily identified, and therefore the resonance frequencies defined, by the off-diagonal peaks with the anomeric proton from each glucose. Furthermore, the glucose H3 positions could also be tentatively located from cross sections through the H2 proton frequencies in the DQFCOSY spectrum.

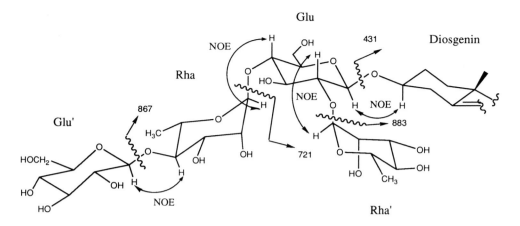

FIGURE 13. Negative-ion FAB-MS fragmentations and interresidue NOEs for sugar sequence and linkage sites of saponin **6**.

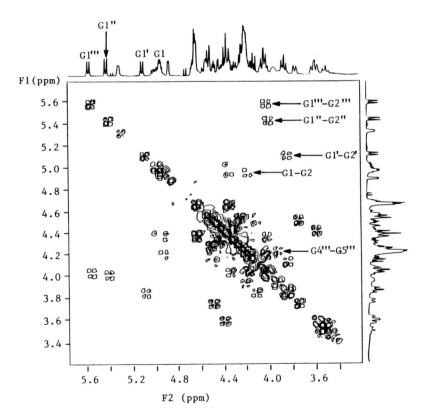

FIGURE 14. DQFCOSY spectrum (phase sensitive) of saponin **9** (93.94 kG, ^{1}H-NMR: 400 MHz), with coherence transfers between the four glucose anomeric 2-position protons, as well as Glu″-H4/H5 coherence transfer indicated by arrows.

The sole exception to locating an H3 glucose resonance occurred with Glu-H3 (i.e., H3 proton for glucose residue attached to the aglycone) where the overlap was too severe. It is important to note that use of the cross sections must be done with caution, since an overlapping off-diagonal element of a slightly different frequency may nevertheless appear in the cross section

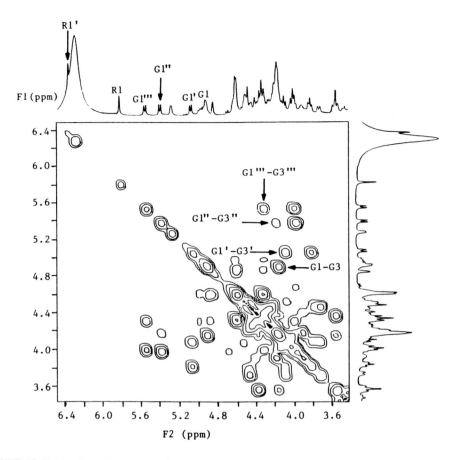

FIGURE 15. Single relay coherence transfer spectrum (RCT) of saponin **9** (93.94 kG, ¹H-NMR: 400 MHz), with relayed coherence transfers between the four glucose anomeric and 3-position protons indicated by arrows.

if the signal has a "tail" at this frequency. Such assignments, therefore, were considered tentative and were always confirmed by the RCT experiment.

The H3 resonances were then unambiguously assigned by the single RCT experiment (Figure 15) where the cross peaks for the relayed coherence for H1/H3 were clearly visible at the H1 cross-section frequencies, with one exception. The overlap between Glu-H2 and Glu-H3 again made the relayed coherence for Glu-H1/H3 impossible to distinguish from the Glu-H1/H2 coherence transfer. An additional feature of concern about this glucose residue was the somewhat smaller coupling between the anomeric proton and H2 ($^1J_{H,H}$ = 5.4 Hz, as measured from the H1 doublet in the one-dimensional spectrum) than anticipated ($^1J_{H1,H2}$ = 7.8 Hz for the remaining glucose residues). The anomeric proton of this residue was confirmed to be axially oriented by measurement of the one-bond anomeric coupling constant ($^1J_{C,H}$ = 159.4 Hz), using a HET2D-J spectrum. Thus, the possibility, albeit remote given the size of $^3J_{H1,H2}$, that this residue was β-mannose with an equatorially oriented H2 was not immediately discounted.

A 2D-NOE spectrum (Figure 16) confirmed the location of the Glu-H3 resonance. In this spectrum, NOEs were revealed between the anomeric proton Glu-H1 and the 3- and 5-position protons (Glu-H3 and Glu-H5, respectively). One of these peaks appeared at the same location as the frequency assigned to Glu-H2 from the DQFCOSY spectrum. That this peak was indeed due to an NOE between Glu-H1 and Glu-H3 (and not with Glu-H2 or Glu-H5, thereby distinguishing the H1/H3 NOE from the H1/H5 NOE and simultaneously locating Glu-H5)

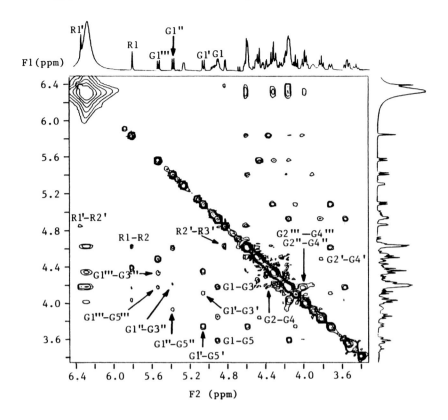

FIGURE 16. 2D-NOE spectrum of saponin **9** (93.94 kG, ¹H-NMR: 400 MHz), with intraresidue NOEs indicated by arrows.

was known by the lack of a new cross peak in the single RCT spectrum corresponding to relayed coherence between H1/H3 (due to Glu-H2/Glu-H3 overlap). An NOE between Glu-H2 and the signal ultimately assigned to Glu-H4 by a double RCT experiment (*vide infra*), and between Glu-H3/Glu-H5, confirmed the axial orientation of the pyranose protons and therefore the sugar stereochemistry as glucose.

In a similar manner the 2D-NOE spectrum confirmed the assignments of the other glucose 3-position protons by NOEs with the respective anomeric protons. Indeed, this spectrum revealed the complete 1,3-diaxial dipolar interactions expected for each glucose residue: H1/H3, H1/H5, H3/H5, and H2/H4 (Table 7, Section IX lists all observed NOEs in the saccharide units of saponins 5–9). The assignments of the glucose 4- and 5-position protons were ultimately confirmed in the double- and triple-RCT experiments, as discussed below. A key feature of the 2D-NOE spectrum at this point was confirmation of the Glu-H3 resonance, which enabled the mapping of the glucose spin systems to continue.

With the glucose 3-position protons assigned, location of the glucose 4-position protons was actually accomplished from the DQFCOSY spectrum, since the H3/H4 coherence for each glucose residue was sufficiently resolved, with the exception of the Glu″-H3/H4 cross peak. Overlap of Glu″-H3 with Glu″-H4, which also overlapped with resonances of Glu-H2, Glu-H3, Glu′-H4, and Glu‴-H4 (about δ 4.16), prevented observation of the Glu″-H3/H4 cross peak. The frequencies of the H4 resonances of the Glu′ and Glu‴ units were confirmed by a double-RCT experiment wherein H1/H4 off-diagonal peaks appeared for the respective sugar residues (Figure 17). Of note in this double-RCT experiment was the lack of a new H1/H4 cross peak for RTC between Glu″-H1 and Glu″-H4. This implied that Glu″-H4 was

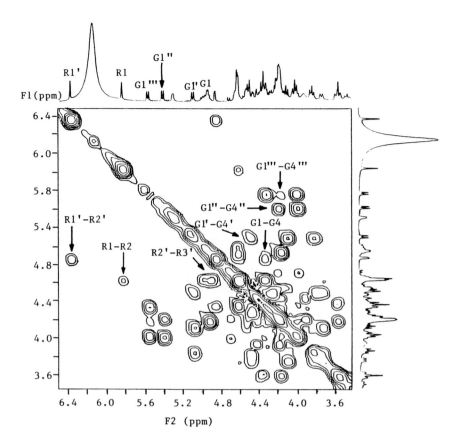

FIGURE 17. Double relay coherence transfer spectrum (RCT) of saponin **9** (93.94 kG, ^{1}H-NMR: 400 MHz), with relayed coherence transfers between the four glucose anomeric and 4-position protons, as well as the coherence transfers between the two rhamnose anomeric and 2-position, and 2- and 3-position protons indicated by arrows.

overlapped with either Glu″-H3 or Glu″-H2. That the former possibility was correct was established by first assigning the Glu″-H5 frequency from the H1/H5 NOE in the 2D-NOE spectrum (Figure 16). With Glu″-H5 located, the DQFCOSY spectrum (Figure 14) revealed the position of Glu″-H4, confirming the overlap with Glu″-H3.

The glucose 5-position protons were most easily assigned from the 1,3-diaxial dipolar couplings (NOEs), with the anomeric protons, observed in the 2D-NOE spectrum (Figure 16). These assignments were corroborated by careful analysis of the single (Figure 15), double (Figure 17), and triple (Figure 18) RCT experiments. For example, the assignment of Glu′-H5 and Glu″-H5 was obtained directly from the triple-RCT spectrum since the H1/H5 off-diagonal peaks linking the H5 resonances with the anomeric protons of their respective sugar residues were well resolved. The cross peak of the latter RTC (Glu″-H1/H5) appeared as a shoulder on the direct coherence transfer between Glu″-H1/H2 in the contour plot, but was clearly resolved in the cross section (Figure 19).

The resonance frequencies of the 6-position protons of the glucose residues were most conveniently determined from the HETCOR spectrum, since these carbons are the only sugar methylene carbons present. For saponin **9**, the multiplicities of the carbon resonances were assigned with a DEPT experiment (Figure 20), the subspectra clearly identifying the methyl, methylene, and methine carbons. The only nonsugar oxygenated methylene carbon in saponin **9** is C26 of diosgenin. This diosgenin resonance was initially identified on the basis of its

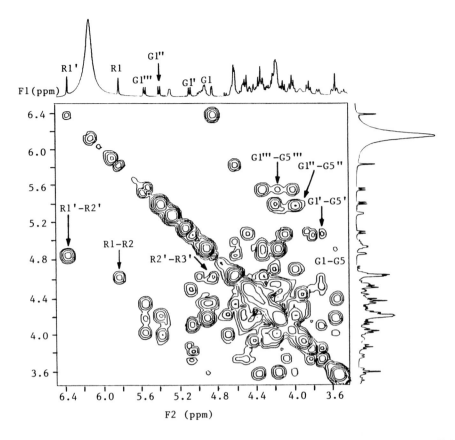

FIGURE 18. Triple relay coherence transfer spectrum (RCT) of saponin **9** (93.94 kG, ^{1}H-NMR: 400 MHz), with relayed coherence transfers between the four glucose anomeric and 5-position protons, as well as the coherence transfers between the two rhamnose anomeric and 2-position, and 2- and 3-position protons indicated by arrows.

chemical shift (δ 66.54 ppm) and consequently confirmed by identifying the methylene protons (H26) in the HETCOR experiment. These protons (H26) showed coupling in the DQFCOSY spectrum to the aglycone H25 proton, while the glucose H6 protons were coupled only to their respective H5 partners (easily distinguished by chemical shift from the diosgenin H25 proton).

After locating the glucose C6 oxygenated methylene carbons from the DEPT experiment, the revelation of the frequencies of their directly bonded, diastereotopic methylene protons should have, in principle, been routine from the HETCOR spectrum. However, the reduced intensity of glucose C6 methylene carbons, with their magnetically nonequivalent methylene protons, rendered these signals difficult to observe in the HETCOR spectrum. To overcome this problem, a fixed evolution HETCOR experiment (see Section III) was used. The 6-position methylene protons were assigned using this sequence, with an evolution time fixed to 18 ms and a refocusing interval fixed to 24 ms. The DQFCOSY and RCT spectra then enabled assignment of the H6 glucose protons to the correct H1 through H5 glucose spin systems (the importance of this modified HETCOR experiment is more clearly illustrated in the structure determination of saponin **7**, Section VI.C).

The assignment of the rhamnose proton resonances to the individual sugar units proceeded analogously, though a few unique features will be pointed out. Given the small values of the homonuclear proton coupling constants $^3J_{1,2}$ and $^3J_{2,3}$, off-diagonal peaks for these connections could not be seen in the DQFCOSY spectrum (Figure 14). Cross peaks for Rha-H1/Rha-H2,

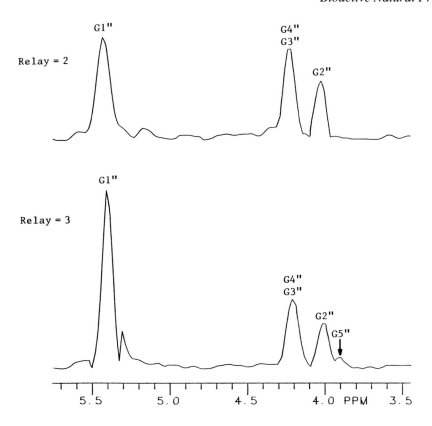

FIGURE 19. Cross sections in double and triple relay coherence transfer spectra (RCT) of saponin **9** (93.94 kG, ¹H-NMR: 400 MHz), at frequency of Glu″-H1, with relayed coherence transfer between Glu″-H1 and Glu″-H5 indicated by arrow in triple RCT cross section.

Rha′-H1/Rha′-H2, and Rha′-H2/Rha′-H3 were revealed in an LRCOSY experiment using a delay of 300 ms (Δ = 300 ms, Figure 21). The near identical chemical shift of Rha-H2 and Rha-H3 prevented this coupling from being directly observed in this experiment (the cross peak was nearly coincident with diagonal peaks and could not be resolved). These assignments were confirmed by the observation of Rha-H1/Rha-H2, Rha′-H1/Rha′-H2, and Rha′-H2/Rha′-H3 NOEs in the 2D-NOE spectrum (Figure 16). The protons within these sets, which have axial/equatorial relationships, lie within 2.5 Å of each other, and so NOEs were expected.

A notable feature of the double- and triple-RCT spectra (Figures 17 and 18, respectively) was the appearance of the Rha-H1/Rha-H2, Rha′-H1/Rha′-H2, and Rha′-H2/Rha′-H3 cross peaks. While these correlations were not observed in the DQFCOSY spectrum due to weak couplings, their appearance in the RCT spectra is understandable by comparison of the basic COSY, LRCOSY, and RCT sequences (Figure 22). In the LRCOSY experiment a delay (Δ) is inserted before and after the 90° mixing pulse of the basic COSY sequence (Figure 22A, Δ = 0), to allow the resolution of weaker couplings. In the RCT sequence the time 2τ (for a single RCT, n = 1) between the second 90° pulse and the third 90° pulse has the same effect on any magnetization not transferred (which is significant!) as the delay Δ in the LRCOSY sequence (i.e., coherence transfer generated by the second 90° pulse which is not further relayed by the third 90° pulse will have weaker couplings resolved by the delay 2τ). In this work an interval (τ) of 20 ms was used in all RCT experiments, which is optimal for 7-Hz coupling, since relayed coherence transfer between proton "A" and "X" via proton "M" is proportional to $[\sin(2\pi J_{AM}\tau)\sin(2\pi J_{MX}\tau)]$.[69] The Rha-H1/Rha-H2, Rha′-H1/Rha′-H2, and Rha′-

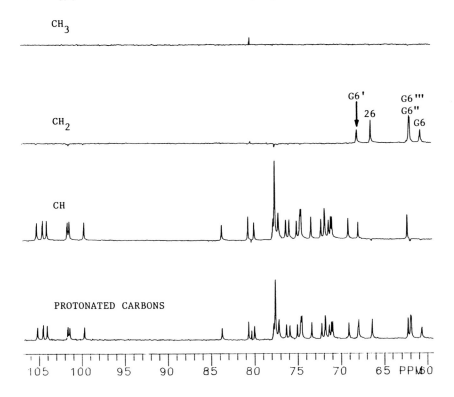

FIGURE 20. DEPT subspectra of sugar carbon region of saponin **9** (93.94 kG, ^{13}C-NMR: 100 MHz), with glucose C6 and diosgenin C26 oxygenated methylene carbons labeled.

H2/Rha′-H3 cross peaks in the double (Figure 22B, n = 2) and triple (Figure 22B, n =3) RCT spectra were observed due to the increased time interval (n × 2τ) between generation of the original coherence transfer (the second 90° pulse) and the final 90° pulse before detection, 80 ms and 120 ms, respectively. The time interval between the second and third 90° pulses in the single-RCT experiment, 40 ms, was not sufficient to allow these weaker couplings to be resolved, and hence, no cross peaks were observed for this spectrum (Figure 15). Note that the generation of coherence transfer, and thus the intensity of cross peaks in a COSY sequence, between two coupled protons "A" and "X" is proportional to $\sin(2\pi J_{AX}t_1)\,e^{(-t_1/T_2)}$, where t_1 is the length of the evolution period.[64,134] Therefore, increasing the number of relays increases the effective t_1, so the sine term becomes observable.

This interpretation of the appearance of the Rha-H1/Rha-H2, Rha′-H1/Rha′-H2, and Rha′-H2/Rha′-H3 cross peaks in the double- and triple-RCT spectra requires that these signals arise from a direct coherence transfer and not from some ill-defined relayed magnetization (which would have required a structural reevaluation of saponin **9**!). This was confirmed by comparing the RCT spectra (single, double, and triple relays) with three new LRCOSY spectra obtained with delays of 40 ms, 80 ms, and 120 ms. In the latter two LRCOSY spectra, the same Rha-H1/Rha-H2, Rha′-H1/Rha′-H2, and Rha′-H2/Rha′-H3 cross peaks as seen in the double- and triple-RCT spectra were observed, but these cross peaks were not visible in the LRCOSY spectrum run with Δ = 40 ms.

While we were able to distinguish the rhamnose H1 and H2 protons for the two separate spin systems in such a fashion, the two 3-position protons as well as the two 4-position protons were coincident (Rha-H3 and Rha′-H3, both δ 4.54; Rha-H4 and Rha′-H4, both δ 4.27). This prevented mapping of the rhamnose spin systems beyond positions 3 and 4, beginning at either H1 or in the reverse direction from H6. Thus, in the double-RCT spectrum, the single relayed

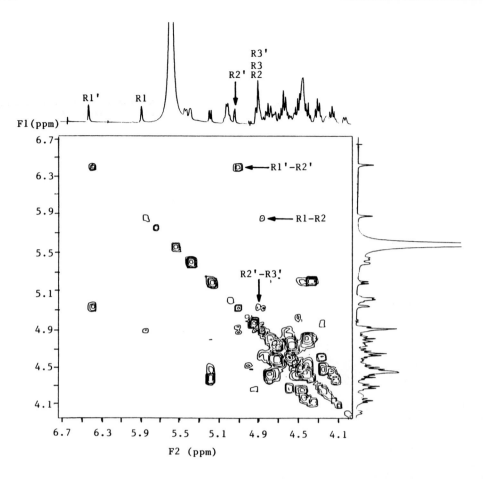

FIGURE 21. LRCOSY spectrum of saponin **9** (93.94 kG, ¹H-NMR: 400 MHz), with long-range coherence transfers between the two rhamnose anomeric and 2-position protons, and the 2- and 3-position protons, indicated by arrows.

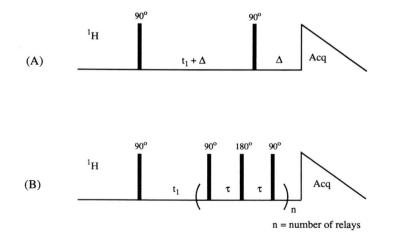

FIGURE 22. Comparison of pulse sequences for COSY (A, $\Delta = 0$), LRCOSY (A, $\Delta \leftarrow 0$), and RCT (B, n = number of relays).

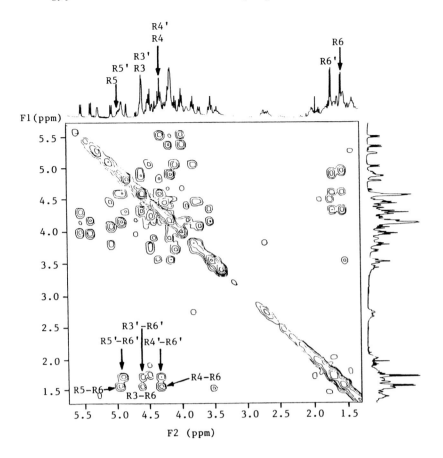

FIGURE 23. Double relay coherence transfer spectrum (RCT) of saponin **9** (93.94 kG, ¹H-NMR: 400 MHz), with relayed coherence transfers from the two rhamnose 6- to 4-position protons, and 6- to 3-position protons, indicated by arrows.

coherence (H4/H6) and double relayed coherence (H3/H6) from the rhamnose methyl groups occurred at the same location along the H4 and H3 frequency cross sections, respectively (Figure 23). The rhamnose 5-position protons were sufficiently distinct (Rha-H5 and Rha′-H5: δ 4.89 and 4.85, respectively) to correlate them with the corresponding H6 methyl doublets (Rha-H6 and Rha′-H6: δ 1.61 and 1.76, respectively). For other saponins with two rhamnose residues, this coincidence of the H3 and the H4 sets of protons was not observed, so mapping of the rhamnose spin systems was not interrupted, and the resonances for the different rhamnose protons could be clearly distinguished in the single- and double-RCT spectra by their H4/H6 and H3/H6 correlations (LRCOSY linked H1, H2, and H3 of these systems).

The problem with mapping the rhamnose spin systems in saponin **9** was matching the H1/H2 vicinal partners with the correct H5/H6 pair on the same rhamnose residue. The carbons of the interior and terminal rhamnose residues were distinguished on the basis of their relative T_1 relaxation times (those of the interior rhamnose being significantly shorter due to their longer correlation times). Thus, the two C6 and C1 carbons were distinguished (T_1 for Rha-C1: 0.194 s; Rha′-C1: 0.274 s; Rha-C6: 0.412 s; Rha′-C6: 0.552; Table 6, Section IX). The respective H6 and H1 resonances were then assigned to the terminal and interior rhamnose residues from the HETCOR spectrum, thereby enabling the distinction of the H2 and H5

TABLE 1
^{13}C T_1 Relaxation Times of Sugar Units and Anomeric Carbons of 9[a]

	50°C		65°C	
	Ave. NT_1	Anomeric carbon	Ave. NT_1	Anomeric carbon
Glu	0.238	0.240	0.243	0.240
Glu′	0.232	0.217	0.244	0.244
Glu″	0.307	0.278	0.358	0.311
Glu‴	0.288	0.265	0.315	0.288
Rha	0.213	0.211	0.230	0.220
Rha′	0.261	0.256	0.290	0.289

[a] Average NT_1 is for all carbons of each sugar where N = number of attached protons.

resonances from the LRCOSY and DQFCOSY spectra, respectively. The two rhamnose C3 and C4 carbons were similarly distinguished and assigned.

2. Linkage Sites and Sugar Sequence

With the ^{13}C-NMR assignments completed, the locations of linkage sites between the sugar residues were indicated by the characteristic downfield shifts (3 to 10 ppm) induced by the formation of the glycosidic bonds. This approach also led to the conclusion that there were two terminal glucose residues and one terminal rhamnose residue, as these sugars showed the typical ^{13}C chemical shifts for unsubstituted β-glucose and α-rhamnose residues. Supporting this conclusion were the T_1 relaxation measurements that also indicated three terminal sugars: two glucose and one rhamnose. In contrast, the Glu residue (the glucose residue with the reduced $^3J_{1,2}$ of 5.4 Hz) appeared to be substituted at C2 (δ 78.1, compared to 75.1 and 75.2 for the terminal glucose residues) and C4 (δ 78.1 compared to 71.6 and 71.8 for the terminal glucose residues), Glu′ was thought to be substituted at C4 (δ 81.2) and C6 (δ 68.6, compared to 62.4 and 62.3 for the terminal glucose residues), and Rha was linked at C4 (δ 84.9, compared to 74.0 in the terminal rhamnose Rha′). The interior nature of these three sugars (Glu, Glu′, and Rha) was also indicated by their T_1 relaxation times (Table 1). Table 5, Section IX lists the ^{13}C-NMR chemical shift assignments for all the saccharide units of the saponins isolated from *A. vineale*.

This chemical-shift-based rationale was conclusively established by the 2D-NOE spectrum (Figure 24) and by analysis of the negative-ion FAB-MS fragmentation patterns (Figure 25). The characteristic glycosidic fragmentations of the oligosaccharide moiety aided greatly in confirming the sugar sequence. Thus, in addition to the [(M-H)]$^-$ molecular ion (*m/z* 1353), loss of each of the individual terminal sugar residues required that at least one glucose ([(M-H)-Glu]$^-$, *m/z* 1191) and one rhamnose ([(M-H)-Rha]$^-$, *m/z* 1207) were terminal. Fragmentations corresponding to loss of any two ([(M-H)-2Glu]$^-$, *m/z* 1029; [(M-H)-Glu-Rha]$^-$, *m/z* 1045) or all three ([M-H)-2Glu-Rha]$^-$, *m/z* 883) terminal sugars were also observed. Loss of the terminal triglucoside unit was observed ([(M-H)-3Glu]$^-$, *m/z* 867), in accord with the chemical shift rationale of the two interior glucose residues, as branching units (two substitutions on each). The two terminal glucose residues identified from the T_1 measurements could therefore be attached to this branching glucose. The remaining branching glucose most likely would bear the terminal rhamnose, since the chemical shifts had indicated that the interior rhamnose is not a branching sugar. The remaining question was therefore whether this second branching glucose residue or the interior rhamnose was attached to the diosgenin aglycone. This latter possibility, which would transpose the Glu residue, along with its terminal rhamnose, and the interior rhamnose unit (Rha), could not be ruled out from the FAB-MS, since a fragment corresponding to the loss of all sugars except the Glu residue (i.e., [(M-H)-3Glu-2Rha], *m/z* 575) was not

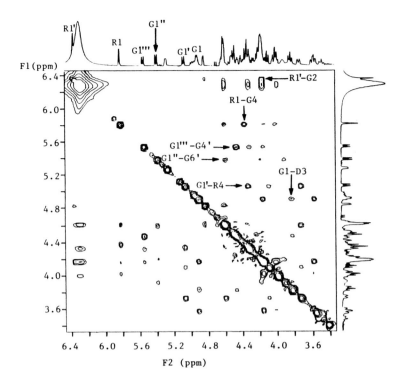

FIGURE 24. 2D-NOE spectrum of saponin **9** (93.94 kG, ^{1}H-NMR: 400 MHz), with interresidue NOEs, that revealed linkage sites indicated by arrows.

observed. The fragment which was observed at *m/z* 721, [(M-H)-3Glu-Rha]⁻, could be due to either the loss of the terminal tetrasaccharide in saponin **9** or loss of the terminal triglucoside unit and the terminal rhamnose unit. This latter possibility left the issue of the sequence of the Rha and Glu residues unresolved.

Unambiguous assignment of the sugar sequence, which also confirmed the linkage sites, therefore required the observation of the NOEs that "crossed" the glycosidic bonds (Figure 24). Thus, the anomeric proton of each sugar showed a prominent NOE in the 2D-NOE spectrum, with the proton bonded to the site of glycosidation on the neighboring sugar. All other NOEs observed in the saccharide unit were intraresidue (1,3-diaxial type). These *trans*-glycosidic NOEs were the only interresidue NOEs detected, even though extensive efforts were made to detect other interresidue NOEs, using both 2D-NOE and DNOE experiments in several of the saponins. The NOE between Glu-H1 and the diosgenin H3 resolved the final ambiguity in the saccharide structure, and the overall structure of saponin **9** was resolved.

3. Concluding Remarks

In summary, the six sugar residues of saponin **9** were identified by establishing the stereochemical relationships of the protons. This was done directly by mapping the individual sugar proton spin systems by the use of DQFCOSY; single, double, and triple RCT; and 2D-NOE spectra, then assigning the relative stereochemistry from the coupling constants as ascertained directly from the one-dimensional proton spectrum when resolution allowed, from the cross sections in the DQFCOSY spectrum, or from cross sections in the HOMO2D-J. This established the all *trans*-diaxial relationship of the protons of the glucopyranose residues and the stereochemistry of the rhamnose units. In cases where the signal overlap was too severe for resolution, axial stereochemistry was readily apparent from the 2D-NOE spectrum,

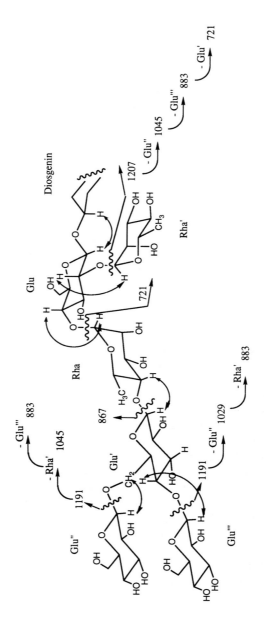

FIGURE 25. Negative-ion FAB-MS fragmentations and interresidue NOEs for sugar sequence and linkage sites of saponin **9**.

which revealed the 1,3-diaxial relationships (H1/H3, H1/H5, H3/H5, and H2/H4 for the glucose residues and H3/H5 for the rhamnose residues) from the intraresidue NOEs (Figure 16).

The carbon assignments were obtained from the HETCOR and fixed-evolution HETCOR experiments once the proton assignments were secured. The latter technique was crucial in establishing the one-bond correlations for the methylene carbons. These assignments were aided by T_1 relaxation measurements that distinguished the carbons of the interior and terminal sugar residues. Indeed, without these T_1 relaxation measurements, distinguishing the C3 and C4 carbons of the rhamnose residues would have been more difficult, since their corresponding protons were indistinguishable. Nevertheless, distinguishing the C3 and C4 rhamnose carbons would have been feasible through any one of several long-range heteronuclear coupling experiments, as described in Section III. The anomeric stereochemistry of each residue was confirmed from the $^1J_{CH}$ values, as determined with a HET2D-J spectrum. The structure of saponin **9** was completed by determining the sugar sequence and their linkage sites by a combination of NMR and MS techniques.

C. THE STRUCTURES OF SAPONINS 5, 7, AND 8

An analogous approach was used to assign the structures of saponins **5**, **7**, and **8**, all of which had been shown by the ^{13}C chemical shifts to have diosgenin as the aglycone. Furthermore, by a count of the anomeric carbon resonances, and confirmed by the FAB-MS, saponin **5** was known to have three sugars, while saponins **7** and **8** each had five sugars in their respective oligosaccharide subunits. The individual spin systems for each sugar were mapped using the DQFCOSY and RCT techniques, and the individual sugar carbon resonances were then assigned with HETCOR and fixed evolution HETCOR spectra. The sugar identities were assigned by defining their relative stereochemistries by coupling constant and NOE analysis, with assignment of the anomeric stereochemistry by measurement of $^1J_{CH}$. Elucidation of the sequences of the sugar units was based upon the relative T_1 relaxation times, mass spectral fragmentations, and, ultimately, by the interresidue NOEs that crossed the glycosidic linkages. These NOEs also confirmed the linkage sites of the sugars, which had been deduced on the basis of the ^{13}C-NMR chemical shifts. The spectroscopic data used to analyze the structures of these saponins are given in the data tables of Section IX (^1H-NMR data: Table 3; ^{13}C-NMR data: Tables 4 and 5; $^1J_{CH}$ for saponins **7** and **8**: Table 9; T_1 relaxation times for saponins **7** and **8**: Table 6; negative-ion FAB-MS fragmentations: Table 8; NOEs: Table 7).

From this data it was concluded that saponin **5** had the same aglycone (diosgenin) and the same three sugars (two glucose and one rhamnose) as found in the known saponin **4**, but with a different sequence or different linkage sites in the trisaccharide unit. From the FAB-MS fragmentation data (Table 8, Section IX), an [(M-H)-Glu]$^-$ fragment was observed at *m/z* 721, but no [(M-H)-Rha]$^-$ fragment appeared. This suggested that the trisaccharide in saponin **5** was linear with only a terminal glucose residue, assuming that the second glucose remained linked to the diosgenin, as had been found in all other saponins. This interpretation was in accord with the ^{13}C-NMR chemical shifts, which also indicated that one glucose was terminal, while the second glucose was substituted at C4 (δ 78.6 ppm compared to 71.3 in the terminal glucose). The rhamnose residue was also thought to be glycosidated at C4 because of the relative downfield position of C4 (δ 85.2 ppm compared to shifts between 73–74 for unsubstituted rhamnosides).

Due to limited amounts of saponin **5**, T_1 measurements and a 2D-NOE spectrum were not undertaken. Nevertheless, the sugar sequence and linkage sites were determined using difference NOE spectra since the three anomeric protons were all well resolved. Thus, NOEs that crossed the glycosidic bonds between Glu-H1 and diosgenin H3, Rha-H1 and Glu-H4, and Glu'-H1 and Rha-H4 were observed. It is interesting to note that due to the relatively long correlation time of saponin **5**, these NOEs were all negative even though only three sugars were present in the oligosaccharide.

The structures of saponins **7** and **8** were determined using the same approach as outlined for saponin **9**, with two unique features. The first problem was the inability to observe the one-bond heteronuclear correlations between the magnetically nonequivalent glucose methylene H6 protons and C6 carbons in the standard HETCOR experiment. While this was also difficult for saponin **9**, the cross sections at the carbon frequencies of the HETCOR spectrum of saponin **9** did enable assignment of these correlations, albeit the cross sections were quite noisy. In the HETCOR spectrum of saponin **7**, even the cross sections did not reveal the correlations to the methylene protons. Numerous attempts were made to optimize the HETCOR experiment by adjusting the polarization transfer and refocusing intervals to match coupling constants expected for oxygenated methylene carbons (the $^1J_{CH}$ couplings of the glucose C6 methylenes in saponins **6–9**, as well as the glucose residues of other unrelated saponins isolated in this laboratory, occurred within the range 144 to 145 Hz[135]), but to no avail. This problem was finally solved using the fixed-evolution HETCOR experiment (Figure 26). All the glucose methylene correlations were observed, enabling assignment of these carbons and protons.

A second noteworthy observation was seen in the 2D-NOE spectra of saponin **8** (Figure 27). When this spectrum was obtained under ambient conditions, no NOEs involving any of the terminal glucose Glu″ protons were observed. The rationalization was that the correlation times of these protons were of the order of $\tau_c \omega_i = 1$ (where ω_i is the precession frequency of the magnetic moments in radians). Under these conditions, the zero and double quantum relaxation pathways will cancel, and no enhancements will be observed.[136,137] Since NOEs for the protons of the other sugars were observed as negative enhancements, the zero quantum relaxation pathway contribution to the NOEs dominated over the double quantum pathway as expected for $\tau_c^2 \omega_i^2 > 1$ (i.e., slower motion than for terminal glucose Glu″). Since the terminal glucose residue was thought to be attached to C6 of interior glucose Glu′, deduced from the FAB-MS fragmentations (Table 8, Section IX) and the relatively low-field ^{13}C chemical shift (δ 69.7 for Glu′-C6 compared to δ 61–63 for C6-unsubstituted glucosides; Table 5, Section IX), more rapid motion for this terminal glucose was considered reasonable. Indeed, the T_1 relaxation time of the anomeric carbon of this residue was the longest found for any of the glucose residues in any of the saponins ($T_1 = 0.418$ s at 35°C, Table 6, Section IX), reflecting the more rapid motion. In contrast, the terminal Glu″ residue of saponin **9** showed negative NOEs even though this residue is also bound to interior glucose Glu′ at C6. The difference with saponin **9**, however, is most likely due to several factors. The adjacent terminal glucose Glu‴ at the Glu′-C4 position may inhibit the motion of Glu″, thereby slowing the correlation time. In addition, the greater viscosity of the hexasaccharide solution in comparison to that of the pentasaccharide, though both samples were run at the same concentration, would also increase the correlation time of saponin **9**.

While a ROESY experiment[138-140] would have enabled observation of these NOEs under ambient conditions, this sequence was unavailable at the time. Therefore, the 2D-NOE experiment was repeated, but at a lower temperature (5°C) in order to increase the correlation times for saponin **8** (Figure 28). Under these conditions the intraresidue and interresidue NOEs involving the terminal Glu″ residue were observed, with the NOE between Glu″-H1 and Glu′-H6 confirming the linkage site. In principle, the 2D-NOE experiment could also have been run at a higher temperature to reduce the correlation times for this terminal glucose residue and give positive NOEs. However, this would have also reduced the correlation times for the protons of the other sugar residues. Since NOEs in these residues were originally all negative, reduced correlation times had the high probability that the NOEs involving these sugars would disappear as $\tau_c \omega_i$ became equal to 1. Since the experiment that would enable observation of all NOEs was most desirable, the temperature was lowered. As expected, all the NOEs observed in saponin **8** at lower temperature were negative.

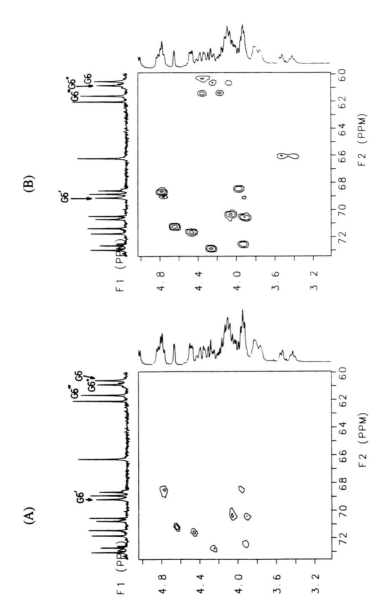

FIGURE 26. (A) Sugar carbon region of standard HETCOR spectrum of saponin **7** (93.94 kG, ^{1}H-NMR: 400 MHz; [13]C: 100 MHz). Correlations for the four glucose C6 (labeled) and diosgenin C26 (δ 66.54, not labeled) were not observed. (B) Same region of fixed evolution HETCOR spectrum of saponin **7**, with correlations to all oxygenated methylene carbons now clearly observed.

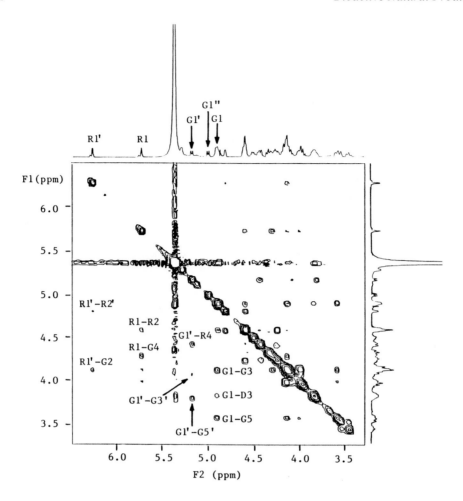

FIGURE 27. 2D-NOE spectrum of saponin **8** (93.94 kG, ¹H-NMR: 400 MHz) at ambient temperature, with NOEs to all anomeric protons indicated by arrows. No NOEs were observed to the Glu″ anomeric proton.

VII. CONFIRMATION OF THE SUGAR IDENTITIES AND DETERMINATION OF THE ABSOLUTE STEREOCHEMISTRIES FOR SAPONINS 6 AND 9

Confirmation of the sugar identities for saponins **6** and **9** by hydrolysis, as well as determination of the absolute stereochemistries for all three aglycones and the two sugars present in saponins **1–9**, was undertaken at the request of a referee following submission of the original manuscript describing this work.[2]

A. GC-FTIR ANALYSIS OF THE SUGAR COMPONENTS

While all nine saponins were identified mainly by the use of NMR techniques, supported by the negative-ion FAB-MS, verification of the sugar identities by GC-FTIR analysis of the persilylated sugars in the acid hydroslates of saponins **6** and **9** removed any doubt. The decision to use GC-FTIR instead of GC-MS in analyzing the sugars was based entirely upon availability. GC-MS is a far more sensitive technique and would have served just as well for the identification of common compounds such as persilylated sugars.[141-144]

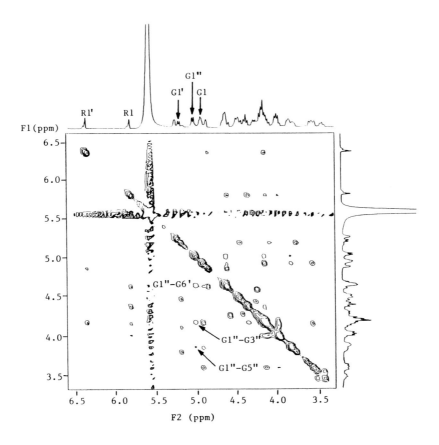

FIGURE 28. 2D-NOE spectrum of saponin **8** (93.94 kG, ^{1}H-NMR: 400 MHz) at 5°C, with NOEs to all anomeric protons indicated by arrows; NOEs were observed to the Glu″ anomeric proton at this temperature.

The saponins were hydrolyzed by refluxing in methanolic HCl for 4 h.[128] After removing the solvent *in vacuo*, the residues were dissolved in water and extracted with methylene chloride to recover the diosgenin. The aqueous layer was neutralized with ammonium hydroxide and lyophilized to dryness, and the residue was trimethylsilylated.[145] Separate D-glucose and L-rhamnose standards, as well as a 1:1 standard mixture of D-glucose and L-rhamnose, were subjected to identical hydrolytic reaction and work-up procedures, and also silylated. The two samples from saponins **6** and **9**, and the glucose and rhamnose standards, were subjected to GC-FTIR analysis (Perkin-Elmer 1800 FTIR equipped with a Sigma 2000 capillary GC) using a methyl silicane capillary column (HP-1, 30 m × 0.32 mm × 0.50 μm film thickness). The persilylated 1-O-methyl-D-glucoside standards appeared as two peaks in a 2.1:1 ratio for the β- and α-anomers, respectively, with the α-anomer having the longer retention time. The persilylated 1-O-methyl-L-rhamnoside standards appeared as three peaks in a 7.8:1.1:1 ratio. The major peak was identified as the α-anomer, after persilylation of α-1-O-methyl-L-rhamnoside. The two minor peaks from the rhamnose hydrosylate were the β-anomer and one unidentified product that always appeared in the rhamnose hydrosylate in the same relative ratio. The persilylated methyl rhamnosides had shorter retention times than the methyl glucosides. Figure 29A shows the GC chromatogram for the standard 1:1 mixture of persilylated methyl glucosides and rhamnosides. Integration of the rhamnose- and glucose-derived peaks a gave a 1:1 ratio, confirming the validity of the technique. The GC-

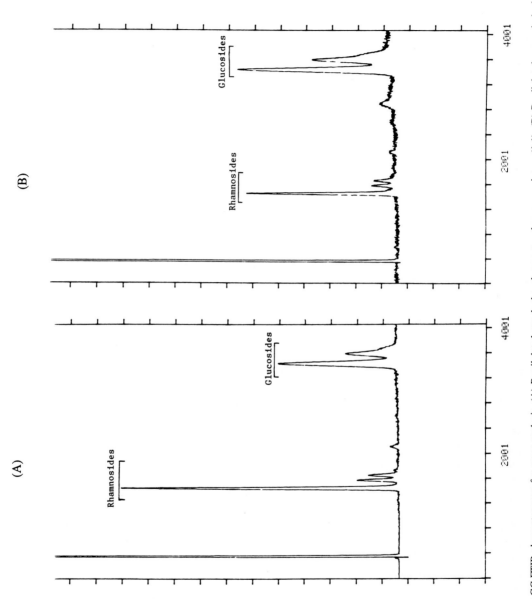

FIGURE 29. Capillary GC-FTIR chromatograms for sugar analysis. (A) Persilylated standard D-glucose + L-rhamnose mixture (1:1). (B) Persilylated sugars in hydrosylate of saponin **9** with D-glucose and L-rhamnose in a 2:1 ratio.

FTIR chromatogram of saponin **9** (Figure 29B) and saponin **6** showed only the presence of the persilylated methyl glucosides and rhamnosides, with the yields of these sugars accounting for all of the original saponin saccharide. Integration of the glucose- and rhamnose-derived peaks in the chromatogram revealed the glucose:rhamnose ratio in saponin **9** to be 2:1, and to be 1:1 in saponin **6**, thereby confirming the NMR studies.

B. ABSOLUTE CONFIGURATIONS OF AGLYCONES AND SUGAR ENTITIES
1. Aglycones
The optical rotations of nuatigenin and isonuatigenin, which were isolated from the ethyl acetate soluble fraction of the original crude methanol extract of *A. vineale* (Section IV, Figure 4), were both negative, i.e., $[\alpha]_D$ –62° (c 0.092, $CHCl_3$) and $[\alpha]_D$ –82° (c 0.177, $CHCl_3$), respectively. Comparison with literature values confirmed the 25 S configuration for these two aglycones.[129] As previously mentioned, diosgenin was not found in the crude methanol extract of *A. vineale*, but was recovered from the acid hydrosylate of saponin **9**. The optical rotation of $[\alpha]_D$ –119° (c 0.350, $CHCl_3$) confirmed the 25 R configuration.[146]

2. Sugar Entities
We had assumed that the absolute stereochemistries of the two sugars would be the "natural" D-glucose and L-rhamnose, since, to the best of our knowledge, the "unnatural" L-glucose and D-rhamnose have never been found in plants. In fact, we were unable to find any reference to D-rhamnose, so this may, in fact, be an unknown sugar. We confirmed this assumption for saponin **7**, since this saponin was the most abundant at the time, by acid hydrolysis and subsequent glycosylation of the monomeric sugars with (S)-(+)-2-butanol, pertrimethylsilylation of the resulting glycosides, and final GC-MS and NMR analysis.[147,148]

Thus, saponin **7** was refluxed in 2N HCl for 4 h and then the reaction mixture was extracted with chloroform to remove the diosgenin. The aqueous layer was neutralized with Amberlite IRA68 ion exchange resin, filtered, and lyophilized. After treatment of the residue with (S)-(+)-2-butanol in trifluoroacetic acid for 12 h at 100°C in a sealed tube, the resulting (S)-2-butyl glycosides were pertrimethylsilylated.[145]

Standard persilylated (S)-2-butyl-glycosides were prepared from commercially available D- and L-glucose, and L-rhamnose under strictly identical conditions (including the hydrolytic step). Since D-rhamnose was not available, both D-glucose and L-rhamnose were subjected to the same hydrolytic and derivatization steps again, but using racemic (±)-2-butanol. This produced two D-glucose and two L-rhamnose diastereomers, i.e., the persilylated (R)- and (S)-2-butyl-D-glucosides, and the persilylated (R)- and (S)-2-butyl-L-rhamnosides. Thus, this reaction generated the enantiomer of (S)-2-butyl-D-rhamnoside, a compound that could not be prepared due to the unavailability of D-rhamnose. Since enantiomers have the same NMR and GC-MS properties, this glycosidation using the racemic 2-butanol enabled the enantiomer of the unnatural rhamnose derivative to be substituted for the NMR and GC-MS analysis. In addition to these standards, a 1:1 mixture of D-glucose and L-rhamnose were also subjected to the hydrolytic and derivatization procedures using both (S)-(+)-2-butanol and (±)-2-butanol as a further control.

a. NMR Determination
Examination of the persilylated (S)-2-butyl glycoside mixture from the hydrosylate of saponin **7** revealed three anomeric signals in the ^{1}H-NMR spectrum (Figure 30A): δ 4.76 (d, J = 3.6 Hz) for 2,3,4,6-tetrakis(trimethylsilyl)-1-O-[(S)-2-butyl]-α-D-glucopyranose, δ 4.24 (d, J = 7.6 Hz) for 2,3,4,6-tetrakis(trimethylsilyl)-1-O-[(S)-2-butyl]-β-D-glucopyranose, and δ 4.56 (bs) for 2,3,4-tris(trimethylsilyl)-1-O-[(S)-2-butyl]-α-D-rhamnopyranose. The α- and β-anomers of the glucose derivatives were present in a 1:1 ratio, and integration revealed the ratio of the glucose-derived anomeric resonances to the rhamnose-derived anomeric

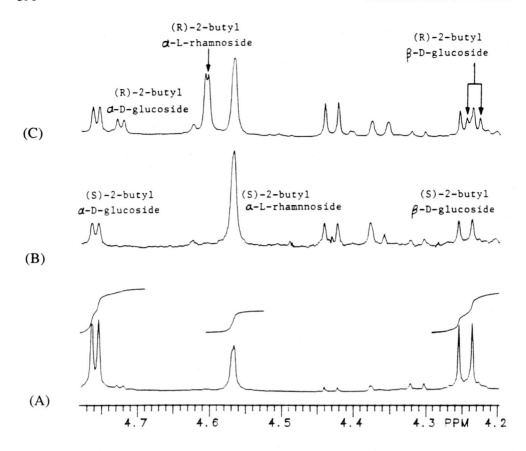

FIGURE 30. ¹H-NMR (93.94 kG, 400 MHz) spectra for sugar absolute stereochemistry assignment. (A) Persilylated (+)-(S)-2-butyl glycosides from hydrosylate of saponin **7** with only D-glucosides and L-rhamnosides. (B) Persilylated (+)-(S)-2-butyl derivatives of D-glucose and L-rhamnose standards. (C) Persilylated (+)-(S)-2-butyl glycosides from hydrosylate of saponin **7** and persilylated (±)-2-butyl glycosides of D-glucose and L-rhamnose standards.

resonances to be 4:1, in agreement with the presence of four glucose residues and one rhamnose residue in saponin **7**. These resonances were identical to the anomeric resonances observed in the spectra of the standard sugar derivatives prepared from D-glucose and L-rhamnose, respectively (Figure 30B), yet distinct from those of the L-glucose derivative. Furthermore, addition of the diastereomeric mixture of 2-butyl-glucosides and rhamnosides, prepared with (±)-2-butanol, to the mixture derived from saponin **7** gave rise to three additional anomeric signals: δ 4.72 (d, J = 3.6 Hz) for 2,3,4,6-tetrakis(trimethylsilyl)-1-O-[(R)-2-butyl]-α-D-glucopyranose, 4.23 (d, J = 7.6 Hz) for 2,3,4,6-tetrakis(trimethylsilyl)-1-O-[(R)-2-butyl]-β-D-glucopyranose, and δ 4.60 (bs) for 2,3,4-tris(trimethylsilyl)-1-O-[(R)-2-butyl]-α-D-rhamnopyranose (Figure 30C). These latter three glycosides would have chemical shifts identical with those found for the α- and β-L-glucose (as confirmed by the derivatization of L-glucose with (S)-(+)-2-butanol) and D-rhamnose derivatives, respectively, prepared with (+)-(S)-2-butanol. Thus, the absolute stereochemistries of the sugars were established (at least for saponin **7**, and reasonably assured for the remaining saponins) as the "natural" D-glucose and L-rhamnose.

b. GC-MS Determination

The GC-MS analysis of the persilylated 1-O-[(S)-2-butyl]glycosides, from the hydrolysis of saponin **7**, in comparison with the diastereomeric mixture prepared from D-glucose and

L-rhamnose with (±)-2-butanol, and with the diastereomerically pure glycosides prepared from D-glucose and L-rhamnose with (+)-(S)-2-butanol, confirmed the absolute stereochemistries of these sugars as well. This work was done using a Finnigan MAT-90, also with the methyl silicane capillary column (HP-1, 30 m × 0.32 mm × 0.50 μm film thickness), as used in the GC-FTIR analysis previously described.

VII. CONCLUSIONS

This chapter is intended to give an overview of the general strategy for the isolation and structure determination of saponins, with the main emphasis of the structure determination placed upon NMR utilization. As isolation and structure determination techniques continue to evolve, newer methods will be applied that, nevertheless, fit within the guidelines of the general strategy. Indeed, as time has progressed since the original publication of this work,[2] several of the sequences used are now considered out of date. For example, the spin lock HOHAHA or TOCSY sequence, and the Rotating Frame nuclear Overhauser Effect Spec–troscopY (ROESY) experiment could be used in place of the multiple RCT and 2D-NOE experiments, respectively. The ROESY technique would have been particularly valuable in observing all the NOEs of saponin **8**, without the necessity of using variable-temperature NOEs.

Once the individual sugar proton spin systems are mapped and the carbons assigned, sequencing and linking of the sugars in the saccharide unit can be accomplished using NMR by the observation of coupling across the glycosidic bonds. In this work dipolar coupling was detected using the 2D-NOE experiment. Detection of heteronuclear scalar coupling ($^3J_{C,H}$) across these glycosidic bonds, using any of the available long-range heteronuclear correlation experiments such as COLOC, FLOCK, or especially HMBC, would also allow for this assignment. A common feature of all these saponins is the relatively small coupling constants ($^3J_{1,2}$) between the anomeric and H2 protons for the glucose residues bonded to the diosgenin. Nevertheless, the heteronuclear coupling constants $^1J_{C,H}$ clearly defined these residues as the β-anomers. The continuously greater resolution offered by ever-advancing NMR capabilities, such as higher-dimensional NMR,[149-151] will make it easier to structurally resolve more complex saccharide units, without the need for degradation studies.

While references have been given to various works on the specific techniques mentioned in the text, the reference list is by no means all inclusive. Rather, only references that were important in our understanding and utilization of the techniques are given. The recent review by Kessler et al.[57] on the numerous NMR techniques provides an excellent overview of modern NMR sequences, while the books by Bax[75,110] and that by Ernst, Bodenhausen, and Wokaun[76,131,134] give thorough explanations of the principles behind the sequences. We have found these works to be particularly invaluable. Finally, the review on the ^{13}C-NMR analysis of steroidal saponins by Agrawal et al. was also particularly valuable.[61]

When work began on these saponins, we were interested in examining the T_1 relaxation times of the sugar carbons not only to distinguish the terminal and interior residues, but also to determine whether or not these measurements could be used to assign the sequence (i.e., the order of the interior sugars). Indeed, successful sequencing of oligosaccharides by this method had been reported in the past.[152-156] Assignment of the sugar sequence would require that the relative mobility of the individual sugars would vary systematically, progressing down the saccharide chain away from the aglycone, and with a variation significant enough to be reflected in the T_1 measurements.

One important feature of the T_1 experiments was the need to raise the temperature for the measurements for the higher saccharides. As more sugars were present in the saponin, the NMR solutions became noticeably more viscous, even though the same concentrations were used, reflecting the greater relative proportion of sugar units. It was also noted that even the

TABLE 2
^{13}C T_1 Relaxation Times of Sugar Units and Anomeric Carbons of 7[a]

| | 50°C | | 70°C | |
	Ave. NT_1	Anomeric carbon	Ave. NT_1	Anomeric carbon
Glu	0.221	0.227	0.260	0.260
Glu'	0.229	0.227	0.272	0.275
Glu''	0.258	0.246	0.300	0.299
Glu'''	0.330	0.336	0.444	0.410
Rha	0.257	0.252	0.297	0.285

[a] Average NT_1 is for all carbons of each sugar where N = number of attached protons.

terminal and interior sugars could not be unambiguously differentiated if the samples were too viscous. Increasing the temperature of the measurement, however, gave a clear distinction between these two sugar types, and this difference in T_1 relaxation times between the interior and terminal sugars continued to increase with increasing temperature, using temperatures up to 70°C. Furthermore, the T_1 relaxation times for all the carbons of the terminal sugar residues were longer than the corresponding carbons of interior sugars, as shown earlier for the glucose residues in saponin **9** (Table 1) as well as those in saponin **7** (Table 2), where Glu'' and Glu''' are the terminal glucose residues.

Analysis of the T_1 relaxation times for all the saponins at various temperatures, however, also indicated that assigning the sequence of the interior sugars would not be unambiguous. Indeed, even those terminal glucose residues linked at C6 to an interior glucose did not always exhibit relatively greater mobility than a terminal glucose linked at a pyranose ring carbon, as might be expected. Thus, while the anomeric relaxation times of the terminal Glu'' residue of saponin **8** was very long and the T_1 relaxation of the Glu'' residue carbons of saponin **9** were not only longer than that of the interior sugars, but also longer than those of the other terminal glucose Glu''' (Table 1, Glu'' bonded to Glu'-C6, Glu''' bonded to Glu' C4), this distinction was not apparent in the T_1 relaxation times for the two terminal glucose residues of saponin **7** (Table 2). While it is often tempting to read more from data than is often there, the conclusion was reached that, in general, T_1 relaxation times of saccharides can really only distinguish terminal and interior sugars under the conditions utilized in this work, particularly when dealing with nonlinear oligosaccharides. Other structural distinctions may exist on a case-by-case basis, but all conclusions drawn from T_1 relaxation measurements should be confirmed by the observations of direct interactions (i.e., couplings that cross the glycosidic bonds). It should be pointed out that in work examining spin-lattice relaxation times as a means to assign sugar sequences, the T_1 measurements had to be performed at elevated temperatures as high as 100°C.[152-155] Other workers, however, have not taken this precaution to minimize intermolecular restrictions on local freedom.

IX. DATA FOR SAPONINS 1–9

In this section are compiled the spectroscopic data used in the structure assignments of saponins **1–9**. All NMR spectra were recorded at 93.94 kG, 400 MHz for ^1H, and 100 MHz for ^{13}C. Tables 3A–C give useful aglycone and complete saccharide proton chemical shifts of most of the saponins investigated in this study. Saponins **1–3** share the same saccharide unit, so only "useful" saccharide shifts are given for saponin **3**, as well as for saponin **4**, which was a known compound. The ^{13}C chemical shifts for the three aglycones of the saponins are given in Table 4. The shifts listed for diosgenin are average values of the seven diosgenin

TABLE 3A
Useful Aglycone and Saccharide ¹H-NMR Chemical Shifts of Saponins 1–3

Saponin	1	2	3
Aglycone, vinyl	5.13 (d, 3.2)	5.29 (d, 3.3)	5.21 (d, 3.2)
methyl	1.22, 1.17 (d, 6.8)	1.37, 1.10 (d, 6.9)	1.12 (d, 6.8), 1.02
	1.03, 0.82	1.04, 0.80	0.80, 0.67 (d, 4.4)
Glu H-1	5.01 (d, 7.3)	5.03 (d, 7.3)	5.00 (d, 7.3)
H-2	4.25 (o)[a]	4.30 (o)	
H-3	4.25 (o)	4.30 (o)	
H-4	4.13 (t, 9.4)	4.17 (t, 9.4)	
H-5	3.90 (o)	3.85 (o)	
H-6	4.48 (d, 11.4), 4.29 (o)	4.51 (d, 11.4), 4.30 (o)	
Rha H-1	6.37	6.38	6.34
H-2	4.79 (br)	4.82 (br)	
H-3	4.60 (dd, 9.2, 3.2)	4.63 (dd, 9.2, 3.2)	
H-4	4.34 (t, 9.2)	4.37 (t, 9.2)	
H-5	4.96 (o)	5.00 (o)	
H-6	1.75 (d, 6.2)	1.78 (d, 6.2)	1.75 (d, 6.1)

[a] Signals unresolved due to overlap are indicated by "(o)."

TABLE 3B
Useful Aglycone and Saccharide ¹H-NMR Chemical Shifts of Saponins 4–6

Saponin	4	5	6
Aglycone, vinyl	5.27 (d, 3.1)	5.27 (d, 3.0)	5.33 (d, 3.4)
methyl	1.13 (d, 6.8), 1.03	1.09 (d, 6.8), 0.84	1.13 (d, 5.8), 0.97
	0.81, 0.68 (d, 3.8)	0.77, 0.64 (d, 4.5)	0.79, 0.69 (d, 4.6)
Glu H-1	4.95 (d, 7.0)	4.94 (d, 7.8)	4.65 (d, 5.3)
H-2		3.93 (t, 8.0)	3.85 (o)[a]
H-3		4.15 (o)	3.85 (o)
H-4		4.35 (t, 9.7)	3.98 (o)
H-5		3.63 (m)	3.41 (d, 8.8)
H-6		4.32, 4.05 (o)	3.86, 4.06 (o)
Glu' H-1	5.04 (d, 7.8)	5.20 (d, 7.8)	4.99 (d, 7.9)
H-2		4.05 (o)	3.57 (t, 8.0)
H-3		4.13 (o)	3.96 (o)
H-4		4.18 (o)	3.84 (o)
H-5		3.71 (m)	3.71 (m)
H-6		4.24 (m)	3.86, 4.06 (o)
Rha H-1	6.25	5.79	5.36
H-2		4.51	4.37 (o)
H-3		4.63 (d, 8.7)	4.37 (o)
H-4		4.42 (t, 8.7)	4.16 (t, 8.9)
H-5		4.93 (m)	4.53 (o)
H-6	1.76 (d, 6.2)	1.69 (d, 6.0)	1.58 (d, 6.0)
Rha' H-1			5.83
H-2			4.60
H-3			4.31 (d, 10.0)
H-4			4.10 (t, 10.0)
H-5			4.58 (m)
H-6			1.60 (d, 5.9)

[a] Signals unresolved due to overlap are indicated by "(o)."

TABLE 3C
Useful Aglycone and Complete Saccharide ¹H-NMR Chemical Shifts
of Saponins 7–9

Saponin		7	8	9
Aglycone,	vinyl	5.23 (d, 3.1)	5.28 (d, 2.8)	5.26 (d, 3.3)
	methyl	1.06 (d, 6.8), 0.93	1.08 (d, 6.8), 0.97	1.13 (d, 6.9), 1.04
		0.74, 0.61 (d, 4.4)	0.76, 0.63 (d, 4.5)	0.82, 0.69 (d, 4.7)
Glu	H-1	4.82 (d, 6.0)	4.88 (d, 5.5)	4.93 (d, 5.4)
	H-2	4.02 (o)[a]	4.08 (o)	4.15 (o)
	H-3	4.05 (o)	4.09 (o)	4.15 (o)
	H-4	4.08 (o)	4.27 (o)	4.36 (o)
	H-5	3.71 (o)	3.54 (o)	3.50 (o)
	H-6	4.38 (o)	4.10, 4.25 (o)	4.13, 3.97 (o)
Glu′	H-1	5.01 (d, 8.0)	5.14 (d, 7.8)	5.11 (d, 7.9)
	H-2	3.89 (o)	3.92 (o)	3.85 (o)
	H-3	3.99 (o)	4.04 (o)	4.12 (o)
	H-4	3.96 (o)	4.10 (o)	4.50 (o)
	H-5	3.77 (o)	3.77 (o)	3.73 (o)
	H-6	3.92, 4.76 (o)	4.56, 4.19 (o)	4.56 (o)
Glu″	H-1	5.04 (d, 7.9)	4.96 (d, 7.9)	5.41 (d, 7.8)
	H-2	3.89 (o)	3.93 (o)	4.00 (o)
	H-3	4.15 (o)	4.15 (o)	4.18 (o)
	H-4	3.89 (o)	4.08 (o)	4.16 (o)
	H-5	4.00 (o)	3.80 (o)	3.91 (o)
	H-6	4.27, 4.10 (o)	4.40, 4.23 (o)	4.41, 4.21 (o)
Glu‴	H-1	4.78 (d, 7.2)		5.56 (d, 7.9)
	H-2	3.89 (o)		4.04 (o)
	H-3	4.21 (o)		4.23 (o)
	H-4	4.05 (o)		4.18 (o)
	H-5	3.78 (o)		4.13 (o)
	H-6	4.18, 4.38 (o)		4.41, 4.21 (o)
Rha	H-1	6.05	5.72	5.76
	H-2	4.64	4.57	4.54
	H-3	4.45 (o)	4.56 (o)	4.54 (o)
	H-4	4.24 (o)	4.40 (o)	4.27 (o)
	H-5	4.76 (o)	4.86 (o)	4.89 (o)
	H-6	1.66 (d, 5.9)	1.68 (d, 6.4)	1.61 (d, 6.2)
Rha′	H-1		6.25	6.40
	H-2		4.78	4.80
	H-3		4.56 (o)	4.55 (o)
	H-4		4.30 (o)	4.26 (o)
	H-5		4.87 (o)	4.85 (o)
	H-6		1.70 (d, 6.7)	1.76 (d, 6.2)

[a] Signals unresolved due to overlap are indicated by "(o)."

saponins **3–9**. The actual diosgenin shifts of saponins **3–9** were within 0.35 ppm of the average values listed in this table. Table 5 lists the complete ¹³C chemical shifts for the saccharide units of saponins **1–9**. Table 6 lists the T_1 relaxation times of the anomeric carbons for saponins **6–9** at the lowest temperature for which the terminal and interior residues were distinguished. Other T_1 relaxation measurements for saponins **9** and **7** were given in Tables 1 and 2, respectively. Table 7 lists all proton homonuclear NOEs for each saccharide unit in **5–9**, with the interresidue NOEs, used to define the linkage sites, and the intraresidue NOEs listed in separate columns. Table 8 lists the mass spectral molecular ions and fragmentations resulting from saccharide cleavages in the negative-ion FAB-MS for saponins **1**, **2**, and **5–9**. Finally, Table 9 lists the heteronuclear one-bond coupling constants between the anomeric protons and the anomeric carbons for each sugar residue of the saccharide units of saponins **6–9**.

TABLE 4
^{13}C NMR Chemical Shifts of Aglycones of Saponins 1–9

Sapogenin	Diosgenin[a]	Nuatigenin	Isonuatigenin
C-1	37.4	37.5	37.8
C-2	30.1	30.2	30.2
C-3	78.2	78.3	78.3
C-4	38.9	40.5	39.0
C-5	140.7	140.8	140.8
C-6	121.7	120.2	121.7
C-7	32.2	32.2	32.3
C-8	31.7	31.6	31.7
C-9	50.2	50.2	50.4
C-10	37.0	37.1	37.0
C-11	21.0	21.1	21.2
C-12	39.8	38.9	40.0
C-13	40.4	39.8	40.5
C-14	56.5	56.5	56.8
C-15	32.0	32.3	32.2
C-16	81.0	80.8	81.4
C-17	62.7	62.6	63.0
C-18	16.3	16.1	16.4
C-19	19.4	19.4	19.6
C-20	41.9	38.4	42.0
C-21	15.0	15.2	15.2
C-22	109.3	121.7	109.6
C-23	31.6	32.6	27.9
C-24	29.2	33.8	33.8
C-25	30.4	85.6	65.9
C-26	66.8	70.1	69.8
C-27	17.3	24.1	27.0

[a] Average values from saponins 3–9 were used, and the differences were within 0.7 ppm.

TABLE 5
^{13}C NMR Chemical Shifts of Saccharide Units of Saponins 1–9

Saponin	1	2	3	4	5	6	7	8	9
Glu C-1	100.3	100.3	100.4	100.0	101.9	99.9	99.6	99.9	100.3
C-2	79.6	79.7	79.7	78.3	76.5	78.7	77.3	78.1	78.1
C-3	77.9	77.9	78.0	76.3	77.2	77.0	76.2	76.9	77.7
C-4	71.8	71.8	71.9	81.8	78.6	77.9	80.5	77.5	78.1
C-5	77.8	77.8	77.9	77.5	76.7	76.3	77.1	75.8	77.0
C-6	62.6	62.5	62.7	61.7	62.5	60.9	60.9	62.2	61.4
Glu' C-1				105.1	106.7	105.2	103.7	105.5	106.2
C-2				75.1	75.6	75.7	73.0	74.8	75.6
C-3				78.2	78.2	77.5	88.2	77.8	76.5
C-4				71.3	71.4	70.8	69.0	71.3	81.2
C-5				77.7	77.4	77.5	77.7	76.6	75.3
C-6				61.7	61.4	61.9	69.5	69.7	68.6
Glu" C-1							104.8	105.1	105.4
C-2							74.6	74.8	75.1
C-3							77.2	77.7	78.2
C-4							71.1	71.3	71.6
C-5							75.7	78.0	78.2
C-6							61.2	60.9	62.4
Glu''' C-1							104.0		104.8

TABLE 5 (continued)

Saponin		1	2	3	4	5	6	7	8	9
	C-2							74.7		75.2
	C-3							77.2		78.3
	C-4							70.9		71.8
	C-5							76.9		78.1
	C-6							62.1		62.3
Rha	C-1	102.1	102.1	102.1	101.7	102.5	101.8	101.5	101.8	102.0
	C-2	72.6	72.6	72.7	72.4	72.1	71.7	71.8	71.6	71.9
	C-3	72.8	72.9	72.9	72.9	72.7	71.7	72.1	71.9	72.8
	C-4	74.2	74.2	74.3	74.3	85.2	83.0	73.4	83.7	84.9
	C-5	69.5	69.5	69.6	69.4	68.5	68.6	69.2	68.4	68.5
	C-6	18.7	18.7	18.9	18.6	18.4	18.0	18.2	18.4	18.6
Rha′	C-1						102.0		101.9	102.2
	C-2						71.4		71.0	72.4
	C-3						72.1		72.3	72.4
	C-4						73.2		73.7	74.0
	C-5						69.5		69.4	69.8
	C-6						18.2		18.3	18.8

TABLE 6
T_1 Relaxation Times of Anomeric Carbons of Saponins 6–9

Saponin	6 (at 30°C)	7 (at 30°C)	8 (at 35°C)	9 (at 45°C)
Glu C-1	0.159	0.203	0.187	0.226
Glu′ C-1	0.234	0.211	0.268	0.234
Glu″ C-1		0.223	0.418	0.256
Glu‴ C-1		0.266		0.305
Rha C-1	0.193	0.262	0.178	0.194
Rha′ C-1	0.247		0.194	0.274
Rha C-6	0.451		0.451	0.412
Rha′ C-6	0.466		0.584	0.552

TABLE 7
Proton NOEs of the Saccharide Units of Saponins 5–9

Saponin		Interresidue NOE	Intraresidue NOE
5	Glu H-1	Dio H-3	Glu H-3, Glu H-5
	Glu′ H-1	Rha H-4	Glu′ H-3, Glu′ H-5
	Glu H-2		Glu H-4
	Glu′ H-2		Glu′ H-4
	Rha H-1	Glu H-4	Rha H-2
6	Glu H-1	Dio H-3	Glu H-3, Glu H-5
	Glu′ H-1	Rha H-4	Glu′ H-3, Glu′ H-5
	Glu H-2		Glu H-4
	Glu′ H-2		Glu′ H-4
	Rha H-1	Glu H-4	Rha H-2
	Rha′ H-1	Glu H-2	Rha′ H-2
7	Glu H-1	Dio H-3	Glu H-3, Glu H-5
	Glu′ H-1	Glu H-4	Glu′ H-3, Glu′ H-5
	Glu″ H-1	Glu′ H-6	Glu″ H-3, Glu″ H-5
	Glu‴ H-1	Glu′ H-3	Glu‴ H-3, Glu‴ H-5
	Glu H-2		Glu H-4
	Glu′ H-2		Glu′ H-4
	Glu″ H-2		Glu″ H-4

TABLE 7 (continued)

Saponin	Interresidue NOE	Intraresidue NOE
	Glu‴ H-2	Glu‴ H-4
	Rha H-1	Glu H-2	Rha H-2
8	Glu H-1	Dio H-3	Glu H-3, Glu H-5
	Glu′ H-1	Rha H-4	Glu′ H-3, Glu′ H-5
	Glu″ H-1ᵃ	Glu′ H-6	Glu″ H-3, Glu″ H-5
	Glu H-2		Glu H-4
	Glu′ H-2		Glu′ H-4
	Glu″ H-2ᵃ		Glu″ H-4
	Rha H-1	Glu H-4	Rha H-2
	Rha′ H-1	Glu H-2	Rha′ H-2
9	Glu H-1	Dio H-3	Glu H-3, Glu H-5
	Glu′ H-1	Rha H-4	Glu′ H-3, Glu′ H-5
	Glu″ H-1	Glu′ H-6	Glu″ H-3, Glu″ H-5
	Glu‴ H-1	Glu′ H-4	Glu‴ H-3, Glu‴ H-5
	Glu H-2		Glu H-4
	Glu′ H-2		Glu′ H-4
	Glu″ H-2		Glu″ H-4
	Glu‴ H-2		Glu‴ H-4
	Rha H-1	Glu H-4	Rha H-2
	Rha′ H-1	Glu H-2	Rha′ H-2

ᵃ NOE observed at 5°C, but not at ambient temperature.

TABLE 8
Mass Spectral Molecular Ions and Saccharide Fragmentations of Saponins 1, 2, 5–9ᵃ

Saponin	1	2	5	6	7	8	9
[M-H]⁻	737	737	883	1029	1207	1191	1353
[(M-H)-Rha]⁻				883	1061	1045	1207
[(M-H)-Glu]⁻			721	867	1045	1029	1191
[(M-H)-Glu-Rha]⁻				721		883	1045
[(M-H)-2Glu]⁻					883	867	1029
[(M-H)-2Glu-Rha]⁻						721	883
[(M-H)-3Glu]⁻					721		867
[(M-H)-3Glu-Rha]⁻							721

ᵃ From negative-ion FAB spectra.

TABLE 9
Heteronuclear Anomeric Coupling Constants ($^1J_{CH}$, Hz) in Saponins 6–9

Saponin	6	7	8	9
Glu	159.4	160.1	159.2	159.4
Glu′	159.4	159.0	157.5	158.6
Glu″		161.6	161.8	158.2
Glu‴		161.6		159.6
Rha	169.8	168.9	174.9	174.6
Rha′	169.8		174.9	172.8

REFERENCES

1. **Chen, S. and Snyder, J. K.,** Molluscicidal saponins from *Allium vineale, Tetrahedron Lett.,* 28, 5603, 1987.
2. **Chen, S. and Snyder, J. K.,** Diosgenin-bearing, molluscicidal saponins from *Allium vineale*: an NMR approach for the structural assignment of oligosaccharide units, *J. Org. Chem.,* 54, 3679, 1989.
3. **Tschesche, R. and Wulff, G.,** Chemie und Biologie der Saponine, in *Fortschr. Chem. Org. Naturst.,* Vol. 30, Herz, W., Grisebach, H., and Kirby, G. W., Eds., Springer-Verlag, New York, 1973, 461.
4. **Price, K. R., Johnson, I. T., and Fenwick, G. R.,** The chemistry and biological significance of saponins in foods and foodstuffs, *Crit. Rev. Food Sci. Nutr.,* 26, 27, 1987.
5. **Shibata, S.,** Saponins with biological and pharmacological activity, in *New Natural Products and Plant Drugs with Pharmacological, Biological or Therapeutical Activity,* Wagner, H. and Wolff, P., Eds., Springer-Verlag, New York, 1977, chap. 6.
6. **Marston, A. and Hostettmann, K.,** Plant molluscicides, *Phytochemistry,* 24, 639, 1985.
7. **Lemma, A.,** Laboratory and field evaluation of the molluscicidal properties of *Phytolacca dodecandra, Bull. World Health Organ.,* 42, 597, 1970.
8. **Swain, T.,** Secondary compounds as protective agents, in *Annual Review of Plant Physiology,* Vol. 28, Briggs, W. R., Green, P. B., and Jones, R. L., Eds., Annual Reviews, Palo Alto, CA, 1977, 479.
9. **Jones, D. A.,** Cyanogenic glycosides and their function, in *Phytochemical Ecology,* Harborne, J. B., Ed., Academic Press, London, 1972, chap. 7.
10. **Schloesser, E.,** Praeformierte Chemische Abwehrstoffe in Pflanzen, *Ber. Deutsch. Bot. Gesell.,* 96, 351, 1983.
11. **Asher, M. J. C. and Shipton, P. J.,** Eds., *Biology and Control of Take-All,* Academic Press, London, 1981.
12. **Crombie, W. M. L. and Crombie, L.,** Distribution of avenacins A-1, A-2, B-1, and B-2 in oat roots: their fungicidal activity toward "Take-All" fungus, *Phytochemistry,* 25, 2069, 1986.
13. **Crombie, W. M. L., Crombie, L., Green, J. B., and Lucas, J. A.,** Pathogenicity of "Take-All" fungus to oats: its relationship to the concentration and detoxification of the four avenacins, *Phytochemistry,* 25, 2075, 1986.
14. **Hufford, C. D., Liu, S., and Clark, A. M.,** Antifungal activity of *Trillium grandifolium* constituents, *J. Nat. Prod.,* 51, 94, 1988.
15. **Takechi, M. and Tanaka, Y.,** Structure–activity relationships of the saponin α-hederin, *Phytochemistry,* 29, 451, 1990.
16. **Nakanishi, K., Goto, T., Ito, S., Natori, S., and Nozoe, S.,** Eds., *Natural Products Chemistry,* Vol. III, University Science Books, Mill Valley, CA, 1983, 255.
17. **Kereselidze, E. V., Pkheidze, T. A., and Kemertelidze, E. P.,** Diosgenin from *Allium albidum, Khim. Prir. Soedin.,* 6, 378, 1970; *Chem. Abstr.,* 73, 117181h, 1970.
18. **Fenwick, G. R. and Hanley, A. B.,** The genus *Allium* — part 2., *Crit. Rev. Food Sci. Nutr.,* 22, 273, 1985.
19. **Khristulas, K. S., Gorovits, M. B., Luchanskaya, V. N., and Abubakirov, N. K.,** A new steroidal sapogenin from *Allium giganteum, Khim. Prir. Soedin.,* 6, 489, 1970; *Chem. Abstr.,* 74, 10356n, 1971.
20. **Gorovits, M. B., Khristulas, K. S., and Abubakirov, N. K.,** Alliogenin and alliogenin β-D-glucopyranoside from *Allium giganteum, Khim. Prir. Soedin.,* 7, 434, 1971; *Chem. Abstr.,* 75, 141102v, 1971.
21. **Kel'ginbaev, A. N., Gorovits, M. B., and Abubakirov, N. K.,** Steroidal saponins and sapogenins of *Allium*. VIII. Structure of gantogenin, *Khim. Prir. Soedin.,* 11, 521, 1975; *Chem. Abstr.,* 84, 74513a, 1976.
22. **Pirtskhalava, G. V., Gorovits, M. B., and Abubakirov, N. K.,** Steroidal saponins and sapogenins of *Allium*. XI. Neoalliogenin from *Allium turcomanicum, Khim. Prir. Soedin.,* 823, 1977; *Chem. Abstr.,* 89, 43942p, 1978.
23. **Pirtskhalava, G., Gorovits, M., and Abubakirov, N.,** Steroid saponins and sapogenins of *Allium turcomanicum,* in *Symp. Pap. IUPAC Int. Symp. Chem. Nat. Prod.,* Marekov, N., Ognyanov, I., and Orahats, A., Eds., Vol. 2, 1978, Bulgarian Academy of Sciences, Sofia, Bulgaria, 351.
24. **Vollerner, Yu. S., Abdullaev, N. D., Gorovits, M. B., and Abubakirov, N. K.,** Steroidal saponins and sapogenins of *Allium*. XIX. Structure of karatavigenin C, *Khim. Prir Soedin.,* 736, 1983; *Chem. Abstr.,* 100, 171548x, 1984.
25. **Peterson, R. T. and McKenny, M.,** *A Field Guide to Wildflowers of Northeastern and North-Central North America,* Houghton-Mifflin, Boston, 1968, 296.
26. **Grieve, M.,** *A Modern Herbal,* Vol. 1, Dover Publications, New York, 1971, 343.
27. **Fenwick, G. R. and Hanley, A. B.,** The genus *Allium* — part I, *Crit. Rev. Food Sci. Nutr.,* 22, 199, 1985.
28. **Britton, N. L. and Brown, A.,** *An Illustrated Flora of the Northern United States and Canada,* Vol. 1, Dover Publications, New York, 1970, 499.
29. **Gerard, J.,** *The Herbal,* revised by Johnson, T., Dover Publications, New York, 1975, 179.
30. **Fenwick, G. R. and Hanley, A. B.,** The genus *Allium* — part 3, *Crit. Rev. Food Sci. Nutr.,* 23, 1, 1985.
31. **Block, E.,** Antithrombotic agent of garlic: a lesson from 5000 years of folk medicine, in *Folk Medicine: The Art and Science,* Steiner, R. P., Ed., American Chemical Society, Washington, D.C., 1986, 125.

32. **Block, E.,** The chemistry of garlic and onions, *Sci. Am.,* 252, 114, 1985.

33. **Block, E., Ahmad, S., Jain, M. K., Crecely, R. W., Apitz-Castro, R., and Cruz, M. R.,** (E,Z)-Ajoene: a potent antithrombic agent from garlic, *J. Am. Chem. Soc.,* 106, 8295, 1984.

34. **Block, E., Ahmad, S., Catalfamo, J. L., Jain, M. K., and Apitz-Castro, R.,** Antithrombic organosulfur compounds from garlic: structural, mechanistic and synthetic studies, *J. Am. Chem. Soc.,* 108, 7045, 1986.

35. **Block, E., Iyer, R., Grisoni, S., Saha, C., Belman, S., and Lossing, F. P.,** Lipoxygenase inhibitors from the essential oil of garlic. Markovnikov addition of the allyldithio radical to olefins, *J. Am. Chem. Soc.,* 110, 7813, 1988.

36. **Natori, S., Ikekawa, N., and Suzuki, M., Eds.,** *Advances in Natural Products Chemistry: Extraction and Isolation of Biologically Active Compounds,* John Wiley & Sons, New York, 1981, chaps. 20, 21, and 28.

37. **Ito, Y. and Bowman, R. L.,** Countercurrent chromatography: liquid–liquid partition chromatography without solid support, *J. Chromatogr. Sci.,* 8, 315, 1970.

38. **Ito, Y. and Conway, W. D.,** Development of countercurrent chromatography, *Anal. Chem.,* 56, 534A, 1984.

39. **Mandava, N. B., Ed.,** Special issue: countercurrent chromatography, *J. Liq. Chromatogr.,* 7, 1984.

40. **McAlpine, J. B. and Hochlowski, J. E.,** Countercurrent chromatography, in *Natural Products Isolation: Separation Methods for Antimicrobials, Antivirals and Enzyme Inhibitors,* Wagman, G. H. and Cooper, R., Eds., Elsevier, New York, 1989, chap. 1.

41. **Snyder, J. K., Nakanishi, K., Hostettmann, K., and Hostettmann, M.,** Applications of rotation locular countercurrent chromatography in natural products isolation, *J. Liq. Chromatogr.,* 7, 243, 1984.

42. **Kubo, I., Marshall, G. T., and Hanke, F. J.,** Rotation locular countercurrent chromatography for natural products isolation, *Chromatogr. Sci.,* 44, 493, 1988.

43. **Ogihara, Y., Inoue, O., Otsuka, H., Kawai, K., Tanuimura, T., and Shibata, S.,** Droplet counter-current chromatography for the separation of plant products, *J. Chromatogr.,* 128, 218, 1976.

44. **Hostettmann, K.,** Droplet countercurrent chromatography and its application to the preparative scale separation of natural products, *Planta Med.,* 39, 1, 1980.

45. **Hostettmann, K., Appolonia, C., Domon, B., and Hostettmann, M.,** Droplet countercurrent chromatography — new applications in natural products chemistry, *J. Liq. Chromatogr.,* 7, 231, 1984.

46. **Hostettmann, K.,** Droplet counter-current chromatography, in *Advances in Chromatography,* Vol. 21, Giddings, J. C., Grushka, E., Cazes, J., and Brown, P. R., Eds., Marcel Dekker, New York, 1982, chap. 5.

47. **Knapp, A.,** Troepfchengegenstromchromatographie und Verwandte Verfahren, *Pharm. Unserer Zeit,* 14, 77, 1985.

48. **Ito, Y.,** Development of high-speed countercurrent chromatography, in *Advances in Chromatography,* Vol. 24, Giddings, J. C., Grushka, E., Cazes, J., and Brown, P. R., Eds., Marcel Dekker, New York, 1985, chap. 6.

49. **Ito, Y.,** Cross-axis synchronous flow-through coil planet centrifuge free of rotary seals for preparative countercurrent chromatography. Part I. Apparatus and analysis of acceleration, *Sep. Sci. Tech.,* 22, 1971, 1987.

50. **Ito, Y.,** Cross-axis synchronous flow-through coil planet centrifuge free of rotary seals for preparative countercurrent chromatography. Part II. Studies on phase distribution and partition efficiency in coaxial coils, *Sep. Sci. Tech.,* 22, 1989, 1987.

51. **Martin, D. G., Biles, C., and Peltonen, R. E.,** Countercurrent chromatography in the fractionation of natural products, *Am. Lab.,* 21, Oct., 1986.

52. **Yang, Y. M., Lloyd, H. A., Pannell, L. K., Fales, H. M., Macfarlane, R. D., McNeal, C. J., and Ito, Y.,** Separation of the components of commercial digitonin using high-performance liquid chromatography and centrifugal countercurrent chromatography. Identification of the products by californium^{-252} plasma desorption mass spectrometry, *Biomed. Environ. Mass Spectrom.,* 13, 439, 1986.

53. **Snyder, J. K., Bruening, R., Derguini, F., and Nakanishi, K.,** Fractionation and proof of structure of natural products, in *Natural Products of Woody Plants ,* Vol. 1, Rowe, J. W., Ed., Springer-Verlag, New York, 1989, chap. 2.

54. **Hanke, F. J. and Kubo, I.,** Increasing the speed of droplet counter-current chromatography separations, *J. Chromatogr.,* 329, 395, 1985.

55. **Kubo, I., Hanke, F. J., and Marshall, G. T.,** Droplet countercurrent chromatography, recent applications in natural products chemistry, *J. Liq. Chromatogr.,* 11, 173, 1988.

56. **Hostettmann, K. and Marston, A.,** Natural products isolation by droplet countercurrent chromatography, *Chromatogr. Sci.,* 44, 465, 1988.

57. **Kessler, H., Gehrke, M., and Griesinger, C.,** Two-dimensional NMR spectroscopy: background and overview of experiments, *Angew. Chem. Int. Ed. Eng.,* 27, 490, 1988.

58. **Doddrell, D. M., Pegg, D. T., and Bendall, M. R.,** Distortionless enhancement of NMR signals by polarization transfer, *J. Magn. Reson.,* 48, 323, 1982.

59. **Patt, S. L. and Shoolery, J. N.,** Attached proton test for carbon-13 NMR, *J. Magn. Reson.,* 46, 535, 1982.

60. **LeCocq, C. and Lallemand, J.-Y.,** Precise carbon-13 nuclear magnetic resonance multiplicity determination, *J. Chem. Soc. Chem. Commun.,* 150, 1982.

61. **Agrawal, P. K., Jain, D. C., Gupta, R. K., and Thakur, R. S.,** Carbon-13 NMR spectroscopy of steroidal sapongenins and steroidal saponins, *Phytochemistry,* 24, 2479, 1985.

62. **Aue, W. P., Bartholdi, E., and Ernst, R. R.,** Two-dimensional spectroscopy. Applications to nuclear magnetic resonance, *J. Chem. Phys.,* 64, 2229, 1976.

63. **Bax, A., Freeman, R., and Morris, G.,** Correlation of proton chemical shifts by two-dimensional Fourier transform NMR, *J. Magn. Reson.,* 42, 164, 1981.

64. **Bax, A. and Freeman, R.,** Investigation of complex networks of spin–spin coupling by two-dimensional NMR, *J. Magn. Reson.,* 44, 542, 1981.

65. **Piantini, U., Sorensen, O., and Ernst, R. R.,** Multiple quantum filters for elucidating NMR coupling networks, *J. Am. Chem. Soc.,* 104, 6800, 1982.

66. **Shaka, A. J. and Freeman, R.,** Simplification of NMR spectra by filtration through multiple-quantum coherence, *J. Magn. Reson.,* 51, 169, 1983.

67. **Rance, M., Sorensen, O. W., Bodenhausen, G., Wagner, G., Ernst, R. R., and Wuthrich, K.,** Improved spectral resolution in COSY ^{1}H-NMR spectra of proteins via double quantum filtering, *Biochem. Biophys. Res. Commun.,* 117, 479, 1983.

68. **Muller, N., Ernst, R. R., and Wuthrich, K.,** Multiple-quantum-filtered two-dimensional correlated NMR spectroscopy of proteins, *J. Am. Chem. Soc.,* 108, 6482, 1986.

69. **Eich, G., Bodenhausen, G., and Ernst, R. R.,** Exploring nuclear spin systems by relayed magnetization transfer, *J. Am. Chem. Soc.,* 104, 3731, 1982.

70. **Bax, A. and Drobny, G.,** Optimization of two-dimensional homonuclear relayed coherence transfer NMR spectroscopy, *J. Magn. Reson.,* 61, 306, 1985.

71. **Braunschweiler, L. and Ernst, R. R.,** Coherence transfer by isotropic mixing: application to proton correlation spectroscopy, *J. Magn. Reson.,* 53, 521, 1983.

72. **Davis, D. G. and Bax, A.,** Assignment of complex ^{1}H-NMR spectra via [2-D] homonuclear Hartman-Hahn spectroscopy, *J. Am. Chem. Soc.,* 107, 2820, 1985.

73. **Schmitz, F. J., Ksebati, M. B., Gunasekera, S. P., and Agarwal, S.,** Sarasinoside A$_1$: a saponin containing amino sugars isolated from a sponge, *J. Org. Chem.,* 53, 5941, 1988.

74. **Ksebati, M. B., Schmitz, F. J., and Gunasekera, S. P.,** Pouosides A–E, novel triterpene galactosides from a marine sponge, *Asteropus* sp., *J. Org. Chem.,* 53, 3917, 1988.

75. **Bax, A.,** *Two-Dimensional Nuclear Magnetic Resonance in Liquids,* D. Reidel, Boston, 1982, chap. 2.

76. **Ernst, R. R., Bodenhausen, G., and Wokaun, A.,** *Principles of Nuclear Magnetic Resonance in One and Two Dimensions,* Oxford University Press, Oxford, 1987, 471.

77. **Bax, A. and Subramanian, S.,** Sensitivity-enhanced two-dimensional heteronuclear shift correlation NMR spectroscopy, *J. Magn. Reson.,* 67, 565, 1986.

78. **Maudsley, A. A. and Ernst, R. R.,** Indirect detection of magnetic resonance by heteronuclear two-dimensional spectroscopy, *Chem. Phys. Lett.,* 50, 368, 1977.

79. **Perpick-Dumont, M., Reynolds, W. F., and Enriquez, R. G.,** Evaluation of pulse sequences combining ^{13}C-^{1}H shift correlation and heteronuclear J spectroscopy with full ^{1}H-^{1}H decoupling, *Magn. Reson. Chem.,* 26, 881, 1988.

80. **Bodenhausen, G. and Freeman, R.,** Correlation of chemical shifts of protons and carbon-13, *J. Am. Chem. Soc.,* 100, 320, 1978.

81. **Bax, A.,** Broadband homonuclear decoupling in heteronuclear shift correlation NMR spectroscopy, *J. Magn. Reson.,* 53, 517, 1983.

82. **Kessler, H., Griesinger, C., Zarbock, J., and Loosli, H. R.,** Assignment of carbonyl carbons and sequence analysis in peptides by heteronuclear shift correlation via small coupling constants with broadband decoupling in t_1 (COLOC), *J. Magn Reson.,* 57, 331, 1984.

83. **Reynolds, W. F., Hughes, D. W., Perpick-Dumont, M., and Enriquez, R. G.,** Pulse sequence which provides rapid, routine ^{13}C-^{1}H shift correlated spectra, *J. Magn. Reson.,* 64, 303, 1985.

84. **Reynolds, W. F., McLean, S., Perpick-Dumont, M., and Enriquez, R. G.,** ^{13}C-^{1}H shift correlation with full ^{1}H-^{1}H decoupling, *Magn. Reson. Chem.,* 26, 1068, 1988.

85. **Reynolds, W. F., McLean, S., Perpick-Dumont, M., and Enriquez, R. G.,** Improved ^{13}C-^{1}H shift correlation spectra for indirectly bonded carbons and hydrogens: the FLOCK sequence, *Magn. Reson. Chem.,* 27, 162, 1989.

86. **Bax, A.,** Structure determination and spectra assignment by pulsed polarization transfer *via* long-range ^{1}H-^{13}C couplings, *Magn. Reson.,* 57, 314, 1984.

87. **Jakobsen, H. J., Bildsoe, H., Donstrup, S., and Sorensen, O. W.,** Simple one-dimensional NMR experiments for heteronuclear shift correlation, *J. Magn. Reson.,* 57, 324, 1984.

88. **Bax, A. and Summers, M. F.,** ^{1}H and ^{13}C assignments from sensitivity-enhanced detection of heteronuclear multiple-bond connectivity by 2D-multiple quantum NMR, *J. Am. Chem. Soc.,* 108, 2093, 1986.

89. **Jeener, J., Meier, B. H., Bachmann, P., and Ernst, R. R.,** Investigation of exchange processes by 2D-NMR spectroscopy, *J. Chem. Phys.,* 71, 4546, 1979.
90. **Macura, S. and Ernst, R. R.,** Elucidation of cross relaxation in liquids by two-dimensional NMR, *Mol. Phys.,* 41, 95, 1980.
91. **Snatzke, G., Ed.,** *Optical Rotary Dispersion and Circular Dichroism in Organic Chemistry,* Heyden, London, 1967.
92. **Crabbe, P.,** *Optical Rotary Dispersion and Circular Dichroism in Chemistry and Biochemistry,* Academic Press, New York, 1972.
93. **Harada, N. and Nakanishi, K.,** *Circular Dichroism Spectroscopy — Exciton Coupling in Organic Stereochemistry,* University Science Books, Mill Valley, CA, 1983.
94. **Kusumi, T., Ohtani, I., Inouye, Y., and Kakisawa, H.,** Absolute configuration of cytotoxic marine cembranolides; consideration of Mosher's method, *Tetrahedron Lett.,* 29, 4731, 1988.
95. **Ohtani, I., Kusumi, T., Ishitsuka, M. O., and Kakisawa, H.,** Absolute configuration of marine diterpenes possessing a xenicane skeleton. An advanced application of Mosher's method, *Tetrahedron Lett.,* 30, 3147, 1989.
96. **Inouye, Y., Ohtani, I., Kusumi, T., Kashman, Y., and Kakisawa, H.,** Absolute configuration of sipholenol-A. Presence of two conformers of MPTA ester moiety in a crystalline state, *Chem. Lett.,* 2073, 1990.
97. **Ohtani, I., Kusumi, T., Kashman, Y., and Kakisawa, H.,** A new aspect of the high-field NMR application of Mosher's method. The absolute configuration of marine triterpene sipholenol-A, *J. Org. Chem.,* 56, 1296, 1991.
98. **Ohtani, I., Kusumi, T., Kashman, Y., and Kakisawa, H.,** High field FT-NMR application of Mosher's method. Absolute configurations of marine terpenoids, *J. Am. Chem. Soc.,* 113, 4092, 1991.
99. **Trost, B. M., Belletire, J. L., Godleski, S., McDougal, P. G., Balkovec, J. M., Baldwin, J. J., Christy, M. E., Ponticello, G. S., Varga, S. L., and Springer, J. P.,** On the use of the O-methylmandelate ester for establishment of absolute configuration of secondary alcohols, *J. Org. Chem.,* 51, 2370, 1986.
100. **Adamczeski, M., Quinoa, E., and Crews, P.,** Novel sponge-derived amino acids. XI. The entire absolute stereochemistry of the bengamides, *J. Org. Chem.,* 55, 240, 1990.
101. **Panek, J. S. and Sparks, M. A.,** Synthesis, resolution and absolute stereochemical assignment of C1-oxygenated allylsilanes and C3-oxygenated vinylsilanes, *Tetrahedron Asymm.,* 1, 801, 1990.
102. **Dorman, D. E. and Roberts, J. D.,** Carbon-13 of some pentose and hexose aldopyranoses, *J. Am. Chem. Soc.,* 92, 1355, 1970.
103. **Gorin, P. A. J.,** Deuterium isotope effect on shifts of ^{13}C magnetic resonance signals of sugars: signal assignment studies, *Can. J. Chem.,* 52, 458, 1974.
104. **Gorin, P. A. J. and Mazurek, M.,** Further studies on the assignments of signals in ^{13}C magnetic resonance spectra of aldoses and derived methyl glycosides, *Can. J. Chem.,* 53, 1212, 1975.
105. **Ritchie, R. G. S., Cyr, N., Korsh, B., Koch, H. J., and Perlin, A. S.,** Carbon-13 chemical shifts of furanosides and cyclopentanols. Configurations and conformational influences, *Can. J. Chem.,* 53, 1424, 1975.
106. **Gorin, P. A. J.,** Carbon-13 nuclear magnetic resonance spectroscopy of polysaccharides, in *Advances in Carbohydrate Chemistry and Biochemistry,* Vol. 38, Tipson, R. S. and Horton, D., Eds., Academic Press, New York, 1981, 13.
107. **Bock, K. and Pedersen, C.,** Carbon-13 nuclear magnetic resonance spectroscopy of monosaccharides, in *Advances in Carbohydrate Chemistry and Biochemistry,* Vol. 41, Tipson, R. S. and Horton, D., Eds., Academic Press, New York, 1983, 27.
108. **Bock, K., Pedersen, C., and Pedersen, H.,** Carbon-13 nuclear magnetic resonance data of oligosaccharides, in *Advances in Carbohydrate Chemistry and Biochemistry,* Vol. 42, Tipson, R. S. and Horton, D., Eds., Academic Press, New York, 1984, 193.
109. **Bodenhausen, G., Freeman, R., Niedermeyer, R., and Turner, D. L.,** High-resolution NMR in inhomogeneous magnetic fields, *J. Magn. Reson.,* 24, 291, 1976.
110. **Bax, A.,** *Two-Dimensional Nuclear Magnetic Resonance in Liquids,* D. Reidel, Boston, 1982, chap. 3.
111. **Batta, G. and Liptak, A.,** Long range ^1H-^1H spin couplings through the interglycosidic oxygen and the primary structure of oligosaccharides as studied by 2D-NMR, *J. Am. Chem. Soc.,* 106, 248, 1984.
112. **Gidley, M. J. and Bociek, S. M.,** Long range ^{13}C-^1H coupling in carbohydrates by selective 2D heteronuclear J-resolved NMR spectroscopy, *J. Chem. Soc. Chem. Commun.,* 220, 1985.
113. **Morat, C. and Taravel, F. R.,** Interglycosidic ^{13}C-^1H coupling constants. An approach to disaccharide conformations, *Tetrahedron Lett.,* 29, 199, 1988.
114. **Allerhand, A. and Doddrell, D.,** Strategies in the application of partially relaxed Fourier transform nuclear magnetic resonance spectroscopy in assignments of carbon-13 resonances of complex molecules. Stachyose., *J. Am. Chem. Soc.,* 93, 2777, 1971.
115. **Doddrell, D. M.,** Structural applications of nuclear spin-lattice relaxation times, *Pure Appl. Chem.,* 49, 1385, 1977.

116. **Lonngren, J. and Svensson, S.,** Mass spectrometry in structural analysis of natural carbohydrates, in *Advances in Carbohydrate Chemistry and Biochemistry,* Vol. 29, Tipson, R. S. and Horton, D., Eds., Academic Press, New York, 1974, 42.

117. **Radford, T. and DeJongh, D. C.,** Carbohydrates, in *Biochemical Applications of Mass Spectrometry,* 1st Suppl. Vol., Waller, G. R. and Dermer, O. C., Eds., John Wiley & Sons, New York, 1980, chap. 12.

118. **Allerhand, A. and Komoroski, R. A.,** Study of internal rotations in gramicidin S by means of carbon-13 spin-lattice relaxation mechanisms, *J. Am. Chem. Soc.,* 95, 8228, 1973.

119. **Lyerla, J. R. and Levy, G. C.,** Carbon-13 nuclear spin relaxation, in *Topics in Carbon-13 NMR Spectroscopy,* Vol. 1, John Wiley & Sons, New York, 1974, chap. 3.

120. **Breitmaier, E., Spohn, K.-H., and Berger, S.,** ^{13}C Spin-lattice relaxation times and the mobility of organic molecules, *Angew. Chem. Int. Ed. Eng.,* 14, 144, 1975.

121. **Barber, M., Bordoli, R. S., Elliott, G. J., Sedgwick, R. D., and Tyler, A. N.,** Fast atom bombardment mass spectrometry, *Anal. Chem.,* 54, 645A, 1982.

122. **Howe, I. and Jarman, M.,** New techniques for the mass spectrometry of natural products, in *Progress in the Chemistry of Natural Products,* Vol. 47, Herz, W., Grisebach, H., Kirby, G. W., and Tamm, Ch., Eds., Springer-Verlag, New York, 1985, 107.

123. **Dell, A.,** F.A.B. mass spectrometry of carbohydrates, in *Advances in Carbohydrate Chemistry and Biochemistry,* Vol. 45, Tipson, R. S. and Horton, D., Eds., Academic Press, New York, 1987, 20.

124. **Bowie, J. H.,** The fragmentations of even-electron organic negative ions, *Mass Spectrom. Rev.,* 9, 349, 1990.

125. **Komori, T., Kawasaki, T., and Schulten, H.-R.,** Field desorption and fast atom bombardment mass spectrometry of biologically active oligosaccharides, *Mass Spectrom. Rev.,* 4, 225, 1985.

126. **Mueller, D. R., Domon, B., Raschdorf, F., and Richter, W. J.,** Applications of tandem mass spectrometry in the structure determination of complex biomolecules. Direct stereochemical assignment of mono- and disaccharide sub-units in larger glycosides by low energy collision induced dissociation, *Adv. Mass Spectrom.,* 11B, 1309, 1989.

127. **Orlando, R., Bush, C. A., and Fenselau, C.,** Analysis of oligosaccharides by tandem mass spectrometry: collisional activation of sodium adduct ions, *Biomed. Environ. Mass Spectrom.,* 19, 474, 1990.

128. **Watanabe, Y., Sanada, S., Ida, Y., and Shoji, J.,** Comparative studies on the constituents of *Ophiopogonis* tuber and its congeners. II. Studies on the constituents of the subterranean part of *Ophiopogon planiscapus* Nakai (1), *Chem. Pharm. Bull.,* 31, 3486, 1983.

129. **Tschesche, R. and Richert, K. H.,** Uber Saponine der Spirostanolreihe. XI. Nuatigenin, ein Cholegenin-Analogon des Pflanzenreiches, *Tetrahedron,* 20, 387, 1964.

130. **Aue, W. P., Karhan, J., and Ernst, R. R.,** Homonuclear broadband decoupling and two-dimensional J-resolved NMR spectroscopy, *J. Chem. Phys.,* 64, 4226, 1976.

131. **Ernst, R. R., Bodenhausen, G., and Wokaun, A.,** *Principles of Nuclear Magnetic Resonance in One and Two Dimensions,* Oxford University Press, Oxford, 1987, 360.

132. **Kasai, R., Okihara, M., Asakawa, J., Mizutani, K., and Tanaka, O.,** ^{13}C-NMR Study of α- and β-anomeric pairs of D-mannopyranosides and L-rhamnopyranosides, *Tetrahedron,* 35, 1427, 1979.

133. **Bock, K. and Pedersen, C.,** A study of ^{13}CH coupling constants in hexopyranoses, *J. Chem. Soc. Perk. Trans. 2,* 293, 1974.

134. **Ernst, R. R., Bodenhausen, G., and Wokaun, A.,** *Principles of Nuclear Magnetic Resonance in One and Two Dimensions,* Oxford University Press, Oxford, 1987, 402.

135. **Bock, K., Lundt, I., and Pedersen, C.,** Assignment of anomeric structure to carbohydrates through geminal ^{13}C-H coupling constants, *Tetrahedron Lett.,* 1037, 1973.

136. **Noggle, J. H.,** *The Nuclear Overhauser Effect; Chemical Applications,* Academic Press, New York, 1971.

137. **Neuhaus, D. and Williamson, M.,** *The Nuclear Overhauser Effect In Structural and Conformational Analysis,* VCH Publishers, New York, 1989.

138. **Bothnerby, A. A., Stephens, R. L., Lee, J., Warren, C. D., and Jeanloz, R. W.,** Structure determination of a tetrasaccharide: transient nuclear Overhauser effects in the rotating frame, *J. Am. Chem. Soc.,* 106, 811, 1984.

139. **Bax, A. and Davis, D. G.,** Practical aspects of two-dimensional transverse NOE spectroscopy, *J. Magn. Reson.,* 63, 207, 1985.

140. **Kessler, H., Griesinger, C., Kerssebaum, R., Wagner, K., and Ernst, R. R.,** Separation of cross-relaxation and J cross-peaks in 2D rotating-frame NMR spectroscopy, *J. Am. Chem. Soc.,* 109, 607, 1987.

141. **Radford, T. and DeJongh, D. C.,** Carbohydrates, in *Biochemical Applications of Mass Spectrometry,* 1st Suppl. Vol., Waller, G. R. and Dermer, O. C., Eds., John Wiley & Sons, New York, 1980, 272.

142. **Dalton, C. G. S.,** Applications of gas–liquid chromatography to carbohydrates: part I, in *Advances in Carbohydrate Chemistry and Biochemistry,* Vol. 28, Tipson, R. S. and Horton, D., Eds., Academic Press, New York, 1973, 12.

143. **Dalton, C. G. S.,** Applications of gas–liquid chromatography to carbohydrates: part II, in *Advances in Carbohydrate Chemistry and Biochemistry,* Vol. 30, Tipson, R. S. and Horton, D., Eds., Academic Press, New York, 1974, 10.

144. **Evershed, R. P.,** Analysis of mixtures by mass spectrometry, part I: developments in gas chromatography/ mass spectrometry, *Mass Spectrom.,* 10, 181, 1989.

145. **Sweeley, C. C., Bentley, R., Makita, M., and Wells, W. W.,** Gas–liquid chromatography of trimethylsilyl derivatives of sugars and related substances, *J. Am. Chem. Soc.,* 85, 2497, 1963.

146. **Windholz, M., Ed.,** *The Merck Index,* 10th ed., Merck, Rahway, NJ, 1983, 481.

147. **Gerwig, G. J., Kamerling, J. P., and Vliegenthart, J. F. G.,** Determination of the D and L configuration of neutral monosaccharides by high-resolution capillary g.l.c., *Carbohydr. Res.,* 62, 349, 1978.

148. **Leontein, K., Lindberg, B., and Lonngren, J.,** Assignment of absolute configuration of sugars by g.l.c. of the acetylated glycosides formed from chiral alcohols, *Carbohydr. Res.,* 62, 359, 1978.

149. **Griesinger, C., Sorensen, O. W., and Ernst, R. R.,** Novel three dimensional NMR techniques for studies of peptides and biological macromolecules, *J. Am. Chem. Soc.,* 109, 7227, 1987.

150. **Fesik, S. W., Gampe, R. T., and Zuiderweg, E. R. P.,** Heteronuclear three-dimensional NMR spectroscopy. Natural abundance ^{13}C chemical shift editing of ^{1}H-^{1}H COSY spectra, *J. Am. Chem. Soc.,* 111, 770, 1988.

151. **Vuister, G. W., de Waard, P., Boelens, R., Vliegenthart, J. F. G., and Kaptein, R.,** The use of 3D NMR in structural studies of oligosaccharides, *J. Am. Chem. Soc.,* 111, 772, 1988.

152. **Yahara, S., Kasai, R., and Tanaka, O.,** New dammararne type saponins of leaves of *Panax japonicus,* chikutsusaponins L_5–L_{9a} and L_{10}, *Chem. Pharm. Bull.,* 25, 2041, 1977.

153. **Neszmelyi, A., Tori, K., and Lukacs, G.,** Use of C-13 spin lattice relaxation times for sugar sequence determination, *J. Chem. Soc. Chem. Commun.,* 613, 1977.

154. **Ishii, H., Kitagawa, I., Matsushita, F., Shirakawa, K., Tori, K., Tozyo, T., Yoshikawa, M., and Yoshimura, Y.,** The configuration and conformation of the arabinose moiety in platycodins, saponins isolated from *Platycodon grandiflorum,* and Mi saponins from *Madhuca longifolia* based on carbon-13 and hydrogen-1 NMR spectroscopic evidence: the total structures of the saponins, *Tetrahedron Lett.,* 22, 1529, 1981.

155. **Hirai, Y., Konishi, T., Sanada, S., Ida, Y., and Shoji, J.,** Studies on the constituents of *Aspiistra elatior* Blume. I. Steroids of underground parts., *Chem. Pharm. Bull.,* 30, 3476, 1982.

156. **Rowan, D. D. and Newman, R. H.,** Noroleanane saponins from *Celmisia petriei, Phytochemistry,* 23, 639, 1984.

Chapter 17

ANTIVIRAL AGENTS FROM HIGHER PLANTS AND AN EXAMPLE OF STRUCTURE–ACTIVITY RELATIONSHIP OF 3-METHOXYFLAVONES

Dirk A. R. Vanden Berghe, Achiel Haemers, and Arnold J. Vlietinck

TABLE OF CONTENTS

I. INTRODUCTION

Although vaccines have been very successful in controlling many viral diseases, some diseases are likely to be controlled only by antiviral chemotherapy. The concept of antiviral drugs has only been accepted slowly, partly because of the toxicity of many of the earlier antiviral agents. The development of compounds useful for the prophylaxis and/or therapy of viral diseases has indeed presented more difficult problems than those encountered in the search for drugs effective in disorders produced by other microorganisms. This is due to the nature of these infectious agents, which depend totally upon the cell they infect for their multiplication and survival. Thus, agents that may inhibit or cause the death of viruses are also likely to injure the host cells that harbor them. As a consequence there are, at the moment, only a few antiviral drugs available for the cure of viral diseases, and even these have a very narrow activity, limited to one or only a few specific viruses, and are not devoid of toxic side effects.

Hence, there is a strong need not only to improve the actual antiviral armamentarium, but also to find an effective therapy for viral infections for which, at present, no clinically useful drugs or vaccines are available. There is also a need for finding new substances with extracellular virucidal activity, since many of the existing disinfectants and antiseptics fail to kill all pathogenic viruses after a 5-min exposure time at room temperature.[1,2]

Therefore, all possible approaches toward the development of new antiviral and virucidal drugs should be pursued. One of the possible methodologies that can be used for the discovery of such drugs is the screening of selected plant extracts for antiviral activity, followed by bioassay-guided fractionation of active plants, leading to the isolation of the pure constituents.

Such a research program is best carried out by a multidisciplinary team consisting of at least a pharmacognocist and a virologist. The antiviral screening system should meet all requirements of any good assay, including validity, lack of ambiguity, accuracy, reproducibility, simplicity, and reasonable cost. Moreover, since we are dealing with plant extracts, the antiviral screen should be highly selective, specific, and sensitive. At this stage it is advisable to discriminate a true antiviral activity from a virucidal activity. Since most of the aforementioned requirements are better met by *in vitro* testing, we not only prefer *in vitro* screening of the plant extracts, but also the use of the same bioassay to guide the isolation of the antivirally active compounds from the plant extracts. The antiviral activity of the pure compounds then has to be confirmed in a later stage by *in vivo* assays.[3,4]

We refer readers to the appropriate literature for an extensive survey of all factors influencing the design of antiviral chemotherapy experiments.[5-12]

II. VIRUS REPLICATION

Before considering agents that have been, or are being, developed for the treatment of virus infections, it may be useful to review how viruses replicate. Since viruses are obligate intracellular parasites, they totally depend on the host cell molecular mechanisms for their replication. A simplified diagrammatic representation of virus replication is shown in Figure 1. The first step consists of the adsorption of a virus particle to the host cell. This step is very specific and leads to the entry of the virus into the cell. A successful collision, by the aid of ions, leads to the early weak attachment of the virus to the cell. For certain viruses, specific virus sites will bind to appropriate cellular receptor sites, although this binding does not always lead to an infection, since attachment might be reversible. The receptors of the plasma membranes of the cells are presumably glycoproteins, whereas protrusions from the outer viral surface, called spikes, are the organs of cell attachment for viruses such as toga-, adeno-, paramyxo-, myxo-, and rhabdoviruses. In viruses lacking these spikes, including picornaviruses, complex binding site polypeptides are involved, the capacity of which depends on the virus architecture.

1. ATTACHMENT (ADSORPTION) OF THE VIRAL PARTICLE TO THE HOST CELL

2. HOST-CELL PENETRATION BY THE INFECTIOUS VIRAL PARTICLE

3. UNCOATING OF THE PARTICLES WITH SUBSEQUENT RELEASE AND TRANSPORT OF VIRAL NUCLEIC ACID AND VIRAL CURE PROTEINS

4. RELEASE AND/OR ACTIVATION OF NUCLEIC-ACID POLYMERASE

5. TRANSLATION OF M-RNA TO POLYPEPTIDES (EARLY PROTEINS)

6. TRANSCRIPTION OF M-RNA

7. REPLICATION OF NUCLEIC ACIDS

8. PROTEIN SYNTHESIS (LATE PROTEINS)

9. CLEAVAGE OF VIRAL POLYPEPTIDES INTO USEFUL POLYPEPTIDES FOR MATURATION

1O. MORPHOGENESIS AND ASSEMBLY OF VIRAL CAPSID AND PRECURSORS

11. ENCAPSIDATION OF NUCLEIC ACID

12. ENVELOPMENT

13. RELEASE

FIGURE 1. Replication of viruses; 13 targets for specific antiviral chemotherapy.

After binding, the virus particle partially or completely enters the cell. This double step is termed "penetration" and "uncoating" and always results in the synthesis of virus-specific proteins or mRNA (steps 2 and 3). A virus may have different mechanisms of penetration into the host cell. For enveloped viruses, fusion of the membrane sometimes occurs, but most viruses are introduced into the cell by a kind of phagocytosis, called viropexis. Following penetration, virus particles are transported, probably along the network of "microtubules" of the cytoplasm, to a specific cell site where subsequent replication should take place. Uncoating results in the liberation of viral nucleic acid into the cell. This event renders virus nucleic acid sensitive to nucleases. For many enveloped viruses, partial or complete uncoating takes place during penetration and transportation, especially when cytoplasm fusion of the viral envelope occurs with cell membranes. Penetration and uncoating finally result in the viral genome together with its virus-associated RNA and DNA polymerase or with a part of the virus structure, which is necessary for the initiation of replication, entering into the cell cytoplasm or nucleus. The remaining part of the virion may stay active outside the cell or in the endocytic vesicles.

There are two types of viruses with a virion polymerase. The first type has a polymerase that is activated when the envelope is already partially disrupted (e.g., vaccinia, RNA tumor viruses, NDV, etc). In the second group the viral capsid has to be partially disrupted by proteolytic digestion, in order to activate the polymerase (e.g., reovirus, cytoplasmic polyhedrosis virus of insects, etc.).

Most viruses probably do not contain a polymerase inside the particle. Penetration, uncoating, and activation of polymerase may occur simultaneously for some viruses. A successful uncoating and penetration always leads to the start of transcription and replication. In these processes viral mRNA(+) plays a major role. Therefore, the relationship between virion nucleic acid and viral mRNA(+) is important. All different types of viruses have to make their own mRNA according to the nature of their virion nucleic acid (e.g., double-stranded DNA, single-stranded DNA(+), single-stranded RNA(+), etc.). The mode of duplication of genetic material to obtain mRNA(+) for protein synthesis, the site in the cell for biosynthesis, the specificity of the replication cycle, and morphogenesis result in different replication patterns. It is difficult to give a general schematic view of these events. Nevertheless, general characteristics are common to most of these replication patterns, such as release and activation of RNA-polymerase (step 4); translation of mRNA to polypeptides (early proteins) for the formation of RNA-polymerase or other polymerases (step 5); transcription of mRNA (step 6), leading to the replication of nucleic acids (step 7); and synthesis of all viral proteins necessary for morphogenesis (late proteins) (step 8). Some newly formed polypeptides have to be cleaved during the generation of the proteins by proteolysis, to yield morphologically and enzymatically active proteins (step 9). Late in the infection, the capsid is built up and assembled (step 10). Therefore, the newly formed proteins have to be transported to the site of morphogenesis (endoplasmic reticulum, nucleus, etc.). The nucleic acid is then associated with the capsid to give a nucleoprotein (step 11) that is released from the cell by envelopment or lysis.

Envelopment requires the synthesis of a specific viral matrix and glycoproteins that bind to the inner surface of the cell membrane so that the glycoproteins are inserted into the lipid bilayer of the cell membrane (step 12). The nucleocapsids bud through the modified membrane and are liberated from the cell if the membrane is a plasma membrane. If the membrane is intracellular (e.g., nuclear membrane for Herpesviridae) or vacuolar (e.g., Togaviridae), the developed particles migrate to the plasma membrane inside the membrane-bound vesicles (step 13).

III. ANTIVIRAL CHEMOTHERAPY

In contrast to the development of other types of chemotherapeutic agents, attempts to develop antiviral drugs have met a variety of problems. Being strictly dependent on cellular metabolic processes, viruses possess only limited intrinsic enzyme systems and building blocks that may serve as specific targets for a drug. Moreover, contrary to a bacteriostatic compound, an effective antiviral drug should not only display considerable specificity in its antiviral action, but should also irreversibly block viral synthesis, in order to stop cell suicide due to the viral infection, and restore normal cell synthesis.

Most known antiviral products indeed have a reversible antiviral activity, which means that virus synthesis will start again once the compounds are omitted. These compounds are consequently unable to cure the infected cells. Furthermore, there exists a problem of latency, so that in an infected host only the multiplication of active virus particles is inhibited, but not the latent viral genomes that are scattered throughout the organism. Consequently, the virus is not eradicated in the infected host. Although the term "eradication" is rather difficult to use in antiviral chemotherapy, eradication can nevertheless occur if irreversibly acting drugs are used and if the viruses present are, at the same time, undergoing the biosynthesis step, which is attacked by the drug. This requires activation of a latent virus simultaneously with antiviral activity itself.

In addition, in order to be effective, an antiviral drug must enter the cells so as to achieve adequate intracellular concentration and must not be incorporated into cellular compounds of infected or noninfected cells, since cell-specific systems must remain intact. On the other hand, it should be possible to develop antiviral drugs that are preferentially activated or transformed by viral-coded enzymes during the active phase in virus-infected cells.

In the literature the term "antiviral" is used in so many contexts as to make it confusing and almost meaningless. Therefore, the many "antiviral agents" must be considered in relation to the specific biologic system in which they act, and in evaluating the systems, the ultimate goal is an agent that can be therapeutically used in animals and man. As indicated in Figure 2, an antiviral may be defined as a product that is able, *in vitro* or *in vivo*, to directly or indirectly reduce the infectious virus in the host cell. As already mentioned, most known antivirals are not useful in antiviral chemotherapy, because they are too toxic and influence or damage the host functions to a certain extent. Consequently, there is still a great need for specific antiviral drugs that show selective antiviral activity.

There are two basic approaches to finding antiviral chemotherapeutic agents. The first is to specifically or nonspecifically stimulate host defense mechanisms that act indirectly on virus multiplication (groups I–III). The best known examples are the interferon system (group Ib) and vaccines (group III), although there are a variety of other nonspecific host defense mechanisms that are important, including macrophages, lymphocyte subpopulations, and other soluble mediators (group Ia). The other approach is to limit the virulence of the invading virus by inhibiting virus replication in the infected host (groups IV and V).

The best candidates for clinically useful antiviral drugs are those substances that act on specific steps of viral biosynthesis. They reversibly or irreversibly inhibit one or more specific genomic processes of one of the replication steps, so that little or no infectious virus is produced by the virus-infected cells. They act at low concentrations that do not influence host cell mechanisms and not only prevent the spread of the virus, but also cure infected cells if the action is irreversible. These substances are termed true antivirals (group V), since they selectively block a virus-specific system in the infected host. The host cells are always involved in this kind of antiviral activity, and the toxicity of the compound for the host is very low.

Many compounds were described as antivirals because a loss of 20% or more infectivity of the virus in tissue culture was observed. Most of these substances, however, belong to the group of antivirals with an activity on extracellular virus, including denaturating agents, detergents, lipid solvents, acids, alcohols, phenolics, urea, polysaccharides, etc. (i.e., group IVb). Their activity is due to an irreversible denaturation of virus proteins or glycoproteins, so that the infectivity of the virus is completely lost. This effect is termed a virucidal activity, since viruses are killed and consequently are no longer able to invade a living cell. These compounds do not act on one of the biosynthesis steps of the virus, and high concentrations of these substances are needed for their activity. The lowering of the virus titer in *in vitro* systems by these substances is mainly due to a toxic effect on the host cell, which influences the virus titer, leading to an inactivation of extracellular virus. It might be possible to use extracellularly virucidal compounds to inhibit enveloped viruses, which are more sensitive to low concentrations of virucidal substances than other viruses, but up until now, no such clinically successful compounds have been reported.

A virucidal drug may not only inactivate the virus extracellularly so that only noninfectious virus particles remain, but it may also attack one of the replication steps in an irreversible way so that, after removal of the drug, the infectious virus is no longer formed (group Vb). In this way, virucidal substances may behave as real antivirals, since they act irreversibly on the viruses and cure the cells completely.

It is difficult to give a good definition of a virustatic compound, since virus growth depends on the host cell, and all substances affecting the host cells might also nonspecifically decrease

ANTIVIRAL

SUBSTANCE ABLE TO REDUCE
OR ERADICATE INFECTIOUS VIRUS IN THE HOST

INDIRECT ACTION

DIRECT ACTION

Group Ia
Antiviral
chemotherapy :
immunomodulators
enhance the antiviral
immune system (antibody
response, cell-mediated
immunity,
macrophage function,
different elements of
the reticulo-endothelial
system, interferon)
Group Ib
Interferon therapy

Group II
In vitro antiviral
viral effect
- action on infected
cells with consequent
reduction of virus
titre
- non-specific activity
virucidal and/or virustatic
activity with delay in
virus spread

Group III
Active immuno-
therapy :
vaccination

Group IV
Action on
extracellular virus

    Group IVa
Passive immunotherapy :
Neutralisation by means
of immune serum; prevents
spread of virus
    Group IVb
Antiviral chemotherapy :
Chemical or physical
damage with consequent
loss of infectivity;
virucidal activity;
prevents spread of virus

Group V
    True antivirals
Antiviral chemotherapy :
Selective action on a step
of the viral biosynthesis
    Group Va
Virustatic activity :
reversible block

    Group Vb
Virucidal activity :
irreversible block

FIGURE 2. Classification of antivirals.

virus growth. Some compounds decrease the number of infectious particles, but do not influence the activity or intensity of viral protein and RNA synthesis. Glucosamine shows such an antiviral activity against *Herpes* virus and measles (group Va).[13] Glucosamine is responsible for the formation of "wrong" glycoproteins, leading to a decrease in the infectivity of the virus particles. It also affects the cells so that picornaviruses are inhibited due to a change in the endoplasmic reticulum that affects the morphogenesis of these viruses (group II).

In conclusion, research in antiviral chemotherapy should be concentrated on the target antivirals (group V) and on virucidal compounds (group IVb).

IV. ANTIVIRAL TEST METHODOLOGY

The viruses to be selected for initial evaluation of plant extracts are obviously of major importance. They must be chosen to represent the different groups of viruses according to their morphology and various multiplication mechanisms and a range of virus diseases for which chemical control would be useful. Besides the need for control, the prevalence of the viral diseases and the resulting projection of the market potential, which are determined by the antigenic abundance of the causative viruses and the problems this represents for vaccine control, are important selection criteria.[14]

In vitro methods are therefore more appropriate, since they allow simultaneous screening of a battery of viruses. In contrast, *in vivo* screening of extracts against a broad array of viruses is not only very expensive, but also extremely laborious. *In vitro* antiviral bioassays utilize thinly confluent monolayers of tissue culture cells with sufficient susceptibility to the infecting viruses that a cytopathogenic effect (CPE), e.g., rounding up, shrinking, or detaching of cells from the monolayer, can be produced and readily observed microscopically. A monolayer of cells consists of animal or human cells (such as chick embryo, rabbit or green monkey kidney cells (Vero cells), or human skin fibroblasts and carcinoma cells [HeLa cells]) grown in culture medium. These continuous cell lines used in virology are mostly "transformed" cells that can be maintained for an indefinite number of generations.

The host cells require an appropriate tissue culture medium in which they can survive for at least 1 week without having to renew the medium. Renewing of the medium causes changes in intra- and extracellular products and alters the virus concentration. The medium must have a stable pH during the whole incubation time and may contain only small amounts of serum, since blood products tend to adsorb many compounds. Usually, a defined synthetic medium, supplemented by some type of serum (such as fetal bovine, calf, or horse), a buffer, and sometimes bacterial and fungal inhibitory antibiotics, is used. According to our experience Vero cells (which allow the growth of many human viruses with visible CPE) grown in a medium without a carbon dioxide buffer are most suitable for antiviral screening of plant extracts.[15]

Many combinations of test viruses are possible, but a battery of six viruses seems to be quite acceptable. Virus types and strains may vary in sensitivity, but they have to be selected as a function of their ability to multiply in the same tissue culture when cell culture models are used as screening systems. In this way an objective comparison of antiviral activities is possible, whereas toxicity tests are minimized. Moreover, virus multiplication must cause a visible CPE within a reasonable period of time, preferably within a week after infection.

The virus battery that we have been using during our screening studies is shown in Figure 3. *Herpes simplex* and adenoviruses are worthy representatives of the DNA viruses, whereas poliomyelitis, vesicular stomatitis, measles, and Semliki forest viruses are good prototypes of the RNA virus group. More specifically, polio represents the picornavirus family, which includes rhinoviruses, which are the causative agents of the common cold, and the enteroviruses, such as coxsackie and echoviruses, which are responsible for many respiratory diseases.

Virus	Family	Morphology	Serotypes	Types of infection induced
Adenovirus	Adenoviridae	Double-stranded DNA, icosahedral, naked, spikes	34	Respiratory, ophthalmic
Herpes simplex	Herpesviridae	Double-stranded DNA, icosahedral, enveloped	2	Ophthalmic, central nervous, skin, oral, upper respiratory, genital
Poliomyelitis or Coxsackie	Picornaviridae	Single-stranded RNA, icosahedral, naked	Polio:3 Coxsackie A:23 Coxsackie B:6	Respiratory, central nervous, cardiovascular
Measles	Paramyxoviridae	Single-stranded and segmented RNA, helical, enveloped	1	Respiratory, skin, nervous
Semliki forest	Togaviridae	Single-stranded RNA, polyhedral, enveloped	2	Central nervous
Vesicular stomatitis	Rhabdoviridae	Single-stranded RNA, helical, enveloped, bullet shaped	1	Respiratory

FIGURE 3. Prototype of a virus battery for screening of plant extracts.

Measles, Semliki forest, and vesicular stomatitis viruses represent the Paramyxoviridae, the Togaviridae, and the Rhabdoviridae, respectively. The latter virus is transmitted by arthropods and is able to multiply both in insect and animal cells. All these viruses cause a visible CPE on monolayers of Vero cells in tissue culture within 3 days after infection. The CPE is spread all over the monolayer, and the infected cells subsequently die. High titers, such as 10^5 TCD$_{50}$ ml^{-1} (50% tissue culture dose end point) or 10^5 PFU ml^{-1} (plaque-forming units per milliliter) or higher, can easily be obtained in Vero cells, using these viruses, which markedly increases the sensitivity of the testing system. Finally, for most of these viruses, suitable animal models are reported that enable the *in vivo* antiviral testing of the pure active compounds isolated from the selected plant extracts.[16]

A. PREPARATION OF SAMPLES FOR ANTIVIRAL TESTING

Ideally, and in contrast to antibacterial screening, no solvents other than physiologic buffer solutions should be used in the *in vitro* antiviral screening of plant extracts, since the samples have to be added to tissue culture cells. Therefore, plant extracts are usually prepared by maceration and/or percolation of fresh green plants or dried, powdered plant material with water or 80% methanol or ethanol. In order to detect antiviral substances present in very small quantities in the plant extracts, testing is carried out on concentrated extracts obtained by evaporation of the solvent *in vacuo*. This often results in the precipitation or coprecipitation of possibly active substances during the testing procedure. If the residue is soluble in water, it is dissolved in sterile 0.01 *M* physiological tris buffer and diluted with tissue culture medium. However, if this is not the case, dimethylsulfoxide (DMSO) may be used. It is our experience that many plant extracts, prepared and evaporated as described, are reasonably soluble in DMSO, especially if little or no water is present in the sample and if the dissolving sample is heated on a water bath. Virucidal and antiviral determinations may then be carried out on test solutions containing not more than 10 and 1% DMSO, respectively. Therefore, dissolved samples of nonpolar plant extracts in DMSO are added dropwise to the maintenance medium in a ratio of 1:10 or 1:100, under stirring. As already mentioned, the maintenance medium may contain antibiotics such as penicillin G (20 μg ml^{-1}), neomycin (1 μg ml^{-1}), and amphotericin B (1 μg ml^{-1}), in order to avoid sterilization of the test solutions. Any contamination by bacteria or fungi would indeed ruin the *in vitro* antiviral bioassay.

B. ANTIVIRAL TEST METHODS

1. Extracellular Virucidal Evaluation Procedures

Most currently used antiseptics and disinfectants kill pathogenic bacteria and fungi at 25°C within 5 min when present in a concentration of about 0.5% (3-log titer reduction). Since we have observed that most of these preparations failed to kill all pathogenic viruses under these circumstances, we developed a method for testing the *in vitro* virucidal effect of plant extracts.

Preincubated (25°C) plant extracts or their twofold dilutions (e.g., 1/2 to 1/16) (1 ml), dissolved in physiological buffer, are thoroughly mixed with the same volume (1 ml) of a preincubated (25°C) virus suspension (e.g., 10^6 PFU ml^{-1} or TCD$_{50}$ ml^{-1}) in physiological buffer. The mixture is incubated at 25°C for 5 min. The incubation is stopped by addition of a tenfold volume (i.e., 20 ml) of ice-cold maintenance medium, and the mixture is immediately filtered through a 0.22-μm filter to eliminate all possible precipitate. The ice-cold filtrate is then filtered through a 0.01-μm filter for enveloped viruses, or a 10,000-MW membrane filter (Amicon, ultrafiltration system) for nonenveloped viruses, to concentrate residual virus on the filter and separate the virus from possibly cytotoxic plant components that pass the filter. A small volume (0.2 ml) of the original sample is left on the filter so that the filter never becomes dry. This residual virus is removed. The filter is washed with maintenance medium supplemented with 5% serum (1 ml), sonicated in an ice-bath for 30 s,

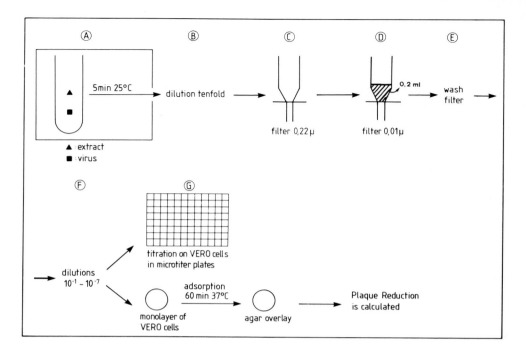

FIGURE 4a. Extracellular virucidal evaluation method.

mixed with the residual virus suspension (0.2 ml), and titrated in tenfold dilutions at 37°C by plaque formation (plaque test, PT) or in microtiter plates, according to the end-point titration technique (EPTT). A virus control in physiological buffer, containing no plant extract, is carried out simultaneously. Figure 4a shows the scheme of this virucidal test methodology.

An essential step of this methodology is the separation of all cytotoxic plant components from the residual virus (filtration), which has to be measured at 37°C. Cytotoxic substances have a greater influence on the activity of extracellular virus at 37°C than at 25°C. This step, however, can be omitted when the plant extracts to be tested are not toxic to the host cells under the conditions of the evaluation procedure.

The final virus titer reduction reflects the extracellular virucidal effect. The initial virus titer has to be high in order to dilute the toxicity of the plant extract and make the CPE visible. A 3-log-titer decrease, in comparison with the viral control, represents a significant virucidal activity, especially if this activity is also found in at least two subsequent dilutions of the extract. Such plant extracts may be investigated further by bioassay-guided isolations of the active components, using the same test method. We have compared both titration methods and can draw the following conclusions. Variabilities in response to the cells used, which greatly influence the precision of virus titer determination, occur to a much higher extent in the EPTT than in the plaque titration (PT) methodology. Consequently, the latter test is more accurate for determining the extracellular virucidal activity of an extract or compound. The EPTT in liquid medium and in microtiter plates, however, is very useful when pronounced virucidal activities must be determined, e.g., more than 2-log reduction of the virus titer, which corresponds with the selection criterion of our method.

2. *In Vitro* Antiviral Evaluation Procedures

The *in vitro* antiviral activity, i.e., the influence of an extract or compound on one or more steps of virus replication, has to be determined by an efficient and economic method. Among

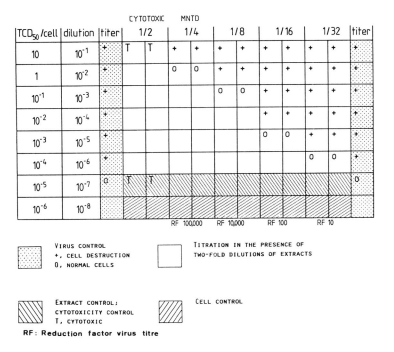

VIRUS CONTROL
+, CELL DESTRUCTION
0, NORMAL CELLS

TITRATION IN THE PRESENCE OF
TWO-FOLD DILUTIONS OF EXTRACTS

EXTRACT CONTROL;
CYTOTOXICITY CONTROL
T, CYTOTOXIC

CELL CONTROL

RF: Reduction factor virus titre

FIGURE 4b. Antiviral evaluation method by means of the end-point titration technique (EPTT) using microtiter plates. Each square represents a microtiter plate hole. MNTD: maximal nontoxic dose.

the different existing *in vitro* systems,[2,17] inhibition of virus plaque formation in susceptible monolayers (plaque inhibition test or PIT) and the EPTT are the only suitable antiviral tests for screening large numbers of plant extracts. In the PIT the development of plaques in the presence of an extract is determined. Thus, paper disks impregnated with serial twofold dilutions of plant extracts in culture medium are placed on the surface of the agar overlayer. After incubation at 37°C for 72–96 h, virus inhibition is evaluated by the measurement in millimeters of the plaque-free zones that appear in the presence of antivirally active compounds. If toxic compounds are also present in the extract, the plaque-free zone of living cells will be found outside a clear zone of cell destruction, which surrounds the disks.

The PIT has several disadvantages. First, the active substance has to diffuse from the disk into the overlayer in order to possibly inhibit virus multiplication. Since viruses have already invaded the cells before the active product has entered the cells, inhibition of the first steps of virus multiplication cannot be detected in the PIT. Second, precipitation of some of the many compounds present in plant extracts may occur in the disk impregnated with the extracts. This might result in an inhibition of the diffusion of possibly active compounds into the agar. In contrast, some toxic compounds, including phenolics and others, may sometimes diffuse faster into the agar than the active substances, so that no effect will be detectable in the toxic zone around the disk, where only dead cells can be observed. This is why we prefer the EPTT to test plant materials.

The EPTT technique (see Figure 4b) is performed on preemptied confluent monolayers of Vero or other cells, grown in the holes (e.g., 96) of microtiter plates, which are infected with serial tenfold dilutions of a virus suspension (100 µl). Starting with monolayers containing 10^4 cells per hole and a virus suspension of, for example, 10^7 TCD$_{50}$ ml^{-1} or PFU ml^{-1}, the first monolayers of cells are infected with a multiplicity of infection (MOI) 10.

By further serial tenfold dilutions of the virus suspension, the MOI decreases from 10 to 10^{-4}. The virus is allowed to absorb for 60 min at 37°C, after which serial twofold dilutions of plant extracts or test compounds in maintenance medium, supplemented with 2% serum and antibiotics, are added. The plates are incubated at 37°C, and the viral CPE is recorded daily by light microscopy during at least 1 week. Cytotoxicity controls (uninfected, but treated cells) and cell controls (uninfected, untreated cells) are run at each treatment concentration, and virus controls (infected, but untreated cells) at each viral dilution. Toxic doses of the extracts (T) are considered to be dilutions that cause destruction and degeneration of the monolayer, so that no virus titer can be determined. The antiviral activity is expressed as the virus titer reduction at the maximal nontoxic dose (MNTD) of the test substance, i.e., the highest concentration (μg ml^{-1}) or lowest dilution (1/n) that does not affect the monolayers under the conditions of the antiviral test procedure.

Viral titer reduction factors (RF, i.e., the ratio of the viral titer reduction in the absence [virus control] and presence of the MNTD of the test sample) of 1×10^3 to 1×10^4 indicate a pronounced antiviral activity and are suitable as selection criteria for further investigation of plant extracts.

Although the PIT gives an accurate idea of the decrease in virus titer, the EPTT is more suitable for testing complex samples such as plant extracts. First, the concentration of the many compounds in the extract remains constant, and consequently the proportion of toxic products vs. active compounds does not change. Second, the exact duration of the antiviral action can be determined by using the EPTT, since the action starts from the moment the extract is added to the cells. By contrast, in the PIT, diffusion of the active compounds occurs after virus multiplication has already started in all infected cells under the agar overlayer. Third, the EPTT, using serial diluted extracts, deals with a dynamic process because the cells are subsequently infected with different MOI. This system allows the correlation of all possible MOI values in the same microtiter plate, with decreasing amounts of plant extracts, so that the noncytotoxic concentration of plant extracts can be determined. At the same time, a correlation between extract toxicity and antiviral activity according to the corresponding MOI can be determined in the same microtiter plate. *It can be stated as a general rule that the detected antiviral activity should be stable in at least two subsequent dilutions of nontoxic concentration of the extract; otherwise, the activity is directly correlated with the toxicity of the extract or is only virucidal.* Moreover, a true antiviral product has to protect the cells that have been infected with low virus dilutions (starting from 0.1 PFU per cell onwards). Finally, it should be stressed that all possible steps of virus replication are included in the EPTT because the cells are inoculated with virus before the plant extracts are added. This means that an antiviral product that protects the cell monolayer at the highest concentrations of virus (10 or 1 TCD$_{50}$ ml^{-1}), under nontoxic conditions, must act on virus replication steps after uncoating. When the cells are completely protected only in the lower MOI (0.1 TCD$_{50}$ ml^{-1}), replication processes before uncoating may be involved.

3. *In Vitro* Anti-HIV Evaluation Procedures

Acquired immune deficiency syndrome (AIDS) is a pandemic immunosuppressive disease that results in life-threatening opportunistic infections and malignancies. Since a retrovirus, designated human immunodeficiency virus (HIV), has been isolated and identified as the etiologic agent of this disease, numerous compounds have been evaluated for their inhibitory effects on HIV replication *in vitro*.[18-29] Several steps in the virus replication cycle offer possible targets for antiviral therapy.[20,30] A possible site of intervention is the inhibition of virus-specific RNA-dependent DNA polymerase (reverse transcriptase), a multifunctional enzyme that transcribes the viral RNA genome to DNA, which can ultimately be integrated as proviral DNA into the cellular genome. Some of the numerous inhibitors of this enzyme have been explored for their inhibitory effects on the infectivity and cytopathogenicity of HIV *in vitro.* The effects of HIV and potential antiviral agents on T cell lines or peripheral blood lymphocytes have been followed by cell survival, syncytia formation, reverse transcriptase activity, or p 24/25 antigen production. Cell viability has been measured by counting cells that include or exclude a dye, such as Trypan Blue, or by measuring incorporation of radioactive nucleotides, such as ^{3}[H]thymidine, during cell proliferation. Each of these dose assays, however, is quite costly, time consuming, and less adaptable to the screening of a large number of samples.[31] Recently, several simple, sensitive, and rapid tests for the screening of anti-HIV agents have been described. Among these tests a microtray assay system using the Trypan Blue exclusion method for monitoring viral antigen expression in parallel with indirect immunofluorescence microscopy and laser flow cytofluorography, and a colorimetric test based upon the transformation of a tetrazolium salt to a colored formazan by living cells, but not by dead cells or culture medium, have shown to be very useful.[32,33] Up until now only one study on the screening of plant extracts for anti-HIV activity has appeared.[26] Twenty-seven medicinal plants reputed in Chinese traditional medicine to have anti-infective properties were tested at their subtoxic concentrations against HIV in the H9 cell line. Anti-HIV activity was tested by means of the indirect immunofluorescence method. The residual HIV infectivity was titrated, and the extracts were also tested for direct virucidal effect against HIV and interferon induction. Not less than 11 of the extracts were found to be active. The extract of *Viola yeodensis* was studied in more detail and was shown to shut off the growth of HIV completely and selectively without inactivating HIV extracellularly or inducing interferon.

V. ANTIVIRALS OF PLANT ORIGIN

Many antiviral agents have been isolated from plant sources and have been partly or completely characterized. Several excellent reviews on this subject have appeared over the last few years.[4,34-37] From these studies it is clear that the chemical structures of these agents belong to the most encountered classes of higher plant metabolites, including alkaloids, mono- and sesquiterpenes, diterpenes, saponins, cardenolides, phenolics, quinones, lignans, tannins, flavonoids, chromones, proteins, lectins, and polysaccharides. A listing of the most interesting antiviral and/or virucidal agents isolated after 1987, and of their antiviral activity, is given in Table 1.

Until now, however, no antiviral compounds from higher plants have yet come into significant clinical use, although several natural substances have been shown to be interesting and promising leads for the development of antiviral and/or virucidal chemotherapeutics. Different natural compounds, including tannins,[38-42] saponins,[43-47] flavonoids,[48] diterpenes,[49-51] alkaloids,[52-55] anthraquinones,[56-60] aminosugars,[13,61] and phenolics[62] exhibit, nevertheless, interesting *in vitro* anti-*Herpes simplex* virus activity. Their *in vivo* activity and bioavailability and other pharmacokinetic parameters are subjects of current investigations that might end up with the development of one or more antivirals with clinically useful properties for the topical treatment of *Herpes* infections.

Some 3-methoxyflavones, and synthetic derivatives thereof, have proven to be promising leads for the development of antirhinovirus drugs.[16,63-68] These compounds also exhibited *in vitro* and *in vivo* antiviral activities against other picornaviruses, including poliomyelitis and coxsackie viruses. Clinical trials with these substances, however, have not yet been published.

Other recently isolated compounds with prominent antirhinovirus properties are sesquiterpene glycosides,[69,70] bufadienolides,[71] and naphthoquinones.[72]

Numerous synthetic and natural compounds (see Figure 5) have been evaluated for their inhibitory effects on HIV replication *in vitro*[30] or on reverse transcriptase of HIV.[24,73] Some

TABLE 1
Products Isolated from Higher Plants with Antiviral Activity
Against Animal or Human Viruses[a]

Compound(s)	Origin	Target(s)	Ref.
Plant-derived compounds	Industry Different Plants	HIV reverse transcriptase	24
Methylgallate	*Sapium sebiferum*	*Herpes simplex*	39
Gallotannins	*Spondias mombin*	Coxsackie B virus Herpes *simplex* virus	42 41
Tannins and related polyphenols	Different plants	*Herpes simplex* virus	40
Tannins (gallo- and ellagitannins)	Different plants	HIV reverse transcriptase	112
Tetragalloyl quinic acids	Turkish and Chinese galls	HIV reverse transcriptase	94
Quinovic acid glycosides (triterpenes)	*Uncaria tomentosa* *Guettarda platypoda*	Vesicular stomatitis virus Rhino virus type 1B	79
Glycyrrhizin (triterpenes)	*Glycyrrhiza radix*	Polypeptide phosphorylation	80
Glycyrrhizin	*Glycyrrhiza radix*	HIV and Vesicular stomatitis virus	76
Glycyrrhizin and modified compounds	*Glycyrrhiza radix*	*Herpes simplex* HIV	47
Castanospermine (alkaloids)	*Castanospermum australe*	Cytomegalo virus HIV	84, 85 83, 86
Different flavonoids	Different plants	Picorna viruses	65
5,7,4′-Trihydroxy- 8-methoxyflavone and others	*Scutellaria baicalensis*	Influenza A virus	99
Isoflavonic glycoside	*Ulex europaeus*	*Herpes simplex* virus Polio virus	48
Flavones and isoflavones	Industry	Avian myelo-blastosis Virus reverse transcriptase	101
4′-Hydroxy-3-methoxyflavones	Synthesis Different plants	Picorna viruses	66
Dammar resin triterpenoids	Plants of the Dipterocarpaceae	*Herpes simplex* virus	43
Triterpenoid saponins	*Wisteria brachybortrys*	Epstein Barr virus	81
Triterpenes	*Euptelea polyandra*	Epstein Barr virus activation	102
Saponins	*Anagallis arvensis*	*Herpes simplex* virus Polio virus	44, 45, 46
Scopadulcic acid B derivatives scopadulin (diterpenoids)	*Scoporia dulcis*	*Herpes simplex* virus	50, 51
Gossypol (polyphenols)	Cotton seed	HIV reverse transcriptase	87, 88
Phototoxins	Different plants	Cytomegalo virus	96
Dextro-odorinol (alkaloids)	*Aglaia roxburghiana*	Ranikhet disease virus	105
Furanochromones (visnagin and khellin)	Species of *Ammi* (Umbelliferae)	Murine cytomegalo virus	97
Alkaloids	Amaryllidaceae	*Herpes simplex* virus	52, 53
Citrusinine I (acridone alkaloid)	Citrus	*Herpes simplex* virus Cytomegalo virus	54
Alkaloids	*Chelidonium majus*	Adenovirus 12 and 5 *Herpes simplex* virus	55
Papaverine (alkaloids)	Industry	HIV reverse transcriptase	89

TABLE 1 (continued)

Compound(s)	Origin	Target(s)	Ref.
Sesquiterpene glycosides	*Calendula arvensis*	Vesticular stomatitis virus Rhinovirus type 1B	69, 70
Aloe emodin (anthroquinones)	*Aloe barbadensis*	Enveloped viruses (virucidal)	103
Lapachol and derivates (naphthoquinones)	Bignoniaceae	Picorna viruses	72
Hypericin and pseudohypericin (naphthobianthrones)	Species of Hypericum family	Retroviruses	59, 60
Hypericin	*Hypericum triquetrifolium*	Cytomegalo virus Sindbis virus HIV	58
Hypericin	*Hypericum triquetrifolium*	*Herpes simplex* virus Influenza A virus	57
Lignins	*Pinus parviflora*	Influenza A virus	98
α-(−)Peltatin (lignans)	*Amanoa oblongifolia*	Sindbis virus Murine cytomegalo virus	108
Lignans	Industry and species of *Amanoa*	Cytomegalo virus Sindbis virus	95
D-Mannose-specific plant lectins	*Galanthus nivalis* *Narcissus pseudonarcissus* *Listeria ovata*	HIV Cytomegalo virus	93
D-Glucosamine and derivatives	Industry	*Herpes simplex* virus and others	13, 61
Bufadienolides (scillarenin and others)	Industry Synthesis	Rhinovirus type 1A Polio virus	71
Plant proteins	*Gelonium multiflorum* *Dianthus caryophyllus*	HIV	90
Trichosanthin and other proteins	*Trichosanthes kirilowii*	HIV	91, 92
Fulvoplumierin (iridoids)	*Plumeria rubra*	HIV reverse transcriptase	24
Allicin (sulfur compounds)	*Allium sativum*	Virucidal activity	104
Prunellin (sulfated polysaccharides)	*Prunella vulgaris*	HIV	75
Phloroglucinol derivates (polyphenols)	*Mallotus japonicus*	*Herpes simplex* virus	62
Catalpol (iridoids)	*Picrorrhiza kurroa*	Hepatitis B virus	100
Epilupeol (triterpenes)	*Vicoa indica*	Ranikhet disease virus	82

[a] All compounds were isolated or studied after 1987. For antivirals studied earlier, see References 4 and 37.

of the compounds showing interesting antiretroviral activities have been isolated from plants. Several sulfated polysaccharides (**1**) have been found to be the anti-HIV active compounds of *Viola yeodensis*[74] and *Prunella vulgaris*,[75] two herbs used in traditional Chinese medicine as anti-infectives. Glycyrrhizin (**2**), one of the main saponins of *Glycyrrhiza glabra*,[19,76] has already been tested in HIV carriers and has delayed the progression of symptoms related to HIV. In another claim, simultaneous administration of the compound appeared to decrease the adverse reactions of zidovudine.[77] Also, other saponins, such as soybean saponin or saponin B1, exhibited strong anti-HIV activity *in vitro*.[78] It should be noted that saponins possess broad spectrum antiviral properties against polio,[46] *Herpes simplex*,[44-47] VSV,[79,80] Epstein-Barr virus,[81] and ranikhet disease virus.[82]

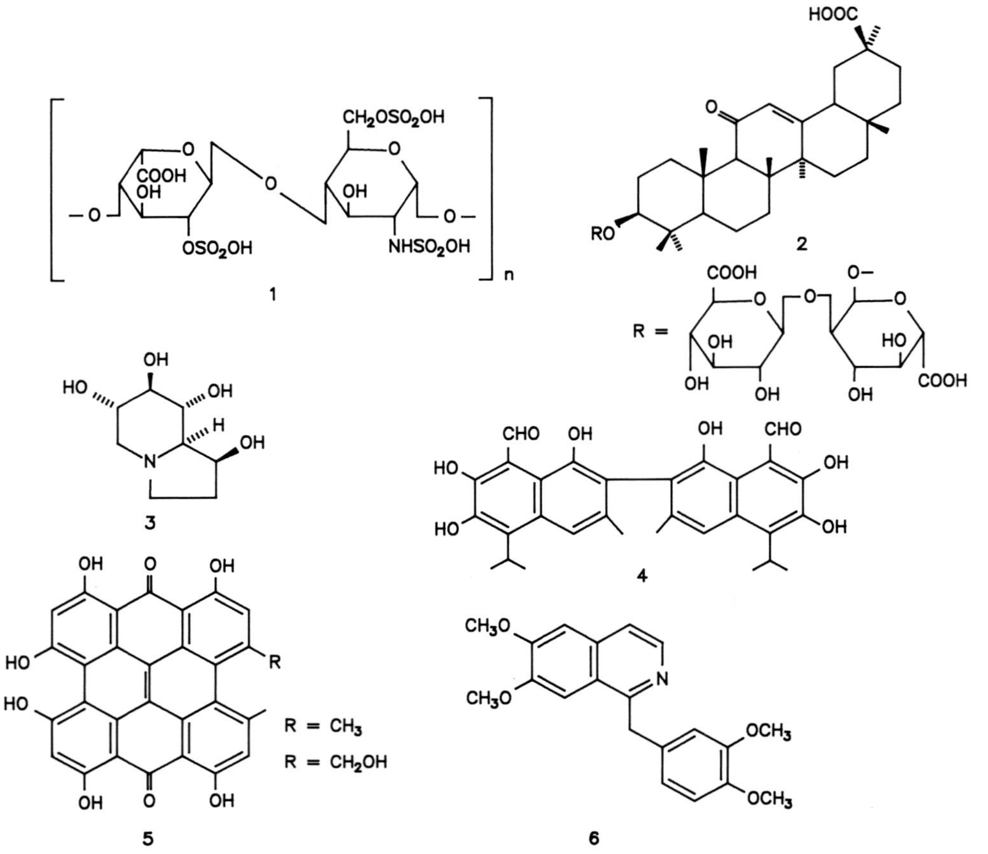

FIGURE 5. Chemical structures of natural products as potential drugs for the treatment of AIDS. **1**: heparin; **2**: glycyrrhizin; **3**: castanospermine; **4**: (–)-gossypol; **5**: hypericin and pseudohypericin; **6**: papaverine.

Castanospermine (**3**), an indolizidine alkaloid from the seeds of *Castanospermum australe*, has been reported to have strong *in vitro* anti-HIV activity[83-86] and to be active *in vivo* when administered orally to mice.

The naphthobianthrones hypericin (**5**) and pseudohypericin (Figure 5), which occur in plants of the *Hypericum* family, are highly effective in preventing viral-induced manifestations that follow infections with a variety of retroviruses *in vitro* and *in vivo*. They also show *in vitro* activity against other enveloped viruses such as *Herpes simplex* virus, influenza virus, and cytomegalovirus.[57-60]

Other plant-derived compounds with inhibitory effect on HIV replication *in vitro* are the (–)-enantiomer of gossypol (**4**), a polyphenolic aldehyde extracted from cotton-seed,[87,88] the well-known alkaloid papaverine,[89] proteins,[90-92] lectins,[93] and tannins (many patent applications).[94] These latter compounds, and some iridoids, also blocked HIV reverse transcriptase.[95]

Some lignans and phototoxins, such as furanochromones, have been found to be active *in vitro* against cytomegalovirus and murine cytomegalovirus,[96-98] whereas influenza and hepatitis B viruses were inhibited by lignins[99] and flavonoids,[100] and by iridoids, respectively.[101]

Other compounds exhibit antiviral or virucidal activity against several enveloped viruses (see also Table 1).[102-109] It should be stressed, however, that the *in vitro* therapeutic index of many of these compounds has not clearly been defined.

We refer to Hudson[37] for a more extensive survey of plant antiviral compounds, especially those described before 1988.

VI. ANTIVIRAL FLAVONOIDS FROM AFRICAN TRADITIONAL DRUGS

A. SCREENING AND BIOASSAY-GUIDED ISOLATION OF ANTIVIRALLY ACTIVE COMPOUNDS FROM AFRICAN MEDICINAL PLANTS

In our antiviral screening program dealing with medicinal plants from central Africa, especially Rwanda, the selection of the plant species to be investigated was mainly based upon their use in the treatment of viral diseases by African traditional healers. This selection method indeed gave a higher percentage of active leads as compared to random selection, although in some cases the same active compounds were isolated from botanically nonrelated active plants.[2,4]

The air-dried plant material was usually extracted with 80% aqueous methanol by maceration and percolation, the solvent was removed, and the residue was solubilized as described earlier in water and/or DMSO, diluted with medium, and screened *in vitro* against a battery of viruses by means of the EPTT. Since traditional healers always prepare their remedies by extraction with water, an aqueous plant extract was prepared and tested as well. Out of a total of 267 crude extracts, corresponding to 100 different plant species of 33 families, 31 extracts (12%) representing 21 plant species (21%) of 16 families (48%) exhibited prominent antiviral properties against one or more of the test viruses (Rf > 10³). Table 2 shows the botanical and vernacular names, the parts of the plants, and the test results of three representative families, including the Asteraceae, Euphorbiaceae, and Fabaceae. Two *Euphorbia* species, i.e., *E. grantii* and *E. hirta* exhibited pronounced activity against picornaviruses. Consequently, a screening of more than 30 different *Euphorbia* species was carried out that resulted in the selection of three of the most promising species for isolation studies of the antiviral compounds, using the same bioassay. Table 3 shows the antiviral screening results of the different *Euphorbia* species (stems are used only) and proves that the antipicornaviral activity is not a common characteristic of all *Euphorbia* species. Despite the fact that, in many African countries, the name *Euphorbia* is considered as being synonymous with poison, many *Euphorbia* species are used for a variety of medicinal purposes, including the treatment of infections and childhood diseases, rheumatism, snake bites, asthma, obstipation, coughs, sores, and skin diseases. It should be noted that the leaves of *E. grantii* (Table 2) are used in Rwandese traditional medicine against poliomyelitis.

All compounds responsible for the pronounced antiviral effects of selected *Euphorbia* species were identified as 3-methoxyflavones. Until now, seven antivirally active 3-methylethers of quercetin (3-MQ) and kaempferol (3-MK) have been isolated from *E. grantii* (0.09%), *E. balsamifera* (0.07%), and *E. hirta* (0.03%) extracts.[4,16,110] Moreover, it was found that the antiviral effects of the other active *Euphorbia* species are also due to the presence of the same, or related, 3-methoxyflavones (Figure 6).

The isolated 3-methoxyflavones from *E. grantii* were highly active in tissue culture, without showing significant cytotoxicity, against all human picornaviruses, including polio, coxsackie, and rhinoviruses, but not against other RNA and DNA viruses, except vesicular stomatitis, vaccinia, and several bunya viruses. The antiviral spectrum of the 3-methylether of quercetin (3-MQ) and the 3,3′-dimethylether of quercetin (3,3′-DMQ) is shown in Table 4. The antiviral activity is expressed as the virus titer reduction in the presence of 5 μg/ml of 3-MQ or 3,3′-DMQ, 3 days and 7 days after infection.

Figure 7 shows the antiviral activity of the same compounds and a naturally occurring mixture of 3,3′-DMQ and the 3,7-dimethylether of quercetin (3,7-DMQ) (3:1) against poliomyelitis virus type 1 and coxsackie B4 virus, grown in Vero cells. Quercetin, rutin, and

TABLE 2
Antiviral Screening of Selected Rwandese Plants[a]

Plant-family botanical names	Vernacular name	Plant part[b]	Polio	Coxsackie	Semliki forest	Herpes	Measles
Asteraceae							
Ageratum conyzoides L.	Karura	L	1	$10^{2.5}$	10^2	1	10
		S	1	$10^{2.5}$	1	1	1
Aspilia africana Pers.(C.A.) Adams	Icyumwa	L	1	1	10	10	1
		S	1	1	10	10	1
		R	1	1	10	10	1
Berkheya spekeana Oliv.	Ikigwarara	L	1	1	1	$10^{0.5}$	1
		S	10	$10^{0.5}$	1	1	10
Bidens pilosa L.	Inyabarasanya	L	1	1	1	1	1
		S	1	1	1	1	1
		R	1	1	1	1	1
		F	1	1	1	1	1
Crassocephalum bumbense S. Moore	Icyunamyi	L	10	10	1	1	1
		S	10	1	1	1	10
		R	1	1	10^2	$10^{0.5}$	10
		F	1	1	10	1	10
Crassocephalum multicorymbosum (Klatt.) S. Moore	Umutagara	L	10	1	1	1	1
		S	10	1	1	1	10
Crassocephalum vitellinum (Benth.)	Umusununu	L	1	10^3	1	10^3	1
		S	1	1	1	1	1
Dichrocephala bicolor (Roth.) Schlech.	Umubuza	L	10	1	1	1	10
		S	10	1	1	1	10
		R	10^2	1	1	1	1
Dicoma anomala Sond.	Umwanzuranya	R	10	1	1	1	1
Erigeron floribundus Sch. Bip.	Wambuba	L	1	1	10	10	10
		S	1	1	1	1	1
		R	1	1	10	1	1
Guizotia scabra (Vis.) Chiov.	Igishikashike	L	1	1	1	1	1
		S	1	1	1	1	1
		R	1	1	1	1	1
Helichrysum hochstetteri (Sch. Bip.) Hook. f.	Umutamatama	L	10	1	1	1	10
		S	10	1	1	1	10
		R	10	1	10	1	10
Laggera brevipes Oliv. et Hiern.	Igitabitabi cy'impyisi	L	10^3	10^4	1	1	1
		S	10^3	10^5	$10^{1.5}$	10	$10^{2.5}$
		F	10^2	10^4	10	1	1
Microglossa pyrifolia O. Kuntze	Umuhe	L	10	1	1	1	1
		S	1	1	1	1	1
		R	10	1	1	1	10
Sonchus exauriculatus (O. et H.) Hoffman	Rurira	L	1	1	1	1	10
		S	1	1	1	1	10
Vernonia aenulans Vatke	Idoma	L	10^4	1	1	1	1
		S	1	1	1	1	1
		R	1	10	1	1	1
		F	1	1	1	1	1
Vernonia amygdalina Del.	Umubilizi	L	1	1	1	1	1
		S	1	1	1	1	1
		R	10	10	1	1	1
		F	10^3	10^2	1	1	1
Vernonia lasiopus O. Hoffm.	Igiriheri	L	1	10^2	1	1	1
		S	$10^{1.5}$	10^4	1	1	1
Vernonia pogosperma Klatt.	Umubimbafuro	L	10	1	1	1	10
		S	1	1	1	1	1

table 2 (continued)

Plant-family botanical names	Vernacular name	Plant part[b]	Polio	Coxsackie	Semliki forest	Herpes	Measles
Euphorbiaceae							
Clutia abyssinica Jaub. et	Unutarishonga	L	10^3	10^3	1	1	1
Spach var. *pedicellaris*		S	$10^{0.5}$	$10^{0.5}$	1	1	1
(Pax) Pax		R	10^2	$10^{1.5}$	1	$10^{1.5}$	1
Clutia paxii Knauf		WP	1	10	1	10	1
Euphorbia candelabrum Trem.	Umuduka	WP	10^2	10	1	1	1
ex Klotschy							
Euphorbia grantii Oliv.	Madwedwe	L	10	1	1	1	10
		S	10^3	10^5	1	1	10
		R	1	1	1	1	10
Euphorbia hirta L.	Nyanduku	WP	10^5	10^4	1	10^3	1
Euphorbia tirucalli L.	Umugenzi	WP	1	1	1	1	1
Macaranga kilimandscharica	Umusekera	L	10^3	10	10^2	10^2	10
Pax							
Tragia brevipes Pax	Isusa	L	1	10	1	10	1
		S	1	1	1	1	1
		F	1	1	1	1	1
Fabaceae							
Cajanus cajan	Umukunde	L	10	10^3	1	1	1
(L.) Millsp.		S	$10^{1.5}$	$10^{4.5}$	1	1	1
		F	$10^{0.5}$	$10^{2.5}$	1	1	10
Crotalaria deserticola Taub.	Utuyogera	L	10	1	1	1	1
ex. Bak. spp. *deserticola*		S	1	10	10	1	10
		F	10	1	1	1	10
Crotalaria incana L. var.	Umuyogera	L	10	10	1	1	10
purpurescens (Lam.) Milne.		S	10	1	1	1	10
Redh.		R	10	1	1	1	1
		F	10	1	1	1	10
Crotalaria mesopontica Taub.	Akayogera	L	$10^{2.5}$	$10^{1.5}$	1	$10^{1.5}$	1
var. *mesopontica*		S	10^3	10	1	10	1
Dalbergia lactea Vatke	Umuhashya	L	1	$10^{2.5}$	1	$10^{0.5}$	10
		S	1	$10^{0.5}$	1	1	10
		R	1	10	1	$10^{0.5}$	10
Erythrina abyssinica	Umuko	S	10^4	10^2	10^4	10^4	1
Lam. ex. A. Rich.							
Glycine javanica L. var.	Umusekerasuka/	L	10	1	1	1	1
claescensii (Willld.) Hauman.	Umucasuka	S	10^2	1	1	10^3	1
Indigofera arrecta Hochst.	Umusororo	L	10^2	10	10^3	10	10^3
ex. A. Rich.							
Sesbania macrantha Welw.	Umunyegenyege	L	1	1	10	1	1
ex. Phillips et Hutch.	wa ruheha	S	1	1	1	1	1
Sesbania sesban (L.)	Umunyegenyege	L	1	1	10	1	1
Merrill var. *nubica* Chiou.	wa nyamabumba	S	10	1	10	1	10
Tephrosia vogelii Hook. f.	Umuruku	L	10	1	1	1	1
		S	1	1	1	1	10
		R	10	1	10	1	10
		F	10	1	1	1	10^2

[a] Virus titer reductions (EPTT method) are represented.

[b] F = fruits; L = leaves; R = roots; S = stems; WP = whole plant.

TABLE 3
Antiviral Screening of *Euphorbia* spp.[a]

Botanical names	Polio	Coxsackie	Semliki forest	Herpes	Measles
Euphorbia abyssinica J.F. Gmel.	10	1	1	1	1
Euphorbia antiquorum L.	10	10	1	1	1
Euphorbia aphylla Brouss.	1	1	1	1	1
Euphorbia balsamifera Ait.	10^6	10^6	1	1	1
Euphorbia barnhartii L. Croiz	10^3	10^3	1	10^3	1
Euphorbia officinarum L. var. *beaumerana* (Hook. f. et Coss.)	10^2	10^2	1	1	1
Euphorbia bivonae Steud.	10	1	1	1	1
Euphorbia x bothae Lotsky et Godd.	10^2	10^3	1	10	10
Euphorbia calycina N.E.Br.	10^2	10	1	1	1
Euphorbia candelabrum Trem. ex. Klotsch.	10^2	10^3	1	1	1
Euphorbia cereiformis L.	1	1	1	1	1
Euphorbia cyparissias L.	1	10^2	1	1	1
Euphorbia esula L.	1	10	1	1	10
Euphorbia franckiana Bgr.	1	1	1	10^4	1
Euphorbia hamata (Haw.) Sweet	10^6	10^6	1	10^4	10
Euphorbia helioscopa L.	1	10	1	10	10
Euphorbia heterophylla L.	1	1	1	10	10
Euphorbia hirta L.	10^5	10^4	1	10^3	1
Euphorbia ingens E. Mey. ex. Boiss.	10^2	1	1	1	1
Euphorbia lathyris L.	1	10	1	10	10
Euphorbia marginata Pursh.	1	10	1	10	10
Euphorbia myrsinites L.	1	1	1	1	1
Euphorbia palustris L.	10^3	10^5	10^2	10^3	10^2
Euphorbia phyloclada Boiss.	10^3	10^3	1	10	1
Euphorbia platyphyllos L.	10	1	10	1	10
Euphorbia ramipressa L. Croiz.	1	10^2	1	1	10
Euphorbia salicifolia DC	1	1	1	1	1
Euphorbia submammilaris (Bgr.) Bgr.	10	1	1	1	1
Euphorbia tirucalli L.	1	1	1	1	1
Euphorbia trigona Haw.	1	1	1	10	1

[a] Virus titer reductions are represented.

apigenin, having respectively a 3-OH, 3-glycosyl, and 3-H function, were inactive under the same circumstances, *proving the importance of the 3-methoxyl group* for potent antiviral activity.

In a follow-up experiment it was shown that the antiviral activity of these compounds was not due to the extracellular inactivation of the virus. As shown in Figure 8, a virucidal activity on extracellular virus was not found at a concentration even 100 times higher than the one that reduces the virus titer by a factor of 10^5. Consequently, *3-methoxyflavones exhibit a "true antiviral" activity* on small human RNA viruses such as picornaviruses. The antiviral activity was not only very pronounced (i.e., no CPE was seen in tissue culture cells that had been infected with more than 1000 virus particles per cell and treated with 2 µg/ml of compound), but the effect was also very stable (i.e., the CPE was not visible even 14 days after infection with a human picornavirus). These two observations reflect the stability of the active substances inside the living tissue culture cells and may explain the difference in results obtained with 3-methoxyflavones and with other flavones that lacked the 3-methoxyl group.

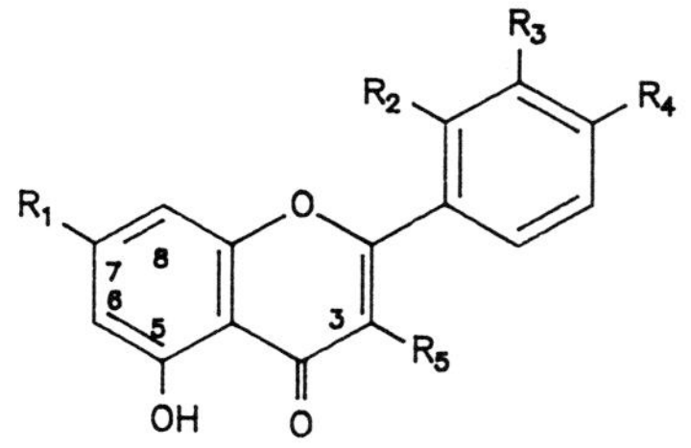

R₁	R₂	R₃	R₄	R₅	
OH	H	OH	OH	OH	QUERCETIN
OH	H	H	OH	OH	KAEMPFEROL
OH	H	H	OH	H	APIGENIN
OH	OH	H	OH	OH	MORIN
OH	H	OH	OH	rutinoside	RUTIN
OH	H	OH	OH	rhamnoside	QUERCITRIN

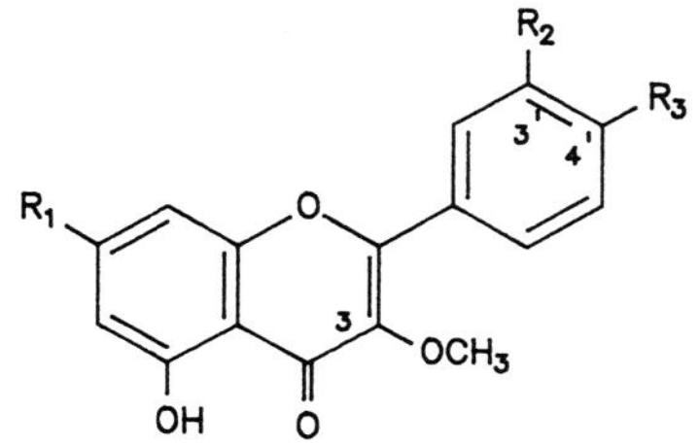

R₁	R₂	R₃	
OH	H	OH	3—METHYLKAEMPFEROL (3—MK)
OCH₃	H	OH	3,7—DIMETHYLKAEMPFEROL
OH	H	OCH₃	3,4'—DIMETHYLKAEMPFEROL
OH	OH	OH	3—METHYLQUERCETIN (3—MQ)
OH	OCH₃	OH	3,3'—DIMETHYLQUERCETIN (3,3'—DMQ)
OCH₃	OH	OH	3,7—DIMETHYLQUERCETIN
OCH₃	OCH₃	OH	3,3',7—TRIMETHYLQUERCETIN

FIGURE 6. Antiviral 3-methoxyflavonoids from selected *Euphorbia* species. The depicted flavones, missing a 3-methoxyl group, were used as references.

When administered intraperitoneally to mice, 3-MQ protected the animals against lethal infections of coxsackie B4, but no signs of toxicity or death were observed in uninfected control mice that were treated with five times the level of 3-MQ used to treat infected mice.[16,111]

It should be noted that Japanese workers have independently found, by a systematic screening of microbial and natural products for antirhinovirus activity, that one of the 3-methoxyflavones isolated from *E. grantii*, namely 3,3',7-trimethylether of quercetin (3,3',7-TMQ), exhibited antirhinovirus effects. This compound was previously isolated from a Chinese medicinal herb and then synthesized.[63]

The cardiovascular effects of 3-MQ, on different isolated organ preparations, have been examined using quercetin and rutin as reference flavonoids. The three compounds were each

TABLE 4
Antiviral Screening[a] of 3,3′-Dimethylether of Quercetin (3,3′-DMQ)
and 3-Methylether of Quercetin (3-MQ) at 5 μg/ml

	3,3′-DMQ		3-MQ	
Polio (10^7)	10^6	10^6	10^7	10^7
Coxsackie B2 (10^7)	10^5	10^5	10^6	10^6
Rhino (10^3)	10	10	10	10
Mengo (10^4)	1	1	1	1
VSV[b] (10^4)	10^3	1–10	10^3	1–10
Semliki forest (10^7)	1	1	1	1
Measles (10^5)	1	1	1	1
Herpes (10^7)	1	1	1	1
Adeno (10^3)	1	1	1	1
Bangin[b] (10^5)	$10–10^2$	1	$10–10^2$	1
Bunyamwera[b] ($10^{6.5}$)	10^3	10	10^3	10
Yellow fever[b] (10^5)	10	1–10	10	1–10

[a] Virus titer reductions are expressed after 3 days (3D) and 7 days (7D). Original virus titers are given in parentheses (TCD_{50}/ml).
[b] A concentration of 25 μg/ml is needed to obtain an antiviral effect.

shown to possess the same pharmacologic properties with regard to platelets and cardiovascular muscle, but only in concentrations that were at least 1000 times higher than the effective antiviral concentration of 3-MQ. Consequently, 3-MQ causes no cardiovascular side effects when used in antivirally active concentrations.[112]

B. MECHANISM OF ANTIVIRAL EFFECTS OF 3-METHOXYFLAVONES

Two classes of antipicornavirus flavonoids can be distinguished according to their antiviral spectrum and mechanism of action.

Some chalcones and flavans selectively inhibit different serotypes of rhinoviruses.[113-116] Compounds such as 4′-ethoxy-2′-hydroxy-4,6′-dimethoxychalcone and 4′,6-dichloroflavan interact directly with specific sites on the viral capsid proteins, thereby preventing uncoating (see Figure 1) and the consequent liberation of viral RNA.[116-120] The sensitivity of the virus depends on the serotype and the compounds, and it has been shown that drug-resistant mutants can be selected readily under *in vitro* conditions.[121]

The second class of flavonoids, consisting of the 3-methoxyflavones, which are active against a wide range of picornaviruses (except mengovirus), have been shown not to interact with the capsid proteins of the viruses, but to interfere with an early stage in the viral RNA synthesis.[122,123] In a detailed investigation it was found that in poliovirus-infected cell cultures, there is a narrow window between 1 and 2.30 h postinfection during which 3-methoxyflavones exert their action and succeed in blocking virus replication. Figure 9 shows the effect of 3-MQ on poliovirus RNA synthesis in HeLa cells when the drug is added together with the virus. During the experiment, depicted in Figure 10, 3-MQ was added 2.30 h and 3 h postinfection. The results indicated that the drug has to be present before 3 h postinfection in order to inhibit completely the polio RNA synthesis. It was also demonstrated that during that period the polio-induced shut off of cellular protein synthesis persisted even when the compound concentration was high enough to completely inhibit viral RNA and protein synthesis. This suggested that the viral factor that caused the shut off did not depend on *de novo* protein synthesis.[124]

Although the exact mode of action is not completely understood yet, it has recently been found that 3-methoxyflavones inhibit the formation of both minus- and plus-strand viral RNA of poliovirus by interacting with the proteins involved in the binding of the virus replication

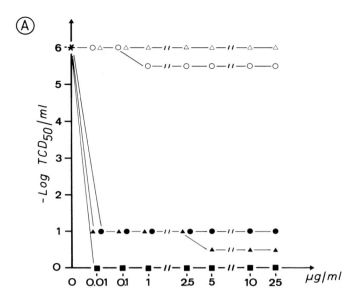

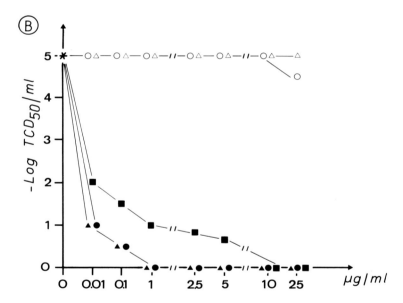

FIGURE 7. Dose–response curve of the activity of 3-methylether of quercetin (3-MQ), 3,3′-dimethylether of quercetin (3,3′-DMQ), a natural mixture (ratio 3:1) of 3-MQ and 3,3′-DMQ, quercetin, rutin, and apigenin on (A) poliovirus, and (B) coxsackie B4 virus, cultivated on VERO cells. Virus titers in the presence of different concentrations of the test compounds were determined using the EPTT technique. ✳: virus control; ●: 3-MQ; ▲: 3,3′-DMQ; ■: natural mixture; △: apigenin and rutin; ○: quercetin.

complex to vesicular membranes where poliovirus replication takes place.[122,125] In contrast to the capsid-binding antiviral flavonoids, no drug-resistant mutants have been detected in the presence of 3-methoxyflavones.

The attractive mechanism of action, the pronounced and broad antiviral activity, and the lack of resistance induction of the 3-methoxyflavones make these compounds good candidates for antivirals against selected picornaviruses.

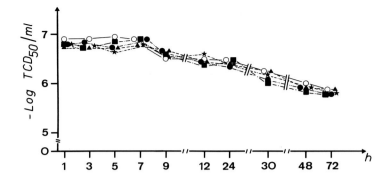

FIGURE 8. Extracellular activity of different concentrations of 3-MQ on poliovirus at 37°C during different incubation periods. EPTT technique was used for the virus titers. ○: 0 μg/ml; □: 100 μg/ml; ●: 50 μg/ml; ▲: 25 μg/ml; ✱: 1 μg/ml.

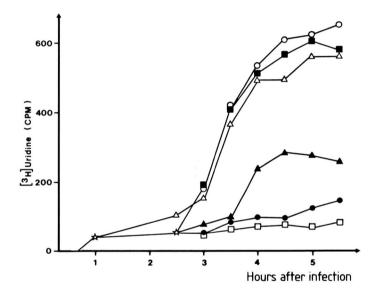

FIGURE 9. Influence of 3-MQ on the poliovirus RNA synthesis in HeLa cells. 3-MQ is added together with poliovirus at time 0. TCA precipitable material was monitored. ○: virus control; ■: 1 μg/ml 3-MQ; ▲: 3 μg/ml 3-MQ; ●: 10 μg/ml 3-MQ; □: cell control (no virus).

VII. STRUCTURE–ACTIVITY RELATIONSHIP STUDIES OF 3-METHOXYFLAVONES: NATURAL COMPOUNDS AND SYNTHETIC DERIVATIVES

The finding of an optimal compound with a high level of antiviral activity and a low degree of toxicity and optimal pharmacokinetic properties is the final goal of structure–activity relationship studies.[126-129]

As already mentioned, 3-methoxyflavones are excellent lead compounds for this kind of study in the antiviral field. Preliminary investigation with natural flavones have shown the 3-methoxyl and 4'-hydroxyl groups to be important for potent antiviral activity.[16] As further indications on structure–activity relationships were lacking, it was decided to synthesize three series of analogs.[66]

INFLUENCE OF 3- METHYLQUERCETIN ON VIRAL

RNA SYNTHESIS

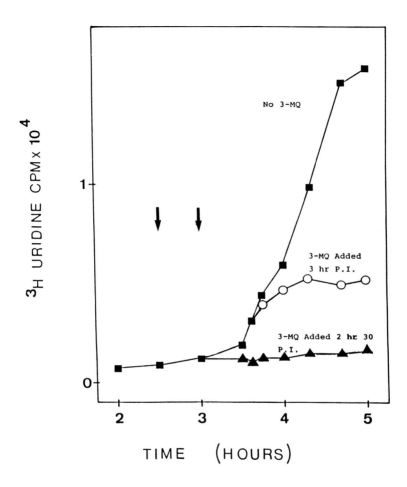

FIGURE 10. Influence of 3-MQ on poliovirus RNA synthesis in HeLa cells. Cells are infected at time 0. The arrows indicate when 3-MQ is added in the different experiments. TCA precipitable material was monitored. ■: virus control (no 3-MQ); ○: 3-MQ added 3 h postinfection; ▲: 3-MQ added 2 h 30 min postinfection.

- A-ring substituted (amino-, halo-, hydroxy-, methoxy-, methyl-, nitro-) 4′-hydroxy-3-methoxyflavones (**1–26**). These were compared with natural 4′-hydroxy-3-methoxyflavones (**27–35**).
- 4′-Chloro- and 4′-methoxy-3-methoxyflavones (**36–42** and **43–45**). In order to confirm the essential role of the 4′-hydroxyl group, these were compared with natural 4′-methoxy- and 4′-acetoxy-3-methoxyflavones (**46–57**).
- 3-Lower alkoxy, 3-amino, 3-chloro-, 3-methyl-4′-hydroxyflavones (**58–62**). The importance of the 3-methoxyl group was confirmed by comparing the activity of these compounds with a 3-methoxyl analog with the same A-ring substitution.

A. SYNTHETIC METHODS

3-Methoxyflavones were prepared by well-known synthetic methods. The 2′-hydroxychalcones were prepared from the corresponding 2′-hydroxyacetophenones and *para-*

substituted benzaldehydes in alkaline medium. These chalcones were cyclized with alkaline hydrogen peroxide to 3-hydroxyflavones (Algar-Flynn-Oyamada oxidation or AFO), which were then transformed into 3-methoxyl derivatives, using dimethyl sulfate. Benzylether protection was used for hydroxyl compounds. Later deprotection was easily achieved by acid hydrolysis.

Since the AFO reaction gave rise to many side products in the preparation of 5-substituted flavones, the latter compounds were prepared by an Allan-Robinson reaction. Thus, 2'-hydroxy-2-methoxyacetophenones were condensed with 4-(benzyloxy)-benzoic acid anhydride in the presence of the potassium salt of 4-(benzyloxy)-benzoic acid as base. Benzyl protecting groups were once again easily removed with acid.

Depending on the target compound, selective methylations, nitroreductions, and nuclear reductions were also carried out.

3-Methyl, 3-chloro, and 3-amino compounds were prepared using specific methods. Full details of all the synthetic methods used have been described elsewhere.[66]

B. ANTIPICORNAVIRUS PROPERTIES

All compounds were tested for their antiviral activity against poliovirus type 1 Brunhilde strain 1A/S3 by the EPTT. The antiviral activity has been expressed as the viral strength reduction factor (RF) after 3 days of incubation in the presence of the maximal nontoxic dose of the compound (MNTD; µg/ml). The minimal dose of the compound that produced an RF_{MNTD} value of 10^3 ($MD_{RF}10^3$) was used as a criterion to compare the antiviral activities of the flavones. This means that the MNTD present in the infected cells results in a minimal titer reduction of 10^3. The results of the antiviral tests of A-ring-substituted 4'-hydroxy-3-methoxyflavones are summarized in Table 5. Interesting compounds were selected, and their therapeutic index 99 (TI_{99}) were determined. The TI_{99} was calculated as the ratio of the maximum drug concentration at which 50% of the growth of normal cells is inhibited (CyD_{50}) to the minimum drug concentration at which 99% of the virus is inhibited (ED_{99}). The compounds were also tested against human rhinovirus type 15. The incubation time, however, was 5 days instead of 3 days, and Vero cells were replaced by human skin fibroblasts. The results are summarized in Table 6. The results obtained with the 4'-substituted-3-methoxyflavones and 3-substituted-4'-hydroxyflavones are presented in Tables 7 and 8.

The results in Tables 5 and 6 reveal 4'-hydroxy-3-methoxyflavones with a monosubstituted A-ring to be less active than the corresponding compounds having a polysubstituted A-ring. In the monosubstituted A-ring series, the hydroxylated analogs (**10**, **11**, and **20**) are the most active substances, whereas substitution of the parent compound (**1**) with a methoxyl, methyl, or chloro function greatly decreases cytotoxicity. The 5-hydroxy, 7-methoxy, and 6-chloro derivatives (**20**, **8**, and **12**) combine a rather low cytotoxicity with a moderate antipolio activity, giving more than ten times higher TI_{99} values (ranging from 16.7 to 50) than that of compound **1**. Monosubstitution of the A-ring of the latter with bromo, iodo, fluoro, amino, and nitro groups affords compounds having low antipolio properties, mostly within the range of the cytotoxic dose.

Within the series of compounds containing a polysubstituted A-ring, 4',7-dihydroxy-3-methoxy-5,6-dimethylflavone (**26**) is the most interesting substance. The ED_{99} value against polio is 0.1 µg/ml, whereas it is not cytotoxic to Vero cells at the highest concentration tested (100 µg/ml), which results in a TI_{99} of more than 1000. Other highly active antipolio substances are the naturally occurring compounds, 4',5-dihydroxy-3,6,7-trimethoxyflavone (penduletin, **31**), 4',5-dihydroxy-3,7,8-trimethoxyflavone (**32**), and 4',5-dihydroxy-3,7-dimethoxyflavone (jaranol, **22**) with TI_{99} values of 500, >400, and 333, respectively. The synthetic compound 4',7-dihydroxy-3-methoxy-5-methylflavone (**24**) has high antipolio activity, but is more toxic, which results in a TI_{99} value of 150.

TABLE 5
Antipoliovirus Activity of 4′-Hydroxy-3-Methoxyflavones 1–35

No.[a]	R_1	R_2	R_3	R_4	R_6	MNTD[b] μg/ml	RF_{MNTD}[c] μg/ml	MD_{RF10^3}[d]
1	H	H	H	H	H	15	10^4	6
2	H	CH_3	H	H	H	12	$10^{4.5}$	10
3	H	H	CH_3	H	H	25	$10^{4.5}$	4
4	H	$CH(CH_3)_2$	H	H	H	5	$10^{1.5}$	–
5	H	H	$CH(CH_3)_2$	H	H	25	$10^{1.5}$	–
6	OCH_3	H	H	H	H	10	$10^{4.5}$	5
7	H	OCH_3	H	H	H	50	$10^{2.5}$	–
8	H	H	OCH_3	H	H	50	$10^{3.5}$	2.5
9	H	H	H	OCH_3	H	50	10^5	2.5
10	H	OH	H	H	H	10	10^4	5
11	H	H	OH	H	H	5	10^4	4
12	H	Cl	H	H	H	100	$10^{3.5}$	1.5
13	H	H	Cl	H	H	100	10^3	20
14	H	H	Br	H	H	30	10	–
15	H	H	I	H	H	10	10^3	10
16	H	H	F	H	H	10	10^2	–
17	H	Cl	H	CH_3	H	25	1	–
18	H	NO_2	H	H	H	50	$10^{2.5}$	–
19	H	NH_2	H	H	H	10	10^3	10
20	OH	H	H	H	H	5	10^3	1.5
21	OH	H	OH	H	H	10	10^5	0.25
22	OH	H	OCH_3	H	H	100	10^5	0.3
23	OCH_3	H	OCH_3	H	H	100	10^4	5
24	CH_3	H	OH	H	H	10	10^4	0.05
25	CH_3	H	OCH_3	H	H	25	10^4	0.5
26	CH_3	CH_3	OH	H	H	>100	10^5	<2.5
27	OH	H	OH	H	OH	20	10^5	0.5
28	OH	H	OCH_3	H	OH	10	10^5	0.5
29	OH	H	OCH_3	H	OCH_3	20	10^5	0.5
30	OH	OCH_3	OH	H	H	20	10^4	1
31	OH	OCH_3	OCH_3	H	H	100	10^5	0.1
32	OH	H	OCH_3	OCH_3	H	25	10^5	0.3
33	OH	H	OCH_3	OCH_3	OH	10	10^4	0.5
34	OH	H	OH	OCH_3	OH	20	10^5	2
35	OH	OCH_3	OH	H	OCH_3	10	10^3	1

[a]   1–26 are synthetic compounds; 27–35 are natural compounds.
[b]   Maximal nontoxic dose.
[c] RF_{MNTD} = reduction factor at MNTD.
[d] MD_{RF10^3} = minimal dose with an RF 10^3.

TABLE 6
Therapeutic Index of 3-Methoxyflavones Against Polio- and Rhinovirus Infections

No.[g]	Polio CyD$_{50}$[a] µg/ml	Polio ED$_{99}$[b] µg/ml	Polio TI$_{99}$[c]	Rhino CyD$_{50}$[a] µg/ml	Rhino ED$_{99}$[b] µg/ml	Rhino TI$_{99}$[c]
1	10	6	1.7	—[d]	—	—
2	75	5	15	75	RF<10^2 [e]	—
3	25	2	12.5	25	25	1
4	10	5	2	10	10	1
5						
6	150	10	15	150	15	10
7	200	25	8	—	—	—
8	>50	2.5	>20	>50	10	>5
9	25	20	1.25	—	—	—
10	3	2	1.5	15	10	1.5
11	15	1	15	25	25	1
12	100	6	16.7	100	RF<10^2	—
13	100	7	14.3	100	25	4
14	30	>30	<1	—	—	—
15	20	4	5	20	RF<10^2	—
16	25	10	2.5	—	—	—
17	50	>50	<1	—	—	—
18	>0.1	NA[f]	—	—	—	—
19	10	5	2	—	—	—
20	25	0.5	50	5	1	5
21	10	0.2	50	—	—	—
22	100	0.3	333.3	100	2.5	40
23	100	10	10	100	20	5
24	15	0.1	150	15	1	15
25	50	1	50	50	25	2
26	>100	0.1	>1000	100	0.5	>200
27	5	0.3	16.7	—	—	—
28	10	0.3	33.3	—	—	—
29	25	2.5	10	—	—	—
30	25	0.5	50	—	—	—
31	100	0.2	500	10	10	1
32	>100	0.25	400	>100	0.5	>200
33	>5	0.2	25	20	2	10
34	10	1.5	6.6	—	—	—
35	1	NA	—	—	—	—

[a] CyD$_{50}$ = 50% cytotoxic dose.
[b] ED$_{99}$ = 99% effective dose.
[c] TI$_{99}$ = therapeutic index 99.
[d] — = not tested.
[e] RF < 10^2 = viral titer reduction factor at CyD$_{50}$ lower than 10^2, no ED$_{99}$.
[f] NA = not active.
[g] No.: for structure see table 5.

These data indicate that substitution of a 5-hydroxylated or methylated A-ring of the parent compound (**1**) with one methoxyl or hydroxyl function, or preferably one hydroxyl and one methyl group or two methoxyl groups, largely increases antipolio potency (at least 20 times) and likewise decreases cytotoxicity to Vero cells (at least ten times), resulting in promising antipolio products. All other compounds of these series have TI$_{99}$ values of 50 or less. It should be noticed that substitution of the 3'-position of the B-ring with hydroxyl or methoxyl functions increases cytotoxicity of the corresponding substance, without influencing antipolio activity very much, indicating that the naturally occurring 3-

TABLE 7
Antipoliovirus Activity of 4′-Chloro, 4′-Methoxy-, and 4′-Acetoxyflavones 36–57

No.[a]	R_1	R_2	R_3	R_4	R_5	R_6	MNTD[b] $\mu g/ml$	RF_{MNTD}[c]	MD_{RF10^3}[d] $\mu g/ml$
36	H	H	CH_3	H	OCH_3	H	100	1	–
37	OCH_3	H	H	H	OCH_3	H	100	1	–
38	H	OCH_3	H	H	OCH_3	H	50	10^2	–
39	H	H	H	OCH_3	OCH_3	H	100	1	–
40	H	OH	H	H	OCH_3	H	100	1	–
41	H	Cl	H	H	OCH_3	H	25	1	–
42	H	H	Cl	H	OCH_3	H	100	1	–
43	H	H	CH_3	H	Cl	H	25	1	–
44	G	Cl	H	H	Cl	H	25	1	–
45	H	H	Cl	H	Cl	H	25	1	–
46	OH	H	OCH_3	H	OCH_3	H	>50	1	–
47	OH	H	OH	H	OCH_3	H	10	10^5	5
48	OH	H	OCH_3	H	OCH_3	OCH_3	>50	1	–
49	OH	H	OH	H	OCH_3	OCH_3	50	10^5	2.5
50	OH	OCH_3	OH	H	OCH_3	H	100	1	–
51	OH	H	OH	OCH_3	OCH_3	H	>10	1	–
52	OH	H	OCH_3	OCH_3	OCH_3	H	7.5	–	–
53	H	OCH_3	OCH_3	H	OCH_3	OCH_3	50	10^3	50
54	OH	OCH_3	OCH_3	OCH_3	OCH_3	H	5	1	–
55	OH	OCH_3	OCH_3	OCH_3	OCH_3	OCH_3	50	10^3	50
56	OCH_3	OCH_3	OCH_3	OCH_3	OCH_3	OCH_3	50	10^4	25
57	$OCOCH_3$	H	$OCOCH_3$	H	$OCOCH_3$	$OCOCH_3$	>50	10^5	2

[a] 36–45 are synthetic compounds; 46–57 are natural compounds.
[b] Maximal nontoxic dose.
[c] RF_{MNTD} = reduction factor at MNTD.
[d] MD_{RF10^3} = minimal dose with an RF 10^3.

O-methylkempferol derivatives possess better TI_{99} values than their corresponding 3-O-methylquercetin analogs.

These results are largely confirmed in the antiviral testing against rhinovirus type 15 and the corresponding cytotoxicity investigation on human skin fibroblasts, as shown in Table 6. For most substances the same degree of cytotoxicity on Vero cells and human skin fibroblasts is observed. In contrast, the antirhinovirus ED_{99} values are usually smaller than the corresponding antipolio ones which results in lower TI_{99} values. Therapeutic ratios of more than 200 are found for compounds **26** and **32**, whereas substances **22** and **24** are ten times less active against rhinovirus than against poliovirus. Surprisingly, penduletin (**31**) is only active against rhinovirus in its cytotoxic concentration. All other compounds tested have TI_{99} values of ten or less.

TABLE 8
Antipoliovirus Activity of 3-Substituted 4'-Hydroxy-7-Methylflavones 58–62

No.[a]	R	MNTD[b] $\mu g/ml$	RF_{MNTD}[c]	MD_{RF10}[3,d] $\mu g/ml$
58	OC_2H_5	25	1	–
59	$OCH(CH_3)_2$	1	1	–
60	Cl	10	10^3	6
61	CH_3	100	10^3	50
62	NH_2	20	1	–

[a] Synthetic compounds.
[b] Maximal nontoxic dose.
[c] RF_{MNTD} = reduction factor at MNTD.
[d] MD_{RF10}[3] = minimal dose with an RF 10^3.

TABLE 9
Antirhinovirus Activity[a] of 4',7-Dihydroxy-3-Methoxy-5,6-Dimethylflavone (26) and 4',7-Dihydroxy-3-Methoxy-5-Methylflavone (24) Against 12 Strains; MIC_{50}, $\mu g/ml$

Compound	2	29	39	85	9	15	59	63	89	41	14	70
26	0.044	0.078	0.088	0.076	0.156	0.088	0.016	0.074	0.5	0.088	0.072	0.097
24	2.4	0.175	12.0	–	0.2	0.4	3.0	–	0.2	0.375	10.8	5.0
Guanidine	>64	>64	109	750	375	438	350	500	1000	1000	7	6
HBB[b]	>50	>50	>50	>50	>50	>50	>50	>50	>50	>50	16	50

[a] The screening was performed by Dr. K. Andries, Janssen Research Foundation, Beerse, Belgium.
[b] 2-(α-Hydroxybenzyl)benizimidazole.

From Table 7 it is clear that a 4'-hydroxyl group is important for antipolio activity. For example, comparison of the antiviral activities of 4'-hydroxy-3-methoxy-7-methylflavone (3), 5,4'-dihydroxy-3,7-dimethoxyflavone (22), and 5,7,4'-trihydroxy-3,6-dimethoxyflavone (30) with those of, respectively, 3,4'-dimethoxy-7-methylflavone (36), 5-hydroxy-3,7,4'-trimethoxyflavone (46), and 5,7-dihydroxy-3,6,4'-trimethoxyflavone (50) shows the superiority of the 4'-hydroxy analogs. Although some of the 4'-methoxy and 4'-acetoxy derivatives show some degree of antipolio activity, viz. compounds 47, 49, and 57, their TI_{99} values are considerably lower than those of the corresponding 4'-hydroxy analogs. None of the 4'-chloro derivatives shows any antipoliovirus activity.

Table 8 indicates the necessity of the 3-methoxyl group. Replacement of this function with different R substitutions results in compounds (58–62) having no, or lower, antipolio activities and/or higher cytotoxicities than the corresponding 3-methoxy derivatives.

TABLE 10
Mutagenicity Testing of 3-Methoxyflavone 26 and Quercetin

		Revertants per plate			
	Salmonella typhimurium	Strain TA 98		Strain TA 100	
Mutagen	**mg per plate**	**No S9**	**With S9**	**No S9**	**With S9**
—	—	12	73	96	87
26	0.25	12	20	110	103
26	2.5[a]	10	9	85	94
Quercetin	0.25	1100	b	589	b
Quercetin	2.5	b	b	b	b

[a] Precipitation occurs after addition of the S9 mix.
[b] Toxic concentration

Substance **26**, being the most interesting compound in both antiviral test systems, was further screened[66] against a series of 12 rhinovirus serotypes that were grown on Ohio HeLa cells. Its activity was compared to that of a less active synthetic flavone, viz. 4′,7-dihydroxy-3-methoxy-5-methylflavone (**24**) and the reference substances guanidine and 2-(α-hydroxybenzyl)benzimidazole (HBB). The data of this screening are presented in Table 9. The results are expressed as the 50%-inhibitory concentration (MIC$_{50}$; µg/ml), i.e., the lowest concentration of compound that protected 50% of the cells from CPE. Whereas the different rhinovirus serotypes widely vary in their susceptibility to the reference substances, the MICs for 50% CPE reduction of **26** range from 0.016 to 0.5 µg/ml. The lower antiviral activity (higher MIC$_{50}$ values) that is found for **24**, against rhinovirus serotype 15, is confirmed for nearly all other rhinovirus serotypes tested. The MIC$_{50}$s vary from 0.175 to 12 µg/ml. It should be pointed out that all rhinovirus serotypes investigated were sensitive to both synthetic flavones. In contrast, it is well known that drug resistance against capsid-binding antiviral drugs (such as flavans and chalcones) is a common feature, so that often cross resistance is exhibited by these antirhinovirus active canyon products.[130, 131]

Since some 3-methoxyflavones, when administered intraperitoneally, have been shown to protect mice from lethal infections from coxsackie B$_4$, the most antivirally active substance of this study, namely compound **26**, should be considered as a promising candidate for antirhinovirus clinical studies in human volunteers. As quercetin and several familiar flavonoids have been reported to be mutagenic in a number of short-term microbial assays,[132-136] mutagenicity experiments were performed with compound **26**, using *Salmonella typhimurium* strains TA 98 and TA 100 with and without S9 mix from rat liver, according to the plate incorporation assay. The results of these tests are given in Table 10. It is clearly shown that the number of revertants is not significantly increased by **26** in concentrations up to 2.5 mg per plate, with or without S9 fraction. On the contrary, the reported mutagenicity of quercetin is confirmed.

VIII. CONCLUSION

This study on the antipicornavirus properties of 3-methoxyflavones has amply shown that this interesting class of natural products is very promising for the development of clinically useful antiviral drugs. The mechanism studies have revealed compounds that interfere with viral RNA and protein synthesis and induce no resistance. Structure–activity relationship studies have given some hope that more active and less toxic analogs might be found, which will be as good as, or better than, the existing antiviral substances. More generally, it has been proven that the potential of higher plants (especially those used in traditional medicine) to provide novel prototype antivirals is considerable and should be exploited more extensively by virologists.

REFERENCES

1. **Springthorpe, V. S., Grenier, J. L., Lloyd-Evans, N., and Sattar, S. A.,** Chemical disinfection of human rotaviruses. Efficacy of commercially available products in suspension tests, *J. Hyg.,* 97, 136, 1986.
2. **Vanden Berghe, D. A. and Vlietinck, A. J.,** Screening methods for antibacterial and antiviral agents from higher plants, in *Methods in Plant Biochemistry,* Vol. 6, Hostettmann, K., Ed., Academic Press, London, 1991, 47.
3. **Ieven, M., Vlietinck, A. J., Vanden Berghe D. A., Totté, J., Dommisse, R., Esmans, E., and Alderweireldt, F.,** Plant antiviral agents. III. Isolation of alkaloids from *Clivia miniata* Regel, *J. Nat. Prod.,* 45, 564, 1982.
4. **Vanden Berghe, D. A., Vlietinck, A. J., and Van Hoof, L.,** Plant products as potential antiviral agents, *Bull. Inst. Pasteur,* 84, 101, 1986.
5. **Hermann, E. C., Jr.,** The detection, assay and evaluation of antiviral drugs, *Prog. Med. Vir.,* 3, 158, 1961.
6. **Kaufman, H. E.,** Problems in virus chemotherapy, *Prog. Med. Vir.,* 7, 116, 1965.
7. **Sidwell, R. W.,** Determination of antiviral activity, *Drugs Pharm. Sci.,* 27, 433, 1986.
8. **Prussof, W. H. and Lin, T. S.,** Experimental aspects of antiviral pharmacology, in *Antiviral Drug Development,* De Clercq, E. and Walker, R. T., Eds., Plenum Press, New York, 1990, 173.
9. **Haseltine, W. A.,** Development of antiviral drugs for the treatment of AIDS: strategies and prospects, *J. Acquir. Immune Defic. Syndr.,* 2, 311, 1989.
10. **Winkler, D. A. and Holan, G.,** Design of potential anti-HIV agents. I. Mannosidase inhibitors, *J. Med. Chem.,* 32, 2084, 1989.
11. **Mitsuya, H. and Broder, S.,** Second generation antiviral therapy against human immunodeficiency virus (HIV), *Mech. Act. Ther. Appl. Biol. Cancer Immune. Defic. Dis. UCLA Symp. Mol. Cell Biol.* (New ser.), 100, 343, 1989.
12. **Allen, L. B., Kehoe, M. J., Hsu, S. C., Barfield, R., Holland, C. S., and Dimitrijevich, S. D.,** A simple method of drying virus on inanimate objects for virucidal testing, *J. Virol. Meth.,* 19, 239, 1988.
13. **Delgadillo, R. A. and Vanden Berghe, D. A.,** Inhibition of the multiplication of enveloped and non-enveloped viruses by glucosamine, *J. Pharm. Pharmacol.,* 40, 488, 1988.
14. **Grunert, R. H.,** Search for antiviral agents, *Ann. Rev. Microb.,* 33, 335, 1979.
15. **Vanden Berghe, D. A., Ieven, M., Mertens, F., Vlietinck, A. J., and Lammens, E.,** Screening of higher plants for biological activities. II. Antiviral activity, *Lloydia,* 41, 463, 1978.
16. **Van Hoof, L., Vanden Berghe, D. A., Hatfield, G. M., and Vlietinck, A. J.,** Plant antiviral agents. V. 3-Methoxyflavones as potent inhibitors of viral induced block of cell synthesis, *Planta Med.,* 50, 513, 1984.
17. **Vlietinck, A. J. and Vanden Berghe, D. A.,** Can ethnopharmacology contribute to the development of antiviral drugs? *J. Ethnopharm.,* 32, 141, 1991.
18. **Ito, M., Baba, M., Sato, A., Pauwels, R., De Clercq, E., and Shigeta, S.,** Inhibitory effect of dextran sulfate and heparin on the replication of human immunodeficiency virus (HIV) in vitro, *Antiviral Res.,* 7, 361, 1987a.
19. **Ito, M., Nakashima, H., Baba, M., Pauwels, R., De Clercq, E., Shigeta, S., and Yamamoto, N.,** Inhibitory effect of glycyrrhizin in the *in vitro* infectivity and cytopathic activity of the human immunodeficiency virus [HIV(HTLV-III/LAV)], *Antiviral Res.,* 7, 127, 1987b.
20. **De Clercq, E.,** Perspectives for the chemotherapy of AIDS, *Anticancer Res.,* 7, 1023, 1987.
21. **De Clercq, E.,** in *Antiviral Drug Development. A Multidisciplinary Approach,* De Clercq, E. and Walker, R. T., Eds., NATO Advanced Science Institutes Series, Series A: Life Sciences, Plenum Press, New York, 1988.
22. **Richman, D. D.,** Antiviral therapy of HIV infection, *Annu. Rev. Med.,* 42, 69, 1991.
23. **Block, T. M. and Grafstrom, R. H.,** Novel bacteriological assay for detection of potential antiviral agents, *Antimicrob. Agents Chemother.,* 34, 2337, 1990.
24. **Tan, G. T., Pezzuto, J. M., and Kinghorn, A. D.,** Evaluation of natural products as inhibitors of human immunodeficiency virus type 1 (HIV-1) reverse transcriptase, *J. Nat. Prod.,* 54, 143, 1991.
25. **Hirsch, M. S.,** Antiviral drug development for the treatment of human immunodeficiency virus infections, *Am. J. Med.,* 85, 182, 1988.
26. **Chang, R. S. and Yeung, H. W.,** Inhibition of growth of human immunodeficiency virus *in vitro* by crude extracts of Chinese medicinal herbs, *Antiviral Res.,* 9, 163, 1988.
27. **Roberts, N. A., Martin, J. A., Kinchington, D., Broadhurst, A. V., Craig, J. C., Duncan, I. B., Galpin, S. A., Handa, B. J., Kay, J., Kröhn, A., Lambert, R. W., Merrett, J. H., Mills, J. S., Parkes, K. E. B., Redshaw, S., Ritchie, A. J., Taylor, D. L., Thomas, G. J., and Machin, P. J.,** Rational design of peptide-based HIV proteinase inhibitors, *Science,* 248, 358, 1990.

28. **Rossmann, M. G.,** Antiviral agents targeted to interact with viral capsid proteins and a possible application to human immunodeficiency virus, *Proc. Natl. Acad. Sci., U.S.A.,* 85, 4625, 1988.

29. **Tochikura, T. S., Nakashima, H., and Yamamoto, N.,** Antiviral agents with activity against human retroviruses, *J. Acquir. Immune. Defic. Syndr.,* 2, 441, 1989.

30. **De Clercq, E.,** Potential drugs for the treatment of AIDS, *J. Antimicrob. Chemother.,* 23, 35, 1989.

31. **Schwartz, O., Henin, Y., Marechal, V., and Montagnier, L.,** A rapid and single colorimetric test for the study of anti-HIV agents, *AIDS Res. Human Retroviruses,* 4, 441, 1988.

32. **Pauwels, R., Balzarini, J., Baba, M., Snoeck, R., Schols, D., Herdewijn, P., De Smijter, J., and De Clercq, E.,** Rapid and automated tetrazolium-based colorimetric assay for the detection of anti-HIV compounds, *J. Virol. Meth.,* 20, 309, 1988.

33. **Takeuchi, H., Baba, M., and Shigeta, S.,** An application of tetrazolium (MTT) colorimetric assay for the screening of anti-*Herpes simplex* virus compounds, *J. Virol. Meth.,* 33, 61, 1991.

34. **Becker, Y.,** Trends in research and development of antiviral substances, *Isr.. J. Med. Sci.,* 11, 1135, 1975.

35. **Becker, Y.,** Antiviral agents from natural sources, *Pharmacol. Ther.,* 10, 119, 1980.

36. **Came, P. E. and Steinberg, B. A.,** Chemotherapy of viral infections, in *Handbook of Experimental Pharmacology,* Springer-Verlag, Berlin, 1982, 61, 479.

37. **Hudson, J. B.,** *Antiviral Compounds from Plants,* CRC Press, Boca Raton, FL, 1990.

38. **Thiel, K. D., Helbig, B., Sprössig, M., Klöning, R., and Wutzler, P.,** Antiviral activity of enzymatically oxidized caffeic acid against *Herpes* virus type 1 and type 2, *Acta Virol.,* 27, 200, 1983.

39. **Kane, C. J. M., Menna, J. H., and Yeh, Y. C.,** Methyl gallate, methyl-3,4,5-trihydroxybenzoate, is a potent and highly specific inhibitor of *Herpes simplex* virus *in vitro*. I. Purification and characterization of methyl gallate from *Sapium sebiferum, Biosci. Rep.,* 8, 85, 1988.

40. **Fukuchi, K., Sakagami, H., Okuda, T., Hatano, T., Tanuma, S., Kitajima, K., Inoue, Y., Inoue, S., Ichikawa, S., Nonoyama, M., and Konno, K.,** Inhibition of *Herpes simplex* virus infection by tannins and related compounds, *Antiviral Res.,* 11, 285, 1989.

41. **Corthout, J., Pieters, L. A., Claeys, M., Vanden Berghe, D. A., and Vlietinck, A. J.,** Antiviral ellagitannins from *Spondias mombin, Phytochemistry,* 30, 1129, 1991.

42. **Corthout, J., Pieters, L., Claeys, M., Vanden Berghe, D., and Vlietinck, A. J.,** Antivirally active gallotannins from *Spondias mombin,* in 36th Annu. Congr. Medicinal Plant Research, 1988, *Planta Med.,* 54, 573, 1988.

43. **Poehland, B. L., Carte, B. K., Francis, T. A., Hyland, L. J., Allaudeen, H. S., and Troupe, N.,** *In vitro* antiviral activity of dammar resin triterpenoids, *J. Nat. Prod.,* 50, 706, 1987.

44. **Amoros, M., Fauconnier, B., and Girre, R. L.,** Effect of saponins from *Anagallis arvensis* on experimental *Herpes simplex* keratitis in rabbits, *Planta Med.,* 54, 128, 1988.

45. **Amoros, M. and Girre, R. L.,** Structure of two antiviral triterpene saponins from *Anagallis arvensis, Phytochemistry,* 26, 787, 1987.

46. **Amoros, M., Fauconnier, B., and Girre, R. L.,** *In vitro* antiviral activity of a saponin from *Anagallis arvensis,* Primulaceae, against *Herpes simplex* virus and poliovirus, *Antiviral Res.,* 8, 13, 1987.

47. **Hirabayashi, K., Iwata, S., Matsumoto, H., Mori, T., Shibata, S., Baba, M., Ito, M., Shigeta, S., Nakashima, H., and Yamamoto, N.,** Antiviral activities of glycyrrhizin and its modified compounds against human immunodeficiency virus type 1 HIV-1 and *Herpes simplex* virus type Q HSV-1 *in vitro, Chem. Pharm. Bull.,* 39, 112, 1991.

48. **Tsuchiya, Y., Shimizu, M., Hiyama, Y., Itoh, K., Hashimoto, Y., Nakayama, M., Horie, T., and Morita, N.,** Antiviral activity of natural occurring flavonoids *in vitro, Chem. Pharm. Bull.,* 33, 3881, 1985.

49. **Radloff, R. J., Deck, L. M., Boyer, R. E., and Vander Jagt, D. L.,** Antiviral activities of gossypol and its derivatives against HIV type II, *Pharmacol. Res. Commun.,* 18, 1063, 1986.

50. **Hayashi, T., Hayashi, K., Uchida, K., Niwayama, S., and Morita, N.,** Antiviral agents of plant origin. II. Antiviral activity of scopadulcic acid B derivatives, *Chem. Pharm. Bull.,* 38, 239, 1990.

51. **Hayashi, T., Kawasaki, M., Miwa, Y., Taga, T., and Morita, N.,** Antiviral agents of plant origin. III. Scopadulin, a novel tetracyclic diterpene from *Scoparia dulcis* L., *Chem. Pharm. Bull.,* 38, 945, 1990.

52. **Renard-Nozaki, J., Kim, T., Imakura, Y., Kihara, M., and Kobayashi, S.,** Effect of alkaloids isolated from Amaryllidaceae on *Herpes simplex* virus, *Res. Virol.,* 140, 115, 1989.

53. **Vrijsen, R., Vanden Berghe, D. A., Vlietinck, A. J., and Boeyé, A.,** Lycorine: an eukaryotic termination inhibitor? *J. Biol. Chem.,* 261, 505, 1986.

54. **Yamamoto, N., Furukawa, H., Ito, Y., Yoshida, S., Maeno, K., and Nishiyama, Y.,** Anti-*Herpes* virus activity of citrusinine-I, a new acridone alkaloid, and related compounds, *Antiviral Res.,* 12, 21, 1989.

55. **Kéry, A., Horvath, J., Nasz, L., Verzar-Petri, G., Kulcsar, G., and Dan, P.,** Antiviral alkaloid in *Chelidonium majus* L., *Acta Pharm. Hung.,* 57, 19, 1987.

56. **Sydiskis, R. J., Owen, D. G., Lohr, J. L., Rosler, K. H. A., and Blomster R. N.,** Inactivation of enveloped viruses by anthraquinones extracted from plants, *Antimicrob. Agents Chemother.,* 35, 2463, 1991.

57. **Tang, J., Colacino, J. M., Larsen, S. H., and Spitzer, W.,** Virucidal activity of hypericin against enveloped and non-enveloped DNA and RNA viruses, *Antiviral Res.,* 13, 313, 1990.

58. **Hudson, J. B., Lopez-Bazzocchi, I., and Towers, G. H. N.,** Antiviral activities of hypericin, *Antiviral Res.,* 15, 101, 1991.

59. **Meruelo, D., Lavie, G., and Lavie, D.,** Therapeutic agents with dramatic antiretroviral activity and little toxicity at effective doses. Aromatic polycyclic diones hypericin and pseudohypericin, *Proc. Natl. Acad. Sci. U.S.A.,* 85, 5230, 1988.

60. **Lavie, G., Valentine, F., Levin, B., Mazur, Y., Gallo, G., Lavie, D., Weiner, D., and Meruelo, D.,** Studies of the mechanisms of action of the antiretroviral agents hypericin and pseudohypericin, *Proc. Natl. Acad. Sci. U.S.A.,* 86, 5963, 1989.

61. **Martin, C. P., Vilas, P., Prieto, S. P., and Martin, A.,** Antiviral activity of a D-glucosamine derivative against herpetic ulcers (HSV type 2) in rabbit cornea, *Acta Ophthalmol.,* 67, 55, 1989.

62. **Arisawa, M., Fujita, A., Hayashi, T., Hayashi, K., Ochiai, H., and Morita, N.,** Cytotoxic and antiherpetic activity of phloroglucinol derivatives from *Mallotus japonicus* Euphorbiaceae, *Chem. Pharm. Bull.,* 38, 1624, 1990.

63. **Ishitsuka, H., Ohsawa, C., Ohiwa, T., Umeda, I., and Suhara, Y.,** Antipicornavirus flavone Ro-09-0179, *Antimicrob. Agents Chemother.,* 22, 611, 1982.

64. **Vlietinck, A. J., Vanden Berghe, D. A., Van Hoof, L., Vrijsen, R., and Boeyé, A.,** Antiviral activity of 3-methoxyflavones, in *Plant Flavonoids in Biology and Medicine,* Cody, V., Middleton, E., Jr., and Harborne, J. B., Eds., Alan R. Liss, New York, 1986; *Progr. Clin. Biol. Res.,* 213, 537, 1986.

65. **Vlietinck, A. J., Vanden Berghe, D. A., and Haemers, A.,** Present status and prospects of flavonoids as antiviral agents, in *Plant Flavonoids in Biology and Medicine II: Biochemical, Cellular and Medicinal Properties,* Cody, V., Middleton, E., Jr., Harborne, J. H., and Beretz, A., Eds., Alan R. Liss, New York, 1988; *Progr. Clin. Biol. Res.,* 280, 283, 1988.

66. **De Meyer, N., Haemers, A., Mishra, L., Pandey, H.-K., Pieters, L. A. C., Vanden Berghe, D. A., and Vlietinck, A. J.,** 4'-Hydroxy-3-methoxyflavones with potent antipicornavirus activity, *J. Med. Chem.,* 34, 736, 1991.

67. **De Meyer, N., Van Hoof, L., Pandey, H. K., Mishra, L., Vanden Berghe, D., Vlietinck, A. J., and Haemers, A.,** Antiviral activity of synthetic 3-methoxyflavones, in Proc. VI Mediterranean Congr. Chemotherapy, *J. Chemother.,* 4 (Suppl.), 1, 1082, 1989.

68. **De Meyer, N., Vlietinck, A., Pandey, H., Mishra, L., Pieters, L., Vanden Berghe, D., and Haemers, A.,** Synthesis and antiviral properties of 3-methoxyflavones, in *Flavonoids in Biology and Medicine III: Current Issues in Flavonoid Research,* Das, N. P., Ed., National University of Singapore, Singapore, 1990, 403.

69. **De Tommasi, N. and Pizza, C.,** Structure and *in vitro* antiviral activity of sesquiterpene glycosides from *Calendula arvensis, J. Nat. Prod.,* 53, 830, 1990.

70. **De Tommasi, N., Conti, C., Stein, M. L., and Pizza, C.,** Structure and *in vitro* antiviral activity of triterpenoid saponins from *Calendula arvensis, Planta Med.,* 57, 250, 1991.

71. **Kamano, Y., Satoh, N., Nakayoshi, H., Pettit, G. R., and Smith, C. R.,** Rhinovirus inhibition by bufadienolides, *Chem. Pharm. Bull.,* 36, 326, 1988.

72. **Pinto, A. V., Pinto, M., Lagrota, M. H., Wigg, M. D., and Aguiar, A. N. S.,** Antiviral activity of naphthoquinones. I. Lapachol derivatives against enteroviruses, *Rev. Latinoam. Microbiol.,* 29, 15, 1987.

73. **Baba, M., Pauwels, R., Balzarini, J., Arnout, J., De Smyter, J., and De Clercq, E.,** Mechanism of inhibitory effect of dextran sulfate and heparin on the replication of human immunodeficiency virus *in vitro, Proc. Natl. Acad. Sci., U.S.A.,* 85, 6132, 1988.

74. **Ngan, F., Chang, R. S., Tabba, H. D., and Smith, K. M.,** Isolation, purification and partial characterization of an active anti-HIV compound from the Chinese medical herb, *Viola yeodensis, Antiviral Res.,* 10, 107, 1988.

75. **Tabba, H. D., Chang, R. S., and Smith, K. M.,** Isolation purification and partial characterization of prunellin, an anti-HIV component from aqueous extracts of *Prunella vulgaris, Antiviral Res.,* 11, 263, 1989.

76. **Ito, M., Sato, A., Hirabayashi, K., Tanabe, F., Shigeta, S., Baba, M., De Clercq, E., Nakashima, H., and Yamamoto, N.,** Mechanism of inhibitory effect of glycyrrhizin on replication of human immunodeficiency virus (HIV), *Antiviral Res.,* 10, 288, 1988.

77. **Hattori, T., Ikumatsu, S., Koivo, A., Matsushita, S., Maeda, Y., Hada, M., Fujimaki, M., and Takatsuki, K.,** Preliminary evidence for inhibitory effect of glycyrrhizin on HIV replication in patients with AIDS, *Antiviral Res.,* 11, 255, 1989.

78. **Nakashima, H., Ohubo, K., Honda, Y., Tamura, T., Matsuda, S., and Yamamoto, N.,** Inhibitory effect of glycosides like saponin from soybean on the infectivity of HIV *in vitro, AIDS,* 3, 655, 1989.

79. **Aquino, R., De Simone, F., and Pizza, C.,** Plant metabolites. Structure and *in vitro* antiviral activity of quinovic acid glycosides from *Uncaria tomentosa* and *Guettarda platypoda, J. Nat. Prod.,* 52, 679, 1989.

80. **Ohtsuki, K. and Iahida, N.,** Inhibitory effect of glycyrrhizin on polypeptide phosphorylation by polypeptide-dependent protein kinase P *in vitro, Biochem. Biophys. Res. Commun.,* 157, 597, 1988.

81. **Konoshima, T., Kozuka, M., Haruna, M., Ito, K., Kimura, T., and Tokuda, H.,** Studies on the constituents of leguminous plants. XII. The structures of new triterpenoid saponins from *Wisteria brachybotrys* Sieb. et Zucc., *Chem. Pharm. Bull.,* 37, 2731, 1989.

82. **Chowdhury, B. L., Hussaini, F. A., and Shoeb, A.,** Antiviral constituents from *Vicoa indica, Int. J. Crude Drug Res.,* 28, 121, 1990.
83. **Taylor, D. L., Fellows, L. E., Farrar, G. H., Nash, R. J., Taylor-Robinson, D., Mobberley, M. A., Ryder, T. A., Jeffries, D. J., and Tyms, A. S.,** Loss of cytomegalovirus infectivity after treatment with castanospermine or related plant alkaloids correlates with aberrant glycoprotein synthesis, *Antiviral Res.,* 10, 11, 1988.
84. **Ruprecht, R. M., Mullaney, S., Andersen, J., and Bronson, R.,** *In vivo* analysis of castanospermine, a candidate antiretroviral agent, *J. Acquir. Immune. Defic. Syndr.,* 2, 149, 1989.
85. **Ruprecht, R. M., Bernard, L. D., Bronson, R., Gama Sosa, M. A., and Mullaney, S.,** Castanospermine vs. its 6'-O-butanoyl analog: a comparison of toxicity and antiviral activity *in vitro* and *in vivo, J. Acquir. Immune. Defic. Syndr.,* 4, 48, 1991.
86. **Rhinehart, B. L., Robinson, K. M., King, C.-H. R., and Liu, P. S.,** Castanospermine-glucosides as selective disaccharidase inhibitors, *Biochem. Pharmacol.,* 39, 1537, 1990.
87. **Lin, T.-S., Schinazi, R., Griffith, B. P., August, E. M., Eriksson, B. F. H., Zheng, D.-K., Huang, L., and Prusoff, W. H.,** Selective inhibition of human immunodeficiency virus type 1 replication by the (−) but not the (+) enantiomer of gossypol, *Antimicrobiol. Agents Chemother.,* 33, 2149, 1989.
88. **Polsky, B., Segal, S. J., Baron, P. A., Gold, J. W. M., Ueno, H., and Armstrong, D.,** Inactivation of human immunodeficiency virus *in vitro* by gossypol, *Contraception,* 39, 579, 1989.
89. **Turano, A., Scura, G., Caruso, A., Bonfanti, C., Luzzati, R., Bassetti, D., and Manca, N.,** Inhibitory effect of papaverine on HIV replication in vitro, *AIDS Res. Hum. Retrov.,* 5, 183, 1989.
90. **Lee-Huang, S., Kung, H. F., Huang, P. L., Li, B. Q., Huang, P., Huang, H. I., and Chen, H. C.,** A new class of anti-HIV agents: GAP31, DAPs 30 and 32, *FEBS Lett.,* 291, 139, 1991.
91. **Ferrari, P., Trabaud, M. A., Rommain, M., Mandine, E., Zalisz, R., Desgranges, C., and Smets, P.,** Toxicity and activity of purified trichosanthin, *AIDS,* 5, 865, 1991.
92. **Lee-Huang, S., Huang, P. L., Kung, H. F., Li, B. Q., Huang, P., Huang, H. I., and Chen, H. C.,** TAP 29: an anti-human immunodeficiency virus protein from *Trichosanthes kirilowii* that is nontoxic to intact cells, *Proc. Natl. Acad. Sci. U.S.A.,* 88, 6570, 1991.
93. **Balzarini, J., Schols, D., Neyts, J., Van Damme, E., Peumans, W., and De Clercq, E.,** Alpha-(1-3)- and alpha-(1-6)-D-mannose-specific plant lectins are markedly inhibitory to human immunodeficiency virus and cytomegalovirus infections *in vitro, Antimicrob. Agents Chemother.,* 35, 410, 1991.
94. **Nonaka, G. -I., Nishioka, I., Nishizawa, M., Yamagishi, T., Kashiwada, Y., Dutschman, G. E., Bodner, A. J., Kilkuskie, R. E., Cheng, Y.-C., and Lee, K.-H.,** Anti-AIDS agents, 2: Inhibitory effects of tannins on HIV reverse transcriptase and HIV replication in H9 lymphocyte cells, *J. Nat. Prod.,* 53, 587, 1990.
95. **Nishizawa, M., Yamagishi, T., Dutschman, G. E., Parker, W. B., Bodner, A. J., Kilkuskie, R. E., Cheng, Y.-C., and Lee, K.-H.,** Anti-AIDS agents. I. Isolation and characterization of four new tetragalloylquinic acids as a new class of HIV reverse transcriptase inhibitors from tannic acid, *J. Nat. Prod.,* 52, 762, 1989.
96. **MacRae, W. D., Hudson, J. B., and Towers, G. H. N.,** The antiviral action of lignans, *Planta Med.,* 55, 531, 1989.
97. **Towers, G. H. N. and Hudson, J. B.,** Potentially useful antimicrobial and antiviral phototoxins from plants, *Photochem. Photobiol.,* 46, 61, 1987.
98. **Hudson, J. B., Graham, E. A., Hudson, L. L., and Towers, G. H. N.,** The mechanism of antiviral phototoxicity of the furanochromones visnagin and khellin, *Planta Med.,* 54, 131, 1988.
99. **Harada, H., Sakagami, H., Nagata, K., Oh Hara, T., Kawazoe, Y., Ishihama, A., Hata, N., and Misawa, Y.,** Possible involvement of lignin structure in anti-influenza virus activity, *Antiviral Res.,* 15, 41, 1991.
100. **Nagai, T., Miyaichi, Y., Tomimori, T., Suzuki, Y., and Yamada, H.,** Inhibition of influenza virus sialidase and anti-influenza virus activity by plant flavonoids, *Chem. Pharm. Bull.,* 38, 1329, 1990.
101. **Mehrotra, R., Rawat, S., Kulshreshtha, D. K., Patnaik, G. K., and Dhawan, B. N.,** *In vitro* studies on the effect of certain natural products against hepatitis B virus, *Indian J. Med. Res. Sect. B,* 92, 133, 1990.
102. **Inouye, Y., Yamaguchi, K., Take, Y., and Nakamura, S.,** Inhibition of avian myeloblastosis virus reverse transcriptase by flavones and isoflavones, *J. Antibiot.,* 42, 1523, 1989.
103. **Konoshima, T., Takasaki, M., and Kozuka, M.,** Studies on inhibitors of skin-tumor promotion. I. Inhibitory effects of triterpenes from *Euptelea polyandra* on Epstein-Barr virus activation, *J. Nat. Prod.,* 50, 1167, 1987.
104. **Sydiskis, R. J., Owen, D. G., Lohr, J. L., Rosler, K. H., and Blomster, R. N.,** Inactivation of enveloped viruses by anthraquinones extracted from plants, *Antimicrob. Agents Chemother.,* 35, 2463, 1991.
105. **Andersen, D., Weber, N., Hughes, B., Lawson, L., Murray, B., and North, J.,** Virucidal activity of allicin from *Allium sativum* garlic, in 89th Annu. Meet. American Society for Microbiology, *Abstr. Ann. Meet. Am. Soc. Microbiol.,* 89, 29, 1989.
106. **Joshi, M. N., Chowdhury, B. L., Vishnoi, S. P., Shoeb, A., and Kapil, R. S.,** Antiviral activity of dextro-odorinol, *Planta Med.,* 53, 254, 1987.
107. **Cheminat, A., Zawatzky, R., Becker, H., and Brouillard, R.,** Caffeoyl conjugates from *Echinacea* species. Structures and biological activity, *Phytochemistry,* 27, 2787, 1988.
108. **De Rodriguez, D. J., Chulia, J., Simoes, C. M. O., Amoros, M., Mariotte, A., and Girre, L.,** Search for *in vitro* antiviral activity of a new isoflavonic glycoside from *Ulex europaeus, Planta Med.,* 56, 59, 1990.

109. **MacRae, W. D., Hudson, J. B., and Towers, G. H. N.,** α-(−)-Peltatin, an antiviral constituent of *Amanoa aff. oblongifolia, J. Ethnopharmacol.,* 22, 223, 1988.

110. **Vanden Berghe, D. A., Vlietinck, A. J., and Van Hoof, L.,** Present status and prospects of plant products as antiviral agents, in *Advances in Medicinal Plant Research,* Vlietinck, A. J. and Dommisse, R. A., Eds., Wissenschaftliche Verlagsgesellschaft mbH, Stuttgart,1985, 4, 47.

111. **Van Hoof, L., Vanden Berghe, D., and Vlietinck, A.,** Antiviral compounds from an African *Euphorbia* species, in *Abstr. 4th Int. Conf. on Comparative Virology,* Banff, Alberta, Canada, 1982, pp. 232, 17–22, 10.

112. **Laekeman, G. M., Claeys, M., Rwangabo, P. C., Herman, A. G., and Vlietinck, A. J.,** Cardiovascular effects of 3-methylquercetin, *Planta Med.,* 6, 433, 1986.

113. **Bauer, D. J., Selway, J. W. T., Batchelor, J. F., Tisdale, M., Cadwell, I. C., and Young, D. A. B.,** 4′,6-Dichloroflavan (BW683C), a new anti-rhinovirus compound, *Nature,* 292, 369, 1981.

114. **Bauer, D. J. and Selway, J. W. T.,** A novel method for detecting the antiviral activity of flavans in their vapour phase, *Antiviral Res.,* 3, 235, 1983.

115. **Ishitsuka, H., Ninomiya, Y. T., Ohsawa, C., Fujiu, M., and Suhara, Y.,** Direct and specific inactivation of rhinovirus by chalcone To 09-0410, *Antimicrob. Agents Chemother.,* 22, 617, 1982.

116. **Ninomiya, Y., Shimma, N., and Ishitsuka, H.,** Comparative studies on the antirhinovirus activity and the mode of action of the rhinovirus capsid binding agents, chalcone amides, *Antiviral Res.,* 13, 61, 1990.

117. **Ninomiya, Y., Uhsawa, C., Aoyama, M., Umeda, I., Suhara, Y., and Ishitsuka, H.,** Antivirus agents, Ro 09-0410, binds to rhinovirus specifically and stabilizes the virus conformation, *Virology,* 134, 269.

118. **Ninomiya, Y., Aoyama, M., Umeda, I., Suhara, Y., and Ishitsuka, H.,** Comparative studies on the modes of action of the antirhinovirus agents Ro 09-0410, Ro 09-0179, RMI-15, 731, 4′,6-dichloroflavan and enviroxime, *Antimicrob. Agent Chemother.,* 27, 595, 1985.

119. **Tisdale, M. and Selway, J. W. T.,** Inhibition of an early stage of rhinovirus replication by dichloroflavan (BW683C), *J. Gen. Virol.,* 64, 795, 1983.

120. **Tisdale, M. and Selway, J. W. T.,** Effect of dichloroflavan (BW683C) on the stability and uncoating of rhinovirus type 1B, *J. Antimicrobiol. Chemother.,* 14 (Suppl. 1), 97, 1984.

121. **Selway, J. W. T.,** Antiviral activity of flavones and flavans, in *Plant Flavonoids in Biology and Medicine: Biochemical, Pharmacological, and Structure–activity Relationships,* Cody, V., Middleton, E., Jr., and Harborne, J. B., Eds., Alan R. Liss, New York, 1986, 521.

122. **Lopez Pila, J. M., Kopecka, H., and Vanden Berghe, D.,** Lack of evidence for strand-specific inhibition of poliovirus RNA synthesis by 3-methylquercetin, *Antiviral Res.,* 11, 47, 1989.

123. **Castrillo, J., Vanden Berghe, D., and Carrasco, L.,** 3-Methylquercetin is a potent and selective inhibitor of poliovirus RNA synthesis, *Virology,* 152, 219, 1986.

124. **Vrijsen, R., Everaert, L., Van Hoof, L. M., Vlietinck, A. J., Vanden Berghe, D. A., and Boeyé, A.,** The poliovirus induced shut-off of cellular protein synthesis persists in the presence of 3-methylquercetin, a flavonoid which blocks viral protein and RNA-synthesis, *Antiviral Res.,* 7, 35, 1987.

125. **Castrillo, J. L. and Carrasco, L.,** Action of 3-methylquercetin on poliovirus RNA replication, *J. Virol.,* 61, 3319, 1987.

126. **Shugar, D.,** Antiviral agents — some current developments, *Pure Appl. Chem.,* 57, 423, 1985.

127. **Flier, J. S. and Underhill, L. H.,** Molecular targets of antiviral therapy, *N. Engl. J. Med.,* 321, 163, 1989.

128. **Harmenberg, J.,** Interactions between virus and antiviral compounds in the host cell, *Med. Biol.,* 62, 299, 1984.

129. **Wiltink, E. H. H. and Janknegt, R.,** Antiviral drugs, *Pharm. Weekbl.* [Sci.], 13, 58, 1991.

130. **Bargar, T., Dulworth, J., Kenny, M., Massad, R., Daniel, J., Wilson, T., and Sargent, R.,** 3,4-Dihydro-2-phenyl-2H-pyrano[2,3-b]pyridines with potent antirhinovirus activity, *J. Med. Chem.,* 29, 1590, 1986.

131. **Tyrrell, D. A. J.,** Hot news on the common cold, *Annu. Rev. Microbiol.,* 42, 35, 1988.

132. **MacGregor, J. T. and Jurd, L.,** Mutagenicity of plant flavonoids: structural requirements for mutagenic activity in *Salmonella typhimurium, Mutat. Res.,* 54, 297, 1978.

133. **Elliger, C. A., Henika, P. R., and MacGregor, J. T.,** Mutagenicity of flavones, chromones and acetophenones in *Salmonella typhimurium, Mutat. Res.,* 135, 77, 1984.

134. **Rueff, J., Laires, A., Gaspar, J., and Rodrigues, A.,** Mutagenic activity in the wine-making process: correlation with rutin and quercetin levels, *Mutagenesis,* 5, 393, 1990.

135. **MacGregor, J. T. and Wilson, R. E.,** Flavone mutagenicity in *Salmonella typhimurium*: dependence on the pKM101 plasmid and excision-repair capacity, *Environ. Mol. Mutagen.,* 11, 315, 1988.

136. **MacGregor, J. T.,** Mutagenic and carcinogenic effects of flavonoids, in *Plant Flavonoids in Biology and Medicine,* Cody, V., Middleton, E., Jr., and Harbone, J. B., Eds., *Pharmacol. Struc. Act. Relationships,* Alan R. Liss, New York, 1986, 411.

Chapter 18

TOXICITY TESTING USING THE BRINE SHRIMP:
ARTEMIA SALINA

Teng Wah Sam

TABLE OF CONTENTS

0-8493-4372-0/93/$0.00+$.50
© 1993 by CRC Press, Inc.

441

I. INTRODUCTION

The study of bioactive compounds from plant sources and extracts in the chemical laboratory is often hampered by the lack of a suitable, simple, and rapid screening procedure. There are, of course, many procedures for bioassay. Whole animals, isolated tissues, or even biochemical systems are available, but they can be quite complicated and expensive. Generally, unless collaborative programs with biologists or pharmacologists are in place, the typical chemical laboratory is not suitably equipped to perform the usual bioassays with whole animals or isolated tissues and organs.

The physiologic or biologic effect to be observed in the screening is also critical. One of the simplest biologic responses to monitor is lethality, since there is only one criterion — either dead or alive. This is a quantal response, and the statistical treatment is also relatively easy. A procedure for general toxicity screening that does not require too much specialization is therefore essential as a preliminary stage in the study of bioactive compounds. A simple animal that has been used for this purpose is the brine shrimp, *Artemia salina* Leach.

The first report of the use of the brine shrimp as a test organism appeared in 1956.[1] Since then there have been many reports on the use of this animal for environmental studies,[2-4] screening for natural toxins,[5,6] and as a general screening for bioactive substances in plant extracts.[7] It has been proposed as a standard test by Vanhaecke and Persoone.[8]

II. LIFE CYCLE OF *ARTEMIA SALINA*

The brine shrimp is a crustacean belonging to the subclass Branchiopoda order Anostraca. It is found worldwide in bodies of water ranging from the brackish to the ultrasaline. This high tolerance to a wide range of salinity (from 10–20 to 180–220 g/l) makes it a relatively easy animal to culture and study.[9] Under ultrasaline conditions, there are few competing organisms or predators, and this probably explains the success of *Artemia* in establishing huge populations in these salt lakes or brine pools. There is considerable variation in form,[10] but generally the life history of this creature may be summarized as follows.[9,11,12]

Seasonally, as the water evaporates from salt lakes or ponds, tiny brown particles measuring about 0.20 mm in diameter rise to the surface and are eventually swept ashore by the wind. These are the inactive, or resting, eggs of the brine shrimp. They can be gathered and separated from sand and other debris by sieving. As long as they remain dehydrated and in diaphase, these eggs have a high resistance to extreme conditions and may be stored for long periods.

When they are returned to a saline solution (for example, seawater), the eggs absorb water. Embryogenesis begins and is completed between 16 to 36 h after immersion. The embryo emerges from the shell still covered in a hatching membrane. However, it soon develops antennae and mandibles, breaks free of the hatching membrane, and becomes an active, free-swimming nauplius. These larvae are considered to be in the instar I or instar II (older) stages. The larva is reddish in color due to the presence of yolk. There are three pairs of appendages (Figure 1): the antennulae, the antennae, and the mandibles. The antennae are constantly moving. The larva grows through about 15 stages. The reddish color persists for up to 3 days, suggesting that these nauplii may survive 72 h on their yolk resource alone. In culture the growing larvae may be fed yeast cells or unicellular algae. In about 20 to 35 days, the animal reaches a length of 8.5 to 9.5 mm and is sexually mature (Figure 2). There is, however, some variation in body length, resulting from different levels of salinity. The adults feed by creating a current, by beating rhythmically with their 11 pairs of thoracic legs, to sweep food particles toward the head. Thus, in all stages of growth of the brine shrimp, active life is indicated by movement of some appendage or other.

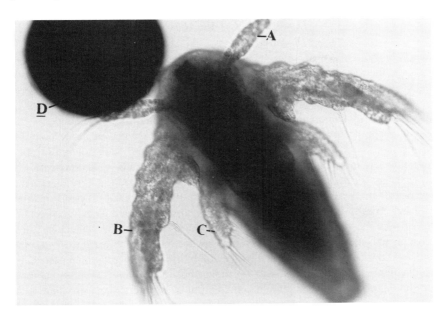

FIGURE 1. Freshly hatched instar I nauplius showing antennulae (A), antennae (B), and mandibles (C), next to a hydrated cyst (D).

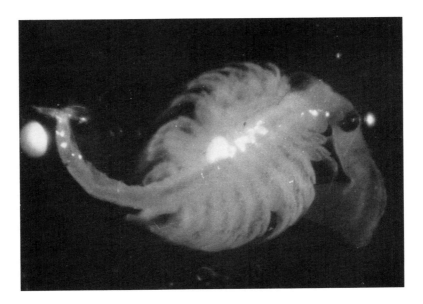

FIGURE 2. Adult male brine shrimp (8–10 mm long) with the antennae transformed into a pair of graspers and showing the stalked lateral eyes and 11 pairs of thoracic legs.

Reproduction can be oviparous or viviparous. This alternation may be governed by the amount of chlorophyll in the diet and of oxygen in the environment. Low oxygen and a high chlorophyll diet favor oviparity. The resting or inactive eggs are produced by oviparous females.

III. ASSAY TECHNIQUE

The availability of the eggs, the ease of hatching them into larvae, the rapid growth of the nauplii, and the relative ease of maintaining a population under laboratory conditions have made the brine shrimp a simple and effective test animal in the biologic sciences and in toxicology. Combined with a reference standard, the brine shrimp offers a bioassay that can be rapid, simple, and, more importantly, inexpensive and reproducible.

All stages in the life cycle of the brine shrimp have been involved in toxicology studies. The hatching rate of the eggs (for example, after 48-h exposure to petroleum oil, pesticides, PCBs, and other environmental contaminants[13] and after exposure to carcinogens[14]) has been used as a criterion for toxicity. More often, the mortality of nauplii at various stages of their development has been used. The most common larval stage is just 24 to 48 h after hatching,[4,5,6,15-17] but older nauplii[18] or the adults[19,20] have also been used as a screening organism. The lethal concentration for 50% mortality after 6 h of exposure, the acute LD_{50}, or after 24 h of exposure, the chronic LD_{50}, is determined as the measure of toxicity of the extract or compound. The choice of time, governed by the solubility of the extract or substance, is largely one of convenience, since the test is to be rapid and kept simple. Thus, polar extracts, or substances fairly soluble in water, should benefit from the shorter exposure, whereas the much lower concentrations that can be achieved with nonpolar or insoluble extracts require the longer times.

The method of choice in our laboratory, a modification of the method proposed by Zillioux et al.[2] makes use of the newly hatched nauplii, for the very simple reason that they do not need maintenance over the period of the bioassay.

A. CYSTS

The eggs are commercially available, since they and the larvae that can hatch from them have become an important food source in aquaculture. For our experiments resting eggs were obtained from the bulk supply in stock at the Fisheries Research Institute, Penang, Malaysia. These were originally San Francisco Bay type. Resting eggs may also be obtained from pet shops specializing in aquarium fishes. Eggs in a desiccator should be stored in a refrigerator (5°C).

B. CULTURE MEDIUM

The medium used is known to influence the hatching of the cysts,[21] but does not affect the experiment with early instar nauplii if a standard formula is used. The medium becomes important if later nauplii stages or adults are used, because their development is dependent upon salinity.[22,23] This can introduce additional variations when these animals are cultured in the laboratory, and thus there are few reports on the use of older nauplii and adults in toxicologic screening.

Natural seawater (about 35 g of salts per liter) is the medium of choice. This should be sterilized by boiling for 30 min, and filtered through a sintered glass funnel. Water loss is made up with distilled water. Filtered air is passed through the cooled solution to reoxygenate the medium. If seawater is not available, then an artificial medium can be prepared according to Table 1.[19] The pH should be adjusted to pH 7 to 8 with sodium bicarbonate. The hatching rate and the viability of the nauplii in natural seawater (NSW) are similar to those in the artificial seawater (ASW).

C. CULTURE METHOD

A large petri dish is divided by a glass plate that does not quite reach to the bottom of the dish (ca. 2-mm gap), and is filled with the sterilized seawater or ASW solution. Cysts (about 100 to 200 mg) are placed in one compartment, and a 60 watt lamp is positioned to

TABLE 1
Composition for an Artificial Seawater Medium
Modified from Morel's Aquil Formula[19]

Salt	g/l
NaCl	24.0
CaCl$_2$.2H$_2$O	1.5
KBr	0.1
KCl	0.7
Na$_2$SO$_4$	4.0
NaHCO$_3$	0.2
MgCl$_2$.6H$_2$O	11.0
Total Salts	41.5

provide direct light and warmth (about 25°C) throughout the embryogenesis.

Free swimming nauplii appear after 16 h, and most of the eggs should have hatched into free-swimming forms by 24 h. The free-swimming nauplii should be ready for collection some 36 to 48 h after sowing. These may be attracted to the other compartment, which is free from emerging embryos, eggs, and egg debris, by a torchlight. The nauplii can be collected using a Pasteur pipette with a nozzle of at least 1 mm diameter.

After 72 h the nauplii will require food. They are transferred to a standard 20 × 20 cm-glass, thin layer chromatography (TLC) developing tank filled with 1 to 2 liters of the same medium used in the hatching process. An aquarium stone aerator provides suitable oxygenation. We have used yeast (disrupted by ultrasound) and *Chlorella* as suitable food sources. *Dunaliella* has also been used.[24]

D. PREPARATION OF TEST SOLUTIONS

In bioassays that require the extract to be in solution, as opposed to being absorbed on a paper disk and immersed in the medium,[5,7] the cosolvent or surfactant can influence the outcome of the assay. The chronic (24 h) median lethal concentrations (LC) for 55 common organic chemicals, among them many solvents, have been estimated. Thus, benzene (LC 66 ppm) and acetic acid (LC 42 ppm) were found to be more toxic than ethanol or methanol (LC 10,000 ppm), whereas butanol and acetone are relatively toxic (3000 ppm and 2100 ppm, respectively).[3] The brine shrimp larvae can tolerate up to 11% of acetonitrile, dimethylformamide, dimethylsulfoxide, dioxane, ethanol, 2-propanol, methanol, and tetrahydrofuran, although acetonitrile and dioxane anesthetized the animals after 1 h, with recovery occurring only 6 h later.[25] Surfactants have also been used as solubilizing aids for the plant extracts. However, care must again be exercized, since it has been shown that several surfactants can be toxic to the brine shrimp larva (Table 2).[26]

This result is surprising, since we have found that the *Artemia* larvae can survive for 24 h, at least in media containing Tween 80 (polyoxyethylenesorbitan monooleate or monolaurate) at concentrations of 50,000 ppm. These conflicting data may simply be due to different experimental protocols, or it may reflect the purity of the Tween. Indeed, it has been shown that a large part of the toxicity of an *Aspergillus sydowii* extract was due to the fatty acid fraction, particularly to the saturated C$_6$ to C$_{12}$ acids, which were most toxic. Among the unsaturated fatty acids, oleic was one of the most toxic.[27] Thus, any Tween 20 that is contaminated with lauric acid would be particularly toxic to the brine shrimp.

In general, weighed amounts of plant extracts, mixtures, or pure compounds are first dissolved in a small amount (0.2 to 1.0 ml) of methanol or ethanol. The alcoholic solution is then added to the saline solution that may contain up to 5% Tween 80 or Tween 20.

Dissolution or emulsification can be assisted by ultrasound or by vigorous stirring with a bar magnet. Dilution of this stock solution with the saline solution gives the series of

TABLE 2
Chronic Median Lethal Concentrations for Some Detergents[a]

Surfactant	LC$_{50}$ (24 h) ppm
Na laurylsulfate	6.9
Tween 20	1447.9
Secosyl	44.6
(Na *N*-laurylsarcosinate)	
Finasol (Na dodecylsulfate)	25.0

[a] Adapted from Castritsi-Catharios, J., Karka, A., and Moraiti, M., *Rev. Trav. Inst. Peches Marit.*, 44, 355, 1980.

concentrations required for the testing. Any insoluble material, especially fibers, must be removed by filtration. Fibers can entrap the swimming nauplii and kill them. Usually five concentrations are obtained for each series of tests. We have always attempted to start with a stock solution of at least 10,000 ppm for an unknown extract (1 g in 100 ml of seawater). If such a concentration is not possible, especially with extracts obtained with nonpolar organic solvents, a stock solution at the highest possible concentration is prepared, and the test is extended to a screening for chronic toxicity at 24 h if necessary.

The negative control solution is simply the same saline solution used to prepare the stock test sample solution. The standard positive control makes use of a heavy metal salt as a toxicant. The use of potassium dichromate is convenient. The acute LC$_{50}$ for dichromate is in the range 500 to 800 ppm, whereas the chronic LC$_{50}$ is in the region of 20 to 40 ppm. Therefore, in a screening series the results are considered acceptable if a concurrent determination with the standard toxicant falls within the expected range and there is a low mortality (<5%) in the negative control.

E. BIOASSAY PROCEDURE

The assay is begun 36 to 48 h after sowing of the cysts (i.e., with larvae that are 20 to 32 h old). Multiwelled culture plates can be used for the bioassay, although any clear glass container with flat bottoms (for example, small beakers or glass vials) will do. Ten nauplii are collected, using a Pasteur pipette, from the hatching dish and are transferred to a well, using the minimum amount of seawater. Two milliliters of the test solution are added, and the time is noted. This is repeated for two additional wells, thereby requiring 30 nauplii for each concentration of test sample. A parallel series of tests with the standard potassium dichromate solution (800 ppm, 600 ppm, 400 ppm) and the blank control are always conducted.

To determine the acute LC$_{50,}$ the number of dead nauplii are counted in every well after 6 h. Counting for the chronic LC$_{50}$ begins 24 h after initiation of the tests. Nauplii are considered dead if they are lying immobile at the bottom of the well. A hand lens is useful to check for inactivity of the appendages (the antennae, antennulae, and the mandibles). Alternatively, the whole tray can be put on top of an overhead projector, and the silhouette, focused on the bottom of the wells, is projected upon a screen. It is then easy to count the dead animals. Live nauplii should then be killed by addition of a few drops of formaldehyde solution, and the total dead counted to confirm the number of animals in each well. The data and the observations are recorded onto a form such as shown in Table 3.

F. ANALYSIS OF RESULTS

There are several simple ways to treat the quantal data derived from a test series, such as the logarithmic-probit method of Miller and Tainter, the DeBeer procedure, and the

TABLE 3
Mortality of the Brine Shrimp Larvae after 6 h of Exposure
to Various Concentrations of Goniothalamin (1) in 10% Tween 20 in Seawater[a]

Dose (ppm)	Dosage (log dose)	Dead	Alive	Accumulated dead	Accumulated alive	Ratio dead:total	Mortality (%)
1000	3.00	15	15	42	15	42/57	73.7
800	2.90	8	22	27	37	27/64	42.2
600	2.78	8	22	19	59	19/78	24.4
400	2.60	7	25	11	84	11/95	11.6
200	2.30	4	26	4	110	4/114	3.5

[a] Number of animals, n, is 30–32. Estimated LC_{50} = 851 ppm

Litchfield and Wilcoxon method.[28] However, the Reed-Muench method described in this chapter is the most convenient.[28,29] This procedure assumes that an animal that survived a given dose would also have survived any lower dose and, conversely, that an animal that died with a certain dose would have also died at any other higher dose. Thus, the information from any one group can be added to that of the other groups in the range of doses tested.

For example, referring to the data (Table 3) for goniothalamin (**1**),[30] 15 animals out of 30 survived a dose of 1000 ppm (log ppm = 3.00). At the next concentration down (800 ppm), 22 animals now survived out of 30. The Reed-Muench method requires that the survivors at 1000 ppm be added to the survivors at 800 ppm to give a total survival of 37 survivors. At the concentration of 600 ppm, there are 22 more survivors, so the total now is 59 survivors, and so on. Deaths are dealt with in a similar manner, working up from the lowest concentration. Thus, the four deaths at 200 ppm are added to the seven at 400 ppm and so on such that the accumulated deaths at 1000 ppm is 42. Therefore, at 1000 ppm there are 42 accumulated deaths and 15 survivors. Consequently, the mortality ratio is 42/(42 + 15) or 73.7%. Similarly, at 200 ppm the accumulated deaths is four and the survivors number 110, and therefore the mortality ratio is 4/114 or 3.5%.

The dose that will kill 50% of the animals may be derived by two graphic procedures, both of which should be used. The first procedure plots the number of the accumulated survivors and the number of accumulated deaths on the same axes (number of animals vs. log dose).[29] The two curves will cross at the dosage (i.e., log dose) where the number of survivors is equal to the number of deaths (Figure 3). This should be used to quickly arrive at an estimate for the median tolerance.

The second method plots dosage against percent mortality and the dosage at 50% mortality is obtained by intersection (Figure 4). In addition, the second method allows the use of a formula to estimate the standard error:[28]

$$SE\ LD_{50} = \sqrt{(0.79\ h\ R/n)}$$

where h = average of the interval between dosages (log dose), R = interquartile range (LD_{75} to LD_{25}) from the cumulative percentages, n = number of animals (or the average). R is obtained from the plot of percent mortality against dosage (Figure 4). If either the LD_{75} or LD_{25} cannot be obtained from the plot, then the interquartile range can be estimated as twice LD_{75} to LD_{50} or twice LD_{50} to LD_{25} (Figure 4).

The 95% confidence limits of the LD_{50} can be derived from the relationship

$$\log LD_{50} \pm 2\ SE\ LD_{50}$$

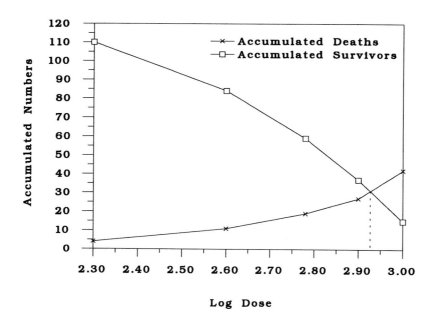

FIGURE 3. Estimation of LC_{50} by plotting the Reed-Muench accumulated deaths and survivors on the same set of axes. The two curves intersect at the 50% lethal dose required for the animal population.

IV. CASE STUDIES

A. THE ACUTE LD_{50} FOR *EUPATORIUM ODORATUM*

Leaves of *Eupatorium odoratum* (250 g) were frozen with liquid nitrogen and then ground to a fine powder. Extraction with methanol (2 × 1 l) for 8 h, followed by evaporation of the solvent under reduced pressure, yielded a dark-green residue (15 g).

A small aliquot of the extract (1.80 g) was dissolved in methanol (0.5 ml) and added to 5% Tween 80 in seawater (80 ml) in a 100-ml volumetric flask. The flask was placed in an ultrasonic bath for 10 min to assist dispersion of the organic material, and the volume was then made up to 100 ml with 5% Tween 80 to form the 18,000 ppm stock solution. Successive dilutions were carried out to obtain concentrations of 16,000 ppm, 14,000 ppm, 12,000 ppm, and 10,000 ppm.

Potassium dichromate, as a standard toxicant, was dissolved in seawater, to give concentrations of 800, 750, 700, 650, and 600 ppm. On the evidence of the brine shrimp assay (Table 4, acute LC_{50} = 12,900 ppm), the leaf extract of *Eupatorium odoratum*, which is traditionally reputed to have anticoagulant properties, is considered nontoxic. Further fractionation of the methanol extract eventually led to the isolation of several flavonoids, two of which were identified as isosakuranetin (**2**) and kempferide (**3**). Both had a chronic (24 h) LC_{50} greater than 1200 ppm,[31] which means that they are well tolerated by the brine shrimp larvae.

B. ACTIVITY-GUIDED SCREENING OF *CERBERA ODOLLAM*

The shrub *Cerbera odollam* is widely used as a roadside ornamental plant and is known to contain cardiac glycosides. Leaves, stem, bark, fruit, and kernels of *C. odollam* (Apocynaceae) were collected and extracted with methanol. The extracts were screened for chronic toxicity against the brine shrimp. The median lethal concentration results are given in Table 5.

On the basis of the preliminary screening tests, the kernels were then subjected to the extraction procedure summarized in Figure 5. The acute median lethal concentrations for each of these extracts toward the brine shrimp are given in Table 6. In this case the larvicidal

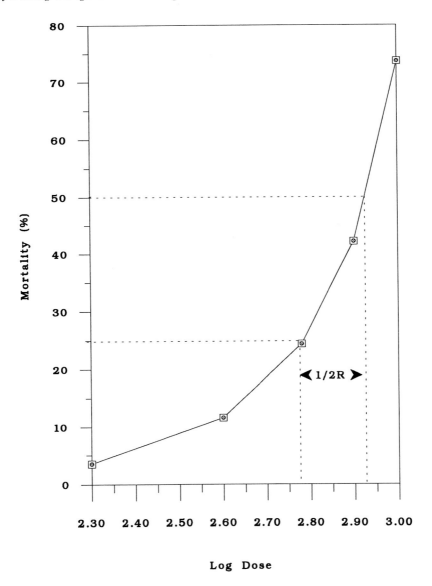

FIGURE 4. Estimation of LC_{50} and standard error by a plot of percent mortality against dosage. In this example half the interquartile range (1/2 R) is taken as Log LC_{50}–Log LC_{25}.

activity coincides with the cardiac activity that is located in the kernel of the fruit.[32] Both cerberin (**4**, LC_{50} (6 h) 650 ppm) and neriifolin (**5**, LC_{50} (6 h) 975 ppm) were isolated from the benzene fraction (Table 6).[33]

C. THE ACUTE LD_{50} FOR A PURE SUBSTANCE: CAMPHOR

Camphor (80 mg in 0.5 ml methanol) was dispersed in 10% Tween 20 in seawater (20 ml) to give a stock solution of 4000 ppm. This stock solution was appropriately diluted with the same saline solution to give concentrations of 2000, 1000, and 500 ppm. The mortality results for brine shrimp are given in Table 7.

The assay was repeated using concentrations of 3000, 2700, 2400, 2100, and 1800 ppm of camphor, and the mortality results are given in Table 8.

TABLE 4
Mortality of Brine Shrimp at Various Concentrations
of the Methanol Extract of Leaves of *E. odoratum*[a]

Dose (ppm)	Dosage (log dose)	Dead	Alive	Accumulated Dead	Accumulated Alive	Ratio dead:total	Mortality (%)
18,000	4.255	33	7	123	7	123/130	94.6
16,000	4.204	34	6	90	13	90/103	87.4
14,000	4.146	24	16	56	29	56/85	65.9
12,000	4.079	22	18	32	47	32/79	40.5
10,000	4.000	10	30	10	77	10/77	11.5

[a] Estimated LC_{50} = 12,900 ppm

TABLE 5
Chronic Median Lethal Concentrations for Plant Parts of *Cerbera odollam*[a]

Plant part	LC_{50} (24 h), ppm	Relative toxicity (RT)
Stem	>10^5	<10^{-4}
Leaves	18,200	1.87×10^{-3}
Fruit	LC_{50} (6 h) >10,000	
Bark	7650	4.44×10^{-3}
Kernel	319	0.11
$K_2Cr_2O_7$	34	1.00

[a] RT = LC_{50} $K_2Cr_2O_7$/LC_{50} test substance.

Estimation of the median lethal concentration by the Reed-Muench method and estimation of the 95% confidence limits by the Pizzi procedure gave an acute LC_{50} of 1820 ppm with 95% confidence limits of 2190 to 1510 ppm.

V. EVALUATION OF ASSAY RESULTS

At this time no single protocol for the determination of lethality to brine shrimp has been accepted as a standard.

Often, the susceptibility of the larvae to various extracts has been tied to a more specific activity. However, there is, at the moment, insufficient data to consistently correlate the toxicity of samples to the brine shrimp, with any relevant type of bioactivity such as antimicrobial or cytotoxic activities. Indeed, Betz and Blogoslawski[20] have noted that it is a poor alternative to a current mouse assay for toxic dinoflagellates. When there is no clear correlation with other biologic activities (which, if present, will confer a distinct advantage), a decision has to be made as to the level of toxicity needed to warrant further evaluation and fractionation of the crude extract. Our laboratory practice is to take a level for the median lethal concentration of 12 times the LC_{50} for potassium dichromate, i.e., about 9000 ppm (relative toxicity, 0.1) for the 6-h test and 450 ppm for the 24-h test. The activity-guided fractionation can proceed from these levels.

The median lethal concentrations are fairly reproducible. For example, the extreme range for goniothalamin (**1**) is 676 to 832 ppm. A probable source of error is the decision whether a nauplius is dead or simply immobilized and dying at the time of scrutiny. Among other factors that can affect the reproducibility are the age of the larva,[34] the temperature,[35-37] and the composition[38] and salinity[39] of the medium. There is therefore reason enough to pay particular attention to consistency in the timing of the sowing of the cysts and the harvesting

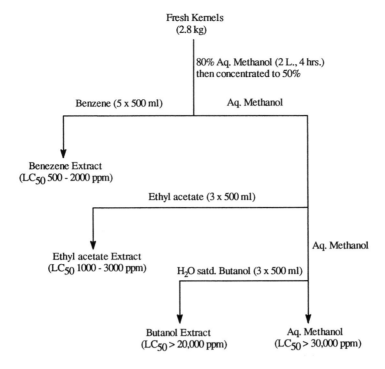

FIGURE 5. Extraction and partition procedure for kernel constituents of *Cerbera odollam*, using the brine shrimp toxicity assay as the guide. Cardiac glycosides, cerberin and neriifolin, are present in the benzene extract.

TABLE 6
Range of Acute Median Lethal Concentrations
for Kernel Fractions of *Cerbera odollam*[a]

Kernel fraction	Acute median lethal concentration (ppm)
Benzene	500–2,000
Ethyl acetate	1,000–3,000
Butanol	>20,000
Aqueous methanol	>30,000

[a] LC_{50} for dichromate = 750–880 ppm.

of the larvae for bioassay. A parallel experiment with potassium dichromate and a negative control is a good practice. If a determination with the control toxicant exceeds the laboratory standard, then the experiment should be rejected and repeated.

Although evaluation is generally by the LC or effective dose (ED) for 50% mortality or by the percent mortality over a fixed period of exposure (e.g., 48 h[5]), the fecundity of *Artemia* has also been evaluated in the presence of metal and pesticide contaminants in the aquatic environment.[40] It was found that the reduced fecundity caused by metal compounds was primarily a consequence of reduced life span.

The evaluation of the DNA profile of the brine shrimp larvae upon exposure to mutagens has been proposed as a potential screen for teratogenic compounds.[41] The respiration rate of adults has been used as another criterion of toxicity.[42] One attempt to investigate the mechanism of the toxicity of food dyes to the brine shrimp larvae suggested that inhibition of lactate dehydrogenase could be a cause of the severe toxicity of rose bengal and

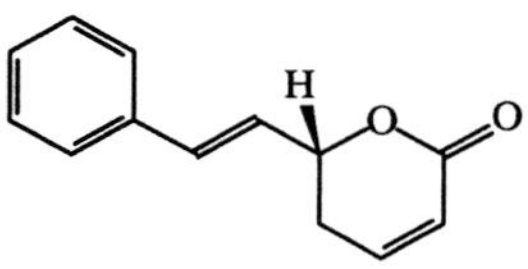

(1)

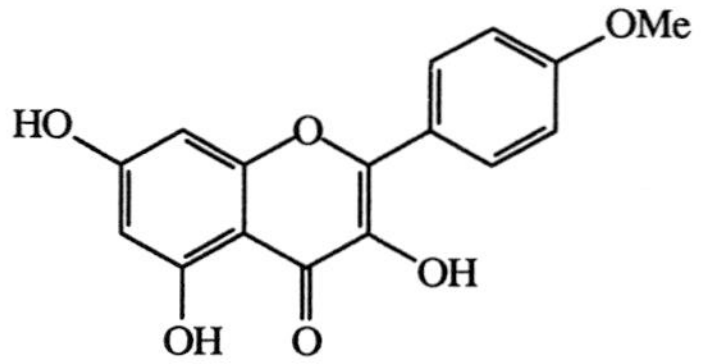

(2)

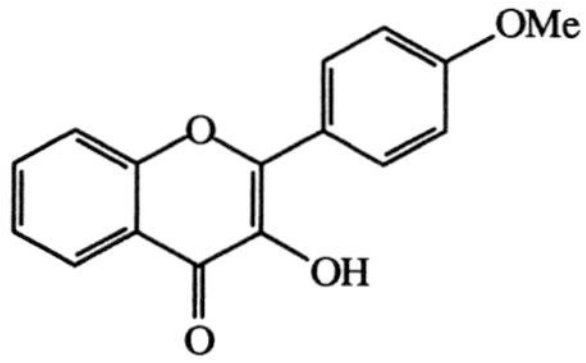

(3)

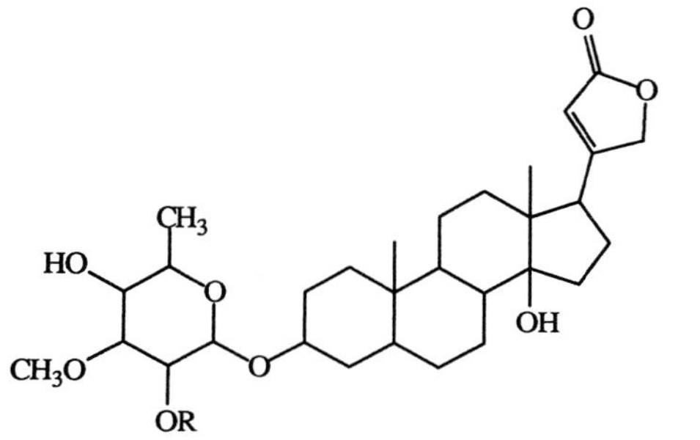

(4) R = Ac

(5) R = H

phloxine.[43] Inhibition of chitin turnover has been suggested as the cause of the toxicity of avermectin to the brine shrimp larvae.[44] As a possible preliminary indicator of anticancer activity, the seed extracts from many species of Euphorbiaceae were screened for brine shrimp lethality, and the results compared with the cytotoxicity of the same extracts toward 9KB- and 9PS-cultured cells.[7] The cytotoxic goniothalenol was isolated from *Goniothalamus giganteus* by this protocol.[45]

The brine shrimp lethality assay has also been used to investigate fungal toxins. A compilation of median lethal doses, estimated by several different groups,[46,47] for sterigmatocystin (STG), diacetoxyscirpenol (DAS), and ochratoxin A (OCT) (Table 9), shows the variations possible under different experimental protocols. Subsequently, Durackova et al.[48] investigated the toxicity of many mycotoxins by a similar method and proposed a model for comparing the data. However, Prior, using the brine shrimp assay to test animal feedstuffs, found a high percentage of false positive results (56%) for three mycotoxins.[49] Bijl et al. compared some bioassay methods for mycotoxins and

TABLE 7
**Mortality of the Brine Shrimp Nauplii after Exposure for 6 h
to Various Concentrations of Camphor in 10% Tween 20 in Seawater[a]**

Concentration (ppm)	Number of dead	Number alive
4000	30	0
2000	8	22
1000	0	30 (weak)
500	0	30

[a] Estimation of the appropriate range for the median lethal concentration.

TABLE 8
**Mortality of the Brine Shrimp Nauplii after Exposure for 6 h
to Various Concentrations of Camphor in 10% Tween 20 in Seawater[a]**

Concentration (ppm)	Number of dead	Number alive
3000	30	0
2700	30	0
2400	29	1
2100	24	6
1800	19	11

[a] Estimated to bracket the median lethal concentration.

TABLE 9
Variations in Reported Toxicity Data for Some Mycotoxins[a]

Measurement	DAS (ppm)	STG (ppm)	OCT (ppm)	Ref.
LC_{50} (24 h)	0.12	0.07	3.9	46
LC_{50} (16 h)	0.47	0.54	10.1	5
LC_{50} (24 h)	0.275	—	—	47
LC_{80}	< 1.0	—	—	6

[a] DAS = diacetoxyscirpenol; STG = sterigmatocystin; OCT = ochratoxin A.

concluded that if sensitivity is not the primary interest, then the brine shrimp method is preferred.[17]

Other studies that have involved a brine shrimp assay include the isolation of the depsipeptide antibiotic beauvericin from *Beauveria bassiana;*[50] screening for ciguatera poisons[51] and other dinoflagellate toxins;[20,24,52] investigation of antifungal polyacetylenes from *Artemisia borelis,*[53] isolation of an antibiotic from several *Fusarium* species;[54] studies with the molluscicidal diterpenoid aldehyde, traversianal;[55] and an investigation into the activity of deoxypodophyllotoxin.[56]

A partial listing of the pure compounds that we have studied appears as Table 10. Cardiac glycosides (cerberin and neriifolin)[33] and embryotoxic compounds (goniothalamin and goniothalamin oxide)[57] can be distinguished from relatively nontoxic compounds such as camphor and menthone. The steam distillate of the aerial parts of the grass *Cyperus zollingerii*, under study as a potential source of insect juvenile hormones,[58] has an LC_{50} (6 h) of 2900 ppm. However, the major component, caryophyllene oxide (LC_{50} > 2300 ppm), is not responsible for this toxicity.

TABLE 10
Average Median Lethal Concentrations of Some Natural Compounds to the Brine Shrimp Larvae[a]

Compound	Av. LC_{50} ppm (exposure time, h)	Relative toxicity
Terpinen-4-ol	880 (6)	0.55
(–)-Carvone	570 (6)	0.88
Camphor	1880 (6)	0.25
Limonene	1470 (6)	0.34
Menthone	1430 (6)	0.33
Citral	500 (6)	0.98
Goniothalamin (1)	845 (6)	0.89
Goniothalamin oxide	608 (6)	1.28
Cerberin (4)	650 (6)	1.30
Neriifolin (5)	975 (6)	0.86
Isosakuranetin (2)	> 1200 (24)	< 0.03
Kaempferide (3)	> 1200 (24)	< 0.03
Caryophyllene oxide	> 2300 (6)	< 0.33
Dihydroartemisin	> 1500 (6)	< 0.48

[a] Toxicity is relative to that of $K_2Cr_2O_7$.

VI. CONCLUSIONS

The brine shrimp larva assay is a simple and rapid toxicity test for bioactive plant products. It is not selective by chemical type and has not as yet been correlated well with other biologic activities. Nonetheless, the simplicity of the procedure should ensure it a place in a natural products laboratory as a preliminary bioassay to screen plant extracts for further study and to guide fractionation.

ACKNOWLEDGMENTS

The Director of the Fisheries Research Institute in Penang, Malaysia, Ong Kah Sing, has given much help and advice in the hatching and culturing of the brine shrimp. He has also given a generous supply of the cysts from his stock. My thanks go also to my colleagues at the Science University of Malaysia (USM), particularly Dr. M. C. Feng and Dr. Y. C. Toong, who have either used the test themselves or have supplied me with plant extracts for testing. This work, which is part of a program on the chemistry of natural products from plants, was funded under USM R&D grant number 123/3104/2502.

REFERENCES

1. **Michael, A. S., Thompson, C. G., and Abramovitz M.,** *Artemia salina* as a test organism for bioassay, *Science,* 123, 464, 1956.
2. **Zillioux, E. J., Foulk, H. R., Prager, J. C., and Cardin, J. A.,** Using *Artemia salina* to assay oil dispersant toxicities, *J. Water Pollut. Control Fed.,* 45, 2389, 1973.
3. **Price, K. S., Waggy, G. T., and Conway, R. A.,** Brine shrimp bioassay and seawater BOD of petrochemicals, *J. Water Pollut. Control Fed.,* 46, 63, 1974.
4. **Sorgeloos, P., Remiche-van der Wielen, C., and Persoone, G.,** The use of *Artemia* nauplii for toxicity tests — a critical analysis, *Ecotoxicol. Environ. Saf.* 2, 249, 1978.

5. **Harwig, J. and Scott, P. M.,** Brine shrimp *Artemia salina* larvae as a screening system for fungal toxins, *Appl. Microbiol.,* 21, 1011, 1971.

6. **Reiss, J.,** Comparing investigations on the toxicities of some mycotoxins to the larvae of the brine shrimp *Artemia salina, Zentralbl. Bakteriol. Parasitenkd. Infektionskr. Erste Abt. Orig. Reihe B. Hyg. Praev. Med.,* 155, 531, 1972.

7. **Meyer, B. N., Ferrigni, N. R., Putnam, J. E., Jacobsen, L. B., Nichols, D. E., and McLaughlin, J. L.,** Brine shrimp: a convenient general bioassay for active plant constituents, *Planta Med.,* 45, 31, 1982.

8. **Van Haecke, P. and Persoone, G.,** Report on an intercalibration exercise on a short term standard toxicity test with *Artemia* nauplii (ARC test), *Colloq. Inst. Natl. Saute Rec. Med.(Test Toxic Aique Milieu Aquat.),* 106, 359, 1982.

9. **Ivleva, I. V.,** *Mass Cultivation of Invertebrates* (transl. Mercado, A.,), Israel Program of Translations, Jerusalem, 1973, 52.

10. **Schmitt, W. L.,** *Crustaceans,* University of Michigan Press, Ann Arbor, 1980, 46.

11. **Sorgeloos, P.,** Life history of the brine shrimp *Artemia,* in *The Brine Shrimp Artemia,* Vol. 1, *Morphology, Genetics, Radiobiology, Toxicology,* Persoone, G., Sorgeloos, P., Roels, O., and Jasper, E., Eds., Universa Press, Wetteren, Belgium, 1980, ix.

12. **Green, J.,** *A Biology of Crustacea,* H. F. & G. Witherby, London, 1967, chap. 5.

13. **Kuwabara, K., Nakamura, A., and Kashimoto, T.,** Effect of petroleum oil, pesticides, PCBs and other environmental contaminants on the hatchability of *Artemia salina* dry eggs, *Bull. Environ. Contam. Toxicol.,* 25, 69, 1980.

14. **Buu-hoi, N. P. and Chanh, P. H.,** Effect of various types of carcinogens on the hatching of *Artemia salina* eggs, *J. Natl. Cancer Inst.,* 44, 795, 1970.

15. **Kinghorn, A. D., Harjes, K. K., and Doorenbos, N. J.,** Screening procedure for phorbol esters using brine shrimp (*Artemia salina*) larvae, *J. Pharm. Sci.,* 66, 1362, 1977.

16. **Stewart, S. and Schurr, K.,** Effects of asbestos on survival of *Artemia,* in *The Brine Shrimp Artemia,* Vol. 1, *Morphology, Genetics, Radiobiology, Toxicology,* Persoone, G., Sorgeloos, P., Roels, O., and Jasper, E., Eds., Universa Press, Wetteren, Belgium, 1980, 235.

17. **Bijl, J., Dive, D., and van Peteghem, C.,** Comparison of some bioassay methods for mycotoxin studies, *Environ. Pollut. Ser. A,* 26, 173, 1981.

18. **Canton, J. H., Wegman, R. C. C., Vulto, T. J. A., Verhoef, C. H., and Van Esch, G. J.,** Toxicity, accumulation and elimination studies with salt water organisms of different trophic levels, *Water Res.,* 12, 687, 1978.

19. **Morel, F. M. M., Rueter, J. G., Andersen, D. M., and Guillard, R. R. L.,** Aquil: a chemically defined phytoplankton culture medium for trace metal studies. *J. Phycol.,* 15, 135, 1979.

20. **Betz, J. M. and Blogoslawski, W. J.,** Toxicity of *Gonyaulax tamarensis* var. *excavata* cells to the brine shrimp *Artemia salina, J. Pharm. Sci.,* 71, 463, 1982.

21. **Clegg, J. S.,** The control of emergence and metabolism by external osmotic pressure and the role of free glycerol in developing cysts in *Artemia salina, J. Exp. Biol.,* 41, 879, 1964.

22. **Ewing, K. D., Conte, F. P., and Peterson, G. L.,** Regulation of nucleic acid synthesis in *Artemia* nauplii by environmental salinity, *Am. J. Physiol.,* 238, R91, 1980.

23. **Reeve, M. R.,** Growth efficiency of *Artemia* under laboratory conditions, *Biol. Bull.,* 125, 133, 1963.

24. **Medlyn, R. A.,** Susceptibility of four geographical strains of adult *Artemia* to *Ptychodiscus brevis* toxin(s), in *The Brine Shrimp Artemia,* Vol. 1, *Morphology, Genetics, Radiobiology, Toxicology,* Persoone, G., Sorgeloos, P., Roels, O., and Jasper, E., Eds., Universa Press, Wetteren, Belgium, 1980, 225.

25. **Blizzard, T. A., Ruby, C. L., Mrozik, H. M., Reiser, F. A., and Fisher, M. N.,** Brine shrimp (*Artemia salina*) as a convenient bioassay for avermectin analogs, *J. Antibiot.,* 42, 1304, 1989.

26. **Castritsi-Catharios, J., Karka, A., and Moraiti, M.,** Toxicity of detergents and a surfactant on *Artemia salina* Leach., *Rev. Trav. Inst. Peches Marit.,* 44, 355, 1980.

27. **Curtis, R. F., Coxon, D. T., and Levett, G.,** Toxicity of fatty acids in assays for mycotoxins using the brine shrimp *Artemia salina, Food Cosmet. Toxicol.,* 12, 233, 1974.

28. **Miya, T. S., Holck, H. G. O., Yim, G. K. W., Mennear, J. H., and Spratto, G. R.,** *Laboratory Guide in Pharmacology,* 4th ed., Burgess Publishing, Minneapolis, 1973, 127.

29. **Ipsen, J. and Feigl P.,** *Bancroft's Introduction to Biostatistics,* 2nd ed., Harper & Row, New York, 1970, chap. 15.

30. **Sam, T. W.,** unpublished results, 1987

31. **Cheah, P. B. and Sam, T. W.,** unpublished results, 1986.

32. **Venkata Rao, E.,** Cardiac glycosides of *Thevetia peruviana* and *Cerbera odollam, Indian J. Pharm.,* 35, 107, 1973.

33. **Ng, A. S. and Sam T. W.,** unpublished results, 1987.

34. **Sleet, R. B. and Brendel, K.,** Homogeneous population of *Artemia* nauplii and their potential use for *in vitro* testing in developmental toxicology, *Teratog. Carcinog. Mutagen,* 5, 41, 1985.

35. **Brown, R. F., Wildman, J. D., and Eppley, R. M.,** Temperature dose relationship with aflatoxin B-1 on the brine shrimp *Artemia salina* bioassay, *J. Assoc. Off. Anal. Chem.,* 51, 905, 1968.

36. **Boothe, P. N.,** Effects of temperature on the chronic toxicity of a pesticide and its metabolites to *Artemia salina, Pharmacologist,* 20, 145, 1978.

37. **Durakovic, S., Galic, J., Rajnovic, P., Pospisil, O., and Stilinovic, L.,** A simple biological reagent for mycotoxin toxicity research: the *Artemia salina* larvae, *Mikrobiologija (Belgr.),* 26, 15, 1989.

38. **Gaonkar, S. N., Karande, A. A., and Rege, M. S.,** SDS-*Artemia* bioassay method for evaluating oil-dispersant toxicity. *Pollut. Res.,* 5, 33, 1986.

39. **Foster, G. D. and Tullis, R. E.,** Quantitative structure-toxicity relationships with osmotically stressed *Artemia salina* nauplii, *Environ. Pollut. Ser. Ecol. Biol.,* 38, 273, 1985.

40. **Grosch, D. S.,** Alterations to the reproductive performance of *Artemia* caused by antifouling paints, algae-cides and an aquatic herbicide, in *The Brine Shrimp Artemia,* Vol. 1, *Morphology, Genetics, Radiobiology, Toxicology,* Persoone, G., Sorgeloos, P., Roels, O., and Jasper, E., Eds., Universa Press, Wetteren, Belgium, 1980, 201.

41. **Sleet, R. B. and Brendel, K.,** Brine shrimp, *Artemia salina*: a potential screening organism for initial teratology screening, *Proc. West. Pharmacol. Soc.,* 26, 169, 1983.

42. **Verriopoulos, G., Moraitou-Apostolopoulou, M., and Xatzispirou, A.,** Evaluation of metabolic responses of *Artemia salina* to oil and oil dispersants as a potential indicator of toxicant stress, *Bull. Environ. Contam. Toxicol.,* 36, 444, 1986.

43. **Kobayashi, N., Taniguchi, N., Sako, F., and Takakuwa, E.,** A screening method for the toxicity of food dyes using *Artemia salina* larvas, *J. Toxicol. Sci.,* 2, 383, 1977.

44. **Calcott, P. H. and Fatig, R. O., III,** Inhibition of chitin metabolism by avermectin in susceptible organisms, *J. Antibiot.,* 37, 253, 1984.

45. **El-Zayat, A. A. E., Ferrigni, N. R., McCloud, T. G., McKenzie, A. T., Byrn, S. R., Cassady, J. M., Chang, C.-J., and McLaughlin, J. L.,** Goniothalenol: a novel bioactive tetrahydrofurano-2-pyrone from *Goniothalamus giganteus* (Annonaceae), *Tetrahedron Lett.,* 26, 955, 1985.

46. **Yamamoto, K.,** Methods for the determination of mycotoxins. III. Investigations of the mycotoxin bioassay method using brine shrimp larvae (*Artemia salina*) and its applications. *Nara Igaku Zasshi,* 26, 264, 1975: *CA,* 84, 174657u.

47. **Eppley, R. M.,** Sensitivity of the brine shrimp (*Artemia salina*) to trichothecenes, *J. Assoc. Off. Anal. Chem.,* 57, 618, 1974.

48. **Durackova, Z., Betina, V., Hornikova, B., and Nemec, P.,** Toxicity of mycotoxins and other fungal metabolites to *Artemia salina* Leach, *Zentralbl. Bakterol. Parasitenkd. Infektionskr. Hyg. Zweite Naturwiss. Abt. Allg. Landwirtsch. Tech. Mikrobiol.,* 132, 294, 1977.

49. **Prior, M. G.,** Evaluation of the brine shrimp *Artemia salina* larvae as a bioassay for mycotoxins in animal feedstuffs, *Can. J. Comp. Med.,* 43, 352, 1979.

50. **Hamill, R. L., Higgens, C. E., Boaz, H. E., and Gorman, M.,** Structure of beauvericin, a new depsipeptide antibiotic toxic to *Artemia salina, Tetrahedron Lett.,* 49, 4255, 1969.

51. **Granada, H. R., Cheng, P. C., and Doorenbos, N. J.,** Ciguatera. I. Brine shrimp (*Artemia salina* L.) larval assay for ciguatera toxins, *J. Pharm. Sci.,* 65, 1414, 1976.

52. **Trieff, N. M., McShan, M., Grajcer, D., and Alam, M.,** Biological assay of *Gymnodinium breve* toxin using brine shrimp (*Artemia salina*), *Tex. Rep. Biol. Med.,* 31, 409, 1973.

53. **Wang, Y., Toyota, M., Krause, F., Hamburger, M., and Hostettmann, K.,** Polyacetylenes from *Artemisia borealis* and their biological activities, *Phytochemistry,* 29, 3101, 1990.

54. **Golinski, P., Wnuk, S., Chelkowski, J., Visconti, A., and Schollenberger, M.,** Antibiotic Y biosynthesis by *Fusarium avenaceum.* Isolation and some physicochemical and biological properties, *Appl. Environ. Microbiol.,* 51, 743, 1986.

55. **Stoessl, A., Cole, R. J., Abramowski, Z., Lester, H. H., and Towers, G. H. N.,** Some biological properties of traversianal, a strongly molluscicidal diterpenoid aldehyde from *Cercospora traversiana, Mycopathologia,* 106, 41, 1989.

56. **Inamori, Y., Kato, Y., Kubo, M., Baba, K., Ishida, T., Nomoto, K., and Kozawa, M.,** The biological actions of deoxypodophyllotoxin (anthricin). I. Physiological activities and conformational analysis of deoxypodophyllotoxin, *Chem. Pharm. Bull.,* 33, 704, 1985.

57. **Sam, T. W., Chew, S.-Y., Matsjeh, S., Gan, E. K., Razak, D., and Mohamad, A. L.,** Goniothalamin oxide: an embryotoxic compound from *Goniothalamus macrophyllus* (Annonaceae), *Tetrahedron Lett.,* 28, 2541, 1987.

58. **Toong, Y. C., Schoolery, D. A., and Baker, F. C.,** Isolation of insect juvenile hormone III from a plant, *Nature,* 333, 170, 1989.

Chapter 19

IN VITRO NEUROTOXICITY BIOASSAY: NEUROTOXICITY OF SESQUITERPENE LACTONES

Richard J. Riopelle and Kenneth L. Stevens

TABLE OF CONTENTS

I. INTRODUCTION

As has been demonstrated in numerous studies,[1-6] the persistent association of Parkinson's disease and rural living suggests that some factors related to rural living contribute to the risk for the disease. If rural factors play a role in causation of certain human neurodegenerative diseases, it might be expected that related disorders could be observed, at times, in domestic animals. One such disorder in horses, called equine nigropallidal encephalomalacia (ENE), with some features of human parkinsonism, was first described in 1954.[7]

ENE was first described as a naturally occurring disease of horses in central and northern California, where *Centaurea solstitialis* (yellow starthistle) is abundant.[7] Subsequently, a similar disorder was observed in horses in western Colorado, where *Acroptilon repens* (Russian knapweed), but not *C. solstitialis*, was found.[8] In two acute feeding experiments, the disease occurred in horses ingesting *C. solstitialis*[7] and *A. repens*.[9] The clinical disorder described by Cordy[7] appeared acutely and was characterized by immobility of facial musculature, idle chewing and tongue-flicking movements, impaired eating and drinking, rhythmical protrusion of the tongue, a posture associated with that seen in drowsiness in the horse, and a wobbly, shuffling gait. Thereafter, the animals became unreactive until sacrificed. All animals showed profound nutritional deficiency, with marked weight loss. Autopsies on six horses in two studies[7,9] showed relatively symmetrical foci of necrosis in *globus pallidus* (four of six animals) and/or in *substantia nigra* (three of six animals). While the number of animals in each study was small, the investigators in one study concluded that the factor in *A. repens* was the most toxic.[9] This interpretation was based upon amount of consumption, exposure time, and bulk of neuropathology.

While the specific toxin or toxins implicated in ENE have not been identified, Merrill and Stevens have speculated that sesquiterpene lactones might be the toxic agents.[10] The sesquiterpene lactones of the genera *Centaurea* and *Acroptilon* are concentrated in the aerial parts of the weeds and are known to have biologic activity producing cytotoxicity and phytotoxicity.[11] To pursue further the hypothesis that sesquiterpene lactones might be neurotoxic, an *in vitro* paradigm using dissociated embryonic neurons has been employed to quantify and to rank order toxicity. Toxicity associated with extracts of the plant family Asteraceae, of which both *Centaurea* and *Acroptilon* are closely related members, may provide clues to support data emerging from analytic epidemiologic studies that suggest a role for rural factors in the production of certain neurodegenerative diseases. Dissociated neuronal tissue culture systems may provide clues to the molecular basis of toxicity of identified compounds of natural origin.

II. NEUROTOXICITY BIOASSAY

To quantify and assign rank order of toxicity to the sesquiterpene lactones, an *in vitro* assay with sensory neurons was utilized. The assay system is based upon a dissociated neuron culture technique initially developed by Sutter et al.[12] and modified by Riopelle and Cameron[13] as a sensitive biologic assay for Nerve Growth Factor (NGF). Eight-day chick embryo[14] dorsal root ganglia (DRG) were freed of meninges, removed aseptically, and kept at 4°C at all times except as mentioned. Ganglia from six embryos (40 to 50 per embryo) were washed in Ca^{2+}- and Mg^{2+}-free Gey's balanced salt solution and exposed to 0.01% trypsin in the same solution for 10 min at 37°C. A half volume of phosphate-buffered Gey's balanced salt solution was added for a further 5 min at 37°C, and the reaction was then stopped with one-third volume of Ham's F12 medium containing 5% fetal calf serum (FCS). The ganglia were then triturated, using a 5-ml narrow-tip pipette, to a single-cell suspension. Following filtration through 37-μM nylon mesh in a millipore chamber to remove clumps, the cell suspension was washed through a 500-µl FCS undercut (700 × g for 5 min at 4°C) and resuspended in 4 ml of Ham's

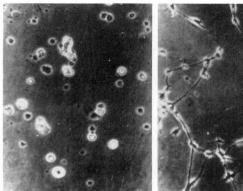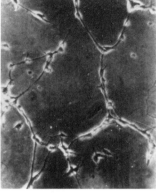

FIGURE 1. Phase contrast photomicrograph of 8-day chick embryo sensory neurons *in vitro* 20 h after plating in the absence (left) and presence (right) of nerve growth factor (NGF). Note that viable (phase bright) cells with processes are present in the absence of NGF at 20 h *in vitro*, while in the presence of NGF, the majority of cells have extended processes.

F12 medium plus 5% FCS. The cell suspension was then preplated on a 100-mm Falcon culture dish and incubated for 45 to 60 min at 37°C in a 5% CO_2 humidified atmosphere. During this period, nonneuronal cells of DRG will preferentially stick to the culture substrate, and a population of cells enriched in neurons can then be decanted for the bioassay. The neuron-enriched cell population was suspended in medium with 4-pM NGF (final concentration) prepared by the method of Mobley et al.[15] Cells were aliquoted in 5-μl volumes at 1000 per well to Terasaki plates previously pretreated with poly-D-lysine (0.1 μg/ml for 24 h). An equal volume of medium containing various concentrations of the sesquiterpene lactones to be analyzed was aliquoted to replicate wells (n = 6 for each concentration). Plates were incubated for 16 to 20 h (37°C, 5% CO_2), at which time wells were scored for neurite-bearing cells (one process in excess of 1.5 cell diameters), using a Leitz Diavert microscope with phase optics (Figure 1). Control wells contained either no NGF or NGF with vehicle (acetonitrile or ethanol).

III. NEUROTOXICITY OF SESQUITERPENE LACTONES

A. EXTRACTION AND ISOLATION OF THE SESQUITERPENE LACTONES

Sesquiterpene lactones were purified from plant material collected in June. The aerial parts of the weeds were air dried, ground, and extracted sequentially with petrol and ether. The ether extract was dissolved in 1 l of 95% ethanol and treated with 1 l of 4% aqueous Pb(OAc)$_2$. After 1 h of stirring, the mixture was filtered, ethanol removed, and then extracted with $CHCl_3$ to give, on evaporation of the solvent, a dark-colored oil from which the sesquiterpene lactones (Figure 2) were subsequently isolated, as described in detail elsewhere.[10,16]

B. NEUROTOXICITY

The biological assay was quantified by counting neurite-bearing cells in each well. For the purposes of the present study, a 4+ or maximal response represented neurite growth from 60 to 70% of plated neurons, while 3+, 2+, and 1+ responses represented 45 to 50, 30 to 40, and 10 to 15% neurite growth, respectively. Figure 3 is a composite representing the means of neurite growth responses in four separate assays over the concentration range depicted for repin and janerin. At concentrations of repin or janerin of 250 nM and greater, the neurite growth response was less than that seen in the absence of NGF (usually 5% neurite-bearing cells), and virtually all cells had lost refractility and were phase dark. The loss of refractility

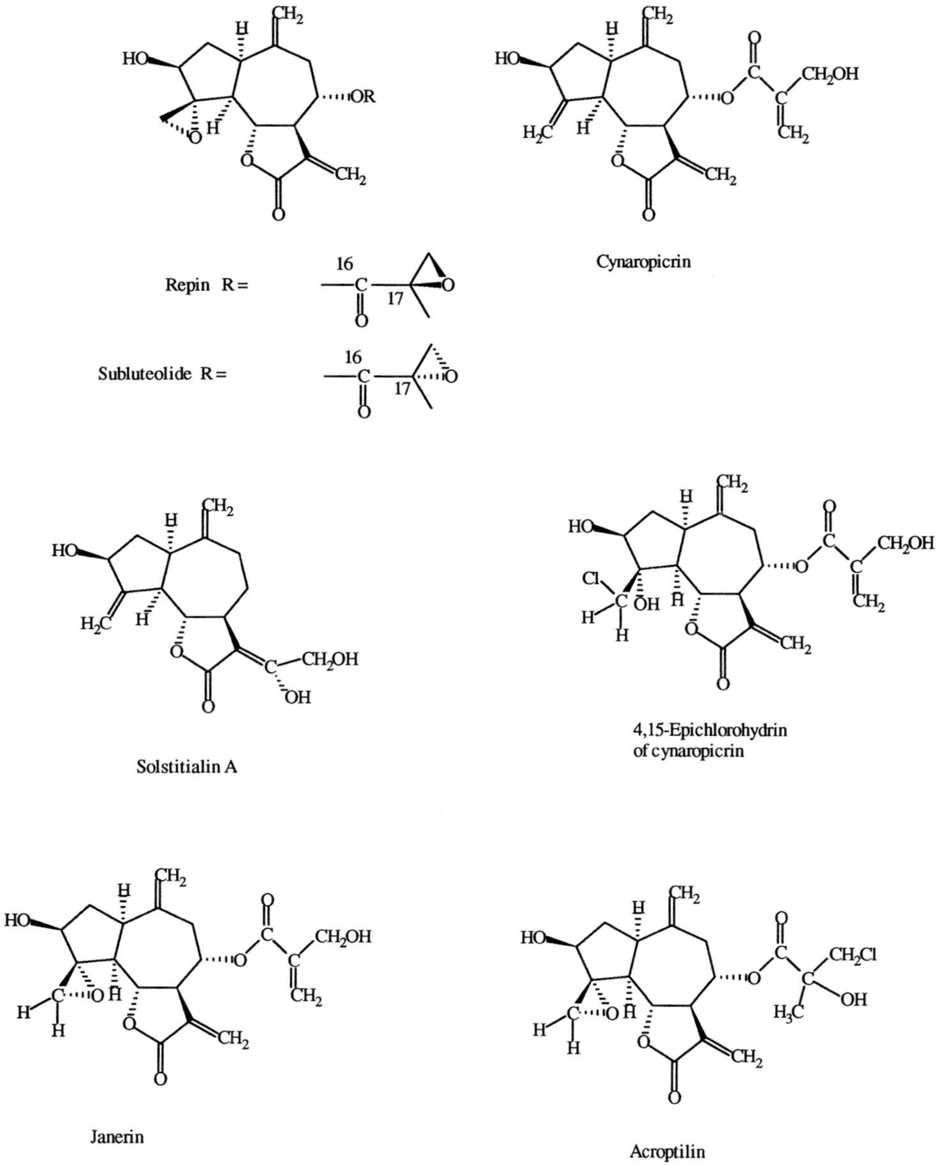

Repin R =

Subluteolide R =

Cynaropicrin

Solstitialin A

4,15-Epichlorohydrin of cynaropicrin

Janerin

Acroptilin

FIGURE 2. Structures of sesquiterpene lactones used in biological assay.

was obvious within 1 h of seeding. Neither the acetonitrile nor ethanol vehicles, used at the appropriate concentrations to account for dilutions of the sesquiterpene lactones, were toxic to the neurons. The 50% toxic dose (TD_{50}) for repin, depicted in Figure 3, was approximately 80 nM, while the TD_{50} for janerin was 350 nM. Using similar assays, TD_{50}s for five of the other sesquiterpene lactones isolated from the plants were established, as listed in Table 1. Repin was the most toxic of the sesquiterpene lactones tested, while subluteolide, which is a C-17 epimer of repin (Figure 2),[16] was some three to four times less toxic, having a TD_{50} of 300 nM. Subluteolide was only slightly more toxic than the other sesquiterpene lactones tested, which had TD_{50}s ranging between 350 and 800 nM (Table 1).

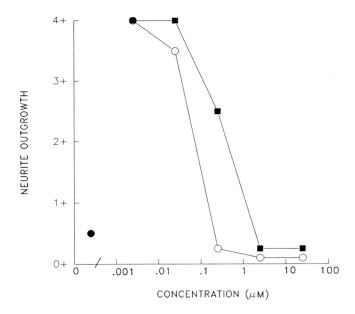

FIGURE 3. *In vitro* toxicity profiles of repin and janerin. Data displayed are a composite of four assays; n = 6 wells for each concentration of sesquiterpene lactone used. ○: repin; ■: janerin; ●: medium without additive and without NGF.

TABLE 1
Rank Ordering of Toxicity of Sesquiterpene Lactones in Dissociated Sensory Neuron Assay System

Sample[a]	TD_{50} (nM)[b]
Repin	80
Subluteolide[c]	300
Janerin	350
Cynaropicrin	550
4,15-Epichlorohydrin of cynaropicrin	650
Acroptilin	650
Solstitialin	800

[a] Vehicle was acetonitrile, except for [c], which was ethanol.
[b] Mean of four assays, n = 6 wells for each concentration of sample, as described in Figure 3.
[c] Vehicle was ethanol.

C. STRUCTURE/ACTIVITY RELATIONSHIPS

The sesquiterpene lactones have long been associated with cytotoxicity, phytotoxicity, and antineoplastic effects and have been suspected as being the allelopathic factors produced by certain genera of the Asteraceae family.[11] The present studies indicate that several of the sesquiterpene lactones purified from *A. repens* and *C. solstitialis* were toxic to neuronal cells *in vitro*. In unpublished studies the most potent of these lactones, repin, was found to produce toxic effects on neurons and glia, when inoculated in nanomole doses into the intact brain of adult rat. All of the sesquiterpene lactones tested displayed neurotoxic effects, which was likely a function primarily of the α,β-unsaturated-γ-lactone moiety with the α-methylene group being exocyclic.[11] The structural configuration around carbon 17 seemed to be of secondary importance, since the rank ordering of toxicity was repin > subluteolide.

IV. CONCLUSIONS

It is of interest that the bulk of pathologic change, the shorter time to onset of symptoms, and the smallest amount of cuttings eaten per day occurred in the group of horses fed *A. repens*, as reported by Young et al.[9] These observations might suggest that the toxic factor or factors in *A. repens* were more potent and/or present in higher concentration than those found in *C. solstitialis*. In this regard, repin, which was the most toxic sesquiterpene lactone in the present *in vitro* studies, is only the third most prevalent sesquiterpene lactone in *A. repens*.

The sesquiterpene lactones may be implicated in the pathogenesis of ENE, a disorder with some clinical features similar to human extrapyramidal disorders. Feeding experiments attempting to reproduce the observations in ENE have been successful only in horses, thereby suggesting a degree of species susceptibility.[17] In the horses where disease has been obvious, the pathologic changes have indicated some selective regional vulnerability in the central nervous system. The present studies of direct exposure *in vitro* and *in vivo* have not demonstrated species, region, or cell specificity, since chick and rat neurons and nonneuronal cells have been affected, and tissue disruption was seen in striatum and cortex. The discrepancies between feeding experiments and the direct exposure studies reported here are numerous and require further study. Specifically, a role of exogenous or endogenous cofactors must be explored. Feeding experiments consistently resulted in signs of nutritional deprivation and weight loss, perhaps suggesting a role for endogenous factors in toxicity. In addition, synergisms between sesquiterpene lactones and other molecular species in the weeds could account for the regional susceptibility seen in ENE.

Rodriguez et al.[11] have correlated cytotoxicity with the degree of oxidation of sesquiterpene lactones. In this regard, a number of oxygenated guaianolides present in species of Asteraceae could possibly be implicated. Stevens has also characterized a group of polyacetylenes from Asteraceae, at least one of which shows phytotoxicity.[18] Synergism between the sesquiterpene lactones and polyacetylenes may be suggested if the latter were found to be neurotoxic.

Correlative studies have yet to be reported on the taxonomy and distribution of species of Asteraceae in regions where analytic epidemiologic data have implicated rural factor(s) in causation of Parkinson's disease. It remains to be determined whether the distribution of the Asteraceae family of weeds could provide partial explanation for the conclusions reached in analytic epidemiologic studies of Parkinson's disease.

ACKNOWLEDGMENTS

These studies were supported by a grant from the Parkinson's Foundation of Canada.

REFERENCES

1. **Rajput, A. H., Uitti, R. J., Stern, W., Laverty, W., O'Donnell, K., O'Donnell, D., Yuen, W. K., and Dua, A.,** Geography, drinking water chemistry, pesticides and herbicides and the etiology of Parkinson's disease, *Can. J. Neurol. Sci.,* 14, 414, 1987.
2. **Tanner, C. M., Chen, B., and Wang, W. Z.,** Environmental factors in the etiology of Parkinson's disease, *Can. J. Neurol. Sci.,* 14, 419, 1987.
3. **Barbeau, A.,** Etiology of Parkinson's disease: a research strategy., *Can. J. Neurol. Sci.,* 11, 24, 1984.
4. **Barbeau, A., Roy, M., Cloutier, T., Plasse, L., and Paris, S.,** Environmental and genetic factors in the etiology of Parkinson's disease, *Adv. Neurol.,* 45, 299, 1987.

5. **Ho, S. C., Woo, J., and Lee, C. M.,** Epidemiologic study of Parkinson's disease in Hong Kong, *Neurology,* 39, 1314, 1989.
6. **Koller, W., Vetere-Overfield, B., Gray, C., Alexander, C., Chin, T., Dolezal, J., Hassanein, R., and Tanner, C.,** Environmental risk factors in Parkinson's disease, *Neurology,* 40, 1218, 1990.
7. **Cordy, D. R.,** Nigropallidal encephalomalacia in horses associated with ingestion of yellow starthistle, *J. Neuropathol. Exp. Neurol.,* 13, 330, 1954.
8. **Larson, K. A. and Young, S.,** Nigropallidal encephalomalacia in horses in Colorado, *JAVMA,* 156, 626, 1970.
9. **Young, S., Brown, W. W., and Klinger, B.,** Nigropallidal encephalomalacia in horses fed Russian Knapweed *(Centaurea repens), Am. J. Vet. Res.,* 31, 1393, 1970.
10. **Merrill, G. B. and Stevens, K. L.,** Sesquiterpene lactones from *Centaurea solstitialis, Phytochemistry,* 24, 2013, 1985.
11. **Rodriguez, E., Towers, G. H. N., and Mitchell, J. C.,** Biological activities of sesquiterpene lactones, *Phytochemistry,* 15, 1573, 1976.
12. **Sutter, A., Riopelle, R. J., Harris-Warrick, R. M., and Shooter, E. M.,** Nerve growth factor receptors, *J. Biol. Chem.,* 254, 5972, 1979.
13. **Riopelle, R. J. and Cameron, D.,** Neurite growth-promoting factors of embryonic chick — ontogeny, regional distribution and characteristics, *J. Neurobiol.,* 12, 185, 1981.
14. **Lillie, F. R.,** *Lillie's Development of the Chick: An Introduction to Embryology,* 3rd ed., revised by Hamilton, H. L., Henry Holt and Company, New York, 1952.
15. **Mobley, W. C., Schenker, A., and Shooter, E. M.,** Characterization and isolation of proteolytically modified Nerve Growth Factor, *Biochemistry,* 15, 5543, 1976.
16. **Stevens, K. L.,** Sesquiterpene lactones from *Centaurea repens, Phytochemistry,* 21, 1093, 1982.
17. **Mettler, F. A. and Stern, G. M.,** Observations on the toxic effects of yellow starthistle, *J. Neuropathol. Exp. Neurol.,* 22, 164, 1963.
18. **Stevens, K. L.,** Allelopathic polyacetylenes from *Centaurea repens* (Russian Knapweed), *J. Chem. Ecol.,* 12, 1205, 1986.

Chapter 20

ULTRASOUND IMAGING: A BIOASSAY TECHNIQUE TO MONITOR FETOTOXICITY OF NATURAL TOXICANTS AND TERATOGENS

Kip E. Panter

TABLE OF CONTENTS

I. INTRODUCTION

Diagnostic ultrasound is one of the great medical advances of the 20th century. Ultrasound imaging is routinely utilized in human health, and its impact with regard to noninvasive diagnostic procedures has been revolutionary. Not only has its use improved the clinical care of obstetric patients, it has permitted the introduction of many new diagnostic and therapeutic procedures. Furthermore, radio-ultrasonography has broadened our research potential in studying human and animal embryonic and fetal physiology and development.

Diagnostic ultrasound has been used extensively in veterinary medicine for diagnostic procedures similar to those used in humans. Its greatest contribution in veterinary medicine has been in pregnancy diagnosis and monitoring follicular development. A recent review covering the last 20 years of veterinary diagnostic ultrasound was published by Christopher et al.[1] Other recent articles have been published for the following individual livestock species: cow, horse, pig, sheep, and goats.[2-6] Recently, ultrasound imaging has been used as a research tool to assess fetotoxic and teratogenic effects of poisonous plants in livestock and to determine normal fetal development and movement patterns in sheep and goats.[7,8]

In this chapter I will briefly review the mechanics and theory of ultrasound imaging, review ultrasound observations of normal fetal growth and development in sheep and goats, discuss research using ultrasound to assess fetotoxic effects of poisonous plants, discuss mechanisms of plant-induced multiple congenital contractures and cleft palate, and discuss the use of ultrasound-measured fetotoxicity as a bioassay tool for measuring the teratogenic potential of individual, isolated plant compounds.

II. MECHANICS AND THEORY OF ULTRASOUND IMAGING

Ultrasonography has been applied to animal studies since the mid-1960s when Lindahl[9] used ultrasound to diagnose pregnancy in sheep. However, its use increased in domesticated animals, following the development of real-time or dynamic imaging scanners in the 1970s,[2] which facilitated the study of anatomy in large and small domestic livestock. Ultrasonography is widely used in research programs and has been integrated into human and veterinary diagnostics.

The physics of ultrasound involves the generation of pulsed electrical energy within a transducer to produce high-frequency sound waves. The imaging system is based on the piezoelectric effect. The transducer contains a piezoelectric crystal that serves as a sending and receiving unit. The crystal changes shape when exposed to an electric potential, and the resulting compression and rarefaction creates a highly directed pulse of sound that passes from the transducer through a coupling agent into the subject. Returning echoes received by the transducer exert pressure on the piezoelectric crystal, and the resulting electrical impulse, generated as a radio-frequency signal, is amplified and displayed as an image on a monitor or television screen. Commonly used transducers in animal research and diagnostics deliver sound frequencies of 3 to 7.5 MHz and consist of either a series of transducers arranged in linear order (linear array) or one or more transducers that rotate or oscillate (sector scanner). The electrical energy of the scanner is dispensed in microsecond impulses, and the focal point of the transducer is determined largely by the number of microsecond impulses generated. The lower the frequency (3 vs. 5 or 7.5 MHz), the greater the focal depth. As the impulses pass through various tissues, some of the waves rebound or are reflected back from the tissue interfaces and are picked up by the transducer. The reflected echoes are displayed as a constellation of spots that form a gray-scaled image of 16 to 64 steps. Nonechogenic structures are black and represent fluid-filled cavities such as ovarian follicles or an embryo yolk sac. At the other end of the gray scale are the highly echogenic dense tissues, such as bone, which are represented as white. Soft tissues such as the uterus and organs are intermediate shades. The resulting picture produced is animated and can be stored on magnetic tapes by a video recorder (VHS or Beta) or viewed in freeze frame for critical analysis, measuring, and/or photography. The two-dimensional image on the monitor allows the ultrasonographer to visualize structures, a process similar to viewing a histologic specimen.

III. NORMAL GROWTH AND DEVELOPMENT OF THE FETAL SHEEP AND GOAT

Ultrasound imaging of normal patterns of fetal growth and development in sheep and goats is not well documented. However, fetal growth in sheep has been studied in animals killed at various stages of gestation or recorded at birth.[10,11] Likewise, there is little information about the normal patterns of fetal activity, fetal cardiac function, and fetal physiology in the sheep and goat. Induction of certain types of malformations by plant toxicants appears to be related to specific periods when fetal movement is first observed, and patterns of inhibited fetal activity. A nonsurgical approach using radio-ultrasound imaging has been used to study embryonic and fetal development in the cow[12] and sheep.[13]

Ultrasound imaging has been shown to be important in assessing fetotoxicity and teratogenicity studies in sheep and goats, because it makes possible the comparison of fetal age to normal patterns of fetal movement and thus relates the latter to certain types of malformations. For example, as will be discussed later in this chapter, there is a direct relationship between the initiation of mouth and tongue movement and the palate closure and cleft palate anomalies induced by poisonous plants.[14,15] Similarly, limb and body movements (strength and frequency) are related to the time during gestation when certain poisonous plants induce multiple congenital contractures. Fetal heart action, strength, and number of beats per minute have been related to fetal death, fluid accumulation, abnormal placental development, and subsequent abortion caused by locoweeds.

Different methods of measuring fetal size relative to gestational age have been reported.[11,16-22] Many of these methods relate fetal growth patterns to the nutritional status of the ewe. Richardson et al.[11] used brain weight, long-bone length, and appendicular ossification centers to estimate developmental age from 50 days gestation until birth. Taylor et al.[22] implanted ultrasound transducers subcutaneously on each side of the fetal skull in late pregnancy, to measure the growth of the fetal skull for 3 weeks. The fetal growth rate during the last 60 to 70 days of gestation was related to nutritional status as reflected in the measurement of fetal girth and crown rump length.[19,20] Ultrasound measurements were compared to those made on fetuses obtained from slaughterhouses. The highest correlations between ultrasound and physical measurements were for thoracic depth ($r = 0.75$) followed by thoracic girth ($r = 0.63$), head width ($r = 0.39$), and thoracic width ($r = 0.17$). These measurements were taken between 83 and 125 days of gestation.[13]

Evans and Sack[23] reviewed extensively the external characteristics of the sheep embryo and fetus by growth features and gestational age. Important points for this discussion include the appearance of the forelimb and hindlimb buds at 20 and 21 days gestation, respectively. Subsequently, the hand and foot plates appear at 22 and 23 days gestation. By 27 days gestation, the third and fourth forelimb digits are prominent, and by day 29, grooves appear between the hindlimb digits. After 29 days, limb development apparently consists of growth and elongation. The facial clefts are closed, and the tongue is visible by 30 days gestation. By day 38 of gestation, the palate is fused.

The relationship between fetal growth and development and teratogenic effects reflects the fact that teratogenic activity is strongly dependent on exposure to an agent (in this case, poisonous plants or alkaloids therefrom) at a specific "sensitive" period in development.[24] Therefore, understanding teratogenesis of a particular agent or xenobiotic requires knowledge not only of the stage of development, but also the patterns of movement, types of movement, heart action, and other parameters.

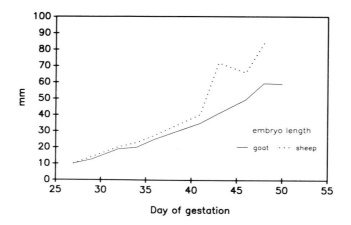

FIGURE 1. Average embryonic length from two sheep and two goats measured weekly from 27 to 50 days gestation.

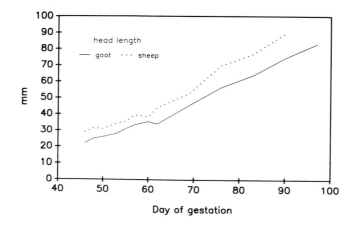

FIGURE 2. Average fetal head length from the tip of the nose to back of the head, from 45 to 90 days gestation.

We speculate that the period of insult in the sheep and goat for cleft palate occurs soon after, or at the time of, initiation of fetal movement and between the time when the tongue is visible and the palate fuses (30 to 38 days gestation). Similarly, during a later period (40 to 60 days) when fetal movement is at its peak and when apparently normal bone elongation, tendon, ligament, and muscle development is dependent on normal fetal movement, multiple congenital contracture (MCC) malformations in sheep and goats may occur.

In order to provide fundamental information on fetal development, two pregnant ewes and two pregnant goats were monitored with ultrasound, and fetal movement and growth patterns compared. Each animal was scanned regularly, beginning 3 weeks after mating and then at weekly intervals to 100 days gestation. Early in gestation, embryonic and fetal length were measured weekly from 27 days gestation to 50 days (Figure 1). Later in gestation only head length, from the tip of the nose to the back of the head (Figure 2), and head width (Figure 3) at the widest region, were measured. Fetal heart rate (Figure 4) and the number of fetal movements (Figure 5) were also recorded.

Results of this preliminary study indicated that fetal movement was initiated near the time of palate closure (35 to 38 days). There was a bimodal pattern of fetal activity in sheep and

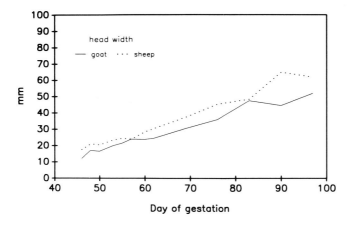

FIGURE 3. Average fetal head width measured from 45 to 96 days gestation.

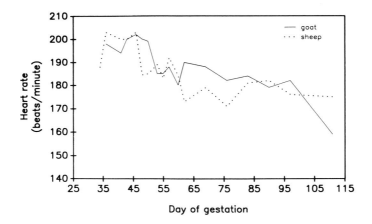

FIGURE 4. Average fetal heart rate in sheep and goats, from 30 to 110 days gestation.

goats (Figure 5). It increased between 38 and 50 days and again between 65 and 90 days. We hypothesize that the early increase in activity is important in normal skeletal development and, if inhibited for a sustained period, will result in MCC malformations. We also speculate that because of small fetal size at this stage of gestation, the fetus is more susceptible to the effects of any circulating toxins.

Fetal heart rate is high (ca. 200 beats per minute; Figure 4) early in gestation, and the first heart beats were observed by ultrasound near 30 days of gestation in both sheep and goats. As gestation progresses and the fetus grows, the heart rate declines to about 150 to 160 beats per minute. Fetal measurements, by the internal calipers of the ultrasound unit, indicate embryonic length (crown rump) increased in a linear fashion until about day 50 (Figure 1). From 45 to 100 days gestation, fetal head width and head length both increase in an almost linear fashion (Figures 2, 3). Further studies must be conducted to confirm these preliminary results. However, early fetal measurements, fetal activity, and fetal heart action must be documented by ultrasound imaging in order to understand normal fetal growth and development and to clarify in teratology the relationship of animal studies to human studies. There is little of this type of information currently available for livestock species.

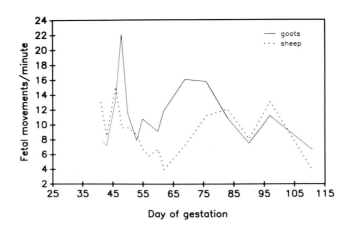

FIGURE 5. Average number of fetal movements per minute from two sheep and two goats, measured from 40 to 110 days gestation.

IV. EFFECT OF LOCOWEED ON FETAL PHYSIOLOGY

Locoweed poisoning, from species of the genera *Astragalus* and *Oxytropis*, occurs in all classes of grazing livestock, and multiple effects have been observed.[25,26] Necropsy studies have shown that locoweed causes fetal malformations, delayed placentation, reduced placental and uterine vascular development, hydrops amnii, abnormal cotyledonary development, interruption of fetal fluid balance, abortion, and various aberrations of male and female reproductive processes.[26-33] Clinical signs of locoweed poisoning include aberrant behavior, ataxia, weakness, emaciation, rough hair coat, and death.[26] The severity and recoverability of locoism varies with the amount and length of time of locoweed ingestion.

Panter et al.[34] monitored, with radio-ultrasound, the effects of locoweed ingestion on fetal and placental development in ewes. Observations were similar to those of others obtained at necropsy,[30,31] i.e., fluid accumulation in the placenta (hydrops allantois, hydrops amnii), altered cotyledonary development, and abortion. Additionally, they reported that fetal heart rate was reduced, and fetal heart contractions were irregular and weak.[34] Fetal cardiac insufficiency and right heart failure, similar to that occurring in calves and cows that ingest locoweed at high elevation, may contribute to the fluid accumulation in the fetus and placenta and thereby cause fetal death and abortion.

Ellis et al.[35] reported locoweed-induced changes in reproductive hormones in pregnant ewes. Serum progesterone values were significantly reduced in a dose-dependent manner, and cotyledonary prostaglandins (6-keto-$PGF_{1\alpha}$, $PGF_{2\alpha}$, and PGE_2) were elevated. 6-Keto-$PGF_{1\alpha}$ and $PGF_{2\alpha}$ were significantly higher than in the controls, while PGE_2 levels tended to be higher. The results of ultrasound studies by Panter et al.[34] suggested that at least one fetus in each ewe fed locoweed died 2 to 3 days before abortion occurred. Ellis et al.[35] reported that myometrial concentrations of 6-keto-$PGF_{1\alpha}$, $PGF_{2\alpha}$, and PGE_2 were significantly higher in ewes bearing dead fetuses than in those bearing live fetuses. Thus, the ultrasound study by Panter et al.[34] and the study reported by Ellis et al.[35] are consistent with the hypothesis that fetal death or necrotic changes in the fetus induce the cotyledonary prostaglandin synthesis that results in abortion. In the case of locoweed-induced abortion, tissue necrosis after fetal death may stimulate prostaglandin production and subsequent abortion.

A condition known as "High Mountain Disease" or congestive right heart failure has been reported in cattle grazing locoweed (*Oxytropis sericea*) at high elevations.[36,37] Based on gross observations and necropsies, the appearance of these animals is similar to those of fetuses from ewes that have ingested locoweed during pregnancy. There was severe edema along the underline, brisket, and jaws and in the thoracic and abdominal cavities. The hearts of calves grazing locoweed at high elevation, and in fetal lambs from ewes fed locoweed, have right ventricular hypertrophy, dilatation, and rounding of the apex.[34,37] High Mountain Disease may result when the effects of locoweed exacerbate the hypoxia of high elevation, thus increasing the vascular resistance and/or vasoconstriction, causing the right ventricle to work excessively, thereby resulting in hypertrophy, cardiac insufficiency, edema, hydrothorax, ascites, and eventual death.

The mammalian fetus is hypoxic and more hypertensive than the adult, because the airways are filled with fluid.[38] Locoweed may cause a similar effect in the fetus as it does in cattle at high elevation, i.e., vasoconstriction or increased vascular resistance. The workload on the fetal heart may increase, causing hypertrophy, dilatation, cardiac insufficiency, fluid accumulation, fetal death, and abortion.

V. TERATOGENIC EFFECTS OF PIPERIDINE AND QUINOLIZIDINE ALKALOID-CONTAINING PLANTS

Conium maculatum (poison hemlock),[39-44] *Nicotiana glauca* (tree tobacco),[45-48] *N. tabacum* (burley tobacco),[49-52] *Lupinus caudatus* (tail cup lupine),[53-56] *L. sericeus* (silky lupine),[53-56] and *L. formosus* (Lunara lupine)[57] have induced congenital birth defects, including cleft palate and skeletal malformations, in livestock. *C. maculatum*[14,43] and *N. glauca*[14,47,58] induced cleft palate in pigs and goats and MCC in pigs,[44,47] goats,[14] sheep,[41,46,59] and cattle.[39-41] *N. tabacum* induced MCC in pigs.[49-51] *L. caudatus*, *L. sericeus*, and *L. formosus* induced cleft palate and MCC in cattle.[53,56,60] *L. formosus* also induced cleft palate and MCC in goats,[61] whereas *L. caudatus* did not.[14] All of these plants contain teratogenic piperidine alkaloids, except *L. caudatus* and *L. sericeus*, both of which contain anagyrine, a quinolizidine alkaloid believed to be the teratogen.[56]

What chemical structural requirements are necessary for piperidines and quinolizidines to be teratogenic? Keeler and Balls[40] orally administered numerous piperidine analogs of coniine (**1**) to cattle. From this research they speculated that for these piperidine alkaloids to be teratogenic they must contain a piperidine ring that was either saturated, as in coniine, or that contained a single double bond, as in γ-coniceine (**2**), and a bulky side chain, three carbons or larger, attached alpha to the nitrogen.

Many plant genera contain piperidine alkaloids that meet these requirements.[48] Some of these include *Conium, Nicotiana, Lupinus, Lobelia, Pinus, Punica, Duboisia, Sedum, Withania, Carica, Hydrangia, Dichroa, Cassia, Prosopis, Genista, Ammondendron, Liparia, Collidium*, and others. The structural similarity to known piperidine teratogens indicates that some of these piperidine alkaloids, such as anabasine (**3**), may be teratogenic.[48]

Why is *Lupinus caudatus*, which contains the quinolizidine alkaloid anagyrine (**5**), teratogenic in cows, but not in sheep and goats? Feeding trials using *L. caudatus* and *L. sericeus* plants (sheep and goats) and anagyrine-rich extracts from these plants (sheep and hamsters) showed that these plants were not teratogenic in sheep,[62] goats,[14] and hamsters.[62] However, they were teratogenic in cattle.[53,56,60] Anagyrine is a quinolizidine alkaloid whose structure is not closely related to that of the known teratogenic piperidines nor to that of the teratogenic piperidine analogs. Why would it induce clinical deformities in cattle, identical to those induced by the piperidines, and yet not do so in sheep, goats, and hamsters? Keeler and Panter[57] hypothesized that cattle metabolize the lupine-derived anagyrine to a complex piperidine, thus meeting the structural requirements of teratogenic piperidines.[40] This mechanism is consistent with the lack of teratogenicity of *L. caudatus* and *L. sericeus* in sheep and goats, and of anagyrine-rich extracts therefrom in hamsters. Keeler and Panter[57] fed *L. formosus* to pregnant cattle and goats and induced cleft palate and MCC in their calves and kids.[61] *L. formosus* contains the piperidine alkaloid ammodendrine (**4**), which has the required structural features of teratogenic piperidine alkaloids proposed by Keeler and Balls.[40] Thus, the proximal teratogens causing cleft palate and MCC may actually be piperidines possessing the structures associated with teratogenicity. Anagyrine (**5**) conversion has yet to be verified in cattle.

In 1983 Panter[42] predicted that the contracture-type birth defects and cleft palate induced in pigs from *Conium maculatum* were associated with the inhibition of fetal movement during critical stages of gestation. This was based on the clinical signs of toxicity (sedation and relaxation) in pregnant sows fed *Conium* and on the appearance of cleft palate and MCC in their offspring. Panter et al.[43] suggested that the alkaloids in *Conium* may cross the placental barrier to the fetus, thus reducing fetal movement due to a similar sedative or anesthetic effect. This fetal immobility could result in the mechanical obstruction by the tongue between the palatine shelves at programmed closure time, resulting in cleft palate. Panter et al.[59] gavaged pregnant ewes twice daily with *C. maculatum*, from gestation days 30 to 60. *Conium* had

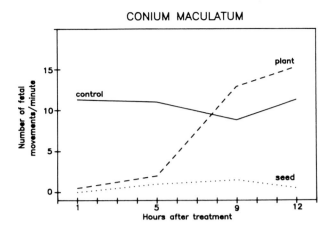

FIGURE 6. Inhibition of fetal movement in goats fed *Conium maculatum* plant and seed, monitored at 45 days gestation. *Conium* seed induced 100% cleft palate and severe MCC-type skeletal defects, whereas *Conium* plant induced minor temporary contracture of carpal joints and no cleft palate.

been shown to cause MCC in cattle[40] and cleft palate and MCC in pigs,[43,44] but results in sheep were not conclusive. Sheep were believed to be resistant to the toxic and teratogenic effects of *Conium*. Nevertheless, they were selected for ultrasound studies because of their size and ease of handling. The pregnant ewes were monitored with ultrasound, at 45, 54, and 60 days gestation, prior to gavage and 1 h after gavage. Fetal movement was significantly ($p < 0.01$) reduced in *Conium*-treated ewes for an undetermined duration, but by 12 to 18 h after treatment, fetal movement had returned to normal. Therefore, inhibition did not persist between doses. None of the lambs from treated ewes had cleft palates, and skeletal malformations were limited to modest carpal flexure of the front limbs, which resolved spontaneously by 8 weeks postpartum. Based on these findings, the duration of reduced fetal movement is apparently a critical factor in permanent limb, spine, and neck deformities and perhaps in cleft palate induction.

Panter et al.[14,15] examined the relationship of the duration of reduced fetal movement and severity of induced terata in goats fed three teratogenic plants, *Conium maculatum, Nicotiana glauca,* and *Lupinus caudatus*. Each of the pregnant goats was monitored with ultrasound at 1, 5, 9, and 12 h after treatment at 45, 51, 55, and 60 days of gestation. The fetal movements in the four groups at 45 days gestation over a 12-hour period were compared (Figures 6, 7, and 8). The significant reduction in fetal movement in goats fed fresh *Conium* was similar to that which occurred in sheep. However, between 5 and 9 h after treatment, fetal movement increased and was almost normal by 12 h after treatment. The kids from these goats had no cleft palates and had modest-to-moderate carpal flexure, which was spontaneously resolved 8 to 10 weeks postpartum, a pattern similar to that described by Panter et al.[59] in sheep. In goats fed *Conium* seed, the fetal movement was reduced over the entire monitoring period (12 h) at all four monitoring times (45, 51, 55, and 60 days of gestation). All kids had bilateral cleft palate and severe skeletal abnormalities, i.e., arthrogryposis, scoliosis, lordosis, torticollis, and rib cage anomalies. Fetal movement was also inhibited for 12 h between doses in the goats fed *N. glauca*. All kids from these goats had bilateral cleft palate and moderate-to-severe skeletal abnormalities, as described for *Conium* seed. Fetal movement was not reduced, and there were no malformations in the goats fed *L. caudatus*.

These results suggest that (1) the duration of reduced fetal movement is a factor in the induction of cleft palate and permanent limb, spine, and neck deformities; and (2) sheep and goats apparently do not metabolize anagyrine (**5**) in *L. caudatus* to a teratogenic piperidine, as suspected in cows.

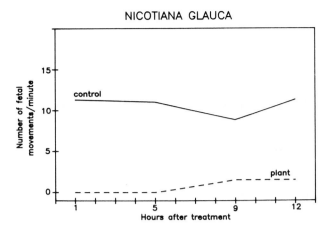

FIGURE 7. Inhibition of fetal movement in goats fed *Nicotiana glauca*, compared to controls, monitored at 45 days gestation. *Nicotiana glauca* induced 100% cleft palate and severe MCC-type skeletal malformations.

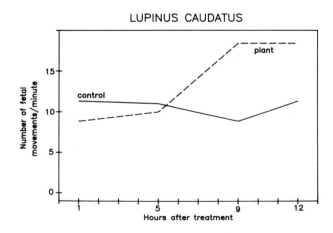

FIGURE 8. Comparison of fetal movement between control and *Lupinus caudatus*-fed goats, monitored at 45 days gestation. *L. caudatus* did not inhibit fetal movement, and there were no cleft palates or MCC-type birth defects induced.

Keeler and Panter[57] fed *L. formosus* to cows and induced severe limb and palate defects. The piperidine alkaloid ammodendrine (**4**), which meets the structural requirement proposed by Keeler and Balls[40] for teratogenic piperidines, was believed to be the teratogen.[57] If piperidine alkaloids must possess certain structural attributes for teratogenicity, then *L. formosus* should be as teratogenic in the goat, as are *Conium* and *Nicotiana*. Goats were gavaged *L. formosus*, and as expected, fetal movement was inhibited, and fetal malformations were the same, and just as severe, as those from both *Conium* and *Nicotiana*.

VI. MECHANISM OF ACTION OF
POISONOUS PLANT-INDUCED CLEFT PALATE
AND MULTIPLE CONGENITAL CONTRACTURES (MCCs)

Recently it was proposed that poisonous plant-induced cleft palate and MCC were caused by a similar physiological mechanism.[15] MCC and cleft palate, of unknown etiology, occur frequently in humans and livestock. Causes proposed for these birth defects include genetic errors, viruses, bacteria, mechanical restriction, chemical teratogens, and plant alkaloids.[63-66] The probable mechanism by which *Conium, Nicotiana,* and *Lupinus* induce cleft palate and MCC is believed to be reduced fetal movement.[14,15] The incidence of MCC and cleft palate in goats (100%), wherein fetal movement was inhibited over the dosage period, strongly suggests that similar mechanisms underlie these two malformations. Cleft palate could be the result of tongue obstruction, and the limb defects could be produced by lack of fetal movement, resulting in postural intrauterine molding or contracture deformities across joints, due to normal long-bone growth, but lack of stretching of tendons and ligaments. Fetal positioning or fetal pressure from a sibling (in the case of multiple births) could result in some of the skeletal defects, i.e., rib cage anomalies, observed in goats.[14]

Other reports support the relationship between sustained reduction of fetal movement and these malformations (cleft palate and MCC). In humans, Pierre Robin syndrome is a birth defect in which cleft palate is associated with micrognathia and glossoptosis.[67] Several workers have observed an association between the Pierre Robin syndrome cleft palate and MCC-type limb defects similar to those we have described. Smith and Stowe[68] reviewed 35 cases of micrognathia associated with defects of the palate and reported that eight of those cases had limb defects, including talipes (club feet), syndactyly, and missing carpal bones. Routledge[69] reported 18 cases of the Pierre Robin syndrome cleft palate, with five of these cases showing limb defects, including talipes, dislocation of the hips, congenital amputations, and multiple deformities of the lower limbs. Crabbe[70] reviewed 232 cases of talipes and found this deformity was associated with cleft palate. Many mechanisms have been proposed for cleft palate in humans. One that fits the inhibited movement theory, and which is supported by many investigators, is the hypothesis of mechanical obstruction by the tongue at programmed palate closure time, first proposed by Poswillo[67] and supported by others.[14,15,42,71,72] Cleft palates have been induced in rats by amniotic sac puncture and subsequent removal of fluid from the amnion.[73,74] Fetal compression and mechanical obstruction by the tongue was believed to be the cause of the cleft palate. Clarke et al.[72] suggested that tongue obstruction during palatogenesis is responsible for cleft palate in chondrodystrophic mice, perhaps due to growth retardation of Meckel's cartilage, which, in turn, prevents the forward displacement of the tongue, thereby leaving it positioned between the fetal palatal shelves and impairing normal shelf elevation and closure. This mandible-tongue-palate relationship is widely accepted as the pathogenetic mechanism underlying cleft palate associated with the Pierre Robin syndrome. Walker and Quarles[75] observed that tongue removal allowed earlier palate closure in fetal mice, which is consistent with the tongue-palate obstruction hypothesis.

Recent observations by Panter and Keeler,[58] using ultrasound, suggest that at 35 to 40 days gestation in the goat, the fetal head is tightly flexed, with intermittent periods of head and neck extension. When fetal movement is inhibited by piperidine-containing plants during this time, there is no head and neck extension observed, and the chin remains tightly pressed against the sternum. Panter et al.[43] proposed that microsurgical removal of the tongue during palate closure would allow normal palate closure even under the influence of the cleft palate-inducing plant *Conium maculatum.* This approach may still help elucidate the mechanism of cleft palate induction by *Conium* and other similar plants.

Similarly, many causes of MCC in humans and animals have been proposed. Swinyard[65] suggested that MCC in livestock is a widespread problem that results in enormous financial losses, and he cited the ingestion of poisonous plants during foraging as the second most frequent cause (viral infections are the first). Several reports of MCC in multiple births are believed to have resulted from intrauterine crowding that may have restricted limb movement.[65] Few specific etiologic agents have been recognized in humans. However, anything that decreases movement or mobility of limbs in a developing fetus may lead to contractures across the joints.[15,62,76-79]

Joint twisting and rigidity have been attributed to a collagenous response around immobile joints.[65] Clinical, anatomical, and biochemical evidence indicate that prenatal fixation of a given joint results from collagenous proliferation, replacing atrophied muscle fibers with connective tissue, and deposition of connective tissue around joint capsules. The joint capsule thickening results in loss of movement and rigidity of the joint.[65]

An early experimental model of xenobiotic-induced arthrogryposis involved the injection of d-tubocurarine (**6**) into developing chicks, causing ankylosis of the joints.[80] Experiments with rats gave similar results, presumably arising from paralysis due to the pharmacologic action of d-tubocurarine.[81] Intra-abdominal pregnancies resulting in arthrogryposis were believed to be the result of fetal compression.[82,83] These studies, together with clinical findings in humans and current research, support the hypothesis that MCC is caused by the loss of fetal movement. Furthermore, where disease conditions are associated with a lack of fetal movement, MCC occurs more frequently than in the general population, i.e., Potter's Syndrome amyoplasia.[76,78,79,84,85] Panter et al.[14,15] hypothesize that any factors that may restrict fetal movement for a sustained period during critical stages of gestation may cause either cleft palate, MCC, or both, depending on the stage of gestation during which inhibited fetal movement occurs. Further ultrasonographic measurements will allow this hypothesis to be tested.

VII. PROPOSED USE OF ULTRASOUND AS A BIOASSAY TECHNIQUE TO TEST ALKALOIDS OR OTHER COMPOUNDS AS TERATOGENS

Ultrasonography to measure fetal movement has potential as a bioassay technique to test many alkaloids, other compounds, and organisms for potential teratogenic activity causing cleft palate and/or MCC. Studies have shown that when fetal movement is severely inhibited for a sustained period of time over the palate closure period and during a longer period for MCC, the result is either cleft palate, which is lethal in livestock, and/or skeletal defects.[14,15] Currently, to test individual compounds, large quantities are needed for oral administration in cows, sheep, or goats for periods not less than twice daily for 7 days. This correlates with amounts ranging from 5 to 10 g or more of pure compound for each animal. Researchers involved in chemical extraction and isolation of natural toxins for testing in livestock species can appreciate this logistic problem. However, based on unpublished data,[61] it appears that if a compound, gavaged a single time, reduces fetal movement for at least 10 to 12 h in the goat, it is a potential cleft palate or MCC inducer. Other possibilities to consider may be the injection of pure compounds into the dam or even direct injection into the allantoic vesicle, thus further reducing the amount of compound needed for testing. The ideal period of gestation in the goat for this testing appears to be about 35 to 40 days. This is when the fetus is very active, yet very small, and probably most susceptible to circulating levels of teratogenic plant compounds of this type.

The following bioassay is proposed and will be tested. Spanish goats (about 30 kg average) will be synchronized in estrus by intravaginal pessary and mated to the same type bucks. Pregnancy will be confirmed by day 30, and the number of fetuses observed and recorded. Normal fetal movement at 35 to 40 days gestation will be determined for 5 min. After the test compound is administered (gavage, sq, im, iv, intrauterine), fetal movement will be monitored for 5 min at 1, 3, 6, 9, and 12 h after treatment. If fetal movement is inhibited at 1 or 3 h after treatment and sustained for 6 to 12 h, then that compound should undergo further testing. The test period would currently include gestation days 35 to 41 in the goat for cleft palate and 41 to 60 for MCC.[61] Treatment from 30 to 60 days induces both cleft palate and MCC, and treatment from 35 to 41 induces cleft palate only. Therefore, 42 to 60 days is thought to be the susceptible period for MCC. However, this specific period has not been tested yet.

VIII. CONCLUSION

Ultrasound imaging has provided new information about the fetotoxic effects of poisonous plants on livestock. The abortifacient action of locoweed involves fetal cardiac insufficiency and probably fetal death prior to abortion. Reduced heart rate during early periods of gestation and reduced strength of cardiac contractions, following ingestion of locoweed, was first observed with ultrasound. These cardiac effects probably lead to fetal congestive right heart failure, fetal death, and subsequent abortion. Ultrasound imaging clarified the critical role of fetal movement in normal growth and development.[8,15] Inhibition of fetal movement by poisonous plants and their alkaloids may result in fetal malformations such as cleft palate and MCC. The poisonous plants *Conium maculatum* (plant and seed) and *Nicotiana glauca* inhibit fetal movement in goats and cause severe birth defects. The alkaloids believed responsible are coniine (**1**) and γ-coniceine (**2**) from *Conium*, and anabasine (**3**) from *N. glauca*, all simple piperidine alkaloids. *Lupinus caudatus*, a plant containing the quinolizidine alkaloid anagyrine (**5**), did not inhibit fetal movement in goats and was not teratogenic.[15] This was the converse of the effect in cattle, perhaps due to differences in the metabolic conversion of anagyrine by cattle and goats.[15] These findings resulted from observations of fetal activity by ultrasound while their mothers were under the influence of poisonous plants.

Ultrasound imaging is a valuable diagnostic tool in human and veterinary medicine, as described in this review, and is also a valuable research tool. Radio-ultrasound is a noninvasive method of observing fetal activity, physiology, growth, and development. Ultrasound may be used in conjunction with biopsy techniques to sample amniotic fluid and fetal blood or to introduce biological material directly to the fetus, thus bypassing the influence of the maternal metabolic or detoxification systems.

Even though ultrasound imaging has only recently been used in poisonous plant research, it has found numerous applications in studies of the fetal physiology, growth, and development of livestock. Ultrasound imaging has increased our knowledge of the effects of poisonous plants on fetal physiology and the mechanism of the subsequent malformations. Ultrasound monitoring as a bioassay will be useful in our efforts to define structure–activity relationships of teratogens and in testing purified alkaloids of the piperidine and quinolizidine types. This research will be accelerated using this bioassay and will further clarify mechanisms involved in cleft palate and MCC induction. With this bioassay the quantities of extractable compounds required for testing fetotoxicity in animals become realistic. The technique may be equally applicable for the evaluation of novel drugs that may be introduced for therapeutic purposes.

REFERENCES

1. **Christopher, R. L., Stowater, J. L., and Pipers, F. S.,** The first twenty-one years of veterinary diagnostic ultrasound, *Vet. Rad.,* 29, 37, 1988.
2. **Pierson, R. A. and Ginter, O. J.,** Ultrasonic imaging of the ovaries and uterus in cattle, *Theriogenology,* 29, 21, 1988.
3. **Kastelic, J. P., Curran, S., Pierson, R. A., and Ginther, O. J.,** Ultrasonic evaluations of the bovine conceptus, *Theriogenology,* 29, 39, 1988.
4. **Squires, E. L., McKinnon, A. O., and Shideler, R. K.,** Use of ultrasonography in reproductive management in mares, *Theriogenology,* 29, 55, 1988.
5. **Madec, F., Marinat-Botte, F., Forgerit, Y., Le Denmat, M., and Vaudelet, J. C.,** Use of ultrasound echotomography in the sow. First trials attempting a codification of some lesions concerning the urogenital tract, *Rec. Med. Vet.,* 164, 127, 1988.

6. **Buckrell, B. C.**, Applications of ultrasonography in reproduction in sheep and goats, *Theriogenology*, 29, 71, 1988.

7. **Panter, K. E., Bunch, T. D., Keeler, R. F., and Sisson, D. V.**, Radio ultrasound observations of the fetotoxic effects in sheep from ingestion of *Conium maculatum* (poison hemlock), *Clin. Tox.*, 26, 175, 1988.

8. **Panter, K. E., Wierenga, T. L., and Bunch, T. D.**, Ultrasonographic studies on the fetotoxic effects of poisonous plants on livestock, in *Handbook of Natural Toxins: Toxicology of Plant and Fungal Compounds*, Vol. 6, Keeler, R. F. and Tu, A. T., Eds., Marcel Dekker, Inc., New York, 1991, 589.

9. **Lindahl, I. L.**, Detection of pregnancy in sheep by means of ultrasound, *Nature*, 212, 642, 1966.

10. **Wallace, L. R.**, The growth of lambs before and after birth in relation to the level of nutrition. II and III, *J. Agric. Sci.*, 38, 243, 1948.

11. **Richardson, C., Hebert, C. N., and Terlecki, S.**, Estimation of the developmental age of the ovine fetus and lamb, *Vet. Rec.*, 99, 22, 1976.

12. **Pierson, R. A. and Ginther, O. J.**, Ultrasonography for detection of pregnancy and study of embryonic development in heifers, *Theriogenology*, 22, 225, 1984.

13. **Kleemann, D. O., Smith, D. H., Walker, S. K., and Seamark, R. F.**, A study of real-time ultrasonography for predicting ovine foetal growth under field conditions, *Aust. Vet. J.*, 64, 352, 1987.

14. **Panter, K. E., Keeler, R. F., Bunch, T. D., and Callan, R. J.**, Congenital skeletal malformations and cleft palate induced in goats by ingestion of *Lupinus, Conium* and *Nicotiana* species, *Toxicon*, 28, 1377, 1990.

15. **Panter, K. E., Bunch, T. D., Keeler, R. F., Sisson, D. V., and Callan, R. J.**, Multiple congenital contractures (MCC) and cleft palate induced in goats by ingestion of piperidine alkaloid-containing plants: reduction in fetal movement as the probable cause, *Clin. Tox.*, 28, 69, 1990.

16. **Thurley, D. C., Revfeim, K. J. A., and Wilson, D. A.**, Growth of the Romney sheep foetus, *N. Z. J. Agric. Res.*, 16, 111, 1973.

17. **Koong, L. J., Garrett, W. N., and Rattray, P. V.**, A description of the dynamics of fetal growth in sheep, *J. Anim. Sci.*, 41, 1065, 1975.

18. **Mellor, D. J. and Murray, L.**, Effects of placental weight and maternal nutrition on the growth rates of individual fetuses in single and twin bearing ewes during late pregnancy, *Res. Vet. Sci.*, 30, 198, 1981.

19. **Mellor, D. J. and Murray, L.**, Effects of long term undernutrition of the ewe on the growth rates of individual fetuses during late pregnancy, *Res. Vet. Sci.*, 32, 177, 1982.

20. **Mellor, D. J. and Murray, L.**, Effects on the rate of increase in fetal girth of refeeding ewes after short periods of severe undernutrition during late pregnancy, *Res. Vet. Sci.*, 32, 377, 1982.

21. **Wenham, G.**, A radiographic study of the changes in skeletal growth and development of the foetus caused by poor nutrition in the pregnant ewe, *Br. Vet. J.*, 137, 176, 1981.

22. **Taylor, M. J., Poore, E. R., Robinson, J. S., and Clewlow, F.**, Measurement of fetal growth in lambs by ultrasound, *Res. Vet. Sci.*, 34, 257, 1983.

23. **Evans, H. E. and Sack, W. O.**, Prenatal development of domestic and laboratory mammals: growth curves, external features and selected references, *Anat. Histol. Embryol.*, 2, 11, 1973.

24. **Shepard, T. H.**, *Catalog of Teratogenic Agents*, 6th ed., Johns Hopkins University Press, Baltimore, 1989, 793.

25. **James, L. F.**, Syndromes of locoweed poisoning in livestock, *Clin. Toxicol.*, 5, 567, 1974.

26. **James, L. F., Hartley, W. J., and Van Kampen, K. R.**, Syndromes of *Astragalus* poisoning in livestock, *J. Am. Vet. Med. Assoc.*, 178, 146, 1981.

27. **Mathews, F. P.**, Locoism in domestic animals, *Tex. Agric. Exp. Stn. Bull.*, 456, 1, 1932.

28. **James, L. F., Shupe, J. L., Binns, W., and Keeler, R. F.**, Abortive and teratogenic effects of locoweed on sheep and cattle, *Am. J. Vet. Res.*, 28, 1379, 1967.

29. **James, L. F., Keeler, R. F., and Binns, W.**, Sequence in the abortive and teratogenic effects of locoweed fed to sheep, *Am. J. Vet. Res.*, 181, 155, 1969.

30. **Van Kampen, K. R. and James, L. F.**, Ovarian and placental lesions in sheep from ingesting locoweed *Astragalus lentiginosus*, *Vet. Pathol.*, 8, 193, 1971.

31. **James, L. F.**, Effects of locoweed on fetal development: preliminary study in sheep, *Am. J. Vet. Res.*, 33, 835, 1972.

32. **James, L. F.**, Effect of locoweed (*Astragalus lentiginosus*) feeding on fetal lamb development, *Can. J. Comp. Med.*, 40, 380, 1976.

33. **McIlwraith, C. W. and James, L. F.**, Limb deformities in foals associated with ingestion of locoweed by mares, *J. Am. Vet. Med. Assoc.*, 181, 155, 1982.

34. **Panter, K. E., Bunch, T. D., James, L. F., and Sisson, D. V.**, Ultrasonographic imaging to monitor fetal and placental developments in ewes fed locoweed (*Astragalus lentiginosus*), *Am. J. Vet. Res.*, 48, 686, 1987.

35. **Ellis, L. C., James, L. F., McMullen, R. W., and Panter, K. E.**, Reduced progesterone and altered cotyledonary prostaglandin levels induced by locoweed (*Astragalus lentiginosus*) in sheep, *Am. J. Vet. Res.*, 46, 1903, 1985.

36. **James, L. F., Hartley, W. J., Van Kampen, K. R., and Nielsen, D.,** Relationship between ingestion of the locoweed *Oxytropis sericea* and congestive right-sided heart failure in cattle, *Am. J. Vet. Res.,* 44, 254, 1983.

37. **Panter, K. E., James, L. F., Nielsen, D., Molyneux, R. J., Ralphs, M. H., and Olsen, J. D.,** The relationship of *Oxytropis sericea* (green and dry) and *Astragalus lentiginosus* with high mountain disease in cattle, *Vet. Hum. Tox.,* 30, 318, 1988.

38. **Brigham, K. L. and Newman, J. L.,** The pulmonary circulation, *Am. Thorac. Soc. News,* 8, 15, 1979.

39. **Keeler, R. F.,** Coniine, a teratogenic principal from *Conium maculatum* producing congenital malformations in calves, *Clin. Tox.,* 7, 195, 1974.

40. **Keeler, R. F. and Balls, L. D.,** Teratogenic effects in cattle of *Conium maculatum* and *Conium* alkaloids and analogs, *Clin. Tox.,* 12, 49, 1978.

41. **Keeler, R. F., Balls, L. D., Shupe, J. L., and Crowe, M. W.,** Teratogenicity and toxicity of coniine in cows, ewes, and mares, *Cornell Vet.,* 70, 19, 1980.

42. **Panter, K. E.,** Toxicity and teratogenicity of *Conium maculatum* in swine and hamsters, Ph.D. thesis, University of Illinois, Urbana, 1983.

43. **Panter, K. E., Keeler, R. F., and Buck, W. B.,** Induction of cleft palate in newborn pigs by maternal ingestion of poison hemlock (*Conium maculatum*), *Am. J. Vet. Res.,* 46, 1368, 1985.

44. **Panter, K. E., Keeler, R. F., and Buck, W. B.,** Congenital skeletal malformations induced by maternal ingestion of *Conium maculatum* (poison hemlock) in newborn pigs, *Am. J. Vet. Res.,* 46, 2064, 1985.

45. **Keeler, R. F.,** Congenital defects in calves from maternal ingestion of *Nicotiana glauca* of high anabasine content, *Clin. Tox.,* 15, 417, 1979.

46. **Keeler, R. F. and Crowe, M. W.,** Teratogenicity and toxicity of wild tree tobacco, *Nicotiana glauca* in sheep, *Cornell Vet.,* 74, 50, 1984.

47. **Keeler, R. F., Crowe, M. W., and Lambert, E. A.,** Teratogenicity in swine of the tobacco alkaloid anabasine isolated from *Nicotiana glauca, Teratology,* 30, 61, 1984.

48. **Keeler, R. F. and Crowe, M. W.,** Anabasine, a teratogen from the *Nicotiana* genus, in *Plant Toxicology, Proc. Australia-U.S.A. Poisonous Plants Symposium,* Seawright, A. A., Hegarty, M. P., James, L. F., and Keeler, R. F., Eds., Queensland Poisonous Plants Committee, Yeerongpilly, Australia, 1985, 324.

49. **Crowe, M. W.,** Skeletal anomalies in pigs associated with tobacco, *Mod. Vet. Pract.,* 50, 54, 1969.

50. **Menges, R. W., Selby, L. A., Marienfield, C. J., Ave, W. A., and Greer, D. L.,** A tobacco related epidemic of congenital limb deformities in swine, *Environ. Res.,* 3, 285, 1970.

51. **Crowe, M. W. and Pike, H. T.,** Congenital arthrogryposis associated with ingestion of tobacco stalks by pregnant sows, *J. Am. Vet. Med. Assoc.,* 162, 453, 1973.

52. **Crowe, M. W. and Swerczek, T. W.,** Congenital arthrogryposis in offspring of sows fed tobacco (*Nicotiana tabacum*), *Am. J. Vet. Res.,* 35, 1071, 1974.

53. **Shupe, J. L., Binns, W., and James, L. F.,** Lupine a cause of crooked calf disease, *J. Am. Vet. Med. Assoc.,* 151, 198, 1967.

54. **Keeler, R. F.,** Lupine alkaloids from teratogenic and nonteratogenic lupins. I. Correlation of crooked calf disease incidence with alkaloid distribution determined by gas chromatography, *Teratology,* 7, 23, 1973.

55. **Keeler, R. F.,** Lupine alkaloids from teratogenic and nonteratogenic lupins. II. Identification of the major alkaloids by tandem gas chromatography-mass spectrometry in plants producing crooked calf disease, *Teratology,* 7, 31, 1973.

56. **Keeler, R. F.,** Lupine alkaloids from teratogenic and nonteratogenic lupins. III. Identification of anagyrine as the probable teratogen by feeding trials, *J. Toxicol. Environ. Health,* 1, 887, 1976.

57. **Keeler, R. F. and Panter, K. E.,** Piperidine alkaloid composition and relation to crooked calf disease-inducing potential of *Lupinus formosus, Teratology,* 40, 423, 1989.

58. **Panter, K. E. and Keeler, R. F.,** Induction of cleft palate in goats by *Nicotiana glauca* during a narrow gestational period and its relation to reduction in fetal movement, *J. Nat. Tox.,* 1, 25, 1992.

59. **Panter, K. E., Bunch, T. D., and Keeler, R. F.,** Maternal and fetal toxicity of poison hemlock (*Conium maculatum*) in sheep, *Am. J. Vet. Res.,* 49, 281, 1988.

60. **Shupe, J. L., James, L. F., Binns, W., and Keeler, R. F.,** Cleft palate in cattle, *Cleft Palate J.,* 1, 346, 1968.

61. **Panter, K. E.,** Unpublished data, 1991.

62. **Keeler, R. F.,** Teratogenicity studies on non-food lupins in livestock and laboratory animals, in *Proc. 2nd Int. Lupine Conf.,* Bellido, L. L., Ed., Publicaciones Agrarias, Madrid, 1984, 301.

63. **Keeler, R. F.,** Plant toxins, in *Survey of Contemporary Toxicology,* Vol. I, Tu, A. T., Ed., John Wiley & Sons, New York, 1980, 285.

64. **Hall, J. G. and Reed, S. D.,** Teratogens associated with congenital contractures in humans and in animals, *Teratology,* 25, 173, 1982.

65. **Swinyard, C. A.,** Concepts of multiple congenital contractures (arthrogryposis) in man and animals, *Teratology,* 25, 247, 1982.
66. **Rousseaux, C. G. and Ribble, C. S.,** Development anomalies in farm animals. II. Defining etiology, *Can. Vet. J.,* 29, 30, 1988.
67. **Poswillo, D.,** Observations of fetal posture and causal mechanisms of congenital deformity of palate, mandible, and limbs, *J. Dent. Res.,* 45, 584, 1966.
68. **Smith, J. L. and Stowe, F. R.,** The Pierre Robin Syndrome, *Pediatrics,* 27, 128, 1961.
69. **Routledge, R. T.,** The Pierre Robin Syndrome. A surgical emergency in the neo-natal period, *Br. J. Plast. Surg.,* 13, 204, 1960.
70. **Crabbe, W. A.,** Aetiology of congenital talipes, *Br. Med. J.,* 2, 1060, 1960.
71. **Seegmiller, R. E. and Fraser, F. C.,** Mandibular growth retardation as a cause of cleft palate in mice homozygous for the chondrodysplasia gene, *J. Embryol. Exp. Morphol.,* 38, 227, 1977.
72. **Clarke, L., Hepworth, W. B., Carey, J. C., and Seegmiller, R. E.,** Chondrodystrophic mice with coincidental agnathia: evidence for the tongue obstruction hypothesis in cleft palate, *Teratology,* 38, 565, 1988.
73. **Trasler, D. G., Walker, B. E., and Fraser, F. C.,** Congenital malformations produced by amniotic sac puncture, *Science,* 124, 439, 1956.
74. **Walker, B. E.,** Effects on palate development of mechanical interference with the foetal environment, *Science,* 130, 981, 1959.
75. **Walker, B. E. and Quarles, J.,** Palate development in mouse foetuses after tongue removal, *Arch. Oral Biol.,* 21, 405, 1976.
76. **Swinyard, C. A. and Bleck, E. E.,** The etiology of arthrogryposis (multiple congenital contracture), *Clin. Orthop.,* 194, 15, 1985.
77. **Hall, J. G.,** Genetic aspects of arthrogryposis, *Clin. Orthop.,* 194, 44, 1985.
78. **Banker, B. Q.,** Neuropathological aspects of arthrogryposis multiplex congenita, *Clin. Orthop.,* 194, 30, 1985.
79. **Hageman, G. and Willemse, J.,** The pathogenesis of fetal hypokinesia: a neurological study of 75 cases of congenital contractures with emphasis on cerebral lesions, *Neuropediatrics,* 18, 22, 1987.
80. **Drachman, D. B. and Coulombre, A. J.,** Experimental clubfoot and arthrogryposis multiplex congenita, *Lancet,* 2, 523, 1962.
81. **Moessinger, A. C.,** Fetal akinesia deformation sequence: an animal model, *J. Pediatrics,* 72, 857, 1983.
82. **Margolis, S. and Luginbeuhl, B.,** Eye abnormalities associated with arthrogryposis multiplex congenita, *J. Pediatr. Ophthalmol.,* 12, 57, 1975.
83. **Guha-Ray, D. K. and Hamblin, M. H.,** Arthrogryposis multiplex congenita in an abdominal pregnancy, *J. Rep. Med.,* 18, 109, 1977.
84. **Thompson, G. H. and Bilenker, R. M.,** Comprehensive management of arthrogryposis multiplex congenita, *Clin. Orthop.,* 194, 6, 1985.
85. **Thomas, I. T. T. and Smith, D. W.,** Oligohydramnios, cause of the nonrenal features of Potter's Syndrome including pulmonary hypoplasia, *J. Pediatr.,* 84, 811, 1974.

Chapter 21

PLANTS AFFECTING LIVESTOCK:
AN APPROACH TO TOXIN ISOLATION

Peter R. Dorling, Steven M. Colegate, and Clive R. R. Huxtable

TABLE OF CONTENTS

0-8493-4372-0/93/$0.00+$.50
© 1993 by CRC Press, Inc.

I. INTRODUCTION

All life forms have evolved strategies for survival. The survival imperative is to stay alive long enough to reach sexual maturity and to reproduce. To do this an individual must maintain a source of nutrients and protect itself from its own predators. Survival is not necessarily a function of physical strength, aggression, speed, or stealth, but also the ability to fight a battle *in situ*. Those species that are rooted to the spot, like plants and fungi, have developed an amazing array of tactics, supreme among which are biologically active chemicals. Most of these chemicals are secondary metabolites in that they do not constitute part of the plant's basic metabolism. Until recently, these compounds were thought to be waste products of little functional importance to the plants. However, recent studies support the notion that many plant and mold secondary metabolites play a vital role in the survival of individual species and, in fact, form part of an ever-escalating chemical arms race between plants and their predators. The major protagonists in this warfare are undoubtedly plants and insects, with herbivorous mammals playing a minor part.

This theater of the ecologic battle was played out in different geographic regions of the world, over many millions of years, with escalation and adaptation between the participants until fairly stable populations and grazing pressures were established. It was the agricultural revolution that destabilized this balance by the rapid domestication of many herbivorous species and, more importantly, transportation of these animals to other geographic regions. In effect, what humans have done is to move well-adapted species from one part of the world and situate them in the middle of someone else's battle, to be damaged in the cross fire of a conflict that had been raging for eons. This is the experience of the Australian animal industries.

Early settlers introduced ruminant herbivores from Europe, where they were well adapted to cope with those toxic plants of their evolutionary conditioning, to Australia, where they were exposed to plant species containing toxins they had never previously encountered. From the earliest years of the colonies, devastating loss of domestic stock was suffered, and the situation only improved following the introduction of pasture plant species. On the other hand, native marsupial herbivores were able to flourish within the natural environment and are far less susceptible to a number of indigenous toxins, for example, monofluoroacetate.[1]

Along with domestic livestock, a number of toxic plants arrived in Australia without their natural predators and thus became uncontrollable weeds in their new environment. Thus, plant species, such as common heliotrope, which have a stable relationship with their predators in Europe, have caused serious stock losses in Australia.

II. INVESTIGATIONS OF POISONINGS BY PLANTS

Estimates of the economic impact of diseases of livestock in Australia, due to poisoning by natural toxins, show that approximately AUST$80 million is lost annually due to animal deaths and production deficits.[2]

A. WHY INVESTIGATE PLANT POISONINGS?

Toxic plants are of particular interest because they cause animal suffering, and hardship and economic loss for farmers and the nation as a whole. For these reasons alone, it is important to study poisonous plants and to understand chemical and clinical aspects of poisoning. The most dramatic effect is the death of animals that have ingested poisonous plants as part of their fodder. In some instances many thousands of animals can be lost in a single outbreak of poisoning, and this can devastate the enterprise of an individual primary producer. However, the major losses are due to reduced wool, meat, and milk production; reduced productive life span; and the cost of bringing affected land into production. These costs include the removal of the offending plants and the loss of real estate values.

Apart from purely veterinary considerations, it is worthwhile taking a closer look at poisonous plants simply because they contain pharmacologically active compounds. After all, the difference between a toxin and a therapeutic drug can simply be the dose rate. Most of our modern drugs are either extracted directly from plants and molds or are synthetic products that are based on, or which were originally derived from, these natural sources. Our modern pharmacopoeia owes much to the early herbalist, and reason demands that there are still many useful compounds awaiting discovery.

Biologically active plant compounds can also be useful tools in the study of metabolic pathways in a variety of living systems. For example, naturally occurring specific enzyme inhibitors and receptor-site ligands have helped elucidate a large number of biologic functions. An excellent example of the latter is the discovery of endorphins as endogenous ligands for morphine receptors.

B. INVESTIGATIVE APPROACH

The investigation of plant poisonings and the isolation of toxic substances from plants is always directed toward a very specific biologic activity. Consequently, many potentially interesting natural products may be overlooked.

1. Experimental Animals

Any investigation should be based on a clear understanding of the toxicosis as it affects the animal species in the field. Therefore, a thorough clinical investigation of the toxicosis, either in natural cases or by means of feeding trials, is essential. Following this clinical investigation, a pathologic study is made. It is crucial that the pathogenesis of the condition is well described so that one can be confident that any later chemical/toxicologic study is dealing with the same syndrome.

Since most domestic herbivores are large animals, they are not ideal subjects for use in a routine bioassay to test for toxicity of extracts of the subject plant. It is therefore desirable to establish an experimental model of the disease in a small laboratory animal species such as rats or mice. In some instances a toxin may induce a different response in the bioassay species. This different response can be used to guide the extraction and purification of the causative compound, which then has to be rigorously checked, in an appropriate species, to ensure that the clinical and pathologic syndrome observed in the field is exactly reproduced.

The use of animals in *in vivo* assays and in the clinico/pathologic studies must be kept to a minimum, and extreme care taken to minimize suffering of experimental subjects. From the early clinico/pathologic studies, the sequence of clinical signs and the pathogenesis of the poisoning are well understood. Therefore, the animals can be humanely sacrificed at the earliest stage in the disease process that is judged to be of value in confirming a positive response.

2. Route of Administration

Since some compounds can have different effects depending on route of administration, it is preferable to mimic the field intoxication conditions as closely as is possible. Therefore, oral administration of plants or extracts of plants is preferred if the experimental animals will voluntarily ingest the plant/toxin. Oral, as opposed to the parenteral administration, also has the advantage that special precautions, such as ensuring absolute sterility and isotonicity, are not necessary. Finely ground whole plant and early extracts can be mixed with the normal diet and presented to the animals in special feeding hoppers. In many cases, depending on palatability, the animals will voluntarily eat such mixtures. In this case they will tend to "titrate" themselves with respect to toxin intake, and therefore across a population there is often a more uniform clinical effect than if the toxin was given on a set dosage regimen. As toxins are progressively purified, it is often more convenient to administer test samples, dissolved in water, mineral paraffin, or some other innocuous carrier, via a stomach tube. Both rats and mice are easy to dose via stomach tube, under very light diethyl ether anesthesia. It is important that anesthesia does not inhibit the swallowing reflex so that the tube is more readily directed down the esophagus rather than the trachea.

3. Isolation of the Toxin

Some plants are always poisonous, some may be seasonally toxic, while others are toxic only when growing in particular soil types or regions. *For this reason it may be important to collect a large supply of the plant under investigation from the site of, and at the time of, a field intoxication.* Once the toxicity of the collected plant has been confirmed, samples of the plant (typically about ten times the toxic dose) are extracted using a series of solvents with a range of polarity. For testing in the bioassay, at least two toxic dose equivalents of the concentrated extracts and the residues are added to normal diet or dissolved in a suitable carrier and administered *per os*. Once a toxic primary extract has been obtained, further purification by routine methods, such as acid/base partition and chromatography in its various forms, is guided by the bioassay system.

The emphasis of this stage of the research is to produce enough pure material for structural analysis. It is also very important to be certain that the pure material elicits the correct pathologic changes in the subject species.

C. STRUCTURAL ANALYSIS OF THE TOXIN

The structural analysis of the pure toxin involves the accumulation of chemical (derivatization, degradation, etc.), physical (melting point, optical activity, crystalline state, etc.), and spectroscopic (mass spectrometry, nuclear magnetic resonance spectroscopy, infrared and ultraviolet absorption spectroscopy, etc.) information about the compound. Analysis and integration of the acquired data provides a chemical structure that should be unambiguous. If any ambiguity persists then synthesis of the proposed structure may be expedient, and the subsequent comparison of the natural and synthetic compounds will be of value. Failing this, and given suitable crystals, X-ray diffraction analysis of the pure compound or of a simple derivative may be possible.

III. SPECIFIC PLANT STUDIES

The following specific studies from the authors' laboratory are offered to illustrate the general principles discussed in the previous sections. The four cases presented here clearly demonstrate how interdisciplinary cooperation has been vital to the elucidation of the problems. The field syndromes are quite different to each other, and the causative toxins vary greatly in structure, including a "cryptic" alkaloid (Section III.A), a phenolic binaphthalene (Section III.B), a novel tetracyclic benzodiazepine (Section III.C), and a very basic guanidino compound (Section III.D).

A. *SWAINSONA CANESCENS*

This investigation commenced with a detailed study of the pathogenesis of the toxicosis, which was facilitated by the fact that the authors were familiar with lysosomal storage diseases in domestic animals: in particular, bovine glycogenosis type II and bovine mannosidosis. This enabled definition of the mechanism of action of the toxin, which was crucial to its subsequent isolation.

1. Clinical Syndrome

Since the days of early European settlement in Australia, certain plants of the genus *Swainsona* (Darling pea) were known to be toxic to grazing livestock. Ingestion of these plants

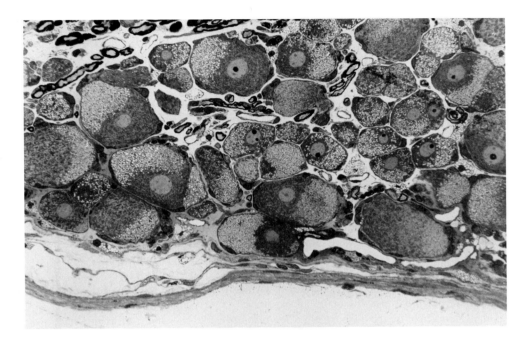

FIGURE 1. Large sensory neuron in a spinal ganglion, showing the intense cytoplasmic vacuolation associated with oligosaccharide storage. (Epon section, Toluidine Blue.)

by cattle, sheep, and horses produces a locomotor disturbance, with affected animals described as being "peastruck". The first confirmed feeding trial showing an association between a locomotor deficit and the ingestion of Darling pea was performed in the colony of New South Wales and reported just prior to the turn of this century.[3]

Clinical signs of poisoning only become apparent after several weeks of significant daily intake of the plant. In cattle there is a progressive loss of condition, followed by head tremor, clumsy gait, increasing incoordination, and weakness, which may lead to death by misadventure. Young cattle appear most susceptible, and cows may have a reduced fertility or abort after maintaining a pregnancy for at least 5 to 7 months.[4] In sheep there may only be a progressive and irreversible loss of condition, leading to death without overt neurologic signs. However, others may show a neurologic disturbance similar to that seen in cattle. There is no documented evidence of abortion or terata in sheep following *Swainsona* ingestion, which is in contrast to the effects of the locoweeds of North America.[5] Emaciation followed by hyperesthesia, excitement, and an ataxic gait are dominant features of poisoning in the horse.[6]

The clinical similarities between *Swainsona*-poisoned animals in Australia and those poisoned by the locoweeds in the U.S. (certain species of *Astragalus* and *Oxytropis*) were recognized in the early 1900s, and a resemblance between the pathological features of peastruck and locoed animals was also noted.[5]

2. Pathogenesis and Biochemical Effects

At postmortem there are no significant gross lesions, apart from signs of emaciation. However, in those animals that had recently ingested *Swainsona*, microscopic examination of tissues reveals widespread, multisystem, neurovisceral, cytoplasmic vacuolation (Figure 1).[4,7-10] On ultrastructural and biochemical grounds, it was suggested that the cytoplasmic vacuolation seen in field and experimental cases of *Swainsona* toxicosis represented swollen secondary lysosomes.[11] It was also suggested that these lysosomes might contain either an endogenous compound, resulting from the inhibition of one of the lysosomal hydrolases, or a compound absorbed from the plant, which could not be broken down within the lysosome.[12] This latter suggestion was supported by the fact that certain exogenous compounds, such as Triton WR 1339, sucrose, and dextran, are indeed taken up by cells and stored within lysosomes, but cannot be degraded by the normal complement of lysosomal enzymes.[13]

The striking similarity between the pathology of *Swainsona* poisoning and mannosidosis in Angus cattle was of considerable comparative interest.[9,14] It became evident that if progress in our understanding of *Swainsona* toxicosis was to be made, it would follow the characterization of the material stored within the cytoplasmic vacuoles in most tissues of affected animals. It was also evident that any new information about *Swainsona* poisoning might be directly applicable to locoism in North America.

3. Characterization of the Storage Product

After feeding rations containing dried, milled *Swainsona canescens* to a Merino ewe for 53 days, a range of tissues were sampled at necropsy for morphologic and biochemical studies.[15] Electron micrographs of vacuolated cells showed that the storage vesicles were essentially clear spaces with very small amounts of dispersed flocculent material. This suggested that the presumptive storage material was highly water soluble, of low molecular weight, and not proteinaceous or lipoid in nature.

Small samples of lymph node from the *Swainsona*-treated ewe and from an untreated sheep were homogenized in water (1 in 5) and deproteinized by precipitation with trichloroacetic acid (5% final concentration). The supernatants, following centrifugation at 100,000 g, were applied to a glass-backed, silica gel thin-layer chromatography (TLC) plate and developed in butanol/acetic acid/water (5:3:2, v/v). The plate was then sprayed with ethanolic sulfuric acid and charred at 200°C. This revealed an intense black spot in the chromatogram of the treated-sheep homogenate, which was not present in the preparation from the normal sheep. Further examination, involving specific visualization reagents, identified the material as an oligosaccharide. It gave a reducing sugar reaction with an Rf approximately one fifth that of glucose. It was also present in urine and other tissues collected from the affected ewe. Similar material was also present in tissues from laboratory animals treated with *Swainsona* extracts (Figure 2).

The stored oligosaccharide was extracted from tissues with ethanol and then precipitated by addition of diethylether. The precipitate was dissolved in water and subjected to Sephadex G25 gel filtration. The oligosaccharide was eluted from the column in a broad band with an average apparent molecular weight of approximately 800 Da. While there was a very small quantity of reducing material of this size in the tissues of the normal sheep, there was no similar compound in extracts of *S. canescens*. Therefore, it seemed likely that tissue vacuolation was caused by the storage of an oligosaccharide originating from the animal's own tissues, rather than by uptake and storage of an oligosaccharide from the plant. As in the genetic lysosomal storage diseases, such an accumulation could occur if the lysosomal enzyme, which normally hydrolyzes the oligosaccharide, was nonfunctional. Inhibition of a lysosomal enzyme by a compound absorbed from the plant therefore became a distinct possibility.

Hydrolysis of the suspected storage product and gas–liquid chromatography (GLC) analysis of the trimethylsilylated hydrosylate demonstrated that the major sugars within the hydrolysate were mannose and glucosamine (in a ratio approximately 3:1). Similar oligosaccharides had previously been isolated from tissues of humans and cattle affected by an inherited deficiency of lysosomal α-mannosidase; that is the lysosomal storage disorder known as α-mannosidosis. In this genetic disease the stored oligosaccharide contained mannose and N-acetylglucosamine, which, under the conditions of acid hydrolysis, would be expected to give glucosamine, as observed in the *Swainsona* intoxication study.

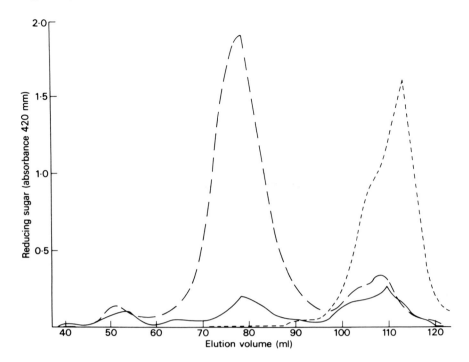

FIGURE 2. Elution, from Sephadex G25, of the crude oligosaccharide preparation from the lymph nodes of an untreated sheep (——), a sheep treated with *Swainsona canescens* (– – –), and a similarly prepared sample from a toxic *S. canescens* extract (- - - -).

This result strongly suggested that *S. canescens* contained an inhibitor of α-mannosidase, which was taken up by tissue lysosomes, thus preventing hydrolysis of mannose-rich oligosaccharides. The pathogenesis of the cellular vacuolation and the development of clinical disease in affected animals was then explicable by comparison with human and bovine mannosidosis.

Extracts of *S. canescens* were subsequently shown to contain an inhibitor that was highly specific for α-mannosidase.

4. Isolation of the Toxin

An *in vitro* assay based on the inhibition of α-mannosidase was developed to guide the extraction and isolation of the toxin.[16]

Extractions and solvent partitions demonstrated that the inhibitor was very water soluble. Association of the activity with a spot on TLC, and subsequent variation of developing solutions, indicated the potential alkaloidal nature of the toxin. However, due to its ready solubility in water, it was not amenable to the usual acid/base extraction procedure for alkaloids. It was not visualized by Dragendorff's reagent, but reaction with ninhydrin indicated, quite falsely as it turned out, the possibility of a secondary amine. Further purification was based on cation exchange techniques.

Thus, powdered plant material was exhaustively extracted with ethyl acetate in a Soxhlet extractor. The extract was evaporated to dryness under reduced pressure, and the residue was suspended in distilled water. Water insoluble materials were removed by filtration, and the pH of the aqueous extract was adjusted to approximately 4 with 1 *M* HCl. The solution was applied to a large column containing Dowex 50W-X8 cation exchange resin in the NH_4^+ form. Following slow application of the sample and extensive washing with distilled water, all α-mannosidase inhibitory activity was retained by the column. The activity was eluted from the Dowex resin by 0.5% aqueous ammonia. Following concentration under reduced pressure, the eluate was adjusted to a pH of 5.0 and was diluted with distilled water until its conductance equaled that of a 10 m*M* acetate buffer, pH 5.0. The solution was applied to a column of CM-Sepharose Cl-6B that had been equilibrated with the acetate buffer, and all inhibitory activity was retained by the column. The column was eluted with a linear NaCl gradient (0 to 1 *M*) in the same acetate buffer. Lyophilization of the inhibitory fractions produced a pale yellow powder that was extracted with dried, ammoniated chloroform. Evaporation of the solvent produced a colorless gel that was crystallized and then recrystallized from chloroform/diethyl ether mixtures to yield sheaves of white needles.[16]

5. Structural Elucidation

Combustion elemental analysis and mass spectrometry yielded a molecular formula, $C_8H_{15}NO_3$, for the toxin.[16] Resonance signals for all eight carbons were observed in the ^{13}C NMR spectrum. The molecular environment and class of these carbons were indicated by the chemical shifts and by the broad-band decoupled NMR experiments. Acetylation of the toxin, with acetic anhydride at room temperature, led to a diacetate or to a triacetate if it was refluxed with acetic anhydride and pyridine. The electron impact mass spectrum (EIMS) of the diacetate and of the triacetate showed the appropriate molecular ion peaks and also showed fragmentations corresponding to losses of two or three acetoxy groups, respectively. The ^{13}C- and ^1H-NMR spectra of the acetates showed the additional carbonyl and methyl carbons and the acetoxy methyl protons expected. The 90 MHz ^1H-NMR spectrum of the toxin, a complex series of multiplets occurring upfield of δ 5 ppm, and the 90 MHz ^1H-NMR spectrum of the diacetate were carefully analyzed using double-resonance decoupling techniques and induced double-resonance (INDOR) techniques and by observing characteristic chemical shift changes when trifluoroacetic acid was added to the NMR solutions. This afforded a nonstereospecific structure for the toxin. Further consideration of the proton–proton coupling constants with respect to the Karplus equation, observation of Bohlmann bands in the IR spectrum, and comparison with molecular models suggested that the relative stereochemistry of the toxin was 1α,2α,8β-trihydroxyoctahydro-8aβ-indolizine (**1**). The toxin was given the trivial name of swainsonine.

Due to the inherent uncertainty in predicting the proton–proton spatial relationships based upon the observed coupling constants, it was considered prudent to examine the structure by X-ray diffraction techniques.[17] Thus, crystals of the 1,2-diacetoxyswainsonine (**2**), which was predicted by NMR and IR examination to have the same relative stereochemistry and the same conformational ring-fusion preference as swainsonine, were of sufficient size and quality for this procedure. In addition to confirming the spectroscopically deduced structure for swainsonine, the X-ray diffraction data indicated that the molecular dimensions of swainsonine diacetate were generally as expected. The six-membered ring was shown to adopt a normal "chair" conformation, with the hydroxyl substituent and the five-membered ring junctions being equatorially oriented. The five-membered ring has a pseudoenvelope conformation, and there was some suggestion of a long hydrogen-bond between the C8 hydroxyl substituent and the C1 oxygen. There does not seem to be any steric reason for nonacetylation of the C8 hydroxyl substituent under the very mild acetylation conditions used.[16] Thus, the partial acetylation observed with the mild reaction conditions may result from the ease of acetylation of the C1 and C2 hydroxyls, rather than any hindrance to the acetylation of the C8 hydroxyl.

The absolute stereochemistry of the molecule was determined by Schneider et al.[18] by reacting the 1,2-acetonide derivative of swainsonine with a racemic mixture of 2-phenylbutanoic anhydride (asymmetric induction method of Horeau[19]). The resulting enantiomeric excess indicated that the 8-hydroxyl substituent was dextrorotatory and therefore was labeled R. Thus, the structure of swainsonine could be absolutely described as (1S,2R,8R,8aR)-1,2,8-trihydroxyoctahydroindolizine. This has subsequently been confirmed by unambiguous, stereospecific synthesis of swainsonine.[20]

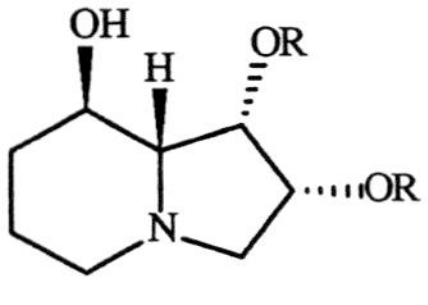

1 R = H

2 R = Acetyl

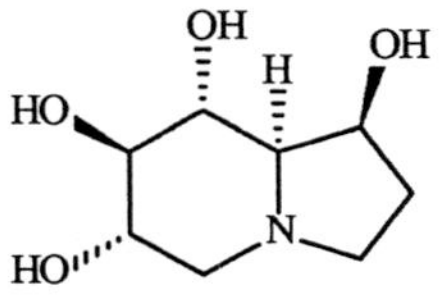

3

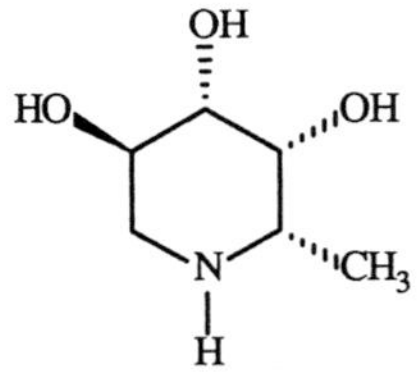

4

6. Conclusions

The most significant advance that led to the isolation of swainsonine was its identification as an α-mannosidase inhibitor and the subsequent development of an *in vitro* assay. Prior to the introduction of this test, only an *in vivo* bioassay was available. This entailed treating mice with plant extracts or preparative fractions for one week and then assessing Kupffer cell vacuolation in liver sections. While this method was satisfactory, it was very slow and often difficult to interpret.

a. Other Sources and Similar Compounds

The isolation of swainsonine from *Swainsona canescens* opened the way for its isolation from other sources. It has now been found in a number of species of *Astragalus* and *Oxytropis* (the locoweeds of America)[21] and from the molds *Rhizoctonia leguminicola*[18] and *Metarhizium anisopliae*.[22]

The identification of swainsonine prompted an awareness for similar small molecular weight polyhydroxylated alkaloids with specific enzyme inhibition activities.[23] Castanospermine (**3**), with its potential for anti-AIDS treatment, ranks among the more notable of these.[24] The purposeful design and synthesis of related compounds has also been pursued, such as that of the fucosidase inhibitor, **4**.[25]

b. Other Biologic Effects of Swainsonine[26,27]

Swainsonine inhibits other α-mannosidases apart from the lysosomal hydrolase. For example, the inhibition of the glycoprotein-processing α-mannosidase II results in the production of hybrid glycosylation rather than the normal complex type.[28] Thus, swainsonine has become an important investigative tool in studies of glycoprotein structure and the role of complex N-linked oligosaccharides in glycoprotein function. Studies to date indicate that this change in glycoprotein processing has little effect on the function of specific glycoproteins.

More recent studies have highlighted the potential of swainsonine and similar compounds as therapeutic agents. Swainsonine is an effective inhibitor of tumor cell metastasis,[29] and it will also inhibit proliferation of certain tumor types in humans.[30] But most interestingly of all, swainsonine is a potent immunomodulator that acts by promoting lymphoproliferation and interleukin-2 production by T-lymphocytes.[31] This may explain much of the antitumor activity of swainsonine, which has been recently reviewed by Olden et al.[32] On this basis, swainsonine may prove to be a very useful therapeutic agent in the management of human and animal malignancies and immune deficiencies.

B. *STYPANDRA IMBRICATA*

Ingestion of *Stypandra imbricata* ("nodding blue lily", "blindgrass") by animals has been implicated in field and experimental intoxications resulting in permanent blindness if the animal survives the acute phase of intoxication.[33,34] *S. imbricata* is a member of the Liliaceae family and has been recently regrouped, along with all other species of this genus, into *S. glauca*.[35] However, as a recent genetic study of the species demonstrates, the taxonomy of the genus is complex and open to review.[36] The plant is not consistently toxic from one area to another or throughout the year, and some plant populations display a white or a patchy-blue white flower rather than the usual blue flower. An investigation of toxic and nontoxic populations of the plant indicated a phytochemical difference that could be genetically based.[37]

1. Clinical Syndrome

The toxicity of *S. imbricata* has been well documented, and the field intoxication has been reproduced in the rat.[33,34,38,39] A toxic dose of plant material taken *per os* induces an acute paretic syndrome with characteristic severe weakness of the pelvic limbs. Apart from remaining blind, animals that survive the acute phase of the intoxication will rapidly return to normal.

2. Pathogenesis

During the acute phase of the intoxication there is diffuse disruption of the structure of central myelin (Figure 3). This results from the separation of myelin lamellae at the intraperiod line,[34,38] a process that effectively reopens the extracellular space obliterated by the formation of the concentrically compacted myelin sheath (Figure 4).

During this acute phase, total brain water is elevated, and plasma cation concentrations are consistent with an increase in normal extracellular fluid. In addition, there is extensive, acute Wallerian degeneration in the optic nerves and tracts. However, no structural lesions are found in the retina during the acute phase.[34]

Over a period of 8 to 10 weeks following acute intoxication, disruption of myelin resolves throughout the central nervous system, and the optic nerves undergo severe atrophy (Figure 5). Concurrently, loss of retinal photoreceptors becomes evident as a multifocal central retinal atrophy (Figure 6). However, no change is seen in the retinal ganglion cells that give rise to those axons that have degenerated in the optic nerves.[34]

3. Isolation and Structural Determination of the Toxin[40]

The extraction of toxic samples of blindgrass and the subsequent purification of the extracts were monitored by the development of characteristic lesions in mice until the toxic response could be associated with a thin-layer chromatography spot. Thus, room temperature extraction of the plant with chloroform, followed by repeated radial chromatography of the petroleum ether insoluble fraction of this extract, yielded crude stypandrol that was recrystallized to greater purity.

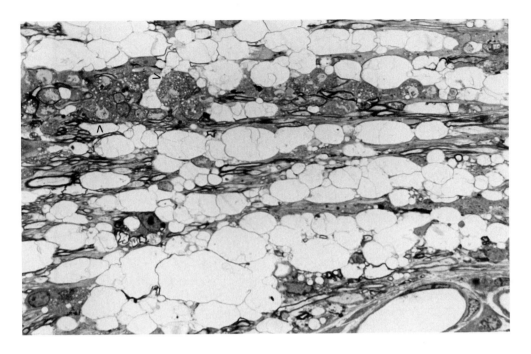

FIGURE 3. Intense myelin vacuolation in the optic nerve of a *Stypandra imbricata*-intoxicated rat. (Epon section, Toluidine Blue.)

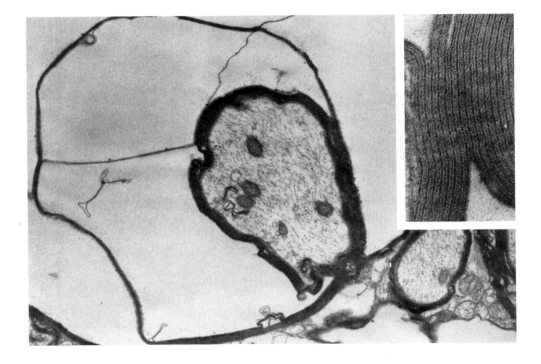

FIGURE 4. Electron micrograph showing vacuolation of central myelin around an axon in *Stypandra imbricata* intoxication. Inset shows separation of lamellae at the intraperiod line.

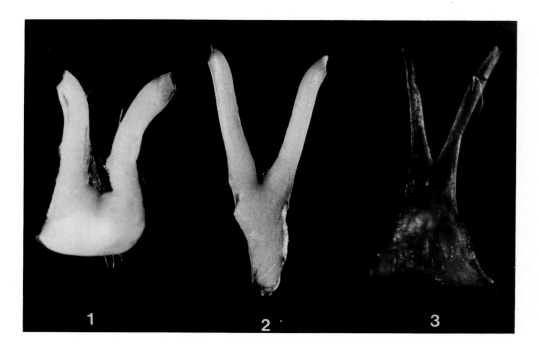

FIGURE 5. Optic nerves from rats showing (1) swelling in acute *Stypandra imbricata* intoxication, (2) normal nerve, (3) marked atrophy 12 weeks postintoxication.

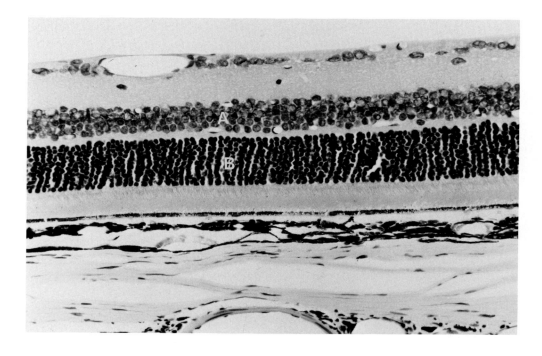

FIGURE 6a. Histology of the retina from a normal rat, to show inner nuclear layer (A) and the outer nuclear layer (B), the latter being the nuclei of the photoreceptor cells.

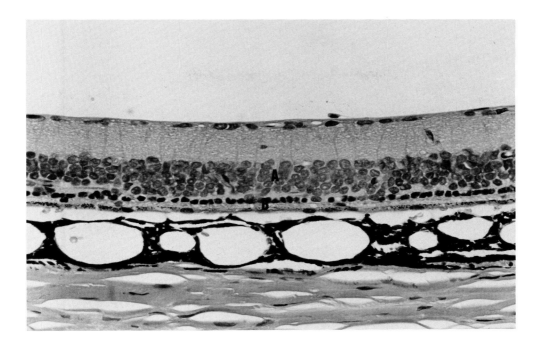

FIGURE 6b. Histology of the retina from a *Stypandra imbricata*-intoxicated rat, 12 weeks after exposure. There is almost total loss of the photoreceptor layer.

The orange powder thereby obtained was only slightly soluble in chloroform and solutions of it were unstable. The phenolic nature of the compound was implied by its ready solubility in aqueous alkali and the formation of an intense blue pigment when potassium ferricyanide was added to the alkaline solution.

The molecular formula, $C_{26}H_{22}O_6$, was established by a combination of mass spectrometry and elemental combustion analysis. Thus, the EIMS of the toxin indicated a molecular weight of 430, and the fragmentation pattern implied the presence of acetyl, hydroxyl, and methyl substituents. Below m/z 216 the EIMS of the toxin was almost identical to that of the naphthalene-diol, dianellidin (**5**), and thus indicated that the toxin could be a dimer of dianellidin. This was supported by an IR spectral absorption at 1625 cm^{-1}, characteristic of an *ortho*-hydroxy-arylketone entity, and by the simplicity of the NMR spectra, which indicated the symmetry of the toxin molecule. An aryl methyl group and an acyl methyl group were indicated by the presence of two three-proton singlets, in the 80 MHz ^{1}H-NMR spectrum, occurring at δ 2.28 and 2.56 ppm, respectively. A broadened, one proton, aromatic singlet (δ 6.98 ppm) was shown to be coupled to the aryl methyl singlet (also broadened) in a double resonance decoupling experiment. The only other signals observed comprised an aromatic AB quartet with a coupling constant consistent with two *ortho*-coupled protons. By comparison, in the 80 MHz ^{1}H-NMR spectrum of dianellidin (**5**), the signals due to the aryl methyl protons and the aromatic proton adjacent to it were observed as a doublet and quartet, respectively, and the remaining aromatic signals were consistent with three adjacent protons.

The broad-band decoupled ^{13}C-NMR spectrum of the toxin comprised 13 signals that, when partial coupling was reintroduced, showed the same multiplicities as in the ^{13}C-NMR spectrum of dianellidin (**5**), except that an extra singlet was observed (in the spectrum of the toxin) at the expense of a doublet (in the spectrum of dianellidin) in the region of 110 to 120 ppm.

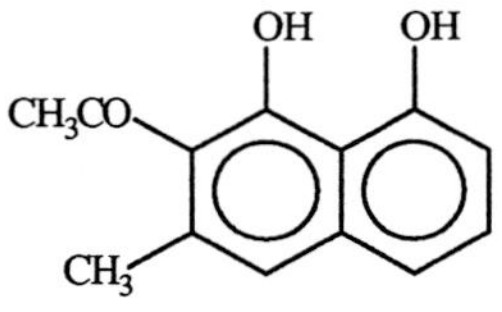

5

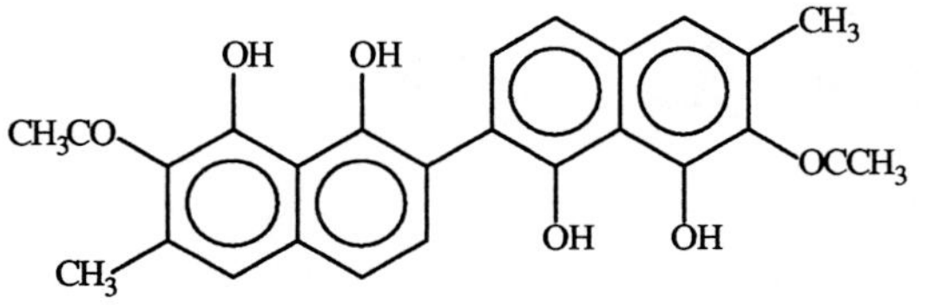

6

This singlet represented the carbon center at which the coupling of two dianellidin moleclues occurs to form the toxin.

The toxin was readily acetylated to form a tetraacetate, as indicated by the EIMS, which confirmed the presence of four hydroxyl substituents. The ^{1}H-NMR spectrum of the acetylated product showed two types of acetoxy methyl protons (instead of four), again demonstrating the symmetry of the molecule.

Therefore, from an examination of the spectroscopic data for the toxin and by comparing it with the data for dianellidin, it was evident that the structure of the toxin could be represented as a dianellidin dimer, with only the position of linkage in doubt. Consideration of the chemical shifts of the aromatic protons in both dianellidin and the toxin led to the conclusion that the two dianellidin molecules dimerized at the 2,2′ position, to yield the toxin that was given the trivial name stypandrol (**6**).

This proposed linkage was confirmed by the X-ray diffraction analysis of the tetraacetate of stypandrol[40] and, more recently, by the synthesis of the toxin involving Fries rearrangement of the dimethyldiacetoxydiospyrol[41] or of the tetraacetoxy derivative of synthetic diospyrol.[42]

4. Pharmacologic and Pathogenic Studies

Utilization of ^{14}C-acetic anhydride in the final steps of the synthesis of the toxin[41] afforded radiolabeled stypandrol, which allowed some preliminary pharmacokinetic studies to be pursued.

Following intravenous or intraperitoneal injection of the radiolabeled toxin, it was determined that stypandrol has a distribution half-life of approximately 1 day and an elimination half-life of about 8 days, in the rat. Following *per os* treatment, these parameters remain the same; however, oral bioavailability is less than 50%.

As a percentage of total dose, 24 h after injection most stypandrol resides in liver, blood, and lungs, while the affected tissues, the brain and eyes, contain a very small proportion of the total counts (Table 1). However, when viewed as counts per gram of tissue, those heavily myelinated tracts, the optic, sciatic, and trigeminal nerves, contain similar concentrations of

TABLE 1
Distribution of Stypandrol Dose 24 h after Treatment

Tissue	Percent of dose
Liver	23
Blood	15[a]
Lungs	11
Muscle	9[a]
Gut	8
Spleen	4
Fat	2[a]
Brain	0.15
Eyes	0.03

[a] Calculated from proportion of body weight.

TABLE 2
Relative Tissue Concentration of Stypandrol

Tissue	dpm/g
Liver	23,850
Cerebrum	2,220
Cerebellum	4,260
Spinal cord	2,910
Optic nerve	20,000
Sciatic nerve	20,830
Trigeminal nerve	15,780
Lung	48,560

label as found in the liver (Table 2). In the blood, more than 96% of stypandrol is bound to plasma proteins, and the label of the total dose is eventually found in the feces and is probably excreted in the bile as a diglucuronide. Insignificant quantities of label appear in the urine.

The pharmacologic data indicate that stypandrol remains in the body for extended periods of time and is present at its site of action in significant concentrations. It is interesting to note, however, that it is also equally concentrated in peripheral myelin in which it has no obvious pathologic effect.

The major lipid classes from whole brain and purified myelin from stypandrol-intoxicated rats were identical to those from control animals. There was no increase in lipid peroxidation, as indicated by the malonyldialdehyde or lipoxidase methods,[43] and purified myelin phospholipids were unaltered in the degree of unsaturation of their fatty acyl chains.[44]

Since lipids appear to be unaffected in stypandrol poisoning, attention must turn to myelin proteins. Myelin of the central nervous system is stabilized by intralamellar protein bridges, while that of the peripheral nervous system is secured by a surrounding basement membrane.[45] It is possible that radical attack on the stabilizing protein within central myelin may account for the observed difference between the effect of stypandrol on central and peripheral myelin. This possibility is currently under investigation.

5. Structure–Activity Studies

The clinicopathologic effects, in rats, of a number of compounds structurally related to stypandrol were examined in the pursuit of structure–activity information.

Dianellidin (**5**), dianellinone (**7**),[46] imbricatonol (**8**),[47] stypandrone (**9**),[48] and diospyrol (**10**)[49] were each tested for stypandrol-like activity in rats. None of these compounds elicited the same clinical and pathologic effects as stypandrol, but if given at a high enough dose rate, they all resulted in peracute death, possibly related to the inhibition of oxidative phosphorylation and the production of methemoglobin.[48,50]

A cyclic voltametric study of the redox properties of some of these compounds demonstrated that only stypandrol formed a stable redox cycle (Figure 7). The other compounds under study were unaffected or were oxidized irreversibly on the first cycle, except for diospyrol (**10**), which displayed a rapidly diminishing reduction peak that lasted for two of three cycles.[51] It is of interest that sporadic cases of blindness have been associated with ingestion of diospyrol as a decoction of the plant *Diospyros mollis*, used as a folk medicinal treatment for internal parasites of humans in Thailand. A clinicopathologic study did not demonstrate any diospyrol-related blindness,[50] but in view of the cyclic voltametric studies, it is conceivable that biochemical circumstances (elevated antioxidant status perhaps?) that impart greater stability to the redox cycle may be a contributing factor to the cause of blindness in these instances.

Apart from its stability, the striking characteristic of stypandrol's redox cycle is that the electric potentials are of the order of the axon potential, that is, the voltage change that passes down the axon with the transmission of each nerve impulse. A mechanism of action of stypandrol might therefore involve damage caused by free radicals (such as short-lived, but reactive, hydrogen radicals) formed as a consequence of a stypandrol redox cycle, such as shown in Figure 7. It should be noted that the stability of this cycle, compared to that of

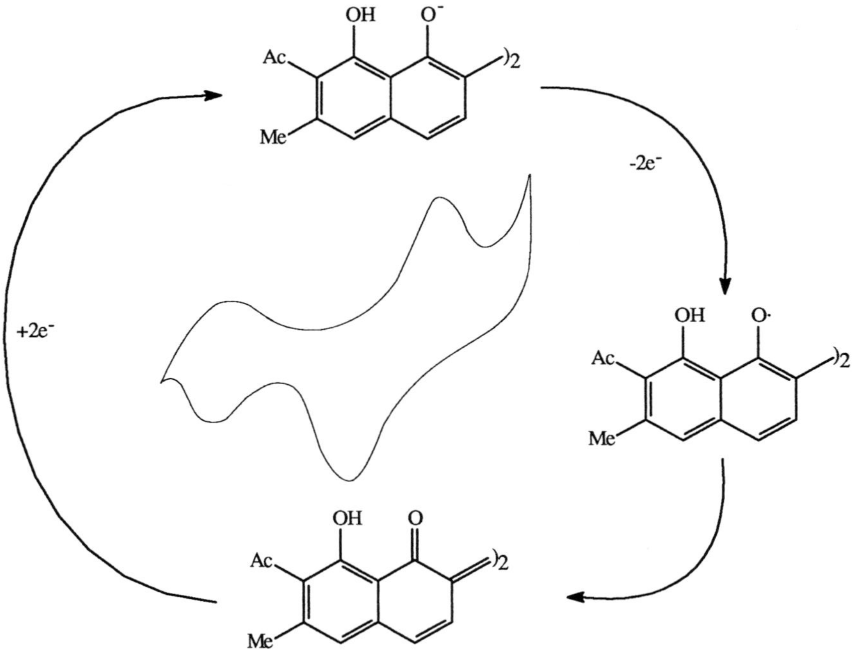

FIGURE 7. The cyclic voltammagram, and a hypothetical explanation for the stability of the redox cycle, observed when stypandrol (Structure 6) is cyclically oxidized and reduced in aqueous basic solution.

diospyrol and dianellidin, may be a consequence of the protected nature of the hydrogen-bonded *ortho*-hydroxy-arylketone entity and the ready delocalization of unpaired electrons from one naphthalene ring system to the other.

6. Conclusions

Purified, natural, and synthetic, stypandrol has been shown to reproduce most, but not all, of the toxic effects of *Stypandra imbricata*. A major problem in working with stypandrol has been the capricious dose–response effect, especially with stypandrol extracted from the plant. This was contributed to by its solubility properties that made accuracy of dosing a difficulty. Synthetic stypandrol was superior in this regard, having a much smaller, and more even, particle size.

What has been clearly shown is that the acute clinical effects of ingestion of *S. imbricata*, the myelin vacuolation and optic axonal degeneration, are due to stypandrol. The latter would account for permanent blindness in animals recovering from the acute phase. Acute death can reasonably be ascribed to cerebral edema associated with severe myelin vacuolation.

No pathogenetic explanation can be offered for the transient paretic syndrome, other than to suggest a functional motor neuron deficit. Axonal degeneration only occurred when vacuolation was severe, and was less marked in immature rats, in which the bones of the growing skull are softer and less rigid. The degenerative process occurred rapidly along the nerve, but did not involve the nerve cell bodies or their most proximal axonal segments. Finally, axonal degeneration has not been seen, to any significant degree, in other regions of the nervous system, in spite of vacuolation of the associated myelin. These results tend to support the idea that optic axonal degeneration is a secondary consequence of myelin vacuolation, swelling of the nerve, and its compression within the optic canals.

As has been previously pointed out, degeneration of retinal photoreceptors is not a consequence of optic axonal degeneration[34] and must be accepted as a separate toxic effect. As the photoreceptor outer segments are a system of compacted membranes, somewhat analogous to myelin,[52] it is attractive to postulate an analogous acute dissociation.

C. *ISOTROPIS FORRESTII*

Isotropis forrestii is one of nine recognized species of this genus that occurs naturally only in Australia. Five species of this genus have been reported to be toxic, while one has been suspected of causing a toxic response in livestock. In western Australia, representatives of the genus are found in both the arid grazing areas and the more intensive agricultural areas of the southwest, where they are referred to as "lamb poisons," "bloom poisons," or "pea-bloom poisons" or by their more benign name of "Granny bonnets." The plants are annual herbs, and it is the emergence of new growth in early spring or late winter or following favorable climatic conditions that has been reported to present the greatest risk to livestock.[39]

1. Clinical Syndrome and Pathologic Effects

Sudden death can occur quietly and with little warning, following ingestion of a relatively small amount of the plant (0.5 to 1 kg by sheep, for example). Animals that survive this peracute reaction may die later as a consequence of renal failure precipitated by acute renal proximal tubular necrosis.[53,54] A study in which sheep were drenched with a single toxic dose of milled plant demonstrated the nephrotoxicity of *I. forrestii*.[55]

The glomerulus and the proximal tubule form the first part of the nephron, the functional unit of the kidney. The proximal tubule receives the glomerular filtrate and begins processing it to produce urine. The specialized cells of the proximal tubules are extremely susceptible to many nephrotoxins, especially at the terminal straight portion where such nephrotoxins may become concentrated.

Clinical signs of intoxication in sheep included weakness, depression, anorexia, diarrhea, and oliguria. At necropsy the kidneys were diffusely pale and, in some cases, slightly swollen. There was intense perirenal edema that extended into the mesenteries and retroperitoneal connective tissue in the more severe cases (Figure 8). Diffuse necrosis of proximal tubular epithelial cells was observed on histopathological examination. The extensive epithelial necrosis resulted in dilated proximal tubules containing cellular debris and proteinaceous material in their lumens (Figure 9). Attempts at epithelial regeneration became evident within the first 36 h following intoxication, and an interstitial lymphocytic infiltrate developed within the renal cortex. The glomeruli and the epithelium of distal tubules appeared normal. Ultrastructurally, changes typical of severe cellular degeneration were observed, such as mitochondrial swelling and degeneration, loss of brush border, and vacuolation of the cytoplasm (Figure 10).

These lesions suggest that tubular obstruction, by proteinaceous casts and necrotic debris, and back leakage of filtrate across the damaged tubular epithelium may account for the initiation and maintenance of renal failure. The pallor of the kidneys further suggests that reduced renal blood flow may also be involved.

2. Isolation and Structural Determination of the Toxin[56]

Extraction of the plant and the subsequent purification procedures were monitored by intraperitoneal injection of aqueous extracts of the test material into mice. Thus, the toxin was extracted from the plant with methanol and then, following precipitation of chlorophylls and other water insolubles, was progressively purified using cation exchange and adsorption chromatography. Recrystallization of the purified extract yielded the toxin iforrestine as fine white needles that, when administered to mice at an intraperitoneal dose of 0.4 mg/kg, elicited the same toxic response as whole plant.

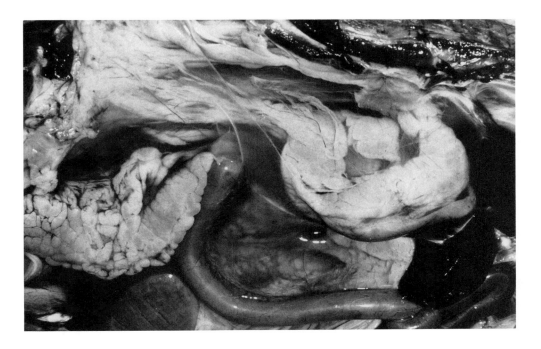

FIGURE 8. Gelatinous edema of the perirenal fat in a sheep with acute *Isotropis forrestii* intoxication.

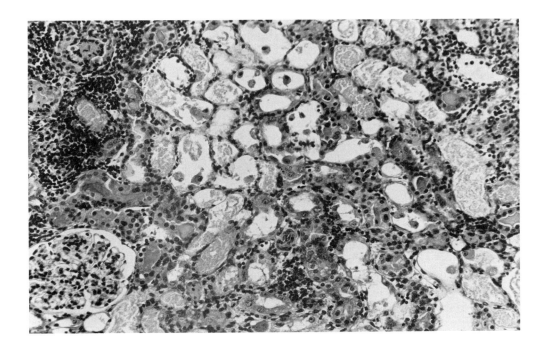

FIGURE 9. Renal tubular necrosis with an interstitial lymphoid infiltrate in a sheep 3 to 4 days after intoxication with a single dose of *Isotropis forrestii*. (Paraffin section, Haematoxylin & Eosin.)

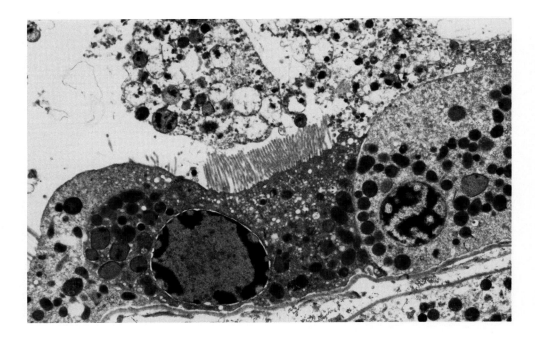

FIGURE 10. Degenerate and necrotic proximal tubular epithelial cells in the early phase of *Isotropis forrestii* intoxication.

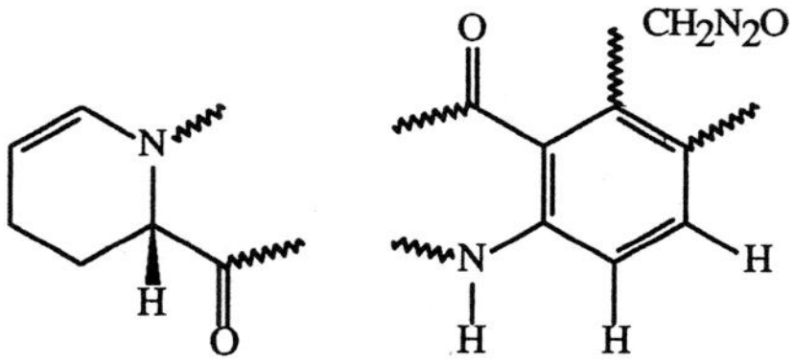

m/z 109   m/z 175

FIGURE 11. Partial structures for the toxin isolated from *Isotropis forrestii*, based upon spectroscopic data.

The chromatographic and solubility properties of the optically active compound ($[\alpha]_D^{20}$ + 362.6°) implied an alkaloidal character. The molecular formula, $C_{14}H_{12}N_4O_3$, was determined by high-resolution EIMS, while data obtained from the ^{1}H-NMR, ^{13}C-NMR, IR, and mass spectra provided a partial structure (Figure 11).

Confirmation of this partial structure and completion of the structural determination was accomplished by a single-crystal X-ray diffraction study of the compound. Thus, the structure and relative stereochemistry of (+)-iforrestine was established as 2-amino-8,9-dihydro-oxazolo[4,5- g]pyrido[2,1-c̲][1,4]benzodiazepine-7,13(6H̲, 7aβH̲)-dione, (**11**; Figure 12).

3. Conclusions

This fused tetracyclic compound appears to represent a novel alkaloidal system, and thus a study of its biosynthesis should prove interesting. Synthesis of the compound from precursors of known stereochemistry will allow for the determination of its absolute stereochemistry and will provide pure compound for pathomechanistic studies and for investigation of other possible bioactivities.

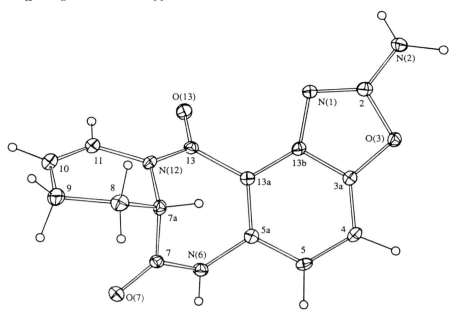

FIGURE 12. X-ray diffraction analysis of the toxin isolated from *Isotropis forrestii.*

D. *SCHOENUS ASPEROCARPUS*

Poison sedge (*Schoenus asperocarpus*) is a short-lived grass-like tussock found only in Western Australia where it is associated with coastal limestone and sands, between the Murchison and Scott Rivers. This plant has not been previously investigated to any great extent, despite the sporadic outbreaks of livestock poisonings that have occurred since the turn of the century.[39]

1. The Field Syndrome

Poisoning has only been reported in sheep and has mainly occurred during the autumn months when the animals have access to regrowth following the first rains of the season.

Ingestion of this plant causes sudden death following respiratory insufficiency. Postmortem findings include a massive pulmonary and pleural exudate that clotted upon exposure to air. This fluid contained large amounts of fibrinogen and other plasma proteins that produced a white, frothy fluid within the trachea and bronchii. No abnormalities were noted in other organs.

Since poison sedge caused sudden death, early investigators assumed that the plant contained cyanogenetic glycosides, like other members of the Cyperaceae.[57,58] It was not until the early 1970s, when 210 sheep from a flock of 1000 died, that a comprehensive case study was conducted and the role of cyanide seriously questioned. Toxic plant material used in the feeding trials did not contain any cyanogenetic material, fluoroacetate, or toxic levels of nitrate, nitrite, oxalate, or alkaloids. In fact, the respiratory lesions alone were considered sufficient to cause death within 24 to 36 h after ingestion of the plant.[59]

No later reports of poisoning have been published, despite many recent outbreaks that have been locally devastating, such as the loss of 2000 sheep in a single paddock 100 km north of Perth.

2. Isolation and Structural Determination of the Toxin[60]

Plant collected from an outbreak of poisoning along the Scott River in Western Australia was air-dried and milled to a fine powder. Extraction and purification of the toxin was guided by the production of respiratory effects in mice, following intraperitoneal injection of test samples.

Thus, the powdered plant material was Soxhlet extracted with ethanol for 24 h. The water soluble portion of the crude extract was desalted by addition of acetone and then evaporated

11

12

to dryness. The residue was chromatographed on silica gel, using CHCl$_3$ and CHCl$_3$/MeOH mixtures as the eluent. The active fractions were repeatedly chromatographed on silica gel (the last steps involving radial chromatography) until TLC indicated a single compound.

The toxin could, with difficulty, be crystallized from MeOH/CHCl$_3$ to yield long colorless needles. The ^{1}H- and ^{13}C-NMR spectra were characteristic of an isoprenyl entity. In addition, a signal at δ 158.5 ppm in the ^{13}C-NMR spectrum was indicative of a iminyl carbon. The mass spectrum implied a molecular formula of C$_6$H$_{13}$N$_3$, which, in conjunction with the NMR data, indicated an allylguanidine structure for the toxin. A review of the literature rapidly identified the toxin as the known butenylguanidine, galegine (**12**). However, the melting point and physical nature of the crystals obtained from the plant were significantly different from that of the free base and any of its reported salts. Therefore, authentic galegine was synthesized, as its sulfate salt, by the method of Desvages and Olomucki.[61] The spectroscopic properties of the synthetic galegine were identical to those of the toxin, and it elicited the same clinical and pathologic effects when administered to sheep either orally or intraperitonealy.

3. Experimental Clinicopathologic Studies

The field syndrome could not be reproduced in mice, rats, or guinea pigs, and therefore sheep, with the consequent larger demands for the toxin, were used for this study. Synthetic galegine administered to sheep as a single intraperitoneal dose of 40 mg/kg induced fatal pulmonary edema approximately 20 h post-dosing in all cases. For the initial 8 to 10 h there was no unequivocal clinical abnormality, but animals developed respiratory distress at about 15 h. When galegine was administered at doses of 30 to 35 mg/kg, approximately 30% of treated animals developed labored breathing at 15 h post-dosing. These animals tended to survive beyond 20 h, and most were subsequently euthanized. One sheep made a complete recovery by 72 h.

Thoracic radiography revealed a caudo-dorsal interstitial pattern several hours prior to the onset of respiratory signs, and a diffuse alveolar pattern by the time dyspnea was evident. This is consistent with the progression of subclinical to clinical pulmonary edema.

By the time respiratory distress was obvious, thoracic ultrasound examination usually revealed hydrothorax and, in some cases, hydropericardium. However, no myocardial abnormalities were evident by ultrasound or electrocardiography. In animals euthanized after the onset of dyspnea, the thoracic cavity contained up to a liter of clear fluid that spontaneously clotted either immediately or after some hours. Mediastinal tissues were markedly edematous, and in some cases hydropericardium was present. The lungs were pale and "wet", and all connective tissue tracts within them were distended and gelatinous.

In animals that died, the above changes were present, but in addition, the lungs were moderately mottled by diffuse congestion, and the airways were filled with stable foam and, in some cases, fibrin casts. Stable foam was present around the nostrils.

Histologically, pulmonary alveolar walls were thickened and hypercellular, due to the accumulation of neutrophils and mononuclear cells in alveolar capillaries. In fatal cases there was also considerable congestion of alveolar capillaries, and proteinaceous exudate in alveolar lumens.

The current interpretation is that the toxin, or a metabolite, directly induces a massive increase in pulmonary alveolar capillary permeability, thereby generating a fibrin-rich lymph that accumulates within the lung and thoracic cavity. The process involves the development of an acute diffuse pneumonitis and terminates fatally when fluid accumulation overwhelms pulmonary respiratory capacity. Limited data suggests that even severe clinical disease may be followed by complete resolution of the pulmonary and thoracic pathology.

4. Conclusions

Galegine was first isolated from *Galega officinalis* (Goat's Rue)[62] and later from *Verbesina encelioides*.[63] Postmortem descriptions of field intoxications by these plants, in contrast to those of poison sedge intoxication, did not mention the high fibrinogen content of the pleural exudate or that this exudate clotted on exposure to air. In addition, microscopic lesions observed in *Verbesina* poisoning included hepatic congestion and centrilobular and tubular necrosis in the kidneys.[64] None of these lesions were noted in poison sedge intoxication.

Galegine, in common with a number of other guanidino compounds, will act as an inhibitor of oxidative phosphorylation.[65] Indeed, at higher doses than those required to reproduce the field syndrome, animals will die suddenly, perhaps as a result of a primary cardiac effect. As a consequence it might be suggested that the pleural and pulmonary exudates associated with galegine intoxication are secondary to pulmonary hypertension due to cardiac failure. However, the results from this study have clearly demonstrated that exudation occurs at a time when heart function is normal. This finding, in conjunction with the occurrence of microscopic lesions in the small arteries, suggests that the primary lesion in sedge intoxication may involve the vascular endothelium rather than a primary insult to the heart.

IV. CONCLUDING COMMENTARY

In the pursuit of causative toxins when investigating animal toxicoses, it is essential to clearly define the clinical syndrome and associated pathology as it occurs in the field. Any putative toxin has to reproduce the field syndrome, with the investigators mindful of possible synergistic, metabolic, and antagonistic effects.

It is also important to realize that different dosages and/or routes of administration can have different effects on the animal under study. For example, a large number of compounds may inhibit or uncouple respiratory chain phosphorylation *in vivo*. These same compounds may even cause peracute poisoning and death due to cardiac arrest, when given in doses of sufficient magnitude to enter the mitochondrial matrix. However, it must not be assumed that, in a given syndrome, a toxin acts via this mechanism simply because it can. A field situation may preclude ingestion of sufficient quantities of the compound for it to act in this way, and in addition, the natural toxicosis may be the result of a metabolite acting at a completely different site, or the parent compound may have a different effect at lower tissue concentrations. For example, both stypandrol and galegine can affect oxidative processes such as ATP production, but it is difficult to see how this action could produce the very different field syndromes described in this chapter.

Investigators should also be aware that the very specific bioactivity that guides the isolation of these natural products should not preclude the possibility of other, potentially very useful, activities. The investigation of *Swainsona* poisoning illustrates this point. The use of swainsonine as a chemotherapeutic agent will undoubtedly prove to be of greater value than its role in the study of a model lysosomal disease. This supports the notion that the search for novel natural products has an intrinsic value that is not always evident at the start of such investigations.

REFERENCES

1. **Twigg, L. E. and King, D. R.,** The impact of fluoroacetate-bearing vegetation on native Australian fauna: a review, *OIKOS,* 61, 412, 1991.
2. **Culvenor, C. C. J.,** Economic loss due to poisonous plants in Australia, in *Plant Toxicology,* Seawright, A. A., Hegarty, M. P., James, L. F., and Keeler, R. F., Eds., Dominion Press-Hedges & Bell, Melbourne, 1985, 3.
3. **Martin, C. J.,** *Agric. Gaz. N. S. Wales,* 8, 363, 1897.
4. **Hartley, W. J. and Gibson, A. J.,** Observations of *Swainsona galegifolia* poisoning in cattle in northern New South Wales, *Aust. Vet. J.,* 47, 300, 1971.
5. **Hartley, W. J.,** A comparative study of Darling pea (*Swainsona* spp.) poisoning in Australia with locoweed (*Astragalus* and *Oxytropis* spp.) poisoning in North America, in *Effects of Poisonous Plants on Livestock,* Keeler, R. F., van Kampen, K. R., and James, L. F., Eds., Academic Press, New York, 1978, 363.
6. **Locke, K. B., McEwan, D. R., and Hamdorf, I. J.,** Experimental poisoning of horses and cattle with *Swainsona canescens* var. *horniana, Aust. Vet. J.,* 56, 379, 1980.
7. **Hartley, W. J. and Kater, J. C.,** Diseases of the central nervous system of sheep, *Aust. Vet. J.,* 41, 107, 1966.
8. **Huxtable, C. R. and Gibson, A. J.,** Vacuolation of circulating lymphocytes in guinea pigs and cattle ingesting *Swainsona galegifolia, Aust. Vet. J.,* 46, 446, 1970.
9. **Hartley, W. J.,** Some observations on the pathology of *Swainsona* spp. poisoning in farm livestock in eastern Australia, *Acta Neuropathol.,* 18, 342, 1971.
10. **Laws, L. and Anson, R. B.,** Neuronopathy in sheep fed *Swainsona luteola* and *S. galegifolia, Aust. Vet. J.,* 44, 447, 1968.
11. **Huxtable, C. R.,** Ultrastructural changes caused by *Swainsona galegifolia* poisoning in the guinea pig, *Aust. J. Exp. Biol. Med. Sci.,* 48, 71, 1970.
12. **Huxtable, C. R.,** The effect of ingestion of *Swainsona galegifolia* on the liver lysosomes of the guinea pig, *Aust. J. Exp. Biol. Med. Sci.,* 50, 109, 1972.
13. **DeDuve, C. R. and Wattiaux, R.,** Functions of lysosomes, *Annu. Rev. Physiol.,* 28, 435, 1966.
14. **Jolly, R. D. and Blakemore, W. F.,** Inherited lysosomal storage diseases: an essay in comparative medicine, *Vet. Rec.,* 92, 391, 1973.
15. **Dorling, P. R., Huxtable, C. R., and Vogel, P.,** Lysosomal storage in *Swainsona* spp. toxicosis: an induced mannosidosis, *Neuropathol. Appl. Neurobiol.,* 4, 285, 1978.
16. **Colegate, S. M., Dorling, P. R., and Huxtable, C. R.,** A spectroscopic investigation of swainsonine: an α-mannosidase inhibitor isolated from *Swainsona canescens, Aust. J. Chem.,* 32, 2257, 1979.
17. **Skelton, B. W. and White, A. H.,** Crystal structure of swainsonine diacetate, *Aust. J. Chem.,* 33, 435, 1980.
18. **Schneider, M. J., Ungemach, F. S., Broquist, H. P., and Harris, T. M.,** (1S,2R,8R,8aR)-1,2,8-trihydroxyoctahydroindolizine (swainsonine), an α-mannosidase inhibitor from *Rhizoctonia leguminicola, Tetrahedron,* 39, 29, 1983.
19. **Horeau, A.,** Principe et applications d'une nouvelle méthode de détermination des configurations dite "par dédoublement partial", *Tetrahedron Lett.,* 506, 1961.
20. **Ali, M. H., Hough, L., and Richardson, A. C.,** A chiral synthesis of swainsonine from D-glucose, *J. Chem. Soc. Chem. Commun.,* 447, 1984.
21. **Molyneux, R. J. and James, L. F.,** Loco intoxication: indolizidine alkaloids of spotted locoweed (*Astragalus lentigenosus*), *Science,* 216, 190, 1982.
22. **Hino, M., Nakayama, O., Tsurumi, Y., Adachi, K., Shibata, T., Terano, H., Kohsaka, M., Aoki, H., and Imanaka, H.,** Studies of an immunomodulator, swainsonine: I. Enhancement of immune response by swainsonine *in vivo, J. Antibiot.,* 38, 926, 1985.
23. **Fellows, L. E.,** The biological activity of polyhydroxylated alkaloids from plants, *Pest. Sci.,* 17, 602, 1986.
24. **Hohenschutz, L. D., Bell, E. A., Jewess, P. J., Leworthy, D. P., Pryce, R. J., Arnold, E., and Clardy, J.,** Castanospermine, a 1,6,7,8-tetrahydroxyoctahydroindolizine alkaloid, from seeds of *Castanospermum australe, Phytochemistry,* 20, 811, 1981.
25. **Fleet, G. W. J., Shaw, A. N., Evans, S. V., and Fellows, L. E.,** Synthesis from D-glucose of 1,5-dideoxy-1,5-imino-L-fucitol, a potent α-L-fucosidase inhibitor, *J. Chem. Soc. Chem. Commun.,* 841, 1985.
26. **Dorling, P. R., Colegate, S. M., and Huxtable, C. R.,** Swainsonine, a toxic indolizidine alkaloid, in *Toxicants of Plant Origin. Vol. 1: Alkaloids,* Cheeke, P. R., Ed., CRC Press, Boca Raton, FL, 1989, chap. 9.
27. **Colegate, S. M., Dorling, P. R., and Huxtable, C. R.,** Swainsonine: a toxic indolizidine alkaloid from the Australian *Swainsona* species, in *Handbook of Natural Toxins, Vol. 6. Toxicology of Plant and Fungal Toxins,* Keeler, R. F. and Tu, A. T., Eds., Marcel Dekker, New York, 1991, chap. 9.
28. **Tulsiani, D. P. R., Harris, T. M., and Touster, O.,** Swainsonine inhibits the biosynthesis of complex glycoproteins by inhibition of Golgi mannosidase II, *J. Biol. Chem.,* 257, 7936, 1982.

29. **Humphries, M. J., Matsumoto, K., White, S. L., and Olden, K.,** Oligosaccharide modification by swainsonine treatment inhibits pulmonary colonization of B16-F10 murine melanoma cells, *Proc. Natl. Acad. Sci. U.S.A.,* 83, 1752, 1986.

30. **Dennis, J. W., Koch, K., and Beckner, D.,** Inhibition of human HT29 colon carcinoma growth *in vitro* and *in vivo* by swainsonine and human interferon-α2, *J. Natl. Cancer Inst.,* 81, 1028, 1989.

31. **Humphries, M. J., Matsumoto, K., White, S. L., Molyneux, R. J., and Olden, K.,** Augmentation of murine natural killer cell activity by swainsonine, a new antimetastatic immunomodulator, *Cancer Res.,* 48, 1410, 1988.

32. **Olden, K., White, S. L., Newton, S. A., Molyneux, R. J., and Humphries, M. J.,** Antineoplastic potential and other possible uses of swainsonine and related compounds, in *Handbook of Natural Toxins, Vol. 6. Toxicology of Plant and Fungal Toxins,* Keeler, R. F. and Tu, A. T., Eds., Marcel Dekker, New York, 1991, chap. 26.

33. **Main, D. C., Slatter, D. H., Huxtable, C. R., Constable, I. C., and Dorling, P. R.,** *Stypandra imbricata* ("Blindgrass") toxicosis in goats and sheep — clinical and pathologic findings in 4 field cases, *Aust. Vet. J.,* 57, 132, 1981.

34. **Huxtable, C. R., Dorling, P. R., and Slatter, D. H.,** Myelin oedema, optic neuropathy and retinopathy in experimental *Stypandra imbricata* toxicosis, *Neuropathol. Appl. Neurobiol.,* 6, 221, 1980.

35. **Macfarlane, T. D.,** Family 132, Phormiaceae, in *Flora of the Perth Region,* Marchant, N. G., Wheeler, J. R., Rye, B. L., Bennett, E. M., Lander, N. S., and Macfarlane, T. D., Eds., Department of Agriculture, Western Australia, 1987, 756.

36. **Russell, J. J. and James, S. H.,** Personal communication, 1989.

37. **Huxtable, C. R., Colegate, S. M., and Dorling, P. R.,** Stypandrol and *Stypandra* toxicosis, in *Toxicants of Plant Origin, Vol. IV. Phenolics,* Cheeke, P. R., Ed., CRC Press, Boca Raton, FL, 1989, chap. 4.

38. **Huxtable, C. R., Dorling, P. R., and Colegate, S. M.,** *Stypandra imbricata* (Blindgrass) toxicosis, in *Plant Toxicology,* Seawright, A. A., Hegarty, M. P., James, L. F., and Keeler, R. F., Eds., Queensland Poisonous Plants Committee, Yeerongpilly, Australia, 1985, 381.

39. **Everist, S. L.,** in *Poisonous Plants of Australia,* rev. ed., Angus and Robertson, Sydney, 1981.

40. **Colegate, S. M., Dorling, P. R., Huxtable, C. R., Skelton, B. W., and White, A. H.,** Stypandrol, a toxic binaphthalenetetrol isolated from *Stypandra imbricata, Aust. J. Chem.,* 38, 1233, 1985.

41. **Rizzacasa, M. A. and Sargent, M. V.,** The synthesis of stypandrol, a toxic binaphthalenetetrol isolated from *Stypandra imbricata:* new syntheses of dianellidin and stypandrone, *Aust. J. Chem.,* 41, 1087, 1988.

42. **Mahidol, C., Tarnchompoo, B., Thebtaranonth, C., and Thebtaranonth, Y.,** Total synthesis of diospyrol, an anthelmintic drug from *Diospyros mollis* Griff., *Tetrahedron Lett.,* 30, 3861, 1989.

43. **Brooksbank, B. W. L. and Balazs, R.,** Superoxide dismutase, glutathione peroxidase and lipoperoxidation in Down's syndrome fetal brain, *Develop. Brain Res.,* 16, 37, 1984.

44. **Cenci di Bello, I., Colegate, S. M., Dorling, P. R., and Huxtable, C. R.,** Unpublished data, 1989.

45. **Waehneldt, T. V. and Linington, C.,** Organization and assembly of the myelin membrane, in *Neurological Mutations Affecting Myelination,* Baumann, N., Ed., Elsevier/North Holland, New York, 1980, 389.

46. **Colegate, S. M., Dorling, P. R., and Huxtable, C. R.,** Dianellidin, stypandrol and dianellinone: an oxidation related series from *Dianella revoluta, Phytochemistry,* 25, 1245, 1986.

47. **Byrne, L. T., Colegate, S. M., Dorling, P. R., and Huxtable, C. R.,** Imbricatonol, a naphthol-naphthoquinone dimer isolated from *Stypandra imbricata* and *Dianella revoluta, Aust. J. Chem.,* 40, 1315, 1987.

48. **Colegate, S. M., Dorling, P. R., and Huxtable, C. R.,** Stypandrone: a toxic naphthalene-1,4-quinone from *Stypandra imbricata* and *Dianella revoluta, Phytochemistry,* 26, 979, 1987.

49. **Borsub, L., Thebtaranonth, Y., Ruchirawat, S., and Sadavongvivad, C.,** A new diglucoside from the anthelmintic berries of *Diospyros mollis, Tetrahedron Lett.,* 105, 1976.

50. **Colegate, S. M., Dorling, P. R., Huxtable, C. R., Tarnchompoo, B., and Thebtaranonth, Y.,** An investigation of possible neurotoxicity of diospyrol, the active principle of *Diospyros mollis* (Maklua), using *Stypandra imbricata* (Blindgrass) induced blindness as a model, *S.E. Asian J. Trop. Med. Public Health,* 21, 139, 1990.

51. **Bailey, S. I., Colegate, S. M., Dorling, P. R., and Ritchie, I. M.,** Unpublished results, 1986.

52. **Towfighi, J., Gonatas, N. K., and McCree, L.,** Hexachlorophene retinopathy in rats, *Lab. Invest.,* 32, 330, 1975.

53. **Gardiner, M. R. and Royce, R. D.,** Poisoning of sheep and cattle in Western Australia due to species of *Isotropis* (Papilionaceae), *Aust. J. Agric. Res.,* 18, 505, 1967.

54. **Huxtable, C. R.,** Toxicity problems associated with *Swainsona, Stypandra* and *Isotropis* spp., in *Veterinary Clinical Toxicology Proceedings No. 103,* University of Sydney Post-graduate Committee in Veterinary Science, Sydney, 1987, 88.

55. **Cooper, T. B., Huxtable, C. R., and Vogel, P.,** The nephrotoxicity of *Isotropis forrestii* in sheep, *Aust. Vet. J.,* 63, 178, 1986.

56. **Colegate, S. M., Dorling, P. R., Huxtable, C. R., Shaw, T. J., Skelton, B. W., Vogel, P., and White, A. H.,** (+)-iforrestine: a novel heterocyclic nephrotoxin from *Isotropis forrestii, Aust. J. Chem.,* 42, 1249, 1989.

57. **Royce, R. D.,** Poison sedge and cyanogenesis, *J. Dept. Agric. W. Aust.,* 1, 497, 1952.

58. **Aplin, T. E. H.,** Cyanogenetic plants of Western Australia, *J. Dept. Agric. W. Aust.,* 4th Ser., 9, 323, 1968.

59. **Nairn, M. E., Aplin, T. E. H., Petterson, D. S., and Brighton, A. J.,** Poison sedge can kill stock, *J. Dept. Agric. W. Aust.,* 4th Ser., 12, 45, 1971.

60. **Colegate, S. M., Dorling, P. R., and Huxtable, C. R.,** Unpublished data, 1992.

61. **Desvages, G. and Olomucki, M.,** Recherches sur les dérivés guanidiques de *Galega officinalis* L.: galégine et hydroxygalégine, *Bull. Soc. Chim. Fr.,* 9, 3229, 1969.

62. **Tanret, G.,** An alkaloid extracted from *Galega officinalis, Compt. Rend. Acad. Sci. Paris Ser. C,* 158, 1182, 1914.

63. **Eichholzer, J. V., Lewis, I. A. S., MacLeod, J. K., Oelrichs, P. B., and Vallely, P. J.,** Galegine and a new dihydroxyalkylacetamide from *Verbesina encelioides, Phytochemistry,* 21, 97, 1982.

64. **King, R. O. C.,** "Crown Beard" (*Verbesina encelioides*), a plant causing pneumonia in sheep, *Agric. Gaz. N. S. Wales,* 48, 364, 1937.

65. **Lotina, B., Tuena de Gomez-Puyou, M., and Gomez-Puyou, A.,** Respiratory changes induced by guanidines and cations in submitochondrial particles, *Arch. Biochem. Biophys.,* 159, 520, 1973.

INDEX

A

Abrusosides, 180, 184–187, see also specific types
Abrus precatorius, 63, 178, 180, 184–186
Abscisic acid, 20
Absorption spectroscopy, 4, 156, 329, 484
Acer saccharum, 178
Acesulfame K, 174
Acetogenins, 197, see also specific types
p-Acetophenol, 310
Acetylenic fatty acids, 269–270, see also specific
    types
O-Acetylisophotosantonin, 152
Acetylvismiones, 233, 238, see also specific types
Acid hydrolysis, 389
Aconitines, 197, see also specific types
Acquired immunodeficiency syndrome (AIDS), 7,
    22, 160, 168, 170, 416, see also Human
    immunodeficiency virus (HIV)
    NCI screens for activity against, 165–168
Acronycines, 197, see also specific types
Acroptilon
    repens, 458, 461
    spp., 458
Actinomycin, 31, 237
Acyclic polyols, 154–155, see also specific types
2-Acyl-1-hydroxy-cyclohexadien-3-ones, 230, see
    also specific types
Adenosine, 283
Adenoviruses, 411, see also specific types
Adsorption chromatography, 34, 38, 44, 352
Affinity chromatography, 288
African medicinal plants, 421–426, see also specific
    types
Agar diffusion, 19
Agar-dilution streak assays, 19
Aglaia
    pyramidata, 210
    spp., 209–210
Aglycones, 188–189, 351, see also specific types
    absolute configurations of, 389
    chemical shifts of, 393–395
    stereochemistry of, 358
    steroidal, 364
Agrobacterium tumefaciens, 18
AIDS, see Acquired immunodeficiency syndrome
Ajugasterone, 148
Alantolacton, 306
Albizia amara, 212–216
Alcohols, 358, see also specific types
Aldehydes, 175, 453, see also specific types
Algae, 20, 45

Alkaline degradation, 288
Alkaloid N-oxides, 66–69
Alkaloids, 11, 15, 59–72, 85, 487, 500, see also
    specific types
    antiviral activity of, 417, 420
    bisbenzylisoquinoline, 46, 208
    bisindole, 196, 198, 222
    chromatography of, 31–32
    complexes of, 60
    conventional, 61–62
    cryptic, see Cryptic alkaloids
    crystallization of, 63
    cytoinhibitory effects of, 222, 238
    cytotoxicity of, 197–199, 214
    detection of, 62
    exciton chirality circular dichroism of, 148
    extraction of, 61, 64–65, 70–71
    fetotoxicity of, 467
    as immunomodulatory compounds, 304–306
    indole, 66
    indolizidine, 420
    insect behavior modification and, 261
    isolation of, 60–62
    isoquinoline, 66, 261, 328
    pipecolic acid-derived, 71
    polar, 32
    polyhydroxy, 69–71
    precipitation of, 63–64, 70
    purification of, 71
    pyridocarbazole, 222
    pyrrolizidine, 30–31, 66, 68, 111
    pyrroloquinoline, 222
    quaternary, 63–66
    quinolizidine, 66, 136, 471–474
    steroidal, 307
    structure of, 60, 71
    teratogenic effects of, 471–474, 476
    tertiary, 44
    tetrahydroxyindolizidine, 11
    tropane, 66
    ultrasonography and, 476
    X-ray crystallography of, 136
1-*O*-Alkyl-2-*O*-acetyl-*ns*-glycero-3-
    phosphorylcholine, 266
Alliogenin, 351
Allium
    cepa, 350
    giganteum, 351
    karataviense, 351
    oleraceum, 351